ADVANCES IN GAS TURBINE TECHNOLOGY

Edited by **Ernesto Benini**

Advances in Gas Turbine Technology
Edited by Ernesto Benini

Published by InTech

Janeza Trdine 9, 51000 Rijeka, Croatia

Publishing Process Manager Alenka Urbancic

Technical Editor Teodora Smiljanic

Cover Designer InTech Design Team

Additional hard copies can be obtained from orders@intechopen.com

Advances in Gas Turbine Technology, Edited by Ernesto Benini
p. cm.
ISBN 978-953-307-611-9

For EU product safety concerns:
IN TECH d.o.o., Prolaz Marije Krucifikse Kozulić 3, 51000 Rijeka, Croatia,
info@intechopen.com or visit our website at intechopen.com.

Meet the editor

Dr. Ernesto Benini is currently a Research Professor of Fluid Machinery at the Department of Mechanical Engineering, University of Padova, Italy, where he first obtained his MSc in Mechanical Engineering and then his PhD in Energy Technology. His research experience is mainly focused on gas turbine engine design and optimization with particular emphasis on aero-thermodynamics of turbomachinery. He authored more than 120 research articles on these topics, the largest part on international refereed journals and congresses.

Contents

Preface

World energy demand is likely to increase over the next 20 years, and it is well ascertained that fossil fuels will still be the dominant source for power generation all over the world. In this scenario, gas turbine (GT) engines will still represent a key technology, either in stand-alone applications or combined with other power generation equipment.

The challenges in GT technology today deal with several issues, such as increased on/off design efficiency, reduced performance degradation over time, and decreased pollutant emissions levels. Major research efforts and investments are being bestowed for the development of new, advanced GT technologies with superior performance, thus helping in the fulfillment of the Kyoto Protocol objectives for greenhouse gases reduction, as well as of many other transnational policies on sustainability and reduced environmental impact of energy technologies.

The book intends to provide an updated picture, as well as a perspective view of some of the major improvements which characterize the GT technology in different applications, from marine and aircraft propulsion to industrial and stationary power generation. Therefore, the target audience for it involves design, analyst, materials and maintenance engineers. Manufacturers, researchers and scientists will also benefit from the timely and accurate information provided.

The book is organized into five main sections, which are comprised of 21 chapters: (I) Aero and Marine Gas Turbines, (II) Gas Turbine Systems, (III) Heat Transfer, (IV) Combustion, and (V) Materials and Fabrication. Starting from a general but evolving vision on GT and their components used in aircraft and marine engines in Section I, a systematic approach is used in Section II to describe the thermodynamic behavior of various GT-based technologies, including combined power, hybrid fuel cell/GT and small-scale GT plants. Then, in Sections III and IV, a total of four specific works (two articles each) are included, which present some of the latest developments in the internal heat transfer and combustion-related phenomena. Finally, a collection of nine papers in Section V. deal with the most recent advances regarding the new materials for GT component fabrication, including interesting suggestions on material damage prevention, diagnosis and repair.

The editor is indebted to all the valuable contributions included in this book from the various GT experts spread worldwide, as well as to InTech Open Access Publisher for giving me the opportunity to edit this volume, and supporting me constantly during its preparation.

Dr. Ernesto Benini
Department of Mechanical Engineering
University of Padova,
Italy

Part 1

Aero and Marine Gas Turbines

1

Future Aero Engine Designs: An Evolving Vision

Konstantinos G. Kyprianidis
Chalmers University of Technology
Sweden

1. Introduction

Public awareness and political concern over the environmental impact of civil aviation growth has improved substantially during the past 30 years. As the environmental awareness increases, so does the effort associated with addressing NO_x and CO_2 emissions by all the parties involved. In the Vision 2020 report made by the Advisory Council for Aeronautical Research in Europe (2001), goals are set to reduce noise and emissions produced by the ever increasing global air traffic. Emissions legislation, set by the International Civil Aviation Organisation (ICAO) and it's Committee on Aviation Environmental Protection (CAEP), is becoming ever more stringent, creating a strong driver for investigating novel aero engine designs that produce less CO_2 and NO_x emissions.

On the other hand, airline companies need to continuously reduce their operating costs in order to increase, or at least maintain, their profitability. This introduces an additional design challenge as new aero engine designs need to be conceived for reduced environmental impact as well as direct operating costs. Decision making on optimal engine cycle selection needs to consider mission fuel burn, direct operating costs, engine and airframe noise, emissions and global warming impact.

CO_2 emissions are directly proportional to fuel burn, and therefore any effort to reduce them needs to focus on improving fuel burn, by reducing engine Specific Fuel Consumption (SFC), weight and size. Reducing engine weight results in a lower aircraft maximum take-off weight, which in turn leads to reduced thrust requirements for a given aircraft lift to drag ratio. Reducing engine size – predominantly engine nacelle diameter and length – reduces nacelle drag and therefore also leads to reduced thrust requirements. For a given engine SFC, a reduction in thrust requirements essentially results in lower fuel burn. Lower engine SFC can be achieved by improving propulsive efficiency and thermal efficiency – either by reducing component losses or by improving the thermodynamic cycle.

Improvements in propulsive efficiency – and hence engine SFC at a given thermal efficiency – can be achieved by designing an engine at a lower specific thrust (i.e. net thrust divided by fan inlet mass flow). This results in a larger fan diameter, at a given thrust, and therefore in increased engine weight, which can partially, or even fully, negate any SFC benefits. Propulsive efficiency improvements at a constant weight are directly dependent on weight reduction technologies such as light weight fan designs and new shaft materials. Increasing engine bypass ratio aggravates the speed mismatch between the fan and the low pressure turbine. Introduction of a gearbox can relieve this issue by permitting the design of these two components at their optimal speeds, and can hence reduce engine weight, as well as improve component efficiency. The first research question therefore rises:

How low can we really go on specific thrust?

Improvements in thermal efficiency – and hence engine SFC at a given propulsive efficiency – can be achieved for conventional cores mainly by increasing engine Overall Pressure Ratio (OPR). At a given OPR there is an optimal level of combustor outlet temperature T_4 for thermal efficiency. However, at a fixed specific thrust and engine thrust, an increase in T_4 can result in a smaller core and therefore a higher engine bypass ratio; in some cases, a potential reduction in engine weight can more than compensate for a non-optimal thermal efficiency. Increasing OPR further than current engine designs is hindered by limitations in high pressure compressor delivery temperature at take-off. Increasing T_4 is limited by maximum permissable high pressure turbine rotor metal temperatures at take-off and top of climb. Increasing turbine cooling flows for this purpose is also fairly limited as a strategy; cooling flows essentially represent losses in the thermodynamic cycle, and increasing them eventually leads to severe thermal efficiency deficits (Horlock et al., 2001; Wilcock et al., 2005). Designing a combustor at very low air to fuel ratio levels is also limited by the need for adequate combustor liner film-cooling air as well as maintaining an acceptable temperature traverse quality (Lefebvre, 1999). The second research question therefore rises:

How high can we really go on OPR and T_4?

Aggressive turbofan designs that reduce CO_2 emissions – such as increased OPR and T_4 designs – can increase the production of NO_x emissions due to higher flame temperatures. The third research question therefore rises:

What is the trade-off between low CO_2 and NO_x?

The research work presented in this chapter will focus on identifying several novel engine cycles and technologies - currently under research - that can address the three research questions raised. These concepts will be evaluated based on their potential to reduce CO_2 and NO_x emissions for engine designs entering service between 2020 and 2025. Design constraints, material technology, customer requirements, noise and emissions legislation, technology risk and economic considerations and their effect on optimal concept selection will also be discussed in detail.

2. An evolving vision

Numerous feasibility studies have been published over the years focusing on future engine and aircraft designs that can reduce fuel consumption; a brief review of some of these publications will be carried out here.

One of the earliest discussions on the subject of improving engine fuel efficiency is provided by Gray & Witherspoon (1976), looking at conventional and heat exchanged cores, as well as non-steady flow combustion processes and open rotor configurations. A similar study focusing on geared and open rotor arrangements as well as heat exchanged cycles is presented by Hirschkron & Neitzel (1976).

An interesting discussion on how specific thrust levels were expected to evolve in the mid-70's based on the economic and technological projections of that time period is given by Jackson

(1976); the author has also provided an update to that discussion based on current economical and technological projections (Jackson, 2009). Wilde (1978), Young (1979), and Pope (1979) provide a good reference on how the future for civil turbofan engines for medium and long range applications was envisaged in the late 70's. Some early discussions on future trends in commercial aviation from the aircraft manufacturer's perspective can be found in Swihart (1970) and Bates & Morris (1983), while Watts (1978) provides an airliner's view of the future. A review on the several technical and economic obstacles that were identified in the late 80's with respect to the realization of the Ultra-High Bypass Ratio (UHBR) turbofan concept is provided by Borradaile (1988) and by Zimbrick & Colehour (1988). Peacock & Sadler (1992) give an update on the subject, focusing further on engine design constraints and the technology advancements required for producing a competitive UHBR configuration. Potential year 2020 scenarios are explored by Birch (2000) while an overview of current aero engine technology and some insight on the future of aircraft propulsion is given by Ruffles (2000). Sieber (1991) and Schimming (2003) provide an excellent discussion on counter-rotating fan designs. Finally, for a review on the development of civil propulsion from the early 50's to recent years the interested reader is referred to Saravanamuttoo (2002). The focus of the next section will be given on recent European research initiatives on enabling technologies relevant to the three research questions that have been set.

3. Enabling technologies and recent research

3.1 Propulsor technologies

Within the EU Framework Program 6 research project VITAL (enVIronmenTALly friendly aero engines, 2009) a number of low pressure system component technologies have been investigated (Korsia, 2009; Korsia & Guy, 2007). The emerging progress will allow the design of new powerplants capable of providing a step reduction in fuel consumption and generated noise.

The VITAL project concentrated on new technologies for the low pressure system of the engine, which enable the development of low noise and low weight fan architectures for UHBR engines. To achieve these objectives, the VITAL project has investigated three different low pressure configurations, leading to low noise and high efficiency power plants. The three configurations are the DDTF (Direct Drive TurboFan) supported by Rolls-Royce, the GTF (Geared TurboFan) by MTU and the CRTF (Counter-Rotating TurboFan) by Snecma.

The DDTF architecture offers a re-optimised trade-off between fan and turbine requirements considering the low weight technologies introduced by the VITAL programme. The GTF combines a fan with a reduction gear train, to allow different rotating speeds for the fan on one hand, and the booster and turbine on the other. The CRTF offers a configuration with two fans turning in opposite directions, allowing for lower rotational speeds, since the two fan rotors split the loads involved.

The technologies being built into the VITAL engines include (Korsia, 2009; Korsia & Guy, 2007):

- New fan concepts with the emphasis on two types: counter-rotating and lightweight fans.
- New booster technologies for different operational requirements; low and high speed, associated aerodynamic technologies, new lightweight materials and associated coating and noise reduction design.

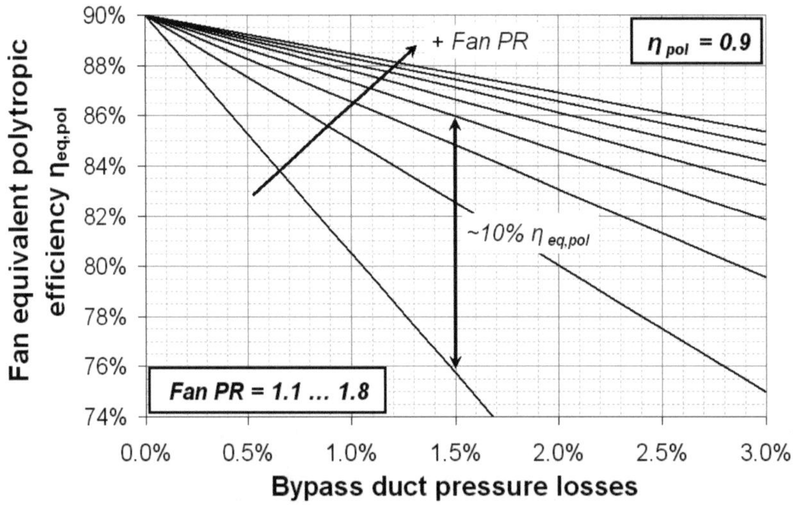

Fig. 1. Effect of fan tip pressure ratio and bypass duct pressure losses on fan equivalent polytropic efficiency

- Polymer composites and corresponding structural design and manufacturing techniques are studied in parallel with advances in metallic materials and manufacturing processes.
- Shaft torque density capabilities through the development of metal matrix composites and multi metallic shafts.
- Low pressure turbine weight savings through ultra high lift airfoil design, ultra high stage loading, lightweight materials and design solutions.
- Technologies for light weight and low drag installation of high bypass ratio engines related to nozzle, nacelle and thrust reverser.

The open rotor engine concept, for high subsonic flight speeds, has also risen as a candidate for improving fuel consumption on several occasions since the advent of the first high bypass ratio turbofan engine. Such engine configurations, often refereed to as propfans in the literature, are direct competitors to ultra high bypass ratio turbofan engines. Their are located at the ultra-low specific thrust region of the design space, where propulsive efficiency benefits for turbofans are negated by very low transfer efficiencies. As illustrated in Fig. 1, this is due to the dominant effect on transfer efficiency that bypass duct pressure losses have when looking at low fan tip pressure ratio engine designs, i.e. low specific thrust. Open rotor engines do not suffer from bypass duct pressure losses and can therefore achieve a very high propulsive efficiency at a good level of transfer efficiency. Compared to turbofans, propfans also benefit from reduced nacelle drag and weight penalties.
Several open rotor programs took place during the 80's, resulting in engine demonstrators and flight tests. The purpose of these projects was to develop propfan concepts that could fly efficiently at speeds comparable to high bypass ratio turbofans, i.e. close to Mach 0.8. General Electric proposed the UDF (UnDucted Fan), a pusher configuration with counter-rotating propellers driven by a counter-rotating low pressure turbine (GE36 Design and Systems Engineering, 1987). The 578-DX, a pusher configuration with counter-rotating

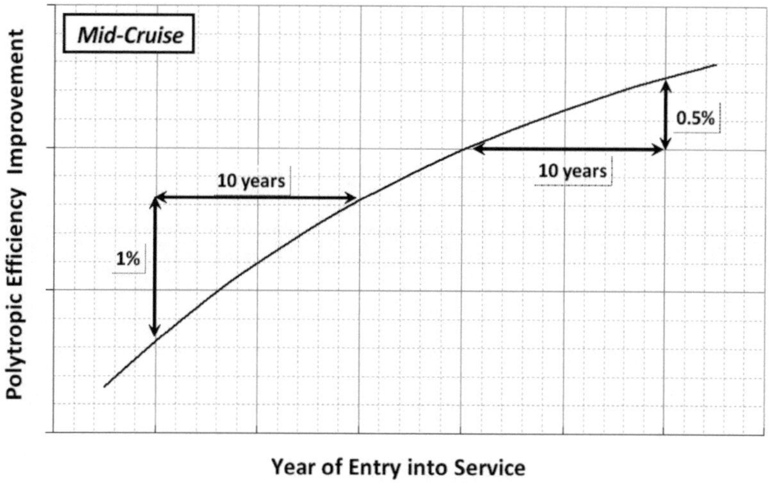

Fig. 2. Compressor efficiency improvement with year of entry into service.

propellers driven by a more conventional low pressure turbine through a differential planetary differential gearbox, was the result of a joint effort by Pratt & Whitney, Hamilton Standard, and Allison.

Both projects were eventually put on hold towards the end of the decade as fuel prices fell significantly. Nevertheless, the open rotor concept has now resurfaced within the EU Framework Program 7 research project DREAM (valiDation of Radical Engine Architecture systeMs, 2011) and the Clean Sky Joint Technology Initiative (2011). Within DREAM, the feasibility of two different open rotor architectures is evaluated including noise. Within Clean Sky, research work is being carried out by some of Europe's largest aero engine manufacturers, such as Rolls-Royce and Snecma, focused on designing, building and testing an open rotor demonstrator.

3.2 Core technologies

Improving core component efficiencies (including reducing losses in the cycle such as duct pressure losses) is one way of improving the engine thermal efficiency. Nevertheless, modern CFD-assisted designs are already quite aggressive and limited benefit may be envisaged by such future advancements (Kurzke, 2003); the increasing effort required to improve an already very good axial compressor design is illustrated in Fig.2.

Within the EU Framework Program 6 research project NEWAC (NEW Aero engine Core concepts, 2011) a number of advanced core component technologies have been investigated that include (Rolt & Kyprianidis, 2010; Wilfert et al., 2007):

- Improved high pressure compressor aero design and blade tip rub management.
- Flow control technologies including aspirated compression systems.
- Active control of surge and tip clearance in compressors.
- Active control of a cooled cooling air system.

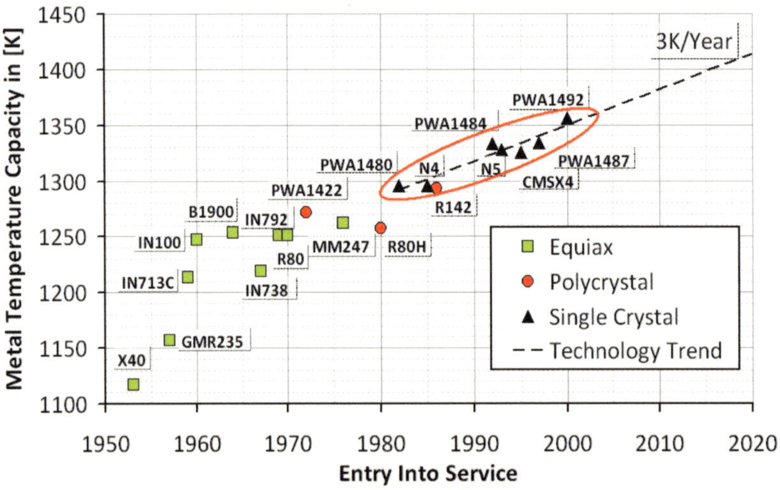

Fig. 3. Evolution of turbine material capability and future trend.

As mentioned earlier another way of improving engine thermal efficiency is to raise the cycle OPR. For conventional cores, increasing OPR and T_4 depends primarily on future advancements in material and cooling technology. The evolution of turbine material capability over a period of 50 years is illustrated in Fig. 3. As can be observed, only mild improvements have been achieved so far and this seems to be a continuing trend; the potential introduction of ceramics would form a major improvement in the field, but substantially more research is still required before realising this. Despite the low improvement rate in turbine material technology (roughly 3 [K/year]) aero engine designs have seen substantial increases in T_4 over the last 60 years (roughly 10 [K/year]); this is illustrated in Fig. 4 for engines designed for long-haul applications. The main reason behind these improvements in T_4 has been the introduction of cooling and Thermal Barrier Coatings (TBC) in turbine designs; the interested reader is referred to Downs & Kenneth (2009) for a good overview of the evolution of turbine cooling systems design.

It is perhaps debatable whether an improvement rate of 10 [K/year] in T_4 can be maintained in the future, and for that reason the design focus for more aggressive thermal efficiency improvements could very well be redirected to the introduction of heat-exchanged cores and advanced compressor technologies for future turbofan designs. In that respect, some of the technologies researched under the NEWAC project can be perceived as intermediate enabling steps for realising new engine core concepts that could improve the core thermal efficiency. These new core concepts comprise of:

- Ultra-high OPR core with intercooling.
- Medium OPR intercooled recuperated core.
- High OPR flow controlled core.
- High OPR active core including active cooling air cooling.

When considering intercooling for an aero engine design, a common textbook misconception is that the thermal efficiency of an intercooled core will always be lower than a conventional

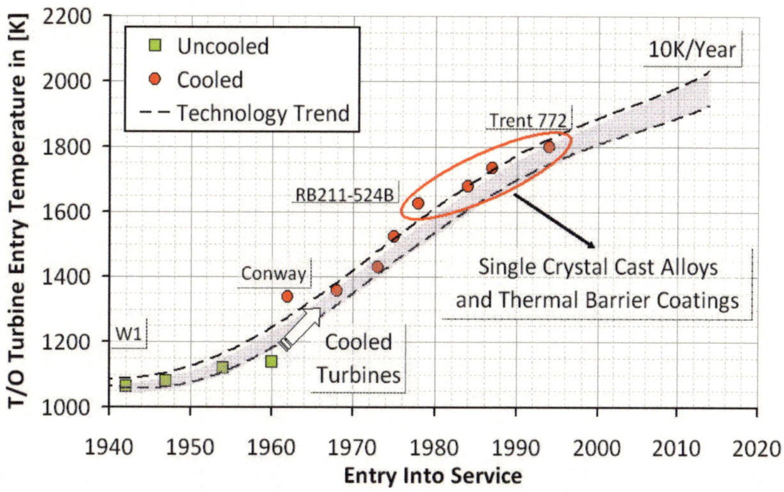

Fig. 4. Evolution of turbine entry temperature and future trend.

core's for a fixed OPR and specific thrust (Saravanamuttoo et al., 2001). The argument behind this is that the heat removed by the intercooler will largely need to be reintroduced in the combustor by burning more fuel, while the reduction in compression work and increase in bypass stream thrust (due to the heat rejection) will only partially compensate for the loss in cycle efficiency, at a fixed specific thrust and T_4. Adding the expected intercooler pressure losses in the cycle calculations would further worsen the SFC deficit and make the increase in specific thrust less marked.

However, cycle calculations based on half-ideal gas properties and no dissociation (i.e. isobaric heat capacity dependent on temperature), presented by Walsh & Fletcher (1998), give a slightly different picture on intercooling. For a given T_4, the optimal OPR for an intercooled core will be much higher than that for a conventional core. Comparing the two concepts at their optimal OPR levels, for a given technology level, can make the intercooled core more attractive with respect to thermal efficiency and not just specific thrust. Canière et al. (2006) and da Cunha Alves et al. (2001) also reached the same conclusion about the thermal efficiency of the intercooled cycle while studying this concept for gas turbines used in power generation. Papadopoulos & Pilidis (2000) worked on the introduction of intercooling, by means of heat pipes, in an aero engine design for long haul applications. Xu et al. (2007) performed a mission optimization to assess the potential of a tubular intercooler. Recent work by Xu & Grönstedt (2010) presents a refined tubular configuration estimating a potential block fuel benefit of 3.4%. The work addresses the limitation that short high pressure compressor blade lengths and related low compression efficiencies may impose on engines designed for short range missions, and suggest a novel gas path layout as a remedy to this constraint. A design study of a high OPR intercooled aero engine is described in Rolt & Baker (2009), while details on the aerodynamic challenges in designing a duct system to transfer the core air into and out of the intercooler are presented by Walker et al. (2009).

The introduction of recuperation in an aero engine, for high thermal efficiency at low OPR, has also been the focus of different researchers. Lundbladh & Sjunnesson (2003) performed a feasibility study for InterCooled (IC) and Intercooled Recuperated Aero engines (IRA) that consider cycle benefits, weights and direct operating costs. Boggia & Rud (2005) provide an extended discussion on the thermodynamic cycle and the technological innovations necessary for realizing the intercooled recuperated core concept. Various aspects of the thermo-mechanical design of a compact heat exchanger have been presented by Pellischek & Kumpf (1991) and Schoenenborn et al. (2006). For a comprehensive review on the development activities for recuperated aero engines since the late 60's the interested reader can refer to McDonald et al. (2008a;b;c).

Finally, three different types of lean-burn combustor technology were also researched within NEWAC with the objective of reducing emissions of oxides of nitrogen (NO_x):

- Lean Direct Injection (LDI) combustor for high and ultra-high OPR cores.
- Partial Evaporation and Rapid Mixing (PERM) combustor for high OPR cores.
- Lean Premixed Pre-vaporized (LPP) combustor for medium OPR cores.

4. Design space exploration

4.1 Methodology, design feasibility and constraints

To effectively explore the design space a tool is required that can consider the main disciplines typically encountered in conceptual design. The prediction of engine performance, aircraft design and performance, direct operating costs, and emissions for the concepts analysed in this study was made using the EVA code (Kyprianidis et al., 2008). Another code, WeiCo, was also used for carrying out mechanical and aerodynamic design in order to derive engine component weight and dimensions. The two tools have been integrated together within an optimiser environment as illustrated in Fig. 5, based on lessons learned from the development of the TERA2020 tool (Kyprianidis et al., 2011). This integration allows for multi-objective optimization, design studies, parametric studies, and sensitivity analysis. In order to speed up the execution of individual engine designs, the conceptual design tool attempts to minimize internal iterations in the calculation sequence through the use of an explicit algorithm, as described in detail by Kyprianidis (2010).

Aero-engine designs are subject to a large number of constraints and these need to be considered during conceptual design. Constraints can be applied within the optimiser environment at the end of the calculation sequence i.e., after the last design module has been executed. During a numerical optimisation, the optimiser will select a new set of input design parameters for every iteration and the resulting combination of aircraft and engine will be assessed. Using user specified objective functions the optimiser will home in on the best engine designs, determining the acceptability/feasibility of each design through the constraints set by the user. Infeasible designs will be ruled out, while non-optimum design values will result in engine designs with non-optimum values for the objective function selected. The optimiser will therefore avoid regions in the design pool that result in infeasible or non-optimum engine designs.

Design constraints set by the user include among others:

- Take-off HPC delivery temperature and other important performance parameters.
- FAR (Federal Aviation Regulations) take-off field length for all engines operating and balanced field length for one engine inoperative conditions.
- Time to height.

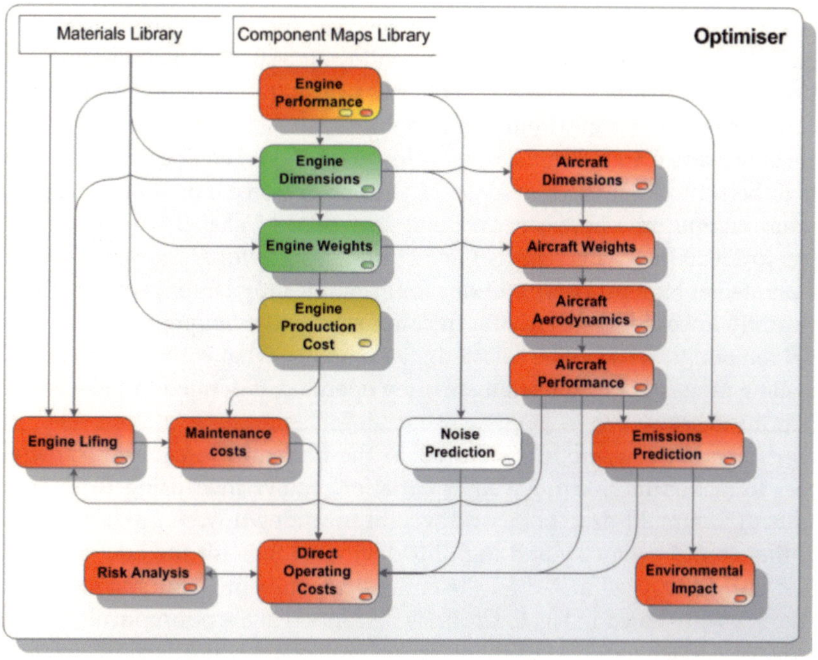

Fig. 5. Conceptual design tool algorithm (Kyprianidis, 2010).

- LTO (Landing and Take-Off) cycle D_pNO_x/F_{oo} vs. ICAO certification limits and CAEP medium and long term goals.
- Cumulative EPNL vs. ICAO certification limits.
- Engine time between overhaul.

Where component design is concerned, for a conventional core the High Pressure Compressor (HPC) delivery temperature, and hence the engine OPR, is typically constrained by the mechanical properties of the HPC disc or HPC rear drive cone or High Pressure Turbine (HPT) disc material (Rolt & Baker, 2009). For an intercooled core, the OPR value is no longer constrained by a maximum allowable HPC delivery temperature. Nevertheless, the intercooling process increases the air density in the gas path and as a result the compressor blades tend to become smaller. Losses from tip clearances become increasingly important and a minimum compressor blade height limitation needs to be applied to maintain state of the art compressor efficiency. Core architecture selections for the conventional core set an upper limit to the HPC design pressure ratio that can achieved when driven by a single-stage HPT. A transonic single-stage HPT design can allow for relatively higher HPC pressure ratios at the expense of a lower polytropic efficiency. A two-stage HPT can offer high HPC pressure ratios at a high polytropic efficiency but a trade-off arises with respect to the need for more cooling air and increased engine length associated with the introduction of a second row of vanes and blades. With respect to the intercooled core, the minimum design pressure ratio for the Intermediate Pressure Compressor (IPC) can in some cases be limited by icing considerations during the descent flight phase. The maximum area variation that may be achieved by the variable area auxiliary nozzle is also constrained by mechanical (and aerodynamic) considerations.

As discussed earlier, designing a combustor at very low air to fuel ratio levels is also limited by the need for adequate combustor liner film-cooling air as well as maintaining an acceptable temperature traverse quality (Lefebvre, 1999); this sets an upper bound on combustor outlet temperature. Furthermore, a maximum permissible mean metal temperature needs to be set to consider turbine blade material limitations. A lower bound on engine time between overhaul also needs to be set to limit the frequency of workshop visits. For short range applications the minimum engine time between overhaul was set to 18000 [hr] while for long range applications to 23000 [hr]. This reflects the fact that designs for short range applications are typically operated at high power conditions for a significantly larger part of their operational life. Significantly lower levels of maximum combustor outlet temperature and turbine blade mean metal temperature had to be selected, compared to what could be selected for engine designs for long range applications that are often operated at derated thrust levels and spend most of their life at cruise.

A rubberised aircraft wing model was used in these studies to capture 'snowball effects' with respect to maximum take-off weight variation, rather than using fixed engine thrust requirements. The aircraft drag polar and weight breakdown were predicted at component level from the aircraft geometry and high lift device settings for the take-off and approach phases. Fuel burned was calculated for the entire flight mission including reserves assuming ISA conditions, as illustrated in Fig. 6. Cruise is performed at the optimum altitude for specific range (fixed cruise Mach number) using a step-up cruise procedure as the aircraft gets lighter. A comprehensive take-off field length calculation is performed for all engines operating and one engine inoperative conditions up to 1500 [ft].

Two baseline aircraft models have been used herein; one model for long range applications and one for short range. The former model is largely based on public domain information available for the Airbus A330-200 while the latter model is based on the Airbus A320-200. The short range aircraft was designed to carry 150 [pax] for a distance of 3000 [nmi] and a typical business case of 500 [nmi]; for long range applications it was designed for 253 [pax], 6750 [nmi] and 3000 [nm], respectively. For the step-up cruise procedure, a minimum residual rate of climb of 300 [ft/min] was set as a constraint for flying at the cruise altitude for maximum specific range.

The maximum values for FAR take-off field length and time to height were set for a load factor of 1 and no cargo. The choice of both is based on customer operational requirements as the aircraft needs to be able to: (i) take-off from a large number of airports around the world and (ii) climb to the initial cruise altitude sufficiently fast to ease operations with local air traffic control (and hence reduce waiting time on the ground). A cumulative distribution of the world's major runway lengths, based on data from Jenkinson et al. (1999), is illustrated in Fig. 7. For short range applications fairly stringent constraints are typically set for the maximum take-off distance and time to height; in this study these were set to 2.0 [km] and 25 [min], respectively. For long range applications a maximum take-off distance of 2.5 [km] was set instead. Stringent constraints result in bigger engines but allow for greater flexibility for engine derating at a smaller block fuel cost.

The choice of load factor and cargo is considered sensible but it does not necessarily constitute a typical airline practice. Validating absolute block fuel predictions with public domain airline data is not a trivial task as different airlines will follow different operational practices. For example for the long range aircraft model, the business case prediction is 10% lower than the published annually-averaged value, given in [lt/(km*pax)], by SwissAir for 2009 for the

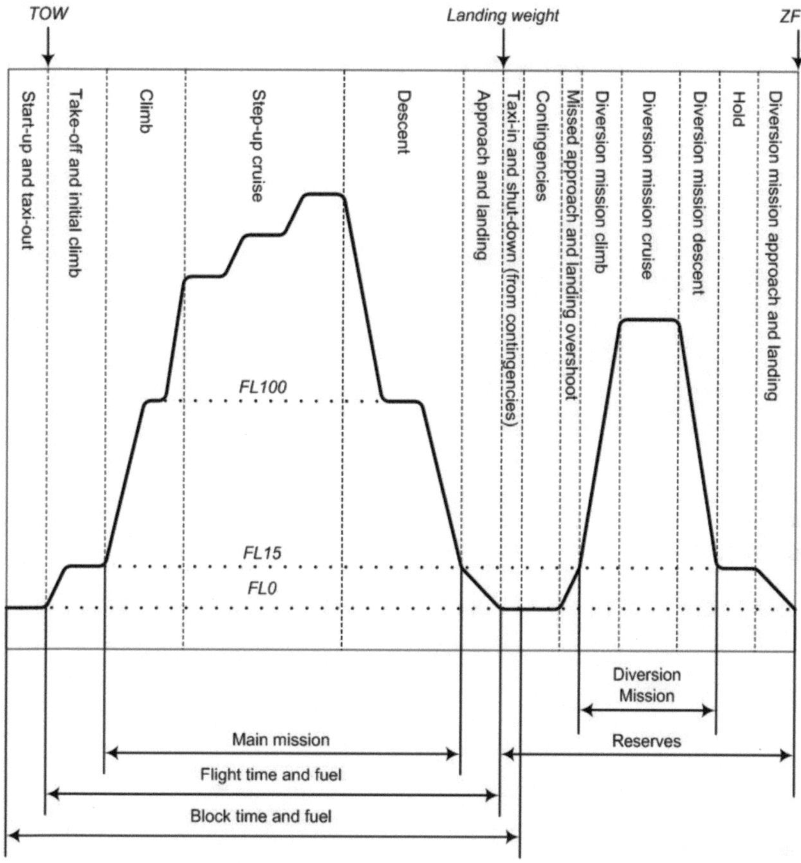

Fig. 6. Typical flight cycle.

Airbus A330-200 (Swiss International Air Lines, 2009). This does not necessarily mean that the model's business case is not a realistic one; nor that it wouldn't fit well with operational practices followed by other airlines. Furthermore, regional Air Traffic Management (ATM) practices can skew available block fuel data, while global ATM regulations may very well change significantly by 2020. It should be noted that fuel planning within the model respects the requirements defined for international flights by Federal Aviation Administration (n.d.) and Joint Aviation Authorities (2008).

Where conceptual design is concerned, exchange rates are perhaps a better type of parameter for evaluating the accuracy of a rubberised wing model, rather than just simply comparing absolute values. Block fuel exchange rates produced with the rubberised wing baseline aircraft models are presented in Table 1 for the business case of the long and short range models and are considered reasonable numbers.

During a block fuel optimization all engine aircraft combinations which do not fulfil the take-off and time to height criteria set will be discarded as infeasible. Due to the underlying physics, this will naturally lead to an optimal engine and aircraft combination for the defined objective function. All large engines will produce heavier aircraft with more drag and thus higher block fuel weight. Engines which are too small will not deliver enough thrust to satisfy the take-off and time to height criteria set.

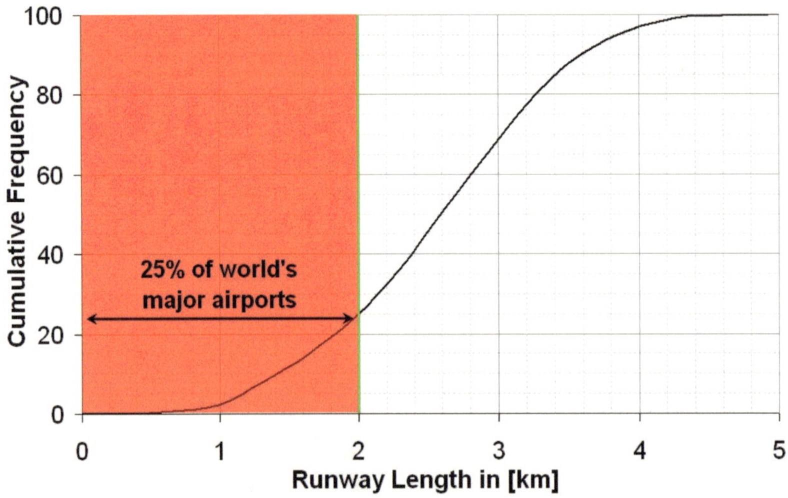

Fig. 7. Cumulative distribution of world's major runway lengths (based on data from Jenkinson et al. (1999)).

	Exchange rate	
Perturbation	Long range	Short range
1000 [kg] weight penalty	0.73%	1.26%
+1% SFC	1.28%	1.09%

Table 1. Block fuel exchange rates using the baseline long range and short range rubberised wing aircraft models.

4.2 Engine design optimality

Whereas optimisation constraints can help ensure the feasibility of an engine design, they do little to help with it's optimality. The optimality of the engine design will depend on the careful selection of the figures of merit used during the optimisation process, such as minimum block fuel, maximum time between overhaul, minimum direct operating costs, minimum noise and LTO NO_x emissions etc.

Determining the optimal aero-engine design is essentially the subject of a multi-objective optimisation, and therefore Pareto fronts need typically be constructed to visualize the region of optimal designs within the design space. A simplified example of utilizing the tool for design space exploration, with active constraints, is illustrated in Fig. 8. In principle, nacelle drag should also be added as a third dimension when plotting design space exploration results that consider varying levels of specific thrust, but this has been omitted here in order to simplify the plot. The aircraft exchange rates for the baseline design were used for plotting a constant block fuel line (ignoring nacelle drag effects and nonlinearities) and this iso-line therefore defines, in a simple manner, the boundaries of trading specific fuel consumption for weight. During a block fuel optimization, the optimizer continuously evaluates different engine designs as it searches for the optimal solution. Designs that fail to meet constraints set by the user are discarded and have been labeled as infeasible in the plot.

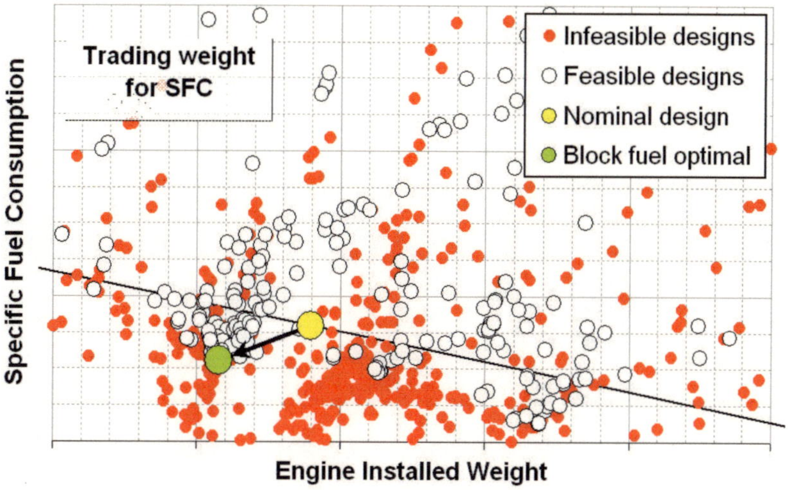

Fig. 8. Visualization example of constrained design space exploration.

4.3 Economic considerations

Safety considerations aside, civil aero engine design has been driven primarily by economic considerations even from its fairly early days. A testament to this has been the advent of the world's first commercial jet-airliner, the de Havilland Comet, powered by 4 Rolls-Royce Avon turbojet engines. Although it burned nearly four times as much fuel compared to piston-driven engines, it's business case was very strong since it permitted significantly higher flight speeds resulting in reduced flight times (i.e. a better airline product) and increased aircraft annual utilization. Furthermore, the excellent power to weight ratio of the turbojet engine meant that it could be used to power aircrafts with significantly higher passenger capacities than what was feasible before. The evolution of aircraft transport efficiency since the late 30's is summarised in Fig. 9 based on data from Avellán (2008).

The aero engine designs proposed herein have been optimized for minimum block fuel for a given aircraft mission (business case), which implies minimum global warming impact if one considers CO_2 emissions alone. The market competitiveness of these fuel optimal designs however is highly dependent on the development of jet fuel prices in the years to come until 2020. The volatility of jet fuel price over the last 10 years is illustrated in Fig. 10. A further economic consideration for European markets may also be the development of the Euro/US$ exchange rate, as well as interest and inflation rates.

For the economic calculations conducted in this study certain assumptions were made. The assumed jet fuel price was 172c$/US gallon. It is worth noting that at the time of writing the average jet fuel price was 320 [c$/US gallon] (International Air Transport Association, 2011; Platts, 2011). Interest and inflation rates were assumed to be 6% and 2%, respectively, while the US$ to Euro exchange rate was assumed to be 0.8222.

It is worth noting that an increase in inflation rates from 2% to 3% can increase the net present cost by as much as 17%, over a period of 30 years. An increase in interest rates from 6% to 7% can increase Direct Operating Costs (DOC) by 2.5% and 4.5% for short and long range applications, respectively.

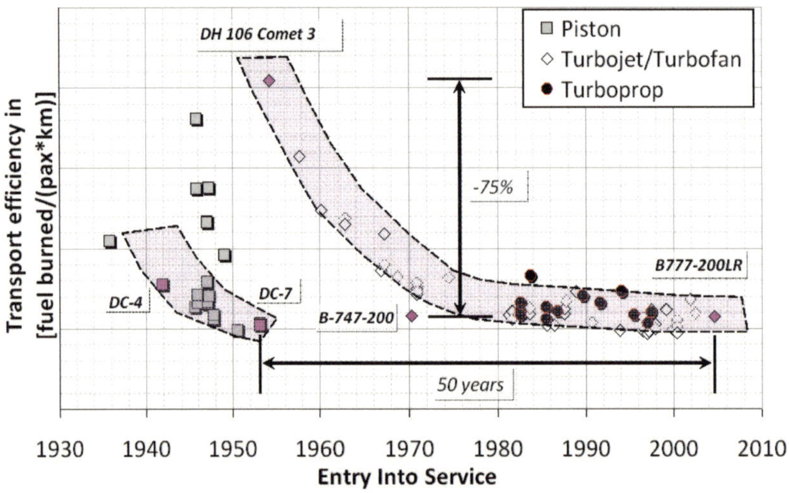

Fig. 9. Evolution of aircraft transport efficiency (based on data from Avellán (2008)).

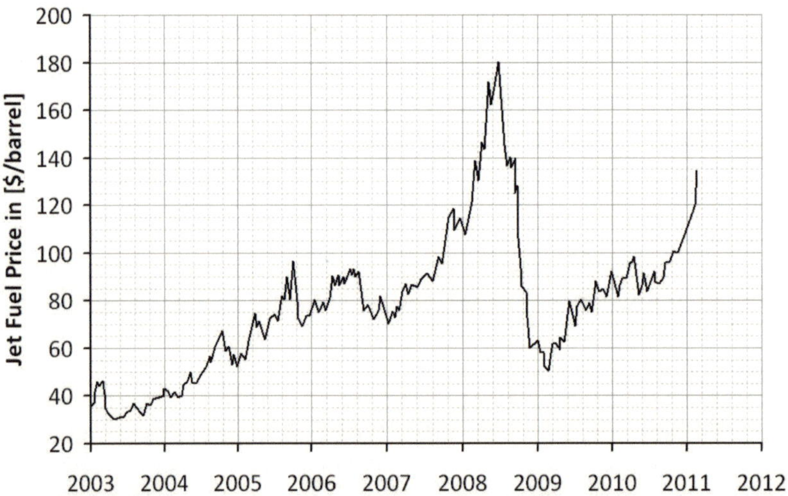

Fig. 10. Long term perspective of jet fuel price movements (based on data from International Air Transport Association (2011) and Platts (2011)).

The cost of fuel as a fraction of the total DOC was predicted to be 13% and 19% for short and long range applications, respectively. An increase in block fuel by 1% translates in an increase of 0.13% and 0.19% in DOC, respectively, and as can be observed it is directly dependent on the ratio of fuel cost over DOC. A doubling of the fuel price would change this ratio to roughly 23% and 32%, respectively, and would also result in 13% and 19% higher DOC levels, respectively.

Higher levels of DOC, as a result of a significant increase in fuel price, would most probably be absorbed by airlines through an increase in fares. This could make fuel efficient designs increasingly market competitive, as the DOC optimal designs would further approach the fuel optimal designs. It would therefore be worthwhile to redirect further research investments towards developing fuel efficient aero engine designs, as has also been the case in the late 70's and through large part of the 80's. The introduction of carbon taxes could also have a similar effect.

5. Summary of design space exploration results

A summary of three different design space exploration case studies using the tools and algorithm described is presented here. This work has looked at the potential block fuel benefits resulting from the introduction of:

- An intercooled core in a direct drive UHBR turbofan configuration.
- An intercooled recuperated core in a geared UHBR turbofan configuration.
- An open rotor propulsor in a geared pusher configuration.

The thrust requirements for the first two concepts are for an engine designed to power the long range aircraft model while the latter concept is centered around powering the short range aircraft model. More details on these studies can be found in Kyprianidis et al. (2011) and Larsson et al. (2011).

For the intercooled core assessment, a year 2020 Entry Into Service (EIS) turbofan engine with a conventional core was set up as the baseline. The intercooled core engine is an ultra high OPR design with also year 2020 EIS level of technology, and features a tubular heat-exchanger, while the fan for both engines has the same diameter and flow per unit of area. Business case block fuel benefits of approximately 3.2% are predicted for the intercooled engine, mainly due to the reduced engine weight and the core's higher thermal efficiency which results in a better SFC. **These intercooling benefits are highly dependent on achieving technology targets such as low intercooler weight and pressure losses**; the predicted lower dry weight, compared to the conventional core engine, can be attributed to various reasons. The intercooler weight penalty is largely compensated by the higher core specific output allowing a smaller core size and hence a higher BPR at a fixed thrust and fan diameter. The high OPR provides an additional sizing benefit, for components downstream of the HPC, by reducing further the corrected mass flow and hence flow areas. The intercooled core Low Pressure Turbine (LPT) was designed in this study with one less stage which reduced both engine weight and length, despite the high cycle OPR requiring a greater number of HPC stages. These observations are summarised in Table 2 with the added components weight group considering the intercooler and its installation standard; this group is not considered in the core weight group which also does not consider the core nozzle or the LPT and its casing.

For the intercooled recuperated core assessment, a year 2000 EIS turbofan engine with a conventional core was set up as the baseline. The intercooled recuperated core configuration is an UHBR design with a year 2020 level of technology. Significant business case block fuel benefits of nearly 22% are predicted for the geared intercooled recuperated core engine due to its higher thermal and propulsive efficiency. The use of HPT cooling air bled from the recuperator exit (Boggia & Rud, 2005; Walsh & Fletcher, 1998) results in a 1.3% SFC improvement due to more energy being recuperated from the exhausts, at a fixed effectiveness level - and despite the considerable increase in cooling air requirements (+3.5% of core mass flow). The predicted dry weight for the intercooled recuperated configuration is higher compared to the conventional core engine. There is a weight benefit from the use of EIS

	Conventional core DDTF LR EIS 2020	Intercooled core DDIC LR EIS 2020
Engine dry weight	Ref.	-5.9%
LPT weight	Ref.	-27.1%
Core weight	Ref.	-32.5%
Added components weight (as % of engine dry weight)	-	7.7%
Block fuel weight	Ref.	-3.2%
Mid-cruise SFC	Ref.	-1.5%
Thermal efficiency	Ref.	+0.007
Propulsive efficiency	Ref.	+0.000

Table 2. Comparison of an intercooled engine with a conventional core turbofan engine at aircraft system level.

	Conventional core BASE LR EIS 2000	Intercooled recuperated core IRA LR EIS 2020
Thrust/weight	Ref.	-12%
Engine dry weight	Ref.	+16.5%
Nacelle weight	Ref.	+29.7%
Fan weight	Ref.	+36.6%
LPT weight	Ref.	-17.1%
Added components weight (as % of engine dry weight)	-	25.4%
Block fuel weight	Ref.	-21.6%
Mid-cruise SFC	Ref.	-18.3%
Thermal efficiency	Ref.	+0.024
Propulsive efficiency	Ref.	+0.120

Table 3. Comparison of an intercooled recuperated engine with a conventional core turbofan engine at aircraft system level.

2020 light-weight materials in most major engine components, as well as from the high speed LPT - due to the reduced stage count. Also, the relatively low engine OPR and the use of an intercooler increases core specific output, resulting in a smaller core. The introduction however of the gearbox, intercooler and recuperator components inevitably results in a significant weight penalty. It should be noted that a lower level of specific thrust, and hence a larger fan diameter, has been assumed for the intercooled recuperated core engine; this results in both a heavier fan and a heavier nacelle. These observations are summarised in Table 3 with the added components weight group considering the intercooler and recuperator and their installation standard, as well as the gearbox.

For the geared open rotor assessment, a year 2020 EIS geared turbofan engine with a conventional core was set up as the baseline. The geared open rotor concept design also assumes year 2020 EIS level of technology, and features two counter-rotating propellers in a pusher configuration powered by a geared low pressure turbine. Significant business case

	Geared turbofan GTF SR EIS 2020	Geared open rotor GOR SR EIS 2020
Engine installed weight	Ref.	+11%
Nacelle weight	Ref.	-88%
Fan/propeller weight	Ref.	+73%
LPT weight	Ref.	+20%
Core weight	Ref.	-31%
Block fuel weight	Ref.	-15%
Mid-cruise SFC	Ref.	-14%
Thermal efficiency	Ref.	-0.013
Propulsive efficiency	Ref.	+0.16

Table 4. Comparison of a geared open rotor engine with a geared turbofan engine at aircraft system level.

block fuel benefits of nearly 15% are predicted for the geared open rotor engine primarily due to its higher propulsive efficiency. Although, the geared turbofan engine benefits from a better thrust to weight ratio it suffers from significantly higher nacelle drag losses, compared to the open rotor design. These observations are summarised in Table 3.

A NO_x emissions assessment of the presented engine configurations has been performed and is illustrated in Fig. 11. The same combustor concept has been considered for both designs i.e., conventional Rich-burn/Quick-quench/Lean-burn (RQL) combustion technology. The results obtained are compared against ICAO Annex 16 Volume II legislative limits (ICAO, 1993), as well as the Medium Term (MT) and Long Term (LT) technology goals set by CAEP (Newton et al., 2007). Balloons have been used to indicate the uncertainty in the NO_x predictions due to the lower technology readiness level associated with the introduction of

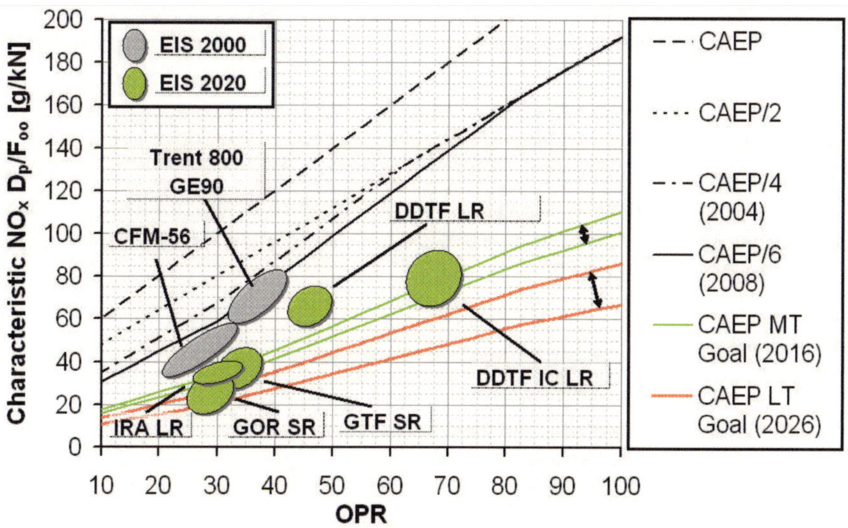

Fig. 11. NO_x emissions assessment for different future aero engine design concepts.

such combustor designs in the proposed future cycles. A sufficient margin against the ICAO CAEP/6 LTO cycle NO_x certification limit may be achieved for all the configurations that have been assessed assuming year 2020 EIS.

6. Conclusions

The research work presented started by reviewing the evolution of the aero engine industry's vision for the aero engine design of the future. Appropriate research questions were set that can influence how this vision may further involve in the years to come. Design constraints, material technology, customer requirements, noise and emissions legislation, technology risk and economic considerations and their effect on optimal concept selection were also discussed in detail.

With respect to addressing these questions, several novel engine cycles and technologies - currently under research - were identified. It was shown that there is a great potential to reduce fuel consumption for the different concepts identified, and consequently decrease the CO_2 emissions. Furthermore, this can be achieved with a sufficient margin from the ICAO NO_x certification limits, and in line with the medium term and long term goals set by CAEP. It must be noted however that aero engine design is primarily driven by economic considerations. As fuel prices increase, the impact of fuel consumption on direct operating costs also increases. The question therefore rises:

> **Can the potential reduction in fuel consumption and direct operating costs outweigh the technological risks involved in introducing novel concepts into the market?**

The answer is left to be given by the choices the aero engine industry makes in the years to come.

7. Acknowledgements

The author is grateful to Richard Avellán (Volvo Aero) for providing the transport efficiency data used in Fig. 9. Stimulating discussions with A.M. Rolt (Rolls-Royce), J.A. Borradaile, S. Donnerhack (MTU Aero Engines), P. Pilidis, (Cranfield University), R. Singh, (Cranfield University), S.O.T. Ogaji, (Cranfield University), P. Giannakakis (Cranfield University), T. Grönstedt (Chalmers University), A. Lundbladh (Volvo Aero) and L. Larsson (Volvo Aero) on advanced concepts and aero engine design are gratefully acknowledged. Finally, the author would like to thank the reviewers of this work for their constructive suggestions to improve the overall quality and clarity of the article.

8. Nomenclature

OPR	Engine overall pressure ratio
SFC	Engine specific fuel consumption
T_4	Combustor outlet temperature

9. References

Advisory Council for Aeronautical Research in Europe (2001). European Aeronautics: A Vision for 2020 – Meeting Society's Needs and Winning Global Leadership. See also URL http://www.acare4europe.org.

Avellán, R. (2008). *Towards Environmentally Friendly Aero Engines*, Licentiate thesis, Chalmers University of Technology, Göteborg, Sweden.

Bates, R. & Morris, J. (1983). A McDonnell Douglas Perspective – Commercial Aircraft for the Next Generation, *AIAA Aircraft Design, Systems and Technology Meeting Proceedings, AIAA-83-2502*, Fort Worth, Texas, USA.

Birch, N. (2000). 2020 vision: the prospects for large civil aircraft propulsion, *RAeS Aeronautical Journal* pp. 347–352.

Boggia, S. & Rud, K. (2005). Intercooled Recuperated Gas Turbine Engine Concept, *Proceedings of 41st AIAA/ASME/SAE/ASEE Joint Propulsion Conference and Exhibit, AIAA 2005-4192*, Arizona, USA.

Borradaile, J. (1988). Towards the optimum ducted UHBR engine, *Proceedings of AIAA/SAE/ASME/ASEE 24th Joint Propulsion Conference, AIAA-89-2954*, Boston, Massachusetts, USA.

Canière, H., Willcokx, A., Dick, E. & De Paepe, M. (2006). Raising cycle efficiency by intercooling in air-cooled gas turbines, *Applied Thermal Engineering* 26(16): 1780–1787.

Clean Sky Joint Technology Initiative (2011). http://www.cleansky.eu.

da Cunha Alves, M., de Franca Mendes Carneiro, H., Barbosa, J., Travieso, L., Pilidis, P. & Ramsden, K. (2001). An insight on intercooling and reheat gas turbine cycles, *Proceedings of the Institution of Mechanical Engineers, Part A: Journal of Power and Energy* 215(2): 163–171.

Downs, J. & Kenneth, K. (2009). Turbine Cooling Systems Design – Past, Present and Future, *ASME TURBO EXPO 2009 Proceedings, GT2009-59991*, Orlando, Florida.

enVIronmenTALly friendly aero engines (2009). http://www.project-vital.org.

Federal Aviation Administration (n.d.). Federal Aviation Regulation Part 121 - Operating Requirement: domestic, flag and supplemental operations, *FAR 121*, Washington, DC, USA.

GE36 Design and Systems Engineering (1987). Full-scale technology demonstration of a modern counterrotating unducted fan engine concept - engine test, *NASA-CR-180869*, GE Aircraft Engines, Cincinnati,, OH, USA.

Gray, D. & Witherspoon, J. (1976). Fuel conservative propulsion concepts for future air transports, *Technical Report SAE-760535*, Society of Automotive Engineers.

Hirschkron, R. & Neitzel, R. (1976). Alternative concepts for advanced energy conservative transport engines, *Technical Report SAE-760536*, Society of Automotive Engineers.

Horlock, J., Watson, D. & Jones, T. (2001). Limitations on Gas Turbine Performance Imposed by Large Turbine Cooling Flows, *ASME Journal of Engineering for Gas Turbines and Power* 123(3): 487–494.

ICAO (1993). International Standards and Recommended Practices - Environmental Protection, Annex 16 to the Convention on International Civil Aviation, Volume II - Aircraft Engine Emissions, *2nd edition plus ammendments*, Montreal, Canada.

International Air Transport Association (2011). http://www.iata.org.

Jackson, A. (1976). Some Future Trends in Aero Engine Design for Subsonic Transport Aircraft, *ASME Journal of Engineering for Power* 98: 281–289.

Jackson, A. (2009). *Optimisation of Aero and Industrial Gas Turbine Design for the Environment*, PhD thesis, Cranfield University, Cranfield, Bedfordshire, United Kingdom.

Jenkinson, L., Simpkin, P. & Rhodes, D. (1999). *Civil Jet Aircraft Design*, 1st edn, Arnold, London, United Kingdom.

Joint Aviation Authorities (2008). Joint Airworthiness Requirements OPS Part 1 – Commercial Air Transportation (Aeroplanes), *Ammendment 14*, Hoofddorp, The Netherlands.

Korsia, J.-J. (2009). VITAL – European R&D Programme for Greener Aero-Engines, *ISABE 2009 Proceedings, ISABE-2009-1114*, Montreal, Canada.

Korsia, J.-J. & Guy, S. (2007). VITAL – European R&D Programme for Greener Aero-Engines, *ISABE 2007 Proceedings, ISABE-2007-1118*, Beijing, China.

Kurzke, J. (2003). Achieving maximum thermal efficiency with the simple gas turbine cycle, *Proceedings of 9th CEAS European Propulsion Forum: "Virtual Engine – A Challenge for Integrated Computer Modelling"*, Rome, Italy.

Kyprianidis, K. (2010). *Multi-disciplinary Conceptual Design of Future Jet Engine Systems*, PhD thesis, Cranfield University, Cranfield, Bedfordshire, United Kingdom.

Kyprianidis, K., Colmenares Quintero, R., Pascovici, D., Ogaji, S., Pilidis, P. & Kalfas, A. (2008). EVA - A Tool for EnVironmental Assessment of Novel Propulsion Cycles, *ASME TURBO EXPO 2008 Proceedings, GT2008-50602*, Berlin, Germany.

Kyprianidis, K., Grönstedt, T., Ogaji, S., Pilidis, P. & Singh, R. (2011). Assessment of Future Aero-engine Designs with Intercooled and Intercooled Recuperated Cores, *ASME Journal of Engineering for Gas Turbines and Power* 133(1). doi:10.1115/1.4001982.

Larsson, L., Grönstedt, T. & Kyprianidis, K. (2011). Conceptual Design and Mission Analysis for a Geared Turbofan and an Open Rotor Configuration, *ASME TURBO EXPO 2011 Proceedings, GT2011-46451*, Vancouver, Canada.

Lefebvre, A. (1999). *Gas Turbine Combustion*, 2nd edn, Taylor & Francis, PA, USA.

Swiss International Air Lines, (2009). Flying Smart, *Swiss Magazine, Issue 12.2009 / 1.2010* pp. 100–105.

Lundbladh, A. & Sjunnesson, A. (2003). Heat Exchanger Weight and Efficiency Impact on Jet Engine Transport Applications, *ISABE 2003 Proceedings, ISABE-2003-1122*, Cleveland, USA.

McDonald, C., Massardo, A., Rodgers, C. & Stone, A. (2008a). Recuperated gas turbine aeroengines, part I: early development activities, *Aircraft Engineering and Aerospace Technology: An International Journal* 80(2): 139–157.

McDonald, C., Massardo, A., Rodgers, C. & Stone, A. (2008b). Recuperated gas turbine aeroengines, part II: engine design studies following early development testing, *Aircraft Engineering and Aerospace Technology: An International Journal* 80(3): 280–294.

McDonald, C., Massardo, A., Rodgers, C. & Stone, A. (2008c). Recuperated gas turbine aeroengines, part III: engine concepts for reduced emissions, lower fuel consumption, and noise abatement, *Aircraft Engineering and Aerospace Technology: An International Journal* 80(4): 408–426.

NEW Aero engine Core concepts (2011). `http://www.newac.eu`.

Newton, P., Holsclaw, C., Ko, M. & Ralph, M. (2007). Long Term Technology Goals for CAEP/7. presented to the Seventh Meeting of CAEP.

Papadopoulos, T. & Pilidis, P. (2000). Introduction of Intercooling in a High Bypass Jet Engine, *ASME TURBO EXPO 2000 Proceedings, 2000-GT-150*, Munich, Germany.

Peacock, N. & Sadler, J. (1992). Advanced Propulsion Systems for Large Subsonic Transports, *ASME Journal of Propulsion and Power* 8(3): 703–708.

Pellischek, G. & Kumpf, B. (1991). Compact Heat Exchanger Technology for Aero Engines, *ISABE 1991 Proceedings, ISABE-91-7019*, Nottingham, United Kingdom.

Platts (2011). http://www.platts.com.

Pope, G. (1979). Prospects for reducing the fuel consumption of civil aircraft, *RAeS Aeronautical Journal* pp. 287–295.

Rolt, A. & Baker, N. (2009). Intercooled Turbofan Engine Design and Technology Research in the EU Framework 6 NEWAC Programme, *ISABE 2009 Proceedings, ISABE-2009-1278*, Montreal, Canada.

Rolt, A. & Kyprianidis, K. (2010). Assessment of New Aero Engine Core Concepts and Technologies in the EU Framework 6 NEWAC Programme, *ICAS 2010 Congress Proceedings, Paper No. 408*, Nice, France.

Ruffles, P. (2000). The future of aircraft propulsion, *Proceedings of the IMechE, Part C: Journal of Mechanical Engineering Science* 214(1): 289–305.

Saravanamuttoo, H. (2002). The Daniel and Florence Guggenheim Memorial Lecture - Civil Propulsion; The Last 50 Years, *ICAS 2002 Congress Proceedings*, Toronto, Canada.

Saravanamuttoo, H., Rogers, G. & Cohen, H. (2001). *Gas Turbine Theory*, 5th edn, Pearson Education Limited, United Kingdom.

Schimming, P. (2003). Counter Rotating Fans – An Aircraft Propulsion for the Future, *Journal of Thermal Science* 12(2): 97–103.

Schoenenborn, H., Ebert, E., Simon, B. & Storm, P. (2006). Thermomechanical Design of a Heat Exchanger for a Recuperated Aeroengine, *ASME Journal of Engineering for Gas Turbines and Power* 128(4): 736–744.

Sieber, J. (1991). Aerodynamic Design and Experimental Verification of an Advanced Counter-Rotating Fan for UHB Engines, *Third European Propulsion Forum*, Paris, France.

Swihart, J. (1970). The Promise of the Supersonics, *AIAA 7th Annual Meeting and Technical Display Proceedings, AIAA 70-1217*, Houston, Texas, USA.

valiDation of Radical Engine Architecture systeMs (2011). http://www.dream-project.eu.

Walker, A., Carrotte, J. & Rolt, A. (2009). Duct Aerodynamics for Intercooled Aero Gas Turbines: Constraints, Concepts and Design Methododology, *ASME TURBO EXPO 2009 Proceedings, GT2009-59612*, Orlando, Florida.

Walsh, P. & Fletcher, P. (1998). *Gas Turbine Performance*, 1st edn, Blackwell Science, United Kingdom.

Watts, R. (1978). European air transport up to the year 2000, *RAeS Aeronautical Journal* pp. 300–312.

Wilcock, R., Young, J. & Horlock, J. (2005). The Effect of Turbine Blade Cooling on the Cycle Efficiency of Gas Turbine Power Cycles, *ASME Journal of Engineering for Gas Turbines and Power* 127(1): 109–120.

Wilde, G. (1978). Future large civil turbofans and powerplants, *RAeS Aeronautical Journal* 82: 281–299.

Wilfert, G., Sieber, J., Rolt, A., Baker, N., Touyeras, A. & Colantuoni, S. (2007). New Environmental Friendly Aero Engine Core Concepts, *ISABE 2007 Proceedings, ISABE-2007-1120*, Beijing, China.

Xu, L. & Grönstedt, T. (2010). Design and Analysis of an Intercooled Turbofan Engine, *ASME Journal of Engineering for Gas Turbines and Power* 132(11). doi:10.1115/1.4000857.

Xu, L., Gustafsson, B. & Grönstedt, T. (2007). Mission Optimization of an Intercooled Turbofan Engine, *ISABE 2007 Proceedings, ISABE-2007-1157*, Beijing, China.

Young, P. (1979). The future shape of medium and long-range civil engines, *RAeS Aeronautical Journal* pp. 53–61.

Zimbrick, R. & Colehour, J. (1988). An investigation of very high bypass ratio engines for subsonic transports, *Proceedings of AIAA/SAE/ASME/ASEE 24th Joint Propulsion Conference, AIAA-88-2953*, Boston, Massachusetts, USA.

2

State-of-Art of Transonic Axial Compressors

Roberto Biollo and Ernesto Benini
University of Padova
Italy

1. Introduction

Transonic axial flow compressors are today widely used in aircraft engines to obtain maximum pressure ratios per single-stage. High stage pressure ratios are important because they make it possible to reduce the engine weight and size and, therefore, investment and operational costs. Performance of transonic compressors has today reached a high level but engine manufacturers are oriented towards increasing it further. A small increment in efficiency, for instance, can result in huge savings in fuel costs and determine a key factor for product success. Another important target is the improvement of rotor stability towards near stall conditions, resulting in a wider working range.

Important analytical and experimental researches in the field of transonic compressors were carried out since 1960's (e.g. Chen et al., 1991; Epstein, 1977; Freeman & Cumpsty, 1992; König et al., 1996; Miller et al., 1961; Wennerstrom & Puterbaugh, 1984). A considerable contribution for the new developments and designs was the progress made in optical measurement techniques and computational methods, leading to a deeper understanding of the loss mechanisms of supersonic relative flow in compressors (e.g. Calvert & Stapleton, 1994; Hah & Reid, 1992; Ning & Xu, 2001; Puterbaugh et al., 1997; Strazisar, 1985; Weyer & Dunker, 1978). Fig. 1 shows the low pressure and high pressure compressors of the EJ200 engine as examples for highly loaded, high performance transonic rotors of an aero engine.

A closer look at the current trend in design parameters for axial flow transonic compressors shows that, especially in civil aircraft engines, the relative flow tip Mach number of the rotor is limited to maintain high efficiencies. A typical value for the rotor inlet relative flow at the tip is Mach ≈ 1.3. The continuous progress of aerodynamics has been focused to the increase in efficiency and pressure ratio and to the improvement in off-design behaviour at roughly the same level of the inlet relative Mach number. Today's high efficiency transonic axial flow compressors give a total pressure ratio in the order of 1.7-1.8, realized by combining high rotor speeds (tip speed in the order of 500 m/s) and high stage loadings ($2\Delta h/u^2$ in the order of 1.0). The rotor blade aspect ratio parameter showed a general trend towards lower values during past decades, with a current asymptotic value of 1.2 (Broichhausen & Ziegler, 2005).

The flow field that develops inside a transonic compressor rotor is extremely complex and presents many challenges to compressor designers, who have to deal with several and concurring flow features such as shock waves, intense secondary flows, shock/boundary layer interaction, etc., inducing energy losses and efficiency reduction (Calvert et al., 2003; Cumpsty, 1989; Denton & Xu, 1999; Law & Wadia, 1993; Sun et al., 2007). Interacting with secondary flows, shock waves concur in development of blockage (Suder, 1998), in corner stall separation (Hah & Loellbach, 1999; Weber et al., 2002), in upstream wakes destabilization (Estevadeordal et al., 2007; Prasad, 2003), and in many other negative flow phenomena. Particularly detrimental is the interaction with the tip clearance flow at the outer span of the rotor, where the compressor generally shows the higher entropy production (Bergner et al., 2005a; Chima, 1998; Copenhaver et al., 1996; Gerolymos & Vallet, 1999; Hofmann & Ballmann, 2002; Puterbaugh & Brendel, 1997; Suder & Celestina, 1996).

Fig. 1. Transonic LPC (left) and HPC (right) of the Eurofighter Typhoon engine EJ200 (Broichhausen & Ziegler, 2005)

As the compressor moves from peak to near-stall operating point, the blade loading increases and flow structures become stronger and unsteady. The tip leakage vortex can breakdown interacting with the passage shock wave, leading to not only a large blockage effect near the tip but also a self-sustained flow oscillation in the rotor passage. As a result, the blade torque, the low energy fluid flow due to the shock/tip leakage vortex interaction and the shock-induced flow separation on the blade suction surface fluctuate with time (Yamada et al., 2004).

Despite the presence of such flow unsteadiness, the compressor can still operate in a stable mode. Rotating stall arises when the loading is further increased, i.e. at a condition of lower mass flow rate. Two routes to rotating stall have been identified: long length-scale (modal) and short length-scale (spike) stall inception in axial compressors (Day, 1993). Modal stall inception is characterized by the relatively slow growth (over 10-40 rotor revolutions) of a small disturbance of long circumferential wavelength into a fully developed stall cell. Spike stall inception starts with the appearance of a large amplitude short length-scale (two to three rotor blade passages) disturbance at the rotor tip, the so-called spike, which grows into a fully developed rotating stall cell within few rotor revolutions.

The following paragraphs give a summary of the possible techniques for limiting the negative impacts of the above reported compressor flow features in aircraft gas turbine engines.

2. Blade profiles studies

For relative inlet Mach numbers in the order of 1.3 and higher the most important design intent is to reduce the Mach number in front of the passage shock. This is of primary importance due to the strongly rising pressure losses with increasing pre-shock Mach number, and because of the increasing pressure losses due to the shock/boundary layer interaction or shock-induced separation. The reduction of the pre-shock Mach number can be achieved by zero or even negative curvature in the front part of the blade suction side and by a resulting pre-compression shock system reducing the Mach number upstream of the final strong passage shock.

Besides inducing energy losses, the presence of shock waves makes transonic compressors particularly sensitive to variations in blade section design. An investigation of cascade throat area, internal contraction, and trailing edge effective camber on compressor performance showed that small changes in meanline angles, and consequently in the airfoil shape and passage area ratios, significantly affect the performance of transonic blade rows (Wadia & Copenhaver, 1996).

One of the most important airfoil design parameter affecting the aerodynamics of transonic bladings is the chordwise location of maximum thickness. An experimental and numerical evaluation of two versions of a low aspect ratio transonic rotor having the location of the tip blade section maximum thickness at 55% and 40% chord length respectively, showed that the more aft position of maximum thickness is preferred for the best high speed performance, keeping the edge and maximum thickness values the same (Wadia & Law, 1993). The better performance was associated with the lower shock front losses with the finer section that results when the location of the maximum thickness is moved aft. The existence of an optimum maximum thickness location at 55% to 60% chord length for such rotor was hypothesized. Similar results can be found in a recent work (Chen et al., 2007) describing an optimization methodology for the aerodynamic design of turbomachinery applied to a transonic compressor bladings and showing how the thermal loss coefficient decreases with increasing the maximum thickness location for all the sections from hub to tip.

Not only the position of maximum thickness but also the airfoil thickness has been showed to have a significant impact on the aerodynamic behaviour of transonic compressor rotors, as observed in an investigation on surface roughness and airfoil thickness effects (Suder et al., 1995). In this work, a 0.025 mm thick smooth coating was applied to the pressure and suction surface of the rotor blades, increasing the leading edge thickness by 10% at the hub and 20% at the tip. The smooth coating surface finish was comparable to the bare metal blade surface finish; therefore the coating did not increase roughness over the blade, except at the leading edge where roughness increased due to particle impact damage. It resulted in a 4% loss in pressure ratio across the rotor at an operating point near design mass flow, with the largest degradation in pressure rise over the outer half of the blade span. When assessed at a constant pressure ratio, the adiabatic efficiency degradation at design speed was in the order of 3-6 points.

The recent development of optimization tools coupled with accurate CFD codes has improved the turbomachinery design process significantly, making it faster and more efficient. The application to the blade section design, with a quasi three-dimensional and more recently with a fully three-dimensional approach, can lead to optimal blade geometries in terms of aerodynamic performance at both design and off-design operating conditions. Such a design process is particularly successful in the field of transonic compressors, where performance is highly sensitive to little changes in airfoil design.

Fig. 2 shows the blade deformation obtained in a quasi 3-D numerical optimization process of a transonic compressor blade section along with the relative Mach number contours before and after the optimization (Burguburu et al., 2004). As shown, no modifications of the inlet flow field occurred after optimization but the flow field structure in the duct is clearly different. The negative curvature of the blade upstream of the shock led to the reduction of the upstream relative Mach number from 1.4 to 1.2. With this curvature change, the velocity slowdown is better driven. Instead of creating a normal shock, the new shape created two low intensity shocks. The new blade gave an efficiency increment of 1.75 points at design condition, without changing the choking mass flow. A large part of the efficiency improvement at the design condition remained at off-design conditions.

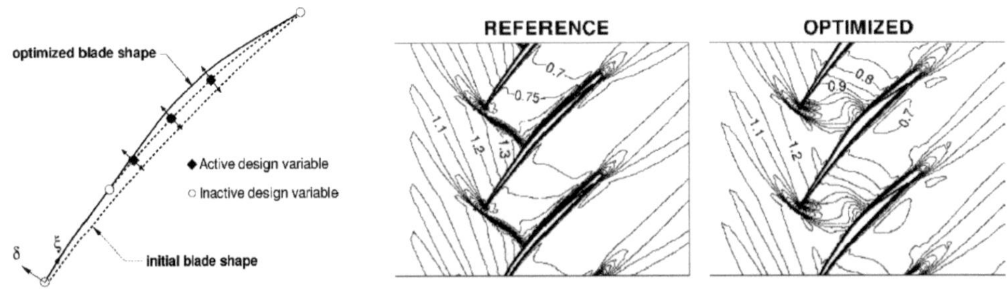

Fig. 2. Blade deformation (left) and relative Mach number contours (right) before and after optimization (Burguburu et al., 2004)

Fig. 3 is related to a both aerodynamic and structural optimization of the well-known transonic compressor rotor 67 (Strazisar et al., 1989), where the aerodynamic objective aimed at maximizing the total pressure ratio whereas the structural objective was to minimize the blade weight, with the constraint that the new design had comparable mass flow rate as the baseline design (Lian & Liou, 2005). The optimization was carried out at the design operating point. Geometric modifications regarded the mean camber line (with the leading and trailing edge points fixed) and thickness distribution of four airfoil profiles (hub, 31% span, 62% span, and tip), linearly interpolated to obtained the new 3-D blade. The chord distribution along the span and the meridional contours of hub, casing, sweep, and lean were maintained.

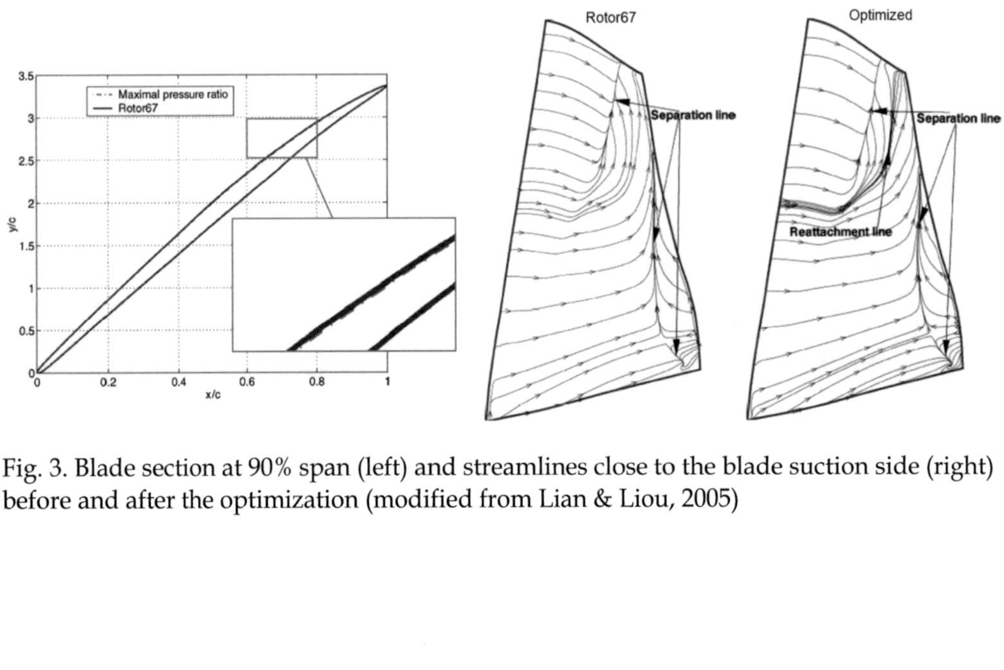

Fig. 3. Blade section at 90% span (left) and streamlines close to the blade suction side (right) before and after the optimization (modified from Lian & Liou, 2005)

At 10% and 50% span (not shown here), the optimization gave a larger camber but lower thickness than the baseline design. The thinner airfoils contributed to reduce the weight of the new design. The calculated difference in the pressure distribution was rather small. At 90% span (see Fig. 3), the new design had a slightly smaller camber and thinner airfoil than the baseline. Nevertheless, the calculated pressure difference was rather large, indicating again that transonic flow is highly sensitive to the profile shape change. One noticeable impact was also in the shock position. The new design showed a more forward passage shock than the baseline.

Such optimized blade gave a decrease of 5.4% in weight and an improvement of 1.8% in the total pressure ratio. The lighter weight came from the thinner blade shape. The higher total pressure ratio was mainly attributed to a reduced separation zone after the shock at the outer span. In Fig. 3, the separation zones are characterized by streamlines going towards the separation lines, whereas reattachment lines look like flow is going away from the separation lines. Compared with the baseline design, downstream of the shock the new design gave a smaller separation zone, which was partially responsible for its higher total pressure ratio.

Fig. 4 is again related to the redesign of rotor 67 using an optimization tool based on evolutionary algorithms (Oyama et al., 2004). Note the particular new design, an improbable design using manual techniques. The optimization gave rise to a double-hump blade shape, especially obvious on the pressure side.

In such new design, the flow acceleration near the leading edge at 33% span diminished because of the decrease of the incidence angle. In addition, at the 90% span, the shock on the suction side moved aft and was weakened considerably because of the aft movement of the maximum camber position. This new blade showed an overall adiabatic efficiency of 2% higher than the baseline blade over the entire operating range for the design speed.

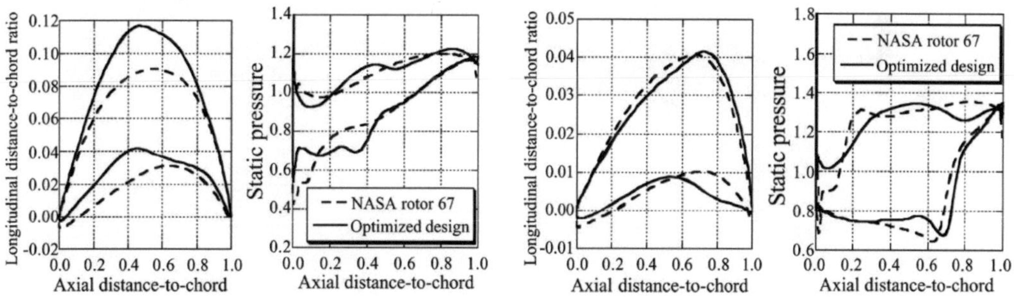

Fig. 4. Comparison between the optimized and baseline design at 33% and 90% span (Oyama et al., 2004)

3. Three-dimensional shaped bladings

The preceding paragraph has shown that a certain maturity in transonic compressors has been reached regarding the general airfoil aerodesign. But the flow field in a compressor is not only influenced by the two-dimensional airfoil geometry. The three-dimensional shape of the blade is also of great importance, especially in transonic compressor rotors where an optimization of shock structure and its interference with secondary flows is required. Many experimental and numerical works can be found in the literature on the design and analysis of three-dimensional shaped transonic bladings (e.g. Copenhaver et al., 1996; Hah et al.,

2004; Puterbaugh et al., 1997). Fig. 5 shows two examples of non-conventional rotors (Rotor2 and Rotor3) derived from the baseline Rotor1 which is conventionally radially-stacked, all developed by TU Darmstadt and MTU Aero Engines. As far as their performance is concerned, Rotor2 gave no real improvement in efficiency and total pressure ratio with respect to the baseline configuration (Blaha et al., 2000; Kablitz et al., 2003a). Rotor3, instead, gave higher performance at design speed (1.5% peak efficiency increment) along with a significantly wider operating range (Passrucker et al., 2003). Information on the favourable impact of Rotor3 blade design on internal transonic flow field is available in the open literature (Bergner et al., 2005b; Kablitz et al., 2003b).

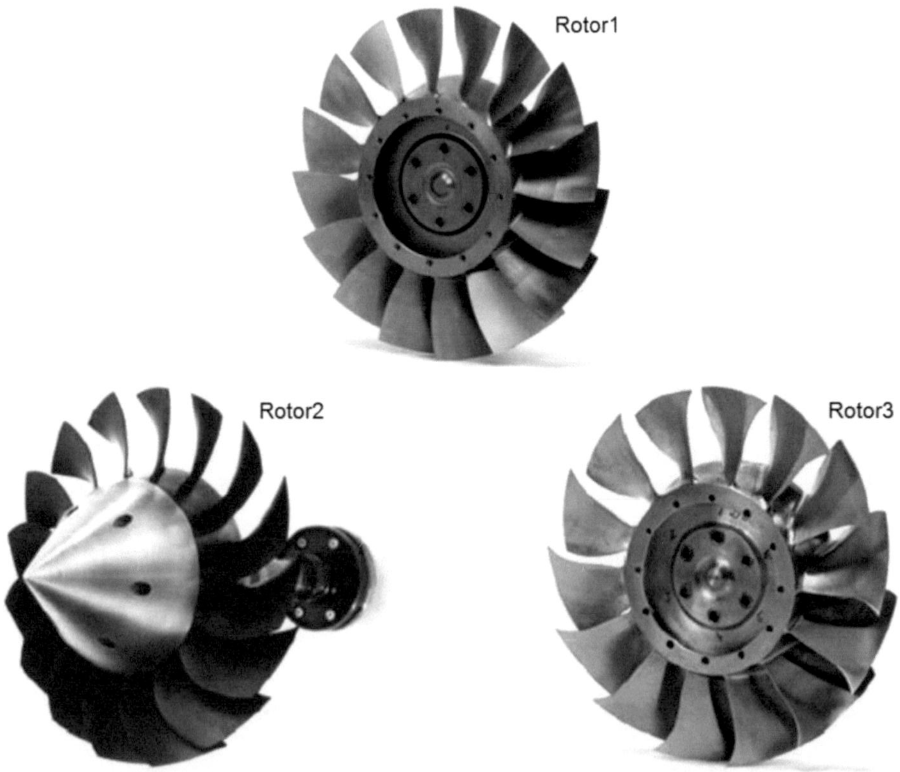

Fig. 5. Transonic compressor test rotors - TU Darmstadt and MTU Aero Engines (Broichhausen & Ziegler, 2005; Passrucker et al., 2003)

A numerical investigation on the aerodynamics of 3-D shaped blades in transonic compressor rotors showed the possibility to have better stall margin with forward sweep (upstream movement of blade sections along the local chord direction, especially at outer span region), maintaining a high efficiency over a wider range (Denton, 2002; Denton & Xu, 2002). This seems to be a general point of view, as confirmed by the following researches.

Numerical and experimental analyses carried out to evaluate the performance of a conventional unswept rotor, a forward swept rotor and an aft swept rotor showed that the forward swept rotor had a higher peak efficiency and a substantially larger stall margin than the baseline unswept rotor, and that the aft swept rotor had a similar peak efficiency with a significantly smaller stall margin (Hah et al., 1998). Detailed analyses of the measured and calculated flow fields indicated that two mechanisms were primarily responsible for the differences in aerodynamic performance among these rotors. The first mechanism was a change in the radial shape of the passage shock near the casing by the endwall effect, and the second was the radial migration of low momentum fluid to the blade tip region. Similar results were obtained in a parallel investigation which identified the reduced shock/boundary layer interaction, resulting from reduced axial flow diffusion and less accumulation of centrifuged blade surface boundary layer at the tip, as the prime contributor to the enhanced performance with forward sweep (Wadia et al., 1998).

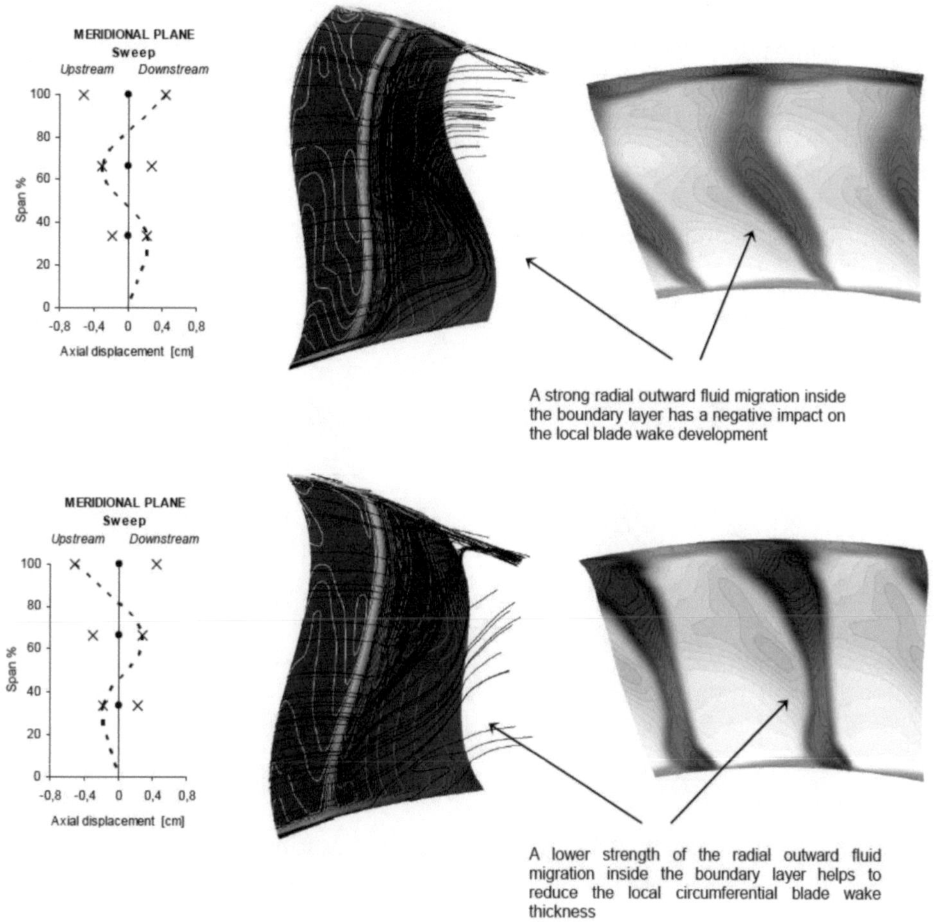

Fig. 6. Blade axial curvature impact on shock, suction side boundary layer and blade wake development (Biollo & Benini, 2008a)

A recent numerical work gave another point of view on the impact of blade curvature in transonic compressor rotors, showing how the movement of blade sections in the axial direction can influence the internal flow field (Benini & Biollo, 2007; Biollo & Benini, 2008a). Such work showed that the axial blade curvature can help to influence the shock shape in the meridional plane, inducing the shock to assume the meridional curvature of the blade leading edge (Fig. 6). In addition, a considerable impact on the radial outward migration of

fluid particles which takes place inside the blade suction side boundary layer after the interaction with the shock has been confirmed. The code predicted a reduction of the strength of such flow feature when the blade is curved downstream and an increment when the blade is curved upstream. Such flow phenomenon is harmful because obstructs the boundary layer development in the streamwise direction, leading to a thickening of blade wakes. A reduction of its strength helped to reduce the entropy generation and the aerodynamic losses associated with the blade wake development. The possibility to increase the peak efficiency of 0.8% at design speed using a proper downstream blade curvature has been showed for the high loaded transonic compressor rotor 37. Details on rotor 37 can be found in the open literature (Reid & Moore, 1978).
The same research group investigated the aerodynamic effects induced by several tangential blade curvatures on the same rotor. It was observed that, when the curvature is applied towards the direction of rotor rotation, the blade-to-blade shock tends to move more downstream, becoming more oblique to the incoming flow. This reduced the aerodynamic shock losses and entropy generation, showing in some cases a peak efficiency increment of over 1% at design speed (Benini & Biollo, 2008). Similar results were previously obtained using a numerical optimization algorithm (Ahn & Kim, 2002). Fig. 7 shows the predicted impact of the optimized design of rotor 37 on the blade-to-blade Mach number.

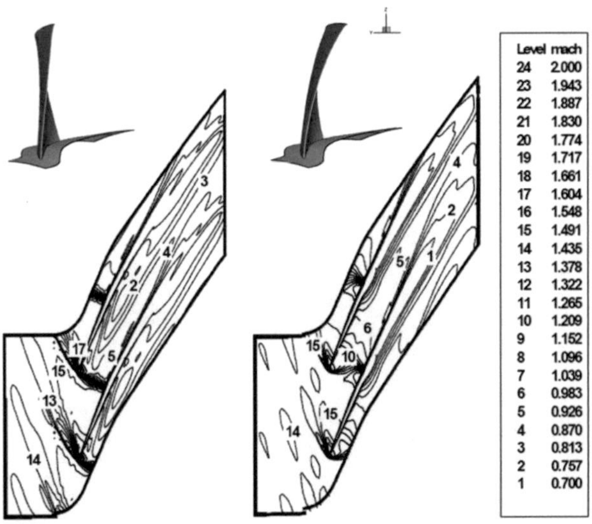

Fig. 7. Baseline (left) and optimized (right) Mach number distributions at 90% span (modified from Ahn & Kim, 2002)

Higher performance can be achieved using a proper combination of two orthogonal blade curvatures, i.e. the use of a blade curved both axially and tangentially, as well as swept and leaned at the same time. Peak efficiency increments from 1% to 1.5% were numerically observed using a blade prevalently curved towards the direction of rotor rotation and slightly backward inclined (Biollo & Benini, 2008b; Jang et al., 2006; Yi et al., 2006).

4. Casing treatments

Hollow structures in the casing to improve the tip endwall flow field of axial flow compressors are commonly referred to as casing treatments. Fig. 8 shows some examples of casing treatments investigated in the 1970's. The interaction of the main flow with the flow circulating in these cavities seems to have a positive impact on rotor stability. However, early studies did not reach a detailed understanding of the phenomenon, since experimental investigations were too expensive and only few configurations could be tested. Only in the past fifteen years numerical simulations made it possible to investigate a larger number of casing treatment solutions and their effects on different compressors. Many researches were carried out on transonic compressor rotors and the potential of this kind of passive devices was revealed: a proper treatment can not only widen the stable working range of a transonic compressor rotor, but also improve its efficiency.

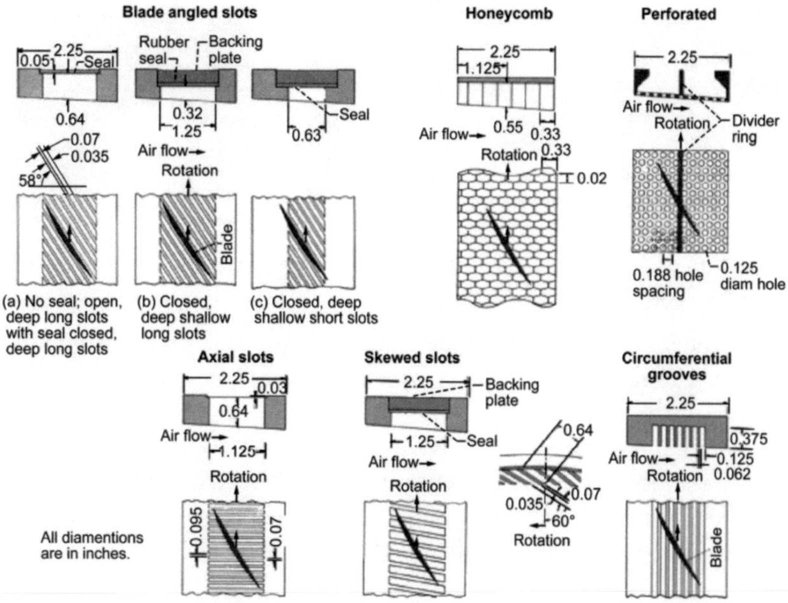

Fig. 8. Various casing treatments investigated in the 1970's (Hathaway, 2007)

4.1 Circumferential groove-type treatments

Recently, the influence of circumferential grooves on the tip flow field of an axial single-stage transonic compressor has been examined both experimentally and numerically (Fig. 9). The compressor stage provided a strongly increased stall margin (56.1%), with only small penalties in efficiency when the casing treatment was applied. Flow analyses showed that at near stall conditions with the smooth casing, the induced vortex originating from the tip clearance flow crossing the tip gap along 20-50% chord length, hits the front part of the adjacent blade, indicating the possibility of a spill forward of low momentum fluid into the next passage, a flow feature considered a trigger for the onset of rotating stall. With the casing treatment applied, the vortex trajectory remained instead aligned to the blade's suction side.

Disadvantages of casing treatments like these are the space they need and the weight increase of the compressor casing. So it is a goal to maintain the positive effects (increased surge pressure ratio in combination with high efficiency) while at the same time reducing the geometric volume of the device. On this regard, an experimental and numerical investigation on the first rotor of a two-stage compressor showed that grooves with a much smaller depth than conventional designs are similarly effective in increasing the stall margin (Rabe & Hah, 2002). The same work also showed that two shallow grooves placed near the leading edge are better than five deep or shallow grooves all over the blade tip. Fewer shallower grooves clearly help to reduce the weight, fabrication costs, and loss generation associated with such a casing treatment.

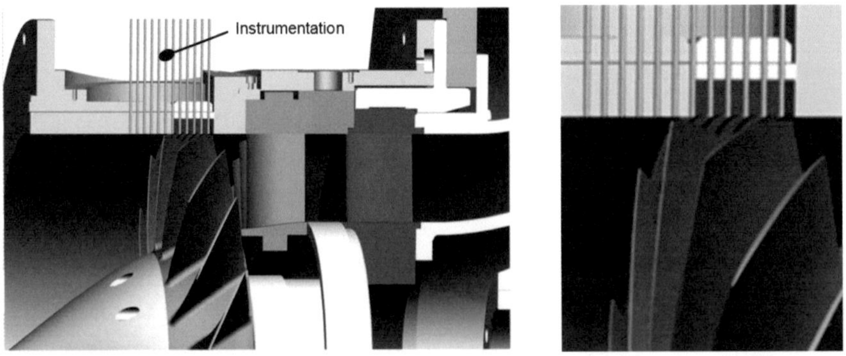

Fig. 9. Cross section of a transonic compressor stage with circumferential grooves (modified from Müller et al., 2008)

Other possible groove-type casing treatment solutions are presented in Figs. 10 and 11. Fig. 10 shows a single extended casing circumferential groove all over the blade tip section. It has been numerically shown that such a casing treatment provides a means for fluid to exit the flow path where the blade loading is severe, migrate circumferentially, and re-enter the flow path at a location where the pressure is more moderate. This can lead to stability improvement since the flow relocation helps to relieve the locally severe blade loading. Using this device, the authors showed the possibility to improve both the efficiency and the stall margin.

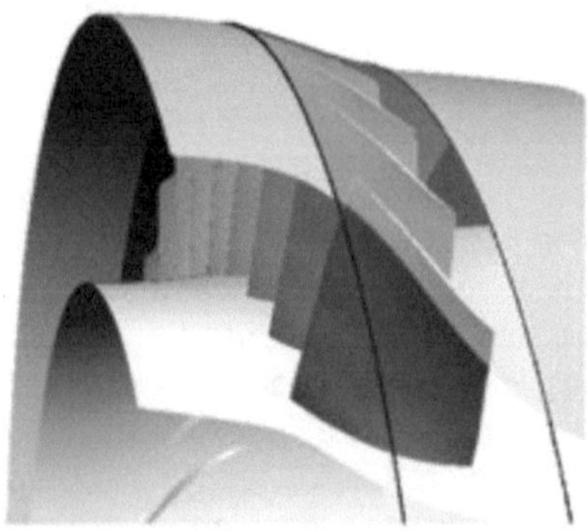

Fig. 10. Single extended casing circumferential groove (Beheshti et al., 2004)

Fig. 11 is related to a numerical investigation of casing contouring effects on flow instability. While the "Type B1" solution gave no improvements in stall margin, the "Type C2" solution extended the calculated stable operating range from 94% to about 92.5% normalized mass flow. With respect to the baseline smooth casing configuration, the successful contouring induced a smaller inflow angle near the leading edge, i.e. a lower incidence, delaying rotating stall inception.

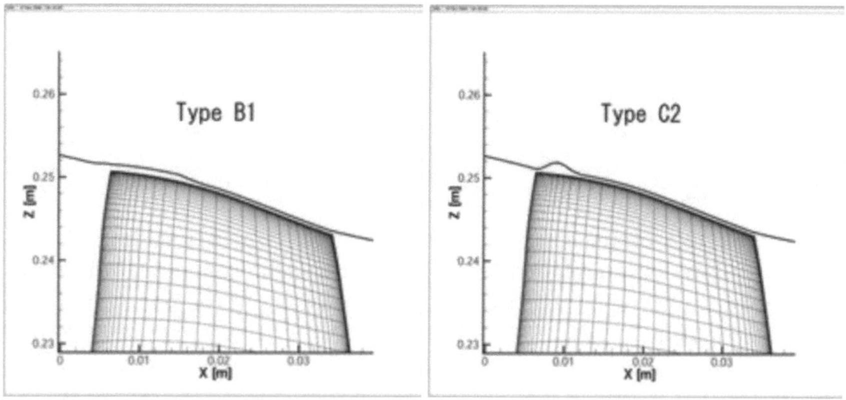

Fig. 11. Endwall casing contouring (Ito et al., 2007)

4.2 Slot-type treatments

Another way to treat the casing with the aim to improve the rotor performance and stability is exemplified in Fig. 12. Here the casing wall is circumferentially treated with a discrete number of axial rectangular slots over the blade tip section.

A similar slot-type casing treatment consisting of four identical axial slots per blade passage and having an open area of 50% in the circumferential direction has been also proposed. (Wilke and Kau, 2004). In this case, the slots are parallel to the rotation axis and inclined by 45° against the meridional plane in the direction of rotor rotation. The slot shape is designed as a semi-circle. Two configurations have been numerically tested. In configuration 1, the position of the slots is centered above the rotor blade tip reaching from 7.5% to 92.5% chord length. In configuration 2, the slots are moved upstream so that only 25% chord length remains covered by the casing treatment.

Fig. 12. EJ200 LPC with axial slots casing treatment (Broichhausen & Ziegler, 2005)

For both configurations, simulations showed a significant increment in flow stability compared to the solid wall, the stalling mass flow passed from 0.95% to 0.75% the design mass flow at the design speed. It was observed that the stabilizing effects are based on the positive impact of casing treatment on the tip clearance flow and its resulting vortex. Configuration 1 led to a massive destruction of the tip leakage vortex, whereas configuration 2 weakened the rolling-up of the tip clearance flow. Configuration 2, gave also a positive impact on the overall efficiency.

5. Air injection and bleeding

One of the first flow control ideas to receive a considerable attention in gas turbine applications is the flow injection and bleeding concept. Air injection near the blade tips has proved to increase compressor stall margin, leading to higher engine operability. Fig. 13 shows a set of successful 12 non-intrusive prototype injectors recently installed in the casing of a transonic compressor test rotor. On the other hand, aspiration (or bleeding) can be used to delay blade separation, which limits the stage work and therefore increases the required number of stages. Doubling the work per stage using aspiration results in a dramatic reduction in the number of stages, though not necessarily an exactly proportional reduction in compressor weight due to the added complexity of blades and added plumbing. For fighter applications this technology looks very attractive due to its potential to improve the thrust/weight ratio of the engine.

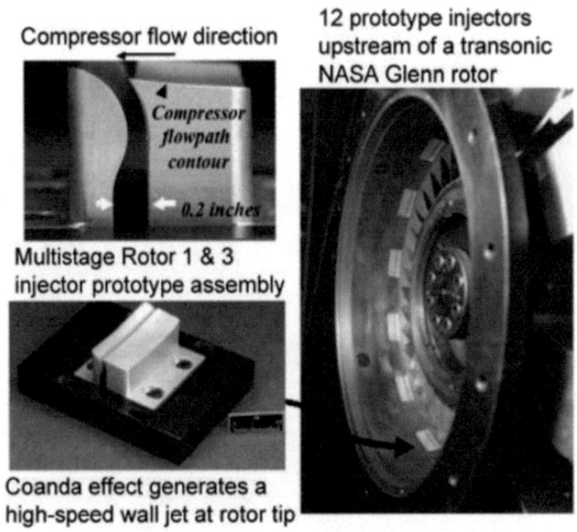

Fig. 13. Discrete tip injection (Hathaway, 2007)

5.1 Air injection

Measurements and simulations for active discrete tip injection have been presented for a range of steady injection rates and distributions of injectors around the annulus of a tip-critical transonic compressor rotor operating in isolation (Suder et al., 2001). In such case, casing-mounted injectors were located at 200% tip axial chord upstream of the rotor. Each injector penetrated 6% span from the casing into the flow field. The simulations indicated that tip injection increases stability by unloading the rotor tip. Tip injection decreases incidence and blade loading at the tip, allowing increased loading at lower blade spans before the blade stalls. With tip injection present, the blade stalls when the loading at the tip reaches the level equal to that for which the blade stalls with no injection. The experiments showed that stability enhancement is related to the mass averaged axial velocity at the tip. For the tested rotor, experimental results demonstrated that at 70% speed the stalling flow coefficient can be reduced by 30% using an injected mass flow equivalent to 1% of the annulus flow. At design speed, the stalling flow coefficient was reduced by 6% using an injected mass flow equivalent to 2% of the annulus flow. Tip injection has also been demonstrated as an effective tool for recovering a compressor from fully developed stall.

In a self-induced (passive) solution, steady discrete tip injection using casing recirculation has been simulated in a single transonic fan rotor (Hathaway, 2002). The idea was to bleed pressurized fluid from downstream of the rotor and properly inject it upstream. Simulations were carried out assuming both a clear and distorted inlet flow. The distortion was circumferentially applied near the casing endwall fixing a lower inlet total pressure. The recirculation model gave 125% range extension without inlet distortion and 225% range extension with inlet distortion, without significant impact on overall efficiency.

Such a solution has been successively tested on a single-stage transonic compressor (Strazisar et al., 2004). Results clearly indicated that recirculation extends the stable operating range of that stage to lower mass flows than that can be achieved without recirculation. With the recirculation activated, the positive change in stalling flow coefficient was 6% at 70% of design speed and 2% at design speed, with a total injected flow of 0.9% of the annulus flow at both operating speeds. In the same work, the potential for using endwall recirculation to increase the stability of transonic highly loaded multistage compressors was demonstrated through results from a rig test of simulated recirculation driving both a steady injected flow and an unsteady injected flow. Unsteady injection increased stability more than steady injection and was capable of changing the unsteady near stall dynamics of the multistage compressor.

5.2 Bleeding

Fig. 14 shows the sketch of a transonic aspirated stage experimentally tested and numerically investigated to demonstrate the application of boundary layer aspiration for increasing the stage work (Schuler et al., 2005). The stage was designed to produce a pressure ratio of 1.6 at a tip speed of 750 ft/s resulting in a stage work coefficient of 0.88. The primary aspiration requirement for the stage was a bleed fraction 0.5% of the inlet mass flow on the rotor and stator suction surfaces. Additional aspiration totalling 2.8% was also used at shock impingement locations and other locations on the hub and casing walls. The stage achieved a peak pressure ratio of 1.58 and through flow efficiency of 90% at the design point. The rotor showed an extremely high efficiency of 97% for a transonic rotor, partially attributed to aspiration and partially to the elimination of the tip clearance flow due to the tip shroud. Aspiration was also effective in maintaining stage performance at off-design conditions. The experimental data showed unstalled stage performance at least 83% of the design mass flow.

The possibility of a very high pressure ratio per single-stage using aspiration has been demonstrated for the fan of Fig. 15. The fan stage has been designed to achieve a pressure ratio of 3.4 at 1500 ft/s. The low energy viscous flow was aspirated from diffusion-limiting locations on the blades and flow path surfaces. Experimental results gave a stage pressure ratio exceeding 3 at design speed, with an aspiration flow fraction of 3.5% of the stage inlet mass flow. CFD simulations showed that aspiration fixes the passage shock position, particularly in the tip region, maintaining good aerodynamic behaviour at off-design operating points.

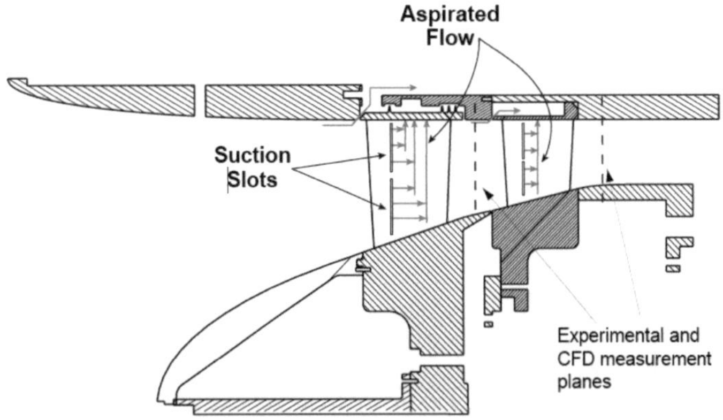

Fig. 14. A tested suction configuration (Schuler et al., 2002)

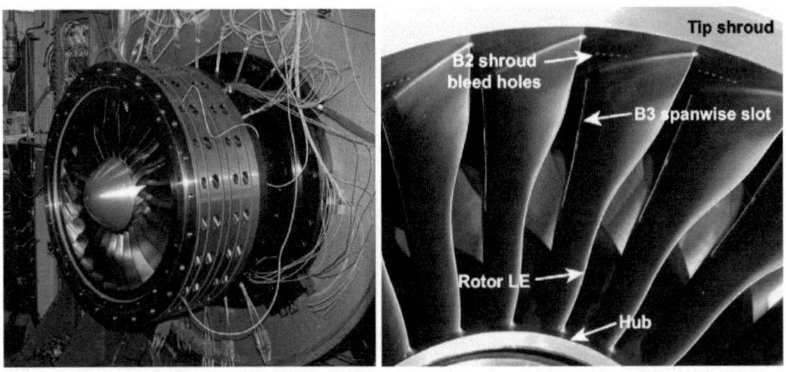

Fig. 15. Test package (left) and rotor aspiration scheme (right) (Merchant et al., 2005)

In a previous work (Dang & Van Rooij, 2003), it was demonstrated the possibility to reduce the amount of aspiration without compromise its benefits. When used as an add-on to an existing design, aspiration can require excessive amounts of suction, whereas with a blade/ aspiration scheme coupled design a significantly lower amount of suction can be needed. A baseline supersonic rotor with 56 blades was used in that work as the starting point. The blade count in that rotor was then reduced to 42 blades, and it was found that 2% of flow suction was needed to pull the shock back into the blade passage for the required back pressure. The aspiration zone was located approximately behind the shock on the suction surface from hub to 95% span. With 42 blades, inspection of the flow field showed that the peak Mach number and loading were significantly higher than in the baseline blade case, resulting in increased shock losses, thickening of the blade suction surface boundary layer, and a large region of low momentum fluid in the tip clearance region.

A new pressure loading shape was developed to mitigate the situation. The new design was shown to have better efficiency potential and a cleaner casing endwall flow using only 0.3%

flow suction. It was also shown that the constant speed throttling characteristic of the new design can be tailored by using varying amounts of suction (up to 2%) to hold the passage shock stationary thereby making it possible to satisfy multiple operating conditions with minimal aerodynamic penalty.

6. Conclusions

Transonic compressors are the state-of-art in the compression system of today's civil and military aero engines. Their capability to provide high pressure ratios maintaining high efficiencies made them preferable to both subsonic (lower pressure ratio) and supersonic (lower efficiency) solutions.
The researches of last decades have greatly contributed to their success. The progress made in optical measurement techniques and the development of computational methods led to a deep understanding of the loss mechanisms associated with their inherent supersonic relative flow, making it possible significant aerodynamic improvements and industrial diffusion.
Nowadays there is still space for further radical improvements and new solutions will be developed in the next future, like highly effective stall/surge control systems and higher pressure ratio configurations, or even new concepts based on new advanced working principles.

7. References

Ahn, C.-S. & Kim, K.-Y. (2002). Aerodynamic Design Optimization of an Axial Flow Compressor Rotor, *Proceedings of ASME Turbo Expo 2002*, GT-2002-30445

Beheshti, B. H.; Teixeira, J. A.; Ivey, P. C.; Ghorbanian, K. & Farhanieh, B. (2004). Parametric Study of Tip Clearance - Casing Treatment on Performance and Stability of a Transonic Axial Compressor, *Proceedings of ASME Turbo Expo 2004*, GT2004-53390

Benini, E. & Biollo, R. (2007). Aerodynamics of Swept and Leaned Transonic Compressor-Rotors, *Applied Energy*, Vol. 84, No. 10, (October 2007), pp. 1012-1027, ISSN 0306-2619

Benini, E. & Biollo, R. (2008). Effect of Forward and Aft Lean on the Performance of a Transonic Compressor Rotor, *International Journal of Turbo and Jet Engines*, Vol. 25, No. 1, (2008), pp. 13-26, ISSN 0334-0082

Bergner, J.; Hennecke, D. K. & Hah, C. (2005a). Tip-Clearance Variations of an Axial High-Speed Single-Stage Transonic Compressor, *Proceedings of 17th Symposium on Air-Breathing Engines*, ISABE-2005-1096

Bergner, J.; Kablitz, S.; Hennecke, D. K.; Passrucker, H. & Steinhardt, E. (2005b). Influence of Sweep on the 3D Shock Structure in an Axial Transonic Compressor, *Proceedings of ASME Turbo Expo 2005*, GT2005-68835

Biollo, R. & Benini, E. (2008a). Impact of Sweep and Lean on the Aerodynamic Behaviour of Transonic Compressor Rotors, *Proceedings of The Future of Gas Turbine Technology - 4th International Conference*, IGTC08_P17

Biollo, R. & Benini, E. (2008b). Aerodynamic Behaviour of a Novel Three-Dimensional Shaped Transonic Compressor Rotor Blade, *Proceedings of ASME Turbo Expo 2008*, GT2008-51397

Blaha, C.; Kablitz, S.; Hennecke, D. K.; Schmidt-Eisenlohr, U.; Pirker, K. & Haselhoff, S. (2000). Numerical Investigation of the Flow in an Aft-Swept Transonic Compressor Rotor, *Proceedings of ASME Turbo Expo 2000*, 2000-GT-0490

Broichhausen, K. D. & Ziegler, K. U. (2005). Supersonic and Transonic Compressors: Past, Status and Technology Trends, *Proceedings of ASME Turbo Expo 2005*, GT2005-69067

Burguburu, S.; Toussaint, C.; Bonhomme, C. & Leroy, G. (2004). Numerical Optimization of Turbomachinery Bladings, *ASME Journal of Turbomachinery*, Vol. 126, No. 1, (January 2004), pp. 91-100, ISSN 0889-504X

Calvert, W. J.; Emmerson, P. R. & Moore J. M. (2003). Design, Test and Analysis of a High-Pressure-Ratio Transonic Fan, *Proceedings of ASME Turbo Expo 2003*, GT2003-38302

Calvert, W. J. & Stapleton, A.W. (1994). Detailed Flow Measurements and Predictions for a Three-Stage Transonic Fan, *ASME Journal of Turbomachinery*, Vol.116, No.2, (April 1994), pp. 298-305, ISSN 0889-504X

Chen, G. T.; Greitzer, E. M.; Tan, C. S. & Marble, F. E. (1991). Similarity Analysis of Compressor Tip Clearance Flow Structure, *ASME Journal of Turbomachinery*, Vol. 113, No. 2, (April 1991), pp. 260-229, ISSN 0889-504X

Chen, N.; Zhang, H.; Xu, Y. & Huang, W. (2007). Blade Parameterization and Aerodynamic Design Optimization for a 3D Transonic Compressor Rotor, *Journal of Thermal Science*, Vol. 16, No. 2, (May 2007), pp. 105-114, ISSN 1003-2169

Chima, R. V. (1998). Calculation of Tip Clearance Effects in a Transonic Compressor Rotor, *ASME Journal of Turbomachinery*, Vol. 120, No. 1, (January 1998), pp. 131-140, ISSN 0889-504X

Copenhaver, W. W.; Mayhew, E. R.; Hah, C. & Wadia, A. R. (1996). The Effect of Tip Clearance on a Swept Transonic Compressor Rotor, *ASME Journal of Turbomachinery*, Vol. 118, No. 2, (April 1996), pp. 230-229, ISSN 0889-504X

Cumpsty, N. A. (1989). Compressor Aerodynamics, Longman Group UK Limited 1989

Dang, T.Q. & Van Rooij, M. (2003). Design of Aspirated Compressor Blades Using Three-Dimensional Inverse Method, NASA Report No. TM-2003-212212

Day, I. J. (1993). Stall Inception in Axial Flow Compressors, *ASME Journal of Turbomachinery*, Vol. 115, No. 1, (January 1993), pp. 1-9, ISSN 0889-504X

Denton, J. D. (2002). The Effects of Lean and Sweep on Transonic Fan Performance: A Computational Study, *Task Quarterly*, Vol. 6, No. 1, (2002), pp. 7-23

Denton, J. D. & Xu, L. (1999). The Exploitation of Three-Dimensional Flow in Turbomachinery Design, *Proc Instn Mech Engrs, Part C: Journal of Mechanical Engineering Science*, Vol. 213, No. 2, (1999), pp. 125-137, ISSN 0954-4062

Denton, J. D. & Xu, L. (2002). The Effects of Lean and Sweep on Transonic Fan Performance, *Proceedings of ASME Turbo Expo 2002*, GT-2002-30327

Epstein, A. H. (1977). Quantitative Density Visualization in a Transonic Compressor Rotor, *ASME Journal of Engineering for Power*, Vol.99, No.3, (July 1977), pp. 460-475, ISSN 0022-0825

Estevadeordal, J.; Gorrell, S. E. & Copenhaver, W. W. (2007). PIV Study of Wake-Rotor Interactions in a Transonic Compressor at Various Operating Conditions, *Journal of Propulsion and Power*, Vol. 23, No. 1, (Jan-Feb 2007), ISSN 0748-4658

Freeman, C. & Cumpsty, N. A. (1992). Method for the Prediction of Supersonic Compressor Blade Performance, *Journal of Propulsion and Power*, Vol. 8, No. 1, (Jan-Feb 1992), pp. 199-208, ISSN 0748-4658

Gerolymos, G. A. & Vallet, I. (1999). Tip-Clearance and Secondary Flows in a Transonic Compressor Rotor, *ASME Journal of Turbomachinery*, Vol. 121, No. 4, (October 1999), pp. 751-762, ISSN 0889-504X

Hah, C . & Loellbach, J. (1999). Development of Hub Corner Stall and Its Influence on the Performance of Axial Compressor Blade Rows, *ASME Journal of Turbomachinery*, Vol. 121, No. 1, (January 1999), pp. 67-77, ISSN 0889-504X

Hah, C.; Puterbaugh, S. L. & Wadia, A. R. (1998). Control of Shock Structure and Secondary Flow Field Inside Transonic Compressor Rotors Through Aerodynamic Sweep, *Proceedings of the 1998 International Gas Turbine & Aeroengine Congress & Exibition*, ASME Paper 98-GT-561

Hah, C.; Rabe, D. C. & Wadia, A. R. (2004). Role of Tip-Leakage Vortices and Passage Shock in Stall Inception in a Swept Transonic Compressor Rotor, *Proceedings of ASME Turbo Expo 2004*, GT2004-53867

Hah, C. & Reid, L. (1992). A Viscous Flow Study of Shock-Boundary Layer Interaction, Radial Transport, and Wake Development in a Transonic Compressor, *ASME Journal of Turbomachinery*, Vol. 114, No. 3, (July 1992), pp. 538-547, ISSN 0889-504X

Hathaway, M. D. (2002). Self-Recirculating Casing Treatment Concept for Enhanced Compressor Performance, *Proceedings of ASME Turbo Expo 2002*, GT-2002-30368

Hathaway, M. D. (2007). Passive Endwall Treatments for Enhancing Stability, NASA Report No. TM-2007-214409

Hofmann, W. & Ballmann, J. (2002). Tip Clearance Vortex Development and Shock-Vortex-Interaction in a Transonic Axial Compressor Rotor, *Proceedings of 40th AIAA Aerospace Sciences Meeting and Exhibit*, AIAA 2002-0083

Ito, Y.; Watanabe, T. & Himeno, T. (2007). Effect of Endwall Contouring on Flow Instability of Transonic Compressor, *Proceedings of the International Gas Turbine Congress 2007 Tokyo*, IGTC2007 Tokyo TS-042

Jang, C.-M.; Samad, A. & Kim, K.-Y. (2006). Optimal Design of Swept, Leaned and Skewed Blades in a Transonic Axial Compressor, *Proceedings of ASME Turbo Expo 2006*, GT2006-90384

Kablitz, S.; Bergner, J.; Hennecke, K.; Beversdorff, M. & Schodl, R. (2003a). Darmstadt Rotor No. 2, III: Experimental Analysis of an Aft-Swept Axial Transonic Compressor Stage, *International Journal of Rotating Machinery*, Vol. 9, No. 6, (2003), pp. 393-402, ISSN 1023-621X

Kablitz, S.; Passrucker, H.; Hennecke, D. K. & Engber, M. (2003b). Experimental Analysis of the Influence of Sweep on Tip Leakage Vortex Structure of an Axial Transonic Compressor Stage, *Proceedings of 16th Symposium on Air-Breathing Engines*, ISABE-2003-1226

König, W. M.; Hennecke, D. K. & Fottner L. (1996). Improved Blade Profile Loss and Deviation Angle Models for Advanced Transonic Compressor Bladings: Part II-A Model for Supersonic Flow, *ASME Journal of Turbomachinery*, Vol. 118, No. 1, (January 1996), pp. 81-87, ISSN 0889-504X

Law, C. H. and Wadia, A. R. (1993). Low Aspect Ratio Transonic Rotors: Part 1-Baseline Design and Performance, *ASME Journal of Turbomachinery*, Vol. 115, No. 2, (April 1993), pp. 218-225, ISSN 0889-504X

Lian, Y. & Liou, M.-S. (2005). Multi-Objective Optimization of Transonic Compressor Blade Using Evolutionary Algorithm, *Journal of Propulsion and Power*, Vol. 21, No. 6, (Nov-Dec 2005), pp. 979-987, ISSN 0748-4658

Merchant, A.; Kerrebrock, J. L.; Adamczyk, J. J. & Braunscheidel, E. (2005). Experimental Investigation of a High Pressure Ratio Aspirated Fan Stage, *ASME Journal of Turbomachinery*, Vol. 127, No. 1, (January 2005), pp. 43-51, ISSN 0889-504X

Miller, G. R.; Lewis, Jr., G. W. & Hartmann, M. J. (1961). Shock Losses in Transonic Compressor Blade Rows, *ASME Journal of Engineering for Power*, Vol. 83, No. 3, (July 1961), pp. 235-242, ISSN 0022-0825

Müller, M. W.; Biela, C.; Schiffer H.-P. & Hah, C. (2008). Interaction of Rotor and Casing Treatment Flow in an Axial Single-Stage Transonic Compressor With Circumferential Grooves, *Proceedings of ASME Turbo Expo 2008*, GT2008-50135

Ning, F. & Xu, L. (2001). Numerical Investigation of Transonic Compressor Rotor Flow Using an Implicit 3D Flow Solver With One-Equation Spalart-Allmaras Turbulence Model, *Proceedings of ASME Turbo Expo 2001*, 2001-GT-0359

Oyama, A.; Liou, M.-S. & Obayashi, S. (2004). Transonic Axial-Flow Blade Optimization: Evolutionary Algorithms/Three-Dimensional Navier-Stokes Solver, *Journal of Propulsion and Power*, Vol. 20, No. 4, (Jul-Aug 2004), pp. 612-619, ISSN 0748-4658

Passrucker, H.; Engber, M.; Kablitz, S. & Hennecke, D. K. (2003). Effect of Forward Sweep in a Transonic Compressor Rotor, *Proc Instn Mech Engrs, Part A: Journal of Power and Energy*, Vol. 217, No. 4, (2003), pp. 357-365, ISSN 0957-6509

Prasad, A. (2003). Evolution of Upstream Propagating Shock Waves From a Transonic Compressor Rotor, *ASME Journal of Turbomachinery*, Vol. 125, No. 1, (January 2003), pp. 133-140, ISSN 0889-504X

Puterbaugh, S. L. & Brendel, M. (1997). Tip Clearance Flow-Shock Interaction in a Transonic Compressor Rotor, *Journal of Propulsion and Power*, Vol. 13, No. 1, (Jan-Feb 1997), pp. 24-30, ISSN 0748-4658

Puterbaugh, S. L.; Copenhaver, W. W.; Hah, C. & Wennerstrom, A. J. (1997). A Three-Dimensional Shock Loss Model Applied to an Aft-Swept, Transonic Compressor Rotor, *ASME Journal of Turbomachinery*, Vol. 119, No. 3, (July 1997), pp. 452-459, ISSN 0889-504X

Rabe, D. C. & Hah, C. (2002). Application of Casing Circumferential Grooves for Improved Stall Margin in a Transonic Axial Compressor, *Proceedings of ASME Turbo Expo 2002*, GT-2002-30641

Reid, L. & Moore, R. D. (1978). Design and Overall Performance of Four Highly Loaded, High-Speed Inlet Stages for an Advanced High-Pressure-Ratio Core Compressor, NASA TP 1337

Schuler, B. J.; Kerrebrock, J. L. & Merchant, A. (2005). Experimental Investigation of a Transonic Aspirated Compressor, *ASME Journal of Turbomachinery*, Vol. 127, No. 2, (April 2005), pp. 340-348, ISSN 0889-504X

Schuler, B. J.; Kerrebrock, J. L. & Merchant, A. (2002). Experimental Investigation of an Aspirated Fan Stage, *Proceedings of ASME Turbo Expo 2002*, GT-2002-30370

Strazisar, A. J. (1985). Investigation of Flow Phenomena in a Transonic Fan Rotor Using Laser Anemometry, *ASME Journal of Engineering for Gas Turbines and Power*, Vol.107, No.2, (April 1985), pp. 427-435, ISSN 0742-4795

Strazisar, A. J.; Wood, J. R.; Hathaway, M. D. & Suder, K. L. (1989). Laser Anemometer Measurements in a Transonic Axial-Flow Fan Rotor, NASA TP 2879

Strazisar, A. J.; Bright, M. M.; Thorp, S.; Culley, D. E. & Suder, K. L. (2004). Compressor Stall Control Through Endwall Recirculation, *Proceedings of ASME Turbo Expo 2004*, GT2004-54295

Suder, K. L. (1998). Blockage Development in a Transonic, Axial Compressor Rotor, *ASME Journal of Turbomachinery*, Vol. 120, No. 3, (July 1998), pp. 465-476, ISSN 0889-504X

Suder, K. L. & Celestina, M. L. (1996). Experimental and Computational Investigation of the Tip Clearance Flow in a Transonic Axial Compressor Rotor, *ASME Journal of Turbomachinery*, Vol. 118, No. 2, (April 1996), pp. 218-229, ISSN 0889-504X

Suder, K. L.; Chima, R. V.; Strazisar, A. J. & Roberts, W. B. (1995). The Effect of Adding Roughness and Thickness to a Transonic Axial Compressor Rotor, *ASME Journal of Turbomachinery*, Vol. 117, No. 4, (October 1995), pp. 491-505, ISSN 0889-504X

Suder, K. L.; Hathaway, M. D.; Thorp, S. A.; Strazisar, A. J. & Bright, M. B. (2001). Compressor Stability Enhancement Using Discrete Tip Injection, *ASME Journal of Turbomachinery*, Vol. 123, No. 1, (January 2001), pp. 14-23, ISSN 0889-504X

Sun, Y.; Ren, Y.-X.; Fu, S.; Wadia, A. R. & Wisler, D. (2007). Numerical Study of the Loss in a Compressor Stage, *Proceedings of the International Gas Turbine Congress 2007 Tokyo*, IGTC2007 Tokyo TS-041

Wadia, A. R. & Copenhaver, W. W. (1996). An Investigation of the Effect of Cascade Area Ratios on Transonic Compressor Performance, *ASME Journal of Turbomachinery*, Vol. 118, No. 4, (October 1996), pp. 760-770, ISSN 0889-504X

Wadia, A. R. & Law, C. H. (1993). Low Aspect Ratio Transonic Rotors: Part 2-Influence of Location of Maximum Thickness on Transonic Compressor Performance, *ASME Journal of Turbomachinery*, Vol. 115, No. 2, (April 1993), pp. 226-239, ISSN 0889-504X

Wadia, A. R.; Szucs, P. N. & Crall, D. W. (1998). Inner Workings of Aerodynamic Sweep, *ASME Journal of Turbomachinery*, Vol. 120, No. 4, (October 1998), pp. 671-682, ISSN 0889-504X

Weber, A.; Schreiber, H.-A.; Fuchs, R. & Steinert, W. (2002). 3-D Transonic Flow in a Compressor Cascade With Shock-Induced Corner Stall, *ASME Journal of Turbomachinery*, Vol. 124, No. 3, (July 2002), pp. 358-366, ISSN 0889-504X

Wennerstrom, A. J. & Puterbaugh, S. L. (1984). A Three-Dimensional Model for the Prediction of Shock Losses in Compressor Blade Rows, *ASME Journal of Engineering for Gas Turbines and Power*, Vol. 106, No. 2, (April 1984), pp. 295-299, ISSN 0742-4795

Weyer, H. B. & Dunker, R. (1978). Laser Anemometry Study of the Flow Field in a Transonic Compressor Rotor, *Proceedings of AIAA Aerospace Sciences Meeting 1978*, AIAA Paper 78-1

Wilke, I. & Kau, H.-P. (2004). A Numerical Investigation of the Flow Mechanisms in a High pressure Compressor Front Stage With Axial Slots, *ASME Journal of Turbomachinery*, Vol. 126, No. 3, (July 2004), pp. 339-349, ISSN 0889-504X

Yamada, K.; Furukawa, M.; Nakano, T.; Inoue, M. & Funazaki, K. (2004). Unsteady Three-Dimensional Flow Phenomena Due to Breakdown of Tip Leakage Vortex in a Transonic Axial Compressor Rotor, *Proceedings of ASME Turbo Expo 2004*, GT2004-53745

Yi, W.; Huang, H. & Han, W. (2006). Design Optimization of Transonic Compressor Rotor Using CFD and Genetic Algorithm, *Proceedings of ASME Turbo Expo 2006*, GT2006-90155

3

Possible Efficiency Increasing of Ship Propulsion and Marine Power Plant with the System Combined of Marine Diesel Engine, Gas Turbine and Steam Turbine

Marek Dzida
Gdansk University of Technology
Poland

1. Introduction

For years there has been, and still is, a tendency in the national economy to increase the efficiency of both the marine and inland propulsion systems. It is driven by economic motivations (rapid increase of fuel prices) and ecological aspects (the lower the fuel consumption, the lower the emission of noxious substances to the atmosphere). New design solutions are searched to increase the efficiency of the propulsion system via linking Diesel engines with other heat engines, such as gas and steam turbines. The combined systems implemented in marine propulsion systems in recent years are based mainly on gas and steam turbines (MAN, 2010). These systems can reach the efficiency exceeding 60% in inland applications. The first marine system of this type was applied on the passenger liner "Millenium". However, this is the only high-efficiency marine application of the combined propulsion system so far. Its disadvantage is that the system needs more expensive fuel, the marine Diesel oil, while the overwhelming majority of the merchant ships are driven by low-speed engines fed with relatively cheap heavy fuel oil. It seems that the above tendency will continue in the world's merchant navy for the next couple of years.
The compression-ignition engine (Diesel engine) is still most frequently used as the main engine in marine applications. It burns the cheapest heavy fuel oil and reveals the highest efficiency of all heat engines. The exhaust gas leaving the Diesel engine contains huge energy which can be utilised in another device (engine), thus increasing the efficiency of the entire system and reducing the emission of noxious substances to the atmosphere.
A possible solution here can be a system combined of a piston internal combustion engine and the gas and steam turbine circuit that utilises the heat contained in the exhaust gas from the Diesel engine. The leading engine in this system is the piston internal combustion engine. It seems that now, when fast container ships with transporting capacity of 8- 12 thousand TU are entering into service, the propulsion engines require very large power, exceeding 50-80 MW. On the other hand, increasing prices of fuel and restrictive ecological limits concerning the emission of NO_x and CO_2 to the atmosphere provoke the search for new solutions which will increase the efficiency of the propulsion and reduce the emission of gases to the atmosphere.
The ship main engines will be large low-speed piston engines that burn heavy fuel oil. At present, the efficiency of these engines nears 45 - 50%. For such a large power output

ranges, the exhaust gas leaving the engine contains huge amount of heat available for further utilisation.

The proposed combined system consisting of a piston internal combustion engine, a gas turbine and a steam turbine can also be used for engines of lower power, ranging between 400 ÷900 kW. For those power ranges a use of low-boiling media of organic-based refrigerant type instead of water (steam) in the steam cycle seems to be a reasonable solution. Piston internal combustion engines of this power range are used on coasting vessels, or in the inland water transport, for instance for driving cargo barges. On the other hand in inland applications the power blocks fired with solid, liquid, or gas fuels are in almost 100% the systems with steam or gas turbines.

In the Central Europe, Poland for instance, the basic fuel in power engineering is coal. Conventional electric power plants have the efficiency of an order of 38-42%, and emit large volumes of CO_2 , NO_x and/or SO_x. In order to decrease the amount of noxious substances emitted to the atmosphere and reduce the cost of production of the electric energy, combined systems are in use - consisting of gas turbines with a steam turbine circuit.

On the other hand, the combined turbine power plants can be complemented by electric power plants with a Diesel engine as the main propulsion. The exhaust gas leaving the engine contains about 30-40% of the heat delivered to the engine in the fuel. Using the heat from the exhaust gas in the gas and steam turbine circuit will increase the efficiency of the entire combined system. For large powers of piston internal combustion engines, the additional gas and steam turbine circuit is a source of measurable economic savings in electric energy production. Moreover, in large-power piston internal combustion engines we can additionally use the low-temperature waste heat, for instance for heating the communal water (Dzida, 2009). In the seaside areas with no large electric power plants, a mobile power plant situated on a platform close to the coast reveals additional advantages:

- increasing production of electric energy in the so-called distributed system,
- diversification of primary energy sources which decreases the consumption of coal in favour of liquid fuels,
- possible combustion of residual heavy fuels from nearby oil refineries,
- reducing large-distance transport of solid fuels, the absence of slag and cinders,
- reducing the emission of CO_2 and NO_x due to the increased system efficiency,
- shorter time of plant erection compared to that of a conventional power plant, and possibility of opening it in stages: first with the Diesel engine alone, and then complementing it, during plant operation, with a combined steam/gas turbine system,
- no problems with the water cooling the condenser, small effect on the environment in water balance aspects,
- mobility of a combined power plant erected on the marine platform.

2. Concept of a combined system

Combined propulsion systems are used in marine engineering mostly in fast special-purpose ships and in the Navy, as the systems being a combination of a Diesel engine and gas turbines (CODAG, CODOG) or solely gas turbines (COGOG, COGAG). The propulsion system of the passenger liner "Millenium" uses a COGES-type system which improved the efficiency and operating abilities of the ship. The system consists of a gas turbine and a steam turbine which drive an electric current generator, while the propeller screws are driven by electric motors. In this system the steam turbine circuit is supplied with the steam generated in the waste heat boiler supplied with the exhaust gas from the gas turbines.

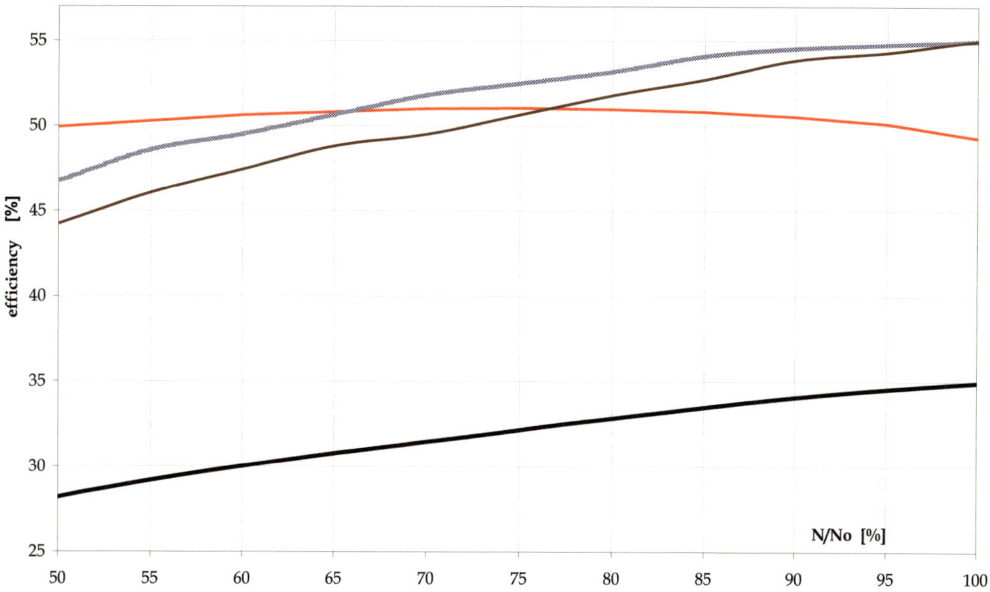

_____ CC Plant with Variable Inlet Guide, _____ CC Plant without Variable Inlet Guide
_____ Simple Gas Turbine, _____ Marine Diesel Engine

Fig. 1. Part - load Efficiency of a Combined - Cycle Plant (GT&ST), Simple Gas Turbine and Marine Diesel Engine

Combined systems used in inland power blocks base on a gas turbine as the main unit and a steam turbine that utilises the steam produced in a waste heat boiler using the heat recovered from the gas turbine exhaust gas. All this provides opportunities for reaching high efficiency of the combined block. The exhaust gas leaving a marine low-speed Diesel engine contains smaller amount of heat, of an order of 30-40% of the energy delivered to the engine.

Figure 1 shows sample efficiency curves of the combined gas turbine/steam turbine systems as functions of power plant load, compared to the gas turbine operating in a simple open circuit and the marine low-speed Diesel engine.

The efficiency curves in Fig. 1 show that the combined cycle gas turbine/steam turbine system has the highest efficiency for maximal loads (maximal efficiency levels for these circuits reach as much as 60%). Gas turbines operating in the simple open circuit have the lowest efficiency (average values of 33÷35%, and maximal values reaching 40%). Low-speed Diesel engines have the efficiency of an order of 47÷50%. It is also noticeable that the Diesel engine curve is relatively flat. This is of special importance in case of marine propulsion systems which operate at heavily changing loads. For the combined cycle gas turbine/steam turbine systems and the gas turbines operating in the simple open circuit the relative efficiency decrease $\Delta\eta/\eta o$ is equal to 15÷20% when the load decreases from 100% to 50%. For the low-speed Diesel engine these numbers are equal to 1÷2%. This property of the Diesel engine, along with the ability to utilise additional heat contained in its exhaust gas, makes the engine the most applicable in marine propulsion systems operating in heavily changing load conditions. The amount of heat contained in the exhaust gas from the gas turbine is approximately equal to 60÷65%, i.e. more than in piston engines, which results from lower exit temperature and less intensive flow of the exit gas leaving the Diesel engine, Figs. 2 and 3.

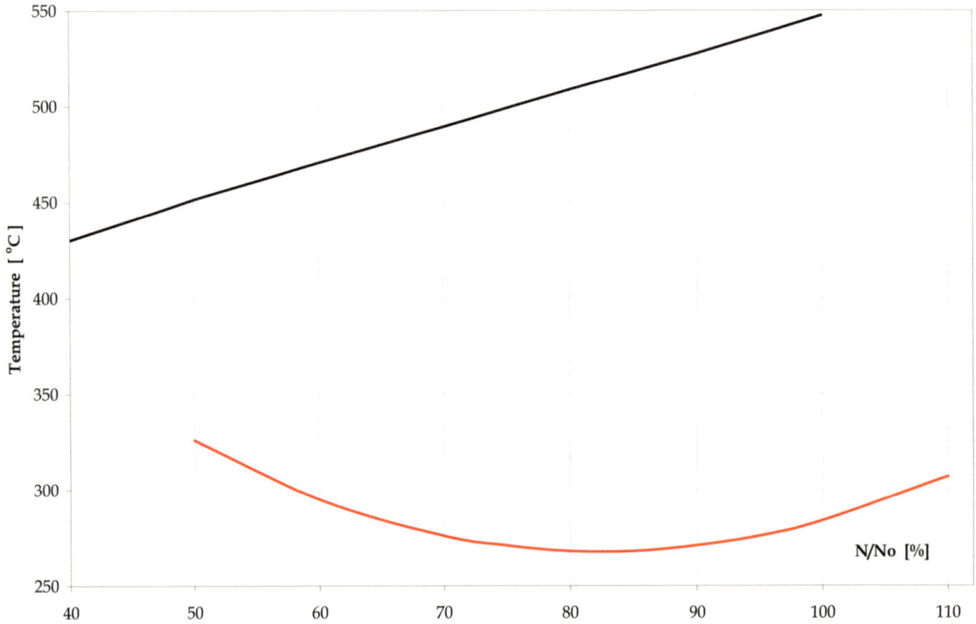

_____ Gas Turbine _____ Marine Diesel Engine

Fig. 2. Temperatures of the exhaust gas from the Diesel engine and the gas turbine as function of power plant load

The exit temperatures of the exhaust gas from the gas turbines range between 450÷600^{0}C, on average, while those from the low-speed Diesel engines are of an order of 220÷300^{0}C. In the gas turbines, decreasing the load remarkably decreases the temperature of the exhaust gas, while in the Diesel engine these changes are much smaller, and the temperature initially decreases and then starts to increase for low loads.

This property of the steam turbine circuit in the combined system with the Diesel engine for partial loads makes it possible to keep the live steam temperature at a constant level within a wide range of load. The related exhaust gas mass flow rate m_g/N [kg/kWh] changes only by about 5% in the Diesel engine when the load changes from 100% to 50%, while in the gas turbine this parameter changes by about 55% for the same load change, Fig. 3.

The combined propulsion system with the low-speed piston internal combustion engine used as the main engine and making use of the heat from the engine exhaust gas is shown in Fig. 4, (Dzida, 2009; Dzida & Mucharski, 2009; Dzida et al., 2009).

The exhaust gas flows leaving individual main engine cylinders are collected in the exhaust manifold and passed to the constant-pressure turbocharger. Due to high turbocharger efficiency ranges (MAN, 2010; Schrott, 1995), the scavenge air can be compressed using the energy contained only in part of the exhaust gas flow. The remaining part of the exhaust gas flow can be expanded in an additional gas turbine, the so-called power turbine, which additionally drives, via a gear, the propeller screw or the electric current generator.

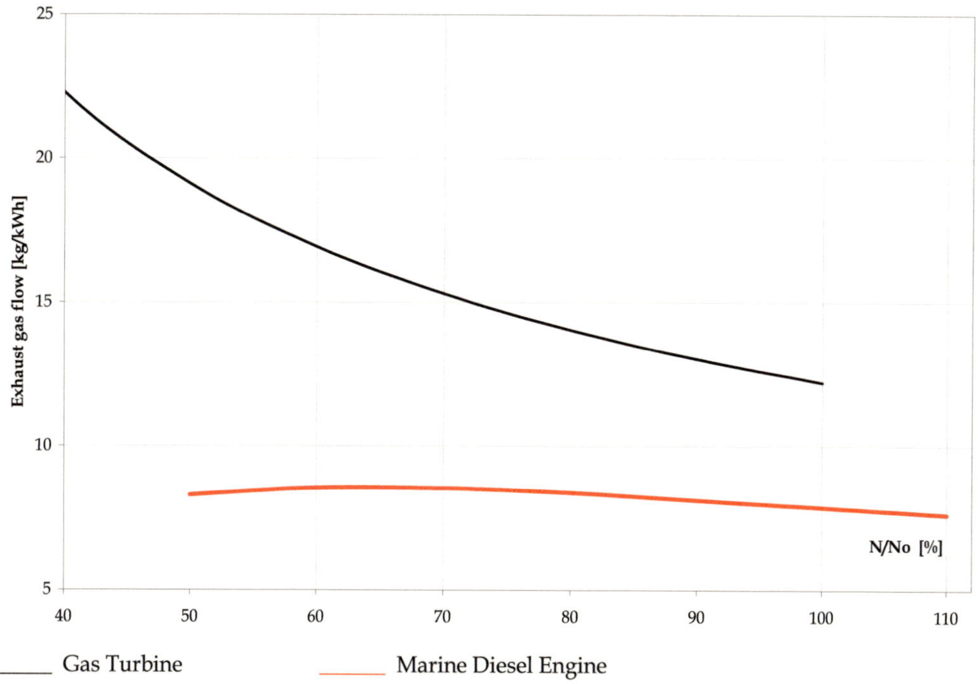

Fig. 3. Related exhaust gas mass flow rate as a function of power plant load

The exhaust gas from the turbocharger and the power turbine flows to the waste heat boiler installed in the main engine exhaust gas path, before the silencer. The waste heat boiler produces the steam used both for driving the steam turbine that passes its energy to the propeller screw, and for covering all-ship needs.

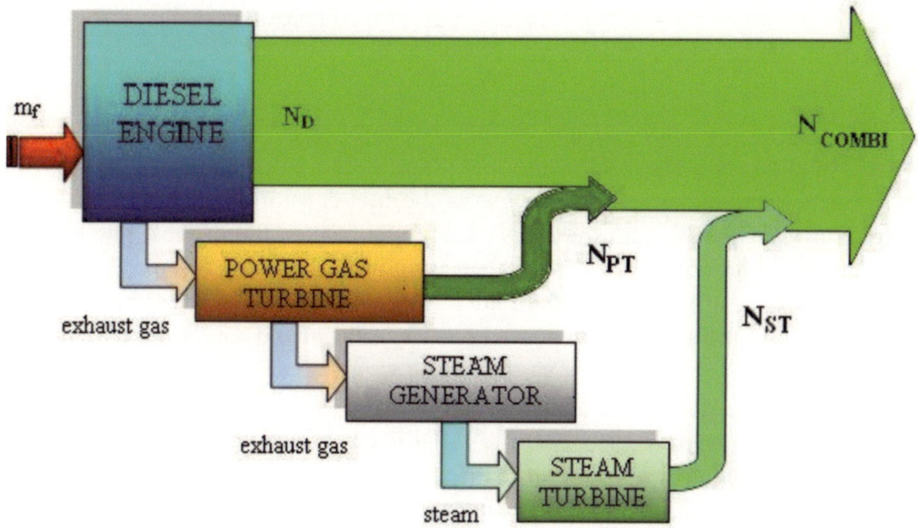

Fig. 4. Concept of the combined propulsion system

In the marine low-speed Diesel engines, another portion of energy that can be used along with the exhaust gas energy is a huge amount of so-called waste heat of relatively low temperature. In the low-speed engines the waste heat comprises the following components (with their proportions to the heat delivered to the engine in fuel):

- heat in the scavenge air cooler (17-20%), of an approximate temperature of about 200⁰C,
- heat in the lubricating oil cooler (3-5%), of an approximate temperature of about 50⁰,.
- heat in the jacket water cooler (5-6%), of the temperature of an order of 100⁰C.

This shows that the amount of the waste heat that remains for our disposal is equal to about 25-30% of the heat delivered in fuel. Part of this heat can be used in the combined circuit with the Diesel engine.

2.1 Energy evaluation of the combined propulsion system

The adopted concept of the combined ship propulsion system requires energy evaluation, Fig. 4. Formulas defining the system efficiency are derived on the basis of the adopted scheme.

The power of the combined propulsion system is determined by summing up individual powers of system components (the main engine, the power gas turbine, and the steam turbine):

$$N_{combi} = N_D + N_{PT} + N_{ST} \tag{1}$$

hence the efficiency of the combined system is:

$$\eta_{combi} = \frac{N_{combi}}{m_{fD} \cdot Wu} = \eta_D \cdot \left(1 + \frac{N_{PT}}{N_D} + \frac{N_{ST}}{N_D}\right) \tag{2}$$

and the specific fuel consumption is:

$$b_{ecombi} = b_{eD} \cdot \frac{1}{(1 + \frac{N_{PT}}{N_D} + \frac{N_{ST}}{N_D})} \quad [g / kWh] \tag{3}$$

where η_D, b_{eD}- is the efficiency and specific fuel consumption of the main engine.

Relations (2) and (3) show that each additional power in the propulsion system increases the system efficiency and, consequently, decreases the fuel consumption. And the higher the additional power achieved from the utilisation of the heat in the exhaust gas leaving the main engine, the lower the specific fuel consumption. Therefore the maximal available power levels are to be achieved from both the power gas turbine and the steam turbine. The power of the steam turbine mainly depends on the live steam and condenser parameters.

2.2 Variants of the combined ship propulsion systems or marine power plants

For large powers of low-speed engines, the exhaust gas leaving the engine contains huge amount of heat available for further utilisation. Marine Diesel engines are always supercharged. Portions of the exhaust gas leaving individual cylinders are collected in the exhaust gas collector, where the exhaust gas pressure $p_{exh\_D} > p_{bar}$ is equalised. In standard solutions the constant-pressure turbocharger is supplied with the exhaust gas from the

exhaust manifold to generate the flow of the scavenge air for supercharging the internal combustion engine.
Present-day designs of turbochargers used in piston engines do not need large amounts of exhaust gas, therefore it seems reasonable to use a power gas turbine complementing the operation of the steam turbine in those cases. Here, two variants of power gas turbine supply with the exhaust gas are possible.

2.2.1 Parallel power gas turbine supply (variant A)

In this case part of the exhaust gas from the piston engine exhaust manifold supplies the Diesel engine turbocharger. The remaining part of the exhaust gas from the manifold is directed to the gas turbine, bearing the name of the power turbine (PT). The power turbine drives, via the reduction gear, the propeller screw or the electric current generator, thus additionally increasing the power of the entire system. Figure 5 shows a concept of this propulsion system, referred to as parallel power turbine supply. After the expansion in the turbocharger and the power turbine, the exhaust gas flowing from these two turbines is directed to the waste heat boiler in the steam circuit.

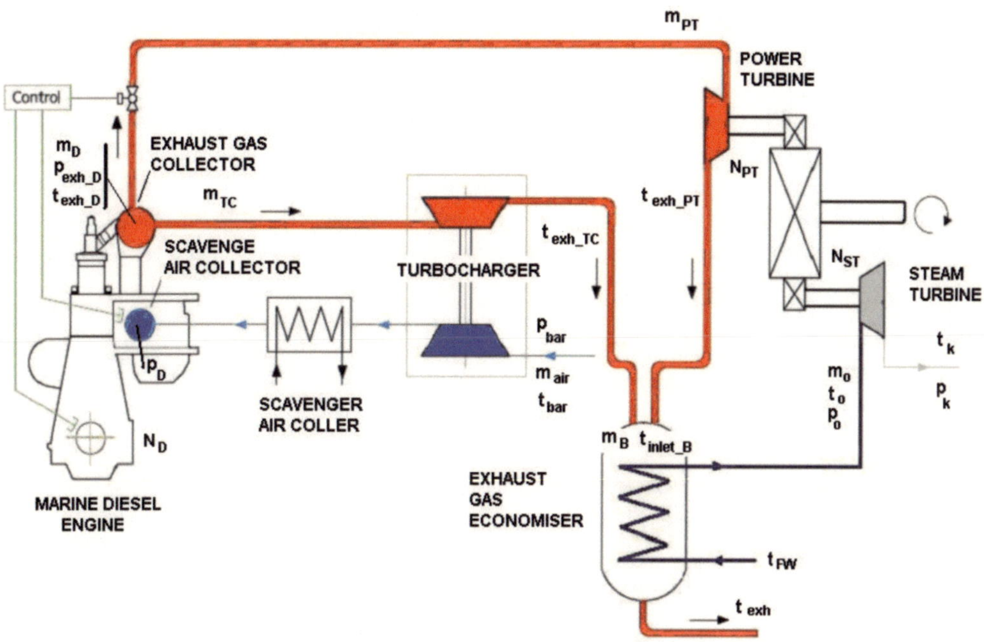

Fig. 5. Combined system with the Diesel main engine, the power turbine supplied in parallel, and the steam turbine (variant A)

In the proposed solution, at low load ranges the amount of the exhaust gas from the main engine is not sufficient to additionally supply the power turbine. In such case a control valve closes the exhaust gas flow to the power turbine, Figure 5. The operation of this valve is controlled by the control system using two signals: the scavenge air pressure signal, and the signal of the propeller shaft angular speed or torque. The waste heat boiler produces the steam which is then used both in the steam turbine and, in case of marine application, to cover the all-ship needs. This system allows for independent operation of the Diesel engine, with the steam turbine or the power turbine switched off. The control system makes it possible to switch off the power turbine thus increasing the power of the turbocharger at partial load, and, on the other hand, direct part of the Diesel engine exhaust gas to supply the power turbine at large load.

Power turbine calculations are based on the Diesel engine parameters, i.e. the temperature of the exhaust gas in the exhaust gas collector, which in turn depends on the engine load and air parameters at the engine inlet. Marine engine producers most often deliver the data on two reference points for the atmospheric air (the ambient reference conditions):

ISO Conditions		Tropical Conditions
Ambient air temperature [^{0}C]	25	45
Barometric pressure [bar]	1	1

2.2.2 Series power gas turbine supply (variant B)

In this variant the exhaust gas from the exhaust manifold supplies first the piston engine turbocharger and then the power turbine, Fig.6.

After leaving the exhaust manifold, the exhaust gas expands in the turbocharger to the higher pressure than the atmospheric pressure, which leaves part of the exhaust gas enthalpy drop for utilisation in the power turbine. The exhaust gas leaving the power turbine passes its heat to the steam in the waste heat boiler, thus producing additional power in the steam turbine circuit.

Also in this combined system, the installed control valve makes it possible to switch off the power turbine at partial piston engine loads, thus increasing the power of the turbocharger by expanding the exhaust gas to lower pressure, Fig. 6. Unlike the parallel supply variant, here the entire mass of the exhaust gas from the piston engine manifold flows through the turbocharger. The exhaust gas pressure at the turbocharger outlet is higher than in variant A.

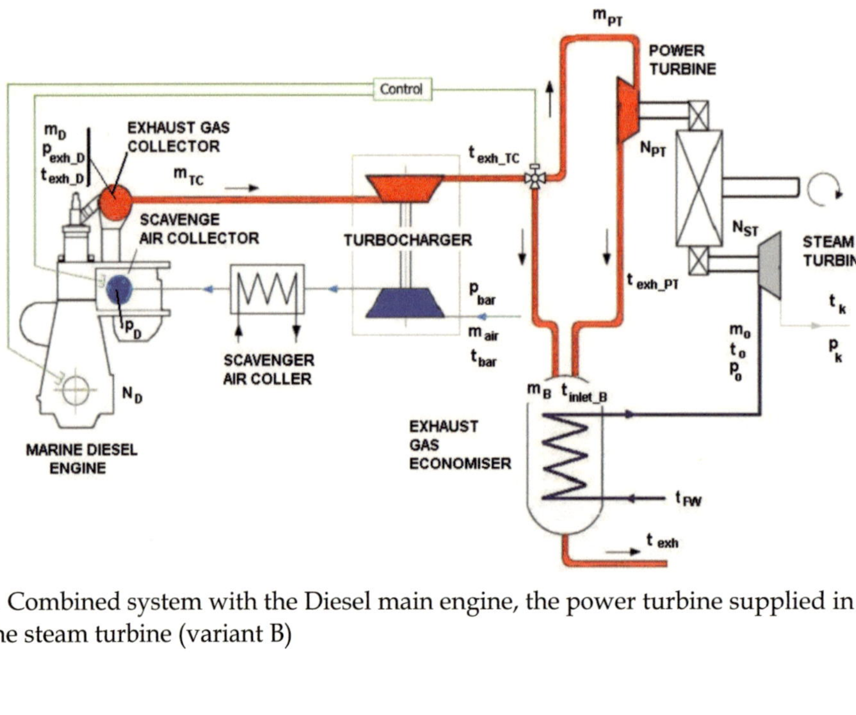

Fig. 6. Combined system with the Diesel main engine, the power turbine supplied in series, and the steam turbine (variant B)

3. Power turbine in the combined system

Calculating the power turbine in the combined system depends on the selected variant of power turbine supply. Usually, piston engine producers do not deliver the exhaust gas temperature in the exhaust manifold (which is equal to the exhaust gas temperature at turbocharger turbine inlet). Instead, they give the exhaust gas temperature at turbocharger turbine outlet (t_{exh_D}). The temperatures of the exhaust gas in the Diesel engine exhaust gas collector are calculated from the turbine power balance, according to the following formula:

$$t_{exh\_D} = \frac{t_{exh\_TC} + 273{,}15}{1 - \eta_T \cdot \left(1 - \frac{1}{\pi_T^{\frac{\kappa g - 1}{\kappa g}}}\right)} \text{ -273,15 [°C]} \quad (4)$$

This formula needs the data on turbocharger turbine efficiency changes for partial loads. These data can be obtained from the producer of the turbocharger (as they are rarely made public), Fig. 7, or calculated based on the relation used in steam turbine stage calculations:

$$\overline{\eta_T} \equiv \frac{\eta_T}{\eta_{To}} = 2 \cdot \overline{v} - \overline{v}^2 \quad (5)$$

where ν - related turbine speed indicator, η_{To}- maximal turbine efficiency and the corresponding speed indicator.

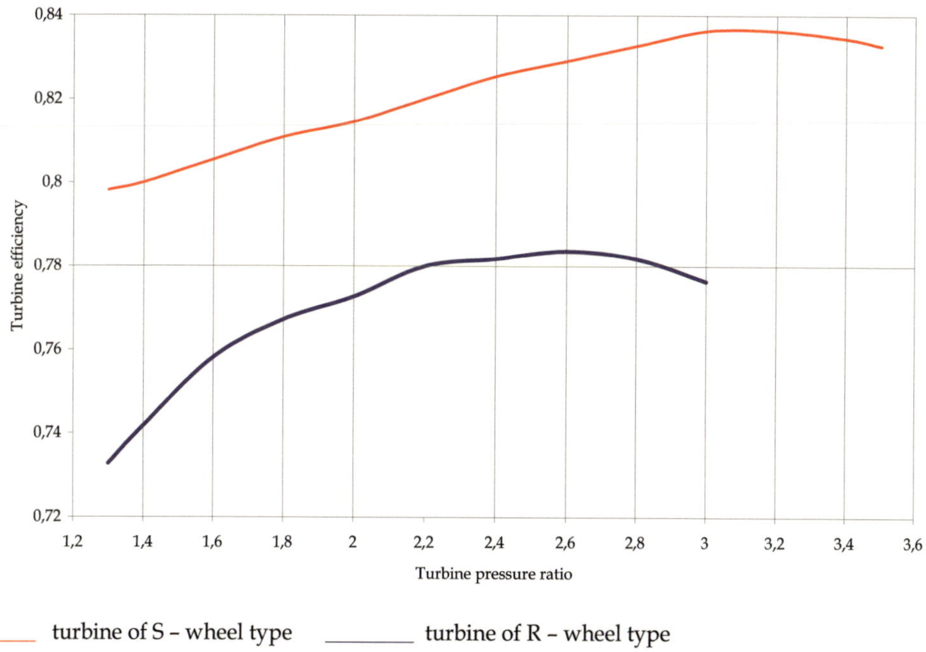

Fig. 7. Turbocharger turbine efficiency as a function of scavenge air pressure, acc. to (Schrott, 1995)

The turbine speed indicator is defined as:

$$\nu = \frac{u}{c_s} = \sqrt{\frac{u^2}{2 \cdot H_T}} \tag{6}$$

where u- circumferential velocity on the turbine stage pitch diameter, H_T- enthalpy drop in the turbine.
The calculations make use of static characteristics of the turbocharger compressor, with the marked line of cooperation with the Diesel engine, Fig.8.
Figure 9 shows the turbocharger efficiency curves calculated from the relation:

$$\eta_{TC} = \eta_T \cdot \eta_C \cdot \eta_m \tag{7}$$

where η_T - the turbocharger turbine efficiency is calculated from relation (5), while the compressor efficiency η_C is calculated from the line of Diesel engine/compressor cooperation, η_m - mechanical efficiency of the turbocharger, Fig. 8. In the same figure a comparison is made between the calculated turbocharger turbine efficiency with the producer's data as a function of the Diesel engine scavenge pressure. The differences between these curves do not exceed 1,5%.
For the presently available turbocharger efficiency ranges, the amount of the exhaust gas needed for driving the turbocharger turbine is smaller than the entire mass flow rate of the exhaust gas leaving the Diesel engine. Fig. 10 shows sample curves of exhaust gas

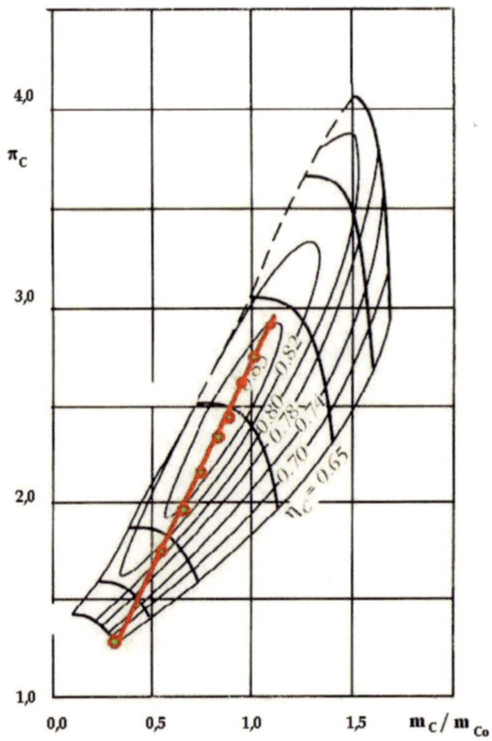

Fig. 8. Diesel engine cooperation line against turbocharger compressor characteristics

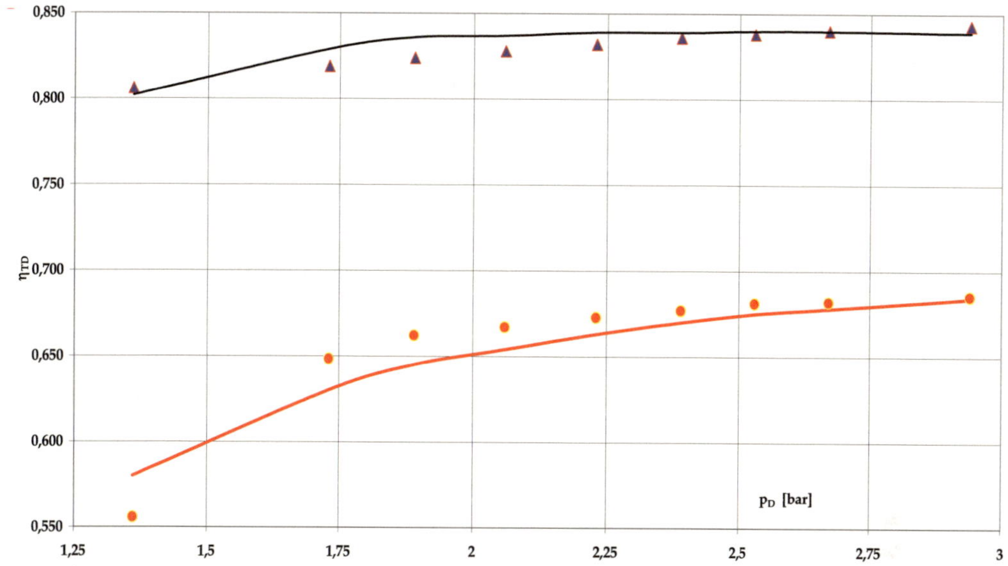

_____ turbocharger efficiency, acc. to producer _____ gas turbine efficiency, acc. to producer ●, ▲calculated efficiency

Fig. 9. Efficiency characteristics of the turbocharger and the turbocharger gas turbine as a function of scavenge air pressure

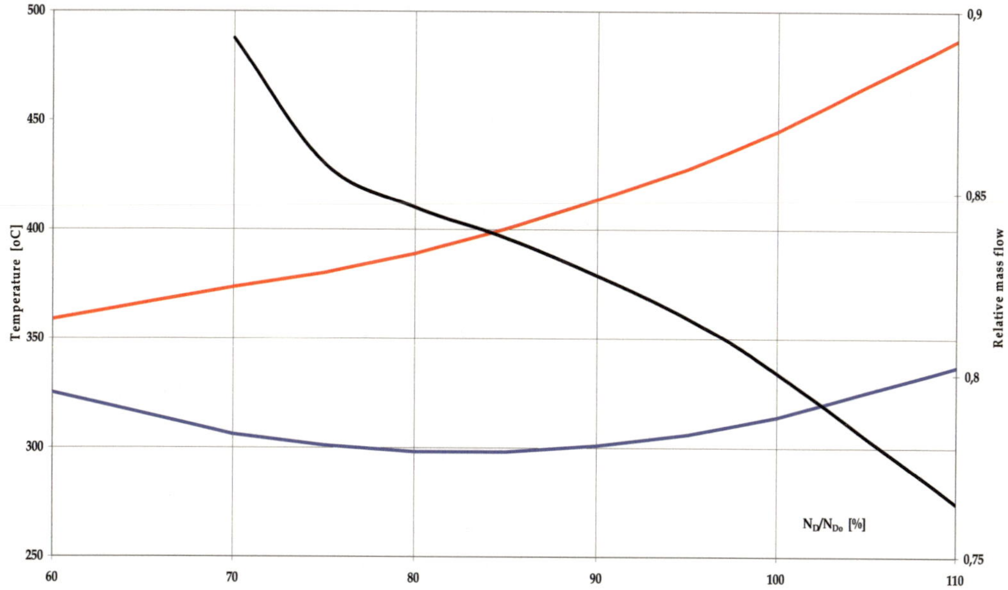

_____ temperature in the Diesel engine exhaust gas collector-calculated curves _____ exhaust gas temperature at turbocharger outlet – producer's data _____ Diesel engine exhaust gas mass flow rate related to the scavenge air mass flow rate

Fig. 10. Sample temperature characteristics of the turbocharger during gas expansion in the turbine to the atmospheric pressure and the related exhaust gas mass flow rates as functions of Diesel engine load temperature changes in the engine manifold (calculated using the relation (4)) and the exhaust gas temperature at the turbocharger outlet (according to the data delivered by the producer) as functions of engine load, when the standard internal combustion engine exhaust gas is expanded to the barometric pressure. The figure also shows the Diesel engine exhaust gas flow rate related to the scavenge air flow rate, as a function of the engine load. This high efficiency of the turbocharger provides opportunities for installing a power gas turbine connected in parallel with the turbocharger (variant A).

The turbocharger power balance indicates that in the power gas turbine we can utilise between 10 and 24% of the flow rate of the exhaust gas leaving the exhaust manifold of the piston engine. The power gas turbine can be switched on when the main engine power output exceeds 60%. For lower power outputs the entire exhaust gas flow leaving the Diesel engine is to be used for driving the turbocharger.

In variant B of the combined system with the power turbine, the turbocharger is connected in series with the power gas turbine. Here, the entire amount of the exhaust gas flows through the turbocharger turbine. Due to the excess of the power needed for driving the turbocharger, the final expansion pressure at turbocharger turbine output can be higher than the exhaust gas pressure at waste heat boiler inlet. In this case the expansion ratio in the turbocharger turbine is given by the relation:

$$\pi_T = \left[\frac{1}{1 - \frac{1}{\eta_{TC}} \cdot \frac{m_a}{m_D} \cdot \frac{c_a}{c_g} \cdot \frac{t_a}{t_{exh\_D}} \cdot \left(\pi_C^{\frac{\aleph_a - 1}{\aleph_a}} \right)} \right]^{\frac{\aleph_g}{\aleph_g - 1}} \tag{8}$$

where: π_C- compression ratio of the turbocharger compressor.

The exhaust gas temperature at turbocharger outlet is calculated from the formula:

$$t_{exh\_TC} = \left(t_{exh\_D} + 273{,}15 \right) \cdot \left[1 - \eta_T \left(1 - \frac{1}{\pi_T^{\frac{\aleph_g - 1}{\aleph_g}}} \right) \right] - 273{,}15 \ [^\circ C] \tag{9}$$

Figure 11 shows sample curves of temperature, compression and expansion rate changes in the turbocharger for variant B: series power turbine supply.

This case provides opportunities for utilising the enthalpy drop of the expanding exhaust gas in the power turbine. The operation of the power turbine is possible when the Diesel engine power exceeds 60%.

3.1 Power turbine in parallel supply system (variant A)

The power turbine (Fig.5) is supplied with the exhaust gas from the exhaust manifold. The exhaust gas mass flow rate m_{PT} and temperature t_{exh_D} are identical as those at turbocharger outlet: the mass flow rate of the exhaust gas flowing through the power turbine results from the difference between the mass flow rate of the Diesel engine exhaust gas and of that expanding in the turbocharger:

$$m_{TD} = m_a \cdot (1 - \overline{m}) + m_{fD} \tag{10}$$

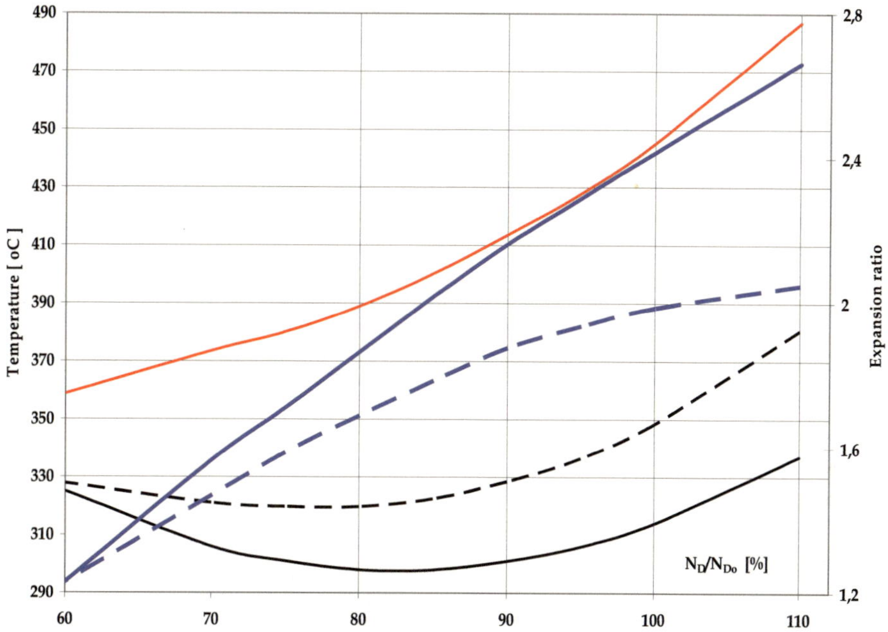

_____expansion ratio in the turbocharger turbine (standard arrangement - without power turbine) _ _ _ expansion ratio in the turbocharger turbine with power turbine ______exhaust gas temperature in the Diesel engine exhaust gas collector ____exhaust gas temperature at turbocharger outlet without power turbine _ _ _ exhaust gas temperature at turbocharger outlet with power turbine

Fig. 11. Changes of temperature and expansion ratio of the turbocharger in the combined system with series power turbine supply (variant B)

The mass flow rate of the exhaust gas needed by the turbocharger is calculated from the turbocharger power balance using the following formula:

$$\bar{m} \equiv \frac{m_{TC}}{m_a} = \frac{1-\dfrac{1}{\pi_T^{\frac{\kappa_g-1}{\kappa_g}}}}{\pi_C^{\frac{\kappa_a-1}{\kappa_a}}-1} \cdot \frac{T_{exh\_D}}{T_a} \cdot \frac{c_g}{c_p} \cdot \eta_{TC} \tag{10.1}$$

The exhaust gas expanding in the power turbine has the inlet and outlet pressures identical to those of the exhaust gas flowing through the turbocharger. The power of the power turbine is given by the relation:

$$N_{PT} = \eta_m \cdot \eta_{PT} \cdot m_{PT} \cdot H_{PT} \tag{11}$$

where η_m- mechanical efficiency of the power turbine, H_{PT} - iso-entropic enthalpy drop in the power turbine.

The power turbine efficiency η_{PT} is assumed in the same way as for the turbocharger turbine, Fig. 9, or using the relation (5). In the shipbuilding, the gas turbines used in combined Diesel engine systems with power turbines are those adopted from turbochargers.

The power turbine system calculations show that the exhaust gas temperature at the power turbine outlet is slightly higher than that at the turbocharger outlet, Fig.12. The increase of the main engine load results in the increase of both the exhaust gas temperature in the exhaust gas collector and the mass flow rate of the exhaust gas flowing through the power turbine. The increase in power of the combined system with additional power turbine ranges from about 2% for Diesel engine loads of an order of 70% up to over 8% for maximal loads, Fig.12.

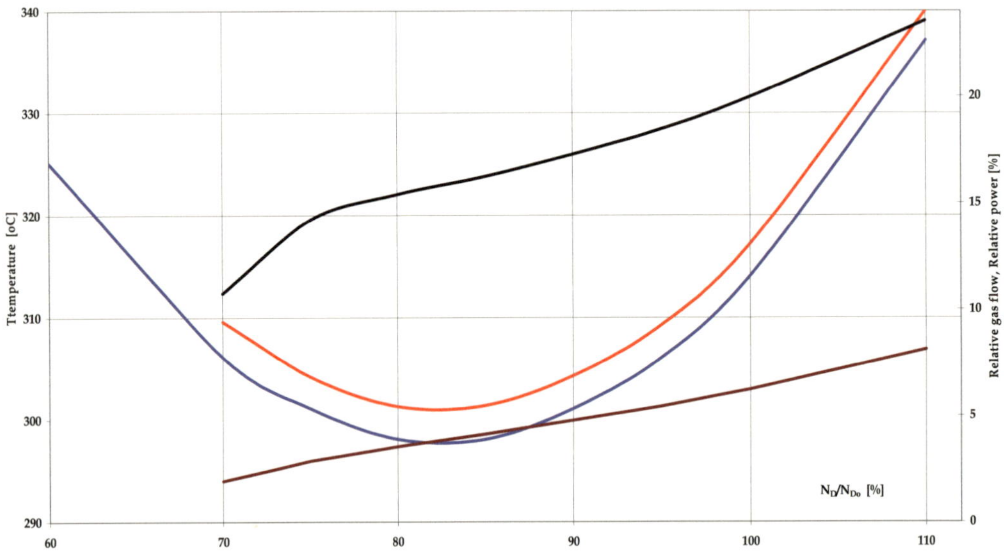

_____ temperature at turbocharger outlet _____ temperature at power turbineoutlet _____ related exhaust gas mass flow rate in power turbine _____ related power turbine power

Fig. 12. Parameters of parallel supplied power turbine as functions of the main engine load – variant A (calculations for tropical conditions)

When the Diesel engine power is lower than 60-70% of the nominal value the entire exhaust gas flow from the exhaust manifold is directed to the turbocharger drive. In this case the control system closes the valve controlling the exhaust gas flow to the power turbine, Fig. 5.

3.2 Power turbine in series supply system (variant B)

In this variant the power turbine is supplied with the full amount of the exhaust gas leaving the Diesel engine exhaust manifold. The power turbine is installed after the turbocharger. The exhaust gas pressure at the power turbine inlet depends on the pressure of the exhaust gas leaving the turbocharger turbine, Fig.11.

In this case the power of the power turbine is calculated as:

$$N_{PT} = \eta_{PT} \cdot m_D \cdot c_g \cdot t_{inl\_PT} \cdot \left(1 - \frac{1}{\pi_{PT}^{\frac{\kappa_g - 1}{\kappa_g}}} \right) \qquad (12)$$

where t_{inl_PT} - exhaust gas temperature at the power turbine inlet, π_{PT}– expansion ratio in the power turbine , η_{PT}- power turbine efficiency. The power turbine efficiency is assumed in the same way as in variant A.
In formula (12) the exhaust gas temperature at the power turbine inlet is assumed equal to that of the exhaust gas leaving the turbocharger, Fig. 13.
The exhaust gas temperature at the power turbine output is calculated from the formula:

$$t_{exh\_PT} = \left(t_{inl\_PT} + 273{,}15\right)\cdot\left[1-\eta_{PT}\left(1-\frac{1}{\pi_{PT}^{\frac{\aleph_g-1}{\aleph_g}}}\right)\right] - 273{,}15[^{\circ}C] \qquad (13)$$

Figure 13 also shows the expansion ratio, the power of the power turbine, and the exhaust gas temperatures at the turbocharger and the power turbine outlets for partial engine loads. The power turbine in this variant increases the power of the combined system by 3% to 9% with respect to that of a standard engine. The turbine power increases with increasing Diesel engine load.

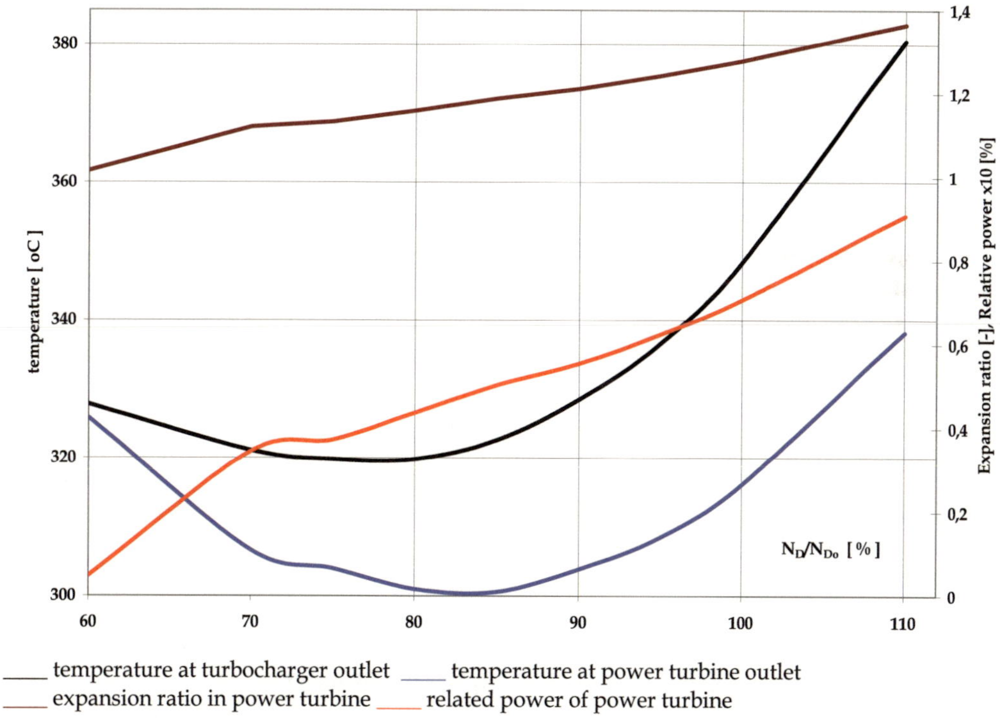

Fig. 13. Parameters of series supplied power turbine as functions of the main engine load - variant B (calculations for tropical conditions)

3.3 Comparing the two power turbine supply variants

The analysis of the two examined variants shows that the power of the combined system increases depending on the Diesel engine load. For both variants the power turbine can be

used after exceeding about 65% of the Diesel engine power. The exhaust gas leaving the power turbine is directed to the waste heat boiler, where together with steam turbine it can additionally increase the overall power of the combined system.
In both cases the temperatures of the exhaust gas leaving the power turbine are comparable. The exhaust gas pressure at power turbine outlet depends on the losses generated when the gas flows through the waste heat boiler and outlet silencers. Following practical experience, the exhaust gas back pressure is assumed higher than the barometric pressure by 300 mmWC, i.e. about 3%. Taking into account powers of the power turbines for the above variants, Fig. 14, it shows that for the same Diesel engine parameters the series supply of the power turbine results in higher turbine power. For lower loads, the power of the series supplied power turbine increases, compared to the parallel supply variant.

4. Steam turbine circuit

The combined system makes use of the waste heat from the Diesel engine. In modern Diesel engines the temperatures of the waste heat are at the advantageous levels for the steam turbine circuit. This circuit makes use of water that can be utilised in a low-temperature process. Adding the steam circuit to the combined Diesel engine/power gas turbine system provides good opportunities for increasing the power of the combined system, and consequently, also the system efficiency, see formula (2).
In the examined combined system the exhaust gas leaving the turbocharger and the power turbine (variant A, Fig. 5) or only the power turbine (variant B, Fig. 6) flows to the waste heat boiler where it is used for producing superheated steam for driving the steam turbine.
The mass flow rate of the exhaust gas reaching the waste heat boiler is equal to that leaving the Diesel engine exhaust gas collector. The exhaust gas temperature at waste heat boiler inlet depends on the adopted solution of power turbine supply. For variant A with parallel supply it is calculated from the balance of mixing of the gases leaving the turbocharger and the power turbine:

$$t_{inl\_B} = \frac{m_{TC} \cdot i_{exh\_TC} + m_{PT} \cdot i_{exh\_PT}}{m_D \cdot c_g} - 273{,}15 \ [^{o}C] \tag{14}$$

while for the series power turbine supply (variant B) it is assumed equal to that at the power turbine outlet, formula (13).
In combined steam turbine systems for small power ranges and low live steam temperatures the single pressure systems are used, Fig. 15, (Kehlhofer, 1991).
Such system consists of a single-pressure waste heat boiler, a condensing steam turbine, a water-cooled condenser, and a single stage feed water preheater in the deaerator.
The main disadvantage of the systems of this type is poor utilisation of the heat contained in the exhaust gas (the waste heat energy). The steam superheater is relatively large, as the entire mass of the steam produced by the boiler flows through it. However, costs of this steam system are the lowest, as poor utilisation of the exhaust gas energy results in high temperature of the exhaust gas leaving the boiler. The deaerator is supplied with the steam extracted from the steam turbine. The application of the single pressure system does not secure optimal utilisation of the exhaust gas energy.

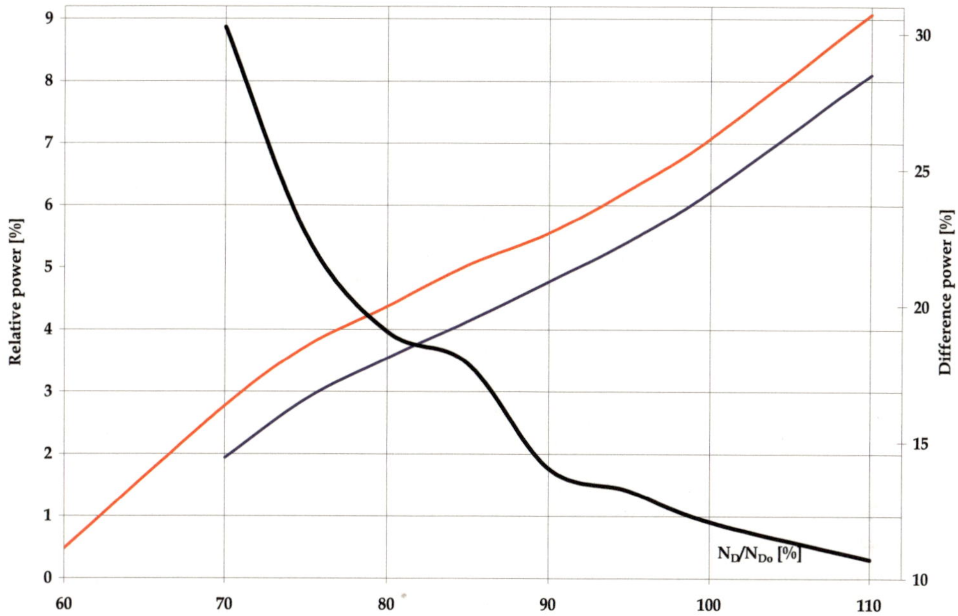

____ parallel power turbine supply (variant A) ____ series power turbine supply (variant B)

Fig. 14. Powers of the power turbine as functions of main engine load

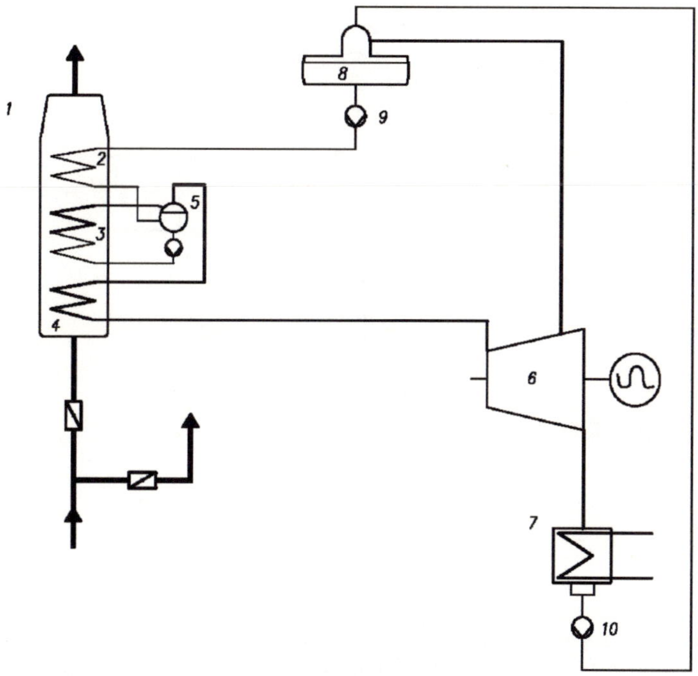

1-Waste Heat Boiler 2-Superheater 3- Evaporator 4-Ekonomizer 5-Boiler drum 6-Steam turbine 7-Condenser 8-Deaerator 9-Feed water pump 10-Condensate pump

Fig. 15. Flow Diagram of the Single Pressure System

Those steam turbine systems frequently make use of an additional low-pressure evaporator, Fig. 16, which leads not only to more intensive utilisation of the waste heat contained in the exhaust gas, but also to better thermodynamic use of the low-pressure steam.
In this solution the high pressure superheater is relatively small, compared to the single pressure boiler. The deaerator is heated with the saturated steam from the low-pressure evaporator. The power of the main high-pressure feeding pump is also smaller. The excess steam from the low-pressure evaporator can be used for supplying the low-pressure part of the steam turbine, thus increasing its power, or, alternatively, for covering all-ship needs.
Figure 16 shows possible use of the temperature waste heat from the scavenge air cooler, the lubricating oil cooler, and from the jacket water cooler in the low-pressure water pre-heater.
The additional low-pressure exchanger in the steam circuit, Fig. 16, makes it possible to increase the temperature of the water in the deaerator. Higher water temperature is required due to the presence of sulphur in the fuel (water dew-point in the exhaust gas) – it is favourable for systems fed with a high sulphur content fuel. If the temperature of the feedwater is low when the system is fed with fuel without sulphur, the heat exchanger 14 in Fig. 16 is not necessary and the waste heat from the coolers can be used in the deaerator. For a low feedwater temperature the deaerator works at the pressure below atmospheric (under the vacuum).

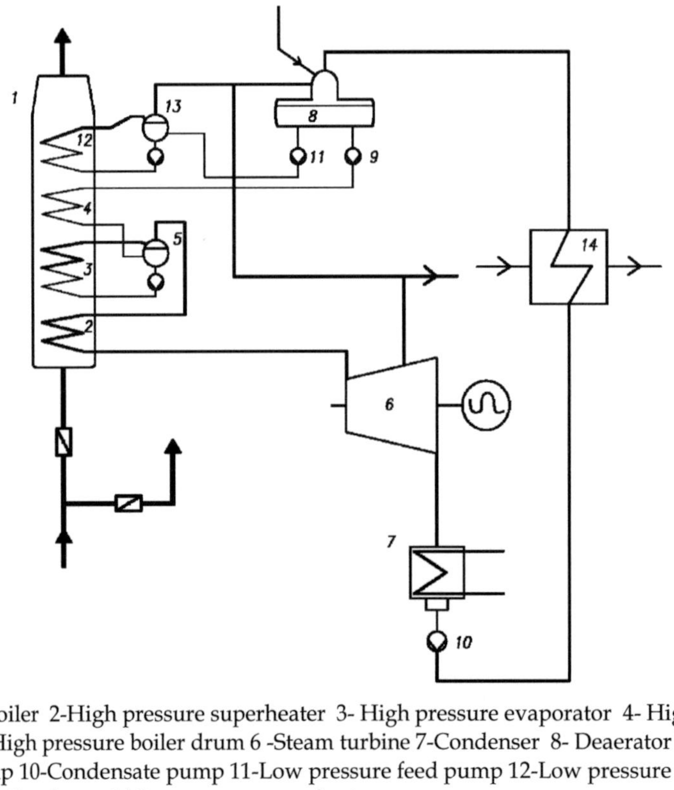

1-Waste Heat Boiler 2-High pressure superheater 3- High pressure evaporator 4- High pressure economizer 5- High pressure boiler drum 6 -Steam turbine 7-Condenser 8- Deaerator 9-High pressure feed water pump 10-Condensate pump 11-Low pressure feed pump 12-Low pressure evaporator 13-Low pressure boiler drum 14-Low pressure pre-heater

Fig. 16. Flow Diagram for a Two – Pressure System

4.1 Limits for steam circuit parameters

The limits for the values of the steam circuit parameters result from strength and technical requirements concerning the durability of particular system components, but also from design and economic restrictions. The difference between the exhaust gas temperature and the live steam temperature, Δt, for waste heat boilers used in shipbuilding is assumed as Δt = 10-15°C, according to (MAN, 1985; Kehlhofer, 1991). The "pitch point" value recommended by MAN B&W (MAN, 1985) for marine boilers is δt = 8-12°C. The limiting dryness factor x of the steam downstream of the steam turbine is assumed as x_{limit}=0,86-0,88. For marine condensers cooled with sea water, MAN recommends the condenser pressure p_K=0,065 bar. This pressure depends on the B&W (MAN, 1985) temperature of the cooling medium in the condenser. Figure 17 shows the dependence of the condenser pressure on the cooling medium temperature. The temperature of the boiler feed water is of high importance for the life time of the feed water heater in the boiler. The value of this temperature is connected with a so-called exhaust gas dew-point temperature. Below this temperature the water condensates on heater tubes and reacts with the sulphur trioxide SO_3 producing the sulphuric acid, which is the source of low-temperature corrosion. That is why boiler producers give minimal feed water temperatures below which boiler operation is highly not recommended. The dew-point temperature is connected with the content of sulphur in the fuel and depends on the excess air coefficient in the piston engine. Figure 18 shows the dew-point temperature as the function of: sulphur content in the fuel, SO_2 conversion to SO_3, and the excess air coefficient in the engine. In inland power installations burning fuels with sulphur content higher than 2%, the recommended level of feed water temperature is t_{FW} > 140-145°C (Kehlhofer, 1991).

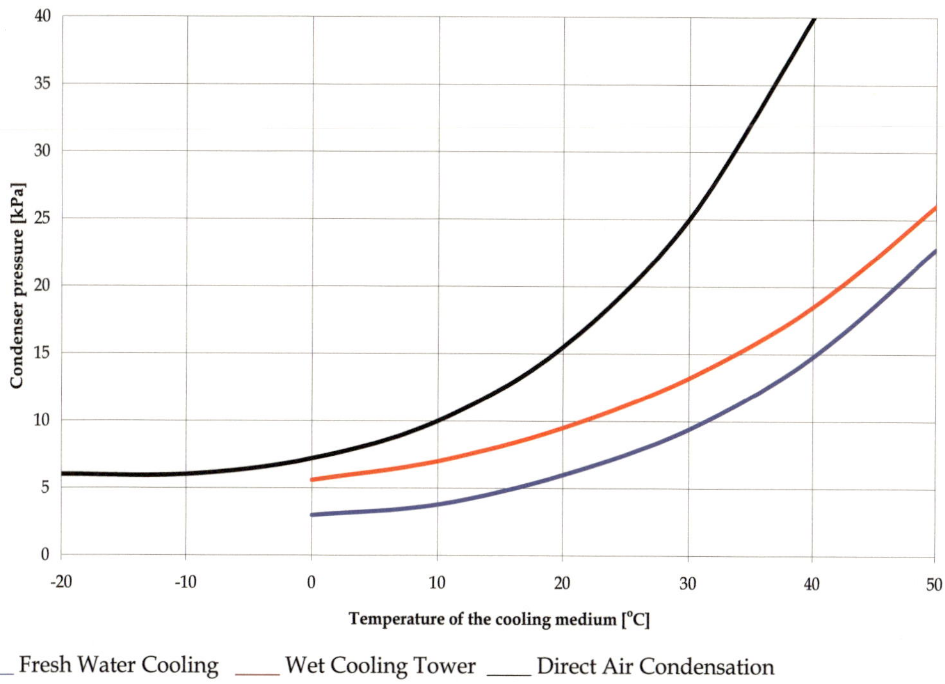

Fig. 17. Condenser pressure as a function of temperature of the cooling medium

In marine propulsion (MAN, 1985) recommends that the feed water temperature should not be lower than 120°C when the sulphur content is higher than 2%. This is justified by the fact that the outer surface of the heater tubes on the exhaust gas side has the temperature higher by 8-15 °C than the feed water temperature, and that the materials used in those heaters reveal enhanced resistance to acid corrosion.

The exhaust gas temperature at the boiler outlet is assumed higher by 15-20°C than the feed water temperature, i.e. $t_{exh} > t_{FW} + (15 - 20°C)$.

Each ship burning heavy fuel in its power plant uses the mass flow rate m_{SS} of the saturated steam taken from the waste heat boiler for fuel pre-heating and all-ship purposes. According to the recommendations (MAN, 1985) the pressure of the steam used for these purposes should range between p_{SS} = 7-9 bar. This pressure is also assumed equal to the pressure in the boiler low-pressure circuit. The back temperature of the above steam flow in the heat box is within 50 - 60°C.

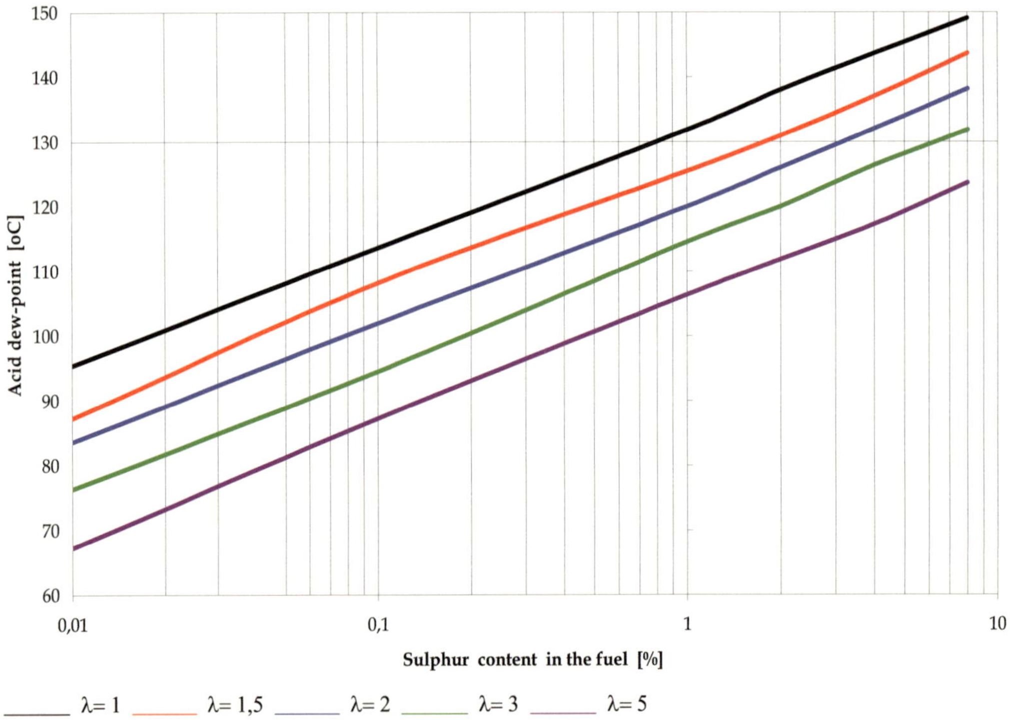

Fig. 18. Acid dew-point as a function of the sulphur content in the fuel and the excess air coefficient λ

4.2 Optimising the steam circuit

Optimisation of the steam system is to be done in such a way so as to reach the maximal possible utilisation of the heat contained in the exhaust gas. In this sense the optimisation is reduced to selecting the steam circuit parameters for which the steam turbine reaches the highest power. The area of search for optimal steam circuit parameters is to be narrowed to the sub-area where the earlier discussed limits imposed on the steam system are met. The use of the steam system with the waste heat boiler increases the power of the propulsion system within the entire range of the main engine load.

Adding a steam turbine to the Diesel engine system increases the power of the propulsion system by $\Delta N_{ST}/N_D$ = 6,5 – 7,5% for main engine loads ranging from 90 to 100%. The power of the steam turbine for both examined variants of power turbine supply are comparable, and slightly higher power, by about 2-4%, is obtained by the steam turbine in the variant with series power turbine supply.

The analysis of the system with an additional exchanger utilising the low-temperature waste heat from the Diesel engine to heat the condensate from the condenser before the deaerator, Fig.16, shows that the steam turbine power increases by 7- 11% with respect to that of the steam turbine without this exchanger.

The requirements concerning the waste heat boiler refer to low loss of the exhaust gas flow (which reduces the final expansion pressure in the power turbine) and small temperature concentrations (pitch points) in the boiler evaporators. There is a remarkable impact of the sulphur content in the fuel on the permissible exhaust gas temperature and the lower feed water temperature limit. In the steam turbine circuit, a minimal number of exchangers should be used (optimally: none). The optimal parameters of this circuit also depend on the piston engine load.

5. Conclusions

It is possible to implement a combined system consisting of a Diesel engine as the leading engine, a power gas turbine, and a steam turbine circuit utilising the heat contained in the Diesel engine exhaust gas. Such systems can reveal thermodynamic efficiencies comparable with combined gas turbine circuits connected with steam turbines.

5.1 Power range of combined systems

Depending on the adopted variant and the main engine load, the use of the combined system makes it possible to increase the power of the power plant by 7 to 15 % with respect to the conventional power plant burning the same rate of fuel. Additional power is obtained by the system due to the recovery of the energy contained in the exhaust gas leaving the piston internal combustion engine. Thus the combined system decreases the specific fuel consumption by 6,4 – 12,8 % compared to the conventional power plant.

In the examined systems the power of the steam turbine is higher than that of the power turbine by 6-29 %, depending on the system variant and the main engine load.

5.2 Efficiency of combined systems

The use of the combined system for ship propulsion increases the efficiency of the propulsion system, and decreases the specific fuel consumption. Additionally, it increases the propulsion power without additional fuel consumption.

Like the power, the efficiency of the combined system increases with respect to the conventional power plant by 7 to 15% reaching the level of 53 - 56% for maximal power ranges. These efficiency levels are comparable with the combined systems based on the steam/gas turbines, Fig. 1. For partial loads the efficiency curves of the combined system with the Diesel engine are more flat than those for the combined turbine systems (smaller efficiency decrease following the load decrease) .

In the combined system the maximal efficiency is reached using particular system components:

- the piston internal combustion engine with the maximal efficiency;
- Turbocharger. The turbocharger with the maximal efficiency should be used as it provides opportunities for decreasing the exhaust gas enthalpy drop in the turbine in case of the series supply variant, or exhaust gas mass flow rate in case of the parallel supply variant, which in both cases results in higher power of the power turbine;
- Power turbine. High efficiency is required to increase its power;
- Steam turbine circuit. The requirement is to obtain the maximal power of the steam turbine from the heat delivered in the exhaust gas flowing through the boiler.

5.3 Ecology

Along with the thermodynamic profits, having the form of efficiency increase, and the economic gains, reducing the fuel consumption for the same power output of the propulsion system, the use of the combined system brings also ecological profits. A typical new-generation low-speed piston engine fed with heavy fuel oil with the sulphur content of 3% emits 17g/kWh NOx, 12g/kWh SOx and 600g/kWhCO_2 to the atmosphere. The use of the combined system reduces the emission of the noxious substances by, respectively, g/kWh NOx, g/kWh SOx and g/kWhCO_2. The emission decreases by % with respect to the standard engine, solely because of the increased system efficiency, without any additional installations.

Depending on the adopted solution, the combined power plant provides opportunities for reaching the assumed power of the propulsion system at a lower load of the main Diesel engine, at the same time also reducing the fuel consumption.

The article presents the thermodynamic analysis of the combined system consisting of the Diesel engine, the power gas turbine, and the steam turbine, without additional technical and economic analysis which will fully justify the application of this type of propulsion systems in power conversion systems.

6. Nomenclature

b_e	- specific fuel oil consumption
c_g, c_a	- specific heat of exhaust gas and air, respectively
i	- specific enthalpy
m	- mass flow rate
N	- power
p	- pressure
T,t	- temperature
Wu	- calorific value of fuel oil
η	- efficiency
κ_g, κ_a	- isentropic exponent of exhaust gas and air, respectively

Indices:

a	- air
bar	- barometric conditions
B	- Boiler
C	- Compressor
combi	- combined system
D	- Diesel engine
d	- supercharging
exh	- exhaust passage
f	- fuel
FW	- feet water
g	- exhaust gas
inlet	- inlet passage
k	- parameters in a condenser
o	- live steam, calculation point
PT	- Power turbine
ST	- Steam turbine
ss	- ship living purposes
T	- Turbine
TC	- Turbocharger
π	- compression ratio in a compressor, expansion ratio in a turbine

7. References

Dzida, M. (2009). On the possible increasing of efficiency of ship power plant with the system combined of marine diesel engine, gas turbine and steam turbine at the main engine - steam turbine mode of cooperation. *Polish Maritime Research,* Vol. 16, No.1(59), (2009), pp. 47-52, ISSN 1233-2585

Dzida, M. & Mucharski, J. (2009). On the possible increasing of efficiency of ship power plant with the system combined of marine diesel engine, gas turbine and steam turbine in case of main engine cooperation with the gas turbine fed in parallel and the steam turbine. *Polish Maritime Research,* Vol 16, No 2(60), pp. 40-44, ISSN 1233-2585

Dzida, M.; Girtler, J.; Dzida, S. (2009). On the possible increasing of efficiency of ship power plant with the system combined of marine diesel engine, gas turbine and steam turbine in case of main engine cooperation with the gas turbine fed in series and the steam turbine. *Polish Maritime Research,* Vol 16, No 3(61), pp. 26-31, ISSN 1233-2585

Kehlhofer, R. (1991). *Combined-Cycle Gas & Steam Turbine Power Plants,* The Fairmont Press, INC., ISBN 0-88173-076-9, USA

MAN B&M (October 1985). The MC Engine. Exhaust Gas Date. Waste Heat Recovery System. Total Economy, *MAN B&W Publication S.A.*, Danish

MAN Diesel & Turbo (2010). Stationary Engine. Programme 4th edition, *Branch of MAN Diesel & Turbo SE,* Germany, Available from www.mandieselturbo.com

Schrott, K. H. (1995). The New Generation of MAN B&W Turbochargers. *MAN B&W Publication S.A.*, No.236 5581E

Part 2

Gas Turbine Systems

4

Exergy Analysis of a Novel SOFC Hybrid System with Zero-CO_2 Emission

Liqiang Duan, Xiaoyuan Zhang and Yongping Yang
School of Energy, Power and Mechanical Engineering, Beijing Key Lab of Energy Safety and Clean Utilization, Key Laboratory of Condition Monitoring and Control for Power Plant Equipment of Ministry of Education, North China Electric Power University, Beijing, People Republic of China

1. Introduction

Now, climate change due to the emission of greenhouse gases, especially the emission of CO_2, is becoming more and more serious. Though many countries have taken all kinds of measures to control and reduce the emission of CO_2, in the short term, CO_2 emission still maintains a rapid growth trend. Power industry is the biggest CO_2 emission sector. So, there exists the greatest CO_2 emission reduction potential in the power industry. Now, many kinds of fossil fuel power generation systems with CO_2 recovery are usually based on the chemical absorption method or the oxygen combustion method. The former demands a chemical absorption and separation unit to recover CO_2 from the flue gas of power systems. The latter demands a special oxygen combustion technology, equipment and a larger ASU (air separation unit). And these technologies all consume great energy and result in the huger equipment investment and higher operating cost. Now, people are eager to develop the high-efficiency power generation technology with the less energy consumption for CO_2 capture. Fuel cell can satisfy the above requirements, with the higher energy conversion efficiency and less energy consumption of CO_2 capture, so it has attracted considerable interest in recent years.

Solid Oxide Fuel Cell (SOFC) is an attractive power-generation technology that can convert the chemical energy of fuel directly into electricity while causing little pollution (Kartha & Grimes, 1994). Because the anode fuel gas is naturally separated from the cathode air by the solid electrolyte, the CO_2 gas with the higher concentration can be obtained in the anode exhaust gas. In addition, SOFC can employ all kinds of fuels, including various hydrocarbon fuels. Compared with the traditional power generation systems, the SOFC hybrid system power plant has the higher system efficiency (net AC/LHV). Even after CO_2 is captured, the efficiency of SOFC hybrid system still can be greater than or equal to that of the traditional power systems without CO_2 capture. In order to further improve the CO_2 concentration of anode exhaust gas, SOFC can employ the O_2/CO_2 combustion mode in the afterburner. Because the required mass flow of pure O_2 is less, the energy consumption is lower. After capturing the CO_2, the SOFC hybrid system does not result in a bigger efficiency reduce. So the SOFC hybrid power system with zero CO_2 emission become a new way which can simultaneously solve the problem of efficient energy utilization and lower pollution emission.

In the last decades, many researchers were involved in study of SOFC stack and the hybrid power system with CO_2 capture. Y.Inui proposed and investigated two types of carbon dioxide recovering SOFC/GT combined power generation systems in which a gas turbine with carbon dioxide recycle or water vapor injection is adopted at the bottoming cycle system (Y.Inui et al, 2005). The overall efficiency of the system with carbon dioxide recycle reaches 63.87% (HHV) or 70.88% (LHV), and that of the system with water vapor injection reaches 65% (HHV) or 72.13% (LHV). A. Franzoni considered two different technologies for the same base system to obtain a low CO_2 emission plant (Franzoni et al, 2008). The first technology employed a fuel decarbonization and CO_2 separation process placed before the system feed, while the second integrated the CO_2 separation and the energy cycle. The result showed that the thermodynamic and economic impact of the adoption of zero emission cycle layouts based on hybrid systems was relevant. Philippe Mathieu presented the integration of a solid oxide fuel cell operating at a high temperature (900°C–1000°C, 55–60% efficiency) in a near-zero emission CO_2/O_2 cycle (Philippe Mathieu, 2004). Takeshi Kuramochi compared and evaluated the techno-economic performance of CO_2 capture from industrial SOFC-combined heat and power plant (CHP) (Takeshi et al, 2009). CO_2 is captured by using an oxyfuel afterburner and conventional air separation technology. The results were compared to both SOFC-CHP plants without CO_2 capture and conventional gas engines CHP without CO_2 capture. B.Fredriksson Moller examined the SOFC/GT configuration with and without a tail-end CO_2 separation plant, and based on a genetic algorithm, selected the key parameters of the hybirid system (Fredriksson et al, 2004). The result of the optimization procedure shows that the SOFC/GT system with part capture of the CO_2 exhibits an electrical efficiency above 60%. Some researchers also studied the performance parameters of the different SOFC hybrid power systems from the thermoeconomic or exergy efficiency point (Bozzolo et al, 2003; Asle & Matteo 2001; Takuto et al, 2007). For example, Ali Volkan Akkaya proposed a new criterion-exergetic performance coefficient (EPC), then applied it in the SOFC stack and SOFC/GT CHP system (Ali et al, 2007, 2009). F. Calisa discussed the simulation and exergy analysis of a hybrid SOFC-GT power system. The result showed that the SOFC stack was the most important sources of exergy destruction (Calisea et al, 2006).

In this paper, a zero-CO_2 emission SOFC hybrid power system is proposed. Using exergy analysis method, the exergy loss distributions of every unit of zero-CO_2 emission SOFC hybrid system are revealed. The effects of different operating parameters on exergy loss of every unit, as well as the overall system performance, are also investigated. The results obtained in this paper will provide useful reference for further study on high-efficient zero emission CO_2 power system.

2. System modelling

The models developed in the paper are all based on the following general assumptions:

1. All components work in adiabatic conditions, pressure drops and refrigerant disclosure are all neglected, and the systems operate at steady-state conditions.
2. The cathode gas consists of 79% nitrogen and 21% oxygen, and all gases are assumed as ideal gases.

3. The mass flow of the input fuel, gas and all the reaction products are stable, the changes of the fluid kinetic energy and potential energy are neglected.
4. The unreacted gases are assumed to be fully oxidized in the after-burner of the SOFC stack, and the after-burner is assumed to be insulation, all the heat exchangers are adiabatic.
5. The temperature of the anode and cathode outlet gases are equal to the cell stack operating temperature, the current and voltage of every cell unit are the same.

2.1 The SOFC stack model and result analysis

2.1.1 SOFC model

The natural gas feed tubular SOFC system process is implemented by using Aspen Plus software. The Aspen Plus contains rigorous thermodynamic and physical property database and provides comprehensive built-in process models, thus offering a convenient and time saving means for chemical process studies, including system modeling, integration and optimization. The simulated SOFC flowsheet is shown in Figure 1. It includes all the components and functions contained in the SOFC stack, such as ejector, pre-reformer, fuel cell (anode and cathode) and afterburner.

Firstly, the preheated fuel (stream 1) mixes with the recycling anode exhausted gas (stream 6), and then the mixed fuel gas (stream 2) is sent to the pre-reformer where the steam reform reaction takes place. After that, the stream (4) enters the anode of SOFC in which the electrochemical reaction of fuel and oxygen from the anode occurs. The reaction product and unreacted flue mixture (stream 5) is separated into two parts. One part (stream 6) is recycled. Another part enters the afterburner and mixes with the nitrogen-rich air (stream 13) from the anode. After the combustion reaction, the exhausted gas from the afterburner (stream 14) is introduced into the regenerator to preheat the air (stream 9) for the anode.

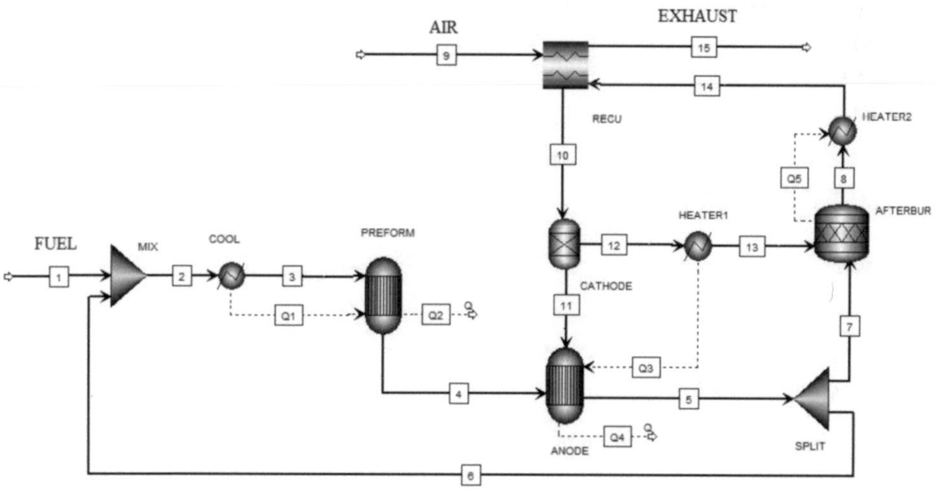

Fig. 1. Aspen Plus SOFC Stack Model Flowsheet

The cell voltage calculation is the core of any fuel cell modeling. The semi-empirical equations from literature (Stefano, 2001) were used to compose the Aspen Plus calculation module to simulate these effects on voltage. Several Design-spec Fortran blocks are used to set the fuel cell system's energy and heat balance. The semi-empirical equations are as follows:

$$\Delta V_P(mV) = C_1 \log(\frac{p}{p_{ref}}) \tag{1}$$

$$\Delta V_T(mV) = K_T(T - T_{ref}) \times i_c \tag{2}$$

$$\Delta V_{an}(mV) = 172 \log \frac{p_{H_2}/p_{H_2O}}{(p_{H_2}/p_{H_2O})_{ref}} \tag{3}$$

$$\Delta V_{cat}(mV) = 92 \log(\frac{p_{O_2}}{(p_{O_2})_{ref}}) \tag{4}$$

By summing the above four correlations, the actual voltage V can be calculated as

$$V_c = V_{ref} + \Delta V_P + \Delta V_T + \Delta V_{cat} + \Delta V_{an} \tag{5}$$

Where i_c is the current density; p_{H_2} / p_{H_2O} is the ratio of H_2 and steam partial pressure; P_{o_2} is the average oxygen partial pressure at the cathode for the actual case; V_c is the actual cell voltage; the subscribe ref is the reference case. P_{ref} = 1 bar; T_{ref} = 1000°C; $(p_{O_2})_{ref}$ = 0.164; $(p_{H_2} / p_{H_2O})_{ref}$ =0.15. ΔV_P stands for the change of SOFC voltage when the operating pressure varies. C_1 is the theoretical Nernst constant. ΔV_T stands for the change of SOFC voltage when the operating temperature varies. K_T is equal to 0.008 when the operating temperature is in the range of 900-1050°C. ΔV_{cat} and ΔV_{an} stand for the voltage change when the cathodic flow composition and anode flow composition vary compared with the reference case, respectively.
The fuel cell power output is the product of the cell voltage and current.

$$Power = current(A) \times V(V) \tag{6}$$

The equivalent hydrogen flow rate $n_{H_2,equivalent}$, can be calculated based on the molar flow rate of H_2 consumed in the electrochemical reaction ($n_{H_2,consumed}$) and the fuel utilization factor (U_f) to be generated:

$$n_{H_2,equivalent}(mol / h) = \frac{n_{H_2,consumed}}{U_f} \tag{7}$$

The value of the equivalent hydrogen flow rate and the known inlet fuel composition (C_i) can determine the amount of fresh fuel required $n_{fresh,fuel}$ from the following equation (8):

$$n_{fresh,fuel}(mol / h) = \frac{n_{H_2,equivalent}(mol / h)}{C_{H_2} + C_{CO} + 4 \times C_{CH_4} + 7 \times C_{C_2H_6} + \cdots} \tag{8}$$

The cell electrical efficiency is calculated according to the following equation (9):

$$\eta = \frac{\text{Power}}{n_{fresh,fuel}(mol / h) \times LHV_{fuel}(J / mol)} \tag{9}$$

2.1.2 SOFC simulation result analysis

In the process of modeling the hybrid power system, the accuracy of the fuel cell stack simulation is critical for the overall system. So this paper firstly checked the accuracy of SOFC stack model. With the same input parameters of the literatures (W.Zhang et al, 2005), the simulation results were compared with those of the literatures (W.Zhang et al, 2005,Veyo, 1996,1998,1999) . As shown in Table 1, the results show that this paper's simulation results are very close to the literature results. So the model of SOFC stack is feasible and reliable, it can be applied to simulate the overall SOFC hybrid system.

Parameters (unit)	Literature Data (W,Zhang et al 2005)	Literature Data (Veyo, 1996,1998, 1999)	Simulation data of this paper
Voltage (V)	0.70	-	0.70
Current Density (mA/cm^2)	178	180	179.1
Air utilization factor	19%	-	18.2%
Pre-reformer Outlet Temperature (℃)	536	550	537.2
Stack Exhaust Composition (EXHAUST)	77.3%N_2 15.9%O_2 4.5%H_2O 2.3%CO_2	77%N_2 16%O_2 5%H_2O 2%CO_2	77.5%N_2 16.4%O_2 4%H_2O 2.1%CO_2
Anode Outlet Composition (stream5)	50.9% H_2O 24.9%CO_2 11.6%H_2 7.4%CO 5.1%N_2	48% H_2O 28%CO_2 14%H_2 5%CO 5%N_2	50.9% H_2O 24.9% CO_2 11.6% H_2 7.4%CO 5.1%N_2
Stack Exhaust Temperature (℃)	834	847	832
Stack Efficiency (AC)	52%	50%	51.99%

Table 1. The Simulation Results of SOFC Stack (120kW DC output)

2.2 Description of the basic SOFC hybrid system

The basic SOFC hybrid system uses a tubular SOFC stack with higher operating temperature. Figure 2 shows the flowsheet of the system. The fuel is compressed and preheated, then is put into pre-reformer to generate the required H_2, CO and CO_2. Air is supplied by a blower and preheated prior to enter the SOFC stack. Then air participates an electrochemical reaction with fuel in the fuel cell stack. Because the pre-reformer needs a larger amount of water vapor, a hot recycle stream from the anode outlet is directed to the pre-reformer. The outlet stream of SOFC anode is mainly composed of H_2O, CO_2 and unconverted fuel (H_2 and CO). Part of this stream is injected into after-burner, then the hot outlet gas with high pressure expands in gas turbine, the rest is recycled and mixed with the compressed and preheated fuel. The recirculation fraction is calculated to obtain a given steam/carbon (S/C) ratio. In this way, the system can prevent from carbon deposition phenomenon, enhance the pre-reformer temperature and get more H_2. Finally, the product gas of SOFC is sent into the after-burner, where the unreacted fuel is burnt with part of the excess air from the cathode. After expansion of the hot gas in gas turbine, it is sent into heat exchangers to heat the inputted fuel and gas.

In order to control the outlet gas temperature and prevent the temperature from exceeding the tolerable inlet temperature of the turbine, water vapor is injected into the afterburner.

After the hot outlet gas with high pressure expands in gas turbine, it is fed to recuperator to heat the inputted fuel and air. Finally it is fed to the heat exchanger 4 for producing the saturated water vapor required for the after-burner.
Although SOFC hybrid power system has realized energy conversion with higher efficiency, the hybrid system still emit some greenhouse gas, which contains a large number of N_2 that will greatly impact the energy consumption of capturing CO_2. The paper proposed a zero CO_2 emission SOFC hybrid power system with water vapor injection and made the detailed exergy analysis on it.

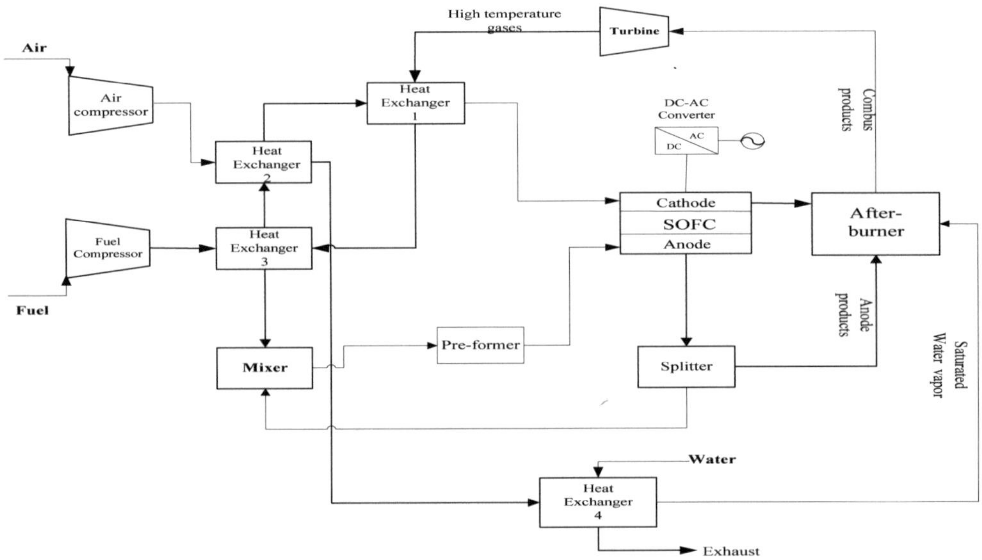

Fig. 2. Flowsheet of the basic SOFC hybrid system

2.3 Description of the new SOFC hybrid system with CO_2 capture

The new system and the basic system's primary processes are the same, but the new system employs O_2/CO_2 combustion mode. Pure oxygen from air separation unit (ASU) is fed into the combustion chamber to burn with the anode exhaust gas. The product gas only contains water vapor and CO_2. As shown in Figure 3, the compressed and preheated air is supplied to enter the SOFC stack and participate in an electrochemical reaction with fuel in the SOFC stack. At the same time, fuel is supplied by a blower and preheated prior to entering the SOFC stack anode. Part of the anode product gas is recycled to the mixer, and then mixed with the input compressed and preheated air; the remaining stream is burnt in the after-burner using oxygen as the oxidizer. After the combustion products expand in gas turbine, they are fed to heat exchangers to heat the inputted fuel and air. Finally the product gases are injected into the heat exchanger 4 and preheat the feed water for the afterburner. Theoretically the produced steam is enough to reduce the combustion gas temperature to the acceptable inlet temperature of turbine. The obtained combustion gas, which is composed of only CO_2 and water vapor, is introduced to gas turbine. After expansion, the cathode products are also injected into the heat exchanger 4 (also called the fourth heat exchange). Generally, capturing CO_2 is a difficult and complicated process, but the system uses O_2/CO_2 combustion mode, as a result, the combustion product gases only contain water vapor and CO_2. CO_2 can be separated by condensation, which does not consume extra energy. Finally it is fed to the multi-stage compressor and makes CO_2 dried from water and liquefied for making the transportation easier.

In order to realize zero CO_2 emission and lower the energy consumption of CO_2 capture, the new system mainly adopts the following measures:

1. In order to prevent the dilution of N_2, only the anode exhaust gas is burned with pure oxygen in the after-burner.
2. In order to enhance the system power, cathode outlet gas of SOFC stack is channeled into turbine to expand. And the product gas is channeled into the heat exchanger 4 to produce enough water vapor for the afterburner.
3. To avoid the great energy consumption, the new system uses multi-stage compression mode. After removing H_2O, the CO_2 concentration of depleted gas is more than 99%.

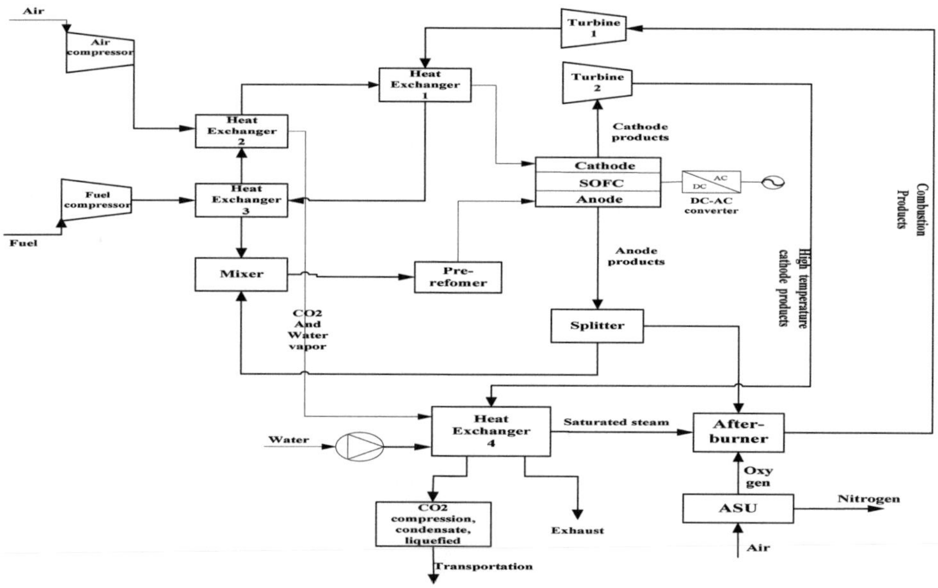

Fig. 3. Flowsheet of the zero CO_2 emission SOFC hybrid system

2.4 Simulation analysis of the basic and the new system

(1) System simulation parameters

The inputted fuel compositions of the basic and the new systems are the same (CH_4 93.6%, C_2H_6 4.9%, C_3H_8 0.4%, C_4H_{10} 0.2%, CO 0.9%). The fuel utilization factor is 85%, and the air utilization rate is 25%. The SOFC operating temperature/pressure is 1123.15K/3atm. The steam turbine isentropic efficiency is 85% and the gas turbine isentropic efficiency is 80%. The DC to AC inverter efficiency is 92%. The efficiencies of CO_2 compressor and air compressor in ASU are 75%.

(2) The performance analysis of the two systems

Table 2 shows the simulation results of the basic and new systems. For the new system, the cathode gas of the new system has a certain pressure; the new system can recover some expansion work of cathode exhaust gas. Though the new system applies the air separation unit and CO_2 capture equipment, which will consume some energy, compared with the cathode gas expansion work, it can be neglected.

Parameters (unit)	Base case	New system
SOFC Voltage (V)	0.64	0.64
SOFC Current Density (mA/cm2)	180.54	180.54
SOFC Stack Efficiency (%)	47.98	47.98
Electrical Efficiency (%) (AC)	59.03	56.26

Table 2. The simulation results of the two systems

3. The exergy analysis of the new hybrid system with CO_2 capture

Exergy analysis method is based on the second thermodynamics Law, considering both the energy quantity and energy quality in every energy conversion process. On the base of the energy balance and mass balance, the overall system is divided into many units and the exergy value of every stream is calculated. Through exergy balance analysis, the distributions of exergy loss of every unit can be acquired. At the same time, the system's largest exergy loss unit can be found. So, the location and direction of reducing exergy loss to improve the efficiency of the system can be identified (Jiaxuan & Shufang 1993).
The total exergy loss ($E_{X_{D,tot}}$) of the hybrid power system is given by the following equation:

$$E_{X_{D,tot}} = E_{X_{D,SOFC}} + E_{X_{D,GT}} + E_{X_{D,ASU}} + ... \tag{10}$$

Where $E_{X_{D,SOFC}}$ is SOFC stack's exergy loss; $E_{X_{D,GT}}$ is the turbine's exergy loss; $E_{X_{D,ASU}}$ is the ASU's exergy loss;
In order to analyze the system exergy loss in detail, the exergy loss rate (σ_i) is designed as the ratio of every component exergy loss ($E_{X_{D,i}}$) to the total system exergy loss.

$$\sigma_i = \frac{E_{X_{D,i}}}{E_{X_{D,tot}}} \times 100\% \tag{11}$$

The exergy efficiency (η_E) of the SOFC hybrid power system is another important exergetic performance criterion, and it can be defined as the ratio of total system effective acquired exergy to fuel input exergy as the following equation:

$$\eta_E = \frac{W + E_Q}{E_F} \times 100\% \tag{12}$$

Where, W is the system power; E_Q is the acquired heat exergy; E_F is the fuel exergy, and it can be calculated as the following equation:

$$E_F = Q_{DW} \times U_{fuel} \tag{13}$$

Where, Q_{DW} is the lower heating value (LHV) of the input fuel; U_{fuel} is the mass flow rate of fuel.
Exergy analysis takes the environmental state as reference state (P_0, T_0). In this paper, the selected temperature/pressure is 1123.15K/1atm as the reference environment state.

3.1 Exergy loss analysis of the system's every unit

Figure 4 shows the every unit exergy loss of the zero CO_2 emission SOFC hybrid power system. The biggest exergy loss unit lies in SOFC stack, which accounts for more than 35% of the total exergy loss. The main reason is that excess air is injected into to the SOFC stack in order to reduce the temperature difference of SOFC stack, and part of the input fuel chemical energy heats the excess air, which will cause a significant irreversible loss. One part of energy generated by electrochemical reaction is directly converted into the electrical energy, while the other part is changed into heat power to ensure that the fuel is reformed into H_2. So it makes the useful work generated by the fuel chemical energy reduce and the exergy loss increase.

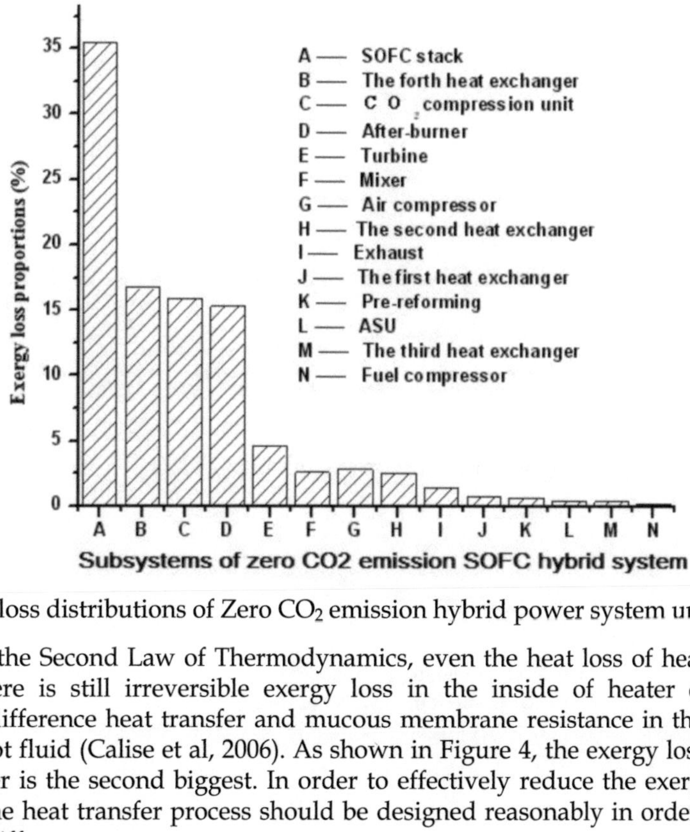

Fig. 4. Exergy loss distributions of Zero CO_2 emission hybrid power system units

According to the Second Law of Thermodynamics, even the heat loss of heat exchanger is neglected, there is still irreversible exergy loss in the inside of heater caused by big temperature difference heat transfer and mucous membrane resistance in the flow process of cold and hot fluid (Calise et al, 2006). As shown in Figure 4, the exergy loss of the fourth heat exchanger is the second biggest. In order to effectively reduce the exergy loss of heat exchangers, the heat transfer process should be designed reasonably in order to reduce the temperature difference.

3.2 Parametric exergy analysis results and discussions

The operating temperature, the operating pressure, the current density and fuel utilization factor of SOFC system are all considered as key variables which greatly influence the overall system performance. In the following discussion, the effects of the above key variables on the exergetic performance of system are respectively discussed.

3.2.1 The operating temperature

When the mass flow of input fuel keeps constant, with the increase of the operating temperature of SOFC, both the fuel cell voltage and system exergy efficiency increase. And then, the required air for cooling fuel cell stack will decrease as shown in Figure 5. In addition, due to the enhancement of cell stack activity, the system exergy loss reduces and the total system output power increases as shown in Figure 6. When the operating temperature is above 920°C, the voltage begins to decrease and system exergy losses increased. Therefore, in the practical situation, the system should operate in the proper temperature.

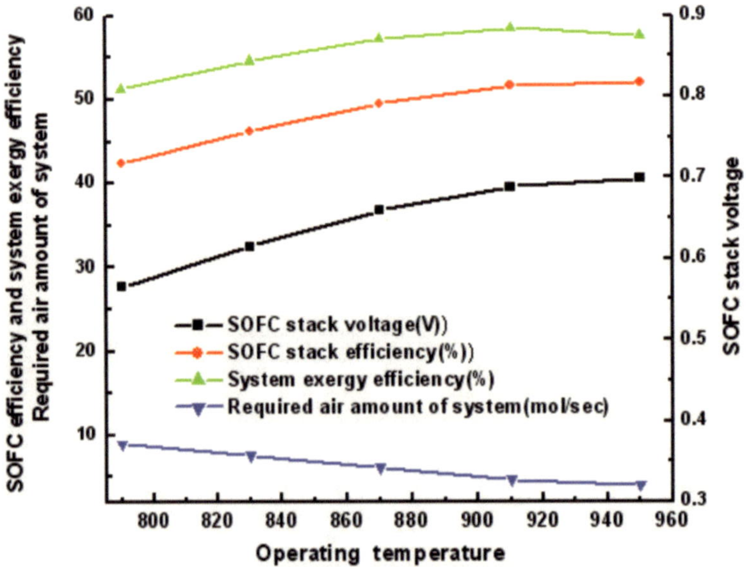

Fig. 5. The effect of operating temperature on system performance

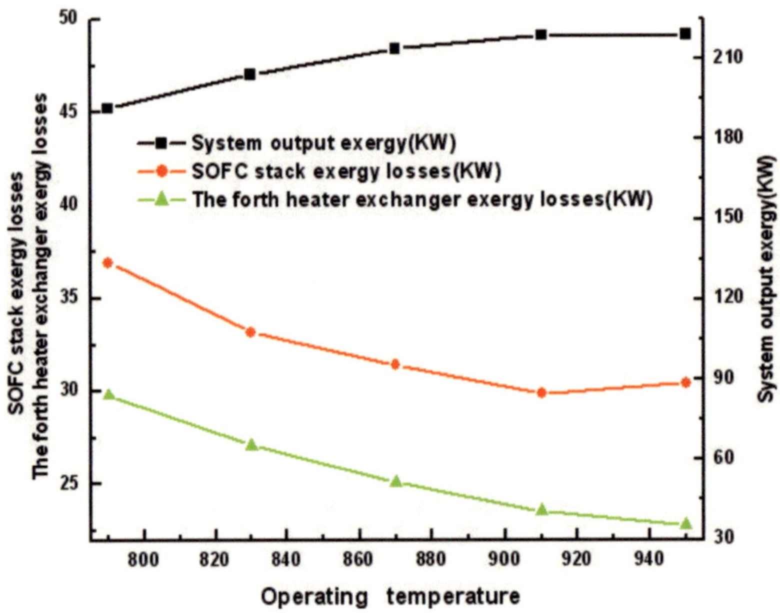

Fig. 6. The effect of operating temperature on system exergy parameters

3.2.2 The operating pressure

The operating pressure is vital to the system performance. Improving the operating pressure of SOFC stack, the SOFC voltage will increase because the H_2 amount in SOFC stack and H_2 partial pressure increase. Figure 7 shows that keeping the current density constant, with the increase of the operating pressure, the voltage increases. However, the growth rate gets smaller.

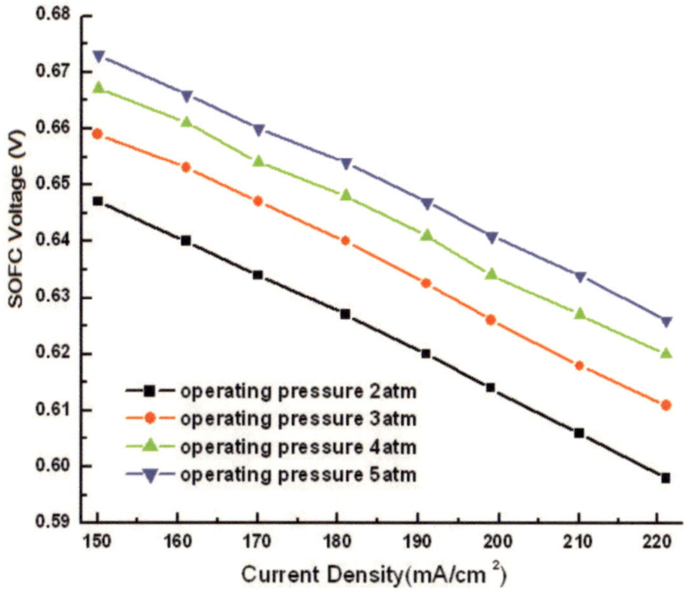

Fig. 7. The effect of operating pressure on SOFC Voltage

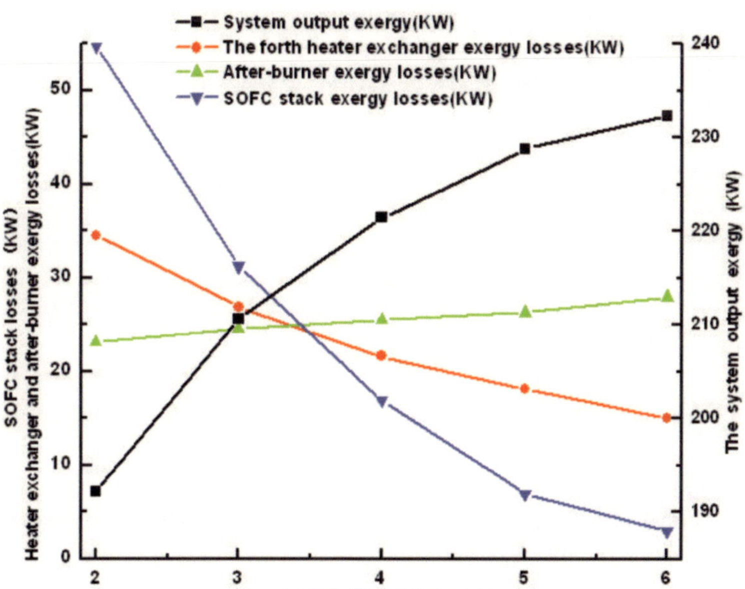

Fig. 8. The effect of operating pressure on system exergy performance

As shown in Figure 8, as the operating pressure increases, the SOFC stack exergy loss decreases and the total system output exergy increases. Because the required air for cooling fuel cell stack slowly increases, the after-burner exergy loss increases and the exergy loss of heat exchanger 4 decreases. In a word, the higher operating pressure is favorable to improving the performance of SOFC hybrid system. However, the higher pressure will increase the cost of system investment. Choosing the appropriate operating pressure should be taken into account when designing the SOFC.

3.2.3 The fuel utilization factor (U_f)

The fuel utilization factor (U_f) has a significant effect on the cell voltage and efficiency. As shown in Figure 9, with the increase of U_f from 0.7 to 0.9, the current density will increase, which will result in the decrease of the cell voltage. At lower values of U_f, when U_f increases, the cell voltage change is not significant, so the system output exergy will increase (as shown in Figure 10). But for higher U_f, the change amount of the cell voltage is bigger than that of the current density, as a result, the system exergy efficiency will reduce as shown in Figure 8. And U_f also has a significant impact on the composition of the anode exhaust stream. The CO_2 concentration at the anode outlet increases when U_f is increased because the fuel is more depleted (less CO and H_2), which will result in the change of the system unit exergy loss as shown in Figure 10.

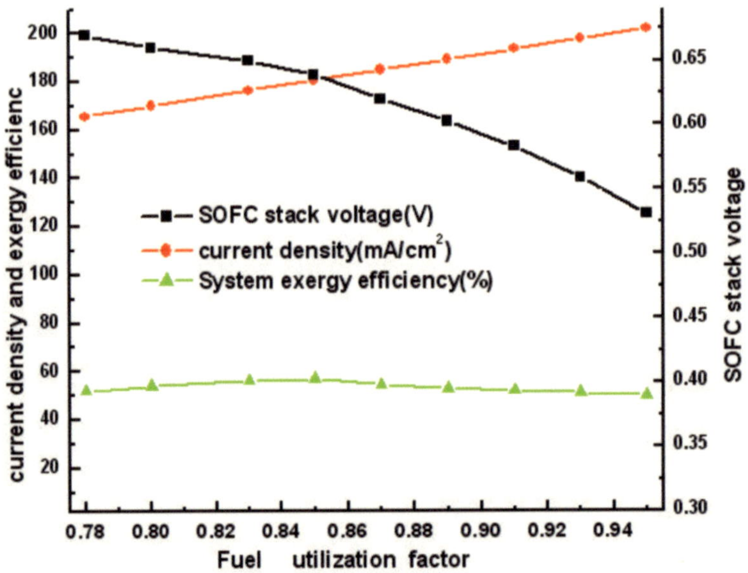

Fig. 9. The effect of fuel utilization factor on system performance

3.2.4 The cathode air input temperature

The cathode air consists of 79% nitrogen and 21% oxygen. As shown in Figure 11, with the increase of the cathode air input temperature, the activity of SOFC stack enhances. At the same time, both the required air for cooling fuel cell stack and the SOFC voltage increase, as a result, the system will produce more power. In addition, the inlet turbine gas temperature also increases, the power output of turbine will boost. But in order to meet the requirement

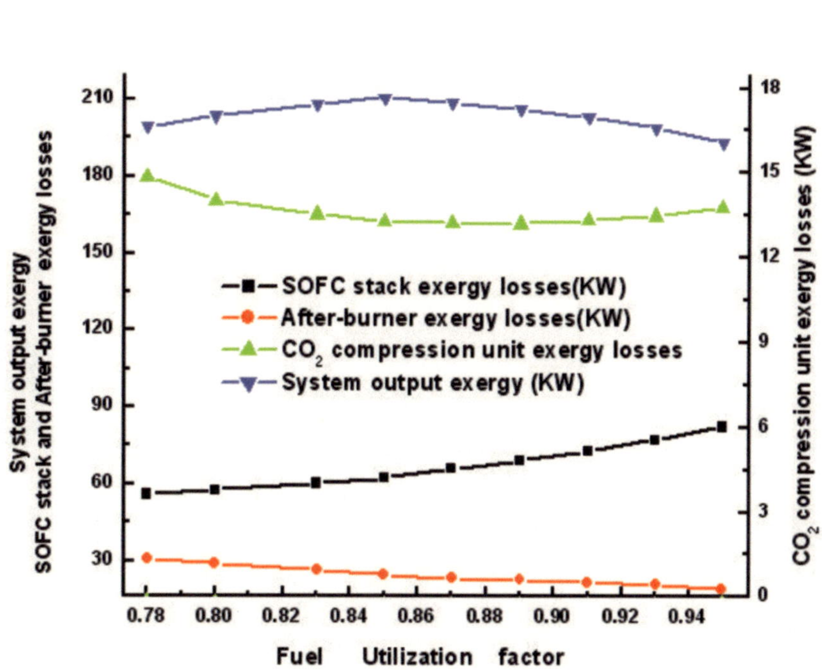

Fig. 10. The effect of fuel utilization factor on system exergy performance

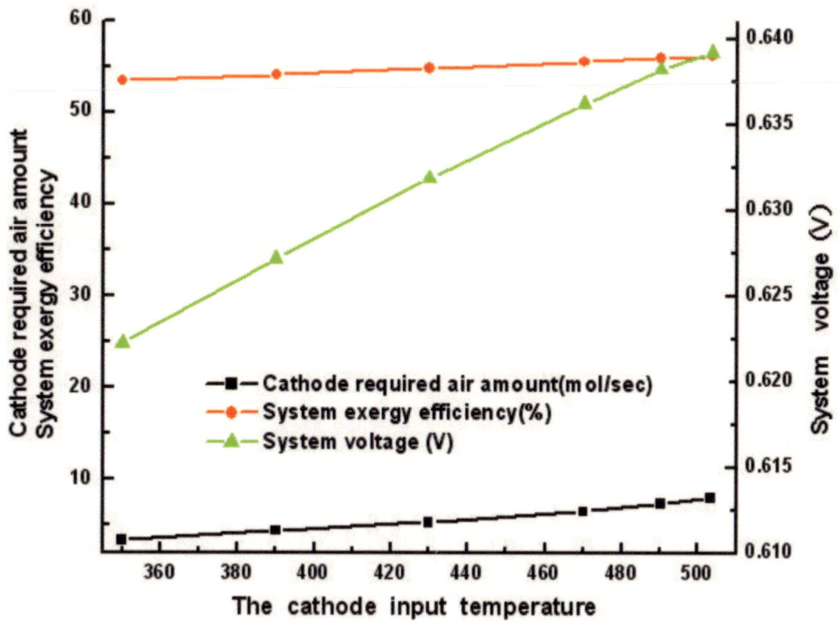

Fig. 11. The effect of cathode air input temperature on system performance

of the inlet air temperature, more heat of the exhaust gas will be consumed. The corresponding exergy loss of heat exchanger increases, so the system exergy efficiency isn't significant increased as shown in Figure 11. And as shown in Figure 12, the input temperature of cathode air also has an important effect on the other system performance parameters. The lower temperature will make the SOFC stack performance deteriorate.

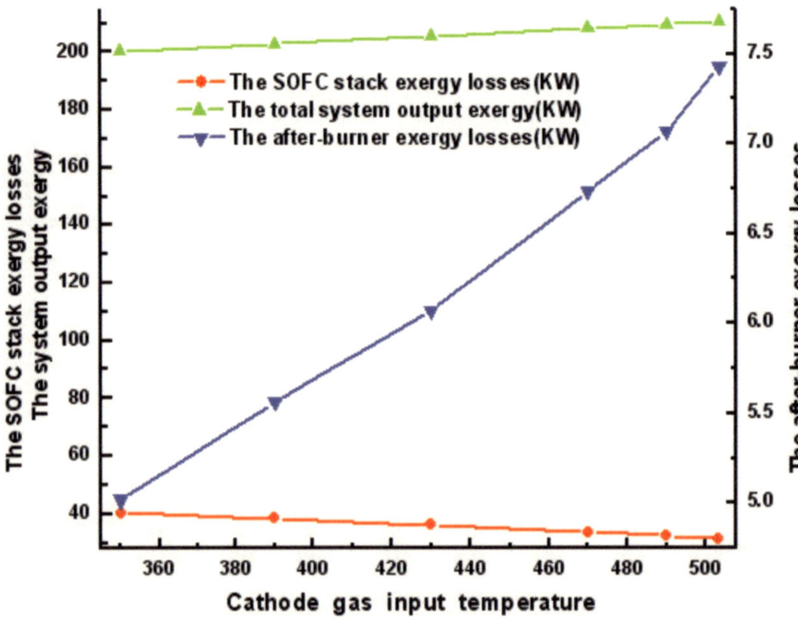

Fig. 12. The effect of cathode air input temperature on system exergy performance parameters

3.2.5 The oxygen concentration effect

As can be seen from Figure 13, when the operating pressure is a constant, as the oxygen purity increases, the O_2 partial pressure of SOFC stack cathode air improves, and this will make the system output exergy and exergy efficiency increase, especially SOFC stack with the lower operating pressure. Because the fuel flow remains unchanged, with the increase of oxygenconcentration, the required air decreases. Due to that the electrochemical reaction is exothermic reaction, it may cause the local area of stack overheat and the battery performance deteriorate. And with the increase of the oxygen concentration the consumed energy for separating the air will become bigger, so the exergy loss of the stack will rise slowly, which will result in the slow rise tend of system output exergy (as shown in Figure 14).

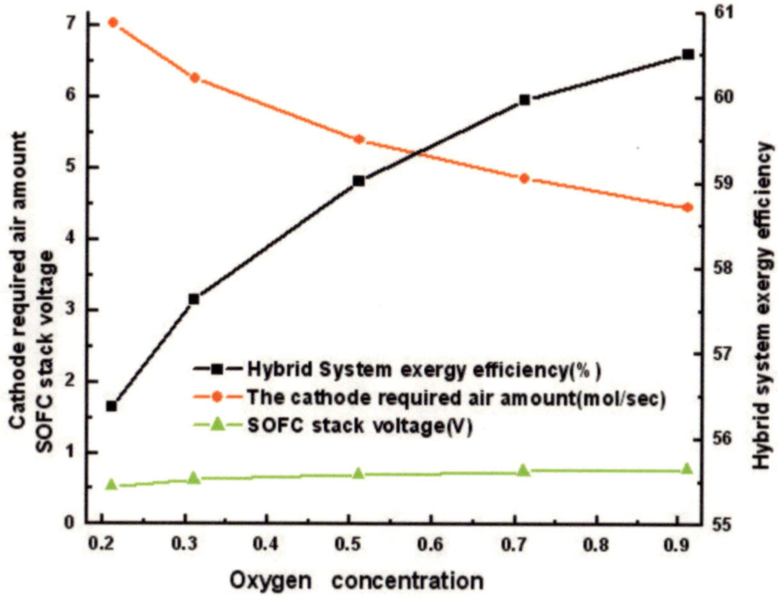

Fig. 13. The effect of oxygen concentration on system performance

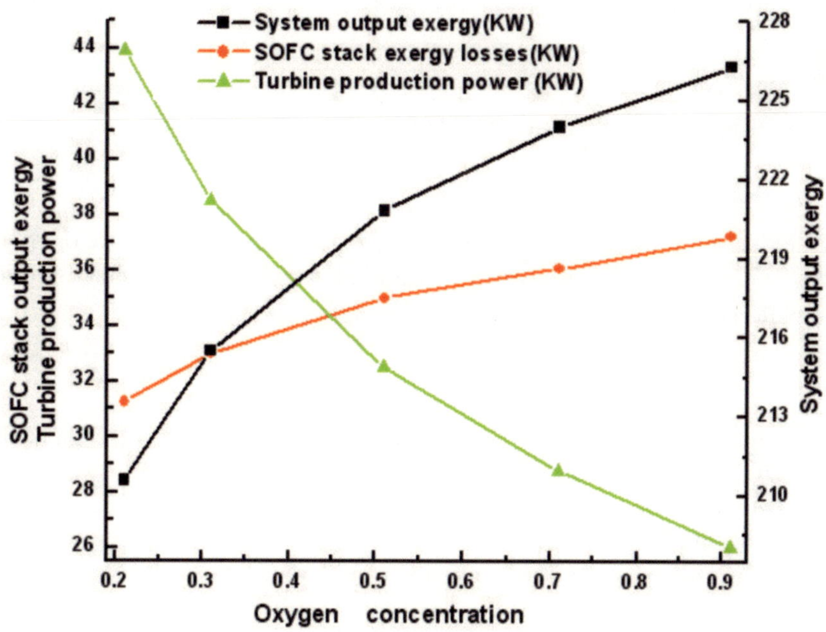

Fig. 14. The effect of oxygen concentration on system exergy performance

4. Conclusions

Based on a traditional SOFC (Solid Oxide Fuel Cell) hybrid power system, A SOFC hybrid power system with zero-CO_2 emission is proposed in this paper and its performance is analyzed. The exhaust gas from the anode of SOFC is burned with pure oxygen and the concentration of CO_2 gas is greatly increased. Because the combustion produce gas is only composed of CO_2 and H_2O, the separation of CO_2 hardly consume any energy. At the same time, in order to maintain the proper turbine inlet temperature, the steam produced from the waste heat boiler is injected into the afterburner, and then the efficiency of hybrid power system is greatly increased. With the exergy analysis method, this paper studied the exergy loss distribution of every unit of SOFC hybrid system with CO_2 capture and revealed the largest exergy loss unit. The effects of main operating parameters on the overall SOFC hybrid system with CO_2 capture are also investigated.

The research results show that the new zero-CO_2 emission SOFC hybrid system still has a higher efficiency. Its efficiency only decreases 3 percentage points compared with the basic SOFC hybrid system without CO_2 capture. The O_2/CO_2 combustion mode can fully burn the anode's fuel gas, and increase the concentration of CO_2 gas; at the same time with the steam injection and the combustion products are channeled into turbine, the efficiency of system greatly increases. The liquefaction of CO_2 by the mode of multi-stage compression and intermediate cooling can also greatly reduce the energy consumption.

The exergy analysis of the zero CO_2 emission SOFC hybrid power system shows that SOFC stack, after-burner and CO_2 compression unit are the bigger exergy loss components. By improving the input temperature of SOFC stack and turbine, the system exergy loss will significantly reduce. The optimal values of the operation parameters, such as operating pressure, operating temperature and fuel utilization factor exist, which make the system efficiency highest. Above research achievements will provide the new idea and method for further study on zero emission CO_2 system with higher efficiency.

5. Acknowledgments

This study has been supported by the National Basic Research Program of China (No. 2009CB219801), National Nature Science Foundation Project (No.50606010), and "the Fundamental Research Funds for the Central Universities" (No.10ZG03)

6. References

Kartha S, Grimes P (1994). Fuel cells: energy conversion for the next century. Physics Today, Vol 47(11) : 54–61,ISSN:0031-9228.

Y.Inui, T.Matsumae, H.Koga, K.Nishiura (2005). High performance SOFC/GT combined power generation system with CO_2 recovery by oxygen combustion method. Energy Conversion and Management, Vol 46(11-12): 1837–1847, ISSN:0196-8904.

A. Franzoni, L. Magistri, A. Traverso, A.F. Massardo (2008). Thermoeconomic analysis of pressurized hybrid SOFC systems with CO_2 separation. Energy, Vol 33(2): 311–320, ISSN: 0360-5442.

Philippe Mathieu (2004). Towards the hydrogen era using near-zero CO_2 emissions energy systems. Energy Vol 29(12-15): 1993–2002,ISSN: 0360-5442.

Takeshi Kuramochi, Hao Wu, Andrea Ramirez, Andre Faaij and Wim Turkenburg(2009). Techno-economic prospects for CO_2 capture from a Solid Oxide Fuel Cell - Combined Heat and Power plant. Preliminary results. Energy Procedia, Vol. 1(1): 3843–3850, ISSN:1876-6102.

B. Fredriksson Möllera, J. Arriagadaa, M. Assadia, I. Pottsb (2004). Optimisation of an SOFC/GT system with CO_2-capture. Journal of Power Sources, Vol. 131(1-2): 320–326, ISSN:0378-7753.

Bozzolo M, Brandani M, Traverso A, Massardo AF(2003). Thermoeconomic analysis of gas turbine plants with fuel decarbonization and carbon dioxide sequestration, ASME transactions. J Eng Gas Turbines Power, Vol. 125(4): 947–53.ISSN:0742-4795.

Asle Lygre, Matteo Ce (2001). Solid Oxide Fuel Cell Power with Integrated CO_2 capture. Second Nordic Minisymposium on Carbon Dioxide Capture and Storage, Gothenburg, Sweden.

Takuto Araki, Takuya Taniuchi, Daisuke Sunakawa (2007). Cycle analysis of low and high H_2 utilization SOFC/gas turbine combined cycle for CO_2 recovery, Journal of Power Sources, Vol 171: 464–470, ISSN:0378-7753.

Ali Volkan Akkaya, Bahri Sahin , Hasan Huseyin Erdem (2009). Thermodynamic model for exergetic performance of a tubular SOFC module. Renewable energy, Vol 34(7):1863-1870, ISSN:0960-1481.

Ali Volkan Akkaya, Bahri Sahin, Hasan Huseyin Erdem (2007). Exergetic performance coefficient analysis of a simple fuel cell system. International Journal of Hydrogen Energy,Vol. 32: 4600- 4609, ISSN:0360-3199.

F. Calisea, M. Dentice d'Accadiaa, A. Palomboa, L. Vanoli (2006). Simulation and exergy analysis of a hybrid Solid Oxide Fuel Cell (SOFC)–Gas Turbine System. Energy, Vol 31 : 3278–3299, ISSN: 0360-5442.

Stefano Campanari (2001).Thermodynamic model and parametric analysis of a tubular SOFC module.Journal of Power Sources, Vol. 92(1-2): 26-34, ISSN:0378-7753.

W. Zhang, E. Croiset, P.L.Douglas, M.W.Fowler, E.Entchev. Simulation of a tubular solid oxide fuel cell stack using AspenPlusTM unit operation models. Energy Conversion and Management 46(2005) 181-196, ISSN: 0196-8904.

Veyo SE. The Westinghouse solid oxide fuel cell program—a status report (1996). In: Proceedings of the 31st IECEC, No. 96570, pp: 1138-43.

Veyo SE, Forbes CA. Demonstractions based on Westinghouse's prototype commercial AES design. In: Proceedings of the Third European Solid Oxide Fuel Cell Forum, 1998, p. 79-86.

Veyo S, Lundberg W. Solid oxide fuel cell power system cycles. ASME Paper 99-GT-356, International Gas Turbine and Aeroengine Congress and Exhibition, Indianapolis, June 1999.

Jiaxuan, Wang, Shufang, Zhang (1993). Exergy method and its application in power plants, China Water Power Press, ISBN: 7-120-01797-7

5

Flexible Micro Gas Turbine Rig for Tests on Advanced Energy Systems

Mario L. Ferrari and Matteo Pascenti
University of Genoa – Thermochemical Power Group (TPG)
Italy

1. Introduction

During the last 15-20 years, microturbine (mGT) technology has become particularly attractive for power generation, especially in the perspective of the development of a distributed generation market (Kolanowski, 2004). The main advantages related to microturbines, in comparison to Diesel engines, are:

- smaller size and weight;
- fuel flexibility;
- lower emissions;
- lower noise;
- vibration-free operation;
- reduced maintenance.

The development of a laboratory based on a large size gas turbine is usually not feasible at the University level for high costs of components and plant management. However, the microturbine (mGT) technology (Kolanowski, 2004) allows wide-ranging experimental activities on small size gas turbine cycles with a strong cost reduction.

Moreover, this technology is promising from co-generative (and tri-generative) application point of view (Boyce, 2010), and is essential for advanced power plants, such as hybrid systems (Massardo et al., 2002), humid cycles (Lindquist et al., 2002), or externally fired cycles (Traverso et al., 2006).

However, if microturbine standard cycle is modified by introducing innovative components, such as fuel cells (Magistri et al., 2002, 2005), saturators (Pedemonte et al., 2007) or new concept heat exchangers (such as ceramic recuperators (McDonald, 2003)), at least two main aspects have to be considered:

- avoiding dangerous conditions (e.g.: machine overspeed, surge, thermal and mechanical stress, carbon deposition);
- ensuring the proper feeding conditions to both standard and additional components.

Moreover, the operation with new devices generates additional variables to be monitored, new risky conditions to be avoided and requires additional control system facilities (Ferrari, 2011).

Experimental support is mandatory to develop advanced power plants based on microturbine technology and to build reliable systems ready for commercial distribution. A possible cheap solution to perform laboratory tests is related to emulator facilities able to generate similar effects of a real system. These are experimental rigs designed to study various critical aspects of advanced power plants without the most expensive components (e.g. fuel cells, high temperature heat exchangers). An important experimental study with mGT based emulators is running at U.S. DOE-NETL laboratories of Morgantown (WV-USA). This activity is based on a test rig designed to emulate cathode side of hybrid systems based on Solid Oxide Fuel Cell (SOFC) technology (Tucker et al., 2009). It is mainly composed of a recuperated microturbine, a fuel cell vessel (without ceramic material), an off-gas burner vessel, and a combustor controlled by a fuel cell real-time model (Tucker et al., 2009). Another emulator facility equipped with a micro gas turbine is under development at German Aerospace Center (DLR), Institute of Combustion Technology, of Stuttgard (Germany) (Hohloch et al., 2008). Its general layout is similar to the NETL test rig. However, this experimental plant includes a fuel cell vessel able to emulate the real stack exhaust gas composition with a water cooling system (coupled with an additional combustor) (Hohloch et al., 2008).

At University of Genoa, TPG researchers developed a new test rig based on a commercial recuperated 100 kW micro gas turbine equipped with a set of additional pipes and valves for flow management. These pipes are essential to perform high fidelity mass flow rate measurements and to connect the machine to an external modular vessel. This additional volume is located between the recuperator outlet (cold side) and the combustor inlet (Ferrari et al., 2009a) to emulate the dimensions of advanced cycle components (such as fuel cells (Ferrari et al., 2009a; Tucker et al., 2009), externally fired gas turbine facilities (Traverso et al., 2006), saturators (Pedemonte et al., 2007)).

This test rig, developed to carry out experimental tests in both steady-state and transient conditions, is designed to have the highest plant flexibility performance. In details, this facility is able to operate in the following configurations:

- simple cycle;
- recuperated and partly recuperated cycle;
- both simple and recuperated cycles coupled with a modular vessel for the emulation of additional component volume (Ferrari et al., 2009a);
- emulation of hybrid systems (Ferrari et al., 2010a) with high temperature fuel cell stack (cathodic and anodic vessels, anodic recirculation (Ferrari et al., 2010a), steam injection for chemical composition emulation, real-time model for components not present in the rig);
- co-generative and tri-generative (with an absorber cooler) systems.

Moreover, on all these plant layouts it is possible to test the influence of the following properties, especially in transient conditions:

- ambient temperature;
- volume size (downstream of the compressor);
- valve fractional opening values;
- bleed mass flow rates;
- grid connection or stand-alone systems;
- control system.

This chapter shows some examples of tests carried out with the rig (using different plant layouts and operative conditions), inside different research projects and during educational activities. In details, the facility was mainly developed inside two Integrated Projects of the EU VI Framework Program (Felicitas and Large-SOFC) and now it is involved in the new EU VII Framework (E-HUB Project) for tests to be carried out with an absorption cooler (tri-generative configuration). Moreover, this experimental plant is essential to introduce undergraduate students to micro gas turbine technology, and Ph.D.s to advanced experimental activities in the same field. With this experimental rig, in addition to learning about thermodynamic cycles and plant layouts, students can also become familiar with their materials, piping, gaskets, technology for auxiliaries, and instrumentation.

2. Nomenclature

CAD	Computer Aided Design
CC	Combustion Chamber
CFD	Computational Fluid Dynamic
Ex	heat Exchanger
FC	Fuel Cell
GT	Gas Turbine
HS	Hybrid System
ISO	International Organization for Standardization
LAN	Local Area Network
mGT	micro Gas Turbine
mHAT	micro HAT cycle
NETL	National Energy Technology Laboratory
REC	RECuperator
RRFCS	Rolls-Royce Fuel Cell Systems
SOFC	Solid Oxide Fuel Cell
UDP	User Datagram Protocol
WHEx	Water Heat Exchanger
Variables	
COP	Coefficient Of Performance
h	heat transfer convective coefficient [W/m^{2}K]
I	current density [A/m^2]
i, j	indexes for Fig. 18
K_p	surge margin
m	mass flow rate [kg/s]
N	rotational speed [rpm]
n	number of moles
p	pressure [Pa]
P	power [W]
q	heat flux [W]
S	surface [m^2]
T	temperature [K]
TIT	Turbine Inlet Temperature [K]
TOT	Turbine Outlet Temperature [K]
U_f	fuel utilization factor
Greek symbols	
β	compressor pressure ratio (total to static)
Δ	variation
ε	recuperator effectiveness
λ	conductivity [W/mK]

Subscripts

0, 1, 2, 3, 4, 5, i	subscripts for Fig. 18
amb	ambient
c	compressor
d	on design
in	inlet
M	measured
s.l.	surge line
t	total

3. The commercial machine

The basic machine is a Turbec T100 PHS Series 3 (Turbec, 2002). It is equipped to operate in stand-alone configuration or connected to the electrical grid. This commercial unit consists of a power generation module (100 kW at nominal conditions), a heat exchanger located downstream of the recuperator outlet (hot side) for co-generative applications, and two battery packages for the start-up phase in stand-alone configuration. Even if the machine is located indoors, the outdoor roof was placed over the machine to use its air pre-filters. Figure 1 shows the plant layout diagram of the micro gas turbine as furnished by the manufacturer in the PHS configuration.

Legend: 1. Generator; 2. Inlet air; 3. Combustion chamber; 4. Air to recuperator; 5. Compressor; 6. Turbine; 7. Recuperator; 8. Exhaust gases; 9. Exhaust gas heat exchanger; 10. Exhaust gas

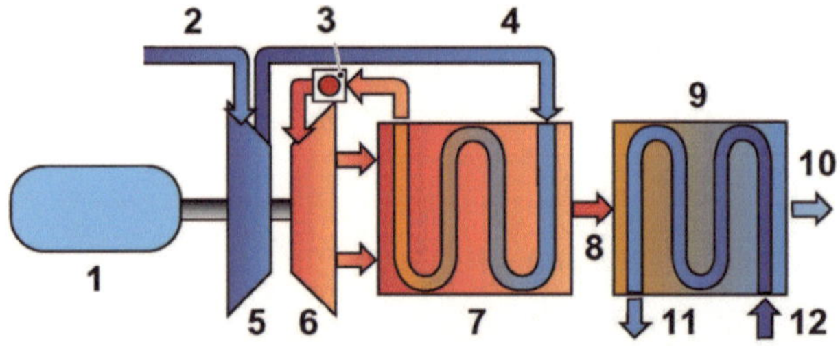

Fig. 1. Turbec T100 machine: PHS standard layout (courtesy of Turbec).

The power module (Fig. 2) is composed of a single shaft radial machine (compressor, turbine, synchronous generator) operating at a nominal rotational speed of 70000 rpm and a TIT of 950°C (1223.15 K), a natural gas fed combustor, a primary-surface recuperator, a power electronic unit (rectifier, converter, filters and breakers), an automatic control system interfaced with the machine control panel, and the auxiliaries. In this test rig the micro gas turbine is operated using its commercial control system. It works at constant rotational speed when the machine is in stand-alone mode. In this mode the control system changes the fuel mass flow rate to maintain the shaft (in steady-state condition) at 67550 rpm. In grid-connected mode this controller works at constant turbine outlet temperature (TOT). So, in this second mode the control system changes the fuel mass flow rate to maintain the machine TOT (in steady-state condition) at 645°C (918.15 K). However, in both modes the power electronic system allows to generate a 50 Hz current output (at each load values).

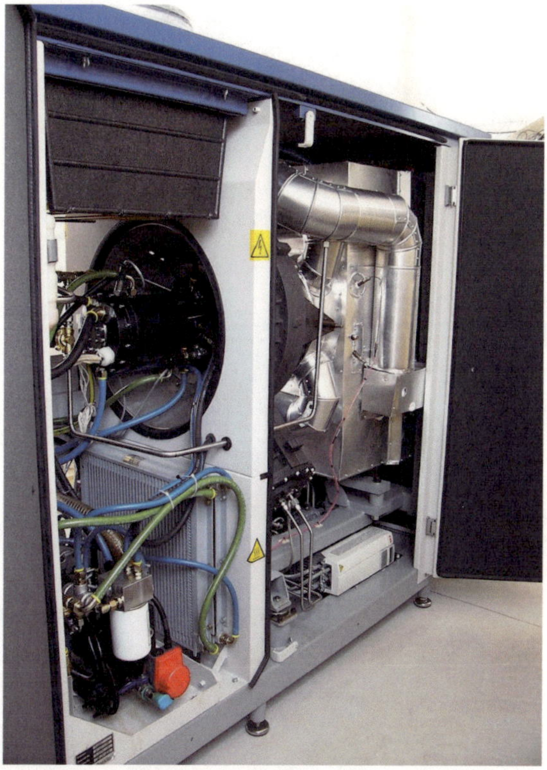

Fig. 2. The T100 power module installed at the laboratory of the University of Genoa (power electronics and control system are not shown).

The heat exchanger for water heating (for cogenerative and trigenerative applications) includes an exhaust gas bypass device to control water temperature values. Each start-up battery package includes thirty 12 Volt batteries connected in series for a nominal voltage of 360 Volt DC.

The machine, in its commercial configuration, is equipped with the following essential probes for control reasons: electrical power (±1%), rotational speed (±10 rpm), TOT (±1.5 K), fractional opening values (pilot and main fuel valves), intake temperature, and heating water temperature meters. Furthermore, the commercial machine is equipped with diagnostic probes (e.g. vibration sensor, filter differential pressure meter, temperature probes for the auxiliary systems).

The TPG laboratory was also equipped with a 100 kW resistor bank (pure resistive load) for turbine operation in stand-alone mode. The bank is cooled through an air fan and continuously controlled by an inverter.

4. Machine modifications and connection pipes

The commercial power unit was modified for coupling with the external connection pipes used for flow measurement and management purposes. These modifications are essential for measuring all the properties necessary for cycle characterization (e.g. air mass flow rate or recuperator boundary temperatures), not available in the commercial layout of the machine. For this reason the following modifications were carried out:

- the two pipes between the recuperator and the combustor inlet were substituted with four pipes for the external connections (Fig. 3);
- at the compressor outlet a check valve (Fig. 3) was introduced to prevent damage and block compressor backflow if surge conditions occur during experimental tests;
- between this valve and the recuperator inlet a T-joint was introduced to have a recuperator bypass line which is necessary for studying non-recuperated cycles or hybrid system start-up and shutdown phases (Ferrari et al., 2010a).

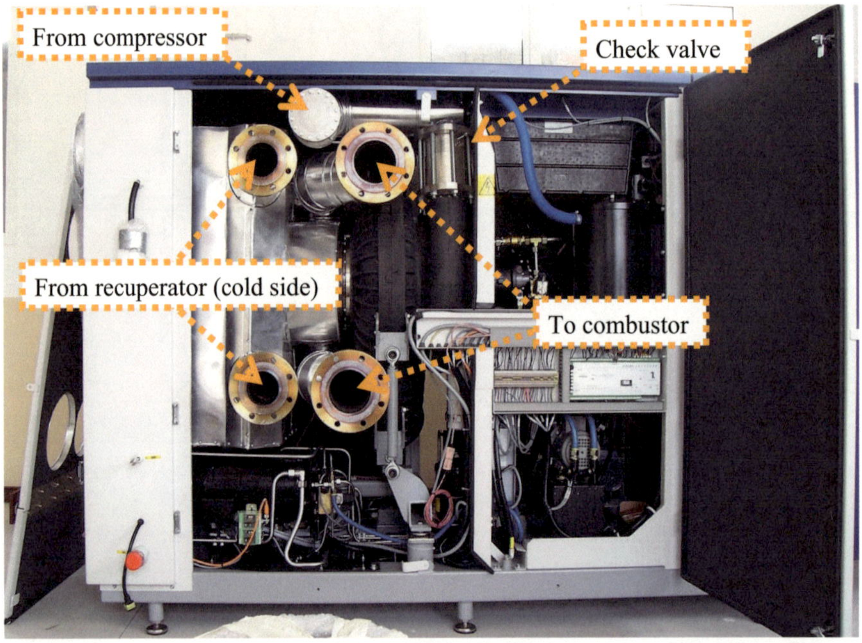

Fig. 3. T100 power module modified for coupling with the external pipe system.

The thermally insulated connection pipes are designed for high fidelity flow measurement, achieved with mass flow, pressure, and temperature probes, and flow management carried out with controlled valves. Moreover, these pipes were designed for easy connection to additional components between the compressor and expander, such as the modular vessel presented in the following paragraph. For this reasons, they merge the two flows at the recuperator outlets and split the combustor inlet flow for the connection to the two combustor pipes (see Figs. 3 and 4). Since this kind of mass flow measurements is carried out through pitot devices, each probe is preceded by a pipe of at least 18 diameter length and followed by a minimum length pipe of 4 diameters. This layout was chosen to obtain flow uniformity essential for high-precision measurements. A cold bypass was included to bypass the recuperator or connect the compressor outlet directly to additional external components (e.g. the modular vessel (Ferrari, 2009) used to emulate additional component dimensions). Moreover, the test rig includes a bleed line (Ferrari, 2009), equipped with a globe valve (VB), located at the compressor outlet. This device is essential to bleed part of flow when operative conditions are too close to the surge region. An additional pipe, equipped with a mass flow rate meter was also installed to directly connect recuperator outlet to combustor inlet (through the VM valve), as in a typical recuperated cycle. These pipes are also essential in changing recuperator mass flow rate values.

The test rig was equipped with additional probes to measure the flow properties (Pascenti et al., 2007) (see Table 1). All additional transducers are connected to a PC, through a FieldPoint™ device, and their signals are acquired via LAN using a software developed in LabVIEW™ environment. These measurement devices are used to measure mass flow rates, pressures and temperatures inside the different lines of the plant. The locations of these probes are shown in the test rig layout diagram (Fig. 4). This plant layout refers to the machine working in the standard recuperated cycle (equipped with a check valve, a recuperator bypass, a bleed line, and additional probes).

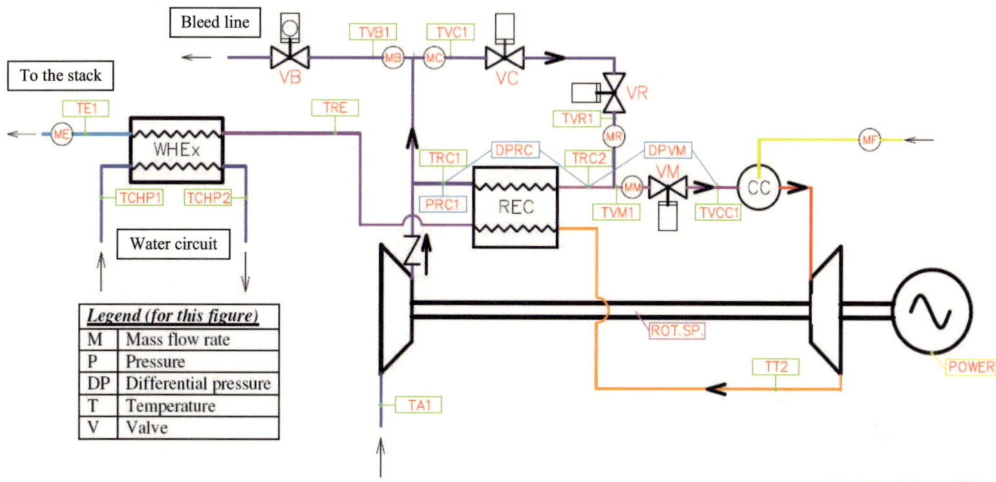

Fig. 4. Plant layout and instrumentation: machine and pipes (see Table 1 for the complete legend).

Since this machine is not designed for connection to additional components between the compressor and expander the connection pipes were developed to incur the lowest possible pressure drops. This is necessary to prevent surge conditions, and to avoid strong machine performance decrease. Therefore, gate valves were chosen (except in the case of the bleed line), and the pipes were designed with wide theoretical support.

A preliminary analysis was carried out with a test rig model in order to evaluate machine performance decrease (due to additional pressure and temperature losses), and to evaluate the operative limits. Then, a CFD analysis (see (Ferrari et al., 2007) for details) was performed with Fluent tool to verify the pipe design.

The test rig model was implemented with TRANSEO tool (in MATLAB®-Simulink® environment). This software (Traverso, 2005) is a simulation program based on an easy access library designed for the off-design, transient and dynamic analyses of advanced energy systems based on microturbine technology. TRANSEO was developed and validated at the University of Genoa in previous studies carried out by both Ph.D. students and TPG researchers (Caratozzolo et al., 2010; Ferrari et al., 2007; Traverso et al., 2005). After an experimental validation in design conditions (Ferrari et al., 2009a), the model was used to calculate the operative curves (at constant TOT) on the compressor map (Fig. 5) at different pressure drop values (Δp) between recuperator and combustor. Each curve was calculated with a valve operating at fixed fractional opening. For this reason, Fig. 5 curves are obtained maintaining constant the ratio between the pressure drop (Δp) and the compressor outlet pressure (p_c). A maximum drop of 485 mbar was calculated at $\Delta p/p_c = 0.108$ to prevent any risk of surge (with the following surge margin limitation: $K_p>1.1$). To have a wide operative range, as required during the tests, a pressure drop limit of 300 mbar was calculated (the curve at $\Delta p/p_c = 0.069$ has a $\Delta p = 290$ mbar at 70000 rpm). These results were necessary to design pipes and valves with wide operative ranges during the tests and manage the rig under safe conditions. The nominal diameter values chosen for these pipes are: (i) 125 mm for the pipes immediately upstream of the combustor, (ii) 100 mm for the other connection pipes. The experimental tests showed good performance in agreement with the design target to obtain the lowest possible additional pressure drop (with the lowest space occupation too). The maximum pressure loss value (DPVM probe) measured during the tests is about 60 mbar.

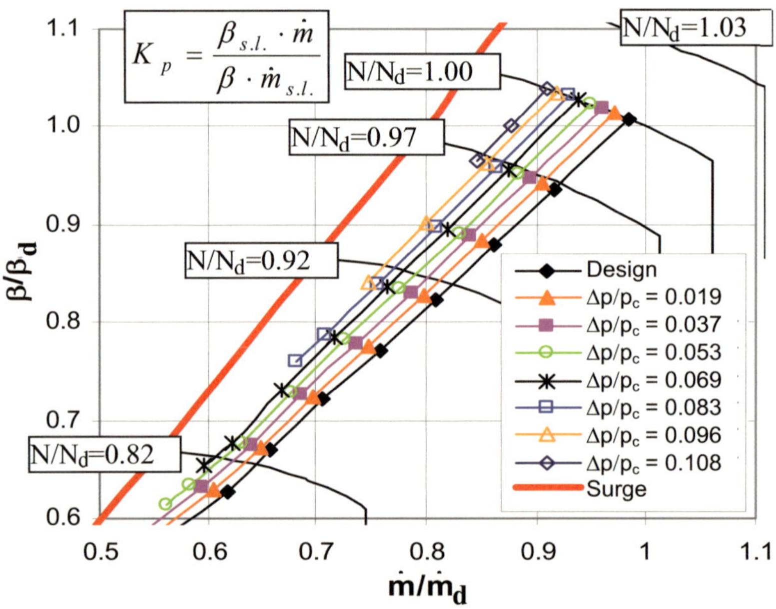

Fig. 5. Machine operation curves at constant $\Delta p/p_c$: TRANSEO model calculations.

4.1 Test example: Machine equipped with the external connection pipes

This paragraph shows an example of possible tests carried out on the machine equipped with the external pipes. For this reason Fig. 6 reports (on manufacturer compressor map) the experimental measurements carried out at different load values in stand-alone mode. These tests were performed using the compressor outlet pressure probe (PRC1 to measure the static wall pressure) and a mass flow meter (MM) located in the direct connection between the recuperator outlet and the combustor inlet (Fig. 4).

It is important to highlight these measurements, because they are essential for a correct theoretical analysis of the machine behaviour at both component and system levels and for a complete model validation. While manufacturer compressor maps are usually essential for commercial machine models, this test rig allows to carry out wide experimental verification activities on these performance curves. Moreover, the experimental compressor maps are essential to prevent surge events during tests operated with external additional components (e.g. the modular vessel).

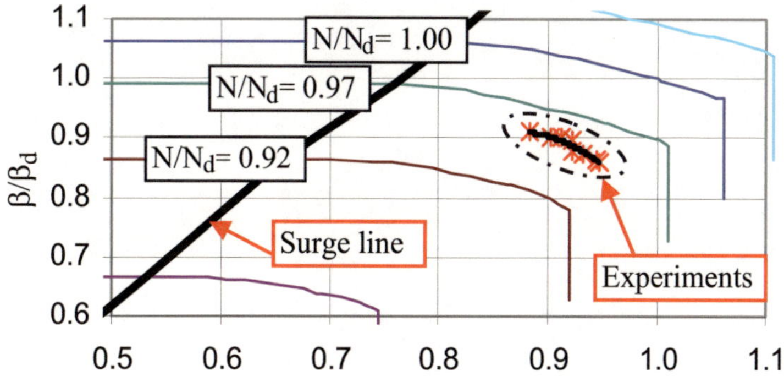

$$\frac{\left(\dot{m}\cdot\sqrt{T_{t\_in}}/p_{t\_in}\right)}{\left(\dot{m}\cdot\sqrt{T_{t\_in}}/p_{t\_in}\right)_d}$$

Fig. 6. Direct-line test: experimental data on compressor map obtained in stand-alone mode.

Since in stand-alone conditions the machine control system operates at constant rotational speed, it is possible to measure part of the curve at N/N_d=96.5. Within the accuracy of the probes utilized (Pascenti et al., 2007), the experimental points (Fig. 6) are well located between the 0.97 and 0.92 curves. The difference between the curve obtained from the manufacturer datum interpolation and the experimental points is due to the ambient temperature, which is around 25°C (298.15 K) instead of 15°C (288.15 K), and to the cited probe accuracy. However, ongoing works will be carried out to measure more map points, by converting the data into ISO conditions or operating at the ISO ambient temperature.

5. Machine connected to an external vessel

The machine operation with an additional volume located between compressor outlet and combustor inlet is typical of different innovative plant layout cycles (hybrid system, mHAT, externally fired cycle). This kind of volume is extremely significant on machine transient behaviour especially for plant component safety reasons (to avoid stress or dangerous conditions (Pascenti et al., 2007)). Therefore, this test rig was equipped with a modular vessel designed for experimental tests on the coupling of the machine with different kinds of components (e.g. saturators, fuel cells of different layouts or technology, or additional heat exchangers). The effect of these innovative components can be emulated through a right number of vessel modules to couple the real volume dimension with the micro gas turbine. The emulation of thermal and flow composition aspects can be carried out through the management of test rig valves (Ferrari et al., 2010a), the hardware/software coupling based on real-time models (Bagnasco, 2011), or the injection of additional flows (e.g. for the chemical composition emulation (Ferrari et al., 2011)).

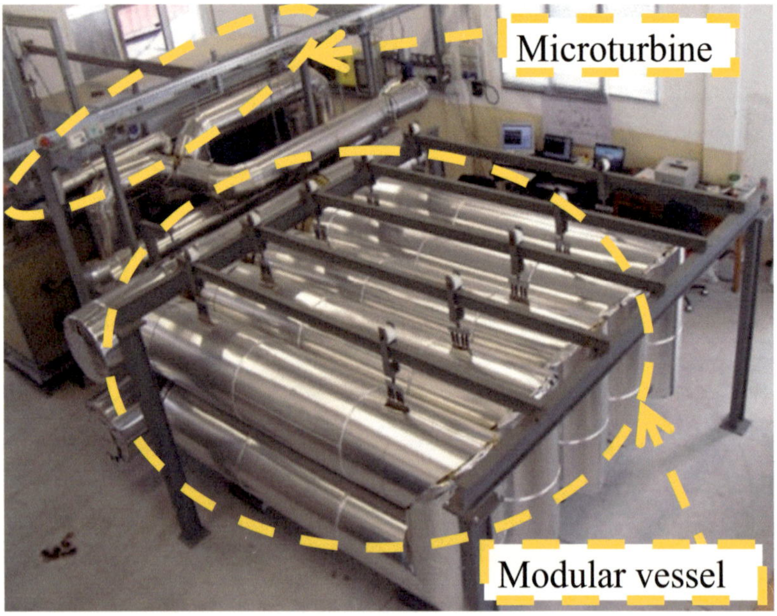

Fig. 7. Modular vessel coupled with the machine.

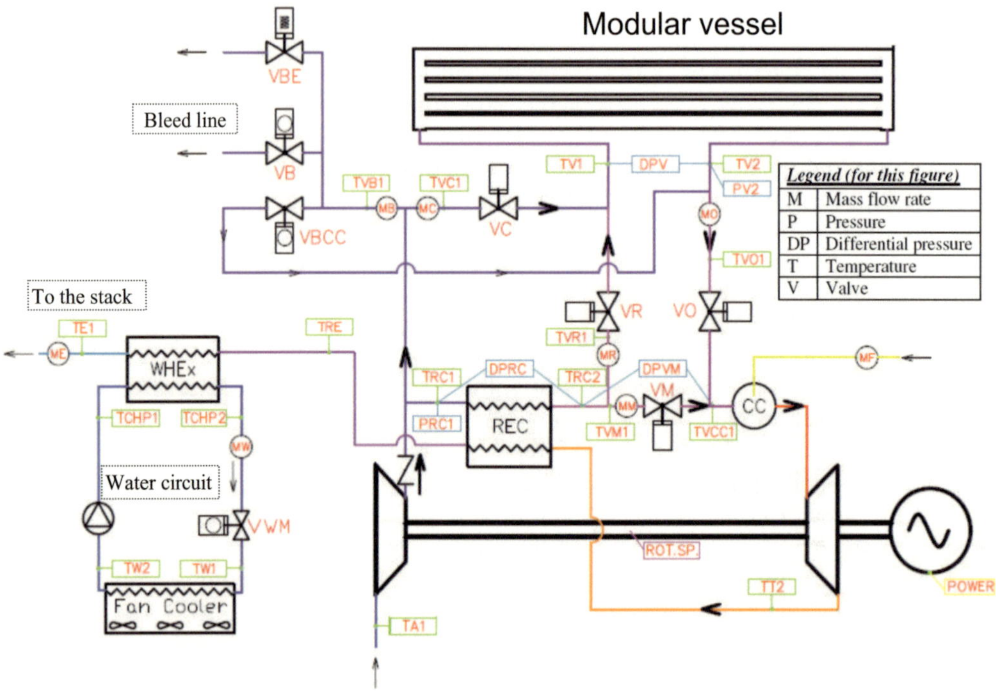

Fig. 8. Plant layout and instrumentation including the cathodic modular vessel (see Table 1 for the complete legend).

Mass flow rates			
Name	Location	Probe type	Accuracy
MM	Main line	Pitot tube	±1%
ME	Plant outlet	Thermal meter	±3%
MF	Fuel inlet	Thermal meter	±1%
MR	Vessel inlet from the recuperator	Pitot tube	±1%
MO	Vessel outlet	Pitot tube	±1%
MC	Vessel inlet from the compressor	Pitot tube	±1%
MB	Bleed outlet	Pitot tube	±1%
MW	Water main line	Magnetic meter	±4%
Static pressures			
Name	Location	Probe type	Accuracy
PA1	Ambient	Ambient sensor	±1%
PRC1	Recuperator inlet	Absolute	±1%
DPRC	Recuperator loss	Differential	±1%
DPVM	Main line loss	Differential	±1%
DPV	Vessel loss	Differential	±1%
PV2	Vessel outlet	Absolute	±1%
Temperatures			
Name	Location	Probe type	Accuracy
TA1	Ambient	Ambient sensor	±1%
TRC1	Recuperator inlet	Thermocouple	±2.5 K
TRC2	Recuperator outlet	Thermocouple	±2.5 K
TVM1	Main line pitot	Thermocouple	±2.5 K
TVCC1	Combustor inlet	Thermocouple	±2.5 K
TT2	Turbine outlet	Thermocouple	±2.5 K
TE1	Plant outlet	Thermocouple	±2.5 K
TVR1	From the recuperator pitot	Thermocouple	±2.5 K
TV1	Vessel inlet	Thermocouple	±2.5 K
TV2	Vessel outlet	Thermocouple	±2.5 K
TVO1	From the vessel pitot	Thermocouple	±2.5 K
TVC1	From the compressor pitot	Thermocouple	±2.5 K
TVB1	Bleed valve inlet	Thermocouple	±2.5 K
TCHP1	WHEx inlet	PT100 RTD	±0.3 K
TCHP2	WHEx outlet	PT100 RTD	±0.3 K
TC1	Compressor inlet	PT100 RTD	±0.3 K
TRE	Recuperator outlet	Thermocouple	±2.5 K
TW1	Cooler inlet	PT100 RTD	±0.3 K
TW2	Cooler outlet	PT100 RTD	±0.3 K

Table 1. Additional probes referred to Fig. 10 layout.

The vessel is composed of two collector pipes, connected to the recuperator outlet and the combustor inlet respectively (see Figs. 7 and 8), and five module pipes connected to both collectors. Both collectors and module pipes have a nominal diameter of 350 millimetres and their total length is around 43 meters for a maximum volume of about 4 m^3. This vessel can emulate the volume of additional components suitable for the machine size.

Figure 8 shows the rig layout with the modular vessel and all the additional probes and valves introduced in this plant configuration (see Table 1 for further instrumentation details). In comparison with the layout shown in Fig. 4, a vessel outlet line equipped with a gate valve (VO) was installed, a bleed emergency valve (VBE) was included to prevent surges during emergency shutdown, and an additional globe pneumatic valve (VBCC) was introduced to connect the compressor outlet directly to the combustor inlet. Moreover, the test rig was further improved with the installation of a water fan cooler located outside of the laboratory. It is based on three 0.7 kW electrical fans used to cool down the water (coming from the WHEx of Fig. 8), and to operate in closed circuit conditions (a 1.5 kW variable speed pump was installed).

6. Hybrid system emulation devices

To perform tests related on high temperature fuel cell hybrid systems a part of the external vessel (collectors and four modules) is used to emulate the cathodic dimension of a fuel cell stack. For this reason, the maximum cathodic volume is about 3.2 m^3. The fifth vessel module is used to emulate the related anodic volume (about 0.8 m^3) and an ejector (Ferrari et al., 2006) based anodic recirculation was included in the rig. Moreover, this facility was equipped with a steam injection system to emulate the turbine inlet composition typical of a hybrid system and a real-time model was connected to the plant for components not physically present in the rig. All these emulation devices were designed to analyse a SOFC based hybrid system of 450 kW electrical load (consistent with the machine size) (Ferrari et al., 2009a) scaled on the basis of the Rolls-Royce Fuel Cell Systems (RRFCS) planar stack (size: 250 kW; fuel utilization: 75%; stack temperatures: 800-970°C; current density: 2940 A/m^2) (Massardo & Magistri, 2003). The layout of the test rig equipped with the anodic recirculation and the steam injection systems is shown in Fig. 9.

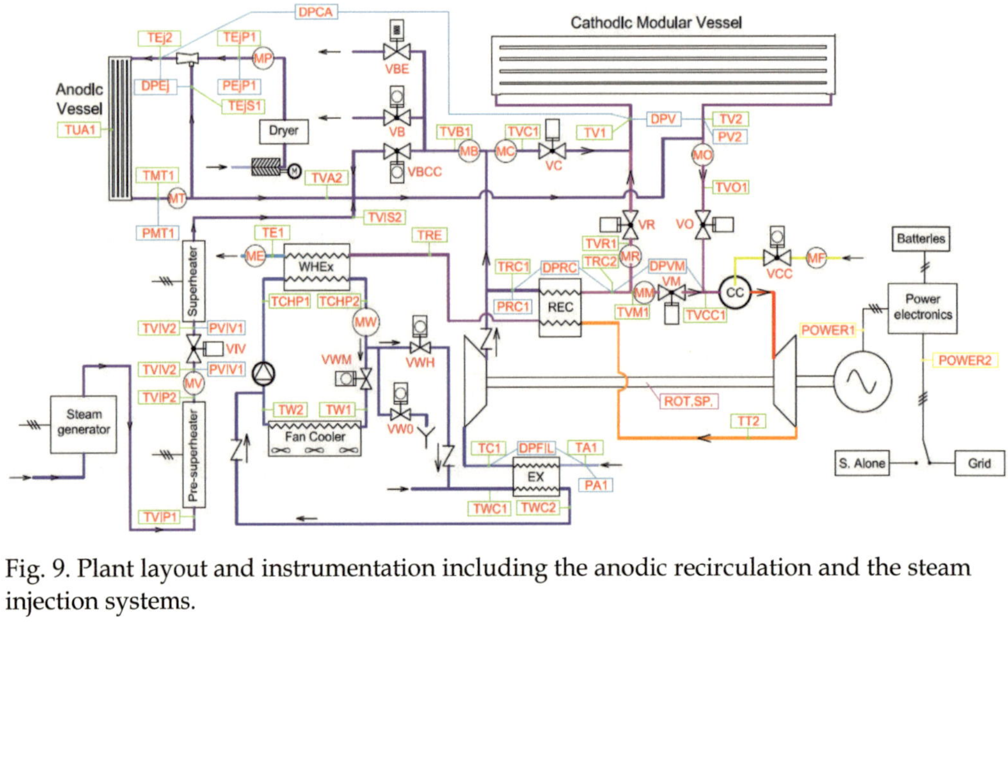

Fig. 9. Plant layout and instrumentation including the anodic recirculation and the steam injection systems.

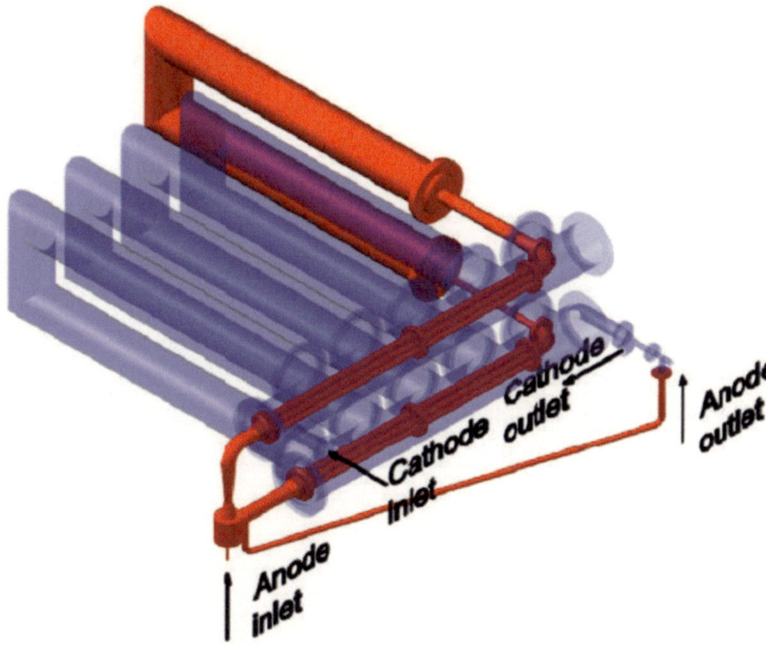

Fig. 10. The anodic loop layout.

6.1 The anodic recirculation and the steam injection systems

The anodic recirculation system (Fig. 10) is composed of a compressed air line (for fuel flow emulation), an anodic single stage ejector, and an anodic vessel. The compressed air line was designed to supply the ejector primary duct with an air mass flow rate up to 20 g/s (Ferrari et al., 2009b). This approach was developed to emulate the fuel mass flow rate at the ejector primary duct with an air mass flow rate. The anodic ejector generates the recirculation flow rate through this system as in a typical SOFC hybrid system. The 0.8 m^3 anodic vessel was designed to heat up the flow in the anodic loop. For this reason, to better emulate the anodic side, a pipe based heat exchanger was developed as shown in (Ferrari et al., 2010b). To better show this layout, a 3-D CAD plot (Fig. 10) was developed: part of the anodic loop was inserted into the cathodic vessel to partially heat the anodic flow.

The steam injection system was designed to obtain at the expander inlet the same c_p value of the reference hybrid system. This similitude approach is completely explained in (Ferrari et al., 2011). For this reason the rig was equipped with a 120 kW electrical steam generator capable of at least 27 g/s, a 40 kW electrical super-heater to increase the steam temperature from the steam generator outlet condition to a temperature suitable for the turbine combustor inlet (around 515°C), and a controlled valve for the mass flow rate management. The measuring of the steam flow rate requires a completely mono-phase steam. So upstream of the mass flow probe, an additional electrical heater (called Pre-superheater) was installed. Furthermore, several thermocouples were added for control and diagnostic purposes. The final layout is shown in Fig. 9.

All the additional probes included in the rig for these additional hardware devices (anodic recirculation and steam injection systems) are reported in Tab. 2.

Mass flow rates			
Name	Location	Probe type	Accuracy
MP	Ejector primary line	Thermal meter	±1%
MT	Anodic volume line	Venturimeter	±3%
MV	Steam generator outlet	Vortex meter	±1%
Static pressures			
Name	Location	Probe type	Accuracy
PEjP1	Ejector primary duct inlet	Absolute	±1%
PMT1	Anodic volume line	Absolute	±1%
DPEj	Anodic ejector	Differential	±1%
DPCA	Cathodic/Anodic circuit pressure difference	Differential	±1%
PVIV1	Steam control valve inlet	Absolute	±1%
PVIV2	Steam control valve outlet	Absolute	±1%
Temperatures			
Name	Location	Probe type	Accuracy
TEjP1	Ejector primary duct inlet	Thermocouple	±2.5 K
TEjS1	Ejector secondary duct inlet	Thermocouple	±2.5 K
TEjS2	Ejector outlet	Thermocouple	±2.5 K
TUA1	Anodic volume (U pipe inlet)	Thermocouple	±2.5 K
TMT1	Anodic flow venturimeter	Thermocouple	±2.5 K
TVA2	Anodic circuit outlet	Thermocouple	±2.5 K
TVIP1	Steam generator outlet	Thermocouple	±2.5 K
TVIP2	Pre-superheater generator outlet	Thermocouple	±2.5 K
TVIV1	Steam control valve inlet	Thermocouple	±2.5 K
TVIV2	Steam control valve outlet	Thermocouple	±2.5 K
TVIS2	Steam system outlet	Thermocouple	±2.5 K

Table 2. Additional probes (not shown in Fig. 8 layout) referred to Fig. 9.

6.2 Fuel cell stack real-time model to be connected to the rig

To complete the emulation of a SOFC hybrid system a real-time model was developed in Matlab®-Simulink® to be coupled to the experimental test rig. This model (developed with components validated in previous works (Ghigliazza et al., 2009a; Ghigliazza et al., 2009b) was based on the simplification of simulation components (cell stack, reformer, anodic loop, off-gas burner, expander) developed in the TRANSEO tool (Traverso, 2005) of the TPG research group. Through the Real-Time Windows Target tool and the UDP interface approach it is possible to study the entire hybrid system using the model for components not physically present in the test rig.

Figure 11 shows how the real-time model and the experimental facility are connected to emulate the entire hybrid system. The real-time model receives (as inputs) the mass flow rate, the pressure and the temperature values at the machine combustor inlet level. Furthermore, it receives the machine rotational speed and the acquisition time values. These input data are transferred from LabVIEW™ software to the real-time model (in Simulink®) through an apt UDP interface. The real-time model is used to calculate the fuel cell system behaviour over time. In detail, this model is composed of a reformer, a SOFC stack, an off-gas burner, an anodic ejector based recirculation system, a cathodic blower based recirculation, and the expander of the T100 machine. So, in this work this real-time model is used to calculate the TOT values coming from an interaction between the SOFC system and the T100 expander. This TOT value is used over time to control the real machine (the electrical load in stand-alone mode) to produce the same TOT value of the model. Also this output interface is managed through the UDP approach. Moreover, the interface includes a port to transfer to LabVIEW™ software the mass flow rate values of anodic ejector primary duct (MP). This values are used to carried out tests with the MP mass flow rate equal to the fuel flow calculated in the model. With this hardware/software interconnection layout it is possible to emulate the stack/turbine interaction from an experimental point of view studying different operative conditions.

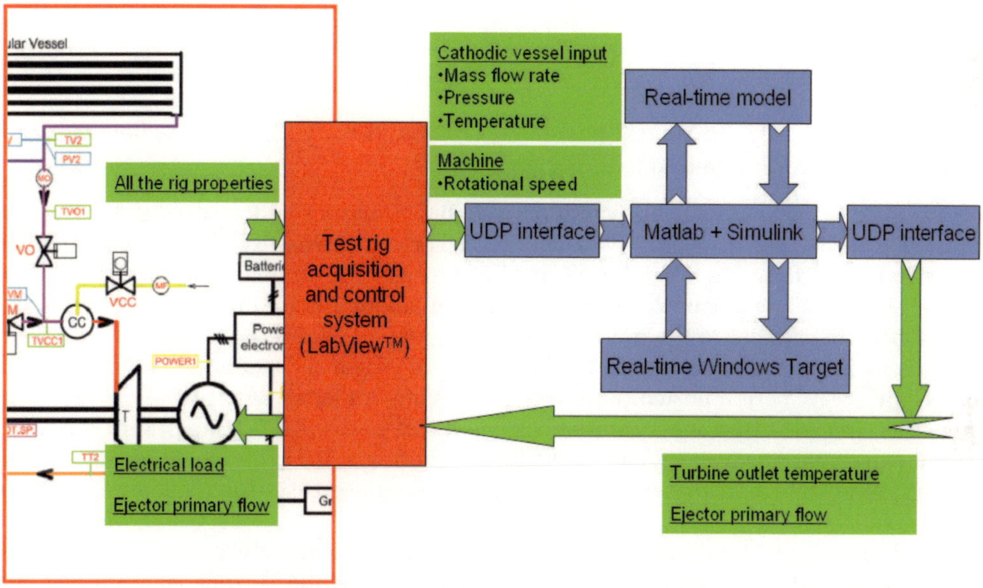

Fig. 11. Hybrid system emulation: hardware/software interconnection layout.

6.2.1 Test example: Hybrid system emulation

This paragraph reports an example of possible tests to be carried out with these emulator devices. This test was performed according to the following procedure:

1. establish connection between the real-time model and the plant;
2. impose requested current and fuel variation on model user interface;
3. the model simulates in real-time mode the evolution of SOFC properties towards a new operative point;
4. the calculated values of TOT and anodic ejector primary flow are continuously fed to the plant control system as set point values;
5. the control system moves machine load until the measured TOT value is equal to the calculated one, and operates on the MP control valve to generate flow values coming from the model.

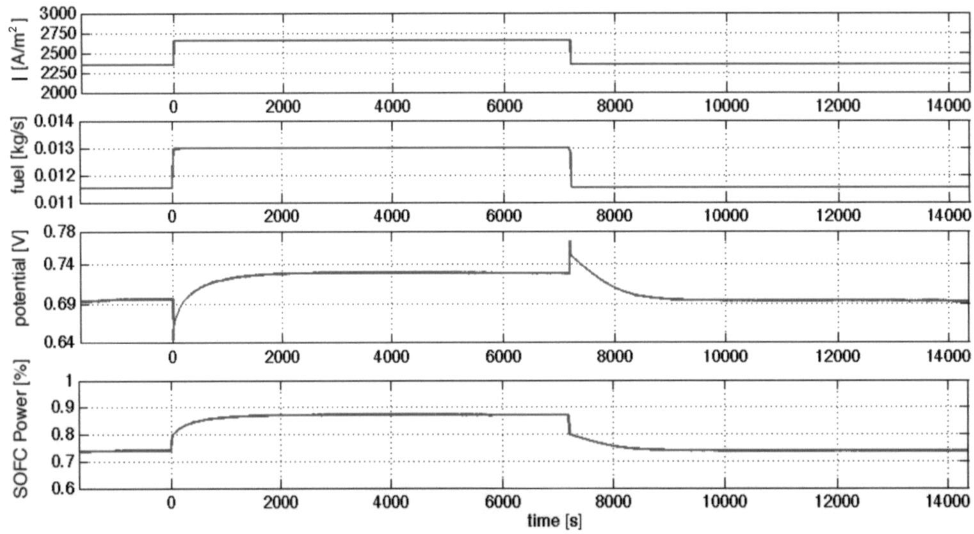

Fig. 12. Emulation test with the real-time model: current density (I), fuel mass flow rate, cell potential, SOFC power (referred to its design value).

The test is based on a concurrent variation of SOFC drawn current and mass flow rate of anodic ejector fuel. From a steady-state condition, corresponding to 80% of nominal current and primary fuel flow, the model performed a step increase in both parameters to 90% of nominal values. When a new steady-state condition (about 2 hours) was reached the model brought back the system to the initial condition. Figure 12 shows the trend of steps related to the input parameters. Moreover, the behaviour of single cell potential, and SOFC total electric power (AC) are reported. Both properties show an initial step variation, due to the input steps, followed by a long duration variation due to the SOFC temperature behaviour over time (high thermal capacitance of the SOFC). In detail, the cell voltage increases when the SOFC temperature increases for the electrical losses decrease (see (Bagnasco, 2011) for further details).

Figure 13 shows the trend of measured turbine outlet temperature and the calculated turbine inlet temperature values: it can be easily noticed how the TOT signal is affected by a constant noise characteristic of values coming from field while the TIT (calculated) shows a smoother behaviour. These values increase with the current and fuel increase (SOFC temperature increase). Figure 13 also shows the variation of the utilization factor (Eq. 1) and hybrid system (HS) global efficiency (Eq. 2). The U_f peak is due to the fluid dynamic delay between the current variation (instantaneous) and the fuel change on the stack. The hybrid system efficiency shows the step effect followed by a long time variation due to thermal aspects. The global efficiency oscillation, with high peaks in the area after the second step (7500-8000 s), are due to unexpected fuel flow discontinuities during the test.

$$U_f = \frac{[nH_2]_{cons}}{[4 \cdot nCH_4 + 6 \cdot nC_2H_6]_{in}} \tag{1}$$

$$\eta_{HS} = \frac{P_{GT} + P_{FC} - P_{blower}}{\dot{m}_{fuel}} \tag{2}$$

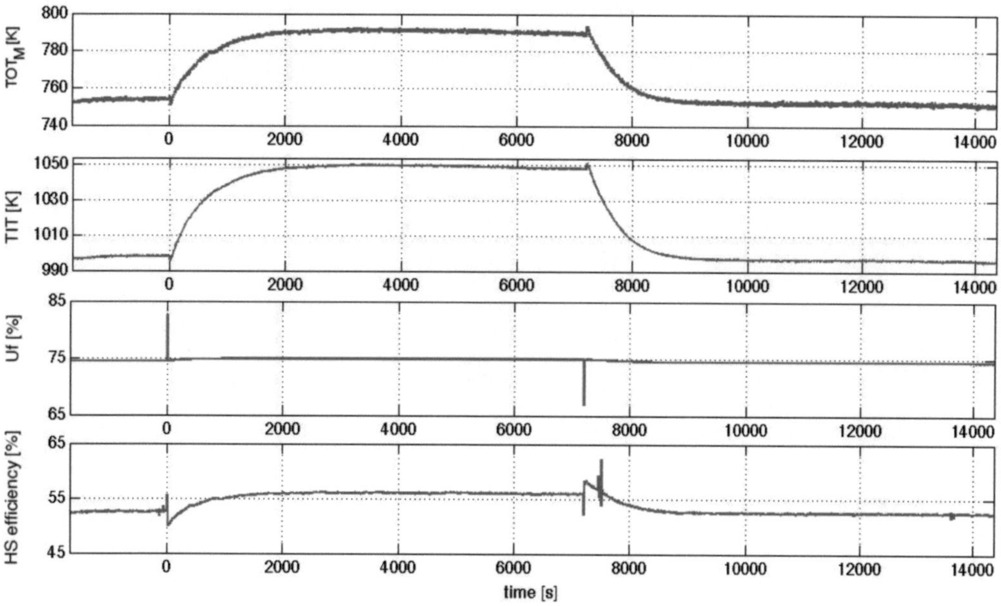

Fig. 13. Emulation test with the real-time model: measured turbine outlet temperature (TOT_M), turbine inlet temperature (TIT), fuel utilization factor, hybrid system efficiency.

7. Compressor inlet temperature control devices

A new water system was designed and installed to control the machine compressor inlet temperature (Fig. 9 layout). This system is composed of three air/water heat exchangers installed at the machine air intakes (Fig. 14) and connected to the water system. Even if the water pipe layout was modified for the connection with an absorption cooler (see the following paragraph), this part of the paper describes the previous rig layout that was used for a wide experimental campaign carried out on the machine recuperator. In this past configuration, the water system was equipped with three controlled electrical valves (VWM, VWH, and VWO of Fig. 9). It was possible to cool down the compressor inlet air by means of cold water from the supply system (opening VWO), and to heat up this air flow (closing VWM, and opening VWH) by means of the hot water coming from the machine co-generation system (WHEx). A new control system (see (Ferrari et al., 2010c) for details) was developed to manage these valves for the required temperature generation. With this past layout, maximum cooling performance depended on the supply water temperature (about 22°C in summer, that is 295.15 K), while the only restriction for heating performance is the maximum compressor inlet air temperature for the machine cooling system, that is 40°C (313.15 K). However, in the following paragraph, it is possible to notice that an absorber cooler connected to the system allows to study tri-generation options and use the produced cold flow for higher compressor inlet temperature cooling.

7.1 Test example: compressor inlet temperature control

As an example of possible tests to be carried out with the compressor inlet temperature control devices, this paragraph shows the experimental data measured on the recuperator of this test rig when operating in the machine standard layout. Since the large influence on the

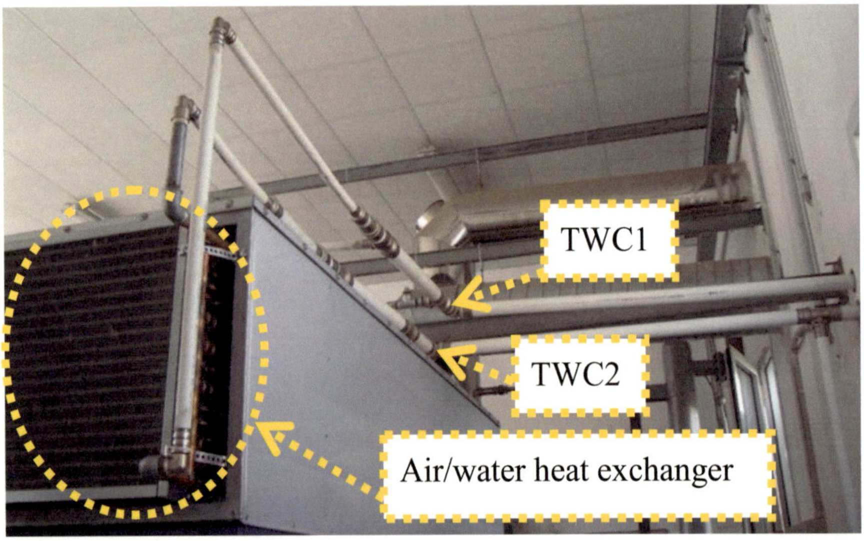

Fig. 14. An air/water heat exchanger with water pipes for compressor inlet temperature control.

recuperator temperatures of the compressor inlet temperature, this new control system was used to maintain this temperature at a fixed value of 28°C (301.15 K, with maximum errors of ±0.3 K during all the tests). These steady-state tests were carried out with the machine connected to the electrical grid to measure recuperator performance at different mass flow rate values. In this configuration the machine control system operates at constant TOT (called TT2 in Fig. 9, that is maintained at 645°C (918.15 K)) and changes the rotational speed (and the air mass flow rate) with load changes. For surge prevention purposes, it is not possible to perform tests below a 20 kW electrical load. After the machine heating phase (during the start-up), the controller does not accept load values below 20 kW.

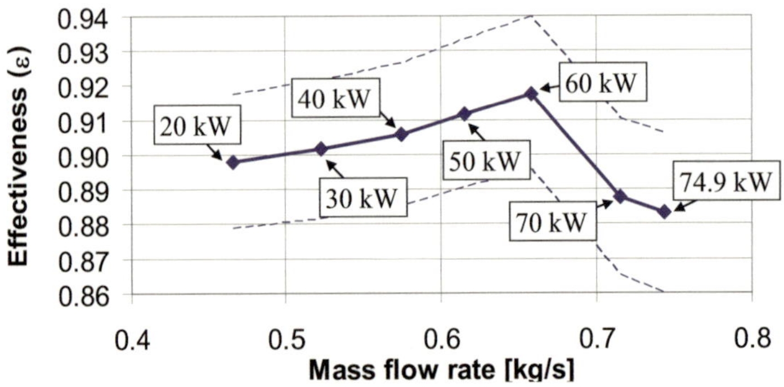

Fig. 15. Steady-state recuperator effectiveness obtained from experimental data at different loads (machine connected to the electrical grid).

Figure 15 shows the recuperator effectiveness (defined in Eq. 3 with Fig. 9 nomenclature) obtained at different electrical load values, i.e. different air mass flow rates. While the continuous line connects the effectiveness values calculated through recuperator boundary temperatures (measured during the tests), the dotted lines show the accuracy values (around ±2%) of this performance parameter. This uncertainty band was calculated through temperature measurements affected by a ±2.5 K accuracy. As shown in previous theoretical works (e.g. (McDonald, 2003)), starting from maximum flow it is possible to observe an effectiveness increase (from 0.883 to 0.918) with the mass flow rate decrease, and an effectiveness maximum followed by a decrease. However, this final trend is not so relevant as in (McDonald, 2003), because machine control system does not enable to operate steady-state tests under 20 kW (under 0.47 kg/s). Further details on all the recuperator temperatures and other tests carried out on this heat exchanger, when operating with the T100 modified machine, are shown in (Ferrari et al., 2010c).

$$\varepsilon = \frac{TRC2 - TRC1}{TT2 - TRC1} \tag{3}$$

8. Test rig integration with an absorber cooler unit

To carry out tests at compressor inlet temperature values under 20°C and to study tri-generative configurations, the facility was equipped with an absorber cooler. This device exploits the thermal content of machine exhaust flow (water at 95°C, that means 368.15 K, produced by the WHEx of Fig. 9) to produce cold water (7-12°C, that means 280.15-285.15 K). This refrigeration energy is essential for the machine inlet and for the laboratory cooling during summer or long time tests.

As shown in Fig. 16 the machine produces, in full load conditions, about 113 kW of thermal power to obtain about 2 l/s of hot water at 95°C (the system operates in closed circuit configuration). With this thermal power the absorber is able to generate about 80 kW of cooling power. This system is based on an absorber inverse cycle (water/lithium bromide) operating (in the test rig conditions) at a 0.7 COP value.

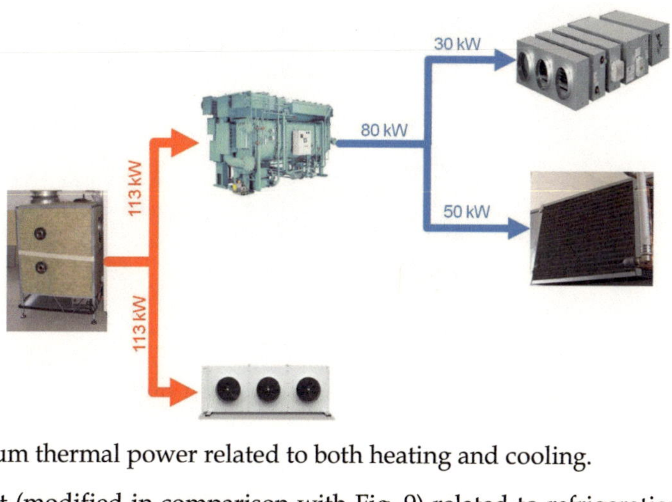

Fig. 16. Maximum thermal power related to both heating and cooling.

The water plant (modified in comparison with Fig. 9) related to refrigeration was designed for a maximum 50 kW cold power for the compressor inlet cooling and 30 kW (at maximum) for the laboratory conditioning. Moreover, with the "Fan Cooler" water/air heat exchanger (already shown in Fig. 9) it is possible to emulate a heating system using a part of the 113 kW thermal power, operating in tri-generative condition. So, While the "Fan Cooler"

is used to emulate a heating system, the heat exchangers used to manage the cold power are essential to emulate a cold thermal load.
Figure 17 shows the plant scheme related to cold water generation (from the absorber unit) and thermal power management. For this new water plant three pumps were installed: 1.5 kW pump for the hot water (2 l/s mass flow rate, 2.45 bar pressure increase), 7.5 kW pump for the cold water (5 l/s mass flow rate, 8.34 bar pressure increase), and a third pump (5.5 kW) to refrigerate the condenser of the absorber unit (12 l/s mass flow rate, 2.74 bar pressure increase). A 260 kW evaporative tower was installed for the refrigeration water (see (Prando et al., 2010) for further details). To operate the laboratory cooling a 20 kW water/air heat exchanger was designed and installed in the rig. Moreover, to perform heating conditions at the compressor inlet level, as already included in the previous facility configuration, (for instance for the emulation of a summer performance during winter) a water/water heat exchanger was included. It is a plate exchanger (power: 80 kW, primary flow: 1.44 l/s, secondary flow: 5 l/s) used to heat the "Ex" water directly with the hot water from the "WHEx" (see Fig. 17 for layout details). The water plant was also equipped with controlled valves for flow management and with mass flow rate (Magnetic meter - accuracy: ±4%) and temperature (PT100 RTD - accuracy: ±0.3 K) probes for measurement of main properties (see (Prando et al., 2010) for further details).

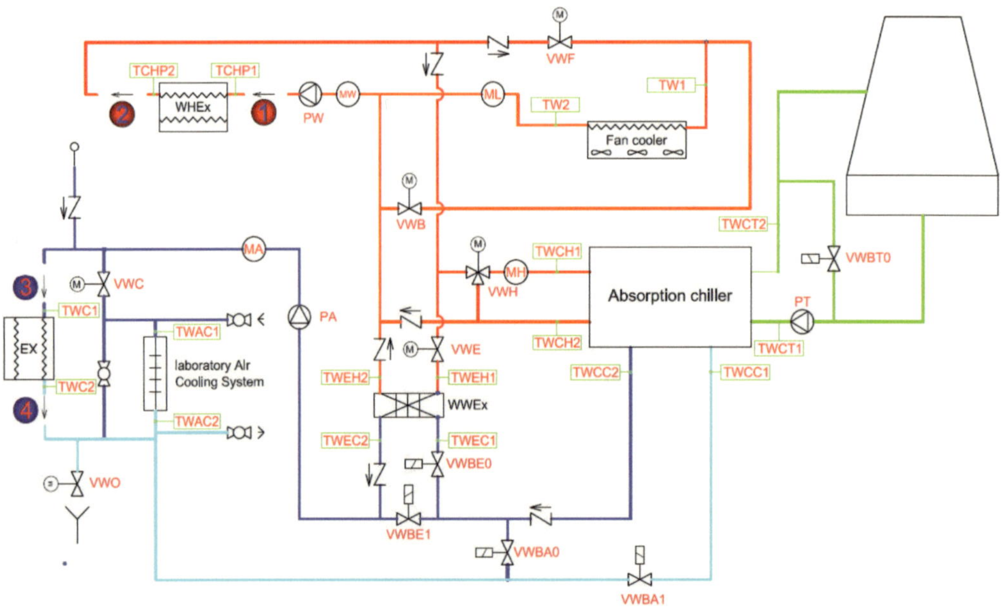

Fig. 17. Water system plant layout for the tests with absorber unit.

9. Model validation activities

Great attention is devoted to this activity to validate time-dependent simulation models at both component (recuperator) and system (the hybrid system emulator test rig) levels. A good level of consistency can be achieved thanks to the complete knowledge of the test rig dimensions, volumes, masses, shaft inertia, thermal capacitances, and operating procedure. Such completeness is difficult to obtain in industrial plants, where details about equipment are often missing or confidential. The following paragraph shows an example of this kind of validation activities focusing the attention on the machine recuperator.

9.1 Test example: The recuperator model

This validation activity regards the primary-surface (cube geometry) recuperator located inside the power case of the Turbec T100 machine (Turbec, 2002). So, a recuperator real-time model was tested against experimental data not in a heat exchanger test rig, but in a real operative configuration, working in a commercial recuperated 100 kWe machine. The recuperator model adopts the lumped-volume approach (Ferrari et al., 2005) for both hot and cold flows. Since momentum equation generates negligible contribution during long-time transients, because it produces quite fast effects (dynamic effects) that are negligible in a component with average flow velocities at around 10 m/s, it is possible to properly represent the transient behaviour of the heat exchanger just using the unsteady form of the energy equation (the actual governing equation (Ghigliazza et al., 2009a) of the system).

The finite difference mathematical scheme (shown in Fig. 18) is based on a recuperator division into four main parts (j = 0, 1, 2, 3). The internal grid is "staggered" to model the heat exchange between each solid cell (j = 1, 3) and the average temperature of the flow (j = 0, 2): M+1 faces correspond to M cells. The resulting quasi-2-D approach is considered a good compromise between accuracy of results and calculation effort. The heat loss to environment and the longitudinal conductivity into solid parts are also included. All the equations and the integration approach of this model are described in (Ghigliazza et al., 2009a). Moreover, this paper reports the main data used for the recuperator model (Table 3) for the results reported here.

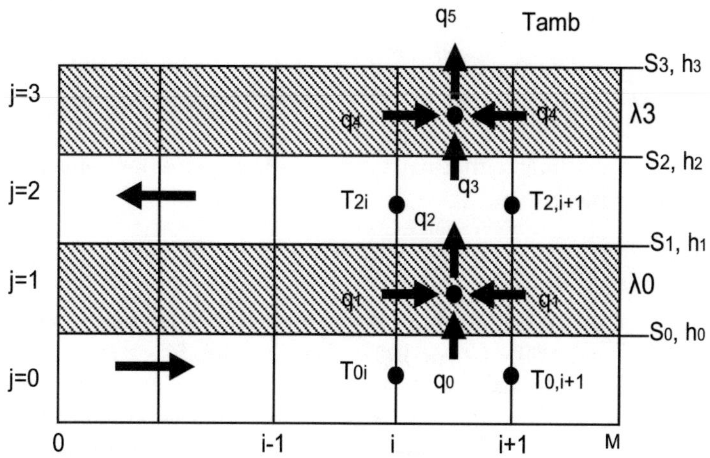

Fig. 18. Real-time model: finite difference scheme.

Figure 19 shows the comparison between experimental data and model results related to recuperator outlet temperature (cold side). The test considered here is a machine start-up phase carried out from cold condition. The results obtained during this test are acceptable, even if same margin of improvement exists. With reference to Fig. 19, the following aspects can be highlighted:

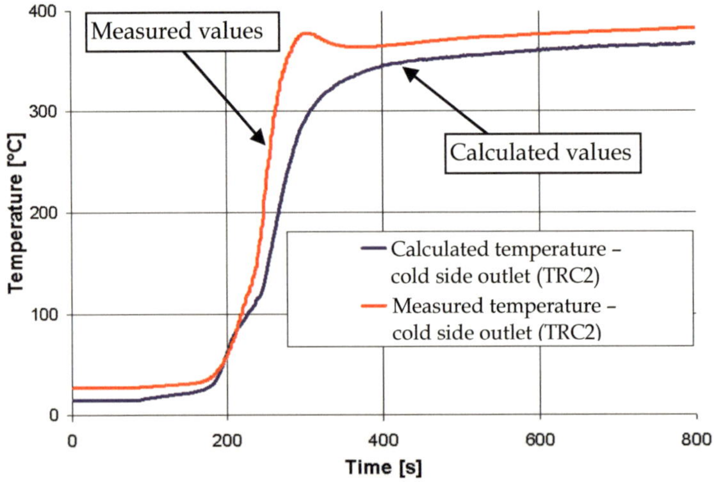

Fig. 19. Recuperator model validation (start-up phase): cold side outlet.

Thermal capacitance	226.05 [kJ/K]
Convective heat exchange (cold side)	500 [W/m^{2}K]
Convective heat exchange (hot side)	250 [W/m^{2}K]
Length	0.35 [m]
Nominal pressure drop (cold side)	0.06 [bar]
Nominal pressure drop (hot side)	0.06 [bar]

Table 3. Recuperator model data.

- the matching between measured data and model predictions is within a difference of 50°C, which can be considered a good result considering real-time simulation performance;
- measurements show a longer thermal delay (likely explanation: effect due to thermal shield of thermocouples).

10. Conclusion

A new test rig based on micro gas turbine technology was developed at the TPG laboratory (campus located at Savona) of the University of Genoa, Italy. It is based on the coupling of different equipments to study advanced cycles from experimental point of view and to provide students with a wide access to energy system technology. Particular attention is devoted on tests related on hybrid systems based on high temperature fuel cells. The main experimental facilities developed and built for both student and researcher activities are:

- A commercial recuperated micro gas turbine (100 kW nominal electrical load) equipped with a hot water co-generation unit and with the essential instrumentation for control reasons and to operate typical tests (start-up, shutdown, load changes) on the machine.
- A set of external pipes connected to the machine for the flow measurement and management. These pipes are used to measure with enough accuracy all the properties necessary for cycle characterization (e.g. the air mass flow rate or recuperator boundary

temperatures), not available in the machine commercial layout. In particular the chapter shows a test example related to the compressor map measuring.

- An external modular vessel to test the coupling of the machine with different additional innovative cycle components, such as saturators, fuel cells of different layouts or technology, or additional heat exchangers.
- Additional devices for hybrid system emulation activities. This part describes the anodic recirculation based on a single stage ejector (coupled to the rig for tests related to the anodic/cathodic side interaction), the steam injection system based on a 120 kW steam generator (used to emulate the turbine inlet composition typical of a hybrid system), and a real-time model used to emulate the components not physically present in the rig (e.g. the fuel cell). As an example of tests carried out with these devices, this chapter reports the main results obtained during fuel and current steps carried out with the real-time model coupled with the rig.
- Compressor inlet temperature control devices (heat exchangers, pipes, pump, and control system) to evaluate performance variations related to ambient temperature changes. Particular attention is focused on tests carried out on the recuperator with the machine operating in grid-connected conditions.
- An absorber unit connected to the plant (the hot water generated by the WHEx is used as primary energy to produce cold water) to carry out tests at compressor inlet temperature values under 20°C and to study tri-generative configurations.
- Great attention is devoted to validation activities for time-dependent simulation models. As an example, this chapter shows the comparison between experimental data and model results related to recuperator outlet temperature (cold side), during a cold start-up phase.

Besides the additional developments and tests on the rig, already planned and presented in (Pascenti et al, 2007; Ferrari et al., 2009a; Ferrari et al., 2010c; Prando et al., 2010), all the different layout configurations will be considered for tests. For instance, in an ongoing work it is planned to use the real-time model for control system development activities related on SOFC hybrid plants and the absorber cooler to carry out tests at lower ambient temperature conditions, also considering tri-generative configurations.

11. Acknowledgment

This test rig was mainly funded by FELICITAS European Integrated contract (TIP4-CT-2005-516270), coordinated by Fraunhofer Institute, by LARGE-SOFC European Integrated Project (No. 019739), coordinated by VTT, and by a FISR National contract, coordinated by Prof. Aristide F. Massardo of the University of Genoa.
The authors would like to thank Prof. Aristide F. Massardo (TPG Coordinator) for his essential scientific support, Dr. Loredana Magistri, (permanent researcher at TPG) for her activities in design point definition, and Mr. Alberto N. Traverso (associate researcher at TPG) for his technological support on absorption cooler installation.

12. References

Kolanowski, B. F. (2004). *Guide to Microturbines*, Fairmont Press, ISBN 0824740017, Lilburn, Georgia (USA).

Boyce, M. P. (2010). Handbook for Cogeneration and Combined Cycle Power Plants, Second Edition, ASME Press, ISBN 9780791859537, New York, New York (USA).

Massardo, A. F., McDonald, C. F., & Korakianitis, T. (2002). Microturbine-Fuel Cell Coupling for High-Efficiency Electrical-Power Generation. Journal of Engineering for Gas Turbines and Power, Vol. 124(1), pp. 110-116, ISSN 0742-4795.

Lindquist, T., Thern, M., Torisson, T. (2002). Experimental and Theoretical Results of a Humidification Tower in an Evaporative Gas Turbine Cycle Power Plant. *Proceedings of ASME Turbo Expo 2002*, 2002-GT-30127, ISBN 0791836010, Amsterdam, The Netherlands, June 3-6, 2002.

Traverso, A., Massardo, A. F., Scarpellini, R. (2006). Externally Fired micro-Gas Turbine: Modelling and Experimental Performance. *Applied Thermal Engineering*, Elsevier Science, Vol. 26, pp. 1935-1941, ISSN 1359-4311.

Magistri, L., Costamagna, P., Massardo, A. F., Rodgers, C., McDonald, C. F. (2002). A Hybrid System Based on a Personal Turbine (5 kW) and a Solid Oxide Fuel Cell Stack: A Flexible and High Efficiency Energy Concept for the Distributed Power Market, *Journal of Engineering for Gas Turbines and Power*, Vol. 124, pp. 850-875, ISSN: 0742-4795, New York, New York (USA).

Magistri, L., Traverso, A., Cerutti, F., Bozzolo, M., Costamagna, P., Massardo, A. F. (2005). Modelling of Pressurised Hybrid Systems Based on Integrated Planar Solid Oxide Fuel Cell (IP-SOFC) Technology. *Fuel Cells*, Topical Issue "Modelling of Fuel Cell Systems", WILEY-VCH, Vol. 1, Issue 5, ISSN 1615-6854.

Pedemonte, A. A., Traverso, A., Massardo, A. F. (2007). Experimental Analysis of Pressurised Humidification Tower For Humid Air Gas Turbine Cycles. Part A: Experimental Campaign. *Applied Thermal Engineering*, Elsevier Science, Vol. 28, pp. 1711–1725, ISSN 1359-4311.

McDonald, C. F. (2003). Recuperator Considerations For Future High Efficiency Microturbines. *Applied Thermal Energy*, Elsevier Science, Vol. 23, pp. 1453-1487, ISSN 1359-4311.

Ferrari, M. L. (2011). Solid Oxide Fuel Cell Hybrid System: Control Strategy for Stand-Alone Configurations. *Journal of Power Sources*, Elsevier, Vol. 196, Issue 5, pp. 2682-2690, ISSN: 0378-7753.

Tucker, D., Liese, E., Gemmen, R. (2009). Determination of the Operating Envelope for a Direct Fired Fuel Cell Turbine Hybrid Using Hardware Based Simulation. *Proceedings of International Colloquium on Environmentally Preferred Advanced Power Generation 2009*, ICEPAG2009-1021, ISBN 3-7667-1662-X, Newport Beach, California, USA.

Hohloch, M., Widenhorn, A., Lebküchner, D., Panne, T., Aigner, M. (2008). Micro Gas Turbine Test Rig for Hybrid Power Plant Application. *Proceedings of ASME Turbo Expo 2008*, GT2008-50443, ISBN 0791838242, Berlin, Germany.

Ferrari, M. L., Pascenti, M., Bertone, R., Magistri, L. (2009a). Hybrid Simulation Facility Based on Commercial 100 kWe Micro Gas Turbine. *Journal of Fuel Cell Science and Technology*, Vol. 6, pp. 031008_1-8, ISSN: 1550-624X, New York, New York (USA).

Ferrari, M. L., Pascenti, M., Magistri, L., Massardo, A. F. (2010a). Hybrid System Test Rig: Start-up and Shutdown Physical Emulation, *Journal of Fuel Cell Science and Technology*, Vol. 7, pp. 021005_1-7, ISSN: 1550-624X, New York, New York (USA).

Ferrari, M. L., Pascenti, M., Magistri, L., Massardo, A. F. (2010b). Analysis of the Interaction Between Cathode and Anode Sides With a Hybrid System Emulator Test Rig, *Proceedings of International Colloquium on Environmentally Preferred Advanced Power Generation 2010,* ICEPAG2010-3435, Costa Mesa, CA (USA).

Turbec T100 Series 3 (2002). Installation Handbook.

Pascenti, M., Ferrari, M. L., Magistri, L., Massardo, A. F. (2007). Micro Gas Turbine Based Test Rig for Hybrid System Emulation, *Proceedings of ASME Turbo Expo 2007,* GT2007-27075, ISBN: 0791837963, Montreal, Canada.

Traverso, A. (2005). TRANSEO Code for the Dynamic Performance Simulation of Micro Gas Turbine Cycles, *Proceedings of ASME Turbo Expo 2005,* GT2005-68101, ISBN: 0791846997, Reno, Nevada (USA).

Traverso, A., Calzolari, F., Massardo, A. F. (2005). Transient Behavior of and Control System for Micro Gas Turbine Advanced Cycles, *Journal of Engineering for Gas Turbine and Power,* Vol. 127, pp. 340-347, ISSN: 0742-4795, New York, New York (USA).

Ferrari, M. L., Liese, E., Tucker, D., Lawson, L., Traverso, A., Massardo, A. F. (2007). Transient Modeling of the NETL Hybrid Fuel Cell/Gas Turbine Facility and Experimental Validation, *Journal of Engineering for Gas Turbines and Power,* Vol. 129, pp. 1012-1019, ISSN: 0742-4795, New York, New York (USA).

Caratozzolo, F., Traverso, A., Massardo, A. F. (2010).Development and Experimental Validation of a Modelling Tool for Humid Air Turbine Saturators, *Proceedings of ASME Turbo Expo 2010,* ASME Paper GT2010-23338, ISBN: 9780791838723, Glasgow, UK.

Bagnasco, M. (2011). Emulation of SOFC Hybrid System With Experimental Test Rig and Real-Time Model, Bachelor Thesis, TPG, Genova, Italy (in Italian).

Ferrari, M. L., Pascenti, M., Traverso, A. N., Massardo, A. F. (2011). Hybrid System Test Rig: Chemical Composition Emulation With Steam Injection, *Proceedings of International Conference on Applied Energy,* pp. 2821-2832, Perugia, Italy.

Ferrari, M. L., Bernardi, D., Massardo, A. F. (2006). Design and Testing of Ejectors for High Temperature Fuel Cell Hybrid Systems, *Journal of Fuel Cell Science and Technology,* Vol. 3, pp. 284-291, ISSN: 1550-624X, New York, New York (USA).

Massardo, A. F., Magistri, L. (2003). Internal Reforming Solid Oxide Fuel Cell Gas Turbine Combined Cycles (IRSOFC-GT) - Part II: Energy and Thermoeconomic Analyses, *Journal of Engineering for Gas Turbines and Power,* Vol. 125, pp. 67-74, ISSN: 0742-4795, New York, New York (USA).

Ferrari, M. L., Pascenti, M., Magistri, L., Massardo, A. F., (2009b). Hybrid System Emulator Enhancement: Anodic Circuit Design, *Proceedings of International Colloquium on Environmentally Preferred Advanced Power Generation 2009,* ICEPAG2009-1041, Newport Beach, California, USA.

Ghigliazza, F., Traverso, A., Pascenti, M., Massardo, A. F. (2009a). Micro Gas Turbine Real-Time Modeling: Test Rig Verification", *Proceedings of ASME Turbo Expo 2009,* GT2009-59124, Orlando, Florida (USA).

Ghigliazza, F., Traverso, A., Massardo, A. F., Wingate, J., Ferrari, M. L. (2009b). Generic Real-Time Modeling of Solid Oxide Fuel Cell Hybrid Systems, *Journal of Fuel Cell Science and Technology,* Vol. 6, pp. 021312_1-7, ISSN: 1550-624X, New York, New York (USA).

Ferrari, M. L., Pascenti, M., Magistri, L., Massardo, A. F. (2010c), Micro Gas Turbine Recuperator: Steady-State and Transient Experimental Investigation, *Journal of Engineering for Gas Turbines and Power*, Vol. 132, pp. 022301_1-8, ISSN: 0742-4795, New York, New York (USA).

Prando, A., Poloni, M. (2010). Absorber Tri-generative System Based on a 100 kWe Micro Gas Turbine: Study and Plant Development, Bachelor Thesis, TPG, Genova, Italy (in Italian).

Ferrari, M. L., Traverso, A., Magistri, L., Massardo, A. F. (2005). Influence of the Anodic Recirculation Transient Behaviour on the SOFC Hybrid System Performance, *Journal of Power Sources*, Elsevier, Vol. 149, pp. 22-32, ISSN: 0378-7753.

6

Biofuel and Gas Turbine Engines

Marco Antônio Rosa do Nascimento and Eraldo Cruz dos Santos
Federal University of Itajubá – UNIFEI
Brazil

1. Introduction

Currently, the interest in using vegetable oils and their derivatives as fuel in primary drives for the generation of electricity has increased due to rising oil prices and concerns over the environmental impacts caused by fossil fuel use. For viability of using biodiesel as a substitute for fossil fuels for power generation, should be considered the emissions of greenhouse gases, i.e., pollutants such as nitrogen oxides (NO_X), sulfur oxides (SO_X), carbon monoxide (CO) and particulates into the atmosphere during the lifetime of the power plant.

According HABIB (2010) the effect of using petroleum-derived fuel in aviation on the environment is significant. Given the intensity of air traffic and civil and military operations, making the development of alternative fuels for the aviation sector is justified, necessary and critical.

Another concern that must be considered is the quality of biofuel to be stored over time, this being an obstacle to be overcome in order to maintain fuel quality and operational reliability of gas turbine installations operating with biofuels.

Biofuels also have the advantage of being renewable and cleaner, this is due in large part because they do not contain sulfur in its composition. The use of distributed generation renewable fuels can be advantageous in isolated regions, far from major urban centers, to generate electricity using the resources available on site.

Among other engines, gas turbines represent one of the technologies of distributed generation, which is characterized by the supply of electricity and heat simultaneously. In principle these machines should operate without major problems by using biofuels, because of similarities with the characteristics of the fuels conventionally used. However, there are few references on the performance of gas turbines operating on biofuels and this is the motivation of this study.

Microturbines are small gas-turbo generators designed to operate in the power range from 10 to 350 kW. Although its operation will also be based on the Brayton cycle, they present their own characteristics that differentiate them from large turbines.

Most gas turbine available today, originated in the military and aerospace industry. Many projects were aimed at applications in the automotive sector in the period between 1950 and 1970. The first gas turbine generation was developed from turbo aircraft, buses and other commercial means of transport (SCOTT, 2000). Interest in stationary generation market has expanded in the years 1980 and 1990, and its use in distributed generation has been accelerated (LISS, 1999).

It is hoped that in future gas turbines of small power is an alternative for power generation for residential and commercial segment, since the operational reliability is one of the main needs in these sectors (WILLIS and SCOTT, 2000). These turbines have various applications such as power generation in the place of consumption (on-site), the uninterrupted supply of electricity, to cover peak loads, cogeneration and mechanical drive, which characterizes the distributed generation (BIASI, 1998).
Gas turbines may use different types of fuel such as diesel, kerosene, ethanol, natural gas and gas obtained from biomass gasification, etc. The shift to gas from biomass has been considered promising, but some changes must be made in supply and combustion systems turbine, aiming to modify the injection and control systems and the volume of the combustion chamber.
The scope of this chapter includes a brief description of the systems of gas turbines, reports the experiences of biofuel use in gas turbines made until today, with emphasis on the experience developed in the Laboratory of Gas Turbines and Gasification the Institute of Mechanical Engineering, Federal University of Itajubá - IEM/UNIFEI aspects of thermal performance and emissions of gases from gas turbines of small power.

2. Biofuels

Biofuels are fuels of biological origin, i.e., not fossil. They are produced from plants such as corn, soy, sugar cane, castor beans, sugar beet, palm oil, canola, babassu oil, hemp, among others. Organic waste can also be used for the production of biofuel. The main biofuels are ethanol (produced from sugar cane and corn), biogas (biomass), bioethanol, biodiesel (from palm oil or soy), among others.
Biofuels can be used on vehicles (cars, trucks, tractors, etc.), turbines, boilers, etc.., in whole or blended with fossil fuels. In Brazil, for example, soy biodiesel is blended with fossil diesel. Is also added to gasoline the ethanol produced from sugar cane.
The advantage of using biofuels is the significant reduction of greenhouse gas emissions. It is also advantageous because it is a renewable source of energy instead of fossil fuels (diesel, gasoline, kerosene, coal).
This section will describe some characteristics and requirements of biofuels that have potential for use in gas turbines.

2.1 Gas turbines operating on liquid fuels

Biofuels have the greatest potential for use in gas turbines are biodiesel and ethanol, due to factors such as availability physical-chemical characteristics similar to fossil fuels such as diesel or jet fuel. Table 1 presents a summary of requirement of liquid fuel as defined by the manufacturers of gas turbines for efficient operations (BOYCE, 2006).

Moisture and sediment	1.0 % (v%) maximum
Viscosity	20 cS at injector
Dew-Point	20 °C at ambient temperature
Carbon Residue	1.0 % (p.) maximum
Hydrogen	11% (p.) maximum
Sulfur	1% (p.) maximum

Table 1. Requirements liquid fuel for gas turbines.

The growing interest in biofuels along with increasing market demand for generators supplied by renewable fuels has led manufacturers to modify the designs of gas turbines and micro-turbines, in order that they can operate on biofuels.
For biodiesel, the supply system is being modified to fit this new biofuel due to some reasons such as higher viscosity, content of acylglycerols and the effects of corrosion. New corrosion-resistant materials, systems control the flow of fuel and improved geometry optimized for the guns are some of the challenges of these new projects.
In scientific literature there is little information about testing of gas turbines for small power, operating on biofuels. To study the impact of biofuel use in the operation and maintenance of gas turbine, one must take the following measures:

- Define the physical and chemical characteristics of both diesel and biofuel used in the tests. Some important characteristics are: density, distillation, viscosity, ash content, phosphorus, iodine and sulfur, water content, cetane number, oxidation stability, flashpoint, freezing point, dew point, volumetric composition of methyl, ethyl and lipids, glycerol, lower calorific value, etc. These values should be compared with the requirements of the standards on diesel and biodiesel to demonstrate that they can be used in the study.
- Once further tests it is possible to define what characteristics of biodiesel are relevant to the determination of changes in engine behavior, considering the performance parameters and emissions. The higher viscosity of biodiesel can lead to difficulties in its injection into the combustion chamber. It is possible reduce the viscosity of the mixture increases its temperature, or by adding alcohol. The lower the flash point of biodiesel could also cause problems in combustion.
- It is possible find accumulation of carbonized material in the inner parts of the gas turbine, after the tests with biodiesel. The biofuel can produce corrosion in fuel supply system.
- It is also recommended to install a filter at least 50 μm at the fuel supply in the gas turbine when using biodiesel.
- It is not advisable to use biofuels in gas turbines without performing a preliminary economic analysis.

Some alternative liquid fuels such as vegetable oil, biodiesel or pyrolysis oil, ethanol and methanol are being tested in gas turbines (GÖKALP, 2004).
As biodiesel has similar properties to diesel, it can be used directly in a gas turbine, blended with diesel in various proportions (usually uses 5 to 30% biodiesel in the blend with diesel). The properties of biodiesel are slightly different to those of diesel in terms of energy content or physical properties.
The Lower Heating Value (LHV) of liquid biofuels such as pure biodiesel (B100), B5 B30 and vegetable oils are between 37,500 and 44,500 kJ/kg, which is close to regular diesel (GÖKALP, 2004).
The viscosities of ethanol, biodiesel and its blends with diesel are lower than the residual oil from the kitchen, making it easier to spray. Vegetable oils and oils derived from pyrolysis have a very high viscosity, which causes problems in its mist inside the combustion chamber of gas turbines. However, these fuels can be heated to reduce its viscosity before being injected into the combustor.
Fuel oil resulting from pyrolysis of wood, vegetable oils and methyl esters has a carbon/hydrogen rate (C/H) higher than that of conventional diesel. As consequence, there

may be an accumulation of soot inside the combustion chamber or turbine blades (GÖKALP, 2004). Another factor that changes as a result of this feature is the transfer of heat by radiation from the flame to the flame tube.

2.2 Gas turbines operating on gaseous fuels

Thermal power plants with gas turbines operating with gas from biomass need to present efficient, simple technology, low cost and operational reliability, in order that these plants could become economically competitive with traditional systems of power generation, such as stationary alternative engines.

The potential of gas turbines for this application is great, although the gas must be subjected to cleaning to remove solid impurities and/or gas that can damage some components of some systems of gas turbine.

Gaseous fuels can be obtained by gasification of biomass, which in addition to the gas generates a set of substances, such as tar which is a compound in gaseous form in the fuel gas, which has an appreciable calorific value, although it shows a tar obstacle to the use of gaseous fuels in internal combustion engines, due to its high corrosive power of components and reduced engine efficiency.

In the case of gas turbines, the tar can be a problem only happens when its condensation. Basically there are two strategies to address the problem of tar, remove it from the fuel gas or burn it in the combustion chamber. In the first case, nickel-based catalysts have shown very promising results. In the second case, the strategy is to keep the fuel temperature above the dew point of tar in the gas supply pipes, and perform your burning at high temperature in the combustion chamber (SCHMITZ, 2000).

Currently, gas turbines are designed for a specific fuel (natural gas or fuel oil). Recent progress has been achieved in the methodologies and tools for the design of combustors for gas turbine. It is possible to perform a clean combustion of fossil fuels by employing low-carbon technologies based on premixed combustion.

There are ongoing projects that aim to harness these advances for applications geared to a wider range of fuels with commercial potential, including those with low calorific value, obtained from biomass gasification. Some procedures should be established for selecting appropriate fuels to be used taking into account the performance of combustion and emissions of soot and NO_X. Furthermore, it should be considered the adaptability of existing burners to use alternative fuels selected (GÖKALP, 2004).

Some gaseous alternative fuels also have potential for use in gas turbines, for example, the synthesis gas from gasification of biomass, the biomass pyrolysis gas, gas from digesters (biogas) and residual gas from industrial processes, which are rich in hydrogen.

Industrial gases such as methane reformed with steam, refinery gas, residual gas from the Fischer-Tropsch gasification gas with oxygen gas and slow pyrolysis of wood, have a LHV comparable to natural gas. This is due to the high hydrogen content of fuel gases, which lies between 19 and 45% of the volume. Rather, the LHV of gas gasification with air and biogas are very low because they are produced at atmospheric pressure, so they must be compressed before being used in gas turbines.

Except reformed methane with steam, all other gaseous fuels mentioned above have a C/H greater than that of natural gas (GÖKALP, 2004).

According BOYCE (2006) in Table 2 presents an overview of the requirements for gaseous fuels that can be used in gas turbines.

Calorific	11,184 - 41,000 kJ/m³
Solid Contaminants	< 30 ppm
Flammability Limits	2.2 : 1
Content of sulfur, sodium, potassium and lithium	< 5 ppm (In the form of meta alkaline sulfate)
H_2O (p.)	< 25%

Table 2. Requirements gaseous fuel for gas turbines.

According KEHLHOFER (2009) the type and composition of the fuel has a direct influence on efficiency and emission of gases from a gas turbine. The LHV of the fuel is important because it defines the mass flow of fuel and consequently its specific consumption. The fuel composition is also important as it influences the performance of the cycle because it determines the enthalpy of gas entering the turbine, and the available enthalpy drop off the engine and the amount of steam generated in the recovery boiler.

2.3 Clean fuel gas

Gaseous fuels obtained through the processes of gasification and pyrolysis produce fuels with low-and medium calorific value. According to Table 1, in some gas turbines require a minimum value for the calorific value of gas, which can be difficult to achieve it.

Manufacturers set very strict parameters regarding the quality of the combustible gases to avoid possible damage to the hot parts of gas turbine. Alkaline components present in the fuel, especially chlorides cause corrosion at high temperatures. The presence of tar should also be considered if the operations of fuel valves occur at temperatures below the dew point of the tar.

Almost all biogenic fuels contain considerable amounts of halogens (Cl, F, Br), alkali (Na, K), alkali oxides and other metals (Zn, Cu, Ca). The sulfur content is usually low. Unless most of these items may be retained in the gasifier or pyrolysis reactor, the requirements of the fuel gas cannot be met when employing biomass. The fuel gas contains impurities mentioned in the gaseous or solid, so it is inevitable that they would perform a deep cleaning of the gas. The chloride concentration should be considerably reduced. The slow pyrolysis produces gases, which are characterized by average values of calorific value, situated in the range between 7,000 and 13,000 kJ/kg, and low levels of chlorides and alkalis.

Should be applied under heating rates and moderate temperatures to produce pyrolysis gas with low alkali chloride. High rates of heating fuel break biogenic structure, which favors the conversion of solid pyrolysis gas, however, compromises the retention of impurities in the vegetable coal. High levels of retention in coal are only achieved with low rates of heating in combination with moderate temperatures.

It is estimated that for slow pyrolysis with a maximum temperature of 350 °C are retained approximately 86% of chloride, and the calorific value of gas reaches 10.9 MJ/kg, which can be considered average. These conditions result in a lower ratio chloride/energy in the gas, although the concentrations required by gas turbines today are even smaller, and therefore, solid and gaseous impurities must be removed (SCHMITZ, 2000).

3. Experiments with biofuels in gas turbine engines

This item will be described some experiences with the utilization of biofuels in gas turbines.

Testing gas micro-turbines operating with biodiesel were performed by LOPP *et. al.* (1995); MIMURA (2003); BIST (2004); SCHMELLEKAMP and DIELMANN (2004); WENDIG (2004) and CORRÊA (2006). LOPP (1995) presented the results of the thermal performance of a gas turbine using a blend of jet fuel (Jet fuel - JF) and soy biodiesel (B). Three different fuels: JF, B10/JF90 and B20/JF80. The turbine efficiency reached its nominal value with B20, with a slightly better performance than with B10. There was an increase in fuel consumption proportional to the addition of biodiesel in the blend. The blends with biodiesel/diesel fuel were compatible enough to allow additional testing and show its potential as an alternative fuel. CO_2 emissions were reduced after the engine was fueled with B20. The two blends B10 and B20 did not show a noticeable increase in emissions of particulate matter compared to the JF. Subsequent inspections in the combustor and turbine blades showed no deposition or degradation of components.

MIMURA (2003) conducted a study of performance and emissions from a micro-turbine supplied with biodiesel from waste food oil regenerated. Operating in cogeneration mode the system had a thermal efficiency of 64%. It was observed emissions of 6 ppm of CO, 23 ppm of NO_X and 1.0 ppm of SO_X.

BIST (2004) performed a feasibility study on the use of methyl esters (biodiesel) derived from soy oil as additives blended with fuel gas turbine aircraft. Several blends were tested to identify which would meet the specifications for this type of engine without the need to change the initial design. Was not noticed a significant increase in fuel consumption for blends of B2, B5 and B10. In cases of B20 and B30 blends the increase in consumption was evident, being 7 and 10%, respectively. It was shown that a decrease in the efficiency of combustion in gas turbine as the percentage of biodiesel in the blend increase. In terms of emissions, an increase of CO content in the gases, due to the increase in the percentage of biodiesel in the blend, was noticed what indicates a reduction in combustion efficiency. This behavior is contrary to that seen in piston engine in which the CO decrease with increasing content of biodiesel blends.

There was also an increase in emissions of NO by increasing the concentration of biodiesel, therefore, NO_2 emissions did not change significantly. The high viscosity of biodiesel is a limiting factor for not testing with larger percentages than B30.

The biodiesel from soybeans contain glycerin. The amount of glycerin in the mixture should be kept as low as possible to avoid problems in combustion process. These measurements indicated that the B30 blend showed a content of glycerin which can be considered negligible. In none of the blends tested has been observed increases in pressure drop through analysis in the fuel filter and therefore none of them produced sediments that could cause blockage in the supply system of the turbine engine (BIST, 2004).

SCHMELLEKAMP and DIELMANN (2004) present the results of using vegetable oil from rape seed in a 30 kW micro-turbine, in blends of 10, 20 and 30%. Fuel consumption increased with increasing biodiesel blend. Using B30 obtained a 12% higher consumption in the range of operation. By using B10 it was verified that CO were lower than in the case of use of fossil diesel. However, when were employed mixtures of B20 and B30 it was observed a higher level of CO emissions. The viscosity of vegetable oil is much greater than that of biodiesel, so it is necessary to make a preheating the blend prior to injection.

The results of operating a 75 kW micro-turbine, burning biodiesel from rapeseed, sunflower, animal fats, were presented by WENDIG (2004). The operation with these three kinds of methyl esters showed a significant increase in emissions of CO and CO_2 at full load. All fuels examined showed a reduction in NO_X emissions in the range of 55%. Were problems related to the corrosive characteristics of biodiesel.

More recently, tests of performance and emissions in a 30 kW micro-turbine, using biodiesel from castor beans, were published by CORRÊA (2006). During the tests it was necessary to preheat the blends to 40 °C to achieve the viscosity values required by the manufacturer. The specific fuel consumption increased nearly 21% when using B100. Throughout the power range studied, it was observed a reduction in emissions of CO and NO_X in the exhaust gases.

WENDIG (2004) by using biogas found a decrease in CO with increasing load, reaching 100 ppm at full load. Since the emission of CO_2 was about 45 ppm, constant throughout the range of operation, and SO_2 emissions decreased. At full load the SO_2 emission was practically zero. The micro-turbine showed the lowest emission of NO_X at full load (20 ppm).

The existing reports on the use of biodiesel in micro-turbines describe increases in fuel flow rate when increasing the proportion of biodiesel in the blend; this is explained in part by lower LHV of biodiesel compared to diesel.

Problems due to viscosity, corrosion and accumulation of foreign material in the turbine blades may be in extended operations. These problems can be mitigated through a rigorous verification of the characteristics of biodiesel, demanding that it complies with the standards and manufacturers recommendations. In Brazil, ABNT published standards NBR 15341, NBR 15342, NBR 15343 and NBR 15344, to specify the properties of biodiesel (ABNT, 2006).

Manufacturers of gas turbines and micro-turbines are currently developing models that can use biofuels such as biodiesel. The changes that are taking place mainly consist in the use of materials resistant to corrosion in fuel and injection systems, and develop systems adjusted to work with injection of fuel physical-chemical characteristics different from diesel.

4. Cycles with potential for use of biofuels

A gas turbine is a set of three components: the compressor, combustion chamber and the turbine itself. This configuration forms a gas thermodynamic cycle accordance with the model ideal Brayton cycle is called. This set can operate in an open cycle, with air as the working fluid, which is admitted to the atmospheric pressure, passes through the turbine and is discharged back into the atmosphere without returning to the admission.

Thus, despite being an open cycle, some energy from the combustion is rejected in the form of heat contained in hot exhaust gases. The heat rejection is a physical limit of the gas turbine, intrinsic to the operation of thermodynamic cycles, even in ideal cases.

The loss of condition ideal cycle in a gas turbine can be quantified by the ratio involving the calorific value of fuel, discounting the power to drive the compressor and power net. Thus, decreasing the losses as it reduces the exhaust temperature, and raises the temperature of turbine inlet. The resistance to high temperature components of gas turbine is a very critical point in building technology such equipment (COHEN, H. *et. al.*, 1996).

Turbines designed to operate in simple cycle, as shown in Figure 1, in view of the thermal efficiency of the cycle, have gas outlet temperature reduced to maximum and have optimized compression ratio. The compression ratio is the ratio between the pressure of air entering and exiting the compressor. For example, if air enters at 1.0 atm, and leaves the compressor at 15.0 atm, the compression ratio is 15:1.

Apart from variation of the simple cycle obtained by the addition of these other components, considerations must be given to two system distinguished by the use of open (Figure1) and close cycles (Figure 2).

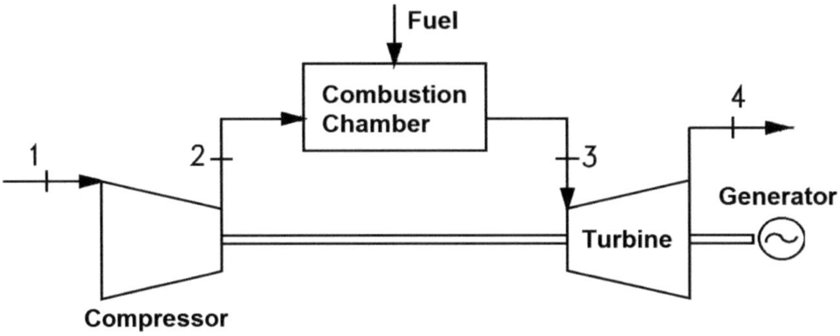

Fig. 1. Scheme of a simple cycle gas turbine system.

In the process of closed cycle operation is the same as the open cycle, the difference is that the working fluid remains within the system and the fuel is burned outside the system.
The biggest advantage of the closed cycle is the possibility of using high pressure throughout the circuit, which results in reducing the size of the turbomachinery, depending on power output, and allows the variation of power output by varying the pressure level of the circuit.
The significant feature is that the hot gases produced in the boiler furnace or reactor core never reach the turbine; they are merely used indirectly to produce an intermediate fluid, namely steam.
In order to produce an expansion through a turbine a pressure ratio must be provided and the first necessary step in the cycle of gas turbine plant must therefore be compression of the working fluid.

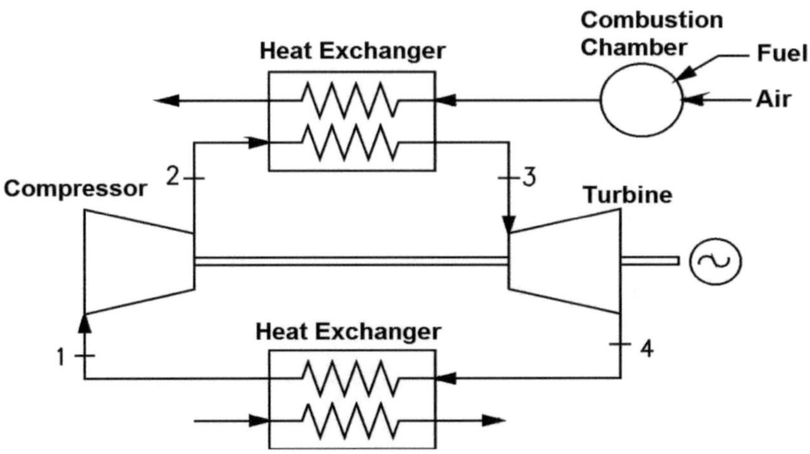

Fig. 2. Scheme of a close cycle gas turbine with a heat exchanger.

If after compression the working fluid was to be expanded directly in the turbine, and there were no losses in either component, the power developed by the turbine would just equal that absorbed by the compressor. Thus if the two were coupled together the combination would do no more than turn itself round. But the power developed by the turbine can be increased by the addition of energy to raise the temperature of the working fluid prior to

expansion. When the working fluid is air a very suitable means of doing this is by combustion of fuel in the air which has been compressed.
One way to increase the thermal efficiency of a closed cycle is the addition of a heat exchanger; however there are limits to the introduction of this equipment because it can cause loss of pressure in the circuit.
A cycle gas turbine with a heat exchanger can also be called a regenerative cycle, as shown in Figure 3, wherein the heat rejected in the exhaust gases of the gas turbine passes through the heat exchanger and heats the air from the compressor before entering the chamber combustion. The pre-heated air reduces the fuel consumption injected into the combustion chamber, increasing the thermal efficiency of the cycle.

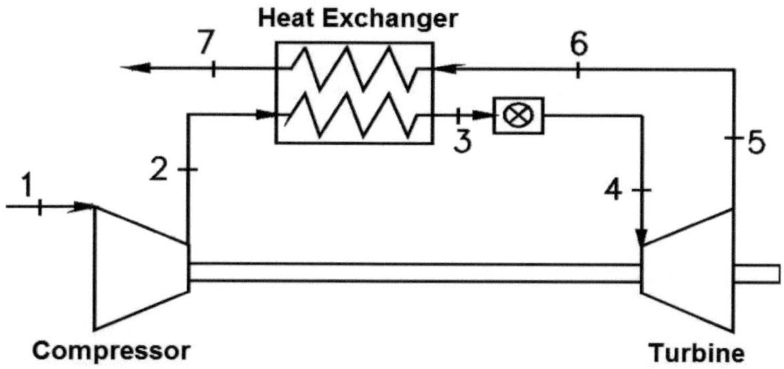

Fig. 3. Scheme of a regenerative cycle gas turbine.

Specific gas turbines to operate in combined cycle, as shown in Figure 4, are developed in order to maximize the thermal efficiency of the cycle as a throughout. Therefore, reducing the temperature of the exhaust gas is not necessarily the most critical point in terms of efficiency, since the gas turbine exit are still used to generate power in other equipment.

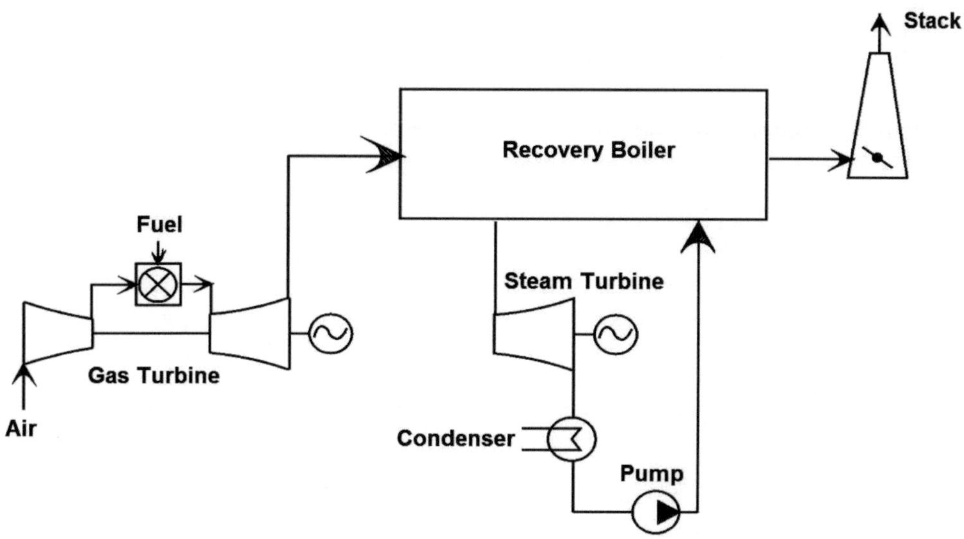

Fig. 4. Scheme of a combined cycle.

Another configuration of installation of gas turbines is the cycle where there is the presence of an intercooler before or between compressors, i.e., compressors among the low and high pressure. In this configuration the air that enters into the first compressor is compressed to a pressure intermediate between the maximum of the cycle and the ambient pressure, as shown in Figure 5.

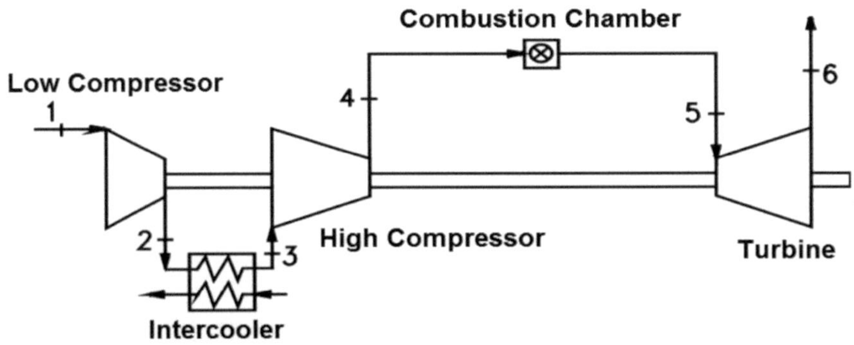

Fig. 5. Scheme of a cycle with intercooler.

In the case of plants with two compressors the air leaving the first compressor (low pressure) enters the intercooler where the heat is removed without major pressure drop. Being that, in practice, the air is returned to the circuit should have a temperature slightly higher than the entrance to the first compressor.

In this type of gas turbine is possible to operate with temperatures above 1450 °C due to the reduction of air temperature of cooling the turbine blades, increasing the thermal efficiency of the cycle, reaching 50%.

The combined cycle consists of one or more gas turbine, whose exhaust gases are injected into a recovery boiler that provides steam to a turbine.

In an open cycle, the thermal efficiency is low, around 30 to 35%. In a combined cycle efficiency can reach 60% (the highest efficiency of all types of driver). In combined cycle occurs the combination of gas turbines with steam turbines.

4.1 Selection cycle gas turbine during operation with gas biofuel

In simple cycle the thermal efficiencies is increased with the increase of pressure ratio and turbine inlet temperature. Therefore, high pressure ratios require high pressure to clean the gas, which results in high costs of equipment, and high compression energy consumption.

In a cycle of regeneration is evident with a peak thermal efficiency at a pressure ratio between 4 and 8. In relation to biomass gasification, the cycle with regenerator is interesting because it is possible to use moderate pressure gasification resulting in lower equipment costs and acceptable thermal efficiencies compared to the simple cycle.

Preheating inlet air aerator does not increase much thermal efficiency, since the output of the reactor; higher temperatures are derived fuel gas. However, this represents a contradiction, since the positive effects of preheating are reversed, it is necessary to perform the cooling of the gas for cleaning.

With gas pyrolysis cycle gas turbine with a regenerator would lower overall efficiency, because the residual heat from the pyrolysis process cannot be used to increase the power of the turbine.

The combined cycle integrated with slow pyrolysis uses a Rankine cycle supplied with charcoal and a cycle gas turbine powered by pyrolysis gas, which results in a high cycle efficiency, because a considerable amount of waste heat can be used to produce Rankine cycle power (SCHMITZ, 2000).

4.2 Adjustments to gas turbine

Due to impurities in the gas or fuel, for example, the synthesis or biofuel, it is necessary redesign the gas turbine combustor. For each type of fuel, run a kind of optimization, with reference to a low value of the LHV of fuel.

To compensate for the lower value of LHV for the fuel gases, the fuel injection system must provide a fuel rate much higher than when the combustor operates fuel with high calorific value. Due to the high rate of mass flow of gas with low LHV, the passage of fuel has a much larger cross section than the section corresponding to natural gas. The fuel pipes and control valves and stop valve have larger diameters and shall be designed to include an additional fuel blend, which consists of the final mixture of the recovered gas with natural gas and steam. The pressure drop and the size of the spiral of air entering the flame tube were adjusted to optimize the combustion process. The system must have high safety standards, so the flanges and gaskets of the combustor and its connections must be good soldiers. The system for low fuel LHV must include:

- Fuel line with a low LHV;
- Natural gas line;
- Steam line to reduce NO_X;
- Line blending of fuel with low LHV;
- Line of nitrogen to purge;
- Lines pilot;
- Compressor;
- Combustion Chamber.

The loading of the gas turbine to the rated load is accomplished through the use of the fuel reserve for security reasons. The procedure for replacing the fuel reserve for the main runs automatically. The characteristics of blends are monitored and analyzed online.

5. Case study with biodiesel and ethanol in gas turbine

For perform the tests in the gas turbine was built a test bench in the laboratory of gas turbines and gasification of the Institute of Mechanical Engineering, Federal University of Itajubá - IEM/UNIFEI. This bench is made of a micro-turbine Capstone C30 model cycle with regenerative power of 30 kW, configured to operate with liquid fuel.

This gas turbine is used mainly for primer power generation or emergency and can work in a variety of liquid fuels. This turbine uses a recovery cycle to improve its efficiency during operation, due to a relatively low pressure what facilitates the use of a single shaft radial compression and expansion (BOLSZO, 2009).

The experimental tests were carried out by using a 30 kW regenerative cycle diesel single shaft gas turbine engine with annular combustion chamber and radial turbomachineries, whose characteristics at ISO conditions are given in Table 3 (CAPSTONE, 2001).

To perform the experiments with different fuel blends and it was implemented a system to preheat the fuel supply aiming to control the viscosity of the fuel used.

For tracking and measuring the parameters of the test was used a supervisory software in test bench (given by the manufacturer of the turbine) and a data acquisition and post processing of data obtained during the tests.

Fuel Pressure	350 kPa
Power Output	29 kW NET (± 1)
Thermal Efficiency	26% (± 2)
Fuel HHV	45,144 kJ/kg
Fuel Flow	12 l/h
Heat Rate (LHV)	14,000 kJ/kWh
Exhaust Temperature	260 °C
Inlet Air Flow	16 Nm^3/min
Rotational Speed	96000 rpm
Pressure Ratio	4

Table 3. Engine Performance data at ISO Condition.

5.1 Tests on gas turbine using biodiesel fuels

To perform the experimental tests were used blends biodiesel/diesel, including the total replacement of diesel with biodiesel in the gas turbine. The blends considered in the experiment were: B10, B20, B30, B50 and B100. Due to low solubility in diesel fuel at low temperature tests with ethanol were performed without pre-mix, and also without the use of additives, which enhance the cost of fuel.
Following the methodology of the measures to be adopted to test gas turbine (item 2), Table 4 shows the physical-chemical properties of biodiesel for testing thermal performance and emissions:

Properties	Soy Biodiesel	Palm Oil Biodiesel	Diesel	Micro Turbine Manufacturer Fuel Limits	ASTM D6751
Cetane Number	58	60	45.8	-	> 47
Sulfur (% mass)	0	0	0.20	0.05 <	< 0.05
Kinematic Viscosity @ 40 °C (mm^2/s)	2.19	2.26	1.54	1.9 – 4.1	1.9 – 6
Density @ 25 °C (g/cm^3)	0.888	0.854	0.838	0.75 – 0.95	-
Flash Point (°C)	136	138	60	38 - 66	> 130
Water (% Volume)	0.05	0.05	0.05	0.05	0.05

Table 4. Biodiesel and diesel physic-chemical characteristics.

Table 4 shows the comparison amongst the characteristics of commonly used types of biodiesel and diesel pure, with the requirements of the fuel made by the manufacturer of gas turbine tested and specifications for biodiesel fuel blend of standard ASTM D6751.
Once obtained the blends, there was the experimental determination of the calorific value, kinematic viscosity and density of different blends (B10, B20, B30, B50, B100) and ethanol, according to the standards ISO 1928-1976 and ASTM D1989-91, respectively. In the case of viscosity were made ten measurements to reduce the standard deviation, and the calorific

value were performed five measurements for each blend. Table 5 shows the LHV of the fuels used in the experiment.

Fuel	Pure (kJ/kg)	B10 (kJ/kg)	B20 (kJ/kg)	B30 (kJ/kg)	B50 (kJ/kg)
Biodiesel (Palm Oil)	37,230.37	41,204.29	40,691.27	39,768.17	38,623.27
Biodiesel (Soy)	28,298.16	41,425.27	40,748.82	39,864.36	39,061.45
Diesel	42,179.27	-	-	-	-
Alcohol	23,985.00	-	-	-	-

Table 5. Lower Heating Value of fuels used.

The use of different fuels implies the need to make adjustments of mass flow rate of them, according to its LHV and your density, because without these adjustments, once established a load, the supply system would feed a quantity of fuel depending on the characteristics of standard fuel (diesel). If the LHV of the new fuel is lower than the standard, the gas turbine power could not reach the demanded.

Initially, the engine was operating with conventional diesel fuel for a period of 20 minutes to reach a steady state condition for a load of 10 kW. After 20 minutes, the mass flow rates were changed to the corresponding values of blends diesel/biodiesel. At this stage, it begins to replace the fuel, in order of increasing content of biodiesel (B10, B20, B30, B50 and B100), closing the inlet valve of diesel and opening the valve of the mixture. In order to ensure that all existing diesel power on the internal circuitry of the engine would be consumed, the engine was left operation for 10 minutes with the same load operation (10 kW).

In order to check if the fuels were able to feed the engine, without experiencing any problems, regarding the fuel injection system, the kinematic viscosity of each fuel was measured. The composition of gas emissions and thermal parameters were also measured in total and average load for each fuel. This whole procedure was performed for the engine operating with loads of 5, 10, 15, 20, 25 and 30 kW in a grid connection mode.

Afterwards it was held on measurement of emissions with gas analyzer, and increased the load of 5 kW, 10 minutes waiting again until it reaches steady state again.

When finished testing with a blend, the engine was left running, in order to accomplish the purging of fuel remaining. After executing the purge the supply system it was loaded with a new mixture. Once completed the tests with biodiesel and blends, was returning to operate the engine with diesel for ten minutes, and then it was disconnected and stopped.

5.2 Thermal performance

The results of performance testing of a 30 kW gas turbine engine supplied with biodiesel from palm oil, soy and ethanol are shown:

Figures 6, 7 and 8 show the relationship between specific fuel consumption and power. The graphs correspond to soy biodiesel, Figure 6, biodiesel from palm oil, Figure 7 and diesel and ethanol, Figure 8.

In the case of biodiesel from soy, Figure 6 observed if an increase in specific fuel consumption, when increasing the fraction of biodiesel in the blend in the range of 10 to 25 kW. The lowest value occurs when using pure diesel oil, the highest value occurs when using pure biodiesel and the difference between the curves of diesel and B100 remains approximately constant at 12.82%. This behavior if repeated when using biodiesel from palm oil as shown in Figure 7. When using ethanol, Figure 8 also shows an increase in specific fuel consumption with respect to diesel.

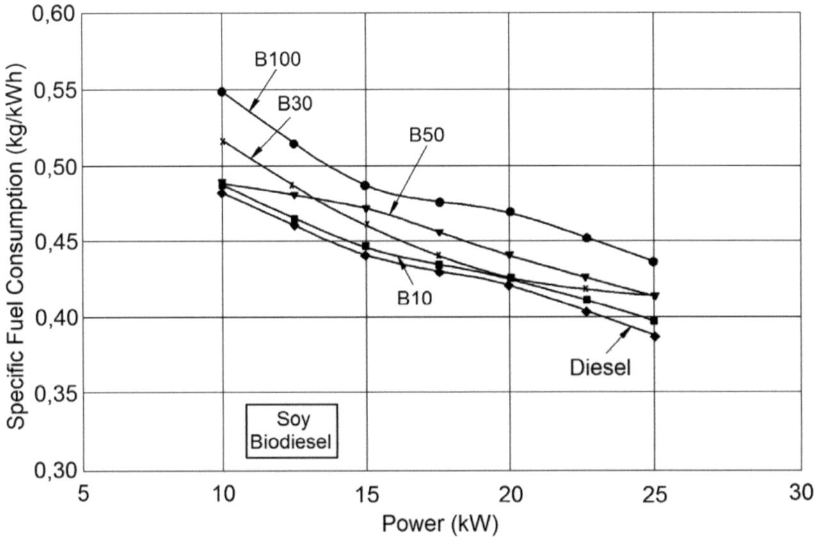

Fig. 6. Specific fuel consumption versus power of the gas turbine for different fuels: soy biodiesel.

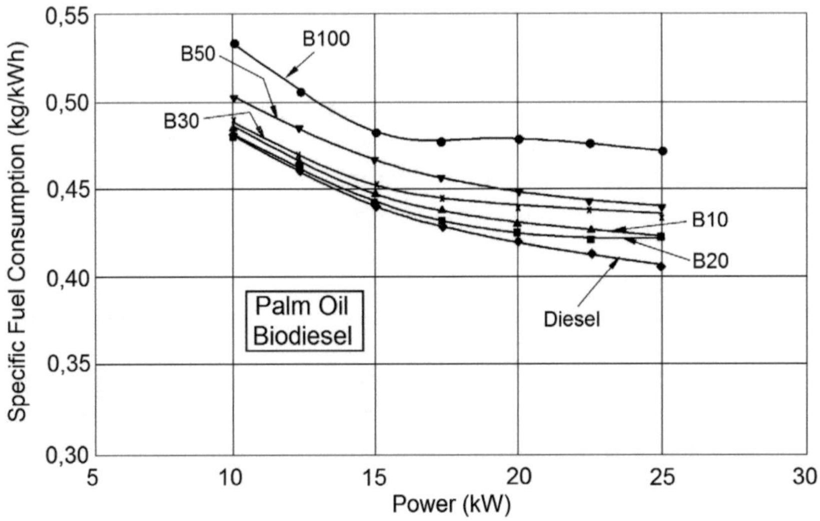

Fig. 7. Specific fuel consumption versus power of the gas turbine for different fuels: palm oil biodiesel.

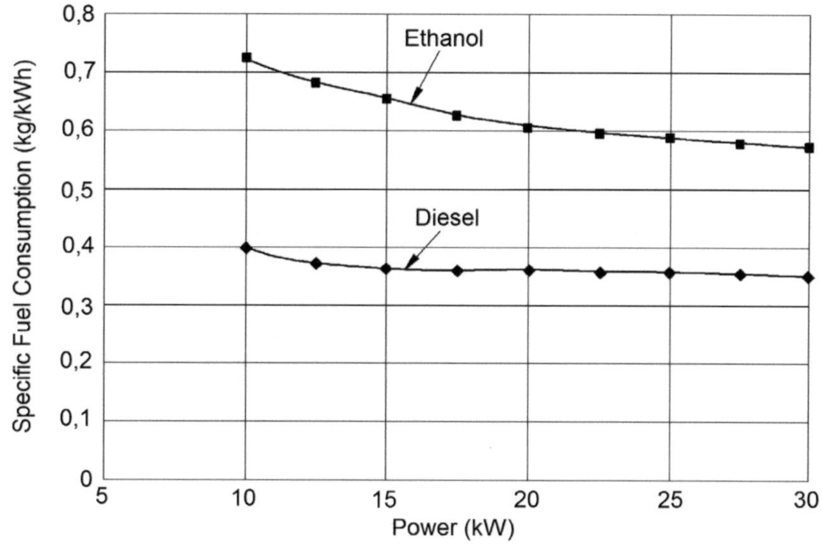

Fig. 8. Specific fuel consumption versus power of the gas turbine for different fuels: diesel and ethanol.

The specific consumption of biodiesel from soy and palm oil were approximately equal. Consumption already of ethanol was higher due to even lower calorific value of fuel compared with the others used fuel.
Figures 9, 10 and 11 show the behavior of power versus Heat Rate of gas turbine for the three different fuels tested.

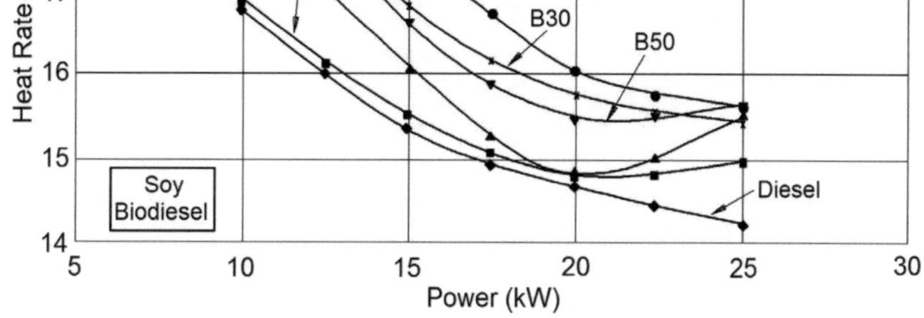

Fig. 9. Heat Rate versus power for different fuels: soy biodiesel.

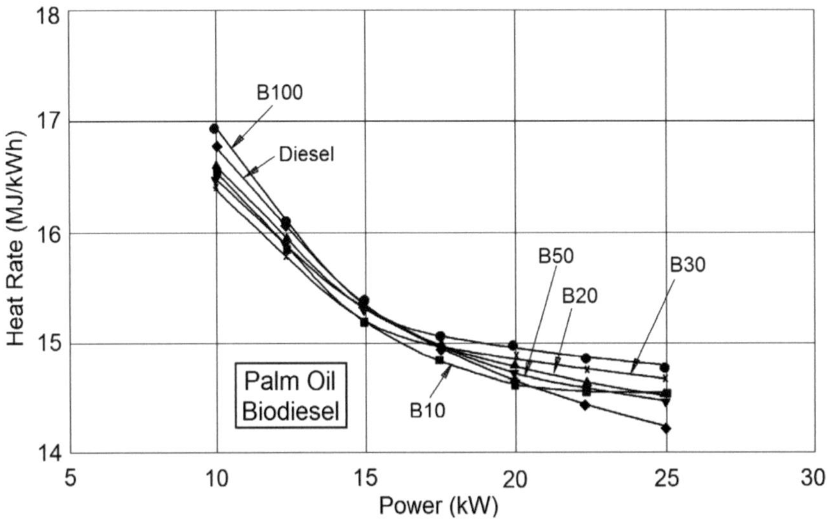

Fig. 10. Heat Rate versus power for different fuels: palm oil from biodiesel.

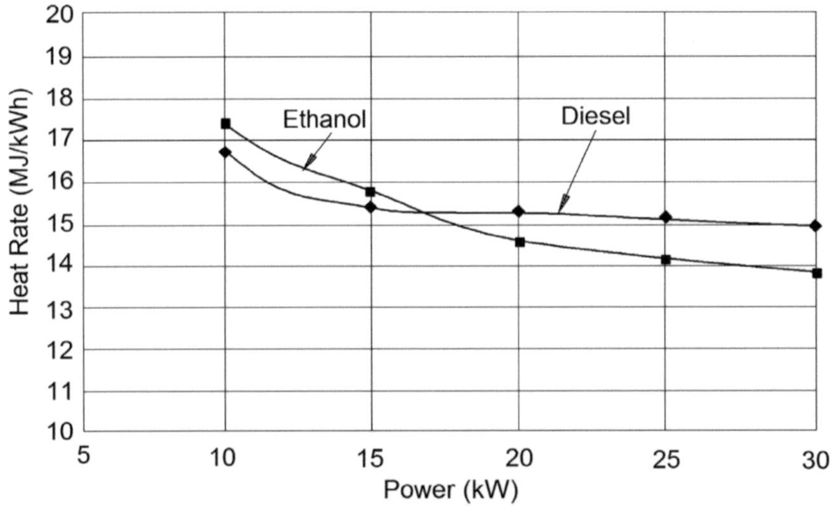

Fig. 11. Heat Rate versus power for different fuels: diesel and ethanol.

In all cases, it was observed a reduction in Heat Rate with the increase in power. Operation with biodiesel from palm oil presented a better performance than soy biodiesel (lower value of the Heat Rate). Similarly to the specific fuel consumption, the differences in the amount of Heat Rate between diesel and B100 remained approximately constant in the range 10 to 25 kW. The lower fuel consumption occurs to the rated power when operating with diesel. This is due to the higher calorific value of diesel.

When a lower LHV of fuel is used, a greater mass of fuel is needed to release in the combustion energy required for a specific power. The mass flow rate of fuel passing through turbine increases and the compressor operating point changes, making their efficiency and, consequently, the cycle efficiency decreases.

Finally, Figures 12, 13 and 14 displays the graphs of the thermal efficiency of the gas turbine for different fuel and power operation. The differences between the efficiency with diesel and biodiesel blends with soy biodiesel, Figure 7 may be due to differences in density, viscosity and LHV of fuel compared with conventional biodiesel. This can cause changes in the process of atomization of the fuel within the combustion chamber, reducing the thermal efficiency of the engine as previously mentioned.
In the case of palm oil biodiesel was not observed differences, and the efficiency values remain equal to those obtained when the engine was tested with diesel, within the entire power range evaluated. Ethanol showed the lowest efficiency among the fuels tested

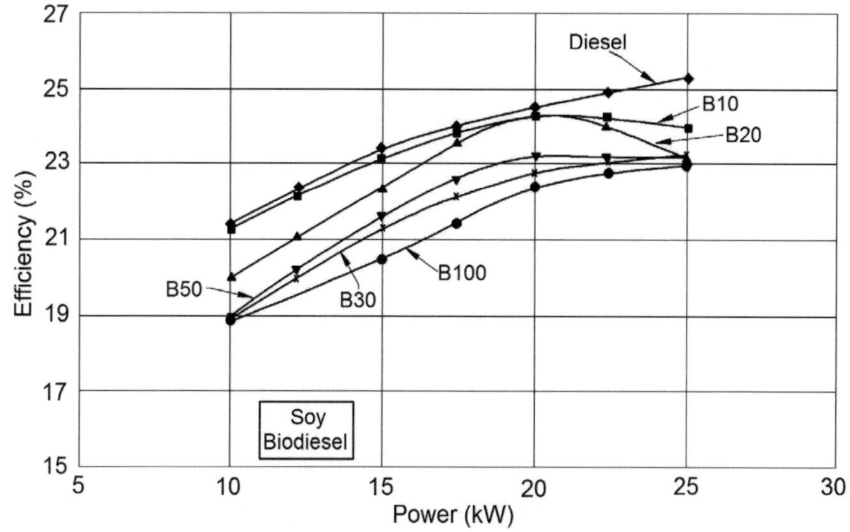

Fig. 12. Efficiency versus power turbine for different fuels: soy biodiesel.

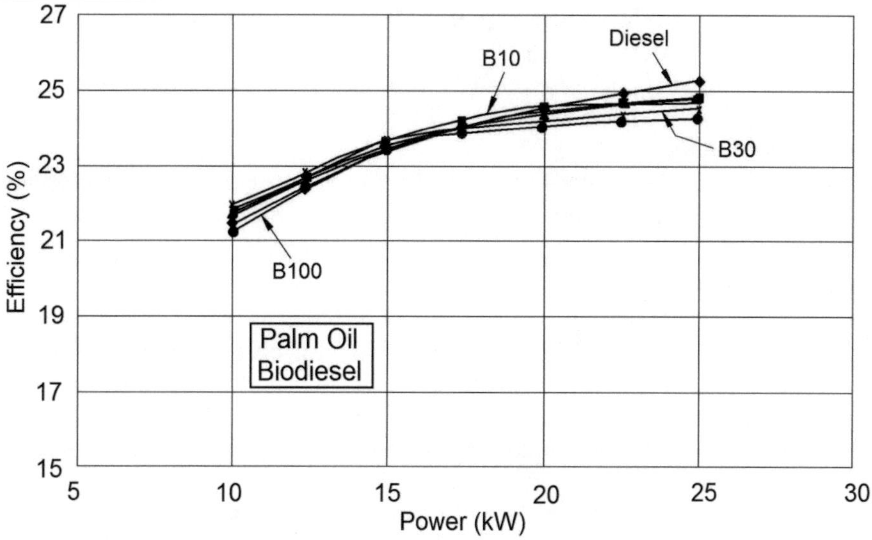

Fig. 13. Efficiency versus power turbine for different fuels: palm oil biodiesel.

As in tests performed by HABIB (2010), the test conducted in the laboratory of UNIFEI with B100 biodiesel resulted in high thermal efficiency compared to other blends, as shown by the graphs of Figures 12 and 13, such performance is attributed to equivalence ratio, which produced the best ratio of air to fuel during firing, resulting in more complete combustion due to the presence of extra oxygen in the biofuel, which resulted in the presence of 16 to 19% oxygen in the gases turbine exhaust.

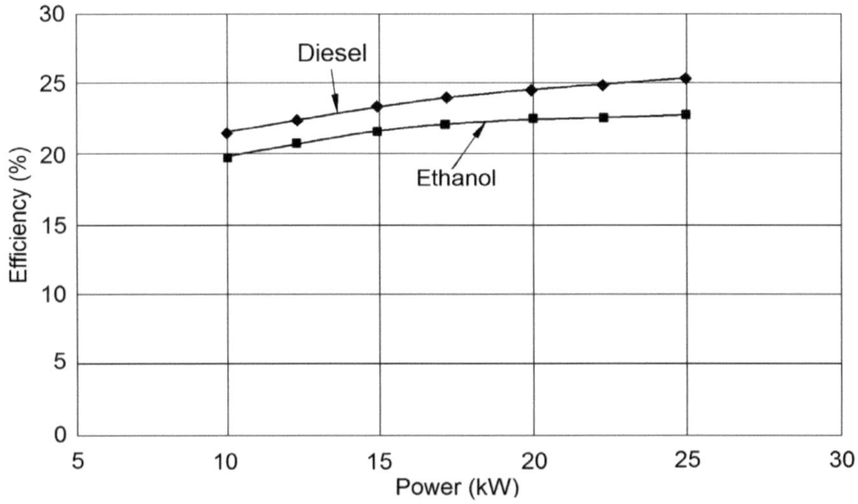

Fig. 14. Efficiency versus power turbine for different fuels: diesel and ethanol.

As mentioned by other authors, it can say that the type of biodiesel is an important factor in the analysis of projects for power generation, as small differences in density, viscosity and LHV cause changes in the parameters of thermal performance of turbines gas, and amongst them when compared with pure diesel.

For the case of palm oil biodiesel, although there is an increase in specific fuel consumption, there were no significant differences in the efficiency of the engine. With soy biodiesel there were differences in the efficiency of around 2% throughout the power range tested.

The results obtained with ethanol were very different for both types of biodiesel tested, due to its lower calorific value.

In practical terms, for distributed generation, ethanol must go through economic and technical assessments in order to detect.

5.3 Emission of pollutants

Publications on experiments with biofuels in gas turbines are not yet sufficient to make definitive conclusions in terms of emissions. However, it is observed that the CO decreases with increasing load. The opposite happens with the NO_X emissions. It is also provided a reduction in the emission of smoke, along with the increase in the emission of NO_X.

The emission results shown in the following are results of experiments with three different biofuels: Soy biodiesel, palm oil biodiesel and ethanol achieved in the laboratory of UNIFEI. The first two were blended with diesel in different proportions and each mixture was tested with a gas micro-turbine operating at full load and partial loads. The pure ethanol was used and compared with the performance of pure diesel, as reported in the results performance.

The analysis focuses mainly on changes in the levels of carbon monoxide (CO) and nitrogen oxides (NO_X), unburned hydrocarbons were not detected in the combustion products, during the tests. This is due to high temperatures and high excess air into the combustion chamber. Likewise there were no emissions of sulfur oxides (SO_X), since biofuels evaluated did not contain sulfur in your composition.

Figures 15, 16 and 17 shows the concentrations of CO in the exhaust gas from the micro-turbine gas to biodiesel blends: soy biodiesel, Figure 15, palm oil biodiesel, Figure 16 and pure diesel and ethanol, Figure 17.

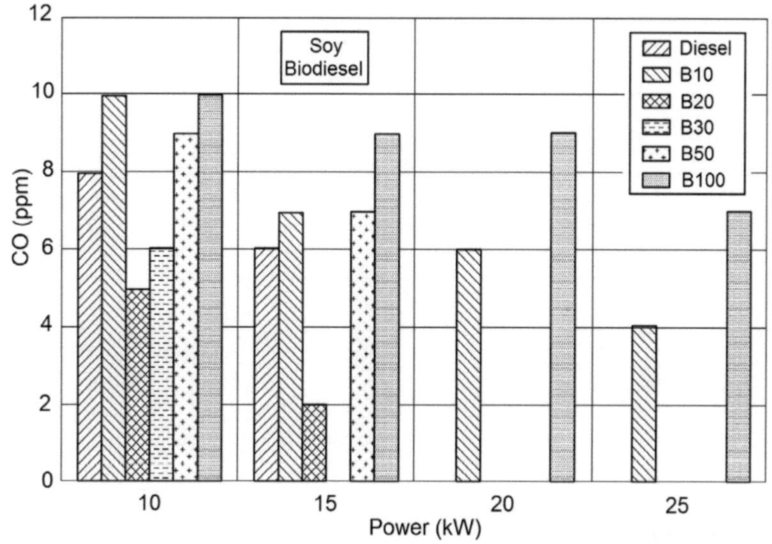

Fig. 15. Concentrations of CO in gas versus the micro-turbine power for different fuels: soy biodiesel.

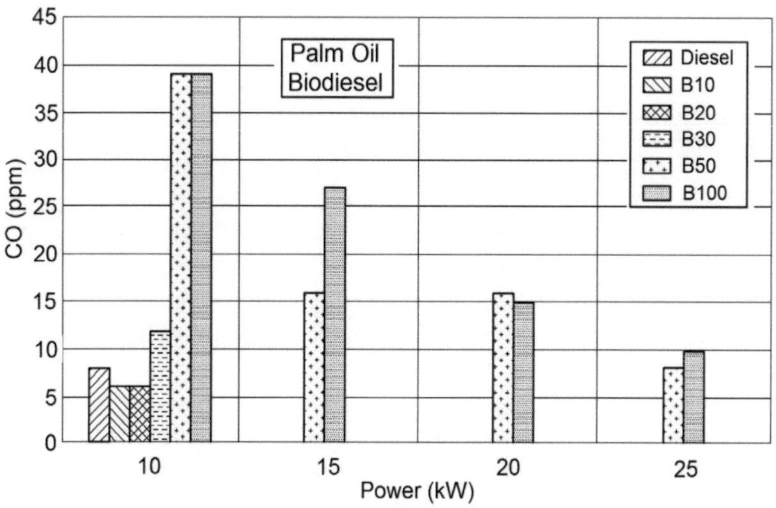

Fig. 16. Concentrations of CO in gas versus the micro-turbine power for different fuels: palm oil biodiesel.

It is observed in Figures 15, 16 and 17 that the concentrations of CO in the exhaust of the micro-turbine operating on pure biodiesel (B100) are higher than when operating with diesel. Emissions decrease as much for diesel as for the blends when the power increases. This if explained by the characteristics of combustion in the combustion chamber.
The soy and palm oil biodiesel has a higher viscosity than the pure diesel. The fuel nozzles not modified in the micro-turbine, must have worsened the quality of atomization of biodiesel compared with diesel, generating higher levels of CO in the exhaust gases as consequence of incomplete combustion. CO emissions at part load are larger than at full load. The lower emission occurs to full load.

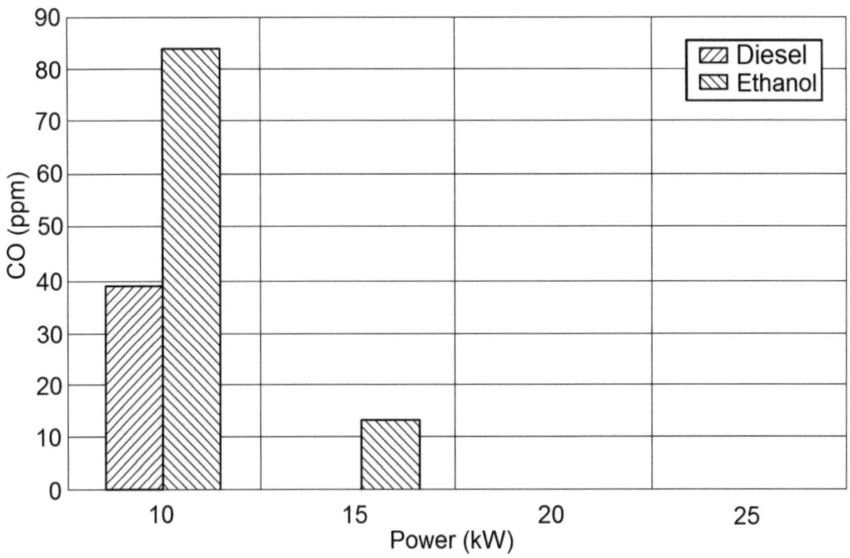

Fig. 17. Concentrations of CO in gas versus the micro-turbine power for different fuels: pure diesel and ethanol.

There are differences in the composition of exhaust gases by using soy and palm oil biodiesel, as can be seen in Figures 15 and 16, as well as been observed previously in thermal efficiency. The operation with soy biodiesel showed no CO in loads exceeding 10 kW, while the operation with palm oil biodiesel (B100) presented CO throughout the power range. The two fuels have different physical-chemical characteristics, which are reflected in a particular behavior in the combustor process.
It is observed in Figure 17 that the emissions of CO to ethanol use are higher than diesel, in all power. This is probably due to lower LHV of ethanol relative to diesel, resulting in higher specific fuel consumption, which reduces the residence time of fuel in the combustion chamber, which may be the cause of greater incomplete combustion.
Figures 18, 19 and 20 show NO_X emissions with different fuels. There are no great differences in the values of NO_X for the fuels.

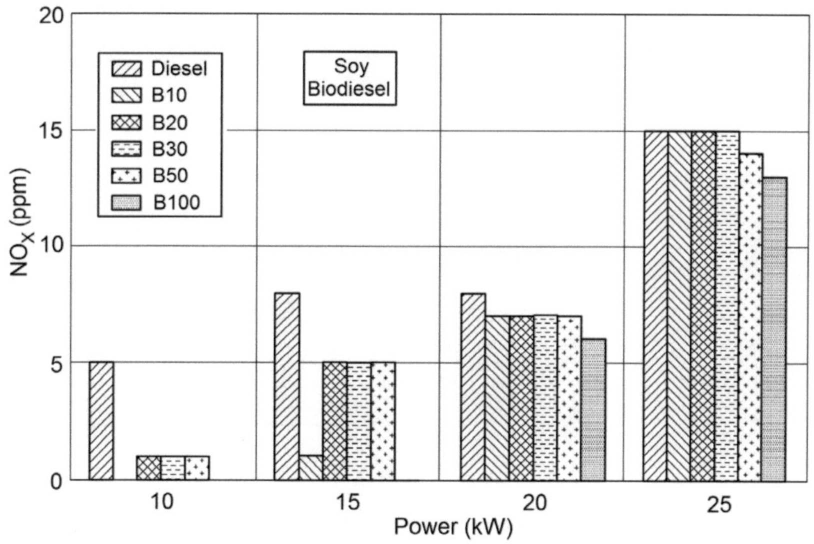

Fig. 18. NO_X emissions versus the micro-turbine power for different fuels: soy biodiesel.

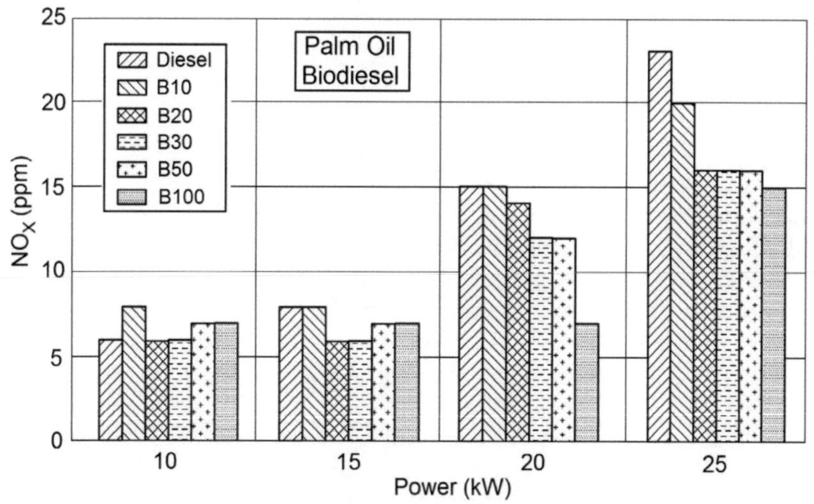

Fig. 19. NO_X emissions versus the micro-turbine power for different fuels: palm oil biodiesel.

NO_X emissions are predominantly of thermal origin and their values are in all cases below 35 ppm, maximum limits set by the engine manufacturer. When used pure biodiesel (B100), the NO_X concentration is lower than the diesel at all loads tested. The results for CO and NO_X show a behavior similar to that presented by PETROV (1999), who also has carried out experiments with a 30 kW micro-turbine.

In the case of ethanol, NO_X emissions showed an inverse behavior to CO, which was expected, because the formation of CO and NO_X is a function of reaction temperature when the CO reduces NO_X increases. Thus, when the amount of CO decreases with increasing power, increases the quantity of NO_X in exhaust gases.

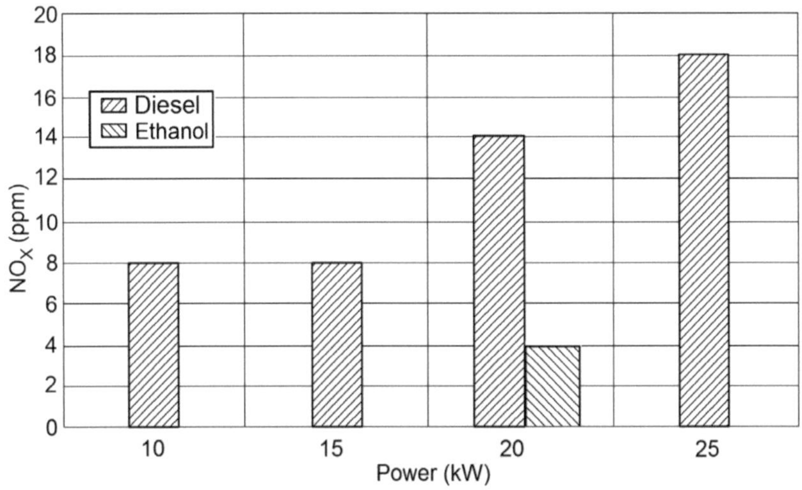

Fig. 20. NO_X emissions versus the micro-turbine power for different fuels: pure diesel and ethanol.

For the three fuels evaluated it was observed a reduction in emissions of NO_X and an increase in CO content compared with diesel, however, all results are within the range indicated by the manufacturer (CAPSTONE, 2001).
It is remarkable that there were no SO_2 emissions and unburned hydrocarbons in any test with biofuels, as mentioned.

6. Conclusions

With increasing industrial development there is the necessity for more refined research on the use of biofuels in gas turbines, covering aspects such as quality of biofuel to be used, form and storage conditions, adjustments in the systems of engines, costs of energy generated in order to maintain high operational reliability of the turbines.
A gas turbine can operate with different types or blends of biofuels with a corrected power loss at around 4.26%, and the corrected heat rate of 8.38% higher than diesel fuel, as shown in this work.
As previously warned care must be taken during the operation of the gas turbine with liquid fuel (from whatever source) and gaseous fuels derived from biomass, because the components of supply systems of gas turbines are very sensitive and the use these biofuels can provide wear or cause loss of efficiency during extended operation.
The physicochemical characteristics of all the fuels evaluated lie within the specifications for their use in gas turbines. The thermal performance tests showed that biodiesel has higher specific consumption than diesel. The reason for that is the lower heat value of the pure biodiesel in comparison with diesel fuels. Agreeing with the findings of other researchers, cited in this work, and verified in tests with biofuels the fuel that presented the smallest difference in terms of heat rate in relation to diesel was the palm oil biodiesel, with a difference of about 17.6 % at full load and less than 1.0 % at medium load.
The tests also found the need to take care with the installations of storage systems and supply of power plants in order to maintain the specifications and properties of mixtures of liquid biofuels within acceptable standards. In the case of gaseous fuels to be careful with

the filter system of gas particles and elimination of harmful substances such as tar and others that, besides causing wear to components of the supply system, undermine the performance of engines.

Emission levels from the experimental tests have shown that CO increases for the palm oil biodiesel and NO_X decreases by approximately 26.6%; SO_X concentration wasn't taken into account when biodiesel was used. Further investigation involving emission has to be carried out for the better understanding of pollutant formation when biodiesel fuels are used.

7. Acknowledgements

We wish to thank CAPES, FAPEMIG, FAPEPE and CNPq, for their financial support.

8. References

ABNT, NBR15431: "Biodiesel - Determinação de glicerina livre em biodiesel de mamona por cromatografia em fase gasosa", 2006.

ABNT, NBR15432: "Biodiesel - Determinação de monoglicerídeos, diglicerídeos e ésteres totais em biodiesel de mamona por cromatografia em fase gasosa", 2006.

ABNT, NBR15433: "Biodiesel - Determinação da concentração de metanol e/ou etanol por cromatografia em fase gasosa", 2006.

ABNT, NBR15434: "Biodiesel - Determinação de glicerina total e do teor de triglicerídeos em biodiesel de mamona", 2006.

ASME performance test code PTC-22-1997. "Gas turbine power plants".

Biasi, V. "Low cost and High efficiency make 30 to 80 kW microturbines attractive", Gas Turbine World, jan.-fev., Southport, 1998.

Bist, S. "Development of Vegetable Lipids Derived Fatty Acid Methyl Esters as Aviation Turbine Fuel Extenders". Master Thesis of Purdue University, 2004.

Bolszo, C. D.; McDonell, V. G. "Emissions Optmization of a Biodiesel Fired Gas Turbine", Proceedings of the Combustion Institute, 32, Science Direct ELSEVIER, 2949-2956 pp., 2009.

Boyce, P. M. "Gas Turbine Engineering Handbook", Third Edition, Gulf Professional Publishing; april 28, 2006.

Capstone Turbine Corporation. "Capstone Microturbine Model 330 system operation manual", Capstone Turbine Corporation, USA, 2001.

Cohen, H.; Rogers, G. F. C.; Saravanamuttoo, H. I. H., Gas Turbine Theory, Fourth edition, 1996.

Correia, P. S. "The use of biodiesel in gas fuel micro turbine: thermal perforamnce and emission testing". Master Theses, UNIFEI, MG, Brazil, 2006.

Gökalp, I.; Lebas, E. "Alternative Fuels for Industrial Gas Turbines (AFTUR)", Applied Thermal Engineering, 24, 2004.

Habib, Z.; Parthasarathy, R.; Gollahalli, S. "Performance and Emission Characteristcs of Biofuel in a Small-Scale Gas Turbine Engine", Applied Energy 87, 24, 1701-179 pp., 2010.

Kehlhofer, R.; Hannemann, F.; Stirnimann, F.; Rukes, B. "Combined-Cycle Gas & Steam Turbine Power Plants", 3rd Edition, PennWell, 2009, 64-69 pp.

Liss, E. W. "Natural Gas Power System For the Distributed Generation Market", Power-Gen International 99 Conference, New Orleans, Louisiana, 1999.

Lopp, D.; Tanley, D.; Ropp, T.; Cholis, J. "Soy-Diesel Blends Use in Aviation Turbine Engines"; Aviation Technology Department of Purdue University, 1995.
Petrov, A. Y.; Zaltash, A.; Rizy, D. T.; Labinov, S. D. "Study of Flue Gas Emissions of gas Microturbine-Based CHP System"; 1999. www.uschpa.org.
Mimura, N. "Biodiesel Fuel: A next Microturbine Challenge"; 2003. www.ornl.gov.
Nwafor, Omi, 2004. "Emission characteristics of Diesel engine operating on rapeseed methyl ester". Renewable Energy, 29:119-29.
Rodgers, G.; Saravanamutto, H. "Gas Turbine Theory", Editora Prentice Hall, março, 2001.
Schmellekamp, Y.; Dielmann, K. "Rapeseed oil in a Capstone C30"; Workshop: Bio-fuelled Micro Gas Turbines in Europe, Belgium, 2004. www.bioturbine.org.
Schmitz, W.; Hein, D. "Concepts for the Production of Biomass Derived Fuel Gases for Gas Turbine Applications", Proceedings of ASME Turbo Expo 2000, Munich, Germany, may 8-11, 2000.
Scott, W. G. "Micro Gas Turbine Cogeneration Applications", International Power and Light Co., USA, 2000.
Sierra, R. G. A. "Experimental test and thermal economical analysis of biofuel used in regenerative micro gas turbine engine"; Master Theses, UNIFEI, MG, Brazil, 2008.
Tan, E. S., Palanisamy, K., 2008. "Experimental and Simulation Study of Biodiesel combustion in a Micro turbine", ASME Turbo Expo 2008, ASME GT2008-51497.
Wendig, D. "Bio fuel in micro gas turbines", Workshop: Bio-Fuelled Micro Gas Turbines in Europe, Belgium, 2004. www.bioturbine.org.
Willis, H. L.; Scott, W. G. "Distributed Power Generation". Planning and Evaluation, Ed. Marcel Dekker, Inc.

7

Afterburning Installation Integration into a Cogeneration Power Plant with Gas Turbine by Numerical and Experimental Analysis

Ene Barbu et al*

National Research and Development Institute for Gas Turbines COMOTI, Romania

1. Introduction

Gas turbines applications represent a continuous challenge for engineers regarding the design, manufacturing and efficient working at nominal and partial loads with respect to environmental and future development requirements. The development of gas turbines in terms of performances as well as of lifetime and safe running has made these installations preferable for cogenerative processes and the lower maintenance costs have lead to particularly tempting recovery terms, down to 3-4 years for a 15-20 years lifetime. The increase in the overall efficiency of the cogenerative group is affected by the degree of use of the heat produced by the gas turbine along with the flue gases. The recovery of the flue gases heat is achieved by using a heat recovery steam generator, usually for producing steam (or hot water). The performances of a heat recovery steam generator depend on the gas turbine's working regime which makes the steam parameters difficult to control. The trend in the field of heat recovery steam generator and afterburning installations are related to the development in gas turbines. The increase in the temperatures of the gas turbine requires new materials that withstand operating regimes in terms of appropriate pollutantsnorms. The oxygen concentration in the gas turbine's flue gases is usually of 11 - 16 % volume. The fact that the combustion process in the gas turbine consumes only a small part of the oxygen from the intake air flow makes possible the application of a supplementary firing (afterburning) for increasing the steam flow rate of the heat recovery steam generator. In aviation the afterburning is used for increasing the thrust of supersonic aircrafts equipped with gas turbines. Introducing the afterburning in cogenerative applications leads to increasing the flexibility and the overall efficiency of the cogenerative group. The burner of the afterburning installation is usually placed between the gas turbine and the heat recovery steam generator, immersed in the flue gases exhausted by the gas turbine, resulting in a complex gas turbine - afterburning - heat recovery steam generator system. This placement results in the afterburning installation affected by the gas turbine and affects the working of the heat recovery steam generator. Usually, when the heat recovery steam generator is used for delivering superheated steam, the working conditions of the afterburning installation

* Valeriu Vilag, Jeni Popescu, Silviu Ionescu, Adina Ionescu, Romulus Petcu, Cleopatra Cuciumita, Mihaiella Cretu, Constantin Vilcu and Tudor Prisecaru
National Research and Development Institute for Gas Turbines Comoti, Romania

influence mainly the superheater of the heat recovery steam generator (with effects on temperature, flow rate etc.). The design of the gas turbine - afterburning installation - heat recovery steam generator system must take into account these variables for insuring the steam parameters required by the technological process. The active control of the combustion is a concept already accepted and the new generation of afterburning installations will need to answer to the requirements of the new "smart" aggregates which automatically take into account the emissions, the energetic efficiency and the process requirements (PIER, 2002). For that purpose the researches conducted at Suplacu de Barcau 2xST 18 Cogenerative Power Plant has focused on the afterburning installation as integral part of the cogenerative group in terms of stack emissions, superficial temperature profile and power quality (Barbu et al., 2010) as well as on its interaction with the heat recovery steam generator.

2. General principles of the mathematical modelling of the thermo-gas-dynamic and chemical processes in the combustion chambers

The classical approach of the combustion chambers study assumes as a general rule the embracing of a steady character of the phenomena taking place in these installations, constituting only a quasi-adequate manner to the problem of analysing the unsteady phenomena generating important collateral effects. The physical-chemical phenomena succeeding in the combustion chamber are extremely complex, each of them (injection, atomization, vaporization, diffusion, combustion) rigorously depending on the physical factors such as air excess, gases pressure, temperature and velocity in the chamber. It may be admitted that the combustion is normal as long as the fluctuations detected in the combustion chamber only depend on the local conditions and they are randomly distributed in the chamber. The high level of complexity of the phenomena, associated to the flow instabilities, the heat transfer and the combustion reactions, makes them inaccurate to model using simplified mathematical models which only globally consider the processes and which are only slightly dependent on the combustion chamber geometry, the combustion configuration, the walls' screening or the intermediary reactions in the flame. Therefore, here is studied the complex and coupled problem of mathematical modelling for pulsating flow (numerical integration of Navier-Stokes equations with a closing model application), the influence on heat transfer (considering the radiation and convection), the combustion reactions (applying complex combustion mechanisms with high number of reactions and intermediary chemical compounds). The generalized model accurately tracking the complex processes in the combustion chamber may be developed as a group of modules, each associated to a phenomenon (flow, heat transfer, combustion reactions and dispersion phase evolution). This modular approach method allows the separate development of several sub-models with higher accuracy for a certain class of problems. Hence the problem of „closing" the equations system describing the studied phenomenon may and has been solved by using several turbulence models: k-ε (standard, realisable or RNG), Reynolds-stress model, LES - large eddy simulation (high scale modelling), or lately, due to the increase in calculation efforts, DNS – direct numerical simulation.

2.1 Mathematical models used for simulating flow, heat transfer and combustion in combustion chambers

There are two fundamentally different manners used for describing the fluid flow equations: the Lagrangian and Eulerian formulations. From the Lagrangian formulation perspective the flow field represents the movement of small, adjacent fluid elements interacting through pressure and viscous forces. The movement of each fluid element is made according to Newton's second law. This method is however impractical because of the high number of mass elements necessary for reaching a reasonable accuracy in describing the flow in a continuous environment. On the other hand the Lagrangian method deserves to be taken into consideration for biphasic flows (gas-droplet type) in describing the dispersion phase because the particles naturally constitute individual mass elements. The complexity of the Eulerian formulation of the biphasic flow does not allow the direct application of the solving schemes existing in the case of monophasic flow. As a consequence of this averaging problem in most numerical models the Lagrangian formulation is used for describing the dispersion phase. The relation between the Lagrangian and the Eulerian formulations is given by the Reynolds transport theorem. Therefore the components on the three axes may be combined in a single vectorial equation:

$$\rho\frac{D\mathbf{u}}{Dt} = \rho\mathbf{F} - \mathrm{grad}\,p + \mu\nabla^2\mathbf{u} + \frac{1}{3}\mu\cdot\mathrm{grad}(\mathrm{div}\mathbf{u}) \tag{1}$$

Equation (1) represents the Navier-Stokes in complete vectorial form describing the movement of a viscous fluid. The Navier-Stokes equation is applied for laminar flows as well as for turbulent flows. However it cannot be directly used in solving the problems associated to turbulent flow because it is impossible to track the minor fluctuations of the velocity associated with the turbulence. In order to determine the flow field, the numerical model solves the mass and impulse conservation equations. For flows involving heat transfer of compressibility additional equations are needed for energy conservation. For flows implying chemical compounds mixing or chemical reactions an equation of chemical compounds conservation is solved or, in the „probability density function" cases (generically called PDF models), conservation equations for the considered mixture fractions as well as equations defining their variations are needed. In flows with turbulent character additional transport equations need to be solved. Mass conservation equation, or continuity equation, may be written:

$$\frac{\partial\rho}{\partial t} + \frac{\partial}{\partial x_i}(\rho u_i) = S_m \tag{2}$$

Equation (2) is the generalized formulation of mass conservation equation and is applicable for incompressible or compressible flows. The source term S_m represents the mass added to the continuous phase, mass resulted from the dispersion phase (due to liquid droplets' vaporization) or a different source. For axi-symmetric bi-dimensional flows, the continuity equation in given by:

$$\frac{\partial\rho}{\partial t} + \frac{\partial}{\partial x}(\rho u) + \frac{\partial}{\partial r}(\rho \mathrm{v}) + \frac{\rho \mathrm{v}}{r} = S_m \tag{3}$$

where x is the axial coordinate, r is the radial coordinate, u is the axial velocity and v is the radial velocity. The impulse conservation on i in an inertial reference plane is described by:

$$\frac{\partial}{\partial t}(\rho u_i) + \frac{\partial}{\partial x_j}(\rho u_i u_j) = -\frac{\partial p}{\partial x_i} + \frac{\partial \tau_{ij}}{\partial x_j} + \rho g_i + F_i \tag{4}$$

where p is the static pressure, τ_{ij} is the tension tensor and ρg_i and F_i are the internal and external gravitational forces (e.g. occurring from the interaction with the dispersion phase) on direction i. F_i also includes other source terms depending on the model (such as for porous environment case). The tension tensor τ_{ij} is given by:

$$\tau_{ij} = \left[\mu \left(\frac{\partial u_i}{\partial x_j} + \frac{\partial u_j}{\partial x_i} \right) - \frac{2}{3} \mu \frac{\partial u_l}{\partial x_l} \delta_{ij} \right] \tag{5}$$

where μ is the dynamic viscosity and the second term in the right side of the relation represents the effect of volume dilatation. For axi-symmetric bi-dimensional geometries the axial and radial impulse conservation equations are given by:

$$\begin{aligned} &\frac{\partial}{\partial t}(\rho u) + \frac{1}{r}\frac{\partial}{\partial x}(r\rho u u) + \frac{1}{r}\frac{\partial}{\partial x}(r\rho \text{v} u) = -\frac{\partial p}{\partial x} + \\ &+\frac{1}{r}\frac{\partial}{\partial x}\left[r\mu\left(2\frac{\partial u}{\partial x} - \frac{2}{3}(\nabla \cdot \vec{\text{v}}) \right) \right] + \frac{1}{r}\frac{\partial}{\partial r}\left[r\mu\left(\frac{\partial u}{\partial r} + \frac{\partial \text{v}}{\partial x} \right) \right] + F_x \end{aligned} \tag{6}$$

$$\begin{aligned} &\frac{\partial}{\partial t}(\rho \text{v}) + \frac{1}{r}\frac{\partial}{\partial x}(r\rho u \text{v}) + \frac{1}{r}\frac{\partial}{\partial x}(r\rho \text{v}\text{v}) = -\frac{\partial p}{\partial r} + \frac{1}{r}\frac{\partial}{\partial x}\left[r\mu\left(\frac{\partial \text{v}}{\partial x} + \frac{\partial u}{\partial r} \right) \right] + \\ &+\frac{1}{r}\frac{\partial}{\partial r}\left[r\mu\left(2\frac{\partial \text{v}}{\partial r} - \frac{2}{3}(\nabla \cdot \vec{\text{v}}) \right) \right] - 2\mu\frac{\text{v}}{r^2} + \frac{2}{3}\frac{\mu}{r}(\nabla \cdot \vec{\text{v}}) + \rho\frac{w^2}{r} + F_r \end{aligned} \tag{7}$$

where $\nabla \vec{\text{v}} = \frac{\partial u}{\partial x} + \frac{\partial \text{v}}{\partial r} + \frac{\text{v}}{r}$ and w is the tangential velocity. The energy conservation equation may be written:

$$\frac{\partial}{\partial t}(\rho E) + \frac{\partial}{\partial x_i}\left[u_i(\rho E + p) \right] = \frac{\partial}{\partial x_i}\left(k_{ef}\frac{\partial T}{\partial x_i} - \sum_{j'} h_{j'} J_{j'} + u_j(\tau_{ij})_{ef} \right) + S_h \tag{8}$$

where k_{ef} is the effective conductivity ($k_{ef} = k + k_t$, where k_t is turbulent thermal conductivity, defined relative to the utilized turbulence model) and $J_{j'}$ is the diffusive flow of the chemical compound j'. The first three members in the left side of equation (8) represent the energy transfer due to conduction, chemical compounds diffusion and respectively viscous dissipation. The term S_h includes the heat exchanged in the chemical reactions or other volume heat sources. In equation (8),

$$E = h - \frac{p}{\rho} + \frac{u_i^2}{2} \tag{9}$$

where the sensible enthalpy h is defined, for ideal gases, by:

$$h = \sum_{j'} m_{j'} h_{j'} \tag{10}$$

and for incompressible flows by:

$$h = \sum_{j'} m_{j'} h_{j'} + \frac{p}{\rho} \tag{11}$$

In equations (10) and (11) $m_{j'}$ is the mass fraction of the chemical compound j' and

$$h_{j'} = \int_{Tref}^{T} c_{p,j'} \cdot dT \tag{12}$$

is the corresponding enthalpy of the compound, and the reference temperature is T_{ref} = *298.15 K*. In combustion studies, when the PDF based nonadiabatic model is used, the model requires solving an equation for total enthalpy, set by the energy equation:

$$\frac{\partial}{\partial t}(\rho H) + \frac{\partial}{\partial x_i}\left[\rho u_i H\right] = \frac{\partial}{\partial x_i}\left(\frac{k_t}{c_p}\frac{\partial H}{\partial x_i}\right) + \tau_{ik}\frac{\partial u_i}{\partial x_k} + S_h \tag{13}$$

In the hypothesis of a unitary Lewis number (*Le* = *1*), the conduction and diffusion terms of the chemical compounds are combined in the first term in the left side of equation (13), while the viscous dissipation contribution in the nonconservative formulation occurs as the second term of the equation. Total enthalpy *H* is defined by:

$$H = \sum_{j'} m_{j'} H_{j'} \tag{14}$$

where $m_{j'}$ is the mass fraction of the chemical compound j' and

$$H_{j'} = \int_{Tref}^{T} c_{p,j'} \cdot dT + h_{j'}^{0}(T_{ref,j'}) \tag{15}$$

$h_{j'}^{0}\left(T_{ref,j'}\right)$ is the enthalpy of formation of chemical compound j' at the reference temperature T_{ref}. Equation (8) includes the terms of pressure and kinetic energy work, terms neglected in incompressible flows. The decoupled solving method for the flow equations does not require including these terms in incompressible flows. However these terms must always be considered when using coupled solving method or for compressible flows. Equations (8) and (13) include the viscous dissipation terms representing the thermal energy created by the viscous tension in the flow. When using the decoupled solving method, the energy equation formulation does not need to explicitly include these terms because the viscous heating is in most cases neglected. The viscous heating becomes important when the Brinkman number, B_r, is close or higher than the unitary value, where

$$B_r = \frac{\mu U_e^2}{k \Delta T} \tag{16}$$

and ΔT represents the temperature difference in the system. The compressible flows usually have a Brinkman number $B_r \geq 1$. In the same time equations (8) and (13) include the enthalpy transport effect due to chemical compounds diffusion. For the decupling solving method the term $\frac{\partial}{\partial x_i}\sum_{j'} h_{j'} J_{j'}$ is included in equation (8), and in the nonadiabatic combustion model (PDF) this term does not explicitly appear in the energy equation, being included in the first term in the right side of equation (13). The energy sources S_h include in equation (8) the energy due to chemical reactions.

$$S_{h,reaction} = \sum_{j'} \left[\frac{h_{j'}^0}{M_{j'}} + \int_{T_{ref,j'}}^{T_{ref}} c_{p,j'} \cdot dT \right] \cdot R_{j'} \tag{17}$$

where $h_{j'}^0$ is the enthaply of formation of compound j', and $R_{j'}$ is the volumetric velocity of creation of compound j'. When using the PDF combustion model, the heat of formation is included in the enthalpy definition so the energy sources of the chemical reaction are no longer included in the formulation of S_h.

3. Suplacu de Barcau 2xST 18 cogenerative power plant

Suplacu de Barcau 2xST 18 Cogenerative Plant (fig. 1), with beneficiary SC OMV PETROM SA, is located in Bihor County, Romania, 75 km from Oradea Municipality. The main technical data are given in table 1. The plant was integrally commissioned in 2004 working in the framework of Suplacu de Barcau Oil Field. The electrical energy is used for driving the reducing gear boxes from the oil wells, the compressors, the pumps, for lighting etc. and the thermal energy (steam) is injected in the deposit being necessary in the oil extraction technological process and/or for other field requirements (heating the buildings or technological pipes). Suplacu de Barcau 2xST 18 Cogenerative Plant comprises two groups (fig. 1, right) which may work together or separately. Each group includes a ST 18 gas turbine (fig. 2, left), an afterburning installation (fig. 2, centre), a heat recovery steam generator (fig. 2, right) and additional installations. The heat recovery steam generator of each cogenerative group is a fire tube type boiler with two flue gas lines - one horizontal and the other vertical, comprising: the uncooled afterburning chamber; the superheater insuring the 300 °C steam temperature; the pressure body producing the saturated steam; the feed water heater assembly - water pre-heater insuring the necessary parameters of the water supplying the pressure body. The superheater is a coil type heat exchanger with 12 coil pipes (ø 38) welded in the steam inlet down-tanks (upper tank - fig. 3, centre) of the pressure body and superheated steam outlet (lower tank - fig. 2, right). The steam in the pressure body enters the upper tank through a PN40 DN150 connector (placed in the middle of the tank) and is distributed to the 12 coil pipes, then enters the lower tank and is delivered to the users.

The gases from the afterburning chamber follow the horizontal line of the heat recovery steam generator (superheater - pressure body) then the vertical one (feed water heater - water pre-heater - stack). Each cogenerative group is able to work in any of the three versions given in table 1, but the basic one is version I. 2xST 18 Cogenerative Power Plant is working automatically, the exploitation personnel being alerted by the command panel, through optical signalling and alarm horns, regarding the deviations of the supervised parameters or the damages occurrence. Certain parameters (pressures, temperatures, flow rates etc) of the equipments are archived and displayed with the help of an acquisition system.

Fig. 1. 2xST 18 Cogenerative Power Plant view (left) and gear placement (right)

No	**Name**	**Version I**	**Version II**	**Version III**
1	**Cogenerative group version**	Gas turbine + afterburning + heat recovery steam generator	Gas turbine + heat recovery steam generator	Heat recovery steam generator + afterburning
2	Fuel	Natural gas		
3	Gas turbine type	ST 18 (Pratt & Whitney - Canada)		
4	Boiler type	Fire tube boiler (SC UTON SA Onesti - Romania)		
5	Electric generator type	GSI-F (Electroputere Craiova - Romania)		
6	Electrical power delivered by the plant	2x1,75 MW (6,3 kV, 50 Hz)		-
7	Superheated steam pressure	20 bar		
8	Superheated steam temperature	300 ^{0}C		

Table 1. Technical specifications of 2xST 18 Cogenerative Power Plant

Fig. 2. ST 18 gas turbine (left), afterburning installation burner (centre) and heat recovery steam generator (right) at 2xST 18 Cogenerative Power Plant

Fig. 3. Superheater of the heat recovery steam generator at 2xST 18 Cogenerative Power Plant

3.1 The afterburning installation at Suplacu de Barcau 2xST 18 cogenerative power plant

The afterburning installation (burner with automatics) at 2xST 18 Cogenerative Power Plant was delivered by Saacke (www.saacke.com - Germany) and has the specifications given in table 2. The burner (fig. 4), produced by Eclipse - Holland, is the "FlueFire" type dedicated to this kind of application. It may be placed directly in the flue gases flow, between the turbine and the recovery boiler, but may work as well on fresh air. The burner has 21 basic modules located on 3 natural gas fuelling ramps and 2 flame propagation modules. The flue gases from the gas turbine are introduced in the "FlueFire" burner through an adaptation section. The mixture with the fuel is obtained through the swirling motion of the flue gases exhausted from the turbine in the fuel jets. This leads to the cooling and the stabilization of the combustion in the burner front allowing downstream high temperatures at a low NO_x content. The air, delivered by a fan (fig. 4, right), is introduced in the adaptation section through a distribution system built to insure an uniform distribution in the transversal section because the emissions depend on the unevenness of the flow, velocity, oxygen concentration etc. The afterburner modules are built in refractory steel, laser cut, for insuring the necessary uniformity.

Each module is fitted in the natural gas fuelling ramps using two gas nozzles in order to allow the free dilatation of the assembly. The ignition is initiated with the help of a pilot burner placed in the lower area of the afterburning burner and the supervision of the flame is insured by a UV type "DURAG D-LX 100 UL" detector placed in the upper area.

No.	Name		Value
1	Natural gas pressure	Before the regulator	0,5-2 bar
		After the regulator	0,4 bar
2	Thermal power	With flue gases (version I)	2,4 MW
		With air at 20 ^{0}C (version III)	6 MW
3	Flue gases maximum mass flow rate		8,75 kg/s
4	Flue gases temperature at the inlet of the burner		524 ^{0}C
5	Flue gases temperature at the end of the afterburning chamber (versions I and III)		770 ^{0}C

Table 2. Technical specifications regarding the afterburning installation at 2xST 18 Cogenerative Power Plant

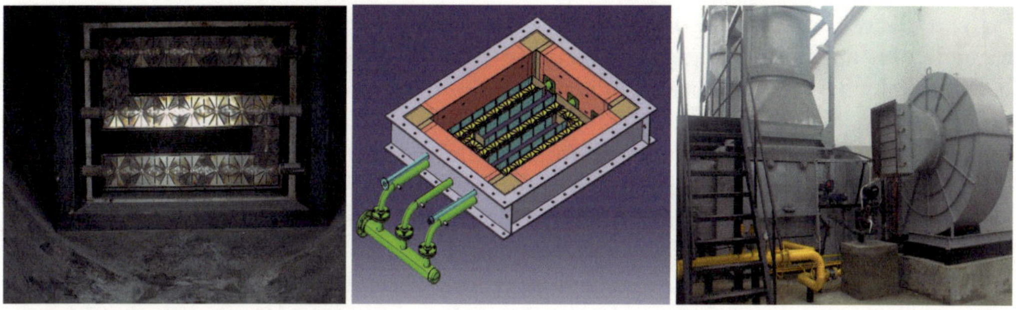

Fig. 4. The burner in the afterburning installation at 2xST 18 Cogenerative Power Plant (left, centre) and the fresh air fan (right)

4. Gas turbine – afterburning interaction

The flue gases flow in the outlet section of the gas turbine is generally turbulent and unevenly distributed. In some areas at the inlet of the afterburning installation backflows may occur. A uniform flow distribution is an important factor concurring to the good working of the afterburning and to the performances of the heat recovery steam generator. Grid type burners are designed to distribute the heat uniformly in the transversal section of the heat recovery steam generator, fact requiring careful oxygen feeding in order to avoid high NO_x emissions and variable length flame. The flow rate, the temperature, the composition of the flue gases exhausted from the gas turbine depend on the fuel type, load, fluid injection in the gas turbine (water, steam), environmental conditions etc. The gas turbines used in industrial applications are fuelled by liquid or gaseous fuels. Regarding the liquid fuels, for economical reasons, there are usually used cheap fuels such as heavy fuels, oil fuels or residual products from different manufacturing processes or chemisation (Carlanescu et al., 1997). Using these types of fuels raises problems concerning: insuring combustion without coating, decreasing the corrosive action caused by the presence of aggressive compounds (sulphur, traces of calcium, lead, potassium, sodium, vanadium) and problems concerning pumping and spraying (heating, filtration etc.). When considering using aviation gas turbines for industrial purposes (existing aviation gas turbines with minimal modifications) the possibilities of using liquid fuels are limited. For each case the technical request of the beneficiary must be analyzed in conjunction with the study of fuel characteristics affecting the processes in the combustion chamber (density, molecular mass, damping limits, burning point, volatility, viscosity, superficial tension, latent heat of vaporization, thermal conductibility, soot creation tendency etc.). For the gaseous fuels the problem is easier considering the high thermal stability, the absence of soot and ashes, and the high caloric power. In this case the problems concern mostly the combustion process in conjunction with the requirements of the used gas turbine. For the valorisation of the landfill gas the TV2 - 117A gas turbine was modified to work on landfill gas instead of kerosene by redesigning the combustion chamber (Petcu, 2010). Numerical simulations and experimentations have been conducted for the gas turbine working on liquid fuel (kerosene) and gasesous fuels (natural gas, landfill gas). The boundary conditions have been either calculated or delivered by the gas turbine manufacturer for three working regimes: take-off, nominal and idle. The results are presented in table 3 with the corresponding temperatures for each regime. The temperature fields are displayed in rainbow with red representing the highest value. The most important result refers to the fact that the numerically obtained temperatures are close enough to the ones indicated by the manufacturer, the differences being explained by the simplifying hypotheses introduced in the simulations. Analyzing the numerical results it may be observed that the flame shortens (column 5) with the decrease in regime, but it fills better the area between two adjacent injectors (column 6). The main criterion validating the numerical results has been the averaged turbine inlet temperature (T_m) in the conditions of fuel and air flow rate imposed by the working regimes of the TV2 - 117A gas turbine. For working on gaseous fuel, the TV2 - 117A gas turbine has suffered adjustments on the fuel system level and particularly on the injection nozzles. The starting point in designing the new injection nozzle was a previous application on the TA2 gas turbine resulted by modifying the TV2 - 117A. Table 4 presents the variation of the CH_4 mass fraction indicating the injected jet shape (left) and the burned gases temperature in the combustion chamber outlet/turbine inlet section (right) for different working regimes. The geometrical parameters of the injection nozzle were set based on the numerical temperature fields in the turbine inlet section and aiming to obtain a compact fuel jet which avoids the combustion chamber walls. It must be noted that a stable combustion process has been obtained using a gaseous fuel in a combustion chamber designed for a different type of fuel (kerosene). The numerical simulations made possible narrowing the variation domains for the geometrical and gas-dynamic parameters in order to establish the constructive solution of the combustion chamber for working on landfill gas. The numerical results have been used for designing and manufacturing the new injection nozzle for the eight injectors of the TV2 - 117A gas turbine, transforming it into the TA2 aero-derivative. From tables 3 and 4 it may be noticed that by changing the fuel and the working regime the temperature distribution in the section of interest is modified, fact that might affect the afterburning installation performances. For reducing the NO_x emissions, reducing the temperature in the combustion area is applied through water or steam injection.

Regime	T_m [K]	Thermal field at the outlet	Thermal field on the walls	Thermal field in the axial-median section	Thermal field on the frontier between two sections
1	2	3	4	5	6
Take-off	1135	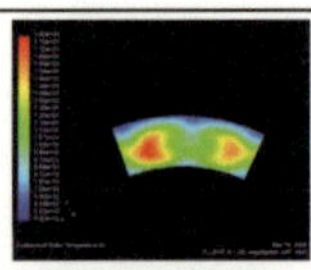	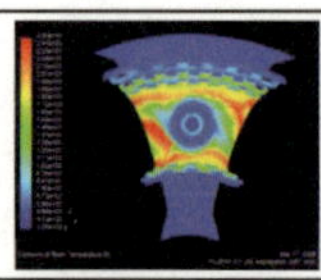	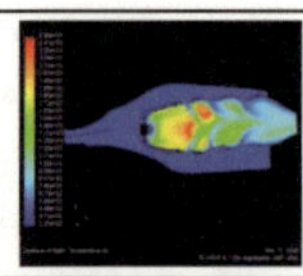	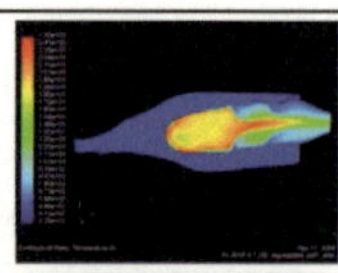
Nominal	1075	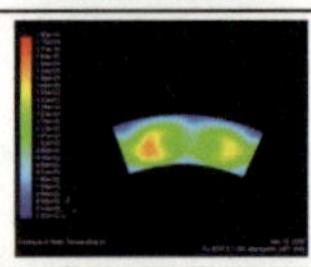	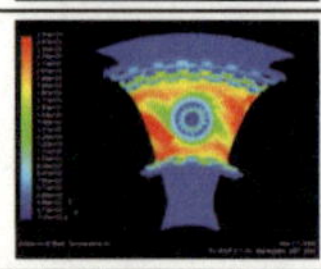	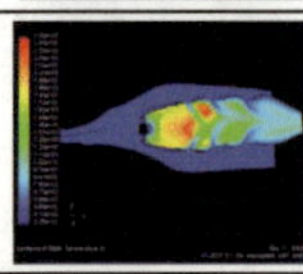	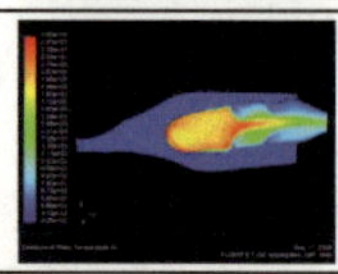
Cruise	1039	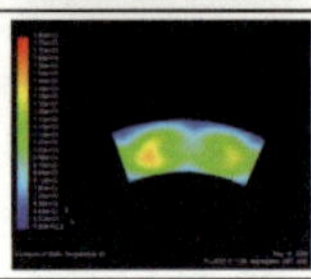	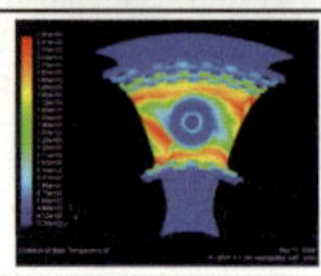	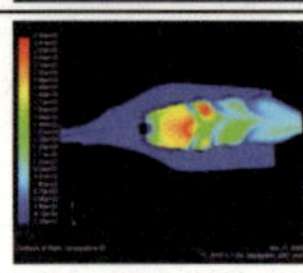	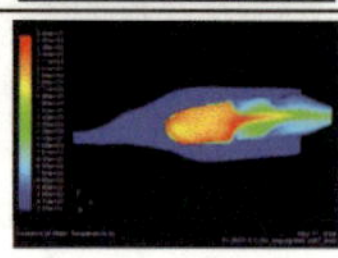

Table 3. Numerical results for the TV2 - 117A gas turbine on liquid fuel (Petcu, 2010)

No.	T_m [K]	Fuel injection jet	Combustion chamber outlet temperature field
1	1053		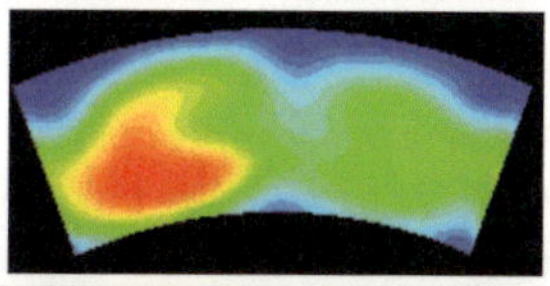
2	952	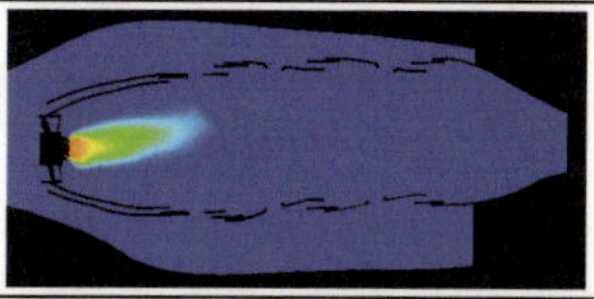	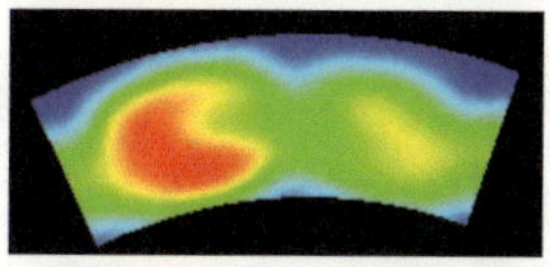
3	944	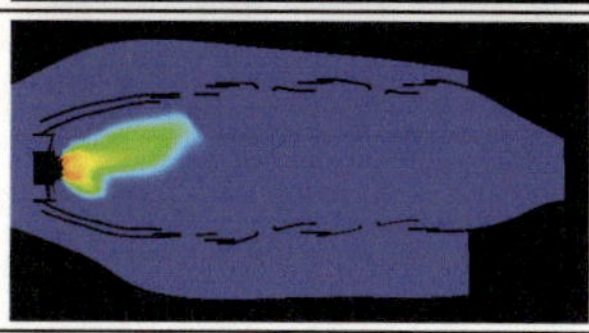	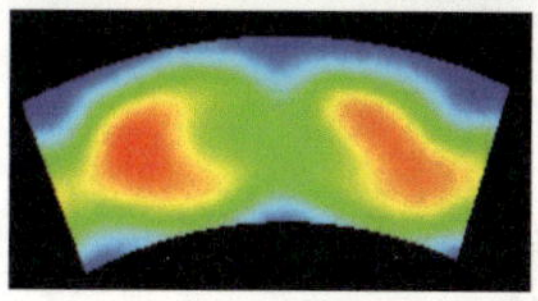

Table 4. Numerical results for the TA2 gas turbine on landfill gas - injection jet and thermal cross section from numerical simulations (Petcu, 2010)

In some cases the water or steam are injected directly in the combustion area through a number of holes in the combustion chamber inlet section or in the fuel injection nozzle. Another solution is injecting the water upstream of the firing tube, usually in the air flow that next passes the turbulence nozzles in its way to the combustion area. This method insures a very good atomization, the small droplets being transported by the air flow while the larger ones form a thin film on the surface of the turbulence nozzles being next atomized by the air passing over the downstream edges of the turbulence nozzle. The efficiency of the water or steam injection in reducing the NO_x emissions has been highlighted by many authors and may be expressed by the following relation (Carlanescu et al., 1998):

$$\frac{NO_{x(wet)}}{NO_{x(dry)}} = \exp-(0.2x^2 + 1.41x) \tag{18}$$

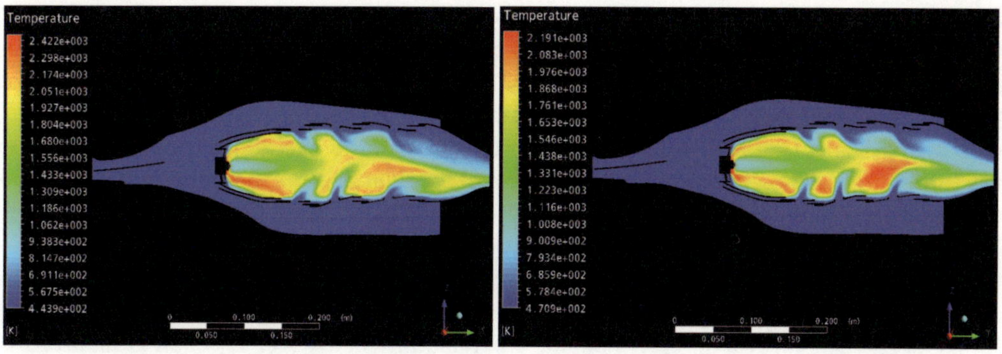

Fig. 5. Temperature field in the axial-median section of the TA2 combustion chamber, T_m =1063 K, without water injection (left) and with water injection (right) (Popescu et al., 2009)

Relation (18) may be applied for liquid as well as for gaseous fuels, showing that approximately 80 % reduction in the NO_x emissions may be obtained at equal water/steam-to-fuel flow rates (x = 1). The water injection is more efficient at higher combustion pressures and temperatures where the NO_x production is higher and less efficient at lower pressures and temperatures. For the independent running of the TA2 gas turbine on methane, without afterburning, the numerical simulations (fig. 5 – 6) regarding the water injection have shown a NO_x reduction of over 50 % (Popescu et al., 2009).

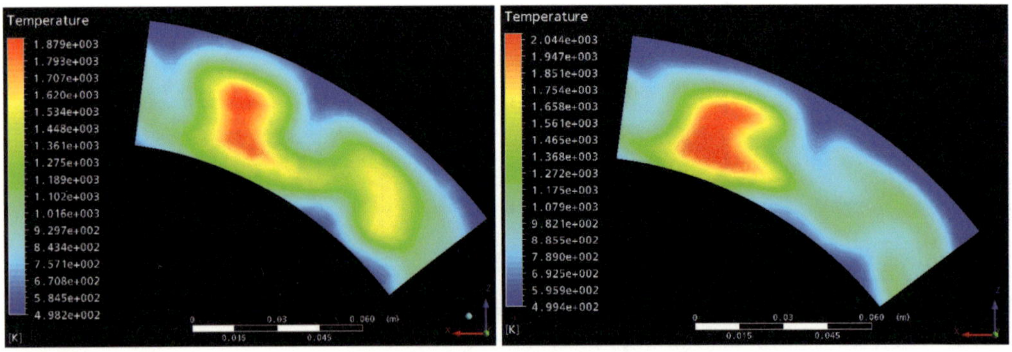

Fig. 6. Temperature field in the combustion chamber outlet section, T_m =1063 K, without water injection (left) and with water injection (right) (Popescu et al., 2009)

However, theoretical and experimental researches on a turbojet have shown that the steam injection reduces the NO emissions up to 16% (mass fractions) when a steam flow which doubles the fuel flow is introduced (Benini et al., 2009). At the same ratio, the NO reduction in the water injection case is approximately 8%. The steam injection slightly reduces the CO level while the water injection raises it with the increase in the injected water quantity. Using the NASA CEA program (McBride & Gordon, 1992; Zehe et al., 2002), the combustion have been analyzed in the TA2 gas turbine for methane, natural gas (the composition at Suplacu de Barcau Cogenerative Plant) and landfill gas (equal volume proportions of methane and carbon dioxide) in the pressure and temperature conditions recommended for different working regimes of the gas turbine (T_m = 1063, 1023 and 873 K). The air excess coefficients have been established for each fuel in the dry working cases in order to obtain the same temperature of the reaction products for a corresponding fuel quantity. Starting from these initial data and increasing the injected water quantity up to 50 % of the fuel quantity, a decrease in temperature has been noticed for each 10 % injected water of approximately 1.46 degrees for methane, 1.62 degrees for natural gas and 1.36 degrees for landfill gas. It was then aimed to establish the dependence of the quantitative water-to-fuel ratio at supplementary fuel injection in order to maintain the maximum temperature in the gas turbine. These analysis have been made only for methane for the same stable working regimes of the TA2 gas turbine (T_m = 1063, 1023 and 873 K). In the theoretical calculations for methane some limitations have been next applied: a ratio between the fuel quantity in the water injection case and the initial fuel quantity of maximum 2; a minimum oxygen concentration in the flue gases from the gas turbine of 11 % volume for afterburning running. The general reaction for methane combustion when water injection is involved is given by equation (19) and the algorythm used in the NASA CEA program for determining the water injection influence is presented in fig. 7:

$$b \cdot CH_4 + 2(O_2 + 3.76N_2) + a \cdot H_2O \rightarrow c \cdot H_2O + d \cdot CO_2 + e \cdot N_2 + f \cdot O_2 + g \cdot CO + h \cdot NO_x \quad (19)$$

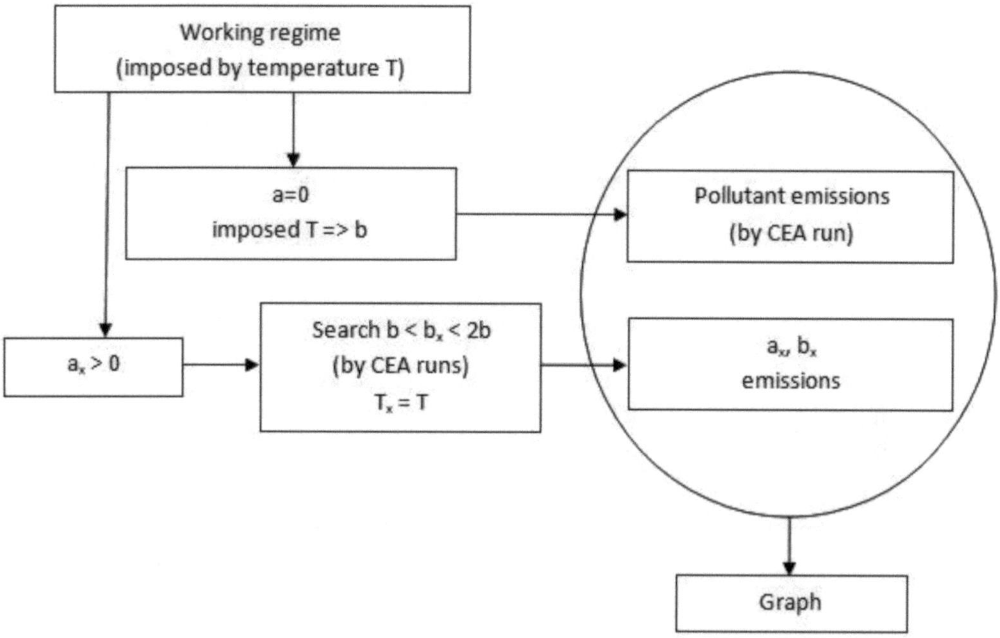

Fig. 7. The algorithm used in NASA CEA program to determine the water injection influence (subscript „x" denominates the seeked coefficients for the desired fixed temperature)

The thermodynamic properties of the system have been tracked with focus on the reaction products concentrations and particularly CO and NO_x. In these conditions the calculations have been made for a water coefficient (denoted a) of maximum 8 (fig. 8). For a higher value than 6.5 an instability of the curves occurs. For a value of the coefficient of approximately 2.8, the gas turbine at the regime T_m = 1063 K is close to the minimum oxygen limit of 11% needed for the afterburning. In fig. 9 and 10 the water coefficient a is limited to 4. Therefore we may have a maximum water coefficient of 2.8 for the regime T_m = 1063 K, 3.1 for the regime T_m = 1023 K and 3.5 for the regime T_m = 873 K, having as result the inaccessible areas of emissions reduction for the TA2 gas turbine with water injection when using the afterburning. The theoretical calculations indicate that the NO_x emissions for the TA2 working at the regime T_m = 1063 K with afterburning may not be lower than 40 ppm. The unevenness of the flow when exiting the combustion chamber (tables 3 and 4, fig. 5 and 6) and the variation in the burned gases composition (fig. 8 - 10) affect the afterburning process and confirm the necessity of a fine control on the injected water quantity particularly when a significant reduction of the emissions is aimed. Therefore the afterburning is affected from the efficiency, emissions, flame stability points of view as well as from the corrosion on the elements subjected to the burned gases action.

The combined actions of water vapours and oxygen concentration and high temperature of the flue gases represent the recipe for an accentuated corrosion (Conroy, 2003).

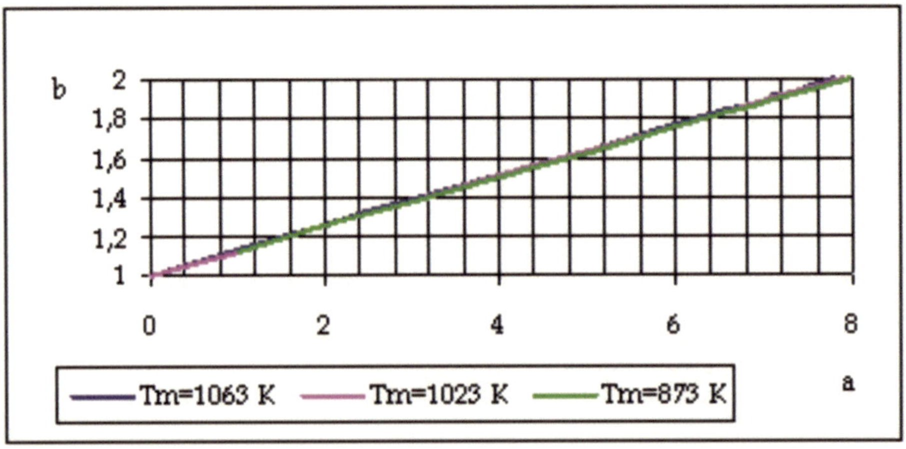

Fig. 8. Fuel coefficient (b) variation depending on injected water coefficient (a)

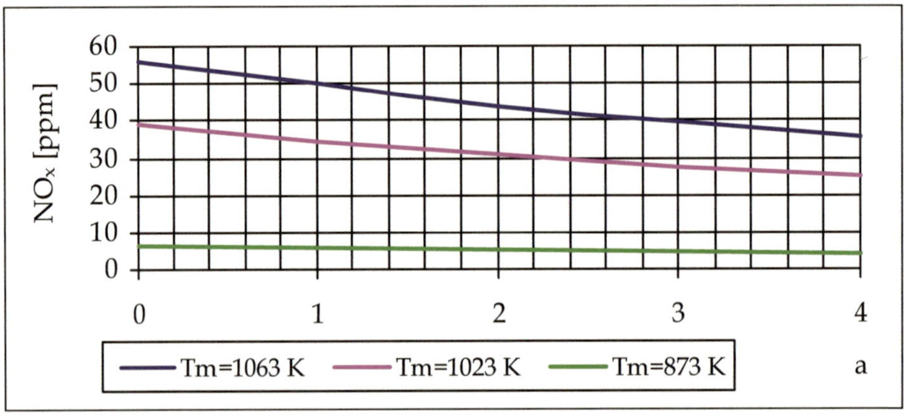

Fig. 9. NO_x concentration variation depending on injected water coefficient (a)

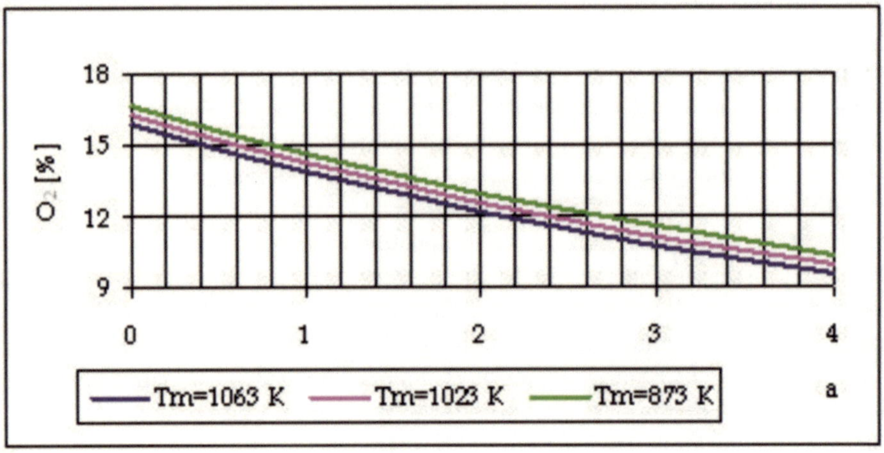

Fig. 10. O_2 concentration variation depending on injected water coefficient (a)

5. Afterburning – Heat recovery steam generator interaction

The heat recovery steam generator in the cogenerative groups is designed to run at certain parameters of the flue gases and superheated steam. The process requirements, the variable environmental conditions (affecting the gas turbine and therefore the afterburning) influence the running of the heat recovery steam generator and the overall efficiency of the cogenerative group. The steam superheater is usually the last heat exchanger in the pressurized thermal circuit water - steam of a heat recovery steam generator. As „end of the line" element it is its duty to maintain the temperature of the superheated steam, imposing mechanical systems and automatics with well defined functions and roles taken into consideration by the designer. Increasing the nominal parameters of modern heat recovery steam generator as a result of the increased performances in gas turbines led to a superheater area larger than that of the vaporizing system, the superheater becoming a large metal consumer as well as the heat exchanger with the highest thermal demand. As consequence, the superheater needs the proper consideration in the design process as well as in activity. If the flow process in the superheater pipes is admitted as isobar, the increase in temperature takes place according to an exponential law. The value of the coefficient of thermal unevenness in the flue gases flow entering the superheater stage depends on the constructive shape and the thermal diagram. This coefficient is defined by (Neaga, 2005):

$$k'_t = \frac{t'_{g\max} + 273}{t'_{g\mathrm{med}} + 273} \tag{20}$$

where the index shows that the temperatures are indicated for the inlet of the stage, $t'_{g\max}$ and $t'_{g\mathrm{med}}$ being the maximum and respectively the averaged temperatures of the gases at the inlet. The ability of the steam to absorb the heat of the flue gases in different areas of the volume occupied by the stage leads to unevenness affecting the safe running of the heat exchanger. The researches show that the highest unevenness on the outlet of the stage is registered in the counter-stream flow of the thermal agents regardless of the number of stages of the superheater and the lowest in the uniflow (Neaga, 2005). The characteristics of the superheated steam are different for the radiation and convection superheaters. The radiation superheater absorbs more heat at low loads while the convection one absorbs more heat at higher loads (Ganapathy, 2001). The superheaters usually have more stages, the radiation and convection combined ensuring a uniform temperature in the steam for a larger range of loads. When the superheater consists in only one stage the problem of controlling the temperature in the superheated steam becomes more complicated. The steam temperature can usually be maintained constant in the 60 - 100 % load range but several factors act on the superheated steam temperature: heat recovery steam generator load, air excess coefficient at the furnace outlet, initial dampness of the fuel, calorific value of the fuel etc. As consequence, the temperature control systems must comply with certain conditions: low inertia, large control range (regardless of the variable parameter leading to variations in the superheated steam temperature), safe running construction, minimal manufacturing and running expenses etc.

6. Researches concerning the integration of the afterburning installation with the gas turbine and the heat recovery steam generator at Suplacu de Barcau 2xST 18 cogenerative plant

6.1 The integration of the afterburning installation with the heat recovery steam generator

The researches for the integration of the afterburning installation with the heat recovery steam generator and the gas turbine at Suplacu de Barcau 2xST 18 Cogenerative Power Plant have been made in several stages. At commissioning, the functional tests at some working regimes of the heat recovery steam generator have shown high temperatures of the superheated steam (table 5) compared to the nominal temperature (300 °C), leading to frequent activations of the heat recovery steam generator. In these conditions the coil pipes have been counted, from 1 to 12 (fig. 11), in the direction of the steam circulation through the lower tank, the outer temperature has been measured (fig. 3, 11) with a contact thermometer (Barbu et al., 2006) and the exchange area of the superheater has been reduced by replacing the first three coil pipes with three L-shaped pipes in order to maintain the steam velocity. Fig. 12 presents the temperature distribution on the outer surface of the 12 coil pipes of the superheater's outlet tank (t_{eSI}) after the replacement, for the averaged temperature of the superheated steam t_{vSI}= 280 °C and the temperature of the gases at the afterburning chamber outlet t_{ca}=690 °C.

Coil pipe	2	4	6	8	10	12	Observations
Outer temperature [°C]	370	380	295	295	245	185	Before replacement; t_{vSI}= 322 ^{0}C
	210	398	366	364	212	198	After replacement; t_{vSI}= 280 ^{0}C

Table 5. Outer temperature of the coil pipes of the superheater's outlet tank

The data in table 5 and fig. 12 (left) shows a modification in temperature distribution after the replacement of the three coil pipes. In the end, the pipes 1 - 4 have been cut and plugged and refractory metal sheets have been applied on each superheater (for a better circulation of the flue gases in the superheater) leading to the temperature distribution presented in fig 12 (right). Fig. 12 (right) illustrates an increase in the thermal field evenness in the coil pipes compared to the initial case at the commissioning of the heat recovery steam generator (table 5). For the process requirements of superheated steam temperature up to 350 °C only the area of the screens has been adequately reduced.

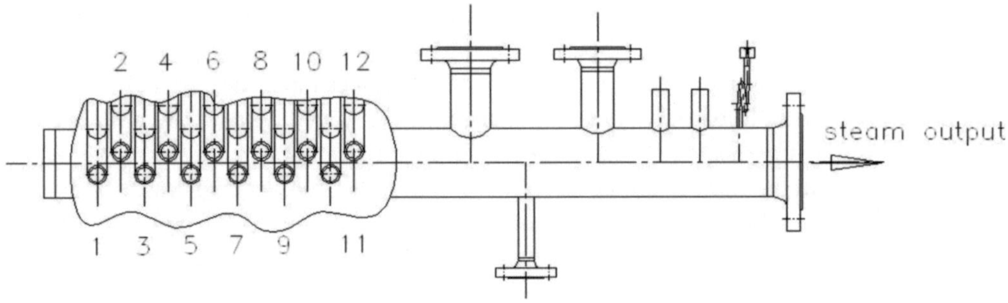

Fig. 11. Counting of the coil pipes of the superheater's outlet tank

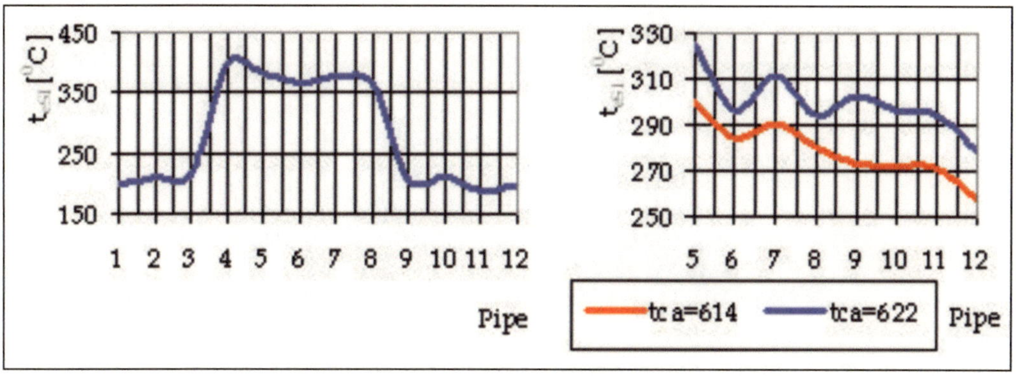

Fig. 12. Outer temperature (t_{eSI}) variation in the outlet tank after replacing the coil pipes 1 - 3 with three L-shaped pipes (t_{vSI}= 280 ^{0}C, t_{ca}=690 ^{0}C - left) and after plugging the coil pipes 1 - 4 and installing the screens on the superheater (t_{ca}=614 ^{0}C s and 622 ^{0}C - right)

6.2 The integration of the afterburning installation with the gas turbine

The data obtained until present resulted from experimentations conducted at 2xST 18 Cogenerative Plant, numerical simulations in CFD environment and experimentations with natural gas and air on the test bench.

6.2.1 The afterburning integrated analysis system

After solving the problem of the superheated steam temperature control, the next step has been simulating in CFD environment the afterburning installation at Suplacu de Barcau 2xST 18 Cogenerative Plant (Barbu et al., 2010). The numerical results have indicated that the air (or flue gases from the gas turbine) distribution in the burner may be improved leading to the installation of a concentrator at group 1. The second cogenerative group has remained unmodified since the commissioning. For determining the influence of the concentrator on the combustion process several aspects have been analysed: emissions at the stack, noise, superficial temperature profile and power quality for different working regimes of the groups. The measurements have been performed in industrial conditions on both cogenerative groups before applying overall optimization solutions in order to not disturb the technological process. For this reason the measurements have been performed for partial loads. For noise analysis have been used three measuring chains and a software application for acoustic prediction according to 2002/49/EC Directive, offering an image of the noise propagation in the area of interest. The noise measurements at the afterburning installation have been performed with a 01dB Metravib SOLO sound level meter. The acoustic field in the station's area has been studied in 50 measuring points with the B&K 2250 sound level meter and the acoustic pressure level of the cogenerative group has been determined with the multi-channel acquisition system 01dB Metravib EX-IF10D/module IEPE with 12 microphones 40AE G.R.A.S.. The measurements concerning the quality of environmental air have been performed with the help of the mobile laboratory (fig. 13, left) especially equipped for the task. For the chemical measurements has been used a Horiba PG 250 gas analyzer with the probe installed at the stack (fig. 13, centre). The outer superficial temperature profile has been determined with the help of a Fluke infrared camera, Ti45FT type, the sighting being in the upper area of the burner (fig. 13, right).

Fig. 13. Mobile laboratory at INCDT COMOTI (left), emissions measurements at the stack (centre) and sighting area of Fluke camera (Ti45FT type) with sound level meter (right)

The process parameters of the gas turbine, heat recovery steam generator and afterburning installation have been displayed in the command room or locally. The correlation of the emissions, noise, superficial temperature profile and power quality has been made related to the time of measurements. The electro-energetic measurements have been made in the electric generator cell through measurement converters, the equipment consisting in devices fixed in panels (ammeters, voltmeters, active and reactive energy counters) and mobile devices (CA 8332B analyser for electro-energetic network and power quality).

6.2.2 Experimental data and numerical simulation of the afterburning at 2xST 18 cogenerative plant

For starters a noise map for cogenerative group 2 has been established (without air concentrator), with group 1 out of work (fig. 14), in order to acquire comparison data for the case of local recording of noise at the burner of the afterburning installation. In the burner area the noise level has been in the 80 - 85 dB range with a significant distortion of noise curves and high values in the area of turbo-generators room. The measurements on group 1 (with concentrator) regarding emissions, noise and outer superficial temperature profile have been performed for the case of the heat recovery steam generator running with the afterburning on fresh air.

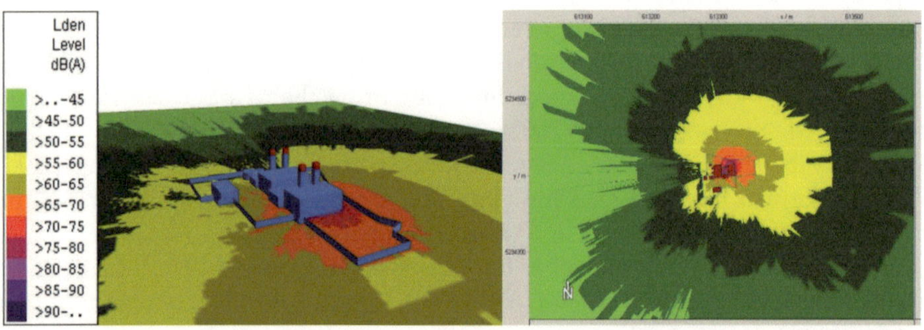

Fig. 14. Suplacu de Barcau 2xST 18 Cogenerative Plant – noise map for group 2, with group 1 out of work

Emissions, noise and outer superficial temperature profile measurements have been performed for group 1 (with concentrator) for the case of the heat recovery steam generator running with the afterburning on fresh air, with group 2 working in cogeneration (gas turbine + afterburning + heat recovery steam generator). The set of experimental data has been obtained for five working regimes defined by the gases temperature at the outlet of the afterburning chamber (t_{ca}): 500 ^{0}C, 552 ^{0}C, 604 ^{0}C, 645 ^{0}C, 700 ^{0}C. The NO_x variation and the noise locally measured at the burner (fig. 13) for group 1 (heat recovery steam generator + afterburning on fresh air) depending on flue gases temperature are given in fig. 15. The flue gases temperature has been measured with a thermocouple at the outlet of the afterburning chamber.

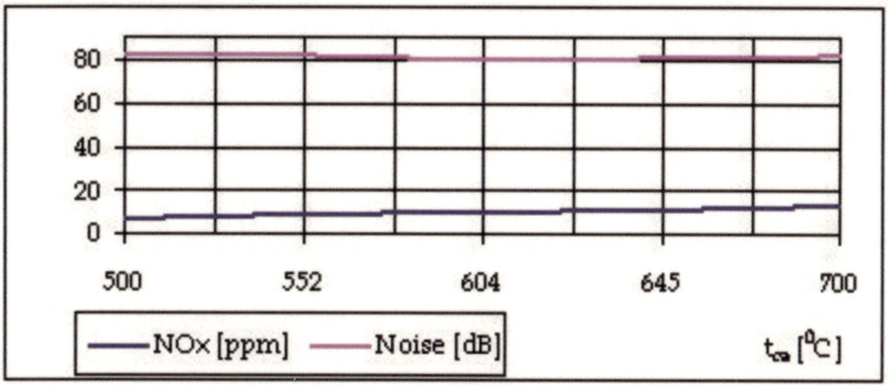

Fig. 15. NO_X and noise variation depending on flue gases temperature - group 1 (heat recovery steam generator + afterburning on fresh air)

Fig. 15 illustrates an increase in NO_x emissions with the flue gases temperature, while the noise level is approximately constant at 80 dB. For the outer superficial temperature profile, according to fig. 13 (right), an increase may be noticed in the area of the isotherms above 200 °C (fig. 16, left) with the increase in the temperature at the outlet of the afterburning chamber from 500 °C to 645 °C. The measurements on group 2 (without concentrator) regarding emissions, noise, outer superficial temperature and power quality have been performed for two cases: without afterburning (gas turbine + heat recovery steam generator) and with afterburning (gas turbine + afterburning + heat recovery steam generator). Another two sets of experimental data have been obtained. One set corresponds to three working regimes of the gas turbine + heat recovery steam generator version, defined by the flue gases temperature at the outlet of the afterburning chamber (t_{ca}): 423 ^{0}C, 437 ^{0}C, 475 ^{0}C. Another set corresponding to the version gas turbine + afterburning + heat recovery steam generator led to the following temperatures (t_{ca}): 536 ^{0}C, 569 ^{0}C, 605 ^{0}C, 645 ^{0}C. The configuration of the isotherms for the gas turbine + afterburning + heat recovery steam generator version is given in fig 16 (right). Working at fresh air rating, more than 645 ^{0}C, the region occupied by the isotherms is reduced. This area is however larger than the one for group 2 (without concentrator - fig 16, right) in gas turbine + afterburning + heat recovery steam generator version or gas turbine + heat recovery steam generator version. The central areas occupied by the isotherms (near the afterburning chamber) for group 2 (without concentrator) are decreasing with the increase in flue gases temperature at the outlet of the afterburning chamber. As opposite, for group 1, as stated above, the areas occupied by the isotherms are increasing with the increase in flue gases temperature.

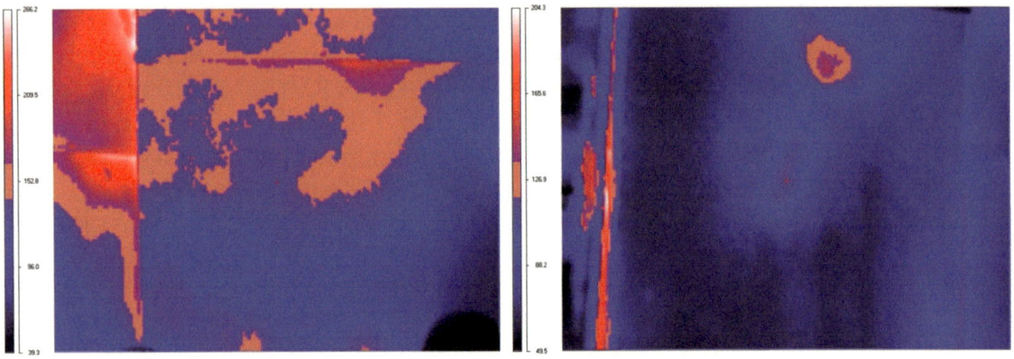

Fig. 16. Isotherms configuration for groups 1 and 2 in infrared; t_{ca} = 645 ^{0}C; left - group 1 (with concentrator): heat recovery steam generator + afterburning on fresh air; right - group 2 (without concentrator): gas turbine + afterburning + heat recovery steam generator

Local noise measurements near the burner confirm the values in fig. 14. The figures 17 - 19 present the electro-energetic measurements, respectively the power variations at the generator's hubs and the distortion coefficients from the fundamental (THD) for current and voltage depending on the flue gases temperature at the outlet of the afterburning chamber.

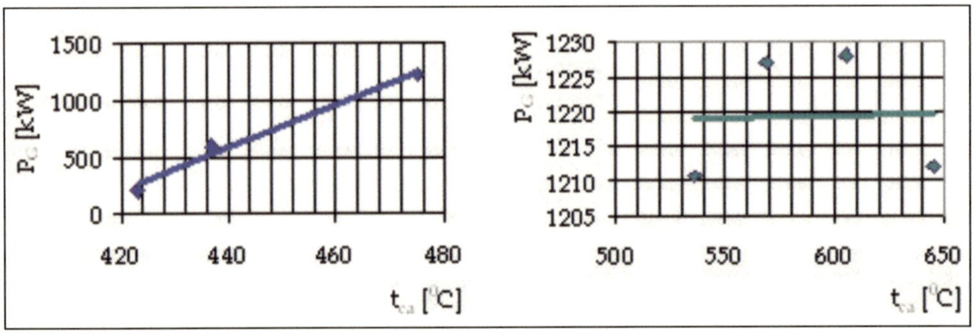

Fig. 17. Electrical power variation for version gas turbine + heat recovery steam generator (left) and version gas turbine + afterburning + heat recovery steam generator (right) depending on the flue gases temperature at the outlet of the afterburning chamber

In version gas turbine + heat recovery steam generator the variation in the heat recovery steam generator load is achieved through the variation of the gas turbine parameters. The electrical power increases with the increase in flue gases temperature (fig. 17, left). For the version gas turbine + afterburning + heat recovery steam generator the heat recovery steam generator load is varied through the afterburning parameters which makes the power quasi-constant (approximately 1220 kW) while the temperature at the outlet of the afterburning chamber increases (fig. 17, right). For the version gas turbine + heat recovery steam generator the value of the distortion coefficients from the fundamental (THD) for current and voltage decreases with the increase in temperature at the outlet of the afterburning chamber (fig. 18, 19 left).

This decrease is more pronounced for the currents (fig. 18, left) indicating an aggravation in the power quality. For version gas turbine + afterburning + heat recovery steam generator the value of the distortion coefficients from the fundamental for current and voltage decreases only slightly (under 1%) with the increase in the flue gases temperature at the outlet of the afterburning chamber (fig. 18, 19 right).

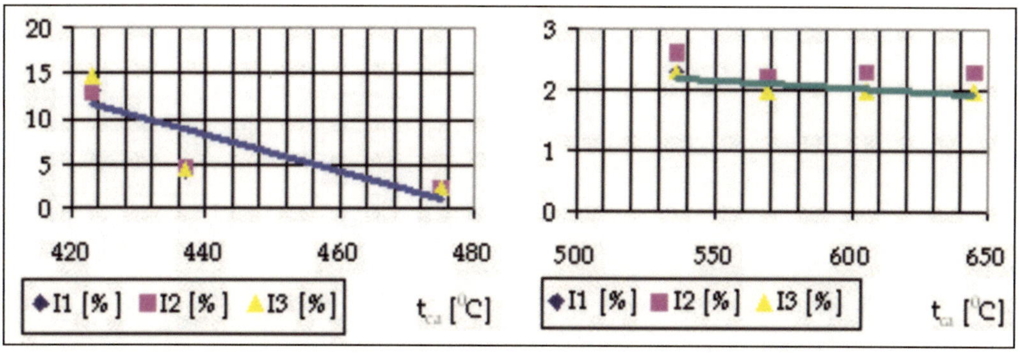

Fig. 18. Variation of distortion coefficients from the fundamental (THD) for current depending on the flue gases temperature at the outlet of the afterburning chamber for version gas turbine + heat recovery steam generator (left) and version gas turbine + afterburning + heat recovery steam generator (right)

For version gas turbine + heat recovery steam generator the variation of the distortion coefficients from the fundamental for current is higher than 5 % (fig. 18, left). This occurs at temperatures at the outlet of the afterburning chamber below 470 °C. The experiments conducted at 2xST 18 Cogenerative Plant have shown an improvement of the flow in the burner section at group 1, particularly in the upper area, but have not allowed an assessment of the performances due to geometric modification of the ST 18 burning module. Based on the experimental data obtained at Suplacu de Barcau 2xST 18 Cogenerative Plant, a new geometry has been obtained for the burning module (ST 18-R, fig. 20 left) analysed through numerical simulations in CFD environment. The new module incorporates an air (flue gases) concentrator and the module ST 18 has been angled to 15° (module ST 18-15) leading to an increased turbulence and a better mixing of the gas fuel with the comburent (Barbu et al., 2010).

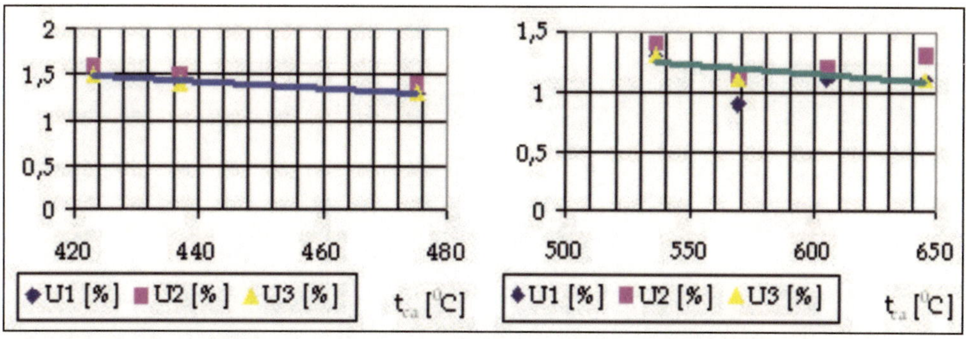

Fig. 19. Variation of distortion coefficients from the fundamental (THD) for voltage depending on the flue gases temperature at the outlet of the afterburning chamber for version gas turbine + heat recovery steam generator (left) and version gas turbine + afterburning + heat recovery steam generator (right)

The numerical simulations (fig. 20, right) indicate NO_x emissions three times lower for the ST 18-R module compared to the old ST 18 model at nominal working regime temperature of the afterburning at 2xST 18 Cogenerative Power Plant (770 ^{0}C).

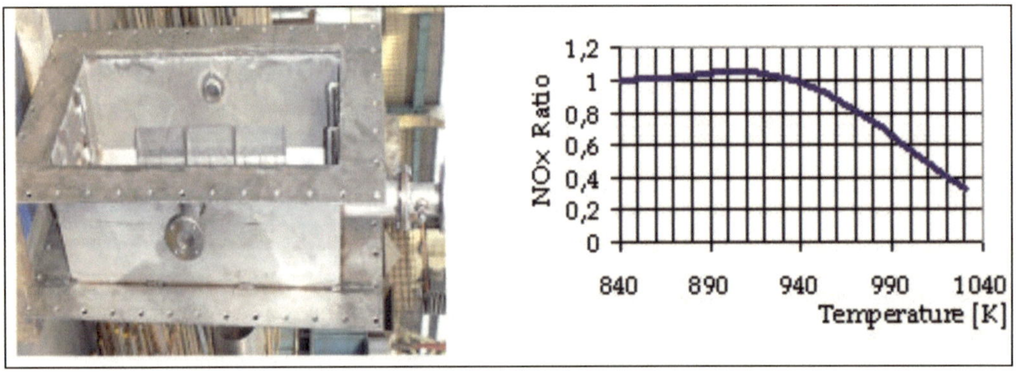

Fig. 20. Burner with three burning modules ST 18-R (left) and the variation of the NO_x ratio depending on the flue gases temperature for modules ST 18 and ST 18-R (right)

6.2.3 Experimental data obtained on test bench

For a thorough investigation of the processes and for eliminating some disturbing factors from the plant test bench examinations were required for the data obtained at 2xST 18 Plant as well as for the numerical results obtained in CFD environment. For this purpose there was designed and manufactured the gas fuel burner INCDT APC 1MGN - UPB (fig. 21, left) with a thermal power of approximately 350 kW, allowing the testing of only one ST 18 or ST 18-15 burning module and adaptable for other geometrical configurations in the same overall dimensions. The natural gas is introduced through a connector in the lower area and the air through a lateral flanged connector. The experiments for the ST 18 (or angled ST 18-15) module took place on the test bench of University Politehnica Bucharest (UPB), Department of Classic and Thermo-mechanical Nuclear Equipment (fig. 21, right). The tests have been made with natural gas and fresh air for different natural gas flow rates (0.5; 0.7 and 0.88 m^3/h). The experimental cell was a rectangular enclosure (890x890x990 mm) with a truncated pyramid segment connected to the exhaust stack. The walls are made of glass allowing the observation of the flame, one side door providing access (fig. 21, centre and right).

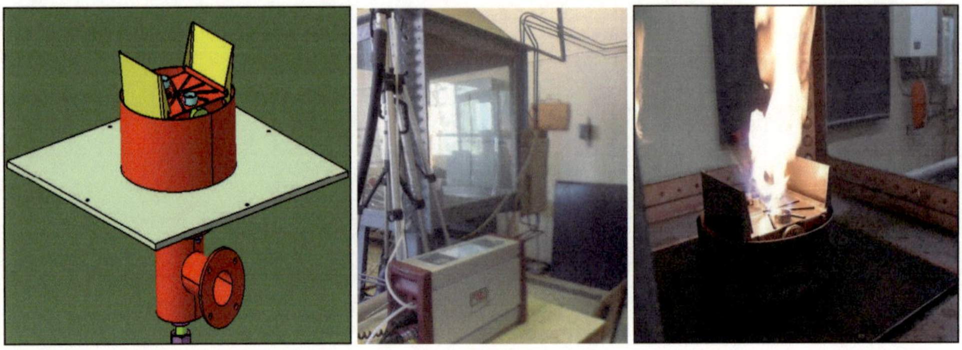

Fig. 21. Burner INCDT APC 1MGN - UPB with module ST 18-15, 3D design (left) and UPB test bench testing (centre and right)

The connection of the burner to the natural gas network was made through a hose and the one to the air fan through a removable flanged assembly. The placement of the burner was achieved through a plate fixed by screws. The emission measurements have been performed with a gas analyser MRU - Analyzer Vario Plus Ind. (fig. 21 centre) and the noise has been monitored with a sound level meter 01 dB Metravib SOLO mounted in the upper area of the enclosure. The outer superficial temperature profile has been determined with the help of a type Ti45FT Fluke infrared camera, the flame being sighted with the access door open (fig. 21, left). The measurements for emissions and noise at the three flow rates have been performed with the access door open. The temperature distribution in the flame has been determined with the help of three thermocouples (type PtRh30% - PtRh46%) placed on a holder on the height estimated for the flame development. The counting of the thermocouples starting from the base was: T_{FL1} , T_{FL2} and T_{FL3}. It was noticed a more homogenous distribution and a slower increase in temperature in the flame on ST 18-15 module, as seen in fig. 22. This also results from table 6 presenting parameters in the infrared recording - module ST 18 (left) and ST 18-15 (right). The areas occupied by the high temperature isotherms are significantly increased for module ST 18-15. Compared to module ST 18, the flame fills more the burning point, it shortens, the temperature distribution is more homogenous and the NO_x and CO emissions decrease. For the flow rate of natural gas of 0.88 m³/h a decrease in the NO_x emissions occurs, over 30 % for module ST 18-15 compared to ST 18 (fig. 23). However the values of noise for the module ST 18-15 are higher than for ST 18 due to increased turbulence (fig. 24). This phenomenon may be better observed particularly for high flow rates (0.88 m³/h). The researches conducted until present have shown the superiority of module ST 18-15, the numerical results being validated by the experimental ones obtained on the test bench.

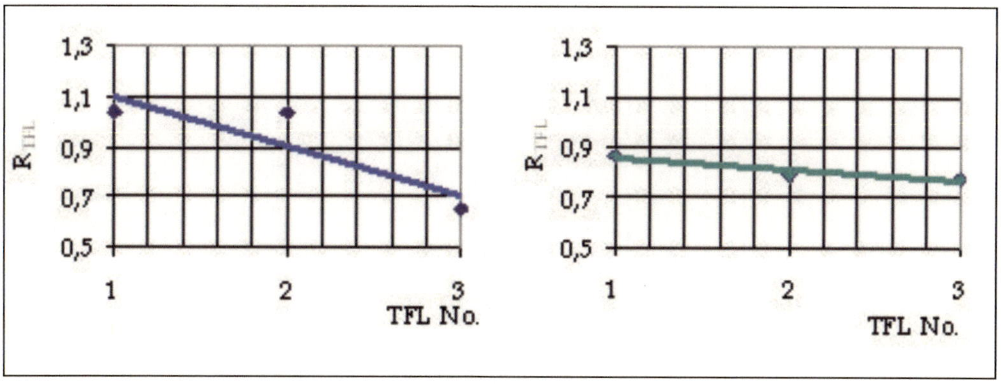

Fig. 22. Variation of the temperature ratios in the flame (for T_{FL1} , T_{FL2} and T_{FL3}), corresponding to modules ST 18-15 and ST 18 at Q_{combs} = 0.5 m³/h (left) and Q_{combs} = 0.88 m³/h (right)

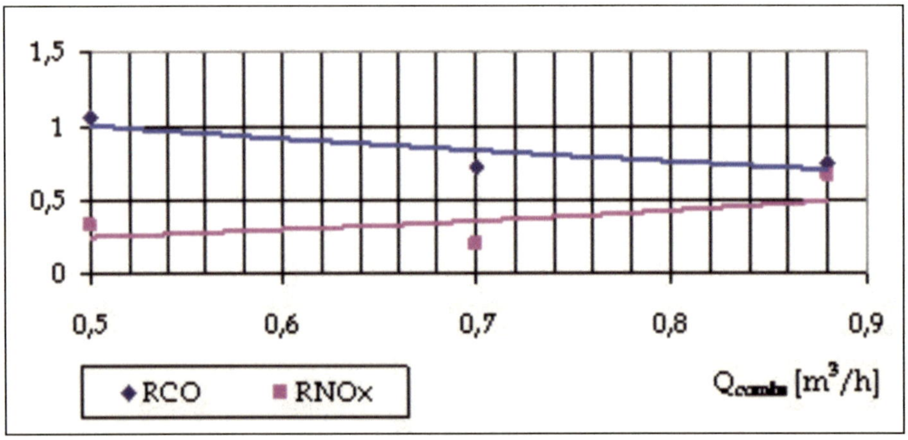

Fig. 23. The CO and NO_x ratios variation corresponding to ST 18-15 and ST 18 modules depending on natural gas volume flow

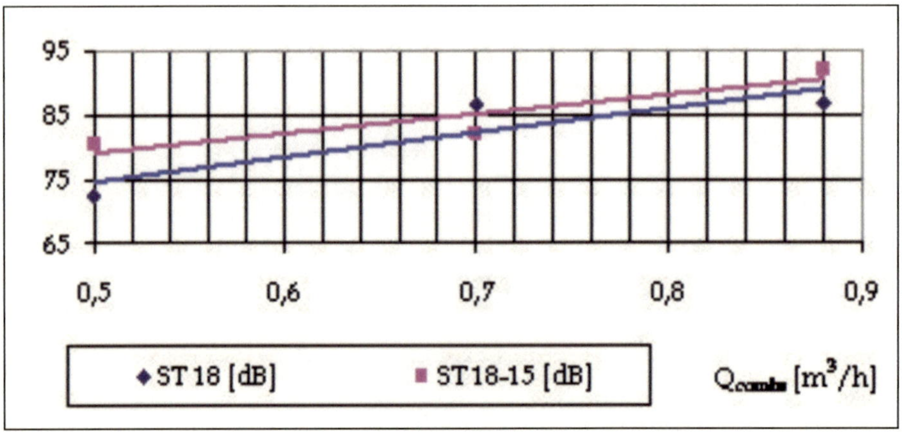

Fig. 24. Noise variation corresponding to ST 18-15 and ST 18 modules, depending on natural gas volume flow

No.	Natural gas flow rate Q_{combs} [m^3/h]	Isotherms (according to the burning module type)	
		Module ST 18	Module ST 18-15
1	0,5	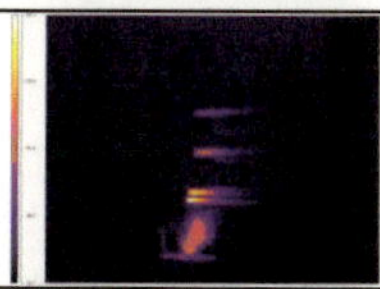	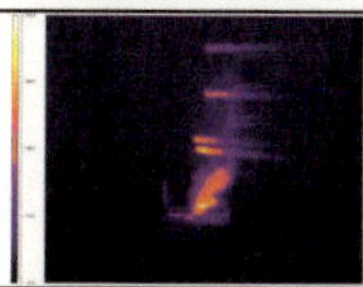
2	0,7	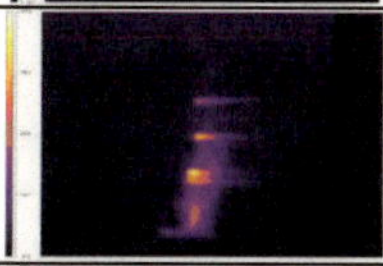	
3	0,88	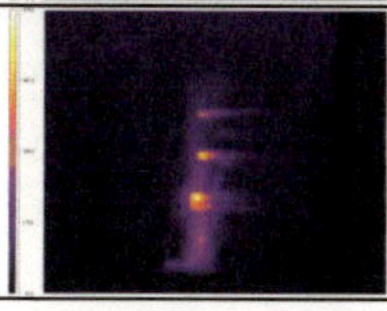	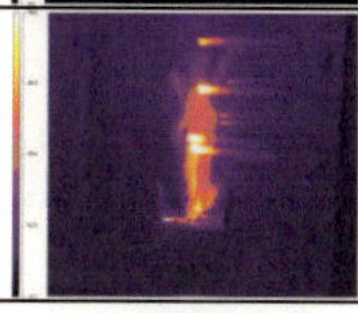

Table 6. Isotherms at the infrared recording for modules ST 18 (left) and ST 18-15 (right) depending on the natural gas flow rate

7. Conclusions

The new generation of afterburning installations will need to respond to the performance requirements of the „smart" aggregates to automaticaly consider emissions, energetic efficiency and process requirements. The researches conducted at Suplacu de Barcau 2xST 18 Cogenerative Power Plant have analyzed the afterburning installation as integrated in the cogenerative group. Based on the measurements and the numerical results, the burning modules have been redesigned, experiments on the test bench have been conducted in order to establish the performances of the new generation of modular burners and the working regimes of the gas turbine have been established for water injection and afterburning cases. For the independent working of the TA-2 gas turbine, the numerical simulations had shown the possibility of 50% decrease in the NO_x emissions. However the modelling of the assembly TA-2 gas turbine with water injection and afterburning has shown that the NO_x emissions decrease (at the working regime defined by T_m = 1063 K) is possible only to 40 ppm (below 30 %). This confirms the necessity of a fine control of the quantity of water injected in the gas turbine particularly when a significant decrease in NO_x emissions is aimed. Future researches will involve test bench experimentations of the gas turbine working on natural gas with water injection, coupled with the multi-module afterburning installation. The experimental data should validate the elaborated numerical model and will constitute the design input data for a new afterburning installation.

8. Acknowledgments

The researches conducted at Suplacu de Barcau 2xST 18 Cogenerative Power Plant have been performed based on contracts 22-108/2008 and 21-056/2007 (Programme "Partnerships in priority fields") financed by Romanian Ministry of Education, Research, Youth and Sports. The consortium involved in the projects includes several Romanian companies: National Research and Development Institute for Gas Turbines COMOTI -

Bucharest, SC OMV PETROM SA, UPB CCT - Bucharest, SC OVM ICCPET SA - Bucharest, SC ICEMENERG SA - Bucharest, SC ERG SRL - Cluj, SC TERMOCAD SRL - Cluj.

9. References

Barbu, E., Rosu, I. & Turcu G. (2006). Supraincalzitorul cazanelor de la centrala cogenerativa 2xST 18 - Suplacu de Barcau, *ETCNEUR - 2006*, pp. 9-12, ISBN 973-7984-49-8, Bucuresti, 6-7 iulie 2006

Barbu, E., Ionescu, S., Vilag, V., Vilcu, C., Popescu, J., Ionescu, A., Petcu, R., Prisecaru, T., Pop E. & Toma T. (2010). Integrated analysis of afterburning in a gas turbine cogenerative power plant on gaseous fuel, *WSEAS Transaction on Environment and Development*, Vol. 6,. Issue 6, June 2010, pp. 405-416, ISSN 1790-5079

Barbu, E., Fetea, Gh., Petcu, R. & Vataman, I. (2010). *Arzator de postardere multimodular pe combustibil gazos*, Dosar OSIM nr. A/00999 din 21.10.2010

Benini, E., Pandolfo, S. & Zoppellari, S. (2009). Reduction of NO emissions in a turbojet combustor by direct water/steam injection: numerical and experimental assessment, *Applied Thermal Engineering*, doi: 10.1016/j.applthermaleng.2009.06.004

Carlanescu, C., Ursescu, D. & Manea, I. (1997). *Turbomotoare de aviatie - Aplicatii industriale*, Editura Didactica si Pedagogica, Bucuresti

Carlanescu, C., Manea, I., Ion, C. & Sterie, St. (1998). *Turbomotoare - Fenomenologia producerii si controlul noxelor*, Editura Academiei Tehnice Militare, Bucuresti

Conroy, J. (2003). Improving duct burner performance through maintenance and inspection, *Energy-Tech*, February 2003, http://www.energy-tech.com/article.cfm?id=17543

Ganapathy, V. (2001). Superheaters: design and performance, *Hydrocarbon processing*, July 2001, http://www.angelfire.com/md3/vganapathy/superhtr.pdf

McBride, B.J. & Gordon, S., (1992) *Computer Program for Calculating and Fitting Thermodynamic Functions*, NASA Lewis Research Center, NASA RP-1271, Cleveland, USA, http://www.grc.nasa.gov/WWW/CEAWeb/RP-1271.pdf

Neaga, C. (2005). *Tratat de generatoare de abur*. vol. III, Editura Printech, Bucuresti, ISBN 973-718-262-6

Petcu, R., (2010). *Contributii teoretice si experimentale la utilizarea gazului de depozit ca sursa de energie*, Teza de doctorat - Decizie Senat nr. 100/12.02.2010, Bucuresti

Popescu, J., Vilag V., Barbu, E., Silivestru, V. & Stanciu, V. (2009). Estimation and reduction of pollutant level on methane combustion in gas turbines, *Proceedings of the „3rd WSEAS International Conference on Waste Management, Water Pollution, Air Pollution, Indoor Climate - WWAI'09"*, pp. 447-451, ISBN 978-960-474-093-2, University of La Laguna, Tenerife, Canary Islands, Spain, July 1-3, 2009

Public Interest Energy Research, [PIER]. (2002). *Active control for reducing the formation of nitrogen oxides in industrial gas burners and stationary gas turbines*, California Energy Commission, http://www.energy.ca.gov/reports/2002-01-10_600-00-009.PDF

Zehe, M.J., Gordon, S. & McBride, B.J. (2002), *CAP: A Computer Code for Generating Tabular Thermodynamic Functions from NASA Lewis Coefficients*, NASA Glenn Research Center, NASA TP—2001-210959-REV1, Cleveland, Ohio, USA, http://www.grc.nasa.gov/WWW/CEAWeb/TP-2001-210959-REV1.pdf

8

Application of Statistical Methods for Gas Turbine Plant Operation Monitoring

Li Pan
Queen's University Belfast
U.K.

1. Introduction

Within a large modern combine cycle gas turbine (CCGT) power station, it is typical for thousands of process signals to be continually recorded and archived. This data may contain valuable information about plant operations. However, the large volume of data accompanied with inconsistencies within the data often limits the ability to identify useful information about the process. Utilising data mining techniques, such as Principal Component Analysis (PCA) and Partial Least Squares (PLS), it is possible to create a reduced order statistical model representing normal plant conditions. Such a model can then be utilised for fault identification and identifying possible improvements in key performance indicators such as thermal efficiency. Moreover, the gas turbine performance can be affected by changes in ambient conditions. A long term nonlinear PLS techniques can be applied here to investigate the seasonal changes in gas turbine.

In this chapter, an approach to establish a long term statistical model for gas turbine will be given, and the application of the model in fault detection and performance analysis will be demonstrated.

2. The data mining techniques

Within the power station, data are continuously collected and archived representing thousands of data points including temperatures, steam flow rates, pressures, etc. Potentially this data may contain valuable information about unit operation. However, collecting a large amount of data does not always equate to a large amount of information, leading to a lot of databases being regarded as data rich, but information poor. The task of extracting information from data is known as data mining, which is defined as the non-trivial process of identifying valid, novel, potentially useful, and ultimately understandable patterns in data. Data mining is the nontrivial process of extracting valid, previously unknown, comprehensible and useful information from large databases (Weiss and Indurkhya, 1998). Also, data mining is a generic term for a wide range of techniques which include intuitive, easily understood methods such as data visualisation to complex mathematical techniques based around neural networks and fuzzy logic (Wang, 1999; Olaru and Wehenkel, 1999). Applications are found within diverse areas such as marketing (Humby et al., 2003), finance (Blanco et al., 2002) and industrial process control (Martin et al., 1996). However, despite being a widely applied technique, it is reported that three

quarters of all companies who attempt data mining projects fail to produce worthwhile results (Matthews, 1997). Unfortunately, this indicates that the potential of data mining techniques, with regard to the available data is often overestimated than the reality.
The act of data mining is itself part of a larger process known as knowledge discovery in data, KDD, which encompasses not only the analysis of data, but the gathering and preparation of data and the interpretation of results. Extracting knowledge from large data sets can be achieved through exploratory data analysis to discover useful patterns in data, in the form of relationships between variables.
Many techniques are applied as classification tools, to categorise new data following the analysis of a historical data set. In this chapter, the first method discussed in Section 2.1 is machine learning techniques which use a logical induction process to categorise a series of examples, resulting in decision tree and rules set which can be implemented in decision making processes. Typical application areas are fault diagnosis in industrial machines (Michalski et a1., 1999) and the assessment of power system security (Voumvoulakis, 2010)
Case based reasoning methods as discussed in Section 2.2, are commonly applied to decision making tasks where previous experience is desirable, but may not be available. Case based reasoning provides an inexperienced user with exposure to experiences from others, through a set of historical 'cases', and has been of particular use in areas such as fault diagnosis (Wang et al. , 2008; Yan et al. , 2007) and system design and planning (Hinkle and Toomey, 1995).
Finally, Section 2.3 discusses multivariate statistical techniques, namely principal component analysis and partial least squares regression, which have been successfully applied to a range of applications areas including chemistry (Wold et a1., 1987), manufacturing (Oliveira-Esquerre et a1., 2004), and power system analysis (Prasad et a1., 2007). It is also extended to finance (Blanco et a1, 2002) and medicine (Chan et a1, 2003) area. Principal component analysis and partial least squares regression are particularly popular in the area of chemometrics, where they are employed in the monitoring of processes which generate large and highly correlated data sets (Yoon and MacGregor, 2001; Kourti et al., 1996).

2.1 Machine learning

Machine learning techniques are those that use logical or binary operations to 'learn' a task from a series of examples, such as symptoms of medical or technical problems, leading to the diagnosis of the problem through the use of decision trees and rule sets which classify data using a sequence of 1ogical steps (Michalski et a1., 1999).
Decision trees are simple top down learning structures, which use Boolean classifiers to 'grow' a tree through recursive partitioning of the sample data using the available attributes. The development of a decision tree starts with the inclusion of all the training data in a root node, resulting in both correctly and incorrectly classified data. In order to 'grow' the tree, the data is recursively split by each attribute until all the attributes in the data have been used. Each node in the final tree, known as a leaf, represents a test on one of the attributes, and the branches from the node are labelled with the Boolean outcomes of the test (Quinlan, 1996).
The basic algorithm of building a decision tree, also known as ID3/C4.5 algorithm, follows a down rule (Quinlan, 1993). In the beginning, all the data is collected in the root node, and the data is recursively subdivided into fewer branches by assessing the information gain of

each attribute in the training data to split the data further, until the terminal node which only contains one attribute is obtained (Quinlan, 1993).

Rule induction is achieved using a bottom up structure, starting with a rule that specifies a value for every available attribute on the decision tree, thereby making the rule as specific as possible. This rule is known as the seed and further rules are developed from it by successfully removing attributes one at a time, until more general rules are acquired. Any rule which includes a counter example is regarded as incorrect and is therefore discarded from the process. The rule learning terminates by saving a set of'shortest'rules. Also a new "RBDT-1" algorithm is devolved for learning a decision tree from a set of decision rules that cover the data instances rather than from the data instances themselves. The method's goal is to create on-demand a short and accurate decision tree from a stable or dynamically changing set of rules (Abdelhalim, A. 2009).

The primary advantage of decision trees is that their simplicity makes them very intuitive to users. However, large data sets call result in vast trees which can be 'needlessly' complex resulting in a largely unusable knowledge base : the ideal tree is as small and linear as possible. Due to their simple nature, decision trees are not suitable for more complex data structures and this is demonstrated by trees that, after pruning, still remain too large to be comprehensible.

2.2 Case based reasoning

Case based learning acquires knowledge from solutions to prior problems and employs it to derive solutions to the current problems. Once a current problem occurs, the similar case and previous solution are retrieved and possibly revised to better fit the current problem. The new solution can be retained into the case base in case to solve future problems. As a result, case based reasoning (CBR) systems are effectively used as lookup tables where 'the system' interrogates an indexed database of relevant cases, and one or more similar cases are retrieved and applied to discover an appropriate solution (Watson, 1999).

A significant issue in CBR is indexing, which limits the search space, thereby reducing case retrieval times. There are many methods for indexing, such as check list based indexing, which identifies predictive features for a case (inductive learning methods may be used) and places them on a list which is then used for indexing, and difference based indexing which selects features as indices that best differentiate one case from another. The user can also manually implement an indexing system, and it has been suggested that selection of indices by the user can be more effective than algorithms for practical applications (Kolodner, 1993). The indexed cases can then either be stored sequentially, making the system easy to maintain but slow to query for larger case sets, or using a hierarchical structure, which will organize cases so that only a small subset are considered during retrieval, thereby reducing search times (Smyth et a1., 2001).

CBR is a self-maintaining system, the database of historical events is updated when new cases occurred and adding to the system's problem solving resources. The advantage of CBR is that it does not require a large number of historical data patterns to achieve satisfactory levels of performance : a CBR model may be created from a small number of cases and the case base can be refined over time (Hinkle and Toomey, 1995). CBR is particularly useful when studying data which has complex internal structures when there is little domain knowledge, enabling the sharing of experience.

Despite these benefits, CBR can be unsuitable for large scale applications as retrieval algorithms are inefficient when faced with handling thousands of cases. Maintenance of the case base, with respect to adding new cases and the removal of out date cases, may also be a problem as it is largely left to human intervention (Watson and Marir, 1994).

2.3 Multivariate statistical techniques

Statistical methods are employed to analyse the relationships between individual points in a data set, determining characteristics such as the average value and distribution of the data. The simple statistical measure represents a univariate approach to data analysis, which lacks the ability to constructively analyse large, multivariate data sets as the interactions between variables are ignored (Martin et a1., 1996). In contrast, multivariate statistical analysis describes methods capable of observation and analysis of the multiple variables required for system monitoring (Kourti and MacGregor, 1995). This section discusses the multivariate techniques, principal component analysis (PCA), least squares regression and partial least squares (PLS), as they are more suitable to the analysis of large data sets than univariate methods.

Principal component analysis (PCA) is a statistical technique useful for identifying underlying systematic structures in data and separating it from noise (Wold et al., 1987). The identification of patterns in data structures allows PCA to be applied to problems requiring a reduction in the dimensionality of a data set, for example image processing(Bharati et a1., 2003), or monitoring of industrial processes including chemical and microelectronics manufacturing processes (Wise and Ricker, 1991 ; MacGregor and Koutodia, 1995). These objectives are achieved by transforming variables, which are assumed to be correlated, into a smaller number of uncorrelated variables called principal components (PCs), providing a simpler description of the data structure. Each successive PC accounts for the most significant variability in the data in a particular direction, with the reduction in dimensionality achieved, the original data set can be represented by few PCs.

PCA is a useful tool when large data sets containing highly correlated variables are to be managed. PCA achieves reductions in data dimensionality, thereby simplifying future observation of variables : plotting a few PCs is significantly more convenient than plotting all original variables. Furthermore, the comparison capabilities between the historical information used to construct the model and newly presented data is a desirable characteristic for system monitoring applications (Martin et a1., 1996). Fault identification can also be undertaken by analyzing the contribution of the independent variables to each PC (MacGregor et a1., 1994).

Projection to latent structures, also known as partial least squares (PLS), is developed to solve the multi-collinearity problem in linear least squares regression (LSR) which can determine the best linear approximation for a set of data points (Freund and Willson, 1997).

The benefit of PLS is achieved by identifying a set of uncorrelated, latent variables. This avoids the co-1inearity problems encountered by likelihood-ratio (LLR) test and utilises some of the techniques associated with PCA, with the new latent vectors composed of scores and orthogonal loading vectors. PLS regression is a robust, multivariate linear regression technique which is considered to be more suitable for the analysis and modelling of noisy and highly correlated data than LLR as parameters do not exhibit large variation when new data samples are included (Otto and Wegscheider, 1985). A high number of variables, with respect to the number of data samples, are also permissible in PLS, which can result in the modelling of noise for LLR(wise and Gallagher, 1996).

In summary, PLS is capable of producing robust, effective models, despite operational data limitations, for example, imprecise measurements and missing data (Oliveira-Esquerre,

2004). The ability to predict dependent data values, especially in the case of product quality data, which is often measured infrequently, is useful in process monitoring (MacGregor, 2005). The proficiency of PLS dealing with the highly correlated and collinear data is also frequently utilized in its application to process monitoring (Kresta, 1994).

2.4 Method selection

A selection of data mining techniques has been presented in this section and the characteristics of each shall now be considered with respect to the problems and available data presented by power plant monitoring and analysis.

Data derived from power plant monitoring can potentially consist of thousands of sensor measurements generated on a second by second basis. The data recorded is highly correlated, due to multiple sensors, which are in place to introduce redundancy into the measurements, and parallel paths within the system, for example, the steam and gas circuits. Data quality is also a factor, with noisy signals and missing data as the common problems. The correlated structure and data quality considerations present in power plant monitoring records bear strong similarities to other application areas discussed in this section, such as chemical process control(Ma et al., 2009; Ahvenlampi and Kortela, 2005), manufacturing (Oliveira-Esquerre et al., 2004; Baharati et a1., 2004; Hisham et al., 2008) and medicine (Chan et a1., 2003).

When monitoring a process which records a vast array of sensor data, individual analysis of each signal by a human operator is clearly not possible. The data analysis techniques discussed in this section can be applied to this problem as they identify essential correlations within the data. This simplifies the monitoring process by identifying the most relevant process signals and thereby reducing the search space.

When undertaking system monitoring, there are three main objectives. The first is the detection of a change in process operation and its nature - should this be a sensor fault, faults within the process or a change in product quality or system performance. Once a change in process behaviour is detected, diagnostic tests are then required to identify the cause of the change, which may require analysis of recent process data and/or consultation with an experienced operator. Finally, with the source of the change ascertained, a solution should then be identified and implementation of corrective action undertaken, if appropriate (Cinar and Undey, 1999). This may require either disabling the source of the problem, or in the event of a faulty sensor reading, reconstruction of the value.

The methods presented here provide different solutions to the process monitoring problem. Clustering, machine learning and CBR are diagnostic tools, which compare current fault conditions to historical examples of faults. While these techniques will identify the nature of the problem, they are not capable of detecting its occurrence or providing signal reconstruction in the event of a sensor fault. Statistical methods offer a model based approach, where process operation data is compared to a system model, based on 'normal' operating conditions. This provides continuous process monitoring, which can supply early warning of a small process change and offers operators the opportunity to take action to prevent the fault becoming more serious. In particular, PCA and PLS can supply the operator with information as to which process variables are outside normal limits (MacGregor et a1., 1994). This serves to focus the operators' attention on the problem area, allowing process knowledge to be used to identify the source of the problem.

Not all of the techniques discussed in this section are capable of providing a solution that enables preventative action to be taken for CCGT power plant.

CBR, providing a historical event that is similar to the currently observed event, is also of little use in this instance, as it again offers classification of new events, however, an explanation of the similarities identified between cases is unavailable.

Cluster analysis and 'rules' may be useful in identifying groups of similar events, as their aim is to suggest correlations in data. Once similarities have been identified for events resulting in groupings with common characteristics further investigation would be required to identify the relationships between variables that cause these groupings. A more complete solution is offered by statistical methods.

Both PCA and PLS are capable of identifying correlations within data, while PLS also offers the ability to extend this to identifying the correlations which are predictive of a dependent quantity. The correlations identified within the data can then be studied using the scores and loading vectors obtained, indicating the contribution of variables, if any, to the variation of the dependent parameter. The historical data is suitable for the development of a system model, by means of PCA and PLS, which can be applied to continuous monitoring.

Archived data is available, detailing sensor data, for example temperatures, pressures, etc, throughout the plant at regular intervals. Once the PCA and PLS models are developed, it can provide a relatively straight forward model which has both the ability for online fault monitoring and offline performance analysis. For practical application, those PCA and PLS models are required for fast online response within 20 seconds, and reasonable prediction accuracy in a wide operation range.

With PCA and PLS identified as possessing properties that are useful in relation to the problems posed by power plant and power system operation, the statistical modelling methods will provide the most suitable approach for operation monitoring and performance analysis of CCGT power station.

3. PCA and PLS algorithm

Give an original data matrix $\mathbf{X}$ $(m \times n)$ formed from m samples of n sensors, and subsequently normalised to zero mean and unit variance, can be decomposed as follows:

$$\mathbf{X} = \mathbf{T}\mathbf{P}^T + \mathbf{E} = t_1 p_1^T + t_2 p_2^T + \ldots + \mathbf{E} = \mathbf{PC}_1 + \mathbf{PC}_2 + \ldots + \mathbf{E} \quad (1)$$

where $\mathbf{T} \in R^{m \times A}$ and $\mathbf{P} \in R^{n \times A}$ are the principal component score and loading matrices, E is the residual matrix (Lewin, 1995).

The principal component matrices can be obtained by calculating eigenvectors of original data. Following the creation of the correlation matrix of original data, the corresponding eigenvalues and eigenvectors are calculated, where an eigenvalue is an eigenvector's scaling factor. As the eigenvectors with the largest eigenvalues correspond to the dimensions than have the strongest correlation in the dataset, the data can then be ordered by eigenvalue, highest to lowest to give the components in order of significance (Jolliffe, 2002). There are a number of methods available to determine the number of ordered PCs. A cross validation which calculates the predicted error sum of squares (PRESS) (Valle et a., 1999) is provide more reliable solutions than a simple scree test (Jackson, 1993).

Partial least square requires two block of data, an $\mathbf{X}$ block (input variables) and $\mathbf{Y}$ block (dependent variables). PLS attempts to provide an estimate of $\mathbf{Y}$ using the $\mathbf{X}$ data, in a similar manner to principal components analysis (PCA). If $\mathbf{T}$ and $\mathbf{U}$ represent the score matrixes for the $\mathbf{X}$ and $\mathbf{Y}$ blocks, and $\mathbf{P}$ and $\mathbf{Q}$ are the respective loadings, the decomposition equations can be presented as:-

$$\mathbf{X} = \mathbf{T}\mathbf{P}^T + \mathbf{E} \quad (2)$$

$$\mathbf{Y} = \mathbf{U}\mathbf{Q}^T + \mathbf{F} \quad (3)$$

where **E** and **F** are the residual matrices. If the relationship between **X** and **Y** is assumed to be linear then the residual matrices **E** and **F** will be sufficiently small, and the score matrices **T** and **U** can be linked by a diagonal matrix **B** such that:

$$\mathbf{U} = \mathbf{B}\mathbf{T} \quad (4)$$

Hence the predicted dependent variable can be translated (Flynn, 2003) as:

$$\hat{\mathbf{Y}} = \mathbf{B}\mathbf{T}\mathbf{Q}^T + \mathbf{E} + \mathbf{F} \quad (5)$$

4. Nonlinear modeling approach

As we discussed previously, PCA and PLS model are powerful linear regression techniques. However, in the real power generation industry, many processes are inherently nonlinear. When applying linear model to a nonlinear problem, the minor latent variables cannot always be discarded, since they may not only describe noise or negligible variance structures in the data, but may actually contain significant information about the nonlinearities. This indicates that the linear model may require too many components to be practicable for monitoring or analyzing the system.

Recognition of the nonlinearities can be achieved using intuitive methods, for example, which apply nonlinear transformations to the original variables or create an array of linear models spanning the whole operating range. More advanced methods have also been proposed including nonlinear extensions to PCA (Li et al. 2000), introducing nonlinear modifications to the relationship between the **X** and **Y** blocks in PLS (Baffi et al., 1999) or applying neural network, fuzzy logic, etc. methods to represent the nonlinear directly.

Transformation of the original variables using nonlinear functions can be introduced prior to a linear PCA and PLS model. For this purpose, the input matrix **X** is extended by including nonlinear combinations of the original variables. However, process knowledge and experience is required to intelligently select suitable nonlinear transformations, and those transforming functions must sufficiently reflect the underlying nonlinear relationships within the power plant. Another problem with this approach is the assumption that the original sets of variables are themselves independent. This is rarely true in practice, which can make the resulting output from the data mining exercise difficult to interpret.

An alternative and more structured approach is the kernel algorithm. The purpose of kernel algorithm is to transform the nonlinear input data set into a subspace with kernel function. In the kernel subspace, the nonlinear relationship between input variables can be transformed into linear relationship approximately. By optimising the coefficients of kernel function, the transformed data can be represented using a Gaussian distribution around linear fitting curve in the subspace. Furthermore, introducing neural network approaches into the kernel structure is generally seen to be more capable of providing an accurate representation of the relationship for each component (Sebzalli and Wang, 2001). In this area, the multilayer perceptron (MLP) networks are popular for many applications. However the initial model training is a nonlinear optimization problem, requiring conjugate gradient and Hessian-based methods to avoid difficulties arising from convergence on local minima. In order to solve this problem, a radial basis function (RBF) network has been selected over other approaches, due to its capability of universal approximation, strong power for input and output translating and better clustering function. A standard RBF network consists of a single-layer feedforward architecture, with the neurons in the hidden layer generating a set of basis functions which are then combined by a linear output neuron. Each basis function is centered at some point in the input space and its output is a function of the distance of the inputs to the centre. The function width should be selected carefully because each neuron should be viewed to approximate a small region of the input surface neighboring its centre. Therefore, the RBF network also has been named localized receptive field network. This localized receptive character implies a concept of distance, e.g. the RBF function is only activated when the input has closed to the RBF network receptive field. For this reason, the performance of RBF network is more dependent on the optimisation of RBF function coefficients rather than the type of function (Jiang et al., 2007).

In order to reduce the neural network dimension, the input data are firstly decomposed into few components, then the output can be reconstructed with nonlinear relationship. Hence, each component will possess its own nonlinear function $f_{non-linear}$, so that

$$\hat{\mathbf{X}} = f^{\mathrm{X}}_{non-linear}(\mathbf{T}) \tag{6}$$

$$\hat{\mathbf{Y}} = f^{\mathrm{Y}}_{non-linear}(\mathbf{T}) \tag{7}$$

In this research, radial basis functions have been selected to represent the non-linearities, since once the RBF centres and widths have been chosen, as described below, the remaining weights can be obtained using linear methods.

4.1 RBF network

The radial basis function network employed in this research is illustrated in Figure 1.

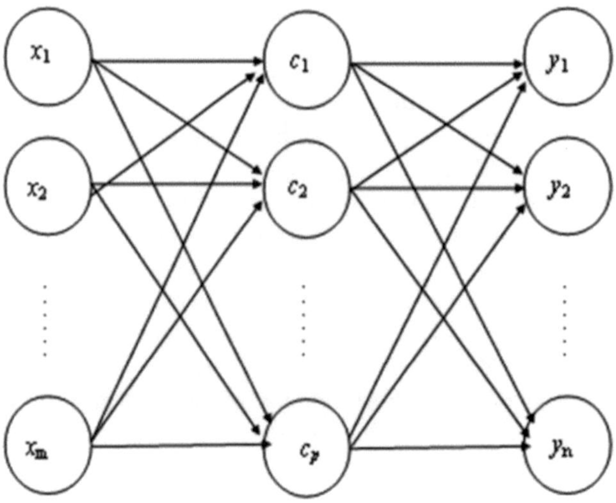

Fig. 1. Radial basis function network

The network topology consists of m inputs, p hidden nodes and n outputs, and the network output, y_i, can be formulated as:-

$$y_i = \sum_{j=1}^{p} w_j^{(i)} \theta_j(\|\mathbf{X} - \mathbf{c}_j\|) \quad i = 1, 2, \ldots n \tag{8}$$

where, $w_j^{(i)}$ are weighting coefficients, and θ_j is the basis function. In this research, a Gaussian base function was selected, which is defined as:-

$$\theta_j(\|\mathbf{X} - \mathbf{c}_j\|) = \exp\left[-\sum_{k=1}^{m}\left(\frac{x_k - \mathbf{c}_j}{\sigma_j}\right)^2\right], \; i = 1, 2, \ldots p \tag{9}$$

The Euclidean distance $\|\mathbf{X} - \mathbf{c}_j\|$ represents the distance between the input space **X** and each RBF centre $\mathbf{c}_j$, where **X** = [$\mathbf{x}_1$ $\mathbf{x}_2$ … $\mathbf{x}_m$], and σ_j is the width coefficient of each RBF node. The coefficient matrix [σ, c, w] is obtained off-line using a suitable training algorithm. Some of the more popular options are least mean squares (LMS) (Moody et al., 1989), orthogonal least squares (OLS) (Li et al,. 2006) and dual-OLS (Billing et al., 1998). These traditional algorithms often employ a gradient descent method, which tends to converge on local minima. In order to address the global optimisation problem, a recursive hybrid genetic algorithm (RHGA) (Li and Liu, 2002, Pan et al., 2007) is employed here to search for valid solutions.

4.2 The genetic algorithm

The typical genetic algorithm (GA) is based upon *survival of the fittest*, and the network framework [σ, c] is coded into the binary genes as illustrated in Table 1. The initial population are selected at random from the entire solution space, with the binary coding denoting whether the training samples are selected as the centers of the hidden neurons (Goldberg, 1989).

All the potential hidden centers	A randomly created gene code	Coded network framework
[σ_1, $\hat{c}_1$]	1	[σ_1, $\hat{c}_1$]
[σ_2, $\hat{c}_2$]	0	---
[σ_3, $\hat{c}_3$]	0	---
[σ_4, $\hat{c}_4$]	1	[σ_4, $\hat{c}_4$]
[σ_5, $\hat{c}_5$]	1	[σ_5, $\hat{c}_5$]

Table 1. Encoding scheme of genes

For each generation, random crossover and mutation is applied to the genes, leading to a new generation of network frameworks being obtained. The fitness, *f*, of the new population is determined using:-

$$\frac{1}{f} = \sum_{j=1}^{n} (\hat{y}_j - y_j)^2 \tag{10}$$

where, $\hat{y}_j$ is the jth RBF output and y_j is the actual value. The most recent framework will be retained if its fitness improves upon previous generations.
Although the genetic algorithm has the capability of wide region searching and efficient global optimizing, it is weak in some local point fitting. This may lead to a decrease in model accuracy. Therefore, the genetic and gradient descent algorithm can be combined in order to obtain both the global and localize optimizing capability (Pan, et al., 2007). In this hybrid algorithm, an initial optimized network can be obtained by the genetic algorithm, and then the structure of network can be further shaped for some specific points with the gradient descent algorithm. The next step is to examine the variate of fitness coefficient. If the fitness reached the preset bound then the regression will be completed, otherwise, the network will be reconstructed for next generation optimisation, and repeat the gradient descent regression, until reach the preset number of generations or meet the request fitness.

5. The auxiliary methods

Once a PCA/PLS model for normal operating conditions has been developed, the real time online DCS data then can be applied into the model to obtain a reconstruction of input data. It can be used to determine whether recorded plant measurements are consistent with historical values and neighboring sensors. A comparison can then be made between the reconstructed value for each variable and the actual measurements. Performed manually this can be a time consuming task. In this section, some efficient auxiliary methods will be discussed for the quality control, sample distribution analysis and fault identification.

5.1 Quality control method

There are two approaches that can quickly help to identify differences between the actual and reconstructed value of a variable, which are the squared prediction error (SPE) and Hotelling's T^2 test.
The SPE value, also know as the distance to the model, is obtained by calculating a reconstruction of each variable, $\hat{x}_i$, from the model, and then comparing it with the actual value, x_i. The SPE for all variables in each data sample can be calculated as

$$\mathbf{SPE} = \sum_{i=1}^{n}(x_i - \hat{x}_i)^2 \tag{11}$$

In order to distinguish between normal and high values of SPE, a confidence limit, known as the Q statistic test is available, which can be determined for α percentile confidence as:

$$\mathbf{Q}_\alpha = \theta_1\left(\frac{c_\alpha\sqrt{2\theta_2 h_0^2}}{\theta_1} + 1 + \frac{\theta_2 h_0(h_0 - 1)}{\theta_1^2}\right)^{1/h0} \tag{12}$$

where c_α is the confidence coefficient for the 1- α percentile of a Gaussian distribution, θ_i is the sum of unused eigenvalues to the ith power and h_0 is a combination of θ as outlined below:

$$h_0 = 1 - \frac{2\theta_1\theta_3}{3\theta_2^2} \tag{13}$$

The T^2 statistic test is designed as a multivariate counterpart to the student's t statistic. This test is a measure of the variation within normal operating conditions. With Tracy- Widom distribution, the T^2 test can be extended to detect peculiar points in the PCA model (Tracy et al., 1993).

Given h components in use, t_i is the ith component score and s_i is its covariance, then the T^2 can be defined as

$$\mathbf{T}^2 = \sum_{i=1}^{h} \frac{t_i^2}{s_i^2} \tag{14}$$

As with SPE, an upper control limit, T_α^2 can be calculated with n training data. This relates the degrees of freedom in the model to the F distribution,

$$\mathbf{T}_\alpha^2 = \frac{h(n^2-1)}{n^2(n-h)} \mathbf{F}_{1-\alpha}(h, n-h) \tag{15}$$

It should be noted that a rise in the SPE or T^2 value does not always indicate a fault, it also may be caused by the process is moving to a new event which is not accounted in the training data. Additionally, both indicators are affected by noise on the system and deviation of measurements from a normal distribution. This can result in nuisance values for both SPE and T^2. However, false alarms can be largely eliminated by simple filtering, and adjustment of the associated threshold (Qin et al., 1997).

5.2 Sample distribution

Both the SPE and T^2 are unlikely to differentiate between a failing sensor and a fault on the power plant. In this case, a plotting of t scores can be combined with the previous methods to distinguish between the two conditions.

The PCA model gives a reduction of data dimension with minimum information less. Therefore, the original m dimension data can be plotted in a plane coordinated by the first two components, and the relative position between each data point is remained the same as the original m dimension space. This character gives a capability to directly observe the similar distribution structure of original sample data, in a 2-dimension plane.

Especially, quoting the T^2 control limit into the 2-dimension plane, we have

$$\mathbf{T}_{2-D}^2 \leq \mathbf{T}_{\alpha\, 2-D}^2 \tag{16}$$

substituting Eq. (14) and (15), the Eq. (16) can be transformed as

$$\left(\frac{t_1^2}{s_1^2} + \frac{t_2^2}{s_2^2} \right) \leq \frac{2(n^2-1)}{n^2(n-2)} \mathbf{F}_{1-\alpha}(2, n-2) \tag{17}$$

Define

$$c = \frac{2(n^2-1)}{n^2(n-2)} \mathbf{F}_{1-\alpha}(2, n-2) \tag{18}$$

then it gives that

$$\frac{t_1^2}{s_1^2}+\frac{t_2^2}{s_2^2}\le c \tag{19}$$

Eq. (19) defines a control ellipse for *t*-score plotting. Score for normal operating conditions should fall within this ellipse. So when a process fault occurs, the individual points on the *t* score plots may be observed drifting away from the normal range into a separate cluster. The relative position of these fault clusters can assist in latter diagnosis.

5.3 Fault orientation

Having confirmed that there is a sensor fault, and not a process condition, the next step is to identify which sensor is failing. If a signal is faulty, a significant reduction in SPE before and after reconstruction would be expected. However, in practice the reduction in SPE can affect all inputs, making the faulty sensor unidentifiable. This situation arises due to a lack of redundancy, or degrees of freedom, among the measurements.

The above difficulties can be overcome by calculating a sensor validity index (SVI) (Dunia et al, 1996). This indicator is determining the contribution of each variable to the SPE value. The SPE value should be significantly reduced by using the reconstruction to replace the faulty input variable. If an adjusted data set z_i represents a input set with the x_i variable being replaced by reconstructed data $\hat{x}_i$, and the adjusted model predicted value being $\hat{z}_i$, then the sensor validity index for i^{th} sensor η_i can be defined as

$$\eta_i^2=\frac{(z_i-\hat{z}_i)^2}{\mathbf{SPE}} \tag{20}$$

The SVI is determined for each variable, with a value between 0 and 1 regardless of the number of samples, variables, etc. The value of SVI close to unity is indicative of a normal signal, while a value approaching zero signifies a fault. It is assumed that a single sensor has failed, and the remaining signals are used for reconstruction. Also, system transients and measurement noise can lead to oscillations in SVI, and possibility of false triggering. Consequently, each signal should be filtered and compared with a user-defined threshold.

6. Application of PCA and PLS model

As these power plants operate in a competitive market place, achieving optimum plant performance is essential. The first task in improving plant operation is the enhancement of power plant operating range. This power plant availability is a function of the frequency of system faults and the associated downtime required for their repair (Lindsley, 2000). As such, availability can be improved through monitoring of the system, enabling early detection of faults. This therefore allows the system working at non-rated conditions, corrective actions, or efficient scheduling of system downtime for maintenance (Armor, 2003).

Monitoring of power plant operations is clearly an important task both in terms of identifying equipment faults, pipe leaks, etc. within the generating units or confirming sensor failures, control saturation, etc. At a higher level, issues surrounding thermal efficiency and emissions production for each generating unit, as measures of plant performance, and the seasonal influence of ambient conditions will also be of interest. Fortunately, the frequency of measurement and distribution of sensors throughout a power

station provides a great deal of redundancy which can be exploited for both fault identification and performance monitoring (Flynn et al., 2006). However, modern distributed control systems (DCSs) have the ability to monitor tens of thousands of process signals in real time, such that the volume of data collected can often obscure any information or patterns hidden within.
Physical or empirical mathematical models can be developed to describe the properties of individual processes. However, there is an assumption that faults are known and have been incorporated into the model. This can be a time-consuming exercise and requires the designer to have extensive knowledge of the application in question (Yoon and MacGregor, 2000). Alternatively, data mining is a generic term for a wide variety of techniques which aim to identify novel, potentially useful and ultimately understandable patterns in data. The most successful applications have been in the fields of scientific research and industrial process monitoring, e.g. chemical engineering and chemometrics (Ruiz-Jimenez et al., 2004), industrial process control (Sebzalli et al., 2000) and power system applications such as fault protection in transmission networks (Vazquez-Martinez, 2003). In the following sections it will be shown how using the principal component analysis (PCA) technique. It is possible to exploit data redundancy for fault detection and signal replacement, as applied to monitoring of a combined cycle gas turbine.
Furthermore, the archived data is used to assess system performance with respect to emissions and thermal efficiency using a partial least square (PLS) technique.

6.1 Raw data pre-process

The PCA and PLS models are trained using historical data to suit the 'normal' plant operating, and the training data have to be selected carefully to avoid failing and over range data from normal power plant operation. The normal power plant operation was defined around the typical output range of 60 MW - 106 MW for single shaft unit and 300 MW - 500 MW for multi-shaft unit. There are severe dynamic conditions existing in the starting up and shutting down period. Therefore, those periods has to be removed from raw data archives. An instance is illustrated in Figure 2, for a single shaft unit operation, approximately one hour operating data was removed after and before system shut down and start up, in order to avoid the transient process.
The DCS normally collects sensor data every second, however, due to the power plant parameters are mainly consisted by temperature and pressure signals, the typical power plant responding time is around minutes. Therefore, consider of the balance of computational complexity and information quality, the sampling interval was determined as 1 minute. Since the raw data sample was archived from DCS, it still contains lots of anomalous signals such as break down process, which the power out suddenly crash down. Noised signal, is a signal disturbed by white noise. And spike, is an instantaneous disturbance which can cause a far deviation from normal signal level. Those data must be pre-filtered before being employed to train a model.
It is generally recognized that CCGT performance, and in particular gas turbine performance, can be affected by changes in ambient conditions (Lalor and O'Malley, 2003). For example, a fall in barometric pressure causes a reduction in air density and hence inlet compressor air flow. Similarly, an increase in ambient temperature causes a reduction in air density and inlet compressor air flow. Since the turbine inlet temperature is maintained as a constant, there is a subsequent reduction in turbine inlet pressure and hence cycle efficiency.

Variations in other external variables such as relative air humidity and system frequency (affecting compressor rotational speed) can also impact on gas turbine performance. Therefore, the training data selection for a widely suitable PCA model has to contain the information of the seasonally changes of ambient condition.

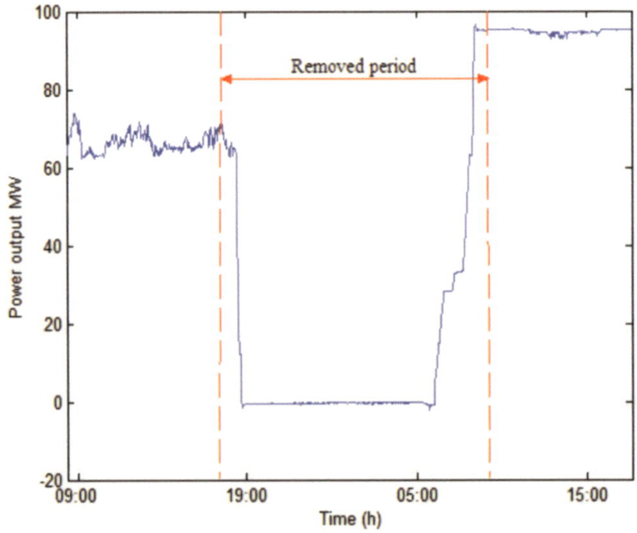

Fig. 2. Removed transient period

In order to obtain a entire seasonal model, the training data sorting process is designed to archive power plant operating data for years, then split all of the ambient variables into many small intervals, and pick up a sample data from each intervals to ensure that the training data contain the operating information for every ambient conditions.

6.2 Sensor data validation

With aging sensors, and the associated performance degradation, inevitable, faulty sensors are a relatively common occurrence in system monitoring. A common example of sensor failure is 'stuck at' signal, as illustrated in Figure 3 (a), which the fault is occurred at 300th data point. The following data is missed and the sensor's output is stuck at the last measurement. Another example is drifting signal, shown as Figure 3 (b), that the original data is disturbed by an increasing interference. Also, a biased signal is a constant noise which biased the sensor's data to other level, as shown in Figure 3 (c).

Univariate limits, i.e. upper and lower bounds are often applied to the detection of these faults. Problems such as biased sensors can be detected when the value eventually exceeds the predefined limits. However, a faulty signal within the univariate limits, such as a drifting sensor, will often go undetected for a long period of time. In order to identify such those faulty sensors, a multivariate approach is required, which will give consideration to the sensor value as part of wider plant operation.

Furthermore, if a sensor is faulty, an operator may choose to disable the sensor, but if the signal is used for feedback/feedforward control, disabling the sensor can only be part of the solution. In this instance, the problem can normally be resolved by signal reconstruction based upon sensor readings from neighboring sensors in the plant. This will require a system model, operating in parallel with the real plant.

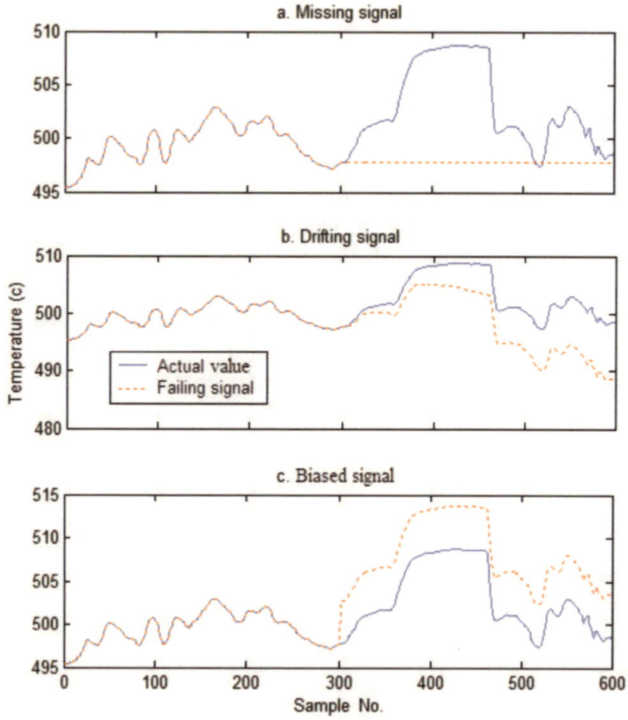

Fig. 3. Sensor faults

Principal component analysis, PCA, as a suitable technique for sensor monitoring and validation as it captures data variability for normal process operation. The development of a PCA model is intended to reduce the dimensionality of a set of related variables, while retaining as much of their variance as possible. This is achieved by identifying new, latent variables known as principal components, PCs, which are linearly independent. A reduced set of these latent variables are then used for process monitoring, with a small number of components normally sufficient to capture the majority of variability within the data.

Monitoring of a system using PCA is a modeling based approach, achieved by comparing observed power plant operation to that simulated by the model from available sensor data. The comparison between model and plant data, resulting in residuals, can then determine if the recorded information is consistent with historical operation and neighboring sensors. Faults are detected by observing deviations from normal operation, which can then be investigated to determine the exact source of the problem.

There are two common automated methods to compare recorded data with the model, as defined in section 5.1, the squared prediction error, SPE, and Hotelling's T2 test. Also, the sensor validity index, SVI, will identify failing sensors, and t score plots, from a cluster representing normal, fault free operation. All of those techniques are detailed in section 5. If an individual sensor is identified as being at fault, it can be replaced with a value reconstructed by the PCA model from other sensor data. However, if the fault is actually with the power plant, corrective maintenance or other necessary action should then be scheduled.

6.3 PCA model performance

In order to demonstrate the monitoring capabilities of this PCA model, a drift signal is introduced to the testing data set. As shown in Figure 4, the drift occurred in the sensor monitoring the steam temperature at 5:00 am. Generally, the lower bound of steam temperature is 500 °C during power plant normal operating period. Consequently, this drift can be detected by under limit indicator approximately 2 hours after the drift was introduced. In contrast to sensors limit indicator, the associated squared prediction error (SPE) monitoring test is illustrated in Figure 5 and shows that the SPE test detects the sensor fault 30 minutes after the introduction of the drift with 95% confidence limit, and 45 minutes with 99% confidence limit. Similarly, the T-squared test detected this sensor fails within 35 minutes using 95% confidence limit, and it crossed the 99% threshold 10 minutes later, as shown in Figure 6. The earlier SPE and T-square fault identification can provide more time for the power plant operator to take actions to solve problems.

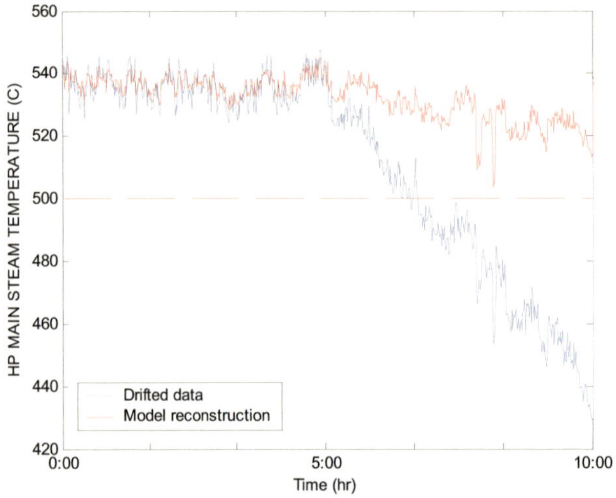

Fig. 4. Sensor drifts for single shaft CCGT unit

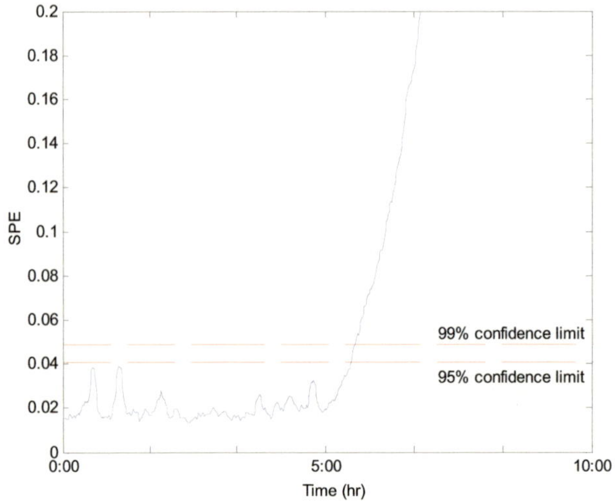

Fig. 5. SPE test for sensor drift in single shaft CCGT unit

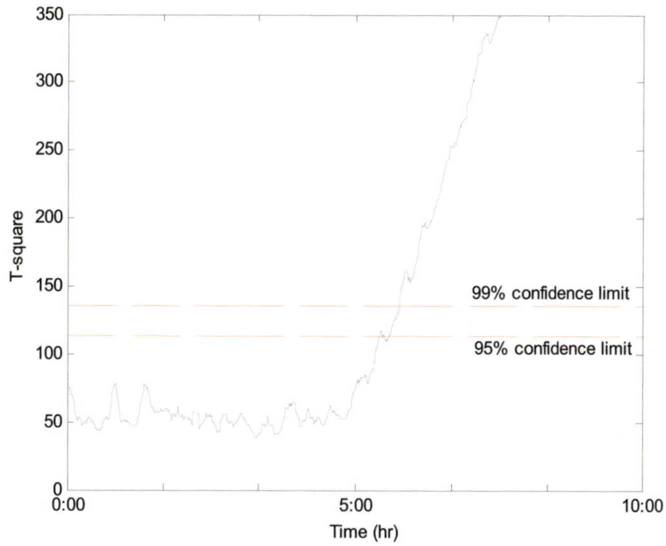

Fig. 6. T-square test for sensor drift in single shaft CCGT unit

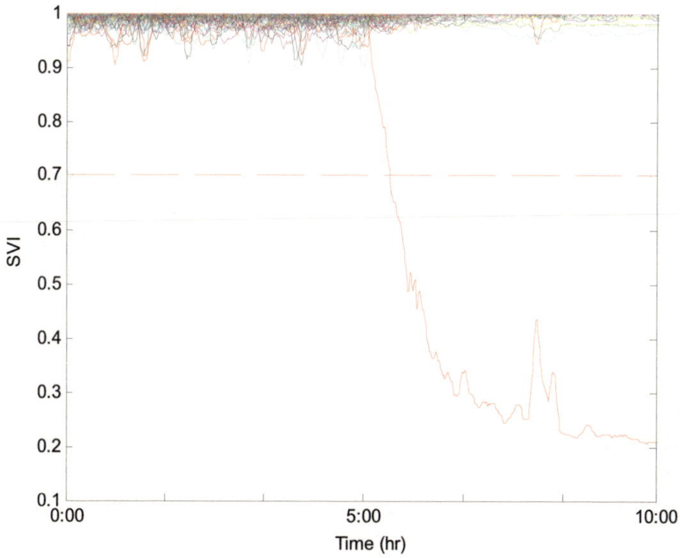

Fig. 7. SVI for sensor drift in single shaft CCGT unit

Following the detection of sensor fault condition, the source of the problem must be identified. Calculation of the sensor validity index (SVI), described in section 5.3, the variations in the SVI for each sensor are illustrated in Figure 7. According to the defined threshold of 0.7, the SVI chart clear identified the faulted sensor at 5:40 am, with the associated index of this signal falling into the range 0.7 to 0.2. Also, system transients and measurement noise can lead to oscillations into the SVI and there is a clearly example of SVI oscillations caused by system transient during 8:00 am to 9:00 am. It should be noted that when the HP main steam temperature signal drifts, the associated indices for the remaining sensors rise toward unity, accentuating identification of the biased sensor. As the fault is with the sensor, not the process, the PCA model can undertake reconstruction of the failed sensor as shown in Figure 4.

6.4 PLS model performance

Having validated the observed sensor data with PCA processor, optimisation of power plant performance should now be addressed. In order to maximize the power generation and simultaneously minimize the fuel consumed and pollution emissions, the performance variables must be able to monitor online and the internal relationship between the performance variables and associated operating parameters should be able to examine through offline analysis.

However, recover some performance variables is a multi-dimensional problem, such as the thermal efficiency, which depends on power demand, supplied fuel type, even the ambient conditions. Due to the expense and complexity of performance variables monitoring, development of online performance monitoring, capable of determine power plant performance from a variety of process variables, is often desirable. Validated and archived plant data can be employed to develop models which are capable of predicting the quality of process operation while providing an insight into the relationship between quality and associated process conditions.

PLS as a suitable technique for plant monitoring and shall be implemented here to demonstrate how system data can be applied to obtain a model of normal plant operation, with respect to a variety of quality variable measures, such as power plant efficiency, emissions and so on.

As with PCA, monitoring of individual fault conditions is not necessary and problems are instead detected as deviations from normal operation. With load cycling of generation plant increasingly common, a wide range of operating conditions are detailed in archived plant data and potentially contain indicators of operating conditions which lead to optimal power plant performance. The availability of operator logs makes it possible to indentify period of generation regarded by operators to be representative of fault-free power plant performance.

6.4.1 Variance explanation contribution

A benefit from the PLS model is that it has the ability to examine the effect of each input variable on the quality variables. Since the PLS model determines the variance explanation contribution of each variable by examining the correlation to the output variables, the PLS model is not only able to find those variables which have the greatest effect on output, but also can find the variables have indirect effect on the quality variables. This function can be applied to research the effect of any variable we interested, such as air temperature, sea water temperature, humidity and so on.

For instance, the variable contributions to the variance explanation of efficiency are charted in Figure 8 for a normal CCGT plant. Since the input variables are selected for highly related to the efficiency, most of them have comparatively high value of variance explanation, and these can be considered to be important variables to be monitored and/or adjusted when attempting to achieve enhanced operating goals. The most important variables are varying similar in both single shaft and multi shaft unit. For example, it can be observed that the No.1 and 2 are outstanding with 87.2% and 86.8% variance explained in the single shaft model, and they pointed to the signals of power output and gas flue flow respectively. Contrasts to multi-shaft model, above variables are identified as parameter No.1 with 85.0% explanation for power output and No.6 with 83% explanation for gas flue flow. Also, a group of sensors measuring the high pressure steam parameters are significant, which is the No.19-21 in the single shaft model with around 85% contributions and No. 27-29, 60-62 for both gas turbines in the multi-shaft model with around 80% contributions.

In addition, variations in ambient conditions is also interested, the last 4 variables in both models represent the effects of humidity, air temperature, barometric pressure and sea water temperature, respectively. It is significant that the sea water temperature has an extremely high effect on the power plant efficiency. The reason is considered of the condensing with sea water. The cooler sea water increases heat transfer from the condensing steam, and hence increase the thermal efficiency.

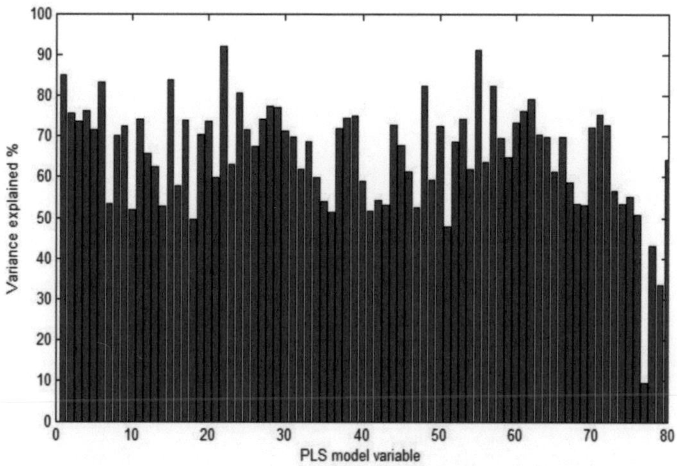

Fig. 8. Variance explanation for CCGT efficiency in multi-shaft unit

6.4.2 Relationship curve

From previous section, the PLS variance explanation suggest that sea water temperature is the most significant ambient condition for thermal efficiency. In order to better appreciate the impact of these environmental variables on the model, we introduce a new technique to study the relationship between input and output variables. It is of interest to lock all model inputs at a normal operating point, e.g. the power output is 90 MW, IGV position is 77% and etc., except the ambient variable being considered. For the simpler structure and closer variables relationship, the instance is chosen to use the single shaft unit, and consequently, Figure 9 illustrate the relative impact of these input parameters on the associated quality output measure for the CCGT plant.

It can be seen that increasing sea water temperature can significantly reduce the efficiency, the linear curve shows that about 50% increase in sea water temperature can cause 8% decrease in efficiency. Observably, the nonlinear curve shows that the relationship between ambient conditions and efficiency is more complicated and non-monotonic. As shown in the red line in Figure 9, the effects of ambient air and sea water temperature on the plant efficiency are represented as two tendency directions. One is the temperature upon the 12 degree, the efficiency is decreased following the increase in the environment temperature, and the reason is well known as we discussed in section 6.1 that the reduced cooling water temperature can enhance the steam cycle efficiency. Another direction present an interesting result where the temperature is lower than 12 degree, the efficiency enhancement seems to be more difficult when the temperature goes down. The reason is considered that the air and fuels inlet temperature will be excessively decreased during chill period and causes the decrease in gas combustion temperature and consequently reduced the gas turbine efficiency; it tends to counteract the effects of decrease in cooling water temperature on efficiency enhancement.

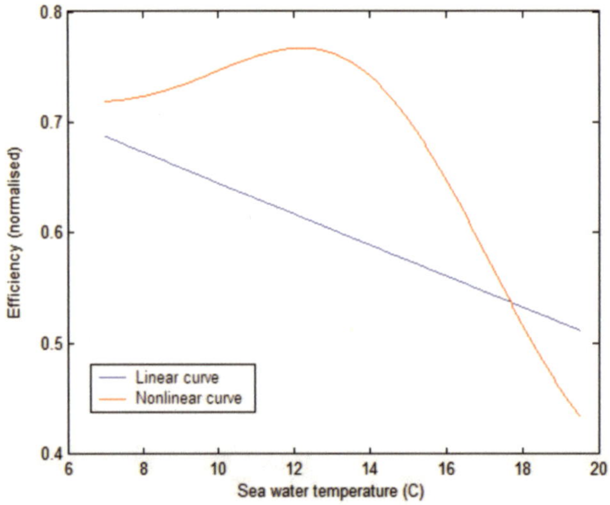

Fig. 9. Relation curve for single-shaft unit: efficiency vs. sea water temperature

7. Conclusion

Distributed control systems provide many advantages in terms of improvements in productivity and plant manoeuvrability when introduced into power plants and other industrial processes. However, the ease of access to a range of plant-wide signals potentially introduces vast problems of scale, since the meaningful information contained within the collected data may be somewhat less than the volume suggests. The task remains, therefore, to identify normal operating regions and relationships within the historical data, and subsequently to apply the collated rules, reference cases, etc. Principal component analysis has received considerable interest as a method of reducing the effective measurement space, and has been considered here for process monitoring of a combined cycle gas turbine.

Traditionally, operator practice has been reactive, whereby actions are taken following the triggering of process alarms, often set over-responsive and mode insensitive - PCA methods enable a more proactive role for the operator, providing early warning of plant irregularities, and identification of instrumentation errors and process faults. The PCA model is identified under normal operating conditions, and subsequently unusual deviations are highlighted and identified.

On the other hand, The PLS model has received considerable interest as a method of analyzing process data and in this instance it has been used for analyzing a combined cycle gas turbine. Analysis of the models variance demonstrates that CCGT performance is affected by changes in ambient conditions. Also a relation curve method can be utilized here to study the impact of these external parameters, from the environment, have on the gas turbine performance.

Future work could extent the quality measurements (efficiency and emissions) to include other important requirements such as plant life, unit flexibility or cost of generation. Where the objective is to optimize power plant quality measurements and consequently enhance plant performance.

Furthermore, it is considered to integrate the physical or empirical mathematical models with the statistical model. Since the statistical models are limited by training range, it is unreliable to be employed under untrained conditions. Compared to physical model, a trained statistical model is difficult to tune in case of power plant renovation or update, due to it requests the input variables remaining a designed internal relationship.

On the other hand, developing the pure physical model can be a time-consuming exercise and requires the designer to have extensive knowledge of the application. Especially for the large scale system, the relationship between each variable is highly correlated and tangled, it is impossible to build a physical model which can provide a detailed and accurate representation of the operation. Therefore, utilizing the statistical model to substitute the complex components in a physical model could be good solution for reduce modeling calculation. Similarly, the adjustability and flexibility of statistical model can be increased by integrating some physical control loop or simulation into the statistical model.

8. References

Abdelhail, A., I. Traore and B. Sayed: 'RBDT-1: A New Rule-Based Decision Tree Generation Technique', *Rule Interchange and Applications*, 58, pp. 108-121, 2009

Ahvenlampi, T. and U. Kortela:' Clustering algorithms in process monitoring and control application to continuous digesters', *Informatica*, 29, pp.101-109, 2005

Armor, A. F.: 'Management and integration of power plant operations', *Thermal power plant simulation and control*, IEE Power and Energy Series 43, pp. 395-416, 2003

Baffi, G., E. B. Martin and A. J. Morris: 'Non-linear projection to latent structures revisited: the quadratic PLS algorithm', *Computers and Chemical Engineering*, 23, (3), pp. 395-411, 1999

Billings S.A and Hong, X.: "Dual-orthogonal radial basis function networks for nonlinear time series prediction". *Neural Networks*, 11 (3), pp. 479-493, April 1998

Blanco, C., D. Soronow and P. Stefiszyn: 'Multi-factor models for forward curve analysis: An introduction to principal component analysis', *Commodities Now*, June, pp. 76-78, 2002

Bharati M. H., J. J. Liu and J. F. MacGregor: 'Image texture analysis: methods and comparisons', *Chemometrics and Intelligent Laboratory Systems*, 72, PP. 57-71, 2004

Chan, C.: 'Application of multivariate analysis to Optimize Function of Cultured Hepatocytes', *Biotechnology Progress*, 19, (2), pp. 580-598, 2003

Cinar, A., and C. Undey: 'Statistical process and controller performance monitoring: A tutorial on current methods and future directions', *Proceedings of the American Control Conference*, San Diego, USA, 4, pp. 2625-2639, 1999

Dunia, R., S. Qin, T.F. Edgar and T.J. MCAVOY: 'Identification of faulty sensors using principal component analysis', *AIChE Journal*, 42, (10), pp. 2797-2812, 1996

Flynn, D., J. Ritchie, M. Cregan and L. Pan: 'Data mining techniques applied to power plant performance monitoring', *16th IFAC World Congress, Prague*, 2006.

Freund, R.J., and W.J. Wilson: 'Regression analysis: statistical modelling of a response variable', Academic Press Ltd., London, UK, 1997

Goldberg, E.D.: " Genetic Algorithms in Search, Optimization, and Machine Learning". *Addision-wesley publishing company*, INC. pp. 27-95, 1989

Jackson, D.A.: 'Stopping rules in principal component analysis: a comparison of statistical and heuristical approaches', *Ecology*, 74, (8), pp. 2204-2214, 1993

Jackson, J.E.: 'A user's guide to principal component analysis', Wiley series in probability and statistics, John Wiley & Sons, Hoboken, New Jersey, 2003

Jolliffe, I.T.: 'Principal component analysis', Springer series in statistics, New York, 2002

Hinkle, D. and C. Toomey: 'Applying case-based reasoning to manufacturing', *AI Magazine*, 16, (1), pp. 65-73, 1995

Kolodner, d.: 'Improving human decision making through case-based decision aiding', *AI Magazine*, 12, (2), pp. 52-68, 1991

Kourti, T., and J. F. Macgregor: 'Process analysis, monitoring and diagnosis, using multivariate projection methods', *Chemometrics and Intelligent Laboratory Systems*, 28, pp. 3-21, 1995

Kourti, T., J. Lee and J.F. Macgregor: 'Experiences with industrial applications of projection methods for multivariate statistical process control', *Computers and Chemical Engineering*, 20, pp. 745-750, 1996

Kresta, J.V., T.E. Martin and J.F. MacGregor: Development of inferential process models using PLS', *Computers and Chemical Engineering*, 18, (7), pp. 597-611, 1994

Lewin, D.R. 'Predictive maintenance using PCA', *Control Engineering Practice*, 3, (3), pp. 415-421, 1995

Li, W., H.H. Yue, S. Valle-Cervantes, and S.J. QIN: 'Recursive PCA for adaptive process monitoring', *Journal of Process Control*, 2000, 10, pp. 471-486

Li, G. and Liu, B.: "RBFNN algorithm based on hybrid hierarchy genetic algorithm and its application". *Control Theory & Applications*. 19(4), pp. 627-630, 2002.

MacGregor, J.F., and T. KOURTI: 'Statistical process control of multivariate processes' *Control Engineering Practice*, 3, (3), pp. 403-414, 1995

MacGregor, J.F., H. YU, S.G. Munoz and J. Flores-Cerrillo: 'Data-based latent variable methods for process analysis, monitoring and control', *Computers and Chemical Engineering*, 29, pp. 1217-1223, 2005

Martin, E.B., A.J. Morris and J. Zhang: 'Process performance monitoring using multivariate statistical process control', *IEEE Proceedings Control Theory and Applications*, 143, (2), pp. 132-144, 1996

Matthews, R.: 'Data miners only strike fool's gold', *New Scientist*, 8th March, pp. 8, 1997

Michalski, R.S., I. Bratko and M. Kubat" 'Machine learning and data mining: Methods and applications', Wiley & Sons, Chichester, England, 1999

Moody, Darken C.: "Fast Learning in Networks of Locally-tuned Processing Units". *Neural Computation*, 1, pp. 281 - 294,1989.

Olaru, C., and L. Wehenkel: 'Data Mining', *IEEE Computer Applications in Power*, 12, pp. 19-25, 1999

Oliveira-Esquerre K.P., D.E. Seborg, R.E. Bruns and M. Mori: 'Application of steady-state and dynamic modelling for the prediction of the BOD of an aerated lagoon at a pulp and paper mill Part I. Linear approaches', *Chemical Engineering Journal*, 104, pp. 73-81, 2004

Otto, M., and W. Wegscheider: 'Spectrophotometric multi-component analysis applied to trace metal determinations', *Analytical Chemistry*, 57, (1), pp. 63-69, 1985

Pan, L., D. Flynn and M. Cregan: 'Statistical model for power plant performance monitoring and analysis', *Universities Power Engineering Conference*, pp. 121-126, 2008

Qin, S., H. Yue, and R. Dunia: 'Self-validating inferential sensors with application to air emission monitoring', *Industrial Engineering Chemical Research*, 36, pp. 1675-1685, 1997

Quinlan, J.R.: 'C4.5: programs for machine learning', 1993, The Morgan Kaufmann series in machine learning, Morgan Kaufmann Publishers, California

Sebzalli, Y.M., and X.Z. Wang.: 'Knowledge discovery from process operational data using PCA and fuzzy clustering', *Engineering Applications of Artificial Intelligence*, 2001, 14, (5), pp. 607-616

Smyth, B., M.T. Keane and P. Cunningham: 'Hierarchical case-based reasoning integrating case-based and decompositional problem-solving techniques for plant-control software design', *IEEE Transactions on Knowledge and Data Engineering*, 13, (5), pp. 793-812, 2001

Valle, S., W. Li and S.J. Qin: 'Selection of the number of principal components: The variance of the reconstruction error criterion with a comparison to other methods', *Industrial Engineering Chemistry Research*, 38, pp. 4389-4401, 1999

Voumvoulakis, E.M. and Hatziargyriou, N.D.: 'A Particle Swarm Optimization Method for Power System Dynamic Security Control', *IEEE Transactions on Power Systems*, pp. 1032-1041, 2010

Watson, I.: 'Case-based reasoning is a methodology not a technology', *Knowledge-Based Systems*, 12, pp. 303-308, 1999

Wang, J.R.: 'Research on web-based multi-agent system for aeroengine fault diagnosis', *The 9th International Conference on Web-Age Information Management*, pp. 195-202, 2008

Weiss, S.M., and N. Indurkhya: 'Predictive data mining, a practical guide', 1998, Morgan Kaufmann Publishers Inc., San Francisco, CA

Wise, B.M. and N.B. Gallagher: 'The process chemometrics approach to process monitoring and fault detection', *Journal of Process Control*, 6, (6), pp. 329-348, 1996

Wold, S., K. Esbensen and P. Geladi: 'Principal component analysis', *Chemometrics and Intelligent Laboratory Systems*, 1987, 2, pp. 37-52

Yang, P.: 'A case-based reasoning with feature weights derived by BP network', *Workshop on Intelligent Information Technology Application, IITA*, pp. 26-29, 2007
Yoon, S., and J.F. Macgregor: 'Fault diagnosis with multivariate statistical models. Part one: using steady state fault signatures', *Journal of Process Control*, 11, pp. 387-400, 2001

Part 3

Heat Transfer

9

Jet Impingement Cooling in Gas Turbines for Improving Thermal Efficiency and Power Density

Luai M. Al-Hadhrami, S.M. Shaahid and Ali A. Al-Mubarak
Associate Professor Center for Engineering Research, Research Institute
King Fahd University of Petroleum and Minerals
Saudi Arabia

1. Introduction

The gas turbine is an engine which produces a great amount of energy depending upon its size and weight. Gas turbines are used for aircraft propulsion and land based power generation. Thermal efficiency and power output (power density) of gas turbines increase with increasing turbine rotor inlet temperatures (RIT). Today there are gas turbines, which run on natural gas, diesel fuel, naphtha, methane, crude, low-Btu gases, vaporized fuel oils, and biomass gases. The last 20 years has seen a large growth in gas turbine technology which is mainly due to growth of materials technology, new coatings, and new cooling schemes. In a simple gas turbine cycle (Figure 1), low pressure air is drawn into a compressor (state 1) where it is compressed to a higher pressure (state 2). Fuel is added to the compressed air and the mixture is burnt in a combustion chamber. The resulting hot products enter the turbine (state 3) and expand to state 4 and the air exhausts. Most of the work produced in the turbine is used to run the compressor and the rest is used to run auxiliary equipment and to produce power. Figure 2 shows schematic of cross section of a small gas turbine.

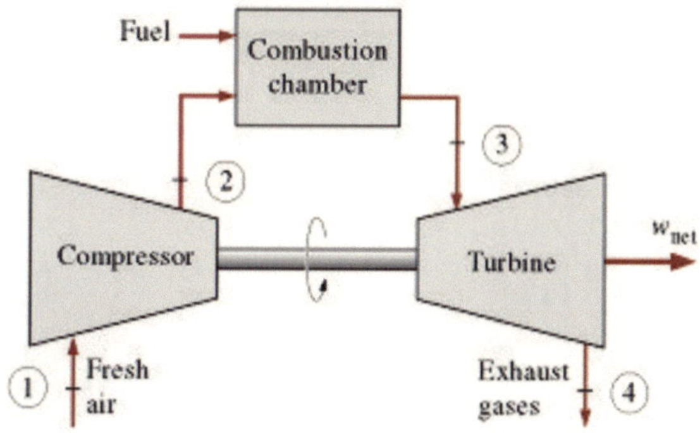

Fig. 1. Schematic of open Gas turbine cycle.

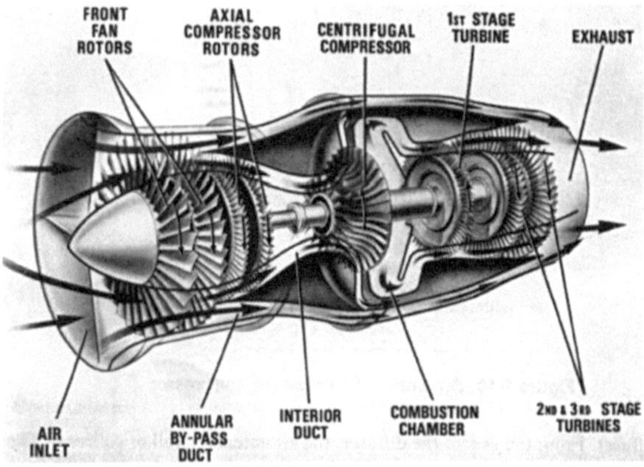

Fig. 2. A schematic of a cutaway of a small gas turbine.

2. Concept and the need for turbine blade cooling

The gas turbine engines operate at high temperatures (1200-1600 °C) to improve thermal efficiency and power output. As the turbine inlet temperature increases, the heat transferred to the turbine blades increases. The above operating temperatures are far above the permissible metal temperatures. Therefore, there is a need to cool the turbine blades for safe operation. The blades are cooled by extracted air from the compressor of the engine. Gas turbine blades are cooled internally and externally. Internal cooling is achieved by passing the coolant through several enhanced serpentine passages inside the blades and extracting the heat from outside the blades. Both jet impingement cooling and pin fin cooling are used as a method of internal cooling. External cooling is also called film cooling. Figure 3 and 4 show different types of turbine blade cooling. The cooling system must be designed to ensure that the maximum blade surface temperatures during operation are compatible with the maximum blade thermal stress.

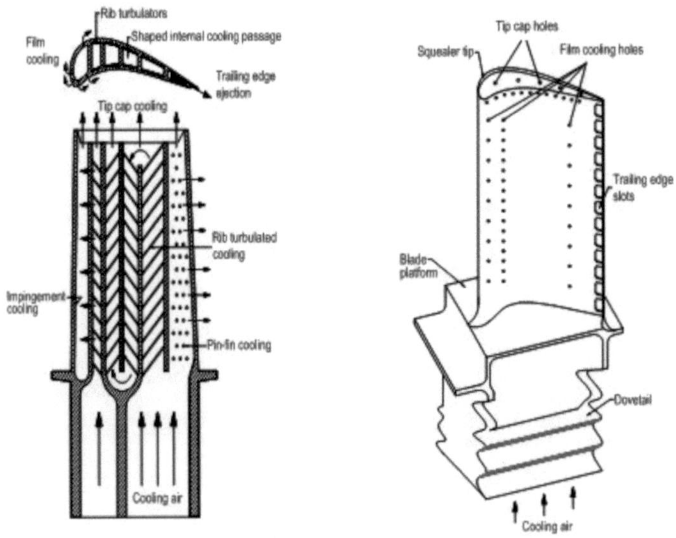

Fig. 3. Schematic of the modern gas turbine with common cooling blade techniques.

3. Typical turbine cooling system

The cooling air is bled from the compressor and is directed to the stator, the rotor, and other parts of the turbine rotor and casing to provide adequate cooling. The effect of coolant on the aerodynamics depends on the type of cooling involved. An example of a typical cooling system is shown in Figure 4.

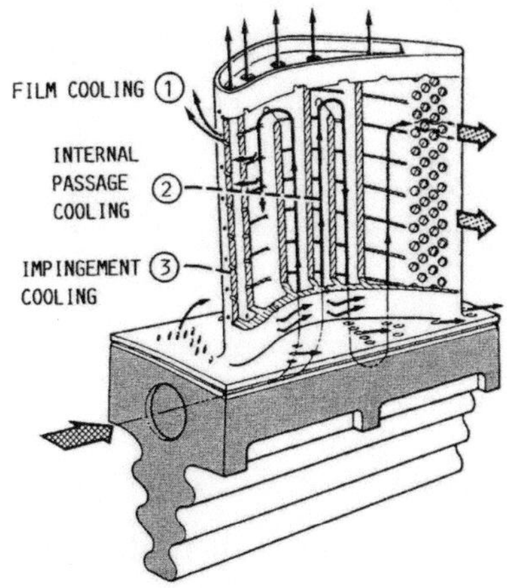

Fig. 4. Typical cooled aircraft gas turbine blade of three dimensions.

4. Jet impingement cooling

Jet impinging on the inner surfaces of the airfoil through tiny holes in the impingement insert is a common, highly efficient cooling technique for first-stage vanes. Impingement cooling is very effective because the cooling air can be delivered to impinge on the hot region. Jet impingement cooling can be used only in the leading-edge of the rotor blade, due to structure constraints on the rotor blade under high speed rotation and high centrifugal loads. Schematic of the impingement jet of the leading edge gas turbine blade is shown in Figure 5.

Several arrangements are possible with cooling jets and different aspects need to be considered before optimizing an efficient heat transfer design. Some of the research studies have focused on the effects of jet-hole size and distribution, cooling channel cross-section, and the target surface shape on the heat transfer coefficient distribution.

Wide range of parameters affect the heat transfer distribution, like impinging jet Re, jet size, target surface geometry, spacing of the target surface from the jet orifices, orifice-jet plate configuration, outflow orientation, etc. Literature indicates that many of these parameters have been studied in appreciable depth [1-20].

Chupp et al. [1] studied the heat transfer characteristics for the jet impingement cooling of the leading edge region of a gas turbine blade. Flourscheutz et al. [2] investigated the heat transfer characteristics of jet array impingement with the effect of initial crossflow. Metzger and Bunker [3] and Flourscheutz et al [4] used the liquid crystal technique to study the local heat transfer coefficients. The authors observed that the jet Nusselt number depends mainly on the jet Re.

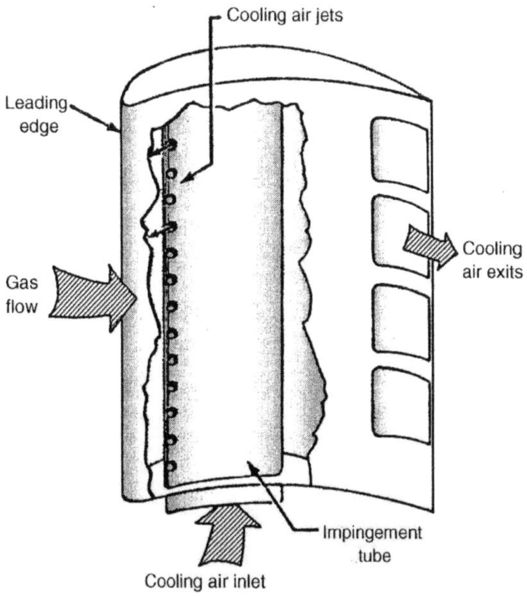

Fig. 5. Schematic of the impingement jet of the leading edge gas turbine blade.

Dong et al [5] determined experimentally the heat transfer characteristics of a premixed butane/air flame jet impinging upwards on an inclined flat plate at different angles of inclination and fixed Reynolds number (Re = 2500) and a plate to nozzle distance of 5d. It was found that decrease in angle of incidence reduced the average heat transfer. Rasipuram and Nasr [6] studied air jet issuing out of defroster's nozzles and impinging on inclined windshield of a vehicle. The overall heat transfer coefficient of the inclined surface for the configuration with one rectangular opening was found to be 16% more than that for the configuration with two rectangular openings. Beitelmal et al [7] investigated the effect of inclination of an impinging air jet on heat transfer characteristics. They found that maximum heat transfer shifts towards the uphill side of the plate and the maximum Nusselt number (Nu) decreases as the inclination angle decreases. Roy and Patel [8] studied the effect of jet angle impingement on local Nu and nozzle to target plane spacing at different Re. They found that heat transfer is the maximum through the shear layer formed near the jet attachment stagnation region. Ekkad et al [9] studied the effect of impinging jet angle ±45 on target surface by using transient liquid crystal technique for single Re = 1.28×10^4. It has been noted that the orthogonal jets provide higher Nu as compared to angled jets.

Tawfek [10] investigated the effect of jet inclination on the local heat transfer under an obliquely impinging round air jet striking on circular cylinder. Their results indicated that with increase in inclination, the upstream side of heat transfer profile dropped more rapidly than the downstream side. Seyedein et.al. [11] performed numerical simulation of two dimensional flow fields and presented the heat transfer due to laminar heated multiple slot jets discharging normally into a converting confined channel by using finite difference method with different Re(600-1000) and angle of inclination (0-20°). Yang and Shyu [12] presented numerical predictions of heat transfer characteristics of multiple impinging slot jets with an inclined confinement surface for different angles of inclination and different Re. Yan and Saniei [13] dealt with measurement of heat transfer coefficient of an impinging circular air jet to a flat plate for different oblique angles (45-90°) and different Re (10000 & 23000) by using transient liquid crystal technique.

Hwang et. al [14] studied the heat transfer in leading edge triangular duct by an array of wall jets with different Re (3000-12600) and jet spacing s/d (1.5-6) by using transient liquid crystal technique on both principal walls forming the apex. Results show that an increase in Re increases the Nu on both walls. Ramiraz et al [15] investigated the convective heat transfer of an inclined rectangular plate with blunt edge at various Re (5600-38500) and angle of inclination (60-70°). The heat transfer distribution over a finite rectangular plate was found to be very much dependent on the orientation of the plate. Stevens and Webb [16] examined the effect of jet inclination on local heat transfer under an obliquely impinging round free liquid jet striking at different Re, angle of inclination, and nozzle sizes. It was found that the point of maximum heat transfer along the x-axis gets shifted upstream. Hwang and Cheng [17] performed an experimental study to measure local heat transfer coefficients in leading edge using transient liquid crystal technique. Three right triangular ducts of the same altitude and different apex angles (30°, 45° & 60°) were tested for various jet Re (3000 ≤Re≤12000) and different jet spacing (s/d=3 and 6). Hwang and Cheng [18] measured experimentally local heat transfer coefficients on two principal walls of triangular duct with swirl motioned airflow induced by multiple tangential jets from the side entry of the duct by using transient liquid crystal technique. Hwang and Chang [19] measured heat transfer coefficients on two walls by using transient liquid crystal technique in triangular duct cooled by multiple tangential jets. The results show that an increase in Re, increases heat transfer of both walls. Hwang and Cheng [20] studied heat transfer characteristics in a triangular duct cooled by an array of side-entry tangential jets.

It is evident from the published literature that no study has been conducted to show the effect of different orifice-jet plate configurations on feed channel aspect ratio with different jet Re, for a given outflow orientation (diameter of jet, d = 0.5 cm) on heat transfer in a channel with inclined heated target surface. Therefore, the following sections include investigation of the above effects by conducting the experimental work. Specifically, the work includes the effect of three orifice-jet plate configurations (centered holes, staggered holes, tangential holes) and three feed channel aspect ratios (H/d=5, 7, and 9) on the heat transfer characteristics for a given outflow orientation (with outflow passing in both the directions) and for a different Reynolds number with inclined heated target surface. The motivation behind this work is that the channel of turbine blade internal cooling circuit at the leading edge is inclined.

4.1 Description of the experimental set-up

The schematic of the experimental set-up is depicted in Figure 6. The test rig used to study the heat transfer characteristics has been constructed using Plexi-glass. The test section consists of two channels, impingement (10) and the feed channel (9). Air enters the test section in the feed channel and is directed onto the heated copper plates in the impingement channel to study the heat transfer characteristics. The target plates (11), made of copper, were heated using a constant flux heater. The other side of the heater was insulated to get the heat transferred only in one direction. The mass flow rate of the compressed air (1) entering the test section was passed through a settling chamber (5) and was controlled with the help of valves (2). The pressure drop was measured using the pressure gauges (4). Gas flow meter (7) was used to measure the mass flow rate entering the test rig which was protected by the air filters (6) of 50μ capacity.

The average surface temperature of each copper plate was determined from the readings of two T-type thermocouples (12) installed in the holes drilled at the back surface of the plates to within 1 mm of the surface in depth. The analog signals generated by these temperature sensors were transmitted to the signal-conditioning unit where they were selectively processed. The resulting analog signals were converted into digital signals by a DAQ (13) card and recorded with application software developed in LabView.

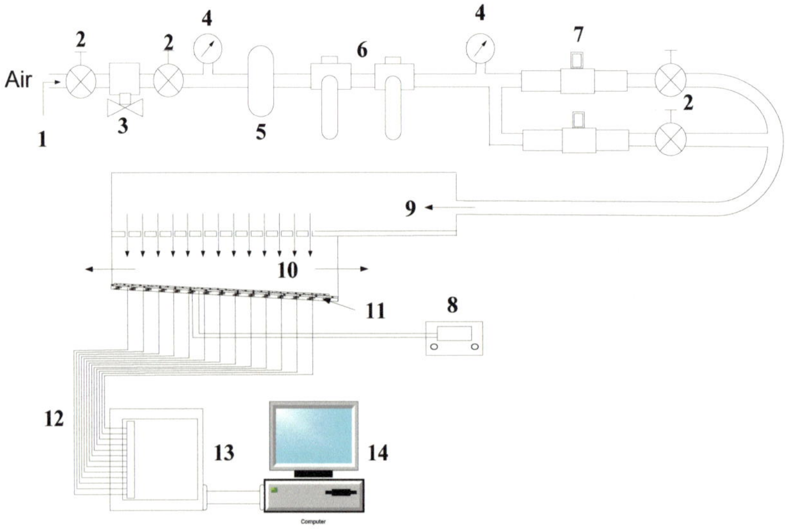

1- Compressed Air.
2- Valves.
3- Discharged Valve.
4- Pressure Gauges.
5- Settling Chamber.
6- Air and Oil Filters.
7- Digital Flow Meters.
8- Transformer.
9- Feeding Channel.
10- Impingement Channel.
11- Target Surface.
12- Thermocouples.
13- Data Acquisition.
14- Computer.

Fig. 6. Schematic of the test section

Figure 7.shows the three-dimensional sketch of the test section. It consists of two channels joined by the orifice plate, which has a single array of equally spaced (centered or staggered or tangential) orifice jets shown in Figure 8. The jet orifice plate thickness is twice the jet diameter. There are 13 jets on the orifice plate. The jet-to-jet spacing is 8 times the jet diameter and the orifice jet diameter $d = 0.5cm$. The length of the test section is 106.5 cm. The width of the feed channel (H) was varied from 2.5 to 4.5 cm (i.e. H/d=5, 7, 9, d=0.5 cm). The impingement target surface constitutes a series of 13 copper plates, each with 4.2 $\times$ 4.1 cm in size, arranged in accordance with the orifice jets such that the impingement jet hits the geometric center of the corresponding plate (however, first and last copper plates are slightly different in sizes). All the copper plates are separated from each other by 1 mm distance to avoid the lateral heat conduction, thus dividing the target surface into segments. The thickness of the copper plate is 0.5 cm. As shown in Figure 9, the length of impingement surface L is 57.3 cm (the target surface is inclined at angle 1.5°, the width of parallel flow side "S2" is 2 cm and the width of the opposite flow side "S1" is 3.5 cm).

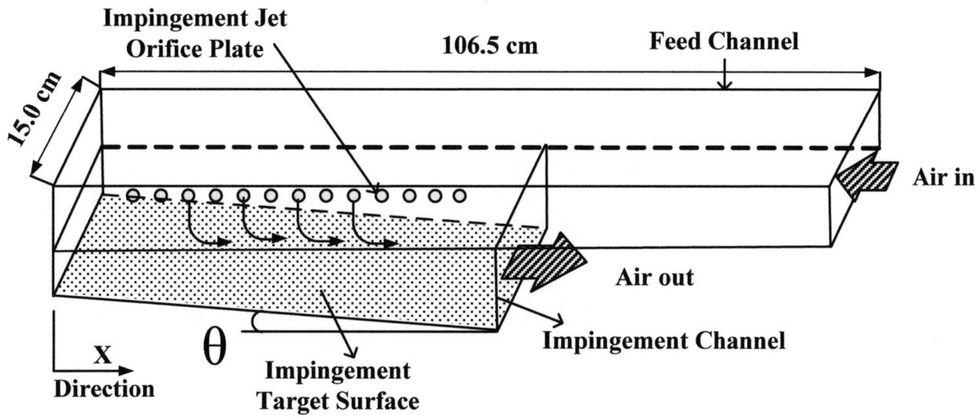

Fig. 7. Three-dimensional view of the test section

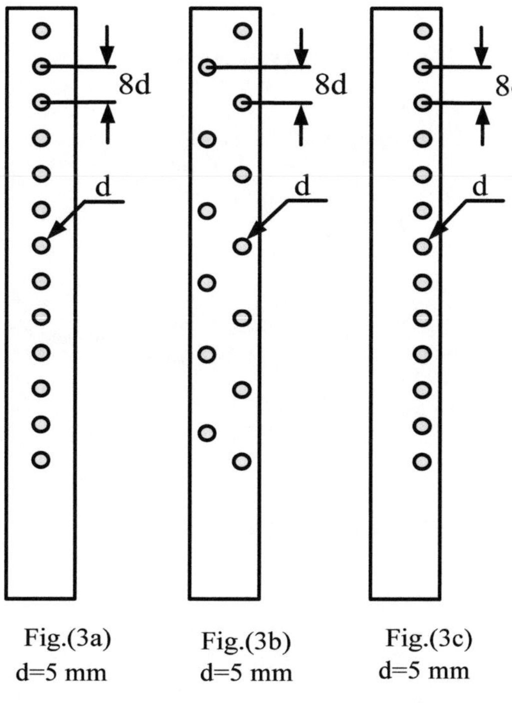

Fig. 8. Illustration of three orifice-jet configurations with single array of jets (d = 5 mm) (Fig. 8a Centered holes, Fig. 8b Staggered holes, Fig. 8c Tangential holes)

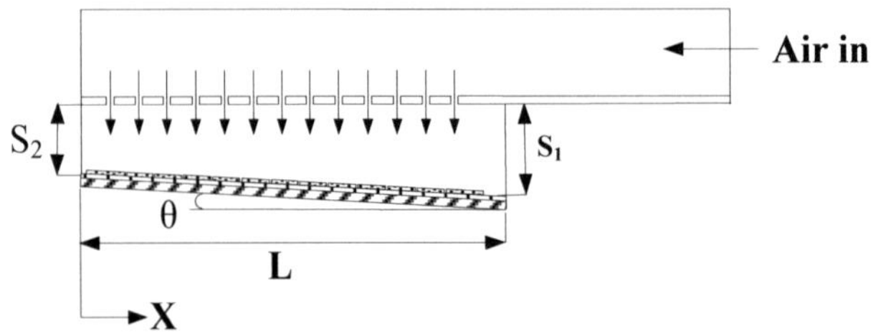

Fig. 9. Inclination angle of the target surface

Figure 10 shows the schematic of the three different outflow orientations. The upper channel is called as the feed channel and the lower channel in which the jets impinge on the target surface is called as the impingement channel. The exit of jets in three different outflow orientations from the impingement channel creates different cross-flow effects as shown in Fig. 10. However, in the present study, attention is focused on Case – 3.

Case - 1 (Outflow coincident with the entry flow),

Case - 2 (Outflow opposing to the entry flow),

Case - 3 (Outflow passes out in both the directions).

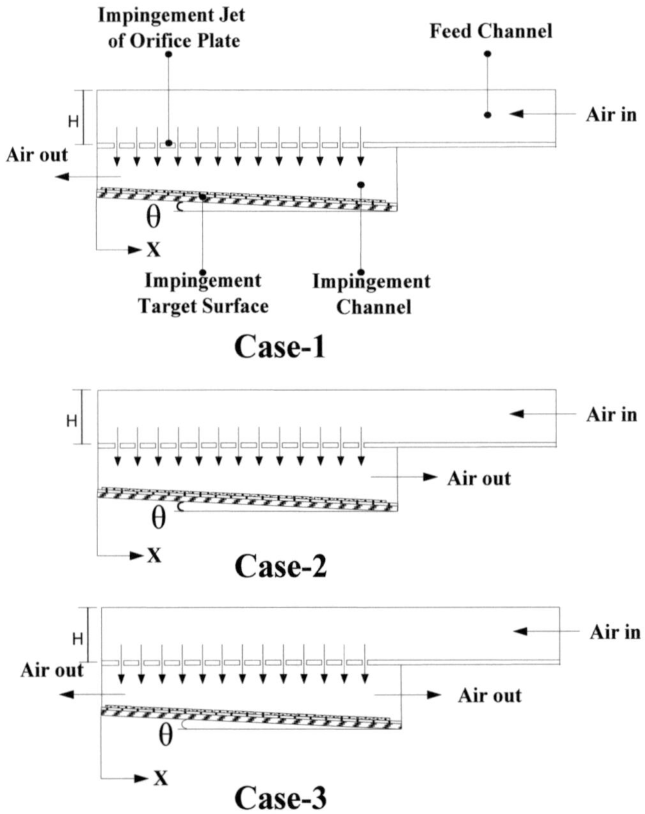

Fig. 10. Illustration of three exit outflow orientations Figure 10a, Case-1 (Outflow Coincident with the inlet flow). Figure 10b, Case-2 (Outflow Opposing to the inlet flow). Figure 10c, Case-3 (Outflow Passes out in both directions).

Figure 11 shows the details of the construction of the target surface. The copper plate is heated with a constant flux heater held between the wooden block and the copper plate by glue (to reduce contact resistance). The ends were sealed with a rubber material to avoid lateral heat losses. The wooden block was 3 cm thick to minimize the heat lost to the surroundings, so that the heat is transferred completely to copper plates only.

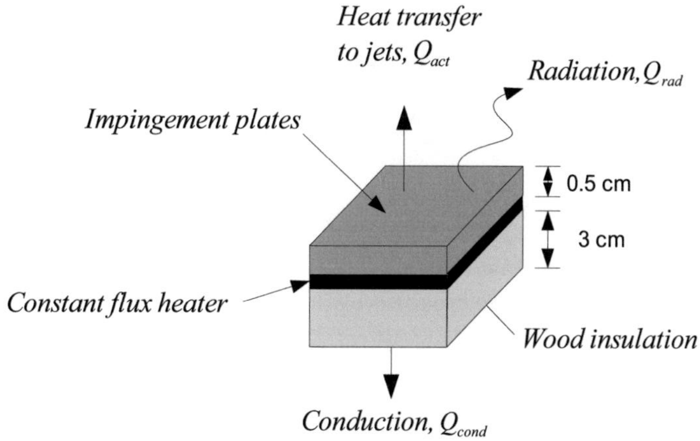

Fig. 11. Overall energy balance over a small element of the impingement plate

4.2 Experimental procedure

To start with, tests were carried out using a given orifice-jet plate (centered or staggered or tangential holes) with jet diameter $d = 0.5cm$ for a given jet Reynolds numbers Re = 18800 (for a given H/d ratio, for outflow passing in both directions) and for a constant heat flux power input. The heated target plate was oriented at a pre-defined angle (1.5°). The mass flow rate was adjusted to the required value for the experiment to be conducted and the air was blown continuously into the test section. Heat was supplied to the copper plates with electric resistive constant flux heaters from backside to provide uniform heat flux. The temperature of the copper plates was measured by two thermocouples mounted in a groove of 2.5 mm on the back of the copper plates. Thus the temperature of a particular plate has been taken as the average of the reading of two thermocouples. The temperature of the copper plates, pressure, temperature of the air at the inlet, and the mass flow rate were continuously monitored. After the temperature of the copper plates reached the steady state condition, all the data was collected with Lab VIEW program. The Nusselt number was then calculated based upon the collected data. The same procedure was repeated for the three different orifice-jet plates described in Figures 8a, 8b & 8c and for different aspect ratios (H/d = 5, 7, 9).

4.3 Data reduction and uncertainty analysis

The collected data was subjected to uncertainty analysis. The method for performing the uncertainty analysis for the present experimental investigation has been taken from Taylor B.N. [21]. The theory for the current uncertainty analysis is summarized in the following discussion:

4.4 Jet reynolds number calculations

The average velocity used to calculate the jet Reynolds number is calculated using the following equation

$$V_{avg} = \frac{\forall}{13 \times \frac{\pi}{4} d^2} \tag{1}$$

The data reduction equation for the jet Reynolds number is taken as:

$$Re = \frac{\rho V_{avg} d}{\mu} = \frac{\rho d}{\mu} \frac{\forall}{13 \times \frac{\pi}{4} d^2} \tag{2}$$

4.5 Uncertainty in jet reynolds number

Taking into consideration only the measured values, which have significant uncertainty, the jet Reynolds number is a function of orifice jet diameter and volume flow rate and is expressed mathematically as follows:

$$Re = f(\forall, d) \tag{3}$$

Density and dynamic viscosity of air is not included in the measured variables since it has negligible error in the computation of the uncertainty in jet Reynolds number. The uncertainty in Reynolds number has been found to be about 2.2 %.

4.6 Nusselt number calculations

The total power input to all the copper plates was computed using the voltage and current, the former being measured across the heater, using the following equation:

$$Q_{total} = \frac{V^2}{R} = VI \tag{4}$$

The heat flux supplied to each copper plate was calculated using:

$$q'' = \frac{Q_{total}}{A_{total}} \tag{5}$$

The heater gives the constant heat flux for each copper plate. The heat supplied to each copper plate from the heater is calculated using the following procedure:

$$Q_{cp,i} = q'' \times A_{cp,i} \tag{6}$$

Where, i is the index number for each copper plate. The heat lost by conduction through the wood and to the surrounding by radiation is depicted in Figure 5 and has been estimated using the following equations for each plate.

$$Q_{cond,i} = k_{wood} A_{cp,i} \frac{(T_{s,i} - T_w)}{t} \tag{7}$$

$$Q_{rad,i} = \varepsilon \sigma A_{cp,i} \left(T_{s,i}^4 - T_{surr}^4\right) \tag{8}$$

The actual heat supplied to each copper plate has been determined by deducting the losses from the total heat supplied to the heater.

$$Q_{actual,i} = Q_{cp,i} - (Q_{cond,i} + Q_{rad,i}) \tag{9}$$

The local convective heat transfer coefficient for each of the copper plate has been calculated using:

$$h_i = \frac{Q_{actual,i}}{A_{cp,i}(T_{s,i} - T_{in})} \tag{10}$$

The average temperature of the heated target surface $T_{s,i}$ has been taken as the average of the readings of the two thermocouples fixed in each copper plate. To calculate h, T_{in} has been considered instead of the bulk temperature or the reference temperature. For a given case (for a given Re, H/d, and orifice-jet plate) T_{in} is fixed. It is measured at the test section inlet, where the air first enters the feed channel. The non-dimensional heat transfer coefficient on the impingement target surface is represented by Nusselt number as follows:

$$Nu_i = \frac{h_i\, d}{k_{air}} \tag{11}$$

The hydraulic diameter has been taken as the diameter of the orifice jet. The data reduction equation for the Nusselt number is considered along with the heat losses by conduction and radiation.

$$Nu_i = \frac{d}{k_{air}} \left(\frac{\dfrac{V^2}{R\,A_{total}} - \dfrac{k_w}{t}(T_{s,i} - T_w) - \varepsilon\sigma(T_{s,i}^4 - T_{Surr}^4)}{(T_{s,i} - T_{in})} \right) \tag{12}$$

4.7 Uncertainty in nusselt number

Temperature of the wood has a very less effect on the uncertainty of heat transfer coefficient due to the large thickness of the wood and also due to the insulation material attached to the wooden block. Temperature of the surroundings and emissivity also has less effect on the uncertainty as the work was carried out in a controlled environment and the temperature of the surroundings was maintained within 21-23 °C through out the experiment. The standard uncertainty in the Nusselt number neglecting the covariance has been calculated using the following equation:

$$\left(U_{c,Nu_i}\right)^2 = \left(\frac{\partial Nu_i}{\partial V} u_V\right)^2 + \left(\frac{\partial Nu_i}{\partial R} u_R\right)^2 + \left(\frac{\partial Nu_i}{\partial T_{s,i}} u_{T_{s,i}}\right)^2 + \left(\frac{\partial Nu_i}{\partial T_{in}} u_{T_{in}}\right)^2 + \left(\frac{\partial Nu_i}{\partial A_{total}} u_{A_{total}}\right)^2 + \left(\frac{\partial Nu_i}{\partial d} u_d\right)^2 \tag{13}$$

Uncertainty propagation for the dependent variable in terms of the measured values has been calculated using the Engineering equation Solver (EES) software. The measured variables x_1, x_2 etc. have a random variability that is referred to as its uncertainty. The uncertainty in Nusselt number in the present study has been found to vary between ± 6 % depending upon the jet velocity.

5. Results and discussions

Jet impingement heat transfer is dependent on several flow and geometrical parameters. The jet impingement Nusselt number is presented in a functional form as follows:

$$Nu_i = \left(\frac{h_i d}{k_{air}}\right) = f\left[\begin{array}{l} Re,\ \left(X/_d, H/_d\right), \\ \textit{outflow orientation} \end{array}\right] \tag{14}$$

Where, Re is the flow parameter, jet spacing to the diameter ratio (X/d) is the geometric parameter. The flow exit direction and target surface geometry are also important parameters having a considerable impact on impingement heat transfer.
The X location starts from the supply end of the channel as shown in Figure 7. For the case 1 shown in Figure 10a, flow enters at X/d = 109.3 and exits at X/d = 0. For case 2 (Figure 10b), flow exits at X/d = 109.3. For case 3 (Figure 10c), flow exits at both ends (X/d = 0 and X/d = 109.3). The flow is fully developed before entering the orifice jets. However, in the present study attention is focused on Case – 3 (out- flow passing out in both directions).

5.1 Effect of orifice-jet-plate configuration on feed channel aspect ratio

Figures 12-14 show the local Nusselt number distribution for three orifice-jet plate configurations and for three H/d ratios as a function of non-dimensional location X/d on the heated target surface (for outflow passing in both directions as shown in Figure 10c, and for a given Re= 18800).
Figure 12 shows the effect of feed channel aspect ratio (H/d) on local Nusselt number for Re=18800 for orifice jet plate with centered holes. It can be observed that, H/d=9 gives the maximum heat transfer over the entire length of the target surface as compared to all feed channel aspect ratio studied. H/d=9 gives 1% more heat transfer from the target surface as compared to H/d=5. Whereas H/d=5 gives of 1% increase in heat transfer as compared to H/d=7.
Figure 13 shows the effect of feed channel aspect ratio (H/d) on local Nusselt number for Re=18800 for orifice jet plate with staggered jets. It can be observed that, H/d=9 gives the maximum heat transfer over the entire length of the target surface as compared to all feed channel aspect ratio studied. H/d=9 gives 1% more heat transfer from the target surface as compared to H/d=5, whereas H/d=5 gives of 6% increase in heat transfer as compared to H/d=7.
Figure 14 shows the effect of feed channel aspect ratio (H/d) on local Nusselt number for Re=18800 for orifice jet plate with tangential holes. It can be observed that, H/d=9 gives the maximum heat transfer over the entire length of the target surface as compared to other feed channel aspect ratio studied. H/d=9 gives 3% more heat transfer from the target surface as compared to H/d=7, whereas H/d=7 gives of 6% increase in heat transfer as compared to H/d=5.

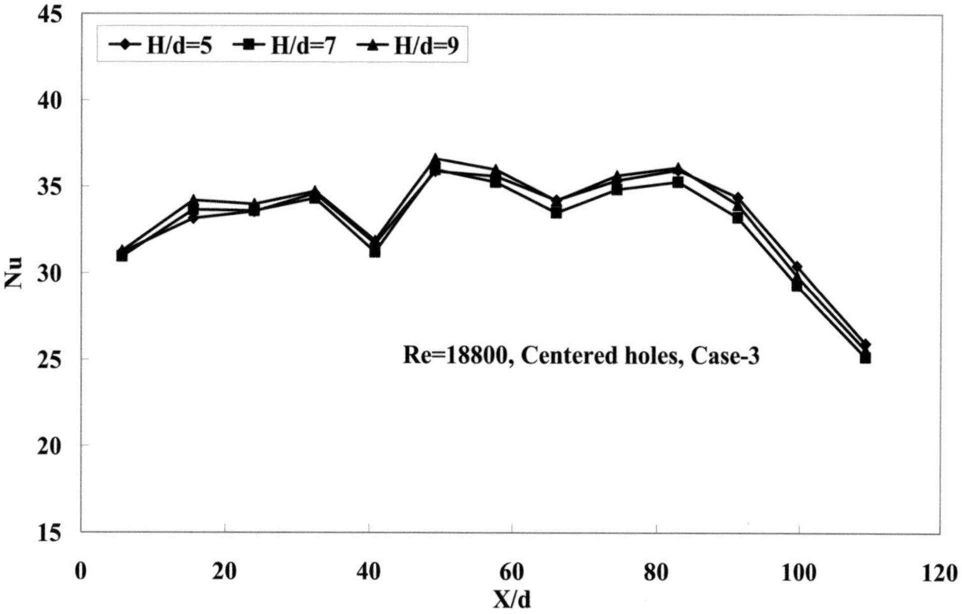

Fig. 12. Nusselt number variation for different aspect ratios and for outflow passing in both directions (for jet-orifice plate with centered holes and for Re =18800)

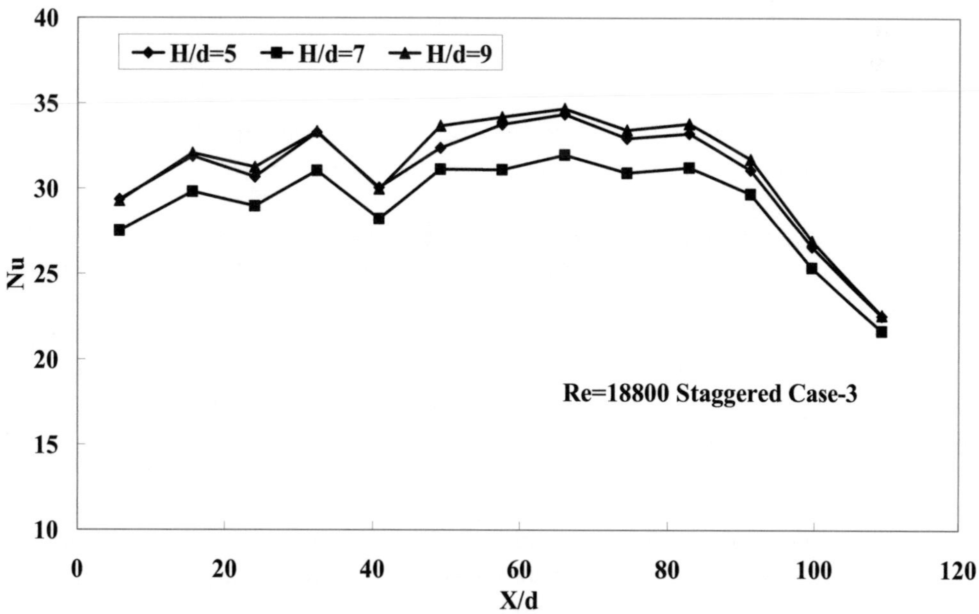

Fig. 13. Nusselt number variation for different aspect ratios and for outflow passing out in both directions (for jet-orifice plate with staggered holes and for Re =18800)

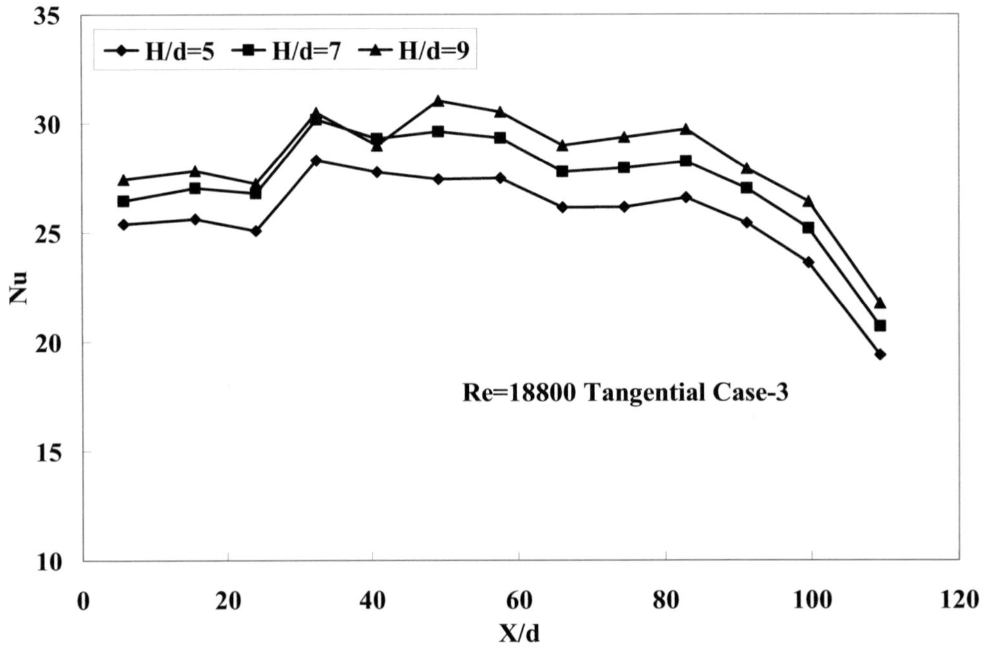

Fig. 14. Nusselt number variation for different aspect ratios and for outflow passing in both directions (for jet-orifice plate with tangential holes and for Re =18800)

5.2 Effect of orifice-jet-plate configuration on local nusselt number

Figures 12-14 also show the effect of the orifice jet plate configurations for different feed channel aspect ratios on local Nusselt number (Nu) along the surface of target surface. Orifice jet plate with centered holes has been found to give better heat transfer characteristics as compared to other plates. For H/d=5, Nu increases in percentage from staggered orifice plate to centered orifice plate by 7% and Nu increases in percentage from tangential orifice plate to staggered orifice plate by 18%. For H/d=7, Nu increases in percentage from staggered orifice plate to centered orifice plate by 11% and Nu increases in percentage from tangential orifice plate to staggered orifice plate by 6%. For H/d=9, Nu increases in percentage from staggered orifice plate to centered orifice plate by 6% and Nu increases in percentage from tangential orifice plate to staggered orifice plate by 10%.

For a given situation (H/d=9, Re=18800 and Case-3) the peak value of local Nusselt number is 36.63 at X/d=49.2 for centered jets. Nu is 34.69 at X/d=66 for staggered jets. Nu is 31.03 at X/d=49.2 for tangential jets.

5.3 Effect of orifice-jet-plate configuration and re on averaged nusselt number

The average Nu is the average of Nu of all 13 copper plates on the target surface for a given situation (i.e. for a given Re, H/d, orifice-jet configuration, outflow orientation).

Figure 15 shows the effect of different orifice jet plate configurations on average Nusselt number for outflow orientation Case-3 (outflow passing out in both directions), for different jet Reynolds numbers and for H/d=9. The Nusselt number has been found to increase with increase in Reynolds number. In general, the percentage increase in average Nusselt number in going from Plate-3 to Plate-2 is 11% and in going from Plate-2 to Plate-1 is 11%. This indicates that Plate-1 (centered orifice-jet configuration) gives higher average Nu as compared to other plates.

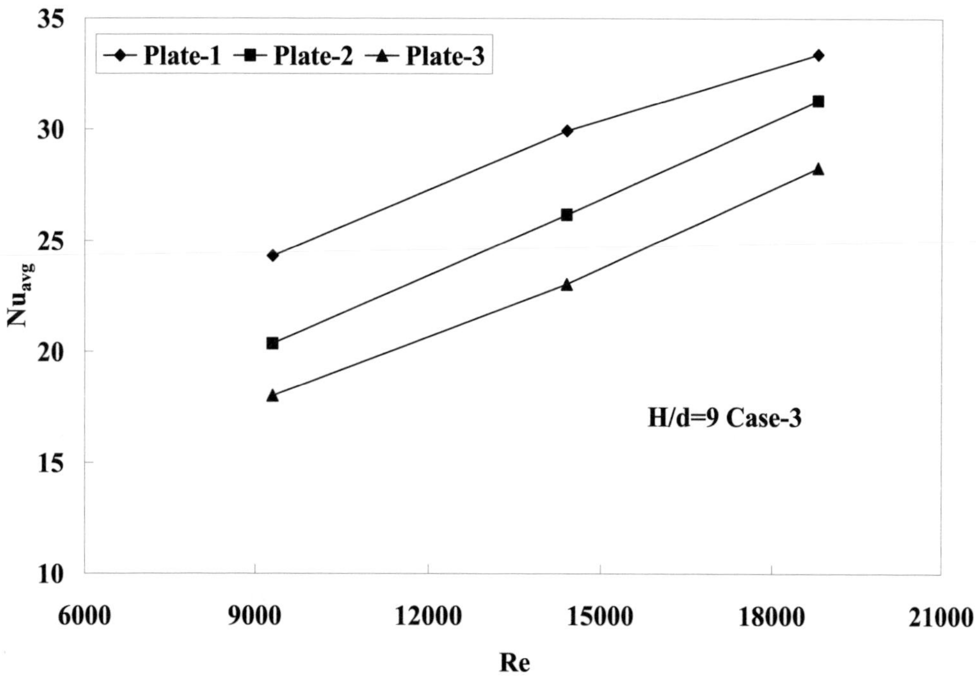

Fig. 15. Average Nusselt number distribution for different jet Re and for different orifice-jet plate configurations (for aspect ratio H/d=9, for outflow passing out in both directions – Case 3)

It is difficult to find out the exact experimental set-up in the literature which has been developed in the present study for comparison of results, however, attempt has been made to make some comparison. Figure 16 compares the results of the present study with archival results of Huang et.al [22] for different jet Re and for different outflow orientations (for a given jet-orifice plate with centered jets). Huang's study focused on multiple array jets, however our study concentrated on single array of centered/staggered/tangential jets (with an inclined target surface). Florschuetz [4] studied experimentally heat transfer distributions for jet array impingement. He considered circular jets of air impinging on heat transfer surface parallel to the jet orifice plate. The air after impingement was constrained in a single direction. Florschuetz presented Nu for centered and staggered hole patterns.

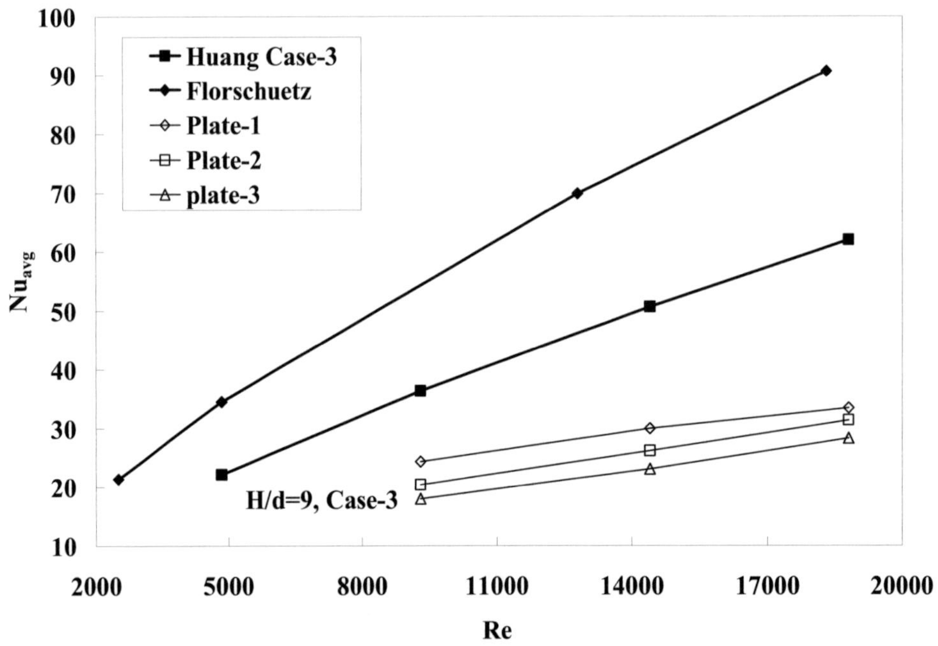

Fig. 16. Comparison of Average Nusselt number of present study with other studies for different jet Re and different orifice-jet plate configurations (for aspect ratio H/d=9, outflow in both directions - Case 3)

6. Conclusions

The above experimental work has discussed in appreciable depth the effect of orifice-jet plate configurations on feed channel aspect ratios (H/d) and on Nusselt number in a channel with inclined target surface cooled by single array of impinging jets (with outflow passing out in both radial directions). In general, it has been observed that Nu is high for higher aspect ratios. For a given plate-1 with single array of equally spaced centered jets and for Re=18800 (outflow passing in both directions), the local Nu for H/d=9 has been found to be greater than Nu of H/d=7 by 5%. The average Nu of plate-1 (centered holes) has been observed to be greater as compared to the Nu of other plate configuration (for a given Re, H/d, and outflow orientation parallel to inlet flow). The averaged Nusselt number has been found to increase with in jet Re regardless of orifice-jet plate configuration. The percentage increase in average Nu has been found to be about 11% with centered holes as compared staggered orifice-jet plate. The percentage increase in average Nu has been found to be about 11% with staggered jet-plate as compared to tangential orifice-jet plate configuration. It can be inferred that from the above results that invariably *(for different combinations impinging jet Re, feed channel aspect ratio, spacing of the target surface from the jet orifices, orifice-jet plate configuration, outflow orientation, etc)* averaged Nu increases with jet impingement cooling. This implies that jet impingement cooling is effective. This eventually results in increase in thermal efficiency and power density of the gas turbines. The observations of the above experimental work offer valuable information for researchers and designers.

7. Acknowledgment

The present work was supported by Research Institute, King Fahd University of Petroleum and Minerals, Dhahran, Saudi Arabia. The authors would like to greatly appreciate the above support. without such support, this work would not have been possible.

8. Nomenclature

$A_{cp,i}$	Area of each copper plate [m²]
A_{total}	Area of all copper plate [m²]
d	Diameter of the orifice jet [m]
h_i	Local convective heat transfer co-efficient [W/m²K]
H	Width of the feed channel [m]
I	Current supplied to heater[Amp]
l	Length of the copper plate [m]
k_{air}	Thermal conductivity of air [W/m.K]
k_{wood}	Thermal conductivity of wood [W/m.K]
Nu_i	Local Nusselt number for each copper plate
Nu_{avg}	Average Nusselt number
q''	Heat flux from the heater [W/m²]
$Q_{cp,i}$	Heat input for each copper plate [W]
Q_{actual}	Actual heat released from target surface [W]
$Q_{cond,I}$	Heat lost due to conduction [W]

$Q_{rad,i}$	Heat lost due to radiation [W]
Q_{total}	Total heat input [W]
Re	Jet Reynolds number
R	Resistance of the heater [ohm]
t	Thickness of wood block behind the heater [m]
T_{in}	Inlet temperature [°C]
$T_{s,i}$	Surface temperature [°C]
T_{surr}	Temperature of the surroundings [°C]
T_w	Wood block temperature [°C]
U	Uncertainty
V	Voltage supplied to the heater [V]
V_{avg}	Average velocity of all jets [m/s]
$\forall$	Volume flow rate [m^3/s]
X	Distance in the x-direction [m]
θ	Inclination Angle [1.5°]

9. Subscripts

cp	Copper plate
i	Index number for each copper plate
j	Jet
w	Wood

10. Greek symbols

ε	Emissivity
σ	Stefan-Boltzman constant [W/(m^{2}K^4)]
μ	Dynamic Viscosity [kg/(ms)]
ρ	Density [kg/m^3]

11. References

Chupp, P. R. E., Helms, H. E., McFadden, P. W. and Brown, T. R. (1969). Evaluation of internal heat-transfer coefficients for impingement-cooled turbine airfoils. *J. Aircraft*, 6, 203-208.

Florschuetz, L. W., Metzger, D. E., Su, C. C., Isoda, Y. and Tseng, H. H. (1984). Heat transfer characteristics for jet array impingement with initial cross flow. *Journal of Heat Transfer*, 106 (1), 34-41.

Metzger, D. E. and Bunker, R. S. (1990). Local heat transfer in internally cooled turbine airfoil leading edge regions: Part I – Impingement Cooling without Film Coolant Extraction. *Journal of Turbo machinery*, 112 (3), 451-458.

Florschuetz, L. W., Metzger, D. E., Su, C. C., Isoda, Y. and Tseng, H. H. (1981). Stream-wise flow and heat transfer distributions for jet impingement with cross flow. *Journal of Heat Transfer*, 103 (2), 337-342.

Dong, L. L., Leung, C. W. and Cheung, C. S. (2002). Heat transfer characteristics of premixed butane/air flame jet impinging on an inclined flat surface. *Heat and Mass Transfer*, 39 (1), pp. 19-26.

Rasipuram, S. C. and Nasr, K. J. (2004). A numerically-based parametric study of heat transfer off an inclined surface subject to impinging air flow. *International Journal of Heat and Mass Transfer*, 47 (23), 4967-4977.

Beitelmal, A. H., Saad, M. A. and Patel, C. D. (2000). Effect of inclination on the heat transfer between a flat surface and an impinging two-dimensional air jet. *International Journal of Heat and Fluid Flow*, 21 (2), 156-163.

Roy, S. and Patel, P. (2003). Study of heat transfer for a pair of rectangular jets impinging on an inclined surface. *International Journal of Heat and Mass Transfer*, vol. 46, no. 3, pp. 411-425.

Ekkad, S. Huang, Yahoo. and Han, Je-Chin (2000). Impingement heat transfer measurements under an Array of Inclined Jets. *Journal of Thermophysics and Heat Transfer*, 14 (2), 286-288.

Tawfek, A. A. (2002). Heat transfer studies of the oblique impingement of round jets upon a covered surface. *Heat and Mass Transfer*, 38 (6), 467-475.

Seyedein, et. al. (1994). Laminar flow and heat transfer from multiple impinging slot jets with an inclined confinement surface. *International Journal of Heat and Mass Transfer*, 37 (13), 1867-1875.

Yang, Y. and Shyu, C. H. (1998). Numerical study of multiple impinging slot jets with an inclined confinement surface. *Numerical Heat Transfer; Part A: Applications*, 33 (1), 23-37.

Yan, X. and Saniei, N. (1997). Heat transfer from an obliquely impinging circular air jet to a flat plate. *International Journal of Heat and Fluid Flow*, 18 (6), 591-599.

Hwang, J. J., Shih, N. C., Cheng, C, S., et. al. (2003). Jet-spacing effect on impinged heat transfer in a triangular duct with a tangential jet–array. *International Journal of Transfer Phenomena*, 5, 65-74.

Ramirez, C., Murray, D. B., and Fitzpatrick, J. A. (2002). Convective heat transfer of an inclined rectangular plate. *Experimental Heat Transfer*, 15 (1), 1-18.

Stevens, J. and Webb, B. W. (1991). Effect of inclination on local heat transfer under an axisymmetric free liquid Jet. *International Journal of Heat and Mass Transfer*, 34 (4-5), 1227-1236.

Hwang, J. J. and Cheng, C. S.(2001). Impingement cooling in triangular ducts using an array of side-entry wall jets. *International Journal of Heat and Mass Transfer*, 44, 1053-1063.

Hwang, J.J. and Cheng, T. T. (1999). Augmented heat transfer in a triangular duct by using multiple swirling jets. *Journal of Heat Transfer*, 121, 683-690.

Hwang, J. J. and Chang, Y. (2000). Effect of outflow orientation on heat transfer and pressure drop in a triangular duct with an array of tangential jets. *Journal of Heat Transfer*, 122, 669-678.

Hwang, J.J. and Cheng, C. S. (1999). Detailed heat transfer distributions in a triangular duct with an array of tangential jets. *Journal of Flow Visalizationa & Image Processing*, 6, 115-128.

Taylor, B. N. and Kuyatt, C. E. (1994). Guidelines for evaluating and expressing the uncertainty of NIST measurement results. *National Institute of Standards and Technology*, 1297-1303.

Hwang, Y., Ekkad, S. V. and Han, J. (1998). Detailed heat transfer distributions under an array of orthogonal impinging jets. *Journal of Thermophysics and Heat Transfer*, 12 (1), 73-79.

10

Influence of Heat Transfer on Gas Turbine Performance

Diango A., Périlhon C., Danho E. and Descombes G.
Laboratoire de génie des procédés pour l'environnement, l'énergie et la santé Conservatoire national des arts et métiers, Paris France

1. Introduction

In the current economic and environmental context dominated by the energy crisis and global warming due to the CO_2 emissions produced by industry and road transportation, there is an urgent need to optimize the operation of thermal turbomachinery in general and of gas turbines in particular. This requires exact knowledge of their typical performance.

The performance of gas turbines is usually calculated by assuming an adiabatic flow, and hence neglecting heat transfer. While this assumption is not accurate for high turbine inlet temperatures (above 800 K), it provides satisfactory results at the operating point of conventional machines because the amount of heat transferred is generally low (less than 0.5% of thermal energy available at the turbine inlet). Internal and external heat transfer are therefore neglected and their influence is not taken into account.

However, current heating needs and the decentralized production of electrical energy involve micro Combined Heat and Power (CHP) using micro-gas turbines (20-250 kW).

In aeronautics, the need for a power source with a high energy density also contributes to interest in the design of ultra-micro gas turbines.

These ultra and micro machines, which operate on the same thermodynamic principles as large gas turbines, cannot be studied with the traditional adiabatic assumption, as has been underlined by many authors such as Ribaud (2004), Moreno (2006) and Verstraete *et al.* (2007). During operation, heat is transferred from the turbine to the outside, bearing oil, casing and compressor, thus heating the compressor and leading to a drop in turbine performance. Consequently, the performances reported on the maps developed under the adiabatic assumption are no longer accurate.

This chapter presents:

- The influence of heat transfer on the performance at an adiabatic operating point of a gas turbine, and a method for determining the actual operating point knowing the amount of heat transfer.
- A study of heat loss versus the geometry scale of the volute and some conclusions concerning the limits of validity of the adiabatic assumption.

2. Relocating an adiabatic operation point subjected to heat transfer on a gas turbine map

2.1 Introduction

The performance of a turbomachine is usually represented graphically with dimensionless coordinates obtained under the assumption of adiabaticity from an existing machine.

These maps are employed by manufacturers and users to determine the overall performance in order to design a new machine or to use the same machine in different operating conditions. The results obtained are not always accurate, however, as this assumption is not valid in all circumstances. Under the influence of heat transfer, the supposedly adiabatic operating point may shift its position. The dimensionless coordinates change, making it necessary to find the actual values for a correct assessment of performance.

In order to simulate the movement of an adiabatic operating point subjected to heat transfer, we consider the single-shaft gas turbine with a simple cycle; the maps are shown in Fig. 13 (Pluviose 2005).

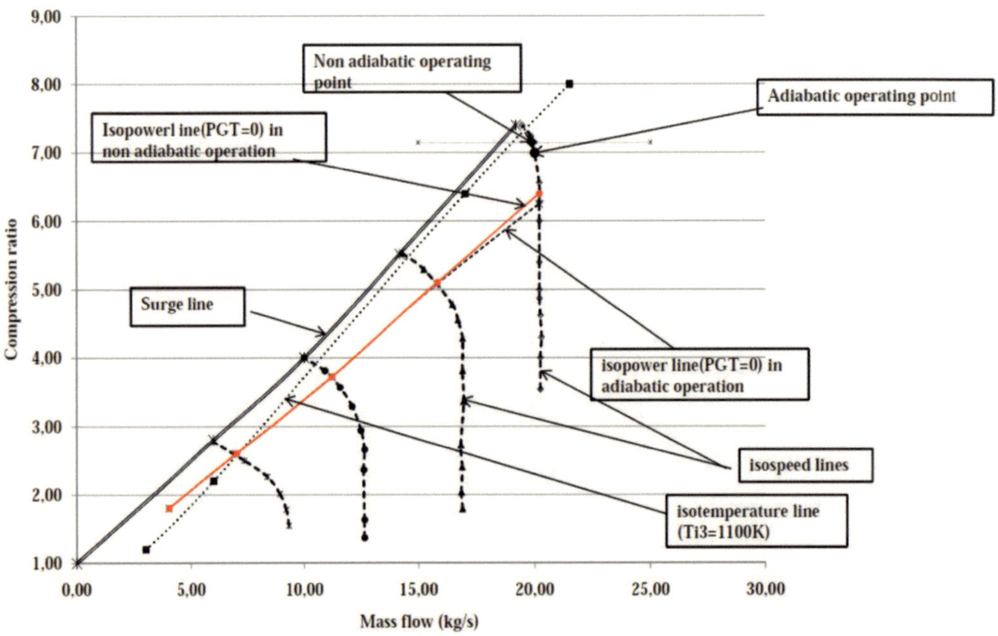

Fig. 1. Adiabatic compressor map of the gas turbine studied (Pluviose, 2005)

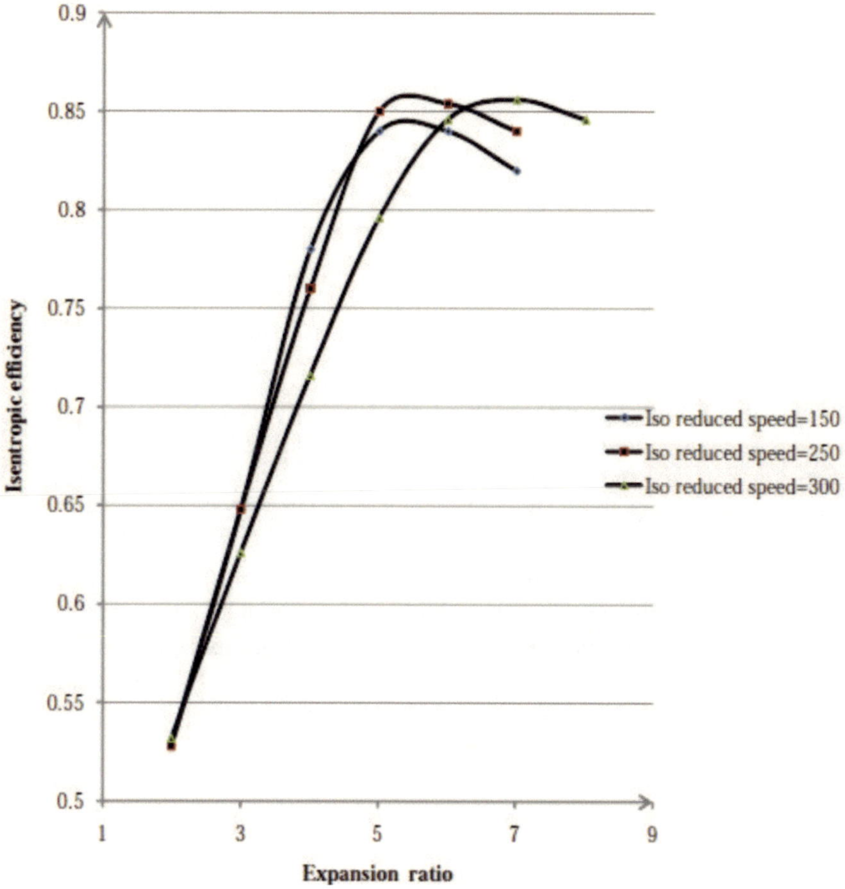

Fig. 2. Isentropic efficiency versus expansion ratio of the turbine (Pluviose, 2005)

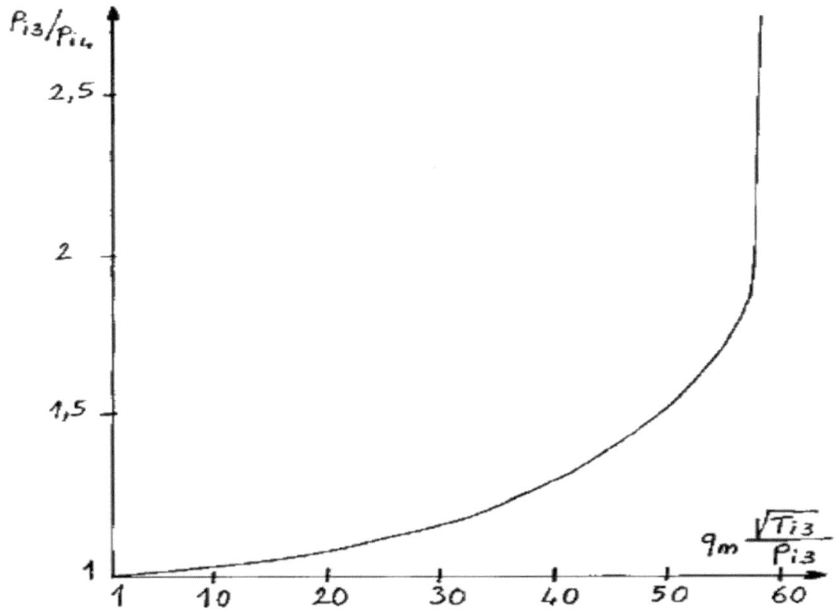

P_{i3}: turbine inlet pressure;
P_{i4}: turbine outlet pressure;
T_{i3}: turbine inlet temperature;
T_{i4}: turbine outlet temperature.

Fig. 3. Expansion ratio versus dimensionless mass flow rate (Pluviose, 2005)

2.2 Adiabatic, insulated and non insulated gas turbine versions

In its simplest form, as shown in Fig. 4, a gas turbine consists of:

- A centrifugal or axial air compressor;
- A combustion chamber in which a mixture of air and fuel is burnt;
- A centripetal or axial turbine;
- A user device (alternator, pumps, etc.).

Neglecting the kinetic and potential energy, the formulation of the first law of thermodynamics in an open system applied to turbomachinery (compressor and turbine) is written:
In transient conditions:

$$dh = \delta w + \delta q \tag{1}$$

In steady conditions:

$$\Delta h = w + q \tag{2}$$

dh: elementary variation of enthalpy;
δw: elementary work exchanged;
δq: elementary amount of heat exchanged with the surroundings;
Δh: specific enthalpy variation;
w: specific work exchanged by the fluid with the moving parts of the machine;
q: heat exchanged by the fluid with its surroundings.

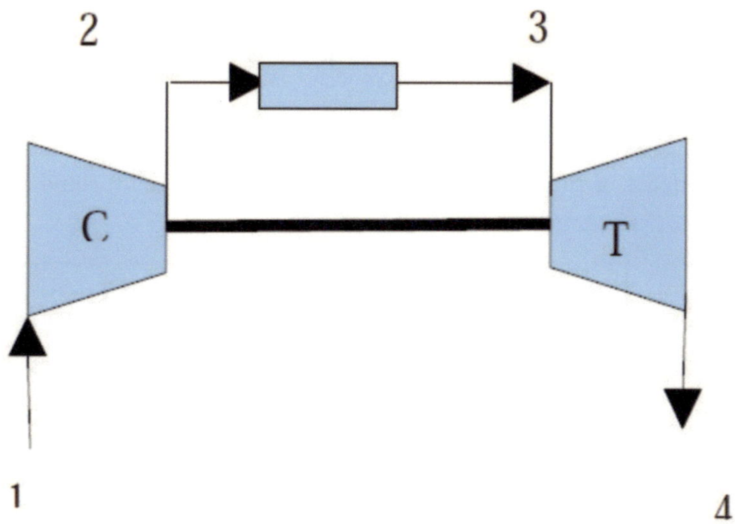

Fig. 4. Simple-cycle, open flow, single-shaft gas turbine

In conventional machines, calculations are usually done by assuming that the gas turbine is adiabatic (q = 0).

The adiabatic version of a gas turbine is one in which the heat exchanged by the fluid with surroundings in the turbomachine is exactly zero (compressor: q_{12}; turbine: q_{34}). This version cannot be obtained in practice because of the difference in temperature between the turbine inlet and the surroundings. In order to approximate this ideal configuration, experimenters introduce some thermal insulation. This leads to the concept of insulated and non-insulated gas turbines.

In an insulated gas turbine, the fluid in the turbomachine is assumed not to exchange thermal energy with the surroundings. In practice, this is achieved by insulating the machines with very low thermally conductive materials. However, because of the external insulation, internal heat exchange (in particular from the turbine to the compressor) is increased and must be taken into account.

The non insulated gas turbine is equivalent to one in which internal and external heat transfer coexist.

2.3 Characteristics of the nominal operating point of an adiabatic gas turbine (Pluviose, 2005)

The assumptions are:

- Power of mechanical losses: P_{ml} = 66 kW;
- Turbine inlet temperature: T_{i3} =973 K;
- Isentropic efficiency of the turbine: η_T = 0.85;
- Rotational speed : N=8000 rpm;
- The compressor mass flow rate: q_m: 20 kg.s^{-1};
- Inlet conditions : p_{i1} =1.013 bar et T_{i1} = 288 K ;
- Specific heat at constant air pressure: c_p=1 kJ/kg
- Specific heat at constant pressure of the burnt gas: c_p=1.13 kJ/kg

- Compression ratio : $\pi_c = 7$;
- Isentropic efficiency of the compressor : $\eta_C = 0.8$;
- Pressure drop in combustion chamber: 5%;
- Specific heat ratios : for air $\gamma = 1.4$; for burnt gas $\gamma = 1.33$;
- Turbine outlet pressure: $p_{i4} = 1.05$ bar.

The characteristics of the operating point of the adiabatic gas turbine are summarized in Table 1.

<table>
<tr><th></th><th>q_m ($kg.s^{-1}$)</th><th>π_C or π_T</th><th>N (rpm)</th><th>P_{GT} (kW)</th><th>T_{i3} (K)</th><th>Q_{cc} (kW)</th><th>η_{GT} %</th></tr>
<tr><td>Adiabatic compressor</td><td>20</td><td>7</td><td>8 000</td><td rowspan="2">1 526.4</td><td rowspan="2">973</td><td rowspan="2">9 431</td><td rowspan="2">16.2</td></tr>
<tr><td>Adiabatic turbine</td><td>20</td><td>6.42</td><td>8 000</td></tr>
</table>

Table 1. Characteristics of the operating point of an adiabatic gas turbine

The energy balance at the operating point is shown in Table 2 (see the detailed calculations in the appendix).

	Energy balance	
	kW	%
Thermal power provided by fuel	9430.7	100
Gas Turbine power	1526.2	16.2
Thermal power loss at exhaust	7838.5	83.12
Mechanical losses	66	0.7
Thermal losses	0	0

Table 2. Energy balance at the operating point of the adiabatic gas turbine

2.4 Characteristics of the nominal operating point of a non-adiabatic gas turbine

As indicated in section 2.2, there are two non-adiabatic versions of the gas turbine: the insulated and the non-insulated version.

2.4.1 Influence of heat transfer on the adiabatic nominal operating point

In order to understand the influence of heat transfer on the nominal operating point, we assume that the turbine is cooled so that the heat losses account for 15% of the adiabatic work. For the non-insulated version, 60% of these losses are considered to contribute to the heating of the compressor (Rautenberg & al. 1981).
In the insulated version, it is assumed that all the heat lost by the turbine is received by the air in the compressor.
In this study, the amount of heat exchanged is assumed known. The internal work depends on the outlet temperature. In practice, during operation, the outlet temperature of the machine can be measured. But, here, we choose $T_{i2} > T_{i2ad}$ (compressor) and $T_{i4is} < T_{i4} < T_{i4ad}$ (turbine).
Table 3 and Table 4 summarize the new performances calculated for the adiabatic gas turbine used in insulated and non-insulated versions at the adiabatic operating point.

	q_m (kg^{-1})	π_C ou π_T	N (rpm)	P_{GT} (kW)	T_{i3} (K)	Qcc (kW)	η_{GT} %
Heated compressor	20	7	8 000	1 128	973	8 521	13.2
Cooled Turbine	20	6.42	8 000				

Table 3. Characteristics of nominal operating point in non insulated version

	q_m (kg^{-1})	π_C ou π_T	N (rpm)	P_{GT} (kW)	T_{i3} (K)	Qcc (kW)	η_{GT} %
Heated compressor	20	7	8 000	1010.38	973	7917	12.8
Cooled Turbine	20	6.42	8 000				

Table 4. Characteristics of nominal operating point in insulated version

The comparison of the results in Table 1, Table 3 and Table 4, leads to the following comments:

- Energy efficiency has dropped from 16.2 to 13.2% (Table 1 and Table 3), and from 16.2 to 12.8% (Table 1 and Table 4);
- Net power of the gas turbine has decreased from 1 526 to 1 128 kW, or by 26% (Table 1 and Table 3); from 1526 to 1010 kW, or by 34% (Table 1 and Table 4);

We can therefore conclude that if the gas turbine operates with heat transfer while maintaining the same parameters as under nominal adiabatic operation, there is a drop in performance.
This significant drop in performance makes it necessary to determine the actual operating point, taking into account heat transfer and the needs of user devices. For example, in a power plant equipped with a gas turbine, meeting the needs of the consumer requires that the power be kept constant. This involves finding the new non-adiabatic operating point which fulfills this criterion (same power at constant rotational speed).

2.4.2 Search for a new operating point able to provide the same power

Search for the new operating point of the compressor

The gas turbine operates under adiabatic or non-adiabatic conditions at 8000 rpm. For this speed, the output power is plotted versus the compression ratio in the three configurations: adiabatic, insulated and non-insulated versions (Figure 5).
For the selected power value, the new compression ratios in insulated and non-insulated operation can be deduced. Then drawing this pressure ratio on the compressor map (Fig. 6), the mass flow rate and the efficiency of this point are deduced.

Comments:

It can be seen on Figure 5 that for the same compression ratio, the net output power is low in the insulated version. The highest output power is obtained in the adiabatic version.
For the same power, the compression ratio is low in the non-insulated version. The lowest value is obtained in the adiabatic version.

Search for the new operating point of the turbine

As the rotational speed is constant and imposed, the required power can be achieved only by means of the quantity of injected fuel which has a direct influence on T_{i3} (turbine inlet temperature).

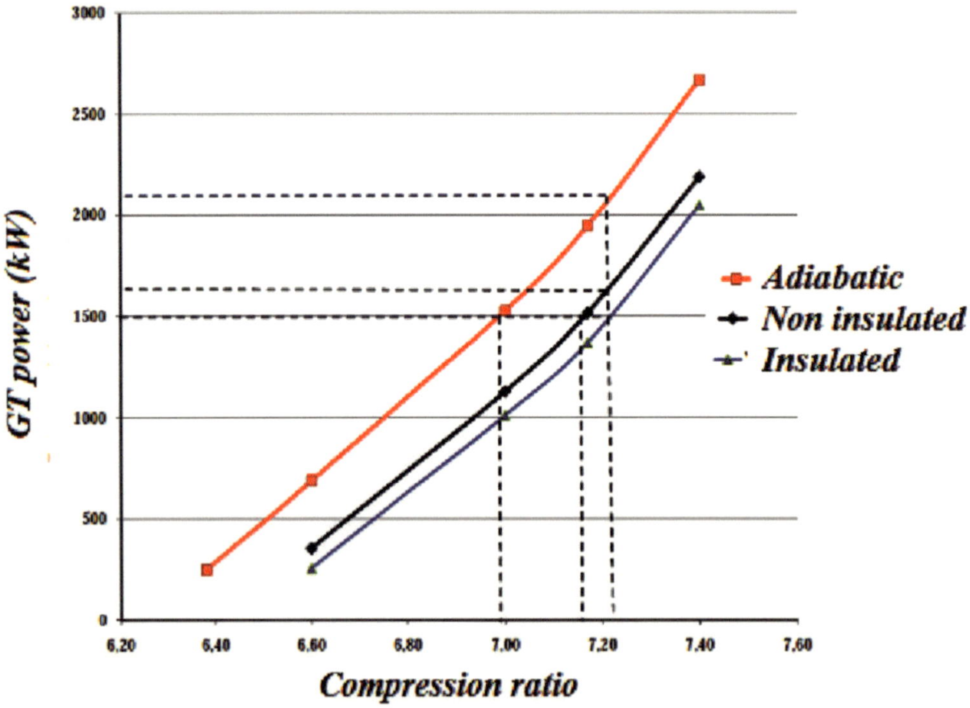

Fig. 5. Output power of the gas turbine versus compression ratio in adiabatic, non-insulated and insulated version

Due to the turbine characteristics, for a pressure ratio above that shown in Fig.3, the reduced mass flowing through the turbine is a constant which was calculated for the nominal operating point in adiabatic conditions (Pluviose, 2005). A reduced mass flow makes it possible to determine the new value of T_{i3} corresponding to the new pressure p_{i3}.

$$q_m \frac{\sqrt{T_{i3}}}{p_{i3}} = cst \quad (3)$$

q_m : mass flow rate ($kg.s^{-1}$);
T_{i3} : turbine inlet temperature (K);
p_{i3} : turbine inlet pressure (p_a).

Non-insulated gas turbine

The characteristics of the new operating points are summarized in table 5 (see calculations in the appendix).
Comparing the results of Table 1 to Table 5, it can be observed that:

- The mass flow rate has decreased. It varies from 20 to 19.8 kg^{-1}. The relative deviation is 1%;
- The compression ratio has increased from 7 to 7.17 with a relative deviation of 2.4%;
- The turbine inlet temperature has risen from 973 to 1 041 K. The maximum is 1100 K;
- The energy efficiency has decreased to 16.2 à 15.6% (the relative deviation is 3.7%).

	q_m ($kg.s^{-1}$)	π_C ou π_T	N (rpm)	P_{GT} (kW)	T_{i3} (K)	Q_{cc} (kW)	η_{GT} %
Heated compressor	19.8	7.17	8 000	1526.2	1040.6	9764	15.6
Cooled turbine	19.8	6.57	8 000				

Table 5. Characteristics of the new operating point *in the non-insulated version*

Insulated gas turbine

In order to simplify calculations, we consider that all the heat lost by the turbine is fully received by the compressor

	q_m ($kg.s^{-1}$)	π_C ou π_T	N (rpm)	P_{GT} (kW)	T_{i3} (K)	Q_{cc} (kW)	η_{GT} %
Heated compressor	19.5	7.22	8 000	1526.2	1088	10253	14.9
Cooled turbine	19.5	6.62	8 000				

Table 6. Characteristics of the new operating point in the *insulated version*

When the results of tables 1 and 6 are compared, it can be seen that:

- The mass flow rate has decreased from 20 to 19.5 $kg.s^{-1}$. The relative deviation is 2.5%;
- The compression ratio has increased from 7 to 7.22. The relative increase is 3.14%;
- The turbine inlet temperature has increased from 973 to 1 088 K. The limit is 1 100 K;
- The energy efficiency has dropped from 16.2 to 14.9% (the relative deviation is 8.02%).

Overall in the two operating configurations, the operating area on the compressor map has slightly narrowed.

However, the temperature increase can be a problem, as this value has a direct influence on the turbine life span.

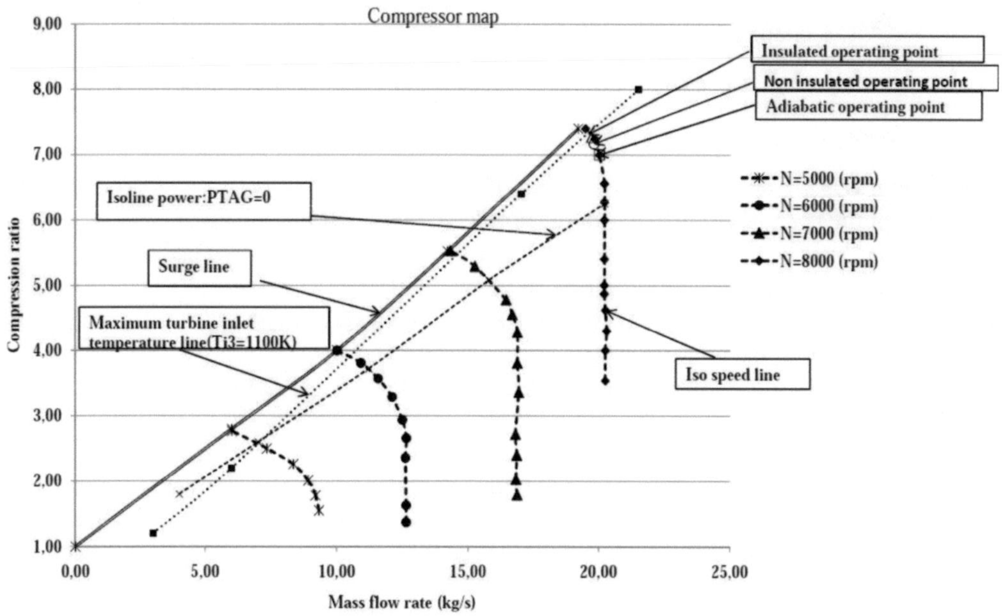

Fig. 6. Adiabatic compressor map with operating points in the three configurations

2.4.3 Energy balance of new operating points

Non-insulated gas turbine

	Energy balance	
	kW	%
Calorific power provided by the fuel	9764	100
GT power	1526.2	15.6
Exhaust power	7129.6	73.0
Mechanical losses	66	0.7
Thermal losses	1042	10.7

Table 7. Energy balance of the new operating point (non-insulated gas turbine)

Insulated gas turbine

	Energy balance	
	kW	%
Calorific power provided by the fuel	10253	100
GT power	1526.2	14.9
Exhaust power	7618.8	74.3
Mechanical losses	66	0.6
Thermal losses	1042	10.2

Table 8. Energy balance of the new operating point (insulated gas turbine)

Comparing tables 7 and 8, we can see that at iso speed and iso net power produced, the efficiency of the gas turbine is better in the non-insulated version.

2.5 Comparison with experimental results

The analysis and the results presented above for the nominal operating point were extended to the other points of the working area.

Figure 8 shows the experimental results obtained by Moreno (2006) on a small gas turbine (75 kW).

The tests were carried out in two versions: an insulated version at 39 000 rpm and a non-insulated one at 40 000 rpm. It may be noted that the speeds are not identical because of the practical difficulties of measurement in testing. But the relative difference of 2.5% between these two speeds can be considered negligible.

Figure 9 shows that, as in the case of our study, iso-speed, iso net power produced by the gas turbine, and energy efficiency are better in the insulated than in the non-insulated version.

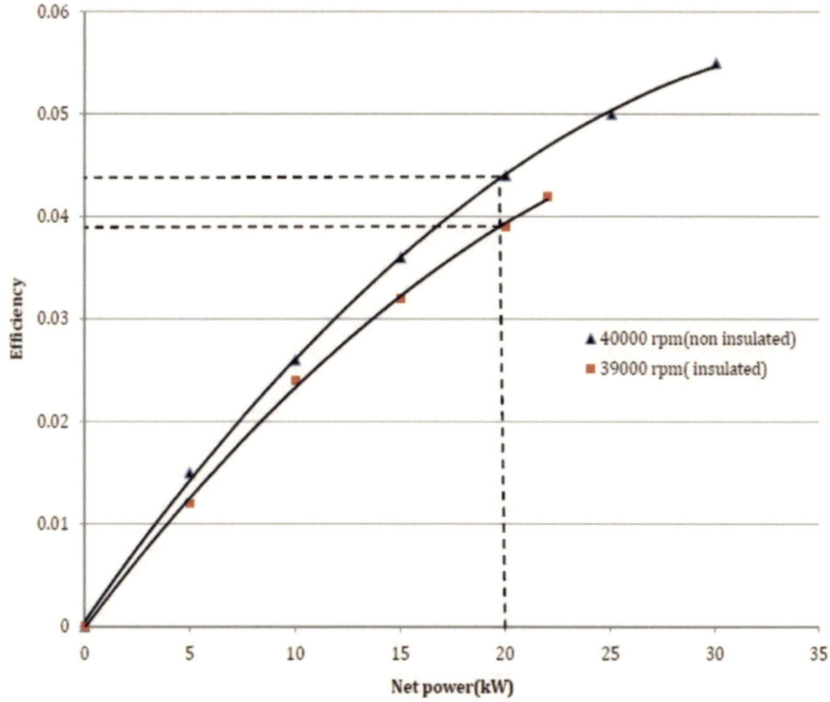

Fig. 7. Energy efficiency versus net power produced (Moreno, 2006)

This study not only confirms the decrease in performance due to heat losses, but also that this drop in performance is proportionally greater with the internal heat transfer.
Compared with the gas turbine studied here (1500 kW), it can be seen that the energy efficiency of the gas turbine used by Moreno (75 kW) is very low: 8% vs 16% at nominal power. This can be attributed to the size of the machine: it is a small machine with a nominal power about approximately twenty times smaller. Heat losses could be the cause of the drop in performance.

3. Heat transfer and geometric scale of gas turbines

As already mentioned in the introduction, the results of the performance calculations carried out in conventional turbomachines remain satisfactory at the full load operating point. In addition, the literature indicates that the impact of heat transfer on the performance of small turbomachines is negative. In these circumstances, it is important to know the characteristic size of the machines in which the assumption of adiabaticity is no longer valid.
This study of heat transfer limited only to the volute of machines studied is conducted in similar operating conditions. It therefore calls on the notion of similarity.

3.1 The similarity of turbomachines: a summary

Similarity makes it possible, when a physical phenomenon for given operating conditions is known, to predict the same phenomenon for other conditions through laws involving dependent and independent dimensionless variables. Similarity generally focuses on two aspects: the geometric aspect that is relative to a family of geometrically similar machines, and the functional aspect that deals with a family of machines with similar operation. These two aspects are simultaneously taken into account.

For adiabatic machines, the dimensionless independent variables used to characterize the similar operating points are (Pluviose, 2005):

- dimensionless mass flow rate
- dimensionless speed.

For adiabatic machines, the dimensionless independent variables used to characterize the similar operating points are (Pluviose, 2005):

- Dimensionless mass flow rate:

$$\frac{q_m\sqrt{rT_{i1}}}{p_{i1}R^2}$$

- Dimensionless rotational speed:

$$\frac{NR}{\sqrt{rT_{i1}}}$$

q_m : mass flow rate ($kg.s^{-1}$);
T_{i1} : turbomachine inlet temperature(K);
p_{i3} : turbomachine inlet pressure (p_a);
R : external radius of the rotor a (m);
r : specific perfect gas constant ($J.kg^{-1}$).

A study of similarity in non-adiabatic turbomachines operating with compressible fluid, conducted by Diango (2010) led to the generalization of Rateau's theorem. The author shows that these two dimensionless variables are also valid when operating with heat transfer.

From the foregoing and for a judicious comparison of heat exchange in different volutes, it is generally assumed that the fluid flows are similar. This leads to the following assumptions:

- Inlet parameters are the same (pressure and temperature);
- Reynolds numbers are equal;
- The inlet dimensionless velocities are identical;
- The mass flow rate is the same in the volutes.

In the heat transfer equations, only the mass flow rate and the inlet conditions are involved. For two machines a and b, the first and last assumptions imply:

$$\frac{q_{ma}\sqrt{rT_{i1}}}{R_a^2 p_{i1}} = \frac{q_{mb}\sqrt{rT_{i1}}}{R_b^2 p_{i1}} \Leftrightarrow q_{mb} = \frac{q_{ma}R_b^2}{R_a^2}$$

$$q_{mb} = \frac{q_{ma}R_b^2}{R_a^2} \tag{4}$$

q_{ma}: mass flow rate of machine a ($kg.s^{-1}$);
q_{mb}: mass flow rate of machine b ($kg.s^{-1}$);
p_{i1}: inlet pressure (p_a);
T_{i1}: inlet temperature (K);

R_a: external radius of the rotor of machine a (m);
R_b: external radius of the rotor of machine b (m);
r: specific perfect gas constant ($J.kg^{-1}$).
Due to the complex geometry of the casing (volute) of turbomachines and the difficulties of calculating heat transfer coefficients, a numerical approach has been adopted.

3.2 Mathematical modeling of heat transfer in the gas turbine volute

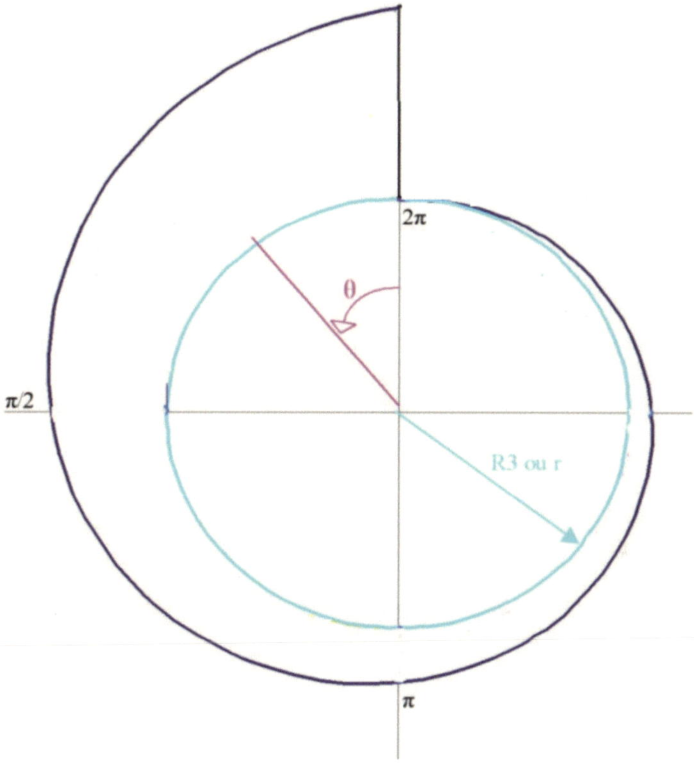

Fig. 8. External shape of the volute

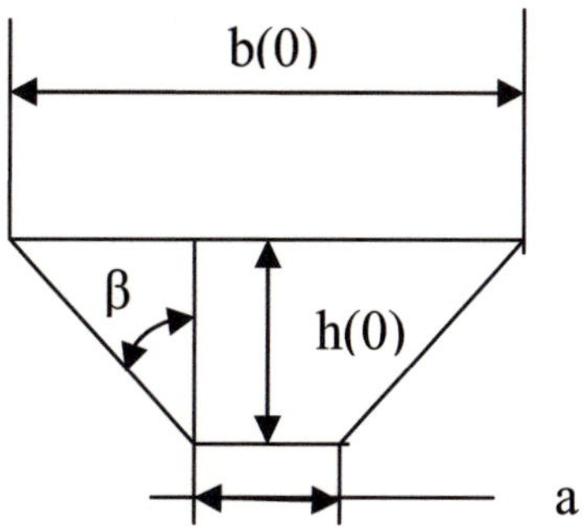

h (0): Inlet height (m)

Fig. 9. Inlet section of the volute (θ = 0)

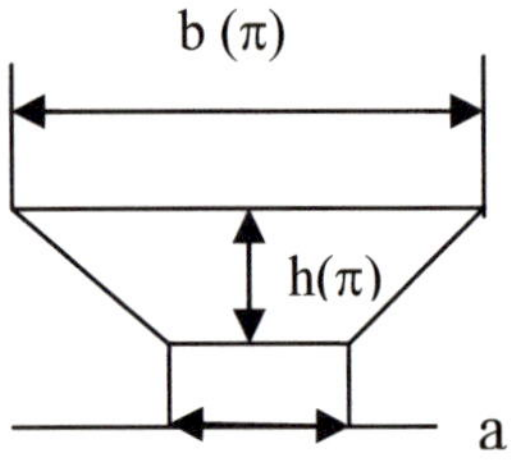

h (π): height at θ = π

Fig. 10. Section at θ = π

The outer shape is a logarithmic spiral.

3.2.1 Energy balance on a mesh volume

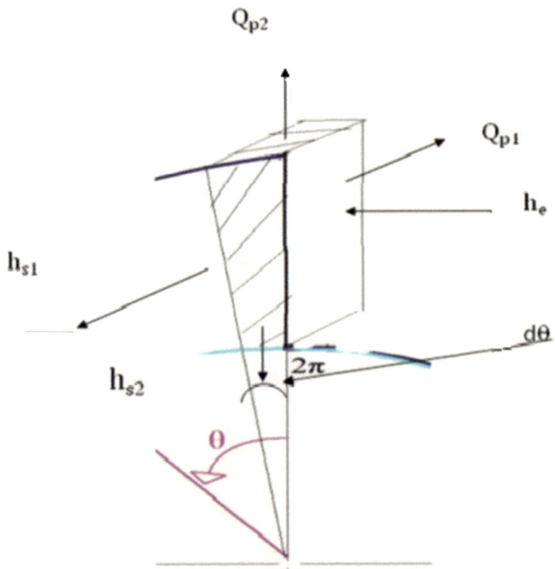

Fig. 11. Mesh volume of the turbine volute

The first law of thermodynamics can be written in any time by the following equation:

$$\frac{dU}{dt} = \sum_i \dot{q}_i + \sum_j \dot{m}_j h_j^* + \sum_k \dot{w}_k \tag{5}$$

$$h_j^* = \left(h_j + \frac{v_j^2}{2} + gz_j \right) \tag{6}$$

$\dot{q}_i$: Heat exchanged by the fluid with surroundings (W);

h_j^* : Total specific enthalpy of component j in the mesh volume (J.kg^{-1});

h_j : Specific enthalpy of component j in the mesh volume (J.kg^{-1});

$\dot{w}_k$: Mechanical power exchanged by surroundings (W)

$\dot{m}_j$: Flow of material exchanged by the component j with surroundings (W);

$\frac{v_j^2}{2}$: Kinetic energy of component j;

gz_j : Potential energy of component j;

In the volute, the fluid does not exchange mechanical energy with the surroundings, so:

$$\sum_k \dot{w}_k = 0$$

Variations in kinetic and potential energy are generally negligible compared to the enthalpy. Equation (5) becomes:

$$\frac{dU}{dt} = \sum_i \dot{q}_i + \sum_j \dot{m}_j h_j \tag{7}$$

In steady state, equation (7) becomes:

$$q_{m\,in} h_{in} - q_{m\,out1} h_{out1} - q_{m\,out2} h_{out2} - \dot{q}_{w1} - \dot{q}_{w2} = 0 \tag{8}$$

$\dot{q}_{w1,2}$: Heat exchanged through the walls;

$h_{in1,2}$: Total specific enthalpy at inlet of the mesh volume (J.kg^{-1});

$h_{out1,2}$: Total specific enthalpy at outlet of control volume (J.kg^{-1});

$\dot{w}_k$: Mechanical power exchanged by the fluid with surroundings (W);

$$\dot{q}_p = \dot{q}_{p1} + \dot{q}_{p2}$$

It is assumed in a first approximation that the specific heat of the burnt gas at constant pressure (cpf) does not vary in the mesh volume and that the inlet temperature of gas in the upstream guide is identical to the outlet of the volume vi.
Knowing that:

$$q_{m_{in}} = q_{m_{out1}} + q_{m_{out2}}$$

Noted henceforth as:

$$q_{m_{in}} = q_{m_{i-1}}$$

The expression (8) becomes :

$$q_{m_{i-1}} c_p T_{i-1} = q_{m_{i-1}} c_p T_i + \dot{q}_p \tag{9}$$

$$\dot{q}_p = h_{eq} S\left(T_{burnt\,gas} - T_a\right) \tag{10}$$

S : Heat exchange surface (m²);
h_{eq} : Equivalent convective heat transfer coefficient (W.m^{-2}.K^{-1});
$T_{burnt\,gas}$: Average temperature of the burnt gas in the control volume (K);
T_a : Average ambient temperature (K)

3.2.2 Heat exchange surface

The mesh volume is considered as a tube in which fluid flows. The elementary exchange surface is the area of the contour of the tube, not including the passage through the upstream guide. It is equal to the perimeter multiplied by the elementary length of the spiral.

The equation of the outer profile of the logarithmic spiral volute ranging from 0 to 2π is given by equation (11), following Moreno (2006):

$$R(\theta) = ae^{-b\theta} + c \tag{11}$$

$R(0)$, $R(\pi)$ and $R(2\pi)$: external radius of the volute at θ = 0 ; π and 2π (m).
Assuming $R(0)$, $R(\pi)$ and $R(2\pi)$ are known, we can write the following relationships, from Moreno (2006), which describe the outer shape of the volute:

$$c = \frac{R(2\pi)R(0) - R^2(\pi)}{R(2\pi) + R(0) - 2R(\pi)} \tag{12}$$

$$b = -\frac{1}{2\pi}\ln\left[\frac{R(2\pi)-c}{a}\right] \tag{13}$$

$$a = R(0)-c \tag{14}$$

a and b: constant real numbers (m)
The arc length of the spiral at the position θ is given by equations (15) and (16), from Berger & Gostiaux (1992):

$$L(\theta) = \int_a^b \sqrt{R'(\theta)^2 + R(\theta)^2}\, d\theta \tag{15}$$

$$L(\theta) = \int_u^v \sqrt{a^2 b^2 e^{-2b\theta} + \left(ae^{-b\theta} + c\right)^2}\, d\theta \tag{16}$$

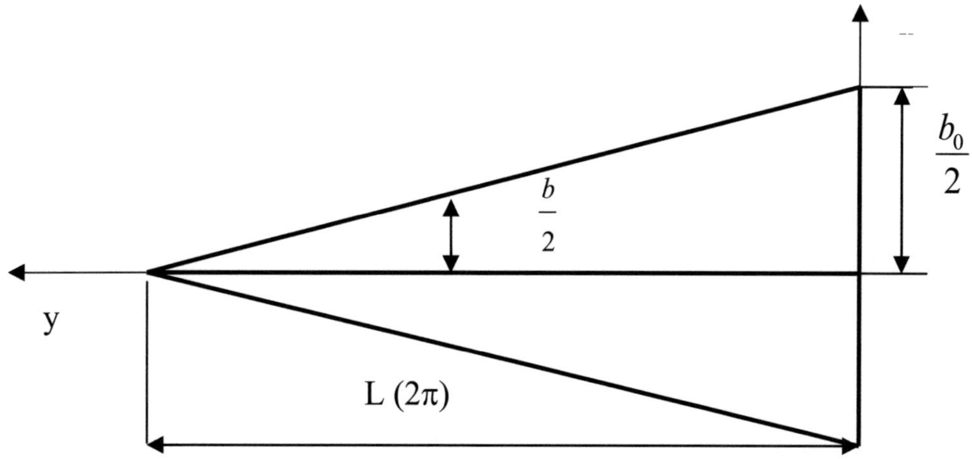

Fig. 12. The width of the upper volute versus the unrolled spiral

The flow section has a trapezoidal shape whose dimensions must be calculated in order to estimate the perimeter.
To determine the large base b (θi), of the trapezium, the equation used is:

$$b(\theta)=b_0\left[1-\frac{L(\theta)}{L(2\pi)}\right] \tag{17}$$

The angle generated by the height and the lateral side (Figure 9, Figure 10) is expressed by:

$$tg\beta=\frac{b(\theta)-l_{tn}}{h(\theta)} \tag{18}$$

l_{tn}: turbine nozzle width.
The length of the lateral side (L_l) is:

$$L_l(\theta)=\frac{h(\theta)}{\cos\beta} \tag{19}$$

Hence the perimeter is:

$$P(\theta)=2L_l(\theta)+b(\theta) \tag{20}$$

The heat exchange surface of the mesh volume is then:

$$dA=PdL=\left[2L_l(\theta)+b(\theta)\right]dL$$

$$A=\left[2L_l(\theta)+b(\theta)\right]\int_u^v\sqrt{a^2b^2e^{-2b\theta}+\left(ae^{-b\theta}+c\right)^2}\,d\theta \tag{21}$$

3.2.3 Heat exchange coefficient

Forced convection inside the volute

The heat transfer coefficient is given by equation (22)

$$h_{fc}(v)=\frac{Nu(v_i)\lambda_{burnt\,gas}}{D_h(v_i)} \tag{22}$$

hfc: coefficient of forced convection ;
λburnt gas: conductivity of burnt gas;
Dh: hydraulic diameter;
vi: mesh volume.
The forced convection model uses the simplified correlation of Sieder & Tate (Kreith 1967) to calculate the Nusselt number (Nu) from the Prandtl (Pr) and Reynolds (Re) numbers:

$$Nu = 0{,}023 Re^{0{,}8} Pr^{0{,}3} \tag{23}$$

$$Pr=\frac{\mu_{burnt\,gas} C_{p\,burnt\,gas}}{\lambda_{burnt\,gas}} \tag{24}$$

$$Re(v_i)=\frac{q_m D_h(v_i)}{\mu_{burnt\,gas} S(v_i)} \tag{25}$$

$\mu_{burnt\,gas}$: dynamic viscosity of burnt gas;

Free convection outside the volute

The Nusselt number is given by the Morgan correlation (Padet 2005) from the Rayleigh number (Ra).

$$\overline{Nu} = A R_a^n \tag{26}$$

$$Ra=\frac{g\beta D_h^3 (T_w - T_a)}{\alpha \nu_a} \tag{27}$$

T_W : wall temperature;
ν_a : Kinematic viscosity of the air;
T_a: air temperature.
The coefficient of external heat exchange is the combination of a coefficient by natural convection and exchange by radiation defined by equation 28.

$$h_{amb} = h_{nc} + h_{rad} \tag{28}$$

h_{amb}: coefficient of heat exchange with ambient air;
h_{nc}: coefficient of heat exchange by natural convection;
h_{rad}: coefficient of heat exchange by radiation.

The coefficient of heat exchange by radiation is given by equation (29):

$$h_{rad}(v_i)=\frac{\varepsilon\sigma\left(T_w^4-T_a^4\right)}{T_w-T_a} \quad (29)$$

The equation of the model of mass flow in the volume is (30):

$$q_{m\theta}=q_{m0}\left(1-\frac{\theta}{2\pi}\right) \quad (30)$$

$q_{m\theta}$: the mass flow at position θ;
q_{m0}: the mass flow at position 0.

3.3 Numerical results of modeling

Calculations are started from a nominal operating point of a turbine of a turbocharger whose dimensions are known (see Table 9).
Geometric similarity requires that the linear dimensions of the similar volutes be multiplied by the same factor. Four other turbines have therefore been considered. The main dimensions and corresponding mass flow are recorded in table 10, which also shows the numerical results of the modeling.

T_{i3} (K)	q_m (kg.s^{-1})	T_w (K)	N (rpm)
761.3	0.848	760	110 000

Table 9. Nominal point of the reference turbine

$$Q_{wdim}=\frac{Q_w}{q_m C_p T_{i3}} \quad (31)$$

Q_w: Thermal power lost through the walls of the volute;
Q_{wdim}: dimensionless thermal power lost through the walls of the volute.

R(0) (m)	R(π) (m)	R(2π) (m)	b (m)	q_m (kg.s^{-1})	$\frac{S}{V}$ (m^{-1})	Q_W (W)	Q_{Wdim} x10^{-4}
0.03104	0.0266	0.020325	0.017	0.0212	463	202	106
0.04656	0.0399	0.03049	0.0255	0.0477	308	273	64
0.06208	0.05321	0.04065	0.034	0.0848	231	341	45
0.0776	0.06651	0.05081	0.0425	0.1325	185	408	34
0.09312	0.07981	0.060975	0.05099	0.1908	154	474	28

Table 10. Heat loss through the volutes of different sizes

Figures 13 and 14 give the modeling results. Figure 13 shows the evolution of the heat exchange surface versus the inlet radius. The greater the volute, the smaller the surface to volume ratio. Small turbomachines therefore have a higher surface to volume ratio.
The necessity of taking into account heat transfer in small turbomachines is largely confirmed by Figure 14: the heat losses in the volute are relatively greater.
In this study, when the inlet radius is halved, the surface to volume ratio doubles and the heat losses are multiplied by about 2.5

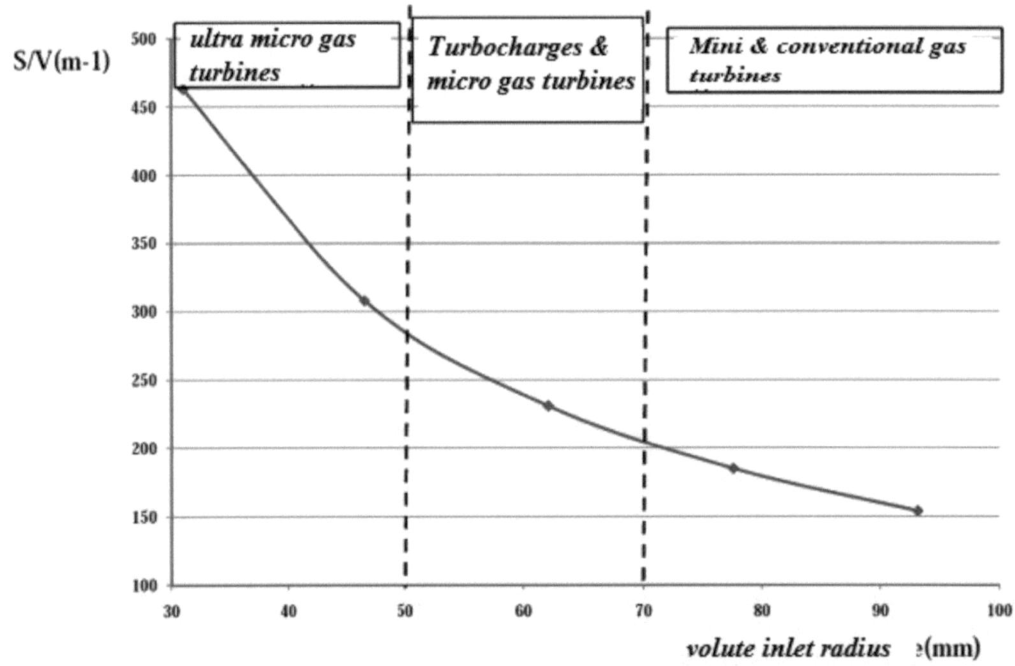

Fig. 13. Ratio of heat exchange surface (S) and the volume (V) of the volute versus the inlet radius

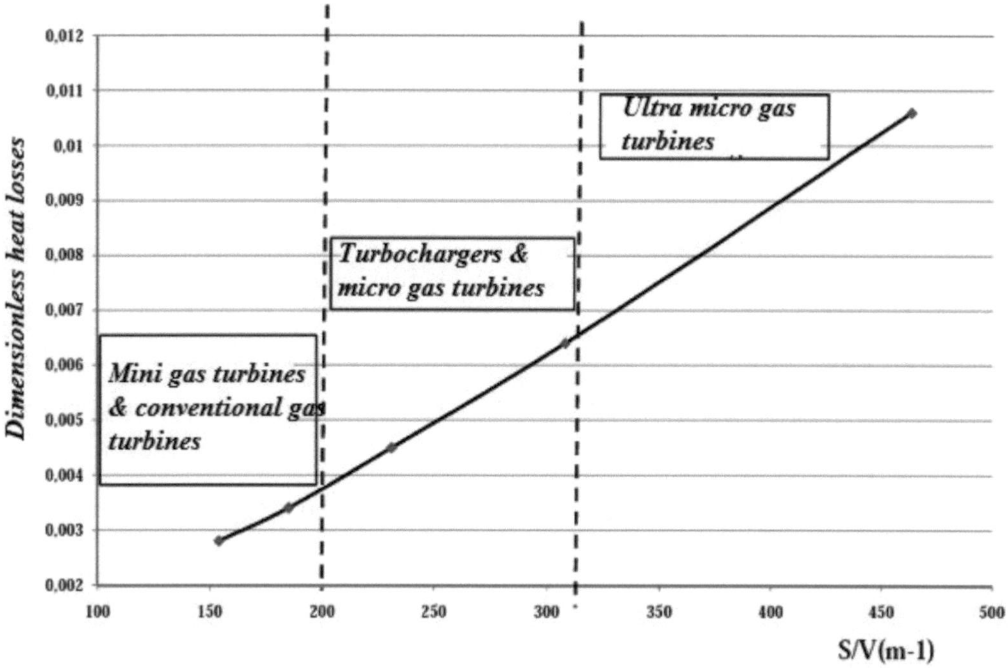

Fig. 14. Heat transfer in the volute versus the size of the machine

4. Conclusion

Internal and external heat transfer induces a drop in the performance of gas turbines. This study shows that the performance of small turbomachines evaluated with the assumption of adiabaticity is not accurate.

For a given operating point, the mass flow and the compression ratio recorded on the maps and the calculated performance do not correspond to the actual characteristics when the machine operates with heat transfer.

The assumption that heat losses represent 15% of the work of adiabatic turbines, of which 60% is received by the compressor (non-insulated), leads to overestimating the power by 35% and the energy efficiency by 23%

Insulation of the turbine, although it seems to be a solution to maintain the operating characteristics of adiabatic turbines, leads in fact to increasing the drop in performance.

For the insulated version, the net power is overestimated by 51% and efficiency by 26.6%. In the absence of an adiabatic gas turbine (ideal machine), which provides the best performance, we must avoid insulating the turbine in order not to decrease performance still further.

To maintain the level of performance, and in particular the net power produced by the gas turbines, despite heat transfer, adjustments are needed. They consist mainly in increasing the fuel flow, resulting in an increase in the turbine inlet temperature. In the case of our study, the fuel flow increase is 3.5% in the non-insulated version and 8.5% in the insulated version. The turbine inlet temperature increase is 6.4% in the insulated version and 11.8% in the non-insulated version.

Finally, this study confirms that the assumption of adiabaticity is not valid in turbochargers, micro and ultra-micro gas turbines. Compared to the available thermal energy at the turbine inlet, heat losses increase with the surface to volume ratio which decreases in small-sized machines. The quality of operation of small turbomachinery cannot be characterized with isentropic efficiency which has no physical meaning because of the relative importance of heat transfer.
The proposal of a new performance indicator and the development of new maps available for any type of thermal turbomachines will therefore be the subject of our forthcoming investigations.

5. Acknowledgment

The authors would like to acknowledge the French Cooperation EGIDE for funding this study.

6. Appendix: Energy balance calculations

1. Adiabatic gas turbine:

Data: (see page 6)
Power of the gas turbine: P_{GT}
Compressor power : Pc

$$P_C = q_m c_p \left(T_{i2} - T_{i1}\right) \; ; \; T_{i2}=T_{i1}+\left[\left(\frac{p_{i2}}{p_{i1}}\right)^{\frac{\gamma-1}{\gamma}}-1\right]\frac{T_{i1}}{\eta_C}=288+\left[(7)^{\frac{0,4}{1,4}}-1\right]\frac{288}{0,8}=555.71\,\text{K}$$

$$P_C=q_m c_p\left(T_{i2}-T_{i1}\right)=20\times 1(555.71-288)=5354.2\,kW$$

Turbine power P_T:

$$P_T=q_m c_p\left(T_{i4}-T_{i3}\right) \; ;$$

$$T_{i4}=T_{i3}+\left[\left(\frac{p_{i4}}{p_{i3}}\right)^{\frac{\gamma-1}{\gamma}}-1\right]T_{i3}\times\eta_T=973+\left[\left(\frac{1}{6.42}\right)^{\frac{0.33}{1.33}}-1\right]\times 973\times 0.85=665.64\,\text{K}$$

$$P_T=\left|q_m c_p\left(T_{i4}-T_{i3}\right)\right|=\left|20\times 1.13\times(665.64-973)\right|=6946.4\,\text{kW}$$

$$P_{GT}=P_T - P_C - P_{ml}=6946.4-5354.2-66=1526.2\text{ kW}$$

Thermal power supplied by combustion chamber: Q_{cc}
The fuel flow is neglected

$$Q_{cc}=q_m c_p\left(T_{i3}-T_{i2}\right)=20\times 1.13\times(973-555.71)=9430.7\,\text{kW}$$

Thermal power lost in the exhaust gas: Q_{exh}

$$Q_{exh}=Q_{CC}-P_{ml}-P_{GT}=9430.7-66-1526.2=7838.5\,kW$$

2. Non insulated gas turbine:

Data: $\pi_c = 7.17$ (Figure 6); $q_m = 19.8$ kg.s^{-1} (From the adiabatic compressor map). $T_{i2} = 604.20$ K; $Q_{12} = 625.2$ kW (thermal power received by the compressor).
Power of the gas turbine: P_{GT}
Thermal power received by the compressor: Q_{12}

$$Q_{12}= 0.15\times0.6\times P_{Tad}=0.15\times0.6\times 6946.4=625.2\ kW$$

P_{Tad} : adiabatic turbine power
Compressor power: Pc

$$h_2\text{-}h_1=\frac{\gamma r}{\gamma\text{-}1}(T_{i2}\text{-}T_{i1})=\frac{0.287\times1.4}{0.4}\times(604.20\text{-}288)=317.62\,kJ.kg^{-1}$$

$$\Delta H_{12}=q_m(\Delta h_{12})=317.62\times19.8=6288.9\ kW$$

$$P_C=\Delta H_{12}\text{-}Q_{12}=6288.9\text{-}625.2=5663.7\ kW$$

Turbine power P_T:

$$P_{GT}=1526.2\ kW$$

$$|P_T|=P_{GT}+P_C+P_{ml}=1526.4+5663.7+66=7256.1\,kW$$

Search for new turbine inlet temperature

The variation in the expansion ratio of the turbine versus the reduced mass flow (Figure 3) shows that when the expansion ratio is greater than two (2), the reduced mass flow remains constant (Pluviose M., 2005). This reduced flow constant calculated in adiabatic conditions enables the new turbine inlet temperature (T_{i3}) corresponding to the new pressure (p_{i3}) to be determined by the following equations.

$$\left(q_m\frac{\sqrt{T_{i3}}}{p_{i3}}\right)_{ad}=\left(q_m\frac{\sqrt{T_{i3}}}{p_{i3}}\right)_{non\,ins}=q_{m(reduced)}=\left(20\times\frac{\sqrt{973}}{6.42\times1.05\times10^5}\right)=92.55\times10^{-5}$$

ad: adiabatic
Non ins: non insulated

$$p_{i3non\,ins}=0.95\times\pi_{Cnon\,ins}\times p_{i1}=0.95\times7.17\times1.01325=6.902\,bars$$

$$T_{i3non\,ins}=\left(\frac{q_{m(reduced)}\times p_{i3non\,ins}}{q_{m\,non\,ins}}\right)^2=\left(\frac{92.55\times10^{-5}\times6.902\times10^5}{19.8}\right)^2=1040.6\,K$$

$$T_{i4isentropic}=T_{i3}\left(\frac{1}{\pi_T}\right)^{\frac{\gamma-1}{\gamma}}=1041.6\times\left(\frac{1}{6.57}\right)^{\frac{0.33}{1.33}}=652.90\,K$$

$$T_{i4non\,ins}=671.23\,K$$

Thermal power supplied by the combustion chamber: Q_{cc}
The fuel flow is neglected

$$Q_{cc}=q_m c_p\left(T_{i3}-T_{i2}\right)=19.8\times1.13\times\left(1040.6-604.2\right)=9764\,kW$$

Thermal power lost in the exhaust gas

$$Q_{exh}=Q_{CC}-P_{ml}-P_{thl}-P_{GT}=9764-66-625.2/0.6-1526.4=7129.56\,kW$$

P_{thl} : power of thermal losses .

3. Insulated gas turbine:

Data: π_c = 7.22 (Figure 6); q_m = 19.5 kg.s^{-1} (from the adiabatic compressor map). T_{i2} = 622.68 K, Q_{12} = 1042 kW (thermal power received by the compressor)
Power of the gas turbine: P_{GT}
Thermal power received by the compressor: Q_{12}

$$Q_{12}=0.15\times P_{Tad}=0.15\times6946.4=1042\text{ kW}$$

P_{Tad} : Adiabatic turbine power
Compressor power: P_C

$$h_2-h_1=\frac{\gamma r}{\gamma-1}\left(T_{i2}-T_{i1}\right)=\frac{0.287\times1.4}{0.4}\times\left(622.68-288\right)=336.19\,kJ.kg^{-1}$$

$$\Delta H_{12}=q_m\left(\Delta h_{12}\right)=317.62\times19.5=6555.6\text{ kW}$$

$$P_C=\Delta H_{12}-Q_{12}=6555.6-1042=5513.6\text{ kW}$$

Turbine power P_T:

$$P_{TAG}=P_T=1526.2\text{ kW}$$
$$\left|P_T\right|=P_{TAG}+P_C+P_{ml}=1526.4+5513.6+66=7105.8\,kW$$

Search for new turbine inlet temperature

$$\left(q_m\frac{\sqrt{T_{i3}}}{p_{i3}}\right)_{ad}=\left(q_m\frac{\sqrt{T_{i3}}}{p_{i3}}\right)_{ins}=q_{m(reduced)}=\left(20\times\frac{\sqrt{973}}{6.42\times1.05\times10^5}\right)=92.55\times10^{-5}$$

ad : adiabatic

$$p_{i3ins}=0{,}95\times\pi_{Cins}\times p_{i1}=0.95\times7.22\times1.01325=6.95\,bars$$

$$T_{i3insulated} = \left(\frac{q_{m(reduced)} \times p_{i3insulated}}{q_{m\,insulated}} \right)^2 = \left(\frac{92.55 \times 10^{-5} \times 6.95 \times 10^5}{19.5} \right)^2 = 1088\,K$$

$$\mathrm{T_{i4isentropic}} = \mathrm{T_{i3}} \left(\frac{1}{\Pi_T} \right)^{\frac{\gamma-1}{\gamma}} = 1088 \times \left(\frac{1}{6.62} \right)^{\frac{0.33}{1.33}} = 680.70\,\mathrm{K}$$

$$\mathrm{T_{i4}} = 690\,\mathrm{K}$$

Thermal power supplied by the combustion chamber: Q_{CC}
The fuel flow is neglected

$$Q_{CC} = q_m c_p (T_{i3} - T_{i2}) = 19.5 \times 1.13 \times (1088 - 622.68) = 10253\,\mathrm{kW}$$

Thermal power lost in the exhaust Q_{exh}

$$Q_{ech} = Q_{CC} - P_{ml} - P_{thl} - P_{TAG} = 10253 - 66 - 1042 - 1526.2 = 7618.8\,\mathrm{kW}$$

7. References

Berger, M., Gostiaux, B., 1992, Géométrie différentielle: variétés, courbes et surfaces *France Presses universitaires de Paris,* ISBN : 2-13-044708-2.

Cormerais, M. 2007, Caractérisation expérimentale et modélisation des transferts thermiques au sein d'un turbocompresseur d'automobile, *Thèse de doctorat de l'école centrale de NANTES,* pp. 1-243.

Diango, A., 2010, Influence des pertes thermiques sur les performances des turbomachines. *Thèse de doctorat du Conservatoire national des arts et métiers,* Paris, pp. 1-244.

Kreith, F., 1967, *Principles of heat transfer,* Masson, [trad.] Kodja Badr-El-Dine, Université d'ALEP (Syrie), Colorado, International textbook Company Scranton, Pennsylvania, 1967. pp. 1-654.

Moreno, N., 2006, Modélisation des échanges thermiques dans une turbine radiale, *Thèse de doctorat de l'École nationale supérieure d'arts et métiers, pp. 158.*

Padet, J., 2005, Convection thermique et massique, *Techniques de l'ingénieur,* BE 8206.

Pluviose, M., 2005. Conversion d'énergie par turbomachines, *Ellipses, pp. 1.277.ISBN 2-7298-2320-4.*

Pluviose, M., 2002, Machines à fluides, *Ellipses, pp. 1-276,* ISBN 2-7298-1175-3.

Pluviose, M., 2005, Similitude des turbomachines à fluide compressible, *Techniques de l'ingénieur,* BM 468007.2005.

Pluviose, M. & Perilhon, C.(2002). Mécanismes de conversion de l'énergie, *echniques de l'ingénieur.* BM 4281, 10-2002..

Pluviose , M., Perilhon, C., 2002, Bilan énergétique et applications, *Techniques de l'ingénieur,* BM 4283, 04.2003.

Rautenberg & Al., 1981, Influence of heat transfer between turbine and compressor on the performance of small turbocharger, *International Gas Turbine Congress, Tokyo,* Asme paper, 1981.

Ribaud, Y., 2004, Overall Thermodynamics Model of an Ultra Micro turbine, *Journal of Thermal Science.* 2004, Vol. 13, 4, pp. 297-301.

Sacadura, J. F., 1993, Initiation aux transferts thermiques, *Lavoisier Tec & Doc,* Vol. 4ème tirage 1993, pp. 1-439, ISBN. 2-85206-618-1.

Verstraete, T. & al., 2007, Numerical Study of the Heat Transfer in Micro Gas Turbines, *Journal of Turbomachinery.* ASME, Octobre 2007, Vol.129, DOI: 10.1115/1.2720874, pp 835-841.

Part 4

Combustion

11

Developments of Gas Turbine Combustors for Air-Blown and Oxygen-Blown IGCC

Takeharu Hasegawa
Central Research Institute of Electric Power Industry
Japan

1. Introduction

From the viewpoints of securing a stable supply of energy and protecting our global environment in the future, the integrated gasification combined cycle (IGCC) power generation of various gasifying methods has been introduced in the world. Gasified fuels are chiefly characterized by the gasifying agents and the synthetic gas cleanup methods and can be divided into four types. The calorific value of the gasified fuel varies according to the gasifying agents and feedstocks of various resources, and ammonia originating from nitrogenous compounds in the feedstocks depends on the synthetic gas clean-up methods. In particular, air-blown gasified fuels provide low calorific fuel of 4 MJ/m^3 and it is necessary to stabilize combustion. In contrast, the flame temperature of oxygen-blown gasified fuel of medium calorie between approximately 9–13 MJ/m^3 is much higher, so control of thermal-NOx emissions is necessary. Moreover, to improve the thermal efficiency of IGCC, hot/dry type synthetic gas clean-up is needed. However, ammonia in the fuel is not removed and is supplied into the gas turbine where fuel-NOx is formed in the combustor. For these reasons, suitable combustion technology for each gasified fuel is important. In this paper, I will review our developments of the gas turbine combustors for the three type gasified fuels produced from the following gasification methods through experiments using a small diffusion burner and the designed combustors' tests of the simulated gasified fuels.

- Air-blown gasifier + Hot/Dry type synthetic gas clean-up method.
- Oxygen-blown gasifier + Wet type synthetic gas clean-up method.
- Oxygen-blown gasifier + Hot/Dry type synthetic gas clean-up method.

Figure 1 provides an outline of a typical oxygen-blown IGCC system. In this system, raw materials such as coal and crude are fed into the gasifier by slurry feed or dry feed with nitrogen. The synthetic gas is cleaned through a dust removing and desulfurizing process. The cleaned synthetic gas is then fed into the high-efficiency gas turbine topping cycle, and the steam cycle is equipped to recover heat from the gas turbine exhaust. This IGCC system is similar to LNG fired gas turbine combined cycle generation, except for the gasification and the synthetic gas cleanup process, primarily. IGCC requires slightly more station service power than an LNG gas turbine power generation.

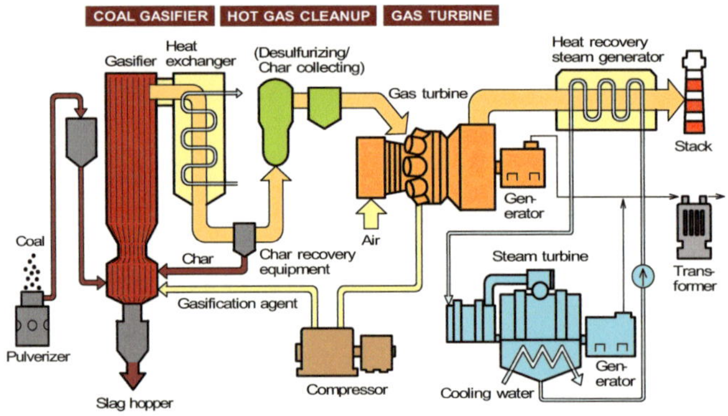

Fig. 1. Schematic diagram of typical IGCC system

1.1 Background of IGCC development in the world

The development of the gas turbine combustor for IGCC power generation received considerable attention in the 1970s. Brown (1982), summarized the overall progress of IGCC technology worldwide up until 1980. The history and application of gasification was also mentioned by Littlewood (1977). Concerning fixed-bed type gasification processes, Hobbs et al. (1993) extensively reviewed the technical and scientific aspects of the various systems. Other developments concerning the IGCC system and gas turbine combustor using oxygen-blown gasified coal fuel include: The Cool Water Coal Gasification Project (Savelli & Touchton, 1985), the flagship demonstration plant of gasification and gasified fueled gas turbine generation; the Shell process (Bush et al., 1991) in Buggenum, the first commercial plant, which started test operation in 1994 and commercial operation in 1998; the Wabash River Coal Gasification Repowering Plant (Roll, 1995) in the United States, in operation since 1995; the Texaco process at the Tampa power station (Jenkins, 1995), in commercial operation since 1996; and an integrated coal gasification fuel cell combined cycle pilot plant, consisting of a gasifier, fuel cell generating unit and gas turbine, in test operation since 2002 by Electric Power Development Co. Ltd. in Japan. Every plant adopted the oxygen-blown gasification method. With regard to fossil-based gasification technology as described above, commercially-based power plants have been developed, and new development challenges toward global carbon capture storage (Isles, 2007; Beer, 2007) are being addressed.

Meanwhile, from 1986 to 1996, the Japanese government and electric power companies undertook an experimental research project for the air-blown gasification combined cycle system using a 200-ton-daily pilot plant. Recently, the government and electric power companies have also been promoting a demonstration IGCC project with a capacity of 1700 tons per day (Nagano, 2009). For the future commercializing stage, the transmission-end thermal efficiency of air-blown IGCC, adopting the 1773 K (1500°C)-class (average combustor exhaust gas temperature at about 1773 K) gas turbine, is expected to exceed 48%(on HHV basis), while the thermal efficiency of the demonstration plant using a 1473 K (1200°C)-class gas turbine is only 40.5%. IGCC technologies would improve thermal efficiency by five points or higher compared to the latest pulverized coal-firing, steam power generation. The Central Research Institute of Electric Power Industry (CRIEPI), developed an air-blown two-stage entrained-flow coal gasifier (Kurimura et al., 1995), a hot/dry synthetic gas cleanup system (Nakayama et al., 1990), and 150MW, 1773K(1500°C)-class gas turbine combustor technologies for low-Btu fuel (Hasegawa et al., 1998a). In order to accept the various IGCC systems, 1773K-class gas turbine combustors of medium-Btu fuels by wet-type or hot/dry-type synthetic gas cleanup methods have undergone study (Hasegawa et al., 2003, 2007).

The energy resources and geographical conditions of each country, along with the diversification of fuels used for the electric power industry (such as biomass, poor quality coal and residual oil), are most significant issues for IGCC gas turbine development, as has been previously described: The development of biomass-fueled gasification received considerable attention in the United States and northern Europe in the early 1980s (Kelleher, 1985), and the prospects for commercialization technology (Consonni, 1997) appear considerably improved at present. Paisley and Anson (1997) performed a comprehensive economical evaluation of the Battele biomass gasification process, which utilizes a hot-gas conditioning catalyst for dry synthetic gas cleanup. In northern Europe, fixed-bed gasification heating plants built in the 1980s had been in commercial operation; the available technical and economical operation data convinced small district heating companies that biomass or peat-fueled gasification heating plants in the size class of 5 MW were the most profitable (Haavisto, 1996). However, during the period of stable global economy and oil prices, non-fossil-fueled gasification received little interest. Then, in the early 2000s when the Third Conference of Parties to the United Nations Framework Convention on Climate Change (COP3) invoked mandatory carbon dioxide emissions reductions on countries, biomass-fueled gasification technology began to receive considerable attention as one alternative. With the exception of Japanese national research and development project, almost all of the systems using the oxygen-blown gasification are in their final stages for commencing commercial operations overseas.

1.2 Progress in gas turbine combustion technologies for IGCCs

The plant thermal efficiency has been improved by enhancing the turbine inlet temperature, or combustor exhaust temperature. The thermal-NOx emissions from the gas turbines increase, however, along with a rise in exhaust temperature. In addition, gasified fuel containing NH_3 emits fuel-NOx when hot/dry gas cleanup equipment is employed. It is therefore viewed as necessary to adopt a suitable combustion technology for each IGCC in the development of a gas turbine for each gasification method.

Dixon-Lewis and Williams (1969), expounded on the oxidation characteristics of hydrogen and carbon monoxide in 1969. The body of research into the basic combustion characteristics of gasified fuel includes studies on the flammability limits of mixed gas, consisting of CH_4 or H_2 diluted with N_2, Ar or He (Ishizuka & Tsuji, 1980); a review of the flammability and explosion limits of H_2 and H_2/CO fuels (Cohen, 1992); the impact of N_2 on burning velocity (Morgan & Kane, 1952); the effect of N_2 and CO_2 on flammability limits (Coward & Jones, 1952; Ishibasi et al, 1978); and the combustion characteristics of low calorific fuel (Folsom, 1980; Drake, 1984); studies by Merryman et al. (1997), on NOx formation in CO flame; studies by Miller et al. (1984), on the conversion characteristics of HCN in H_2-O_2-HCN-Ar flames; studies by Song et al. (1980), on the effects of fuel-rich combustion on the conversion of the fixed nitrogen to N_2; studies by White et al. (1983), on a rich-lean combustor for low-Btu and medium-Btu gaseous fuels; and research of the CRIEPI into fuel-NOx emission characteristics of low-calorific fuel, including NH_3 through experiments using a small diffusion burner and analyses based on reaction kinetics (Hasegawa et al, 2001). It is widely

accepted that two-stage combustion, as typified by rich-lean combustion, is effective in reducing fuel-NOx emissions (Martin & Dederick, 1976; Yamagishi et al, 1974).
On the other hand, with respect to the combustion emission characteristics of oxygen-blown medium calorific fuel, Pillsbury et al. (1976) and Clark et al. (1982) investigated low-NOx combustion technologies using model combustors. In the 1970s, Battista and Farrell (1979) and Beebe et al. (1982) attempted one of the earliest tests using medium-Btu fuel in a gas turbine combustor. Concerning research into low-NOx combustion technology using oxygen-blown medium calorific fuel, other studies include: Hasegawa et al. (1997), investigation of NOx reduction technology using a small burner; and studies by Döbbeling et al. (1994), on low NOx combustion technology (which quickly mixed fuel with air using the double cone burner from Alstom Power, called an EV burner); Cook et al. (1994), on effective methods for returning nitrogen to the cycle, where nitrogen is injected from the head end of the combustor for NOx control; and Zanello and Tasselli (1996), on the effects of steam content in medium-Btu gaseous fuel on combustion characteristics. In almost all systems, surplus nitrogen was produced from the oxygen production unit and premixed with a gasified medium-Btu fuel (Becker & Shetter, 1992), for recovering power used in oxygen production and suppressing NOx emissions. Since the power to premix the surplus nitrogen with the medium-Btu fuel is great, Hasegawa et al. studied low-NOx combustion technologies using surplus nitrogen injected from the burner (Hasegawa et al, 1998b) and with the lean combustion of instantaneous mixing (Hasegawa et al, 2003). Furthermore, Hasegawa and Tamaru(2007) developed a low-NOx combustion technology for reducing both fuel-NOx and thermal-NOx emissions, in the case of employing hot/dry synthetic gas cleanup with an oxygen-blown IGCC.

1.3 Subjects of gas turbine combustors for IGCCs

The typical compositions of gasified fuels produced in air-blown or oxygen-blown gasifiers, and in blast furnaces, are shown in Tables 1. Each type of gaseous mixture fuel consists of CO and H_2 as the main combustible components, and small percentages of CH_4. Fuel calorific values vary widely (2–13 MJ/m^3), from about 1/20 to 1/3 those of natural gas, depending upon the raw materials of feedstock, the gasification agent and the gasifier type.
Figure 2 shows the theoretical adiabatic flame temperature of fuels which were: (1) gasified fuels with fuel calorific values (HHV) of 12.7, 10.5, 8.4, 6.3, 4.2 MJ/m^3; and (2) fuels in which methane is the main component of natural gas. Flame temperatures were calculated using a CO and H_2 mixture fuel (CO/H_2 molar ratio of 2.33:1), which contained no CH_4 under any conditions, and the fuel calorific value was adjusted with nitrogen. In the case of gasified fuel, as the fuel calorific value increased, the theoretical adiabatic flame temperature also increased. Fuel calorific values of 4.2 MJ/m^3 and 12.7 MJ/m^3 produced maximum flame temperatures of 2050 K and 2530 K, respectively. At fuel calorific values of 8.4 MJ/m^3 or higher, the maximum flame temperature of the gasified fuel exceeded that of methane, while the fuel calorific value was as low as one-fifth of methane. Furthermore, each quantity of CO and H_2 constituent in the gasified fuels differed, chiefly according to the gasification methods of gasifying agents, raw materials of feedstock, and water-gas-shift reaction as an optional extra for carbon capture system. However, it could be said that the theoretical adiabatic flame temperature was only a little bit affected by the CO/H_2 molar ratio in the case of each fuel shown in Tables 1. That is to say, in air-blown gasified fuels, fuel calorific values are so low that flame stabilization is a problem confronting development of the combustor.

Fuel	BFG	COG	Gasified fuel									
Resource			Waste	RDP	Coal					wood		Heavy residue
Gasifier			Fluidized		Fixed	Fluidized	Entrained			Fluidized		Entrained
Coal supply					Dry	Dry	Dry	Dry	Slurry			
Developer					BGL	BC	IGC	Shell	Texaco	Lurgi		Texaco
Oxydizer			Air	Air+O_2	O_2	Air	Air	O_2	O_2	Air	O_2	O_2
Composition												
CO[%]	20	6	6	30	56.4	7.9-14.7	25.9-27.6	65.2-69.5	40.9	20	46	51.7
H_2[%]	3	56	1.6	22	25.6	13.2-15.0	10.9-9.4	28.8-31.0	29.9	10	13	43.1
CH_4[%]	-	30	0.9	0.4	6.6	1.5-2.8	1.4-0.5	0.01-0.03	0.1	5	10	0.2
CO_2[%]	20	-	12.4	4.1	2.8	10-12	6.7-5.4	1.0-2.8	9.5	14	23	3.2
H_2O[%]	-	-	23.4	5.9	-(a)	11.5-18.4	-(a)	-(b)	12.3	-	-	-(b)
NH_3[ppm]	-	-	-	-	-(a)	500-1000	1000	100-600	-(a)	-	-	-(a)
H_2S+COS[%]	-	-	-	-	20ppm	-(a)	404-714ppm	0.14-1.1	-(a)	-	-	1.6
Others[%]	N_2	C_2H_2 etc.	N_2	C_2H_2 etc.	8.6	45.9-47.3	54.2-56.1	-(a)	7.3	51	8	0.2
CO/H_2	7	0.1	3.8	1.4	2.2	-(a)	2.4-3.0	2.1-2.4	1.4	2	3.5	1.2
HHV[MJ/m³]	2.9	20	1.8	6.8	13.0	3.6-4.1	4.9-5.2	12.2-12.5	9.0	5.8	11.4	12.1

BFG:Blast furnace gas, COG:Coke-oven gas, RDF:Refuse derived fuel, Waste:Municipal solid waste, (a):No description, (b):Drybase

Table 1. Various gasified fuels

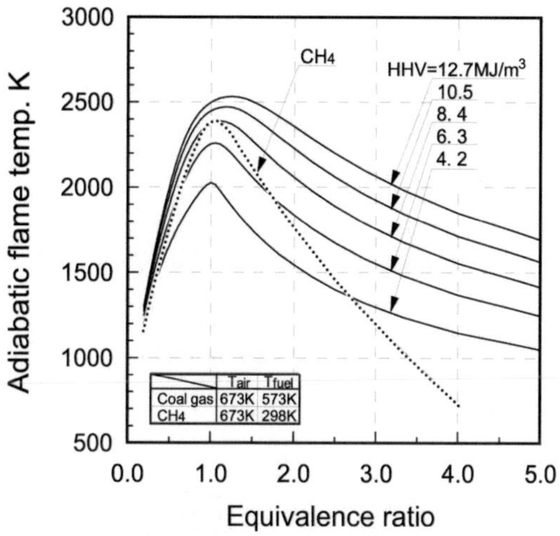

Fig. 2. Relationship between equivalence ratio and adiabatic flame temperature for gasified fuels and CH_4.

On the other hand, in the case of oxygen-blown gasified fuels, flame temperature is so high that thermal-NOx emissions must be reduced. Therefore, in oxygen-blown IGCC, N_2 produced by the air separation unit is used to recover power to increase the thermal efficiency of the plant, and to reduce NOx emissions from the gas turbine combustor by reducing the flame temperature. Furthermore, when hot/dry synthetic gas cleanup is employed, ammonia contained in the gasified fuels is not removed, but converted into fuel-NOx in the combustor. It is therefore necessary to reduce the fuel-NOx emissions in each case of air-blown or oxygen-blown gasifiers.

Because fuel conditions vary depending on the gasification method, many subjects arose in the development of the gasified fueled combustor. Table 2 summarizes the main subjects of combustor development for each IGCC method.

		Synthetic gas cleanup	
		Wet type	Hot/Dry type
Gasification agent	Air	• Combustion stability of low-calorific fuel	• Combustion stability of low-calorific fuel • Reduction of fuel-NOx
	O_2	• Surplus nitrogen supply • Reduction of thermal-NOx	• Surplus nitrogen supply • Reduction of thermal- and fuel-NOx emissions

Table 2. Subjects for combustors of various gasified fuels

2. Test facilities and method for gasified fueled combustors

This chapter indicates a typical example of a test facility and method for a single-can combustion test using simulated gasified fuels.

2.1 Test facilities

The schematic diagram of the test facilities is shown in Figure 3. The raw fuel obtained by mixing CO_2 and steam with gaseous propane was decomposed to CO and H_2 inside the fuel-reforming device. A hydrogen separation membrane was used to adjust the CO/H_2 molar ratio. N_2 was added to adjust the fuel calorific value to the prescribed calorie, and then simulated gases derived from gasifiers were produced.
This facility had another nitrogen supply line, by which nitrogen was directly injected into the combustor. Air supplied to the combustor was provided by using a four-stage centrifugal compressor. Both fuel and air were supplied to the gas turbine combustor after being heated separately with a preheater to the prescribed temperature.

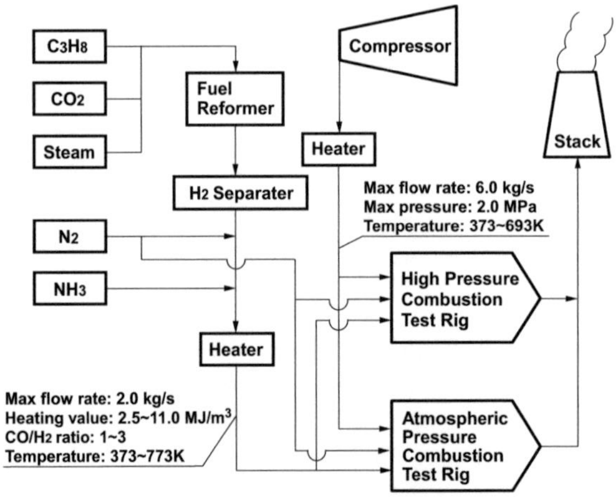

Fig. 3. Schematic diagram and specifications of test facility

The combustion test facility had two test rigs, each of which was capable of performing full-scale atmospheric pressure combustion tests of a single-can for a "several"-hundreds MW-class, multican-type combustor as well as half-scale high-pressure combustion tests, or full-scale high-pressure tests for around a 100MW-class, multican-type combustor. Figure 4 shows a cross-sectional view of the combustor test rig under pressurized conditions. After passing through the transition piece, the exhaust gas from the combustor was introduced into the measuring section where gas components and temperatures were measured. An automatic gas analyzer analyzed the components of the combustion gases. After that, the gas temperature was lowered through a quenching pot, using a water spray injection system.

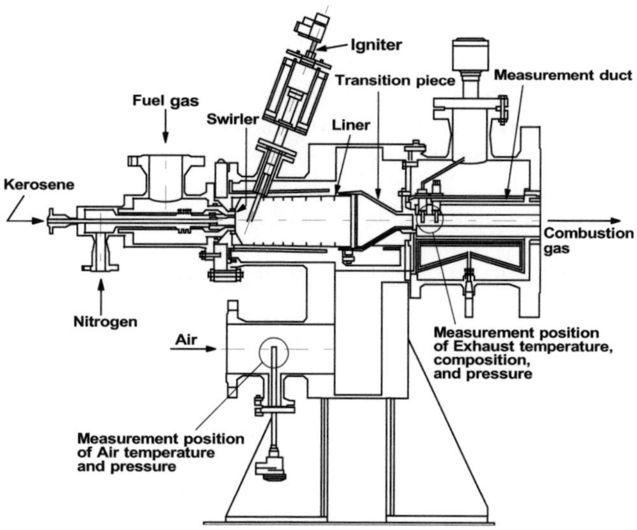

Fig. 4. Combustion test rig

2.2 Measurement system

Exhaust gases were sampled from the exit of the combustor through water-cooled stainless steel probes located on the centerline of a height-wise cross section of the measuring duct. The sample lines of exhaust gases were thermally insulated with heat tape to maintain the sampling system above the dew point of the exhaust gas. The exhaust gases were sampled from at an area averaged points in the tail duct exit face and continuously introduced into an emission console which measured CO, CO_2, NO, NOx, O_2, and hydrocarbons by the same methods as the test device for basic studies using the small diffusion burner. The medium-Btu simulated fuel were sampled from the fuel gas supply line at the inlet of combustor, and constituents of CO, H_2, CH_4, H_2O, CO_2 and N_2 were determined by gas chromatography. Heating values of the simulated gaseous fuel were monitored by a calorimeter and calculated from analytical data of gas components obtained from gas chromatography.

The temperatures of the combustor liner walls were measured by sheathed type-K thermocouples with a diameter of 1mm attached to the liner wall with a stainless foil welding. The temperature distributions of the combustor exit gas were measured with an array of three pyrometers, each of which consisted of five type-R thermocouples.

3. Gas turbine combustors for the gasified fuels

This chapter indicates the characteristics of the combustion technologies being applied to the gasified fuels classified into four types in Table 2. Based on the knowledge through experiments using a small diffusion burner and numerical analyses, prototype combustors were constructed, tested and their performances were demonstrated.

3.1 Combustor for air-blown gasification system with hot/dry type synthetic gas cleanup

3.1.1 Design concept of combustor

Figure 5 shows the relation between the combustor exhaust temperature and the air distribution in the gas turbine combustor using low-calorific gasified fuel. To calculate air distribution, the overall amount of air is assumed to be 100 percent. The amount of air for combustion is first calculated at 1.2 times of a theoretical air (ϕ =0.83), 30 percent of the total air is considered as the cooling air for the combustor liner wall, and the remaining air is considered as diluting air. According to this figure, as the gas turbine temperature rises up to 1773K, the ratio of cooling and diluting air decrease significantly, and the flexibility of the combustor design is minimized. To summarize these characteristics, it can be said that the design concept of the gas tur-bine combustor utilizing low-calorific fuel should consider the following issues when the gas turbine temperature rises:

- Combustion stability; it is necessary to stabilize the flame of low-calorific fuel.
- Low NOx emission technology to restrain the production of fuel NOx.
- Cooling structure to cool the combustor wall efficiently with less amount of air.

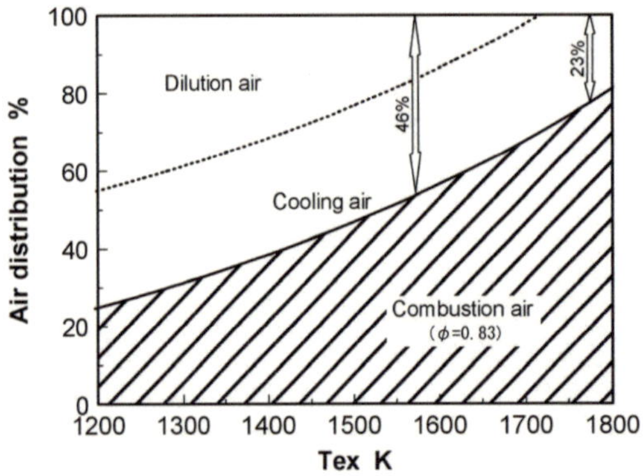

Fig. 5. Air distribution design of a gas turbine combustor that burns low-Btu gasified fuel

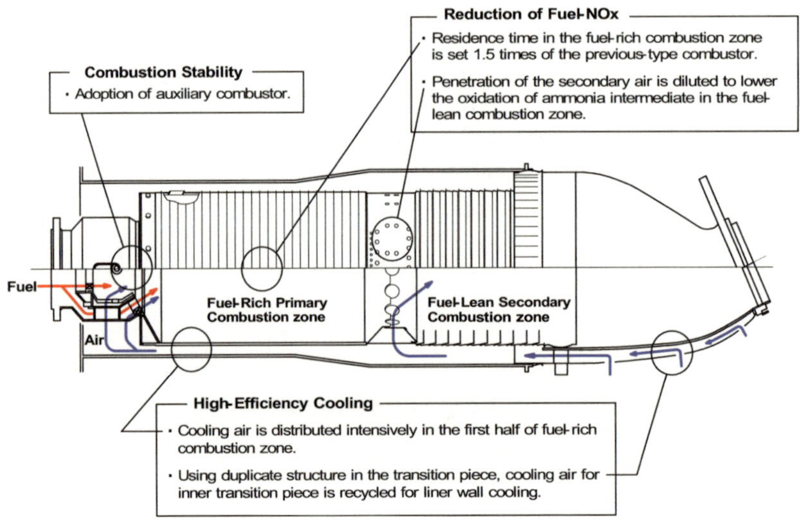

Fig. 6. Design concept of 1773K-class low-Btu fueled combustor

Burner   Combustion Liner

Fig. 7. Tested combustor

Figure 6 presents characteristics of the designed and tested 1773K-class combustor. Figure 7 illustrates the external view of the burner of the combustor. The main design concept of the combustor was to secure stable combustion of a low-calorific fuel in a wide range of turn-down operation, low NOx emission and enough cooling-air for the combustor liner. The combustor is designed for advanced rich-lean combustion which is effective in decreasing fuel NOx emissions resulting from fuel bound nitrogen.

3.1.1.1 Assurance of flame stabilization

In order to assure flame stability of low-calorific fuel, an auxiliary combustion chamber is installed at the entrance of the combustor. The ratio of the fuel allocated to the auxiliary combustion chamber is 15 percent of the total amount of fuel. The fuel and the combustion air are injected into the chamber through a sub-swirler with a swirling angle of 30 degree. By setting the stoichiometric condition in this chamber under rated load conditions, a stable flame can be maintained. The rest of the fuel is introduced into the main combustion zone from the surrounding of the exit of the auxiliary combustion chamber.

3.1.1.2 Fuel-NOx reduction

To restrict the production of fuel NOx that is attributable to NH_3 contained in the fuel, a two-stage combustion method (rich-lean combustion method) is introduced. The tested combustor has a two chamber structure, which separates the primary combustion zone from the secondary combustion zone. In addition, the combustor has two main design characteristics for reducing fuel NOx as indicated below:

3.1.1.2.1 Air to fuel ratio in primary combustion zone

The equivalence ratio of the primary combustor is determined setting at 1.6 based on the combustion tests previously conducted using a small diffusion burner (Hasegawa et al., 2001).

Figure 8 shows an outline of the experimental device of the small diffusion burner. The combustion apparatus consists of a cylinder-style combustion chamber with an inner diameter, 'D', of 90mm and a length of 1,000mm, and a primary air swirler and fuel injection nozzle. The combustion chamber is lined with heat insulating material and the casing is cooled with water. In order to simulate two-stage combustion, secondary air inlets at a distance from the edge of the fuel injection nozzles of 3×'D' are used. The diameter of the secondary air inlets at the entry to the combustion chamber is 13mm, and six inlets are positioned on the perimeter of one cross-section. The tested burner consists of a fuel injection nozzle and a primary air swirler. There are twelve injection inlets with a diameter of 1.5mm on the fuel injection nozzle with an injection angle, θ, of 90-degree. The primary air swirler has an inner diameter of 24.0mm, an outer diameter of 36.4mm, and twelve vanes with a swirl angle, θ_a, of 45-degree. Swirl number, S, which is calculated from the following equation, is 0.84.

$$S = \frac{2}{3} \times \frac{1 - B^3}{1 - B^2} \times \tan\theta_a \tag{1}$$

Where B (boss ratio of swirl vane)=0.66.

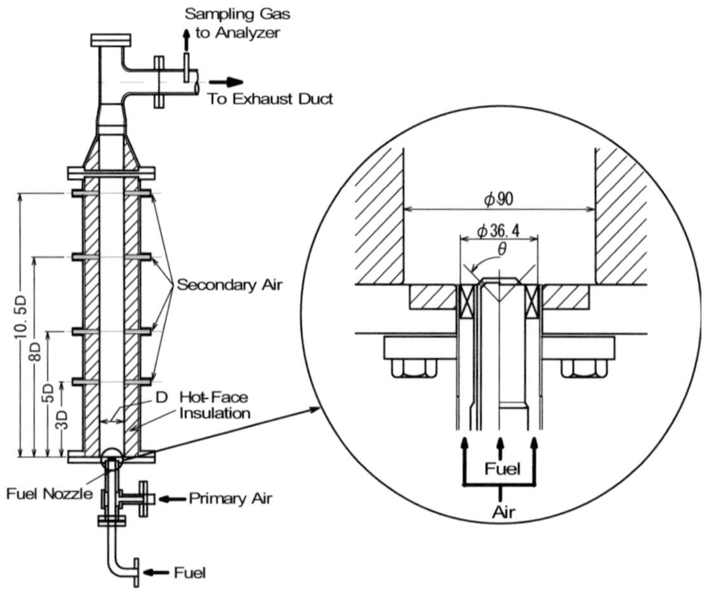

D : inner diameter of cylinder-style combustion chamber, 90mm
θ : injection angle of fuel nozzle, 90 degrees

Fig. 8. Combustion chamber and diffusion burner of basic experimental device

Figure 9 presents an example of the test results which indicates the influence of the equivalence ratio of the primary combustion zone to the conversion rate of NH_3 to NOx, C.R., at the exit of the secondary combustion zone. It also indicates the influence of the CH_4 concentration in the fuel.

$$C.R. = \frac{([\mathrm{NOx}] - [\mathrm{NOx}_{th}]) \times (\text{volume flow rate of exhaust})}{[\mathrm{NH}_3] \times (\text{volume flow rate of fuel})} \tag{2}$$

To obtain the conversion rate of NH_3 to NOx, the concentration of thermal-NOx, '[NOxth]', was first measured after stopping the supply of NH_3, then the concentration of total NOx, '[NOx]', was measured while NH_3 was supplied, and finally fuel-NOx was calculated by deducting the concentration of thermal-NOx from that of total NOx. In the tests investigating fuel-NOx emissions, 1000ppm of NH_3 is contained in the low-Btu fuel which consists of CO, H_2 (CO/H_2 molar ratio of 2.33:1), and small amount of CH_4. In the case of changing CH_4 concentration, fuel calorific value was adjusted by N_2 dilution.

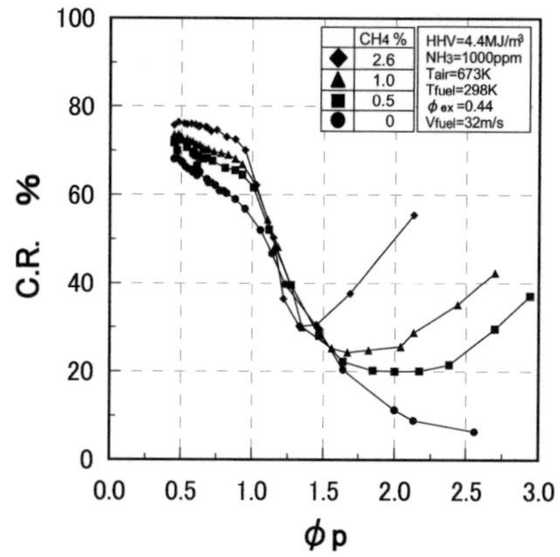

Fig. 9. Effect of methane content on conversion rate of ammonia in the fuel to NOx, defining by the experiments using a small diffusion burner (Hasegawa et al., 2001)

From the test results, it is known that the conversion rate of NH_3 to NOx is affected by both the equivalence ratio in the primary combustion zone using the two-staged combustion method and CH_4 concentration. When the fuel contains CH_4, HCN produced in the primary combustion zone is easily converted to NOx in the secondary combustion zone along with the decomposition of NH_3. Therefore, there is a particular equivalence ratio, which minimizes the NOx conversion rate. Since the low-calorific fuel derived from the IGCC subject to development contained approximately 1.0 percent of CH_4, the equivalence ratio in the primary-combustion zone was set at 1.6. The fuel and the primary combustion-air are injected from the burner, which has 30 degree swirl angle and 15 degree introvert angle.

3.1.1.2.2 Introduction method of secondary air

An innovative idea was applied for secondary air introduction. With the decomposition of fuel N, a large portion of the total fixed nitrogen produced in the primary combustion zone, including NO, HCN and NHi, is converted to NOx in the secondary combustion zone. The influence of secondary air mixing conditions on the NOx production was examined from the viewpoint of reaction kinetics with modular model where each combustion zone means a perfect stirred reactor, neither the effect of diffusion nor that of radiant heat transfer of the flame are taken into account. As a result, it was found that the slower mixing of the secondary air made the conversion rate of NH_3 to NOx decline further (Hasegawa et al., 1998a). Based on this result, an exterior wall was installed at the secondary-air inlet section in the tested combustor to make an intermediate pressure zone of the dual structure. By providing this dual structure, the flow speed of the secondary air introduced to the combustor decreased to 70m/s, compared to 120m/s without an exterior wall, thus the secondary air mixing was weakened.

3.1.1.2.3 Cooling of combustor liner wall

In order to compensate for the declined cooling air ratio associated with the higher temperature of the gas turbine, the tested combustor is equipped with a dual-structure transition piece so that the cooling air in the transition piece can be recycled to cool down the combustor liner wall. The cooling air that flowed into the transition piece from the exterior wall cools the interior wall with an impingement method, and moves to the combustor liner at the upper streamside.

For the auxiliary combustor and the primary combustion zone in which temperatures are expected to be especially high, the layer-built cooling structure that combined impingement cooling and film cooling was employed. For the secondary combustion zone, the film cooling method was used.
In addition to the above design characteristics, the primary air inlet holes are removed in order to maintain the given fuel-rich conditions in the primary combustion zone. Also, the overall length of the combustor, including the auxiliary chamber, is 1317mm, and the inside diameter is 356mm.

3.1.2 Test results

Combustion tests are conducted on under atmospheric pressure conditions. Concerning the pressure influence on the performance of the combustor, a half scale combustor, which has been developed by halving in dimension, was tested under pressurized conditions. Supplied fuels into the combustor were adjusted as same components as that of air-blown entrained-flow gasified coal fuel shown in Table 1. The standard rated conditions in the combustion tests are summarized in Table 3. Combustion Intensity at the design point is 2.0×10^{2} W/(m^{3}•Pa).

T_{air}	700K
T_{fuel}	633K
T_{ex}	1773K
P	1.4MPa
ϕ_{ex}	0.62
Combustion Intensity	2.0×10^{2} W/(m^{3}•Pa)

Table 3. Rated test conditions

3.1.2.1 Combustion emission characteristics

Figure 10 shows the combustion emission characteristics, under the gas turbine operational conditions. When the gas turbine load was 25 percent or higher, which is the single fuel firing of gasified fuel, the conversion rate of NH_3 to NOx was reduced as low as 40 percent (NOx emissions corrected at 16 percent O_2 was 60ppm), while the combustion efficiency shows around 100 percent in each gas turbine load.

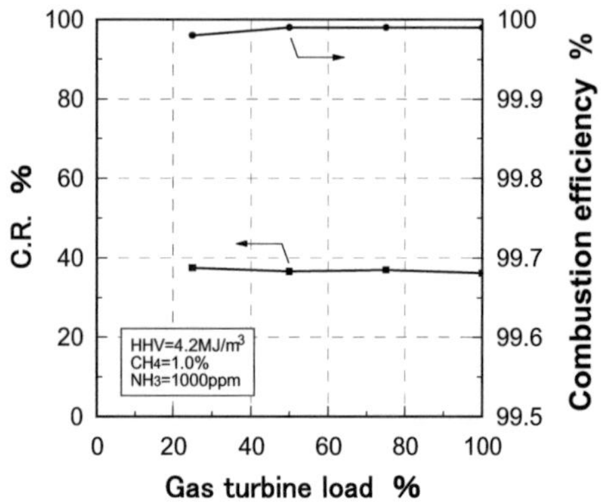

Fig. 10. Combustion emission characteristics

3.1.2.2 Thermal characteristics of combustor liner wall

Figure 11 shows the temperature distribution of the combustor liner wall at the rated load condition. From this figure, it could be said that the overall liner wall temperature almost remained under 1123K (850°C), the allowable heat resistant temperature, while the wall temperature increased to an adequate level and a stable flame was maintained in both the auxiliary-combustion chamber and the primary combustion zone.

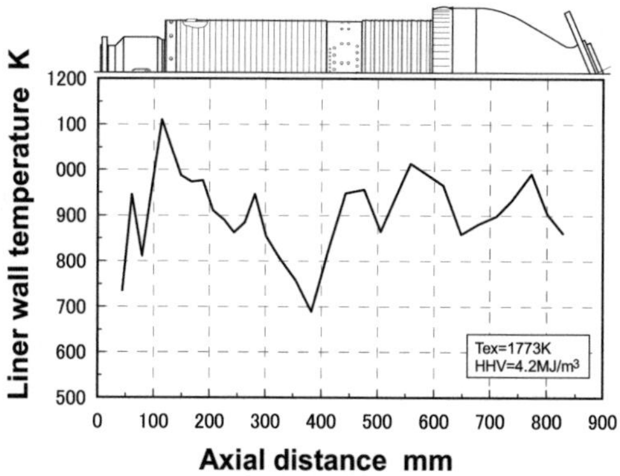

Fig. 11. Combustor wall temperature distribution

3.2 Combustor for oxygen-blown gasification system with wet type synthetic gas cleanup

3.2.1 Subjects of combustor

In the case of oxygen-blown IGCC, which has an air-separation unit to produce oxygen as gasification agent, medium-Btu gasified fuels are produced compared with the case of the air-blown gasified low-Btu fuels. That is, the maximum flame temperature of medium-Btu fuel is higher than that of each low-Btu fuel or high-calorie gas such as natural gas. Thermal-NOx emissions are expected to increase in the case of medium-Btu fueled combustors.

Furthermore, in the oxygen-blown IGCC system, large quantity of nitrogen is produced in the air separation unit. In almost all of the systems, a part of nitrogen is used to feed raw material such as coal into the gasifier and so on, gasified fuels are premixed with the rest of the nitrogen and injected into the combustor to increase electric power and to decrease thermal-NOx emissions from the gas turbine. It is necessary to return a large quantity of the surplus nitrogen (as much as the fuel flow rate) to the cycle from the standpoint of recovering power for oxygen production. So, we intend to inject the surplus nitrogen directly into higher temperature regions from the burner and to decrease thermal-NOx emissions produced from these regions effectively. Analyses confirmed that the thermal efficiency of the plant improved by approximately 0.3 percent absolutely by means of nitrogen direct injection into the combustor, compared with a case where nitrogen is premixed with gasified fuel before injection into the combustor.

3.2.2 Design concept of combustor

Figure 12 presents characteristics of the designed, medium-Btu fueled 1573K (1300°C)-class combustor based on the above considerations. The main design concept for the tested combustor was to secure a low-NOx and stable combustion of medium-Btu fuel with nitrogen injection in a wide range of turn-down operations. The overall length of the combustion liner is 650mm and the inside diameter is 230mm.

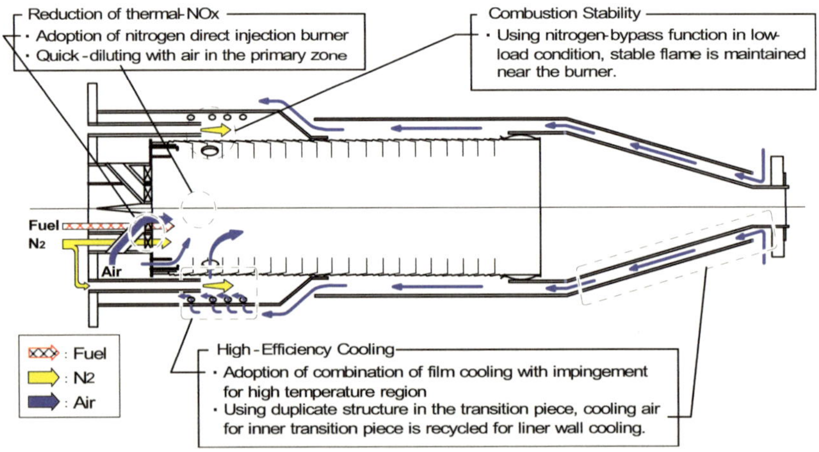

Fig. 12. Design concepts of medium-Btu fueled combustor for wet-type synthetic gas cleanup

According to the combustor cooling, a convection method was employed in the transition piece, and moves to the combustor liner on the upstream side. For the primary combustion zone where temperatures are expected to be especially high, the dual-cooling structure was employed, in which the cooling air was impinged from the air flow guide sleeve to the combustion liner and used as film cooling air for the combustion liner. For the secondary combustion zone, the film-cooling method was used.

To restrict thermal-NOx production originating from nitrogen fixation and CO emissions, the burner was designed with nitrogen injection function, based on combustion tests previously conducted using a small diffusion burner (Hasegawa et al., 2001) and a small model combustor (Hasegawa et al., 2003).

Figure 13 presents an example of the test results using the small diffusion burner shown in figure 8, which indicate the influence of the primary equivalence ratio on NOx emission characteristics in two-staged combustion for comparing three cases: 1) a fuel calorific value (*HHV*) of 12.7MJ/m^3, without nitrogen injection; 2) a fuel calorific value of 12.7MJ/m^3, where nitrogen is blended with the combustion air from the burner; 3) a fuel blended with nitrogen of the same quantity as case 2), or low-Btu fuel of 5.1MJ/m^3. From figure 13, we know that nitrogen blended with fuel or air injected from the burner has a great influence over decreasing NOx emissions from nitrogen fixation. On the other hand, not shown in here, in the case where nitrogen blended with air was injected into the combustor, CO emissions decreased as low as medium-Btu gasified fuel not blended with nitrogen, while CO emissions significantly increased when fuel was blended with nitrogen. That is, in the medium-Btu fuel combustion with nitrogen injection, all of the surplus nitrogen should be injected into the primary combustion zone to reduce the thermal-NOx emissions and should not be blended with fuel, or the primary zone should be fuel lean condition for a low NOx and stable combustion in a wide range of turn-down operations.

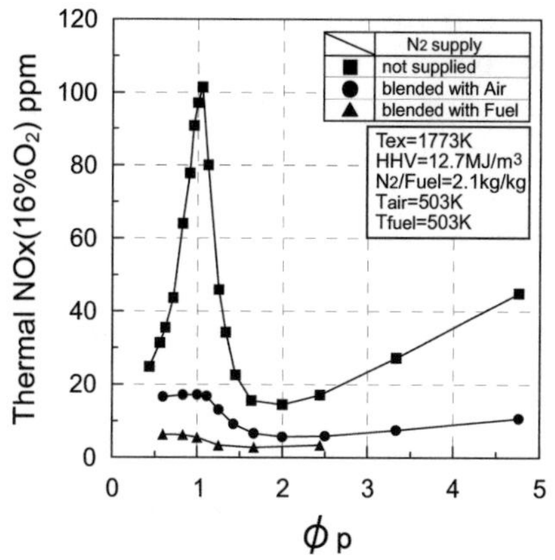

Fig. 13. Effect of nitrogen injection on thermal-NOx emission characteristics in two-stage combustion, using a small diffusion burner

Figure 14 shows the combustion gas temperature distribution in the both cases of no nitrogen injection and of nitrogen injection of 1.0kg/kg N_2/Fuel from the burner under atmospheric pressure condition, using a model combustor. In tests, the combustor outlet gas temperature is set at 1373K. From figure 14, we know that nitrogen injection from the burner has a great influence over decreasing hot regions by around 200K in this test conditions. So, in this way of nitrogen injection, thermal-NOx production was restrained one fifth that of the case no nitrogen injection.

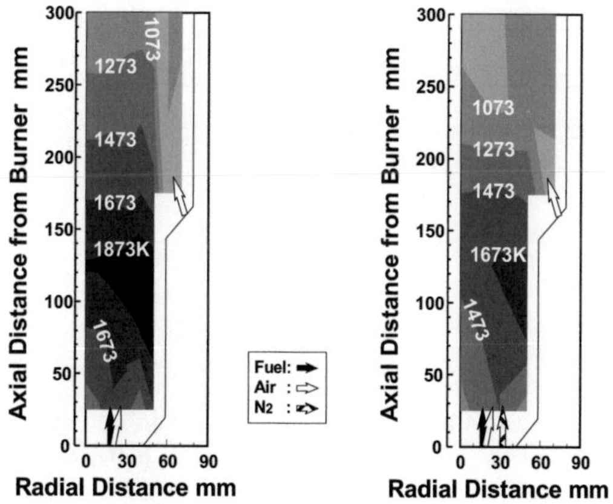

(1)No injection of nitrogen (2)Nitrogen injected

Fig. 14. Effect of nitrogen injection on combustion gas temperature distribution using a model combustor

Based on these basic test results, we arranged the nitrogen injection intakes in the burner and adopted the lean primary combustion, as shown in figure 12. The nitrogen injected directly into a combustor has the effect of decreasing power to compress nitrogen, compared with the case where the nitrogen was blended with fuel or air evenly. And it is possible to control the mixing of fuel, air and nitrogen positively by way of nitrogen being injected separately into the combustor. The nitrogen direct injection from the burner dilutes the flame of medium-Btu fuel. Furthermore we intended to quench the flame as soon as possible, both by sticking the combustion air injection tubes out of the liner dome and by arranging the secondary combustion air holes on the upstream side of the combustion liner. Design of the combustor was intended for the medium-Btu fuel, the nitrogen injection function was combined with the lean combustion technique for a low NOx combustion. By setting the primary combustion zone to fuel lean state under the rated load condition, the NOx emissions are expected to decrease, and by bypassing nitrogen to premix with the combustion air under partial load conditions, a stable flame can be maintained in a wide range of turn-down operations.

3.2.3 Test results

Table 4 and 5 show the typical properties of the supplied fuel and the standard test conditions, respectively. Higher heating value of the supplied fuel, HHV, was set at 10.1MJ/m^3, CH_4 was contained higher concentration of 6.8 percent. A part of surplus nitrogen produced from the air-separation unit was used to feed coal or char into the gasifier and the flow rate of the rest was about 0.9 times the fuel flow in the actual process. Since the density of the supplied fuel is higher than that of the commercial gasified fuel and temperature of supplied nitrogen is lower in the case of the test conditions than in the actual operations, the combustor performance is investigated in the case of 0.3kg/kg N_2/Fuel ratio, in which firing temperature of the burner outlet corresponds to the case of actual operations. The rated temperature of combustor-outlet gas, T_{ex}, is around 1700K.

Composition CO	30.4 vol%
H_2	27.5 vol%
CH_4	6.8 vol%
CO_2	35.3 vol%
HHV	10.1 MJ/m^3

Table 4. Typical conditions of supplied fuel

T_{air}	603K
T_{fuel}	583K
T_{N_2}	333K
T_{ex}	1700K
N_2/Fuel ratio	0.3kg/kg
P	1.4MPa
Combustion Intensity	2.2×10^2 W/(m^3•Pa)

Table 5. Rated test conditions

Figure 15 shows the relationship between the gas turbine load and the combustion emission characteristics, under the condition where the pressure in the combustor is set to 1.0MPa of slightly less than that of the practical operation at the equivalent, rated load. When the gas turbine load was 25 percent or higher, which is the single fuel firing of gasified fuel, the NOx emission was reduced as low as 11ppm (corrected at 16 percent O_2), while the NOx emission tends to increase slightly with the rise in the gas turbine load. Considering the effects of pressure, it could be said that NOx emission was surmised as low as 12ppm (corrected at 16 percent O_2) at any gas turbine load. On the other hand, combustion efficiency shows around 100 percent in the case where the gas turbine load was 25 percent or higher, by bypassing nitrogen to premix with the combustion air at low load conditions.

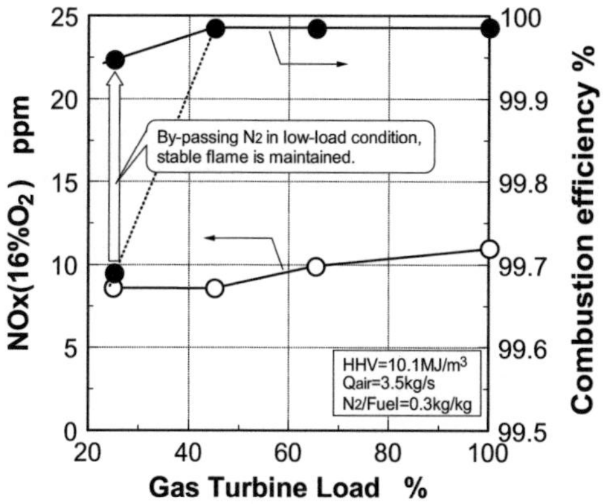

Fig. 15. Effect of the gas turbine load on combustion emission characteristics

3.3 Combustor for oxygen-blown gasification system with hot/dry type synthetic gas cleanup

In order to improve the thermal efficiency of the oxygen-blown IGCC, it is necessary to adopt the hot/dry synthetic gas cleanup. In this case, ammonia contained in the gasified fuels could not be removed and fuel-NOx is emitted from the gas turbine. It is necessary to develop to low NOx combustion technologies that reduce fuel-NOx emissions originating from ammonia in the fuel at the same time as reducing thermal-NOx ones.

3.3.1 Subjects of combustor

From the characteristic of medium-Btu, gasified fuel as mentioned above, it could be said that the design of a gas turbine combustor with nitrogen supply, should consider the following issues for an oxygen-blown IGCC with the hot/dry synthetic gas cleanup:

1. Low NOx-emission technology: Thermal-NOx production from nitrogen fixation using nitrogen injection, and fuel-NOx emissions originating from ammonia using a two-stage combustion must be simultaneously restrained.
2. Higher thermal efficiency: Nitrogen injection must be tailored so as to decrease the power to compress nitrogen, which is returned into the gas turbine in order to recover a part of the power used for the air-separation unit.

3.3.2 Design concept of combustor

Figure 16 presents the configuration and its function of a designed, medium-Btu fueled 1773K (1500°C)-class combustor based on the above considerations. The main design concepts for the tested combustor were to secure stable combustion of medium-Btu fuel with nitrogen injection in a wide range of turn-down operations, and low NOx combustion for reducing fuel-NOx and thermal-NOx emissions. In order to secure stable combustion, we installed an auxiliary combustion chamber at the entrance of the combustor. To reduce thermal-NOx emissions, the nitrogen injection nozzles was set up in the main-swirler, which is installed at exit of the auxiliary combustion chamber. The overall length of the combustion liner is 445mm and the inside diameter is 175mm.

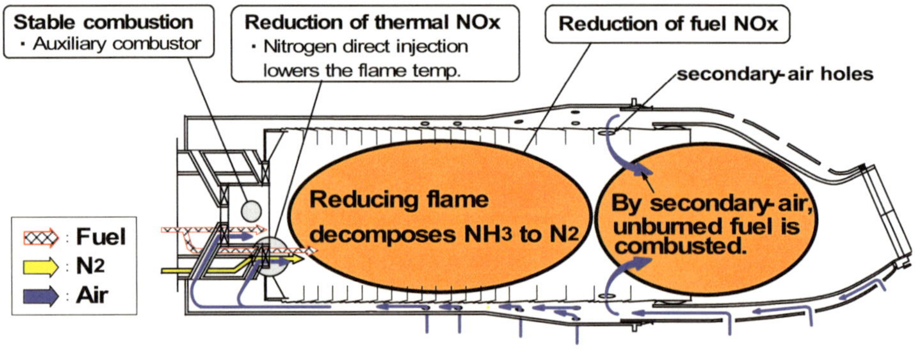

Fig. 16. Design concept of a medium-Btu fueled gas turbine combustor for hot/dry-type synthetic gas cleanup

Figure 17 illustrates the axial distribution of equivalence ratio at the rated load condition. In order to reduce the fuel-NOx emissions, we adopted the two-stage combustion, in which a fuel-rich combustion was carried out in the primary zone maintaining the equivalence ratio of 0.66 at exit of the combustor. And the designed combustor has following characteristics.

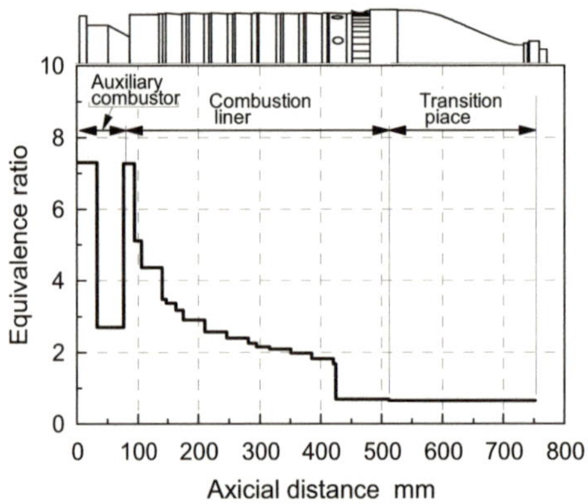

Fig. 17. Axial distribution of equivalence ratio at the rated load condition

3.3.2.1 Assurance of flame stabilization

The ratio of the fuel allocated to the auxiliary combustion chamber is 30 percent of the total amount of fuel. The fuel and air are injected into the chamber through a sub-swirler with a swirling angle of 30-degree. By setting the mean equivalence ratio in the auxiliary chamber at 2.4 under rated load conditions, a stable flame can be maintained in the fuel-rich combustion zone and reduction of NH_3 to N_2 could proceed in lower load conditions. The rest of the fuel is introduced into the main combustion zone from the surrounding of the exit of the auxiliary combustion chamber.

3.3.2.2 Nitrogen injection

From figure 13, we just noticed that nitrogen supply, which is blended with fuel or primary air, drastically decreases thermal-NOx emissions, and also NOx emissions decreases with rises in the primary equivalence ratio, ϕ_p, in the case of using the two-stage combustion. That is, thermal-NOx emissions decrease significantly by setting a fuel-rich condition when

ϕ_p is 1.3 or higher in the case of nitrogen premixed with fuel, and by setting ϕ_p at 1.6 or higher in the case of nitrogen premixed with primary combustion air.

With regard to fuel-NOx emissions on the other hand, figure 18 indicates the effects of nitrogen injection conditions on the conversion rate of NH_3 in the fuel to NOx, C.R. in the same conditions with figure 13 except for fuel containing NH_3. In the tests investigating fuel-NOx emissions, 1000ppm of NH_3 is contained in the medium-Btu fuel. In the case of a fuel blended with nitrogen, fuel was diluted, or fuel calorific value decreased to 5.1MJ/m^3 and NH_3 concentration in the fuel decreased to 400ppm. From figure 18, whether with or without nitrogen supplied, the staged combustion method effectively decreased the fuel-NOx emissions, or C.R. drastically decreased as the primary equivalence ratio, ϕ_p, become higher than 1.0, which is a stoichiometric condition, and shows the minimum value at the appropriate ϕ_p. Those optimum ϕ_p become lower when the medium-Btu fuel was blended with nitrogen, while the optimum ϕ_p was in a wide range in the case of nitrogen blended with the primary combustion air injected from the burner, and C.R. showed a tendency to become a little higher than in the other two cases. Furthermore, under lean-lean combustion conditions with a lower ϕ_p than 1.0, in the case of nitrogen premixed with medium-Btu fuel, C.R. becomes higher than in the case of nitrogen premixed with the primary combustion air.

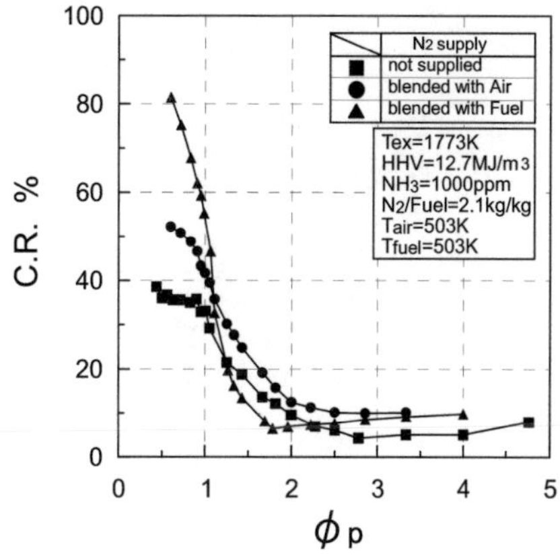

Fig. 18. Effect of nitrogen injection on the conversion rate of NH_3 to NOx in two-stage combustion, using a small diffusion burner

From the above, it was shown that the supply method of nitrogen premixed with medium-Btu fuel possibly decreases total emissions of thermal-NOx and fuel-NOx, but careful attention must be paid to the homogeneity of mixture of fuel and nitrogen, or thermal-NOx emissions will increase. In the case of nitrogen premixed with the primary combustion air, total NOx emissions grow slightly higher than the case of nitrogen premixed with fuel, and the power to compress nitrogen increases a little or the thermal efficiency of the plant decreases. That is, it is necessary to blend nitrogen with medium-Btu fuel evenly in the combustor, in which the lowest power to compress nitrogen is needed for nitrogen supply into the gas turbine, and not to collide the medium-Btu fuel with combustion air directly.

Based on these basic experimental results, we arranged the nitrogen injection intakes between fuel and air intakes in the main-swirler surrounding the primary-flame from the auxiliary combustion chamber for low thermal-NOx emissions. Additionally the fuel, the combustion air, and the nitrogen from the burner are separately injected into the combustor through a swirler, (which has a 30-degree swirl angle and a 15-degree introverted angle), to collide medium-Btu fuel with air in an atmosphere where nitrogen is superior in amount to both fuel and air.

3.3.2.3 Fuel-NOx/Thermal-NOx reduction

In order to decrease fuel-NOx emissions, we adopted fuel-rich combustion in the primary zone and set the equivalence ratio in the primary-combustion zone is determined based on the combustion test results using a small diffusion burner shown in figure 8. Figure 19 presents a relation between the primary equivalence ratio, ϕ_p, and the conversion rate of NH_3 to NOx, C.R., with CH_4 concentration as a parameter in two-staged combustion. In test, the average temperature of the exhaust, T_{ex}, is set to 1773K and fuel calorific value is 11.4MJ/m^3 for fuel containing 1000ppm of NH_3, CO and H_2 of 2.33 CO/H_2 molar ratio. In the same way as low-Btu fuels, the primary equivalence ratio that minimizes the conversion rate of NH_3 to NOx is affected by CH_4 concentration in the fuel. Because the supplied fuel contains 3 percent of CH_4, the equivalence ratio in the primary-combustion zone was set around 1.9 and the equivalence ratio in the auxiliary-combustion chamber was around 2.4 to maintain the flame stabilization and to improve reduction of NH_3, simultaneously.

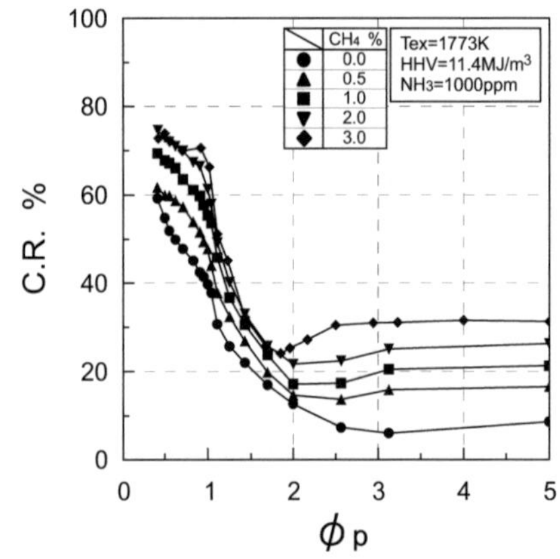

Fig. 19. Effect of the CH_4 concentration on conversion rate of NH_3 to NOx in two-stage combustion of medium-Btu fuel

The effect of the CH_4 concentration on the fuel-NOx produced by NH_3 in gasified fuel was studied using the elementary reaction kinetics (Hasegawa et al., 2001). The model of the flow inside the combustor introduced the Pratt model (Pratt et al., 1971) and each stage combustion zone is assumed to be a perfectly stirred reactor. The reaction model employed here was proposed by Miller and Bowman(1989), values for thermodynamic data were taken from the JANAF thermodynamics tables(Chase et al., 1985) or calculated based on the relationship between the Gibbs' standard energy of formation and the chemical equilibrium constant. The values of Gibbs' standard energy of formation were obtained from the CHEMKIN database (Kee et al., 1990). The GEAR method (Hindmarsh, 1974) was used for the numerical analysis. Also, it is assumed that the species are evenly mixed, and diffusion and stirring processes are not taken into consideration in the reaction process. The appropriateness of the model for reaction NH_3 with NO in the gasified fuels (Hasegawa, 1998c) has been confirmed by comparison with test results.

The nitrogen of NH_3 in the fuel has weaker bonding power than N_2. In the combustion process, NH_3 reacted with the OH, O, and H radicals and then easily decomposed into the intermediate NHi by the following reactions (Miller et al., 1983).

$$NH_3 + OH\ (O, H) \Leftrightarrow NH_2 + H_2O\ (OH, H_2) \tag{3}$$

$$NHi\ (i{=}1, 2) + OH\ (H) \Leftrightarrow NHi\text{-}1 + H_2O\ (H_2) \tag{4}$$

When hydrocarbon is not contained in the fuel, NHi is converted into N_2 by reacting with NO in the fuel-rich region. If fuel contains CH_4, HCN is produced by reactions 5 and 6 in the fuel-rich region and the HCN is oxidized to NO in the fuel-lean zone (Heap et al, 1976) and (Takagi et al, 1979).

$$CHi\ (i{=}1,2) + N_2 \Leftrightarrow HCN + NHi\text{-}1 \tag{5}$$

$$R\text{-}CH + NHi \Leftrightarrow HCN + R\text{-}Hi,\ (R\text{-} : \text{Alkyl group}) \tag{6}$$

Some HCN is oxidized into NO by reactions 7 and 8, and the rest is decomposed into N radical by the reaction 9. NH radical is decomposed into the NO by reactions 10, 11, and 12. With the rise in CH_4 concentration in gasified fuel, the HCN increases, and NOx emissions originated from HCN in the fuel-lean secondary combustion zone increase.

$$HCN + OH \Leftrightarrow CN + H_2O \tag{7}$$

$$CN + O_2 \Leftrightarrow CO + NO \tag{8}$$

$$CN + O \Leftrightarrow CO + N \tag{9}$$

$$NH + OH \Leftrightarrow N + H_2O \tag{10}$$

$$N + O_2 \Leftrightarrow NO + O \tag{11}$$

$$N + OH \Leftrightarrow NO + H \tag{12}$$

On the other hand, some NH radical produced by the reactions 3, 4 and 5 are reacted with Zel'dovich NO, Prompt NO and fuel-N oxidized NO, which produced by above reactions, and decomposed into N_2 by the reaction 13.

$$NO + NH \Leftrightarrow N_2 + OH \tag{13}$$

That is, it is surmised that each of increase in thermal-NOx concentration and fuel-NOx affected the alternative decomposition reaction of intermediate NH radical with NO, so the each of NOx emissions originated from the nitrogen in the air or fuel-N decreased.

These new techniques those adopted the nitrogen direct injection and the two-stage combustion, caused a decrease in flame temperature in the primary combustion zone and the thermal-NOx production near the burner was expected to be controlled. On the contrary, we were afraid that the flame temperature near the burner was declined too low at lower load conditions and so a stable combustion cannot be maintained. The designed combustor was given another nitrogen injection function, in which nitrogen was bypassed to premix with the air derived from the compressor at lower load conditions, and a stable flame can be maintained in a wide range of turn-down operations. Also, because the nitrogen dilution in the fuel-rich region affected the reduction characteristics of NH_3, the increase in nitrogen dilution raised the conversion rates of NH_3 to NOx. This tendency showed the same as that of the case where nitrogenous compounds in fuel increased, indicated by Sarofim et al.(1975), Kato et al.(1976) and Takagi et al.(1977). That is, it is necessary that the nitrogen bypassing technique is expected to improve fuel-NOx reduction in the cases of higher concentration of NH_3.

3.3.3 Test results

Supplied fuels into the combustor were adjusted as same propertied as that of the slurry-feed coal gasified fuel. In tests, the effects of the CH_4 concentrations, etc. in the supplied fuels on the combustion characteristics were investigated and the combustor's performances were predicted in the typical commercial operations. Figure 20 estimates the combustion emission characteristics under the simulated operational conditions of 1773K-class gas turbine for IGCC in the case where gasified fuel contains 0.1 percent CH_4 and 500ppm NH_3. Total NOx emissions were surmised as low as 34ppm (corrected at 16 percent O_2) in the range where the gas turbine load was 25 percent or higher, which is the single fuel firing of gasified fuel, while the NOx emissions tend to increase slightly with the rise in the gas turbine load. In the tests of the simulated fuel that contained no NH_3, thermal-NOx emissions were as low as 8ppm (corrected at 16 percent O_2). On the other hand, we can expect that combustion efficiency is around 100 percent under operational conditions of the medium-Btu fueled gas turbine.

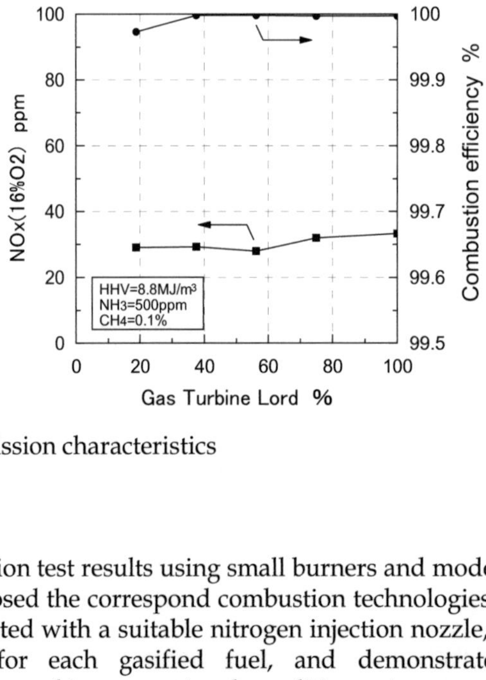

Fig. 20. Combustion emission characteristics

4. Conclusion

Based on basic combustion test results using small burners and model combustors, Japanese electric industries proposed the correspond combustion technologies for each gasified fuels, designed combustors fitted with a suitable nitrogen injection nozzle, two-stage combustion, or lean combustion for each gasified fuel, and demonstrated those combustors' performances under gas turbine operational conditions. As summarized in Table 6, the developed combustors showed to be completely-satisfied with the performances of 1773K-class gas turbine combustor in the actual operations. That is, these combustion technologies reduced each type of NOx emissions for each gasified fuel, while maintaining the other

combustor's characteristics enough. Furthermore, developed technologies represent a possible step towards the 1873K-class gas turbine combustor.
To keep stable supplies of energy and protect the global environment, it will be important that human beings not only use finite fossil fuel, such as oil and coal, but also reexamine unused resources and reclaim waste, and develop highly effective usage of such resources. The IGCC technologies could have the potential to use highly efficient resources not widely in use today for power generation.

		Synthetic gas cleanup	
		Wet type	Hot/Dry type
Gasification agent	Air	• 1573K-class gas turbine combustor for BFG • thermal-NOx ≦20ppm*	• 1773K-class combustor • NOx emissions ≦60ppm* • thermal-NOx ≦ 8ppm* • P.F.(rated) ≦ 8%
	O_2	• 1573K-class combustor • thermal-NOx ≦11ppm* • P.F.(rated) =10～13%	• 1773K-class combustor • NOx emissions ≦34ppm* • thermal-NOx ≦8ppm* • P.F.(rated) ≦ 7%

* : Concentration corrected at 16% oxygen in exhaust

Table 6. Performances of gasified fueled combustors

5. Acknowledgment

The author wishes to express their appreciation to the many people who have contributed to this investigation.

6. Nomenclature

CO/H_2	Molar ratio of carbon monoxide to hydrogen in fuel [mol/mol]
C.R.	Conversion rate from ammonia in fuel to NOx [%]
HHV	Higher heating value of fuel at 273 K, 0.1 MPa basis [MJ/m^3]
N_2/Fuel	Nitrogen over fuel supply ratio [kg/kg]
NOx(16%O_2)	NOx emissions corrected at 16% oxygen in exhaust [ppm]
P	Pressure inside the combustor [MPa]
T_{air}	Temperature of supplied air [K]
T_{ex}	Average temperature of combustor exhaust gas [K]
T_{fuel}	Temperature of supplied fuel [K]
T_{N_2}	Temperature of supplied nitrogen [K]
ϕ_{ex}	Average equivalence ratio at combustor exhaust
ϕ_p	Average equivalence ratio in primary combustion zone

7. References

Battista, R.A. & Farrell, R.A. (1979). Development of an Industrial Gas Turbine Combustor Burning a Variety of Coal-Derived Low Btu Fuels and Distillate, *ASME paper,* No.79-GT-172, San Diego, California, USA, March 11-15, 1979.

Becker, B. & Schetter, B. (1992). Gas Turbines Above 150 MW for Integrated Coal Gasification Combined Cycles (IGCC), *Trans. ASME: J Eng. Gas Turbines Power*, Vol.114, pp.660-664, ISSN 0742-4795.

Beebe, K.W.; Symonds, R.A. & Notardonato, J. (1982). Evaluation of Advanced Combustion Concepts for Dry NOx Suppression with Coal-Derived, Gaseous Fuels, DOE/ NASA/ 13111-11, NASA TM, 82985.

Beer, J.M. (2007). High Efficiency Electric Power Generation: The Environmental Role. *Prog. Energy Combust. Sci.* , Vol.33, pp.107–134, 2007, ISSN 0360-1285.

Brown, T.D. (1982). Coal Gasification: Combined Cycles for Electricity Production. *Prog. Energy Combust. Sci.*, Vol.8, pp.277–301, ISSN 0360-1285.

Bush, W.V.; Baker, D.C. & Tijm, P.J.A. (1991). Shell Coal Gasification Plant(SCGP-1) Environmental Performance Results, *EPRI Interim Report* No. GS-7397, Project 2695-1.

Chase, Jr.M.W.; Davies, C.A.; Downey, Jr.J.R.; Frurip, D.J.; McDonald, R.A. & Syverud, A.N. (1985). *JANAF Thermodynamical tables 3rd Edition.*, J. Phys. Chem. Reference Data, Vol.14, ISSN 0047-2689.

Clark, W.D.; Folsom, B.A.; Seeker, W.R. & Courtney, C.W. (1982). Bench scale testing of low-NOx LBG combustors, *Trans. ASME: J Eng. Power*, Vol.104, pp.120-128, ISSN 0742-4795.

Cohen, N. (1992). Flammability and Explosion Limits of H_2 and H_2/CO: A Literature Review; Aerospace Report No.TR-92(2534)-1; the Aerospace Corporation: El Segundo, CA, USA, September 10, 1992.

Consonni, S.; Larson, E.D. & Berglin, N. (1997). Black Liquor-Gasifier/Gas Turbine Cogeneration, *ASME paper,* No.97-GT-273, Orlando, Florida, USA, June 2-5, 1997.

Cook, C.S.; Corman, J.C. & Todd, D.M. (1994). System evaluation and LBtu fuel combustion studies for IGCC power generation, *ASME paper,* No.94-GT-366, The Hague, The Netherlands, June13-16, 1994.

Coward, H.F. & Jones, G.W. (1952). Limits of Flammability of Gases and Vapors, Bulletin 503 Bureau of Mines, United States Government printing office, Washington.

Dixon-Lewis, G. & Willians, D.J. (1969).The Oxidation of Hydrogen and Carbon Monoxide, chapter 1. In: *Comprehensive Chemical Kinetics,* Bamford, C.H., Tipper, C.F.H., Eds., pp.1-248, Elsevier Pub. Co.: Amsterdam, The Netherlands, ISBN 0444416315.

Döbbeling, K.; Knöpfel, H.P.; Polifke, W.; Winkler, D.; Steinbach, C. & Sattelmayer, T. (1994). Low NOx premixed combustion of MBtu fuels using the ABB double cone burner (EV burner), *ASME paper,* No.94-GT-394, The Hague, The Netherlands, June13-16, 1994.

Drake, M.C.; Pitz, R.W.; Correa, S.M., & Lapp, M. (1984). Nitric Oxide Formation from Thermal and Fuel-bound Nitrogen Sources in a Turbulent Nonpremixed Syngas Flame, *Proc. 20th Symp.(Int.) Combust.*, The Combust. Inst., pp.1983-1990, ISSN 0082-0784, Ann Arbor, MI, USA, August 1984.

Folsom, B.A.; Courtney, C.W. & Heap, M.P. (1980). The Effects of LBG Composition and Combustor Characteristics on Fuel NOx Formation, *Trans. ASME: J. Eng. Power*, Vol.102, pp.459-467, ISSN 0742-4795.

Haavisto, I. (1996). Fixed Bed Gasification of Solid Biomass Fuel, In: *Power Production from Biomass II with Special Emphasis on Gasification and Pyrolysis R&DD*, Sipila, K., Korhonen, M., Eds.; Technical Research Centre of Finland: Espoo, Finland, Vol.164, pp. 127–132, ISBN 9513845559.

Hasegawa, T. & Sato, M. (1997). Study on NOx Formation Characteristics of Medium-Btu Coal Gasified Fuel, *Trans. Japan. Soc. Mech. Eng.*, B, Vol.63, pp.3123–3130, ISSN 0387-5016. (in Japanese).

Hasegawa, T.; Sato, M. & Ninomiya, T. (1998a). Effect of Pressure on Emission Characteristics in LBG-Fueled 1500℃-Class Gas Turbine, *Trans. ASME: J. Eng. Gas Turbines Power*, Vol.120, pp.481-487, ISSN 0742-4795.

Hasegawa, T.; Hisamatsu, T.; Katsuki, Y.; Sato, M.; Yamada, M.; Onoda, A. & Utsunomiya, M. (1998b). A Study of Low NOx Combustion on Medium-Btu Fueled 1300 °C-Class Gas Turbine Combustor in IGCC, *ASME Paper*, No.98-GT-331, Stockholm, Sweden, June 1998.

Hasegawa, T. & Sato, M., (1998c). Study of Ammonia Removal from Coal-Gasified Fuel, *Combust. Flame*, Vol.114, pp.246-258, ISSN 0010-2180.

Hasegawa, T.; Sato, M. & Nakata, T. (2001). A Study of Combustion Characteristics of Gasified Coal Fuel, *Trans. ASME: J. Eng. Gas Turbines Power*, Vol.123, pp.22-32, ISSN 0742-4795.

Hasegawa, T.; Hisamatsu, T.; Katsuki, Y.; Sato, M.; Koizumi, H.; Hayashi, A. & Kobayashi, N., (2003). Development of Low NOx Combustion Technology in Medium-Btu Fueled 1300℃-Class Gas Turbine Combustor in IGCC, *Trans. ASME: J. Eng. Gas Turbines Power*, Vol.125, pp.1-10, ISSN 0742-4795.

Hasegawa, T. & Tamaru, T. (2007). Gas Turbine Combustion Technology Reducing both Fuel-NOx and Thermal-NOx Emissions in Oxygen-Blown IGCC with Hot/Dry Synthetic Gas Cleanup. *Trans. ASME: J. Eng. Gas Turbines Power*, Vol.129, pp.358–369, ISSN 0742-4795.

Heap, M.P.; Tyson, T.J; Cichanowicz, J.E.; Gershman, R.; Kau, C.J.; Martin, G.B. & Lanier, W.S. (1976). Environmental Aspects of low BTU gas combustion, *Proc. 16th Symp. (Int.) on Combust.*, The Combust. Inst., pp.535-545, ISSN 0082-0784, Cambridge, Massachusetts, USA, August 15-20, 1976.

Hindmarsh, A.C. (1974). GEAR: Ordinary Differential Equation System Solver, Lawrence Livermore Laboratory, Univ. California, Report No. UCID-30001, Rev.3.

Hobbs, M.L.; Radulovic, P.T. & Smoot, L.D. (1993). Combustion and Gasification of Coals in Fixed-beds, *Prog. Energy Combust. Sci.*, Vol.19, pp.505–586, ISSN 0360-1285.

Ishibasi, Y.; Oomori, T. & Uchiyama, Y. (1978). Experimental Study on Swirl Flame of Low-Calorific Gas, *Proc. the 6th Annual Conf. Gas Turbine Soc. Jpn.*, pp.7-11, Tokyo, Japan (in Japanese).

Ishizuka, S. & Tsuji, H. (1981). An Experimental Study of Effect of Inert Gases on Extinction of Laminar Diffusion Flames, *Proc. 18th Symp. (Int.) on Combust.*, The Combust. Inst., pp.695-703, ISSN 0082-0784, Waterloo, Canada, August 17-22, 1980.

Isles, J. (2007). Europe Clean Coal Power Priorities are on Carbon Capture and Storage. In: *Gas Turbine World*, DeBiasi, V., Ed.; Pequot Publishing: Fairfield, CT, USA, Vol.37, pp. 20–24, ISSN 0746-4134.

Jenkins, S.D. (1995). Tampa electric company's polk power station IGCC project, *Proc. 12th. Annual Int. Pittsburgh Coal Conference*, p.79, Pittsburgh, PA, USA.

Kalsall, G.J.; Smith, M.A. & Cannon, M.F. (1994). Low Emissions Combustor Development for an Industrial Gas Turbine to Utilize LCV Fuel Gas, *Trans. ASME: J. Eng. Gas Turbines Power* Vol.116, pp.559-566, ISSN 0742-4795.

Kato, K.; Fujii, K.; Kurata, T. & Mori, K. (1976). Formation and Control of Nitric Oxide from Fuel Nitrogen : 1st Report, Experimental and Modelling Studies of Fuel NO in Premixed Flat Flames, *Trans. Japan. Soc. Mech. Eng.*, Series 2, Vol.42, pp.582-591, ISSN 0029-0270. (in Japanese)

Kee, R.J.; Rupley, F.M. & Miller, J.A. (1990). The CHEMKIN Thermodynamic Data Base, Sandia Report, SAND 87-8215B.

Kelleher, E.G. (1985). Gasification fo kraft black liquor and use of the pruducts in combined cycle cogeneration, phase 2 final report, DOE/CS/40341-T5, prepared by Champion Int'1. Co. for U.S. Dept. of Energy, Wash., D.C.

Kurimura, M.; Hara, S.; Inumaru, J.; Ashizawa, M.; Ichikawa, K. & Kajitani, S. (1995). A Study of Gasification Reactivity of Air-Blown Entrained Flow Coal Gasifier, *Proc. 8th. Int. Conference on Coal Science*, Elsevier Science B.V., Amsterdam, Vol.1, pp.563-566, ISBN:9780444822277.

Littlewood, K. (1977). Gasification: Theory and Application. *Prog. Energy Combust. Sci.*, Vol.3, pp.35–71, ISSN 0360-1285.

Martin, F.J. & Dederick, P.K. (1977). NOx from Fuel Nitrogen in Two-stage combustion, *Proc.16th Symp. (Int.) on Comb.*/The Comb. Institute., pp.191-198, ISSN 0082-0784, Cambridge, Massachusetts, USA, August 15-20, 1976.

Merryman, E.L. & Levy, A. (1997). NOx Formation in CO Flames; Report No.EPA-600/2-77-008c; Battelle-Columbus Laboratories: Columbus, OH, USA, January 1997.

Miller, J.A.; Smooke, M.D.; Green, R.M. & Kee, R.J. (1983). Kinetic Modeling of the Oxidation of Ammonia in Flames, *Combust. Sci. Technol.*, Vol.34, pp.149-176, ISSN 0010-2202.

Miller, J.A.; Branch, M.C.; McLean, W.J.; Chandler, D.W.; Smooke, M.D. & Kee, R.J. (1984). The Conversion of HCN to NO and N_2 in H_2-O_2-HCN-Ar Flames at Low Pressure, *Proc. of the 20th Symp.(Int.) Combust.*, The Combust. Inst., pp.673–684, ISSN 0082-0784, Ann Arbor, MI, USA, August 1984.

Miller, J.A. & Bowman, C.T. (1989). Mechanism and modeling of nitrogen chemistry in combustion, *Prog. Energy Combust. Sci.*, Vol.15, pp.287-338, ISSN 0360-1285.

Morgan, G.H. & Kane, W.R. (1962). Some Effects of Inert Diluents on Flame Speeds and Temperatures, *Proc. 4th Symp.(Int.) on Combust.*, The Combust. Inst., pp.313-320, ISSN 0082-0784.

Nagano, T. (2009), Development of IGCC Demonstration Plant, *Journal of the Gas Turbine Society of Japan*, Vol.37, No.2, pp.72-77, ISSN 0387-4168 (in Japanese).

Nakayama, T.; Ito, S.; Matsuda, H.; Shirai, H.; Kobayashi, M.; Tanaka, T. & Ishikawa, H. (1990). Development of Fixed-Bed Type Hot Gas Cleanup Technologies for Integrated Coal Gasification Combined Cycle Power Generation, Central Research Institute of Electric Power Industry Report No.EW89015.

Paisley, M.A. & Anson, D. (1997). Biomass Gasification for Gas Turbine Based Power Generation, *ASME paper*, No.97-GT-5, Orlando, Florida, USA, June 2-5, 1997.

Pillsbury, P.W.; Cleary, E.N.G.; Singh, P.P. & Chamberlin, R.M. (1976). Emission Results from Coal Gas Burning in Gas Turbine Combustors, *Trans. ASME: J Eng. Power*, Vol.98, pp.88-96, ISSN 0742-4795.

Pratt, D.T.; Bowman, B.R. & Crowe, C.T. (1971). Prediction of Nitric Oxide Formation in Turbojet Engines by PSR Analysis, *AIAA paper* No.71-713, Salt Lake City, Utah, USA, June 14-18, 1971.

Roll, M.W. (1995). The construction, startup and operation of the repowered Wabash River coal gasification project, *Proc. 12th. Annual Int. Pittsburgh Coal Conference*, pp.72-77, Pittsburgh, PA, USA.

Sarofim, A.F.; Williams, G.C.; Modell, M. & Slater, S.M. (1975). Conversion of Fuel Nitrogen to Nitric Oxide in Premixed and Diffusion Flames, *AIChE Symp. Series.*, Vol.71, No.148, pp.51-61.

Savelli, J.F. & Touchton, G.I. (1985). Development of a gas turbine combustion system for medium-Btu fuel, *ASME paper*, No.85-GT-98, Houston, Texas, USA, March 17-21, 1985.

Song, Y.H.; Blair, D.W.; SimisnskiV.J. & Bartok, W. Conversion of Fixed Nitrogen to N_2 in Rich Combustion, *Proc. of the 18th Symp.(Int.) on Combust.*, pp. 53–63, ISSN 0082-0784, Waterloo, Canada, August 17-22, 1980.

Takagi, T.; Ogasawara, M.; Daizo, M. & Tatsumi, T. (1976). NOx Formation from Nitrogen in Fuel and Air during Turbulent Diffusion Combustion, *Proc. 16th Symp. (Int.) Combust.*, The Combust. Institute., pp.181-189, ISSN 0082-0784, Cambridge, Massachusetts, USA, August 15-20, 1976.

Takagi, T.; Tatsumi, T. & Ogasawara, M. (1979). Nitric Oxide Formation from Fuel Nitrogen in Staged Combustion: Roles of HCN and NHi, *Combustion and Flames*, Vol.35, pp.17-25, ISSN 0010-2180.

White, D.J.; Kubasco, A.J.; LeCren, R.T. & Notardonato, J.J. (1983). Combustion Characteristics of Hydrogen-Carbon Monoxide Based Gaseous Fuels, *ASME paper*, No.83-GT-142, Phoenix, Arizona, USA, March 27-31, 1983.

Yamagishi, K.; Nozawa, M.; Yoshie, T.; Tokumoto, T. & Kakegawa, Y. (1974). A Study of NOx emission Characteristics in Two-stage Combustion, *Proc.15th Symp. (Int.) on*

Comb., The Comb. Institute., pp.1157-1166, ISSN 0082-0784, Tokyo, Japan, August 25-31, 1974.

Zanello, P. & Tasselli, A. (1996). Gas Turbine Firing Medium Btu Gas from Gasification Plant, *ASME paper*, No.96-GT-8, Birmingham, England, June10-13, 1996.

12

Characterization of a Spray in the Combustion Chamber of a Low Emission Gas Turbine

Georges Descombes
Laboratoire de génie des procédés pour l'environnement, l'énergie et la santé
France

1. Introduction

The use of a turbo-alternator in Lean Premixed Prevaporized combustion (LPP) for hybrid vehicles is beneficial in reducing pollutant emissions at the nominal operating point. The electric thermal hybrid demonstrator studied here consists of a low-emission gas turbine and an alternator which provides the electric power to an electric propulsion motor and a storage battery.
The combustion chamber of the gas turbine is adapted to the nominal operating point so as to function in pre-vaporized combustion, premixed and lean mixtures. A problematic point, however, is the emission of smoke and unburnt hydrocarbons during start-up because the geometry of the combustion chamber is not adapted to moderate air flows.
In the transitional stages of start, an air-assisted pilot injector vaporizes the fuel in the combustion chamber. The jet is ignited by a spark, the alternator being used as an electric starter. This starting phase causes, however, the formation of a fuel film on the walls which can be observed as locally rich pockets.

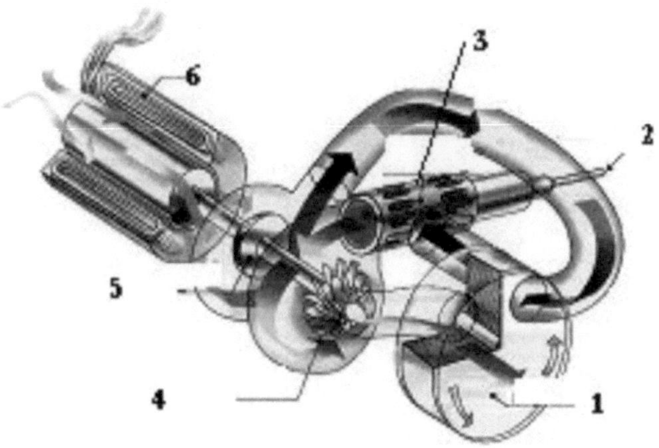

1	2	3	4	5	6
Exchanger	Fuel	Ignition	Turbine	Compressor	Alternator

Fig. 1. Diagram of the turbo alternator

2. The turbo alternator

The turbo alternator has a single-shaft architecture on which the wheels of the compressor and turbine, as well as the high speed alternator, are fixed. The turbine is a single-stage compression/expansion, radial machine with a heat exchanger, as shown in Figure 1. At the nominal operating point, the supercharging air is preheated upstream of the combustion chamber by recovering heat from exhaust gases, thus improving the output of the cycle while decreasing the compression ratio. The exchanger consists of a ceramic heat storage matrix rotated around its axis by a hydraulic engine.

The turbo-alternator delivers an electric output of 38 kW at full load at 90000 rpm. The acceptance tests provide the cartography of the stabilized performance of the turbo-alternator from the turbine inlet temperature and the number of revolutions. The power and the output increase naturally with the temperature, and the optimal operating range is between 70000 and 85000 rpm; the temperature is between 975°C and 1025°C.

3. The combustion chamber

The Lean Premixed Pre-vaporized (LPP) combustion chamber is divided into three zones (Figure 2). First of all, the fuel is injected and vaporized in a flow of hot air with which it mixes. In this zone, complete evaporation and a homogeneous mixture must be achieved before the reaction zone preferably just above the low extinction limit in order to limit the formation of NO_x emissions (Leonard and Stegmaïer, 1993, Ripplinger et al., 1998). The flame is then stabilized with the creation of re-circulation zones, and combustion proceeds with a maximum flame temperature generally lower than 2000K (Poeschl et al., 1994, Ohkubo et al., 1994). The third area is the dilution zone which lowers the temperature below the threshold imposed by the temperature limit of the turbine blades (Turrell et al., 2004).

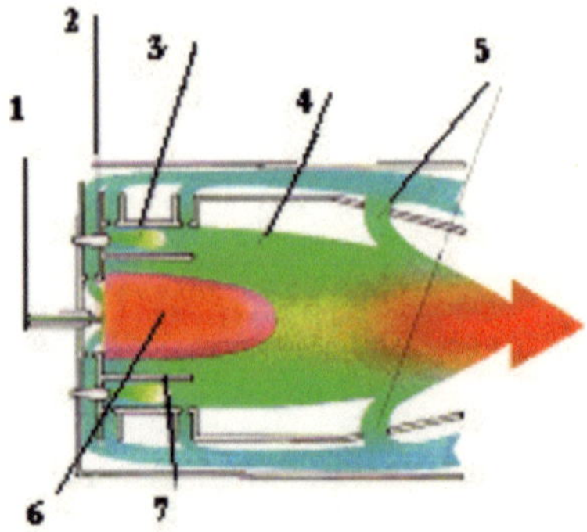

1	2	3	4	5	6	7
Pilot injector	Main injectors	Lean mixture	Lean combustion	Dilution zone	Pilot flame	Mixture pipe

Fig. 2. Diagram of the LPP combustion chamber

The geometry of this combustion chamber is optimised for nominal operation. As modification of the aero-thermodynamic characteristics of the air flow at partial load and at start-up is not conducive to flame stability (Schmidt, 1995), a pilot injector is therefore used; this also serves as a two-phase flame whose fuel spray does not burn in premixed flame.

4. The pilot injector

During the starting phase, the low compression ratio and thermal inertia of the exchanger means that the inlet air cannot be preheated, making LPP operation impossible. The main injectors do not intervene during this phase and are used only when a temperature above 800°C is reached at the turbine inlet.

A pilot injector is used to vaporize the fuel during start-up. The jet is ignited by the spark and a turbulent two-phase flame ensures the temperature increase of the machine. Additional fuel is also provided by the pilot injector to stabilize the flame in weak combustion modes and at low power.

The coaxial injector is characterized by a central fuel jet surrounded by a peripheral high-speed gas flow. The system provides the injector with predetermined and adjustable quantities of liquid fuel and air flow. It is composed of two parts, an air-assisted circuit and a pressurized fuel circuit.

It is observed that the maximum fuel flow, which is about 8 kg/h of fuel for a pressure of 12 bar, remains insufficient to obtain correct vaporization of the fuel. A complementary air-assisted circuit is therefore necessary to interact with the fuel swirl of the pilot injector where atomisation begins. Fuel atomisation is intensified by the counter-rotating movement of the two fluids (Figure 3).

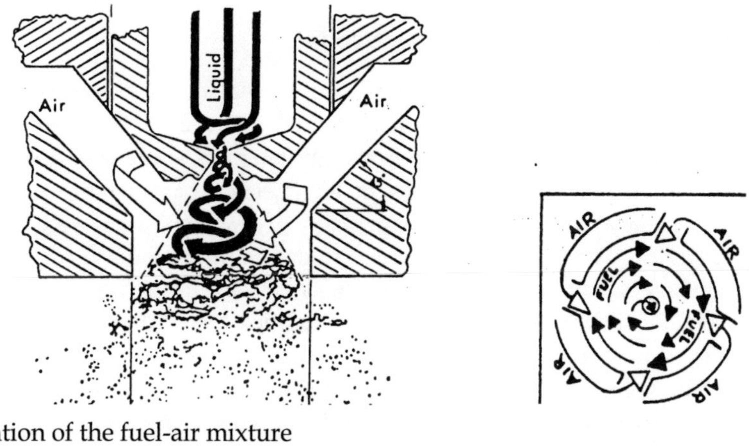

Fig. 3. Formation of the fuel-air mixture

The tests carried out in the laboratory on a turbo-alternator test bench also showed the need for a variable air flow in the pilot injector because the fuel jet of the pilot injector does not always ignite correctly. When a significant increase in temperature is detected in the exhaust, smoke is emitted and its concentration varies significantly depending on the injection parameters . The evolution of the air flow acts directly on the ignition timing and the temperature, as shown by the curves on figure 4.

The ignition timing increases with the increase in the air pressure and the temperature increases more rapidly when the air pressure rises. It is observed that smoke appears approximately thirty seconds after the start-up of the turbine, but vanishes more quickly when the air pressure is higher. Increasing the temperature velocity setting of the turbine made it possible to optimise the burnt fuel fraction and to reduce smoke emissions (Pichouron, 2001).

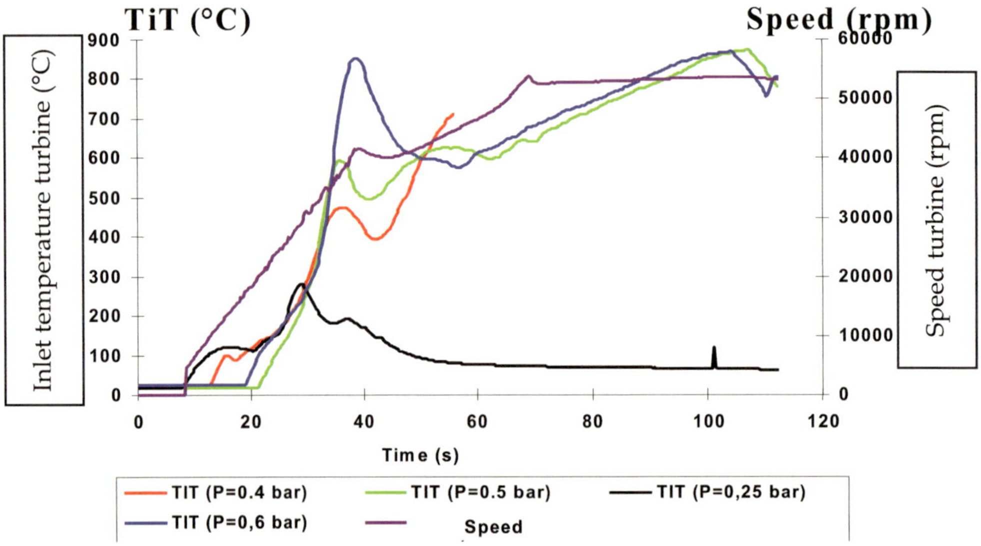

Fig. 4. Evolution of the turbine inlet temperature (TiT) and turbine speed (rpm) at start up of the gas turbine as a function of time and air pressure

5. Experimental study of the non-reactive jet

The preliminary start tests and the analytical study revealed the existence of a correlation between the ignition and the combustion of a fuel spray as a function of its physical characteristics (Pichouron, 2001). The vaporization dynamics of the pilot injector were first studied in the starting phase. The influence of the injection parameters were controlled as was the quality of the jet in terms of drop size, law of distribution as well as jet angle and mass fuel distribution. This cartography aimed to define the optimised operating points as well as the boundary conditions which were then used in the numerical study of the jet.
The air flow of the pilot injector significantly modifies the structure of the jet which is characterized by the spray angle, the fragmentation length, the size distribution of the droplets inside the spray and the penetration. Photographs of the jet taken on the injection bench in the laboratory show the effect of the air flow on the structure of the jet (Figure 5).

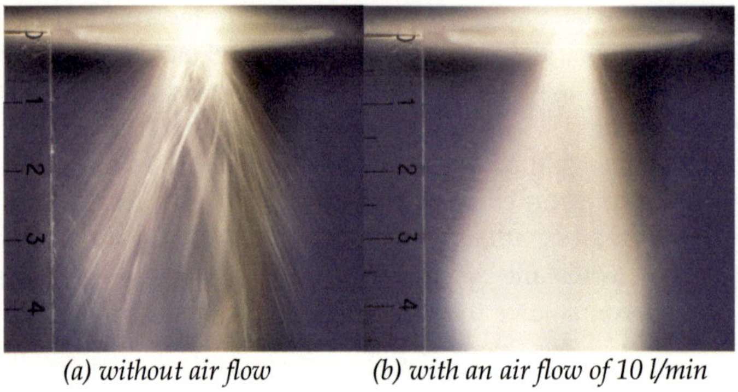

(a) without air flow *(b) with an air flow of 10 l/min*

Fig. 5. Cartography of the spray (liquid flow: 7.3 kg/h)

A granulometric study conducted with the participation of the laboratory CORIA (Rouen, France) also made it possible to measure the distribution of the drop diameters of the injector as a function of the air pressure, the viscosity and the fuel pressure (table 1). The drop sizes were measured by the optical diffraction of a laser beam which passes through the cloud of drops. By measuring the thickness of the cloud of drops in the path of the laser beam and the attenuation of the direct beam, the volume concentration can be obtained (Figure 6). These results made it possible to give the initial conditions of the jet and its dispersed phase.
The geometry of the jet was experimentally investigated in order to measure the angle formed by the jet, to determine the mass distribution of the fuel in the jet and to study axial symmetry. The test bench is composed of a feeding circuit of fuel and air (Figure 7). The fuel jet which develops with the free air is studied and the air mass fuel rates of air flow for the operating points are given in table 1.

Fuel flow (kg/h)	4,4	5,7	6,6	7,7
Air flow (l/min)	10-16-24	10-16-24	10-16-24	10-16-24

Table 1. Operating points for the geometrical study of the spray

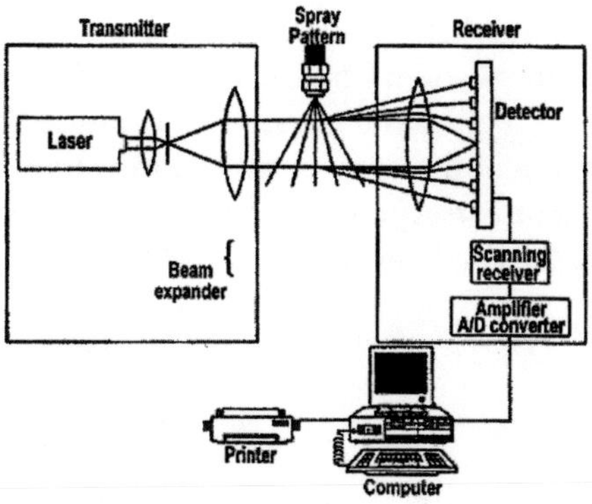

Fig. 6. Diagram of the drop size measurements

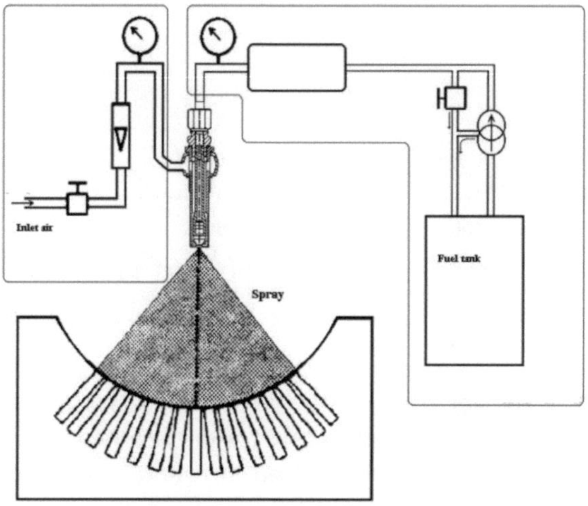

Fig. 7. Diagram of the test rig for characterization of the spray

6. Modelling of the jet

6.1 Identification of a volume law of distribution

The most widely used expression is that originally developed for powders by Rosin and Rammler, where Q is the fraction of total volume contained in drops with a diameter lower than D, X and Q are two parameters which characterize the drops composing the jet (Eq. 1).

$$1 - Q = \exp^{-\left(\frac{D}{X}\right)^q} \quad (1)$$

By identifying X and Q using the experimental results of the granulometric study (Ohkubo and Idota, 1994), the distribution of the drop sizes of the injector must be checked by the Rosin-Rammler law where X is the diameter when 63.2% of the liquid volume is dispersed in drops smaller than X, Q being calculated starting from the Rosin-Rammler law (Eq. 2).

$$q = \frac{\ln(-\ln(1 - Q))}{\ln(D / X)} \quad (2)$$

Figure 8 shows the experimental distribution curve and the associated Rosin Rammler law. The measurements were made at the centre of the spray. The air and fuel mass flows are respectively 16 l/min and 7.7 kg/h. The curves are cumulative distributions of the drop sizes and represent the fraction of the total spray volume in drops larger than the diameter considered. Each measurement corresponds to an operating point of the injector to which corresponds a calculation of the coefficients X and Q of the Rosin-Rammler law.

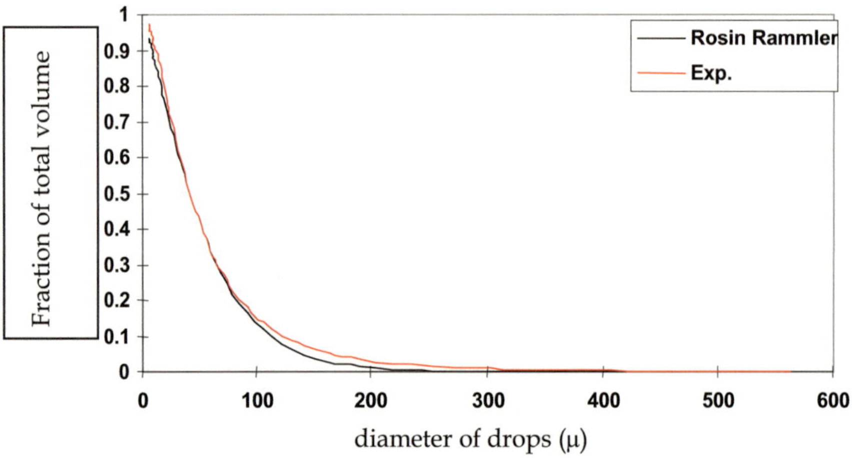

Fig. 8. Experimental distribution of the drop sizes and of the associated Rosin-Rammler law

The Rosin-Rammler law correctly describes the drop size distribution at the centre and the periphery of the jet, in particular when the air flow is low. The validity of the law was then checked for all the injector operating points and for the two fuels: diesel fuel and kerosene. The modeling of the fuel jet in terms of drop size and volume distribution was thus validated by the Rosin-Rammler law in which coefficients are given starting from the granulometry results.

6.2 Cartography of the jet

The effect of the air flow can be very clearly observed on figure 9 when the mass fuel rate of flow is maintained constant. For an air flow of 24 l/min, 50% of the volume of fuel injected is vaporized in drops with a diameter less than 50 microns. If the air flow is reduced to 3.5 l/min, the maximum drop size required to vaporize the same volume of fuel reaches 150 microns.

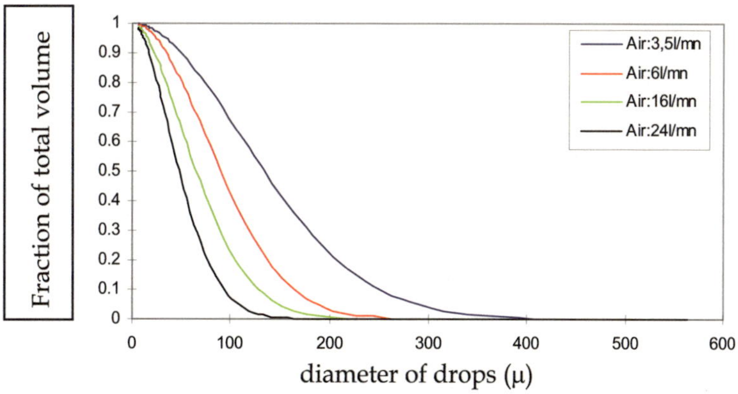

Fig. 9. Evolution of the spray granulometry as a function of the air flow (measurements made 10 mm from the spray centre, mass fuel flow 7.7 kg/h)

The study also shows that the increase in the mass fuel flow rate makes the jet less uniform by producing a significant number of large drops. The increase in the mass fuel flow rate from 4.4 to 7.7 kg/h causes an increase in the maximum drop size from 150 to 250 microns in the centre of the jet. The effects related to the increase of the mass fuel flow rate are also greater at the periphery than in the centre of the jet. These results confirm that the level of atomisation in the jet can be estimated by calculating the mass ratio of the mass fuel flow rate and the air flow.

6.3 Angle of the spray

The jet angle has a value ranging between 30 and 35° on both sides of the longitudinal axis of the injector when the air flow is 24 l/min and it is the same for a flow for 10 l/min. This shows that the geometry of the jet is independent of the mass fuel flow rate when the air flow is 24 l/min. Finally, in agreement with Lefebvre (1989), it can be concluded that the jet angle is only slightly influenced by the air flow.

6.4 Mass distribution of the fuel in the jet

The air flow strongly influences the mass distribution of the fuel in the jet, since increasing the air flow concentrates a high proportion of the fuel in the centre of the jet. Only a small quantity of fuel is then located beyond 30° from the injector axis. The tests show conclusively that the axial symmetry of the jet is respected for the operating conditions, in particular with air flows above 20 l/min.

6.5 Correlations of the SAUTER average diameter

The lack of a consolidated theory on vaporization processes meant that empirical correlations had to be used to evaluate the relation between a representative diameter, the

average diameter and the injection conditions which relate to the physical properties of the liquid, the geometrical characteristics of the injector as well as the outputs of liquid and air flow.
Several definitions of the average diameter have been established depending on the processes observed, but the SAUTER average diameter is generally used to describe vaporization in a medium in which mass and heat transfer phenomena dominate, such as the combustion of a fuel jet (Inamura and Nagai, 1985, Simmons, 1979, Elkotb et al., 1982, Faeth, 1983).
The evolution of the properties of the pilot injector jet is estimated starting from the correlation of Elkotb et al., 1982. It takes into account a geometrical parameter (the diameter of the injector exit), the physical properties of the fluid to be vaporized (surface stress, density and viscosity) and the operating conditions (relative velocities of the liquid and the ambient air, and ratio of the air flow to the liquid flow).
The correlations studied make it possible to better understand the operation of the air-assisted injector used at turbine start-up. The SAUTER average diameter grows with the increase in viscosity and the surface tension of the liquid spray. The use of kerosene, which is less viscous than diesel fuel, makes it possible to decrease the SAUTER average diameter, and the air flow contributes very significantly to vaporization. It is indeed necessary to obtain a high relative speed between the liquid spray and the ambient conditions to ensure good atomisation. This speed is obtained by maintaining the ratio of the mass throughput of the air flow to the mass throughput of liquid spray close to a value of 0.4.

7. Numerical study of the non-reactive jet

Modelling is based on the concept of average size but the aim is not to seek the spatial and temporal evolution of the instantaneous sizes, rather to study their average behaviour. The instantaneous flow field is therefore replaced by an average part and a fluctuating part. These definitions are applied to the conservation equations and the "average temporal" operator is then applied to the resulting equations.
The non-linearity of the convection terms reveals additional terms which represent the correlations of the fluctuations in the physical sizes of the flow. These unknown factors are approximated using an isotropic k-ε model both for the study of the non-reactive jet and for the later study of turbulent combustion in gas phase. The concept of turbulent viscosity proposed by Boussinesq shows that it is possible to approach the additional terms (Pichouron, 2001).

7.1 Liquid phase

The spray is modelled according to a Lagrangian description by a particle unit and it is assumed that the dispersed phase is sufficiently diluted to neglect interactions between the drops (Zamuner, 1995). In practice, the volume fraction occupied by the drops in the jet should not exceed 10 to 12%. Primary disintegration, coalescence and collisions between drops can therefore be neglected. The jet is thus modelled by a set of drops grouped in layers with initial conditions relating to the position, velocity, size, temperature and number of drops represented.
The drops are assumed to be spherical and non-deformable, without clean rotation or interaction (Zamuner, 1995, Wittig et al., 1993). The flow around a drop is assumed to be

homogeneous and the particle density much higher than that of gas. Gravity, the Archimedes force, the added mass term, the force due to the pressure gradient , the Basset force and the Saffman force, are neglected.
The initial conditions of the calculation of the drop trajectories in the dispersed phase result from the experimental study of the jet. The initial drop diameters were determined by the value of the diameters measured 30 mm downstream from the injector nozzle and it was verified that the droplets did not undergo secondary vaporization outside the path of the laser beam.
With low relative speeds, the spherical shape of the droplets is preserved by the combined action of surface stress and the viscous forces of the fuel. When the speed increases, the aerodynamic loads acting on the surface of the drop cause deformation, oscillation, and finally disintegration of the liquid particle.
Two groups of parameters make it possible to distinguish the various modes from secondary disintegration (Schmel et al., 1999, Yule and Salters, 1995), namely Weber numbers *We,* and Ohnesorge numbers *On* (Eqs. 3 and 4) which respectively determine the relationship between the aerodynamic loads exerted on the drop and the surface stress, and the relationship between viscous friction in the drop and surface stress (Krzecskowski, 1995, Pilch and Erdman, 1987). ρ is the density of surrounding gas, u_{rel} is relative speed between gas and the particle, and D, σ_g, μ_g, ρ_g are respectively the diameter, surface stress, viscosity and density of the fuel.

$$We = \frac{\rho u_{rel}^2 D}{\sigma_g} \tag{3}$$

$$On = \frac{\mu_g}{\sqrt{\rho_g D \sigma_g}} \tag{4}$$

No deformation, or oscillation is observed when the Weber number is lower than a breaking value W_{ec}. Beyond this breaking value, three different mechanisms are observed which control the disintegration of the droplets in the case of typical Weber numbers of the flows in a gas turbine combustion chamber. For an Onhesorge number higher than 0.1, a significant influence of viscosity is observed and the transition between the various modes is given by the Weber number. Correlation (5) can then be used to assess the degree of vaporization in the two-phase flows measured. For the relative speeds studied, the drops must have a minimum diameter of 100 μm to undergo secondary vaporization.

$$W_{ec}=12\ (1+1.077\ On^{1.6}) \tag{5}$$

The characterization of the jet shows that for an air flow of 24 l/min, the pilot injector emits a jet made up mainly of drops with a diameter lower than 70 μm. In this configuration, only a very small quantity of the drops is subjected to secondary vaporization. On the other hand, when the air flow is 10 l/min, a maximum diameter of drops of about 180 μm is reached.
Taking into account the ejection speeds estimated for the drops, it can be noted that only the drops with diameters larger than 100 μm are likely to reach the disintegration mode. This indicates that secondary vaporization may therefore occur only over 1.5% in mass of the total fuel flow. Lastly, even if the relative speed increase between the drops and gas favours

secondary vaporization, this physical phenomenon will never be very important within the present framework. Secondary vaporization was therefore be neglected, as was the behaviour of the drops after rebound from the walls.

7.2 Gas phase

The fuel drops warm up and evaporate during their trajectory in the gas phase. The evaporation process of a drop composed of a mixture of hydrocarbons can be divided into three fields for modelling mass and heat transfer (Prommersberger et al., 1999, Aggarwal and Peng, 1994).

The most fully developed approaches (model DLM, Diffusion Limit Model) take account of the heterogeneous temperature field in the droplet, of the influence of the drop and the multi-component composition of the hydrocarbon (Hallmann et al., 1995, Li, 1995). Certain models treat drop heating and vaporization simultaneously, while others assume that the droplet warms up initially without evaporating, and that when it reaches a sufficient temperature, it vaporizes (Schmehl et al., 1999). It is the latter approach which is adopted here, following three successive behavior laws (Pichouron, 2001). This involves calculating reheating of the droplet without exchange of mass with the surrounding medium from the ejection temperature until the vaporization temperature.

Beyond the vaporization temperature, the mass and heat transfer between the drop and the surrounding medium is calculated, up to a boiling point. The convective boiling of the drop at iso-temperature is then predicted.

Calculation proceeds in a fixed geometry with motionless walls, entries for the dilution and air for combustion and an exit for the combustion products. For the entries and the exit, the boundary conditions are imposed in flow in the study of the non-reactive jet and in pressure in the later study of the turbulent combustion of the jet.

The limiting conditions of flow and pressure resulting from the experiment are obtained on the test bench.

7.3 Coupling of the liquid and gas phases

The drops act on gas by the source terms introduced into the equations. The source terms are determined by summing the exchanges along the trajectory of the particles which pass through the control volume. The momentum transfer from the continuous phase to the dispersed phase is obtained by calculating the variation in momentum of the particle traversing the control volume. The heat exchanged between the continuous and dispersed phases is deduced from the thermal variation in energy of the drop which passes through the control volume. The mass transfer of the dispersed phase towards the continuous phase is obtained by calculating the mass variation of the drop traversing the control volume (Reitz and Bracco, 1982).

8. Limiting conditions of calculation

8.1 Space distribution of the drops at the injector outlet

For the 3D representation, the jet is described by a hollow cone. By defining several hollow cones of identical origin and axis, but with a different ray R and angle ϑ, it is possible to represent the jet of the pilot injector. The fuel drops initially form crowns, and taking into account the secondary assumption of non-disintegration, the origin of the crowns is located at the injector nozzle (Litchford and Jeng, 1991).

Initially, the fuel drops are thus divided regularly and into an identical number on concentric crowns (Figure 10). Many numerical tests were carried out to determine the optimal number of crowns. The best representation is obtained with ten crowns for vaporization with an air flow of 24 l/min, and with five crowns for 14 l/min, the crown having a constant external diameter of 2 mm.

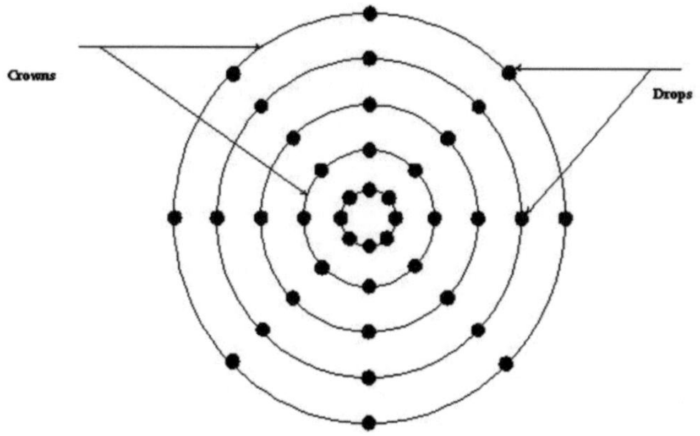

Fig. 10. Example of space distribution of the drops on the injector outlet

8.2 Jet angle

The jet is represented by five or ten hollow cones for which it is necessary to define an angle corresponding to the initial direction of the drops distributed on the crowns. From experimental measurements of the jet angle, the external taper angleϑ_{ext}, was defined; the hollow taper angle i[th] is then given by Equation (6), where N is the number of crowns.

$$\theta_i^n = i.\frac{\theta_{ext}}{n} \tag{6}$$

8.3 Initial diameters

To determine the initial Lagrangian diameters, two series of diameters were selected, based on the granulometry measurements. It was observed that when the air flow in the injector was 24 l/min, the drop sizes are less widely dispersed and that drops with a diameter above 100 µm are rare. Similar observations were made for an air flow of 14l/min, with a maximum diameter of 200 µm. Five (respectively 10) classes of drops were therefore defined for an air flow of 24 l/min (respectively 14 l/min). Each class corresponds to an initial diameter in the calculation of the droplet trajectory.

8.4 Mass flow by trajectory

As the liquid phase flow has to be respected whatever the number of drops injected into the calculation field, it is assumed that a fraction of this flow is allocated with each calculated trajectory and the summation of the flows for each trajectory is equal to the fuel flow in the injector. To attribute a flow to each trajectory, the Rosin-Rammler law was used, as it satisfactorily characterizes the drop size distribution in the pilot injector jet . This representation assumes that several identical drops forward per unit of time on each trajectory (Zamuner, 1995).

8.5 Initial particle speed

It is difficult to determine the initial particle speed given that the fuel is injected via the tube into the injector envelope and that the action of the air flow on the liquid jet takes effect at the very end of the injector envelope, in a zone very close to the section considered as the modelling injection surface. Lay (1997) suggests choosing the initial speed of the drops randomly in a range which evolves by 0 m/s at the local speed of the gas phase. It is further assumed that the initial speed of the drops is constant whatever the diameter and the spatial position of the particle. The estimated speed is based on the speed characteristics of the liquid exiting the injector channel. Knowing that the effective bypass section at the nozzle of the tube is 96.3.10^{-3} mm^2, a speed V_l was determined for the liquid phase. The initial speed of the drops V_g is given by applying a corrective factor whose value depends on the air flow.

8.6 Grid of the geometry

The geometry selected to validate the injection model is a cylinder 300 mm in diameter and 200 mm in length; the medium is composed of air at rest under atmospheric pressure. At one extremity of the cylinder, a centred disc with a diameter of 2.35 mm represents the injector nozzle exit where the initial jet conditions and boundary conditions are imposed to model the air flow. The other end of the cylinder is the simulation exit.

The grid of the geometry consists of 68000 hexahedral cells with a total of 70520 nodes; the quality of the cells and nodes is monitored by a tracer. In the zone downstream from the injector nozzle, the mesh size was fixed at a value ranging between 0.2 and 0.5 mm, based on the results of Sugiyama et al., (1994) cited by Levy (1997), which indicate care should be taken that the mesh does not exceed 1 mm on side in the delicate zones. The modelled field is the top of the combustion chamber made up of the premixing tube and the air inlet ducts. The fuel is vaporized in the premixing tube by the injector, the extremity of which is placed in inlet duct.

8.7 Dispersed phase

Three reference test cases were used to compare the results, and the initial conditions of the dispersed phase are defined following description given previously (table 2).

Case n°1	Air flow =0.285g/s (14l/min) Fuel flow 1=1.84g/s (6,6 kg/h)
Case n°2	Air flow =0.5g/s (24l/min) Fuel flow=1.22g/s (4,4 kg/h)
Case n°3	Air flow =05g/s (24l/min) Fuel flow=1.84g/s

Table 2. Operating points for the study of the spray

8.8 Gas phase

A speed condition is imposed for the injector air flow which makes it possible to combine the requirements of flow (experimental measurement of 14 l/min or 24 l/min) and of flow direction (imposed by the geometry of the injector nozzle). A fraction of the total flow measured on the test bench is imposed on the inlet ducts, based on the division of the discharge between the air intake according to their effective surface.

9. Validation of calculations

9.1 Concentration

The calculated volume concentration of the liquid phase is compared with the experimental results. In the post-processing phase, the concentration value calculated along lines registered on a fictitious cylinder representing the laser beam and located 30 mm downstream from the origin of the injection was recorded (Figure 11).

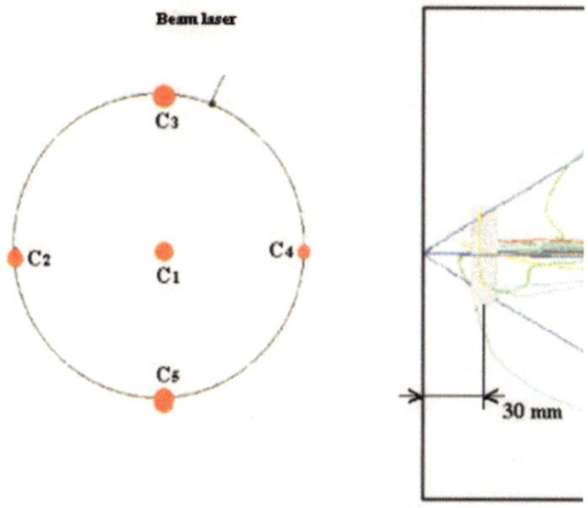

Fig. 11. Calculation of the average concentration of the liquid phase

Figure 12 represents the instantaneous concentrations of the liquid phase along the 5 radial lines (C1 to C5) and the configurations of the concentrations remain similar to those described in the literature for the three cases tested. A deficit in fuel drops on the axis of the injector can be observed, followed by a very clear increase while moving away from the axis, before a second deficit which corresponds to the physical limit of the jet. Note that this same tendency was observed in the experimental study of the jet angle.

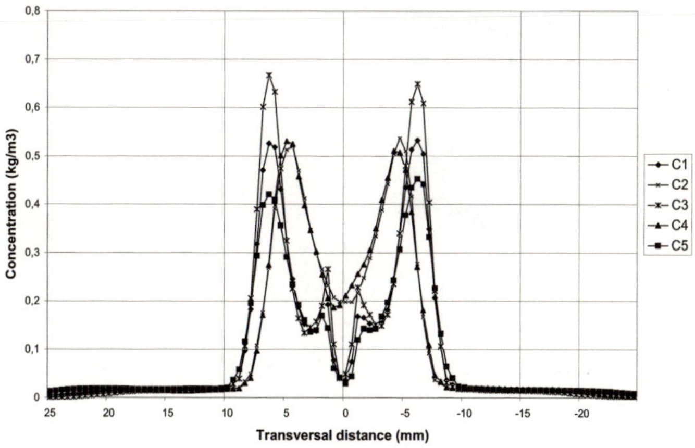

Fig. 12. Instantaneous concentration of the liquid phase as a function of the transverse position. *(mass fuel flow rate: 1.22 g/s and air flow: 0.5 g/s)*

The numerical and experimental average concentrations for the three studied cases were compared. The agreement is very good in the case of the lowest air flow (0.285g/s). For the air flow of 0.5 g/s, the difference between the numerical and experimental results shows the limits of the injection model used here: the experimental study revealed the very strong impact of the air flow on the disintegration of the jet and correlatively on the initial particle speed, which significantly influences the trajectory and thus the concentration of the droplets. The injection model however is kept to study the evolution of the two-phase flow in the ignition zone.

9.2 Simulation of the ignition zone

The experiment shows that the explosion limits depend on the average diameter of the droplets, and that an optimal drop size minimizes the ignition energy required. It was as highlighted as in fact the smallest droplets govern the behavior of the spray to the ignition. The injection model is thus used to numerically study the particle behavior in the ignition zone as a function of their diameter, their initial position in relation to the injector nozzle, and the air flow in the inlet ducts. In the simulation field, the end of the spark which penetrates the ignition zone was not reproduced.

9.3 Aerodynamic field of the jet

Figure 13 shows the field rate of the gas flow for test case n°3. It is observed (13A) that the gas has a high speed at the beginning of the jet due to the air flow, confirming the influence of the air flow on the droplet trajectory that was observed in experiments. Downstream, inside the premixing tube, it can be seen that the field speed is characterized by an intense zone on the injector axis (again evidencing the influence of the air flow) and a calmer area when deviating from the axis. Along the wall of the premixing tube, a return flow is observed (13B), though the speeds reached locally (less than 7m/s) remain low compared to the speeds reached on the injector axis.

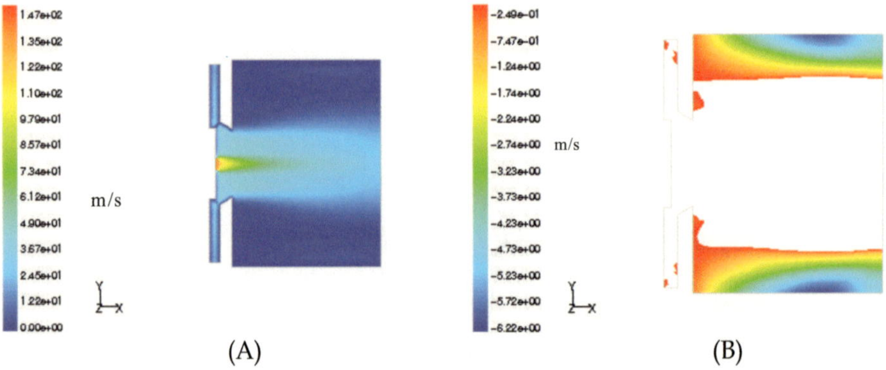

Fig. 13. Gas flow rate in a diametrical plane Field speed (A) and negative axial speed along the walls (B)

The evolution of the radial profiles of the components speed is shown on Figure 14 for 4 positions on the axis of the premixing tube geometry. A strong decrease in axial speed can be observed when the flow deviates from the axis, as along the axis. Observation of the evolution of the radial and tangential components highlights the axi-symmetry of the flow (14A and 14B). It can be seen that the maximum intensity of each component is reached in a zone located 5-6 mm from the axis of the geometry, the dimensions of this corridor being conditioned by the throttling diameter (12 mm) at the end of the convergent one. Beyond this corridor, marginal recirculation zones with weak kinetic energy are formed under the action of the central flow shearing forces.

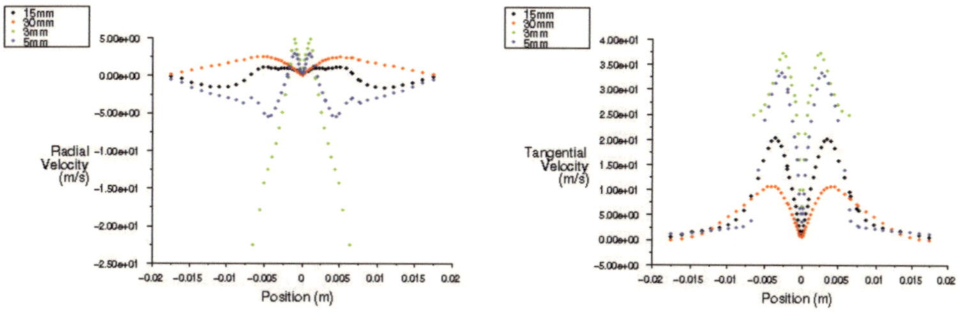

(A) (B)

Fig. 14. Evolution of the profiles of the radial (A) and tangential (B) velocity components for 4 positions on the axis of the geometry

The simulation results for two positions on the axis (3 mm and 15 mm) and for each component speed (Figure 15) were also compared. These test cases differ by their air flows. The variation in the air flow significantly modifies the axial speed in the zone of the axis of geometry (15A and 15B), has a more moderate influence on radial speed (15C and 15D), and has no impact on tangential speed (15E and F).

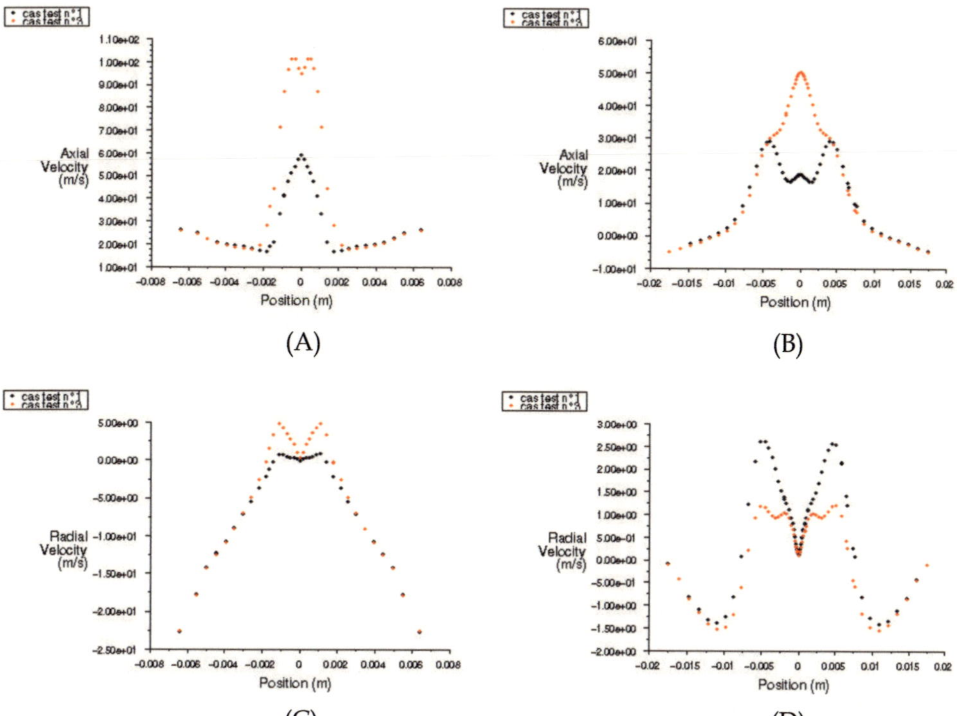

(A) (B)

(C) (D)

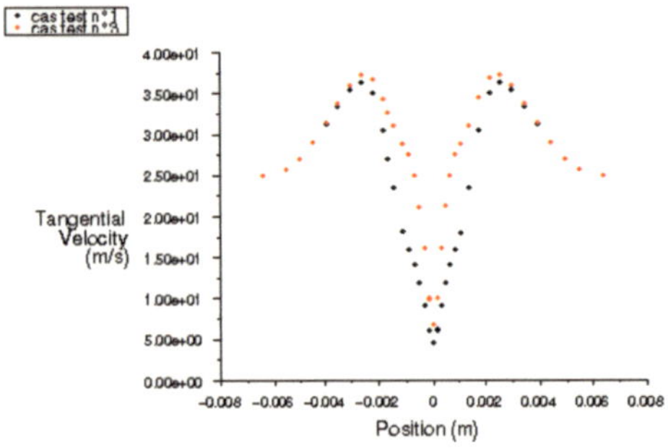

(E)

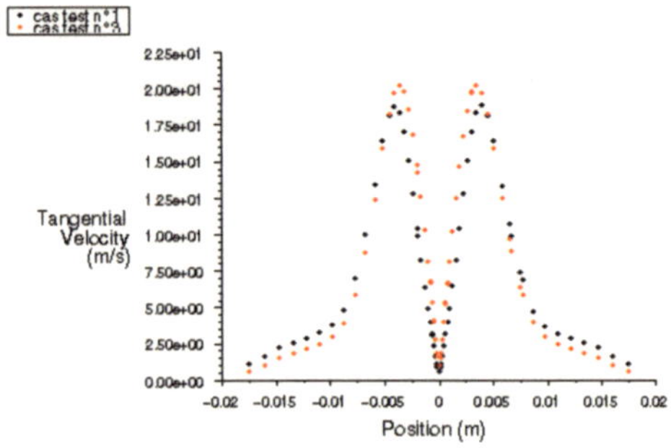

(F)

(left-hand column: calculation with the position 3mm, right-hand column: calculation with the position 15mm)

Fig. 15. Comparison of the axial (A and B), radial (C and D) and tangential (E and F) velocity components of gas for test cases 1 and 3.

The field velocity simulated close to the spark is shown on Figure 16. The mean velocity in this zone is influenced primarily by the air flow entering the inlet ducts. It can be seen that the re-circulation zone increases upstream with the increase in the air flow of combustion (16A and B), and that the tangential component is significantly increased (16C and D). The studies by Snyder et al., 1994, and Yamada et al., 1995 showed that the increased mean velocity increases the minimal ignition energy and this effect will have to be taken into account to define the ignition delay in the combustion chamber.

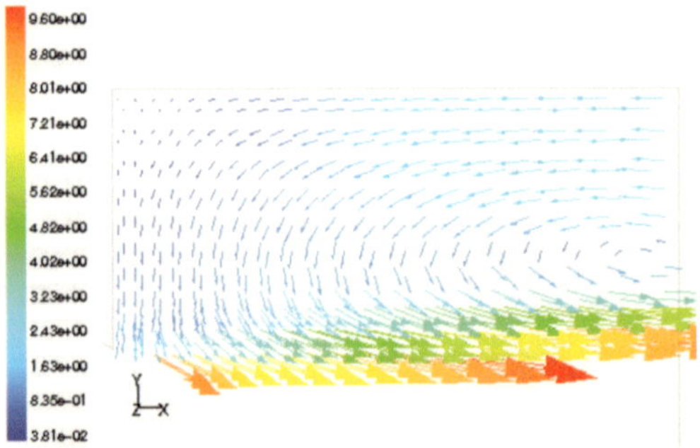

(A)

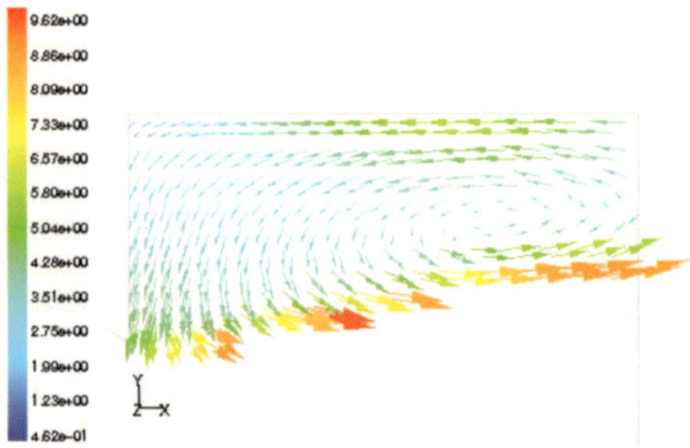

(B)

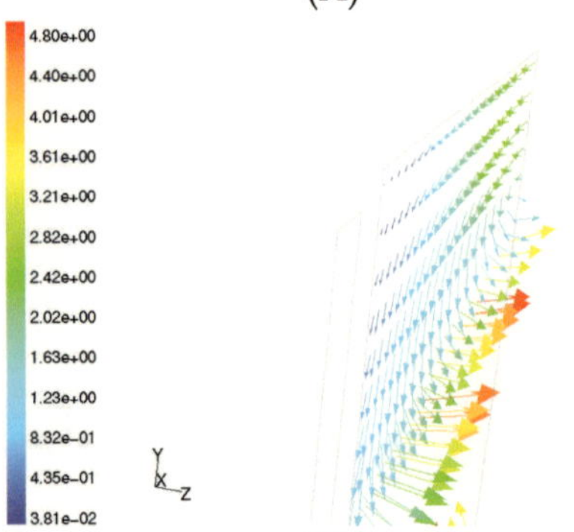

(C)

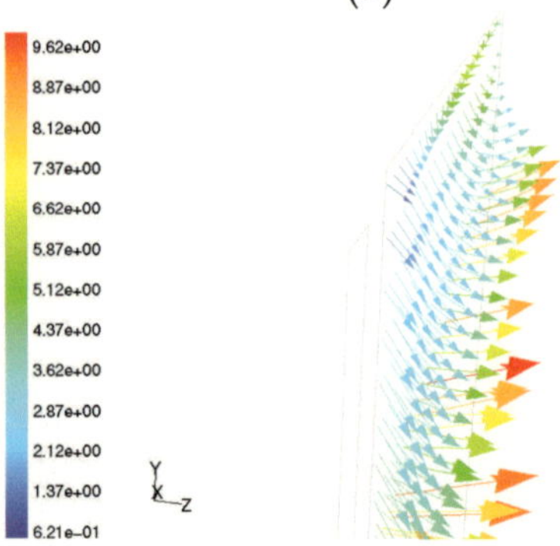

(D)

Fig. 16. Comparison of the field velocity in the ignition zone for two rotation speeds: 5000 rpm (A) and (C) and 40000 rpm (B) and (D)

9.4 Trajectory of the fuel drops

The trajectory of the drops is studied according to three principal parameters. The injector air flow determines the granulometry of the jet and influences the aerodynamics along the simulation axis. The rotation speed of the turbine determines the air flow entering the inlet ducts and influences the swirl intensity in the premixing tube . The penetration depth of the injector nozzle modifies the geometry of the simulation field.

The numerical study of the trajectories shows that drops with a diameter equal to or higher than 40 µm traverse the field, do not impact the walls and that their trajectories are helicoidal. This effect is imposed by the swirl movement in the premixing tube. The increase in the air flow (14 l/min to 24 l/min), by increasing the initial speed of the drops and the axial speed of the gas phase along the axis of symmetry, decreases their residence time in the ignition zone. For lower diameters, the drop-gas interaction has a stronger effect.

Figure 17 compares the trajectories of the 20 µm drops injected from an identical origin and with the same angle. In the case of vaporization at 14 l/min, the drop has less kinetic energy and undergoes the surrounding aerodynamic effects more. The left-hand column on figure 17A shows the case with 14 l/min, and that on the right-hand side the case with 24 l/min (figure 17B). The comparison shows that for the lowest flow, the trajectory of the drops deviates more and that for the maximum angle, the return flow along the walls collects more drops.

It can thus be considered that the damping of the walls is due to the smallest drops. They have the lowest speed and are located in the zone outside the jet, and consequently, the jet angle and particle speed need to be considered to study this phenomenon. The numerical study also shows that damping of the walls can be carried out by direct impact of the largest drops in the case of an open jet angle. In this case, the large drops have a rectilinear trajectory from injection to the wall.

$\phi = 20\mu m,\ \alpha = 12°,\ r = 0{,}3mm,\ q_{assistance} = 14l/\min$	$\phi = 20\mu m,\ \alpha = 12°,\ r = 0{,}3mm,\ q_{assistance} = 24l/\min$

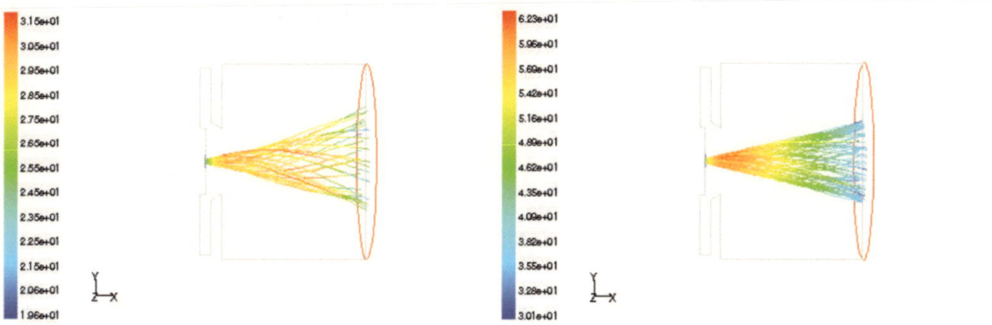

(A) (B)

Fig. 17. Trajectory of the 20-micron drops as a function of the air flow

9.5 Influence of the turbine rotation speed

The turbine rotation speed modifies the air flow in the entry ducts. We saw that this flow does not have a significant influence on the gas flow in the zone close to the field axis where the effects of the air flow of the injector dominate. Beyond this zone, on the contrary, the swirl intensity increases with the air flow in the entry ducts and modifies the field gas speed and the trajectory of the drops. Figure 18 shows that the increase in the swirl of the flow increases the jet angle and contributes to its ventilation. We tracked the trajectories of 20 μm and 40 μm drops, launched with same initial speed, and for two air flows in the inlet ducts, 1.25g/s (15000 rpm) and 3g/s (40000 rpm, 18A and 18B). The results show the influence of the air flow on the particle trajectory and the jet angle. This is true for the small-sized drops, but the larger the drops, the less they are deviated. Drops with an initial diameter above 60 μm have a practically identical trajectory for the two air flows.

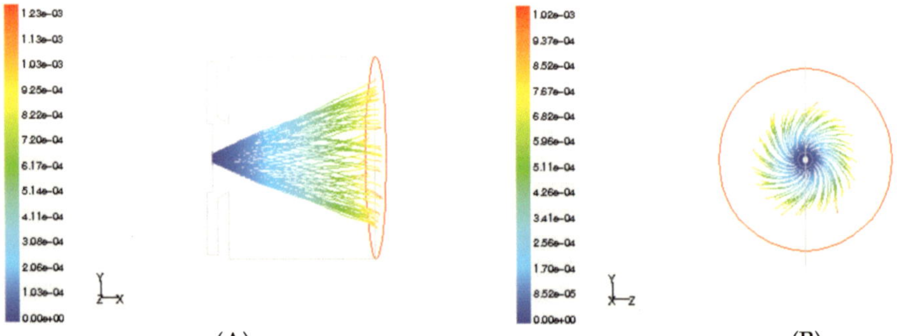

(A) (B)

Fig. 18. Trajectories of the fuel drops as a function of the air flow in the entry ducts *for the same batch of drops (20 and 40 microns)*

9.6 Influence of the injector position

The position of the injector modifies the gas flow along the walls of the premixing tube (Figure 19). It can be seen that the return flow zone is wider and it is the negative axial speed of stronger intensity close to the wall which develops as a function of the penetration depth of the injector (19A). This phenomenon influences especially the trajectory of drops with a diameter below 40μm located at the periphery of the jet. Figure 19B shows the evolution of a batch of drops (20μm) for the three injector positions and identical injection conditions. It can be observed that for the most advanced injector position, drops interact with the return flow and undergo a change of trajectory.

$$(\dot{m}_l = 6{,}6kg/h \text{ - } q_{air\,assistance} = 24l/\min \text{ - } N = 40000tr/\min).$$

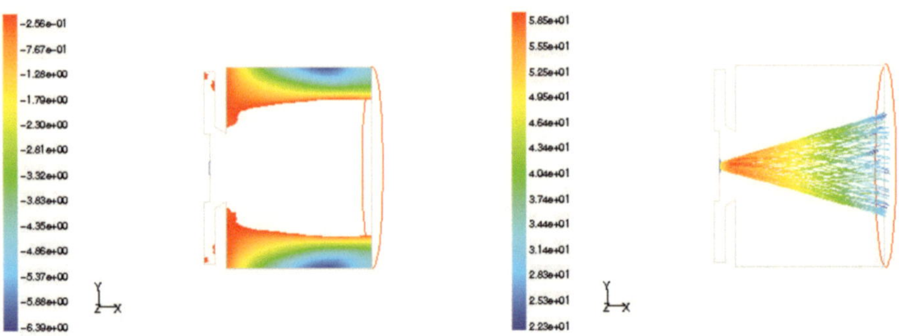

(A) (B)

Fig. 19. Influence of the injector position on the two-phase flow in the premixing tube

9.7 2D Simplification of the geometry

In order to reduce the grid size and the computing time, the digital simulation of combustion is restricted to a 2D approach, which requires simplifications of the geometry and a new definition of the boundary conditions on the air inlet in the combustion chamber. Simplifications consist in making the geometry axisymmetric by removing the entry ducts which induce the overall swirl movement in the flow, then redefining new boundary conditions on the entry surfaces of the air. This operation reveals a new surface located at the base of the inlet ducts which have just been removed and where it is necessary to define entry conditions for the air.

The post-processing of calculations on the unsimplified geometry makes it possible to define the boundary conditions to apply to the entry surface of the simplified geometry. The sizes necessary are measured at the base of the inlet ducts, and are then applied to the simplified geometry. The computation results of the field speed of the gas phase and the particle trajectory make it possible to validate the simplification.

The three components of the aerodynamic field obtained with the two geometries are then compared. On Figure 20, the axial component is calculated along the axis of the geometry and at a transversal distance of 10 mm downstream from the air flow intake. It can be seen that there is a variation along the axis (20B). This variation is also observed in the results calculated along the axis (20A). This phenomenon results from the fact that a uniform field velocity is imposed as boundary condition for the air intake in the simplified geometry, whereas the corresponding field in the complete geometry presents space variations.

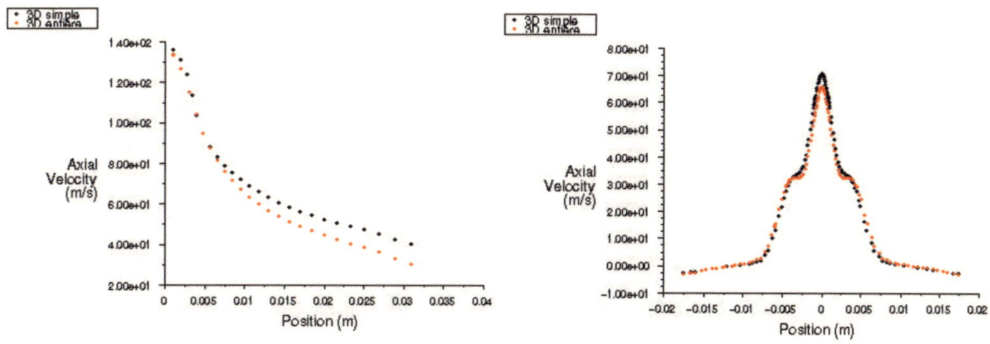

(A) (B)

Fig. 20. Gas flow axial velocity for the two geometries

On Figure 21, the difference between the two field velocities is still observed, either along the axis where the contribution of the radial (21A) and tangential (21B) components is very weak (quasi 1D flow), but in a zone located between 2.5 mm and 10 mm from the axis.
Lastly, the evolution of the same drop in the two simulation fields (complete and simplified geometry) is compared. The particle behavior is very similar in the two cases, with a weak drift on the axial velocity which is higher in simulation with the simplified geometry, which involves the variation noted on the axial co-ordinate of the particle.

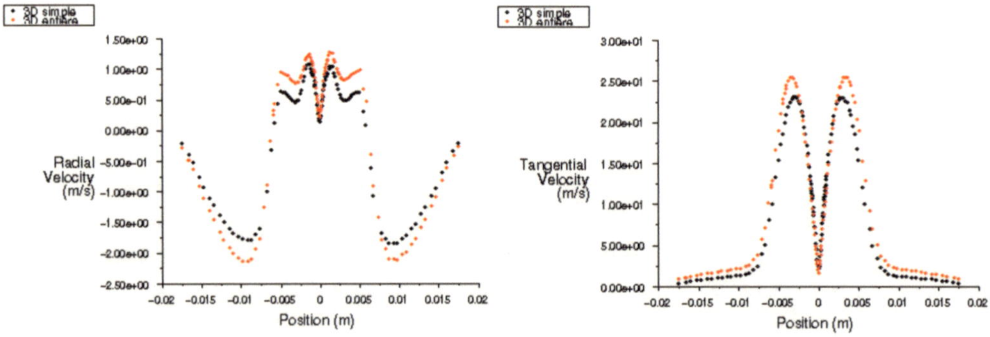

(A) (B)

Fig. 21. Radial velocity (A) and tangential speed (B) for the two geometries

10. Conclusion

This study has investigated the optimization of the start phase of a gas turbine coupled to an alternator used as an electric generator in a hybrid electric vehicle. During this phase, the turbo alternator is actuated electrically and the fuel is vaporized by a pilot injector in the combustion chamber, then ignited by a spark to ensure the rise in temperature of the gas turbine.
The laboratory bench tests made it possible to identify and process hierarchically the dominant parameters which act on the opacity of the smoke. They revealed the paramount role of the air flow of the pilot injector, which is confirmed by the state of the art on spray ignition phenomena.
The experimental characterization tests of the injector controls made it possible to define the granulometry and the geometrical characteristics of the jet as a function of the variations in the injection parameters. We have thus been able to construct the cartography of start-up, starting from the physical properties of the spray versus the air-fuel mass ratio, thus opening up the possibility of optimization without any major technological modifications to the injection system. All the tests provided results which also made it possible to define the initial conditions and the limits of calculation, and to validate the digital simulations of the jet.
The numerical study of the jet aimed to widen the exploratory field concerning the influence of the start parameters on the phenomenon of damping of the walls, on the evaporation of the liquid phase and the interaction between the fuel drops and air in the

combustion chamber. Having observed, in preliminary tests, occurrences of start failure, either by complete absence of ignition of the jet, or by a very slow rate of temperature increase at the turbine inlet, we decided to study the jet under non-reactive conditions initially.

The digital simulation of the non-reactive jet focused on in the ignition zone located inside the premixing tube with a view to defining very precisely the initial conditions of the fuel jet. In agreement with many authors, we assumed that the jet was completely atomized at the injector outlet. The model selected was validated by comparing the measured and calculated liquid phase concentration downstream from the point of injection. Confronted with the difficulty which modeling the atomization of the jet represented, and without precise information on the initial particle speed, we performed more numerical tests to evaluate it while being guided by the experimental results. In this way, we validated our model in experiments by acknowledging in advance that it was not completely exportable to other configurations.

The experimental data relating to the concentration of the liquid phase of a spray vaporized in atmosphere made it possible to validate an injection model which correctly describes the initial conditions of the dispersed phase. This model shows that the gas flow consists of two distinct zones in the premixing tube . The first zone is located along the axis of the tube and is dominated by the effects of the air flow with a high axial component velocity.

In the second zone, located at the periphery, it is the velocity of the tangential component which dominates because it is influenced by the air flow entering the inlet ducts depending on the rotation speed of the turbine. in the trajectory of the droplets whose diameter is lower or equal to 40 µm, when the air flow in the conduits of admission corresponds to a minimal mode of turbine of 15000 rpm.

This modeling was based on simplifications of representation of the jet such as the possibility of various drops occupying the same space, uniform initial speed for the drops, and the nonrandom variation in the jet angles. It is noted that it is necessary to define a high number of injections for the wide distribution of the drop diameters.

The results of this work show that increasing the injector air flow contributes to the formation of a better ventilated jet. The field rate of the flow in the ignition zone was also studied to evaluate the convective effects ; an experimental study of the influence of spark length on ignition remains to be done. The injection model finally made it possible to validate the assumption of axisymmetry of the geometry in order to reduce the calculation of the trajectories to a 2D approach and to thus limit the computation time.

An improvement in the model could be obtained by adopting a surface injection model instead of the conical model, and by introducing a probability function for the initial droplet velocity and jet angle. Estimating the distribution of the diameters and initial speeds of the drops with a model of jet disintegration could also be considered.

A later stage will relate to the choice and the development of a model of turbulent combustion using a semi-detailed chemical mechanism in a multi-phase medium. We will show that it is possible to reproduce the experimental observation of the temperature levels at the entry of the turbine as well as the evolution of the unburnt residues. A parametric analysis will also be conducted to evaluate the influence of the injection parameters and the rotation speed of the turbine on the trajectory of the dispersed phase, these observations being directly applicable to optimize the cartography of the start phase of gas turbines.

11. Acknowledgements

The author is grateful to the grant-in-aid from RENAULT SA research program for providing support for this research project and the PhD of J.F. Pichouron.

12. References

Aggarwal S.K. and Peng F., 1994, A Review of Droplet Dynamics and Vaporization Modeling for Engineering Calculation, ASME 94-WP-215, 1994.

Elkotb, Mr., Mahdy Mr. A., Montaser, Mr. E., 1982, Investigation off external mixing air assist atomizer, International Proceeding off the 2nd conferences one liquid atomization and sprays, pp. 107-115, Madison, 1982.

Faeth, G., 1983, Evaporation and combustion off sprays, Program Energy combustion Sciences, Vol.9, pp. 1-76, 1983.

Hallmann, Mr. Scheurlen, Mr., Wittig, S., 1995, Computation off turbulent evaporating sprays, Eulerian versus lagrangian approach, Transactions off the ASME, vol. 117, pp. 112-119, 1995.

Inamura T. and Nagai NR., 1985, The relative performance off externally and internally-mixed twin fluid atomizers, Proceedings off the 3rd International conference one liquid atomization and sprays, London, July 1985.

Krzecskowski, S.A., 1995, Atomization and sprays, Shorts race 1995, Institute für Thermische Strömungsmaschinen, Universität Karlsruhe, 1995.

Lay, Mr. K., 1997, CFD analysis off have liquid spray combustion in gas turbine combuster, International gas turbine and aeroengine congress, Orlando, Florida, june 2-5, 1997, ASME Paper 97-WP-309, 1997.

Lefèbvre, H.A., 1989, Atomization and sprays, Hemisphere publishing corporation, 1989.

Leonard G. and Stegmaier J., 1993, Development off year aero-derivative gas turbine dry low emissions combustion system, ASME 93-WP-288, 1993.

Levy, NR., 1997, numerical Study of a reactive Diesel jet, Thesis of the central School of Lyon, 1997.

Li X., 1995, Mechanism off atomization off liquid jet, Atomization and Sprays, 5:89-105, 1995.

Litchford, R.J. and Jeng, S.J., 1991, Efficient statistical turbulent transport model for particle dispersion in spray, AIAA Newspaper, Vol.29, n°9, 1991.

Ohkubo, Y. and Idota, Y., 1994, Evaporation Characteristics off Spray in Lean-Premixed-Prevaporisation Combuster for has 100 kw automotive Ceramic Turbine, ASME Paper n°94-WP-401, 1994.

Pichouron J.F., 2001, Contribution to the study of the performances of a gas turbine low emissions in launching phase, thesis of doctorate of the University Paris 6, supported on March 28, 2001.

Pilch, Mr. and Erdman, A.C., 1987, Use off station-wagon-up time dated and velocity history dated to predict the off maximum size stable fragments for acceleration induced station-wagon-up off has liquid drop, International newspaper off multiphase flow 13 (6), pp. 741-757, 1987.

Poeschl G., Ruhkamp W., Pfost H., 1994, Combustion with low pollutant emissions off liquid fuels in gas turbines by premixing and prevaporization, ASME, 94-WP-43, 1994.

Prommersberger, K., Maier, G., Wittig, S., 1999, Validation and application off droplet evaporation model for real aviation fuel, Research and technology organization meeting Proceedings, vol. 14, 1999.

Reitz R.D. and Bracco, F.V., 1982, Mechanism off atomization off has liquid jet, Physics off fluids, Vol.25, pages 1730-1742, 1982.

Ripplinger Th., Zarzalis NR., Meikis G., Hassa C., 1998, Nox Reduction by Lean Premixed Prevaporized Combustion, Applied Vehicle Technology Panel Symposium, Lisbon, Portugal, 12-16 October 1998.

Schmehl, R., Klose, G., Maier, G., Wittig, S., 1999, Efficient numerical calculation off evaporating sprays in combustion chamber flows, Research and technology organization meeting proceedings, vol. 14, 1999.

Schmel, R., Klose, G., Maier, G., Wittig, S., 1999, Efficient numerical calculation off evaporating sprays in combustion chamber flows, Research and technology organization meeting proceedings-14, 1999.

Schmidt S.E., 1995, Bus temperature and LDA velocity measurements in A turbulent swirling premixed propane/air fuel model gas turbine combuster, ASME 95-WP-64, 1995.

Simmons H.C., 1979, The prediction off To jump mean diameter for gas turbine fuel nozzles off different types, ASME Paper 79-WA/GT-5, 1979.

Snyder T.S., Rosfjord T.J., Mc Vey J.B., 1994, Emissions and performances off has lean-premixed gas fuel aeroderivative injection system for gas turbine engine, ASME 94-WP-234, 1994.

Sugiyama, G., Ryu, H., Kobayashi, S., 1994, Computational simulation off Diesel combustion with high presses fuel injection, International symposium COMODIA, pp. 391-396, Yokohama, 11-14 july 1994.

Turrell, Mr., Stopford, P., Syed, K., Buchanan, E., 2004, CFD simulation off the flow within and downstream off has high-swirl lean premixed gas turbine combuster, Proceedings off ASME Turbo Expo 2004, Power for Land, sea and air, June 14-17, 2004, Vienna, Austria, ASME Paper GT2004-53112.

Wittig, S., Hallmann, Mr., Scheurlen, Mr., Schmehl, S., 1993, A new eulerian model for turbulent evaporating sprays in recirculating flows, AGARD-CP-536, 1993.

Yamada H, Shimodaira K., Hayashi S., 1995, There engine evaluation off emissions characteristics off is variable geometry lean-premixed combuster, ASME 95-WP-48, 1995.

Yule A.J. and Salters D.G., 1995, One the diesel distance required to atomize sprays injected from nozzles opening-type, Volume 209, pages 217-226, Proceeding off ImechE: Newspaper off Automobile Engineering, 1995.

Zamuner, B., 1995, Experimental and numerical Study of a spray in combustion resulting from a coaxial injector liquid-gas, Thesis of the central School of Paris, 1995.

Part 5

Materials and Fabrication

13

Materials for Gas Turbines – An Overview

Nageswara Rao Muktinutalapati
VIT University
India

1. Introduction

Advancements made in the field of materials have contributed in a major way in building gas turbine engines with higher power ratings and efficiency levels. Improvements in design of the gas turbine engines over the years have importantly been due to development of materials with enhanced performance levels. Gas turbines have been widely utilized in aircraft engines as well as for land based applications importantly for power generation. Advancements in gas turbine materials have always played a prime role – higher the capability of the materials to withstand elevated temperature service, more the engine efficiency; materials with high elevated temperature strength to weight ratio help in weight reduction. A wide spectrum of high performance materials - special steels, titanium alloys and superalloys - is used for construction of gas turbines. Manufacture of these materials often involves advanced processing techniques. Other material groups like ceramics, composites and inter-metallics have been the focus of intense research and development; aim is to exploit the superior features of these materials for improving the performance of gas turbine engines.

The materials developed at the first instance for gas turbine engine applications had high temperature tensile strength as the prime requirement. This requirement quickly changed as operating temperatures rose. Stress rupture life and then creep properties became important. In the subsequent years of development, low cycle fatigue (LCF) life became another important parameter. Many of the components in the aero engines are subjected to fatigue- and /or creep-loading, and the choice of material is then based on the capability of the material to withstand such loads.

Coating technology has become an integral part of manufacture of gas turbine engine components operating at high temperatures, as this is the only way a combination of high level of mechanical properties and excellent resistance to oxidation / hot corrosion resistance could be achieved.

The review brings out a detailed analysis of the advanced materials and processes that have come to stay in the production of various components in gas turbine engines. While there are thousands of components that go into a gas turbine engine, the emphasis here has been on the main components, which are critical to the performance of the engine. The review also takes stock of the R&D activity currently in progress to develop higher performance materials for gas turbine engine application. On design aspects of gas turbine engines, the reader is referred to the latest edition of the Gas Turbine Engineering Handbook (Boyce, 2006).

2. Compressor parts for aircraft engines – Titanium alloys

Titanium, due to its high strength to weight ratio, has been a dominant material in compressor stages in aeroengines. Titanium content has increased from 3 % in 1950s to about 33% today of the aeroengine weight. Unlike predictions made for requirements of ceramic and metal matrix composites for aeroengines, predictions made for titanium alloys have come true or even surpassed. High temperature titanium alloys have found extensive application in aeroengines. Ti-6Al-4V is used for static and rotating components in gas turbine engines. Castings are used to manufacture the more complex static components. Forgings are typically used for the rotating components. For example, the alloy is used for fan disc and low pressure compressor discs and blades for the Pratt and Whitney 4084 engine. The alloy is used in the cooler compressor stages up to a maximum temperature of about 315 °C. Ti-8Al-1Mo-1V is used for fan blades in military engines (Bayer, 1996). The alloys 685 (Ti-6Al-5Zr-0.5Mo-0.25Si) and 829 (Ti-5.5Al-3.5Sn-3Zr-1Nb-0.25Mo-0.3Si) are used in many current European aeroengines such as RB2111, 535E4 in fully beta heat treated condition to maximize creep resistance (Gogia, 2005). Alloy 834 (Ti-5.8Al-4Sn-3.5Zr-0.7Nb-0.5Mo-0.35Si-0.06C), a relatively recent grade, in contrast is used in $\alpha+\beta$ condition, with a 5-15% equiaxed α in the microstructure to optimize both creep and fatigue strength (Gogia, 2005). The alloy was aimed at replacing the Alloys 685 and 829 preferred in European jet engines. Alloy 834 is used as a compressor disc material in the last two stages of the medium-pressure compressor, and the first four stages of the high pressure compressor in variants of the Rolls-Royce Trent series commercial jet engine. The Ti-1100 (Ti-6Al-2.8Sn-4Zr-0.4Mo-0.4Si), a competitive alloy to IMI834, is designed to be used in the β heat treated condition. The alloy is under evaluation by Allison Gas Turbine Engines for higher thrust versions of their 406/GMA3007/GMA2100 family of engines, primarily for castings (Gogia, 2005). The alloy has a claimed use temperature of 600 °C. IN US, Ti6-2-4-2 (Ti-6Al-2Sn-4Zr-2Mo) is the preferred high temperature alloy for jet engine applications. A variant of this alloy, Ti6-2-4-2S is also commercially available. The 'S' denotes addition of 0.1-0.25 % Si to improve the creep resistance. It is used for rotating components such as blades, discs and rotors at temperatures up to about 540 °C (Bayer, 1996). It is used in high pressure compressors at temperatures too high for Ti-6-4, above about 315 °C, for structural applications.

Today, the maximum temperature limit for near-α alloys for elevated temperature applications is about 540 °C. This temperature limitation for titanium alloys mean the hottest parts in the compressor, i.e. the discs and blades of the last compressor stages, have to be manufactured from Ni-based superalloys at nearly twice the weight. Additionally, problems arise associated with the different thermal expansion behavior and the bonding techniques of the two alloy systems. Therefore enormous efforts are underway to develop a compressor made completely of titanium. Titanium alloys are required that can be used at temperatures of 600 °C or higher. This has been the impetus for extensive research and development work in the area of elevated temperature titanium alloys.

Table 1 gives the chemical composition and the maximum service temperature of various grades of titanium alloys mentioned above. Figure 1 shows schematically the relative creep capability of these grades in the form of a Larson Miller plot. The reader is referred to some excellent reviews on use of titanium alloys in gas turbine engines (Bayer, 1998; Gogia, 2005). The technical guide on titanium published by ASM International (Donachie, 2000) also gives much information on titanium as a gas turbine material.

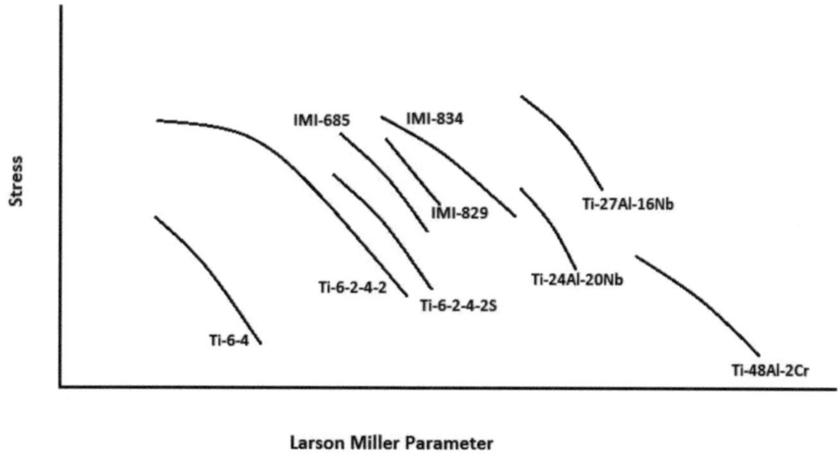

Fig. 1. Relative creep capability of titanium alloys used for compressor parts in the form of a Larson Miller plot (Schematic).

3. Compressor blading materials for land based gas turbines – Special steels

Until recently, all production blades for compressors are made from 12% chromium containing martensitic stainless steel grades 403 or 403 Cb (Schilke, 2004). Corrosion of compressor blades can occur due to moisture containing salts and acids collecting on the blading. To prevent the corrosion, GE has developed patented aluminum slurry coatings for the compressor blades. The coatings are also meant to impart improved erosion resistance to the blades. During the 1980's, GE introduced a new compressor blade material, GTD-450, a precipitation hardened martensitic stainless steel for its advanced and uprated machines (Schilke, 2004). Without sacrificing stress corrosion resistance, GTD-450 offers increased tensile strength, high cycle fatigue strength and corrosion fatigue strength, compared to type 403. GTD-450 also possesses superior resistance to acidic salt environments to type 403, due to higher concentration of chromium and presence of molybdenum (Schilke, 2004).

Grade designation	Nominal chemical composition	Maximum service temperature (°C)
Ti64	Ti-6Al-4V	315
Ti811	Ti-8Al-1Mo-1V	400
Alloy 685	Ti-6Al-5Zr-0.5Mo-0.25Si	520
Alloy 829	Ti-5.5Al-3.5Sn-3Zr-1Nb-0.25Mo – 0.3Si	550
Alloy 834	Ti-5.8Al-4Sn-3.5Zr-0.7Nb-0.5Mo-0.35Si-0.06C	600
Ti1100	Ti-6Al-2.8Sn-4Zr-0.4Mo-0.4Si	600
Ti6242	Ti-6Al-2Sn-4Zr-2Mo	
Ti6242S	Ti-6Al-2Sn-4Zr-2Mo-0.2Si	540

Table 1. Titanium alloys used for compressor parts in aircraft engines - chemical composition and maximum service temperature

Table 2 gives the chemical composition of the different steel grades used for compressor blading.

Grade	Chemical composition	Remarks
AISI 403	Fe12Cr0.11C	Martensitic stainless steel
AISI 403+Nb	Fe12Cr0.2Cb0.15C	Martensitic stainless steel with Nb addition
GTD-450	Fe15.5Cr6.3Ni0.8Mo0.03C	Precipitation hardening stainless steel

Table 2. Compressor blade materials for land based gas turbines

4. Combustion hardware for aircraft and industrial gas turbines (IGTs)

Driven by the increased firing temperatures of the gas turbines and the need for improved emission control, significant development efforts have been made to advance the combustion hardware, by way of adopting sophisticated materials and processes.
The primary basis for the material changes that have been made is improvement of high temperature creep rupture strength without sacrificing the oxidation / corrosion resistance. Traditionally combustor components have been fabricated out of sheet nickel-base superalloys. Hastelloy X, a material with higher creep strength was used from 1960s to 1980s. Nimonic 263 was subsequently introduced and has still higher creep strength (Schilke, 2004). As firing temperatures further increased in the newer gas turbine models, HA-188, a cobalt base superalloy has been recently adopted for some combustion system components for improved creep rupture strength (Schilke, 2004). Coutsouradis et al. reviewed the applications of cobalt-base superalloys for combustor and other components in gas turbines (Coutsouradis et al., 1987). Nickel base superalloys 617 and 230 find wide application for combustor components (Wright & Gibbons, 2007). Table 3 gives the chemical composition of combustor materials.

Grade	Chemical composition	Remarks
Hastelloy X	Ni22Cr1.5Co1.9Fe0.7W9Mo0.07C0.005B	Nickel-base superalloy
Nimonic 263	Ni20Cr20Co0.4Fe6Mo2.1Ti0.4Al0.06C	Nickel-base superalloy
HA188	Co22Cr22Ni1.5Fe14W0.05C0.01B	Cobalt-base superalloy
617	54Ni22Cr12.5Co8.5Mo1.2Al	Nickel-base superalloy
230	55Ni22Cr5Co3Fe14W2Mo0.35Al0.10C0.015B	Nickel-base superalloy; values for Co, Fe and B are upper limits.

Table 3. Combustor materials

In addition to designing with improved materials, combustion liners and transition pieces of advanced and uprated machines involving higher firing temperatures are given a thermal barrier coating (TBC). The coating serves to provide an insulating layer and reduces the underlying base metal temperature. Section 9 deals with the subject of TBC in detail.

5. Turbine disk applications

5.1 Aircraft engines – Superalloys

A286, an austenitic iron-base alloy has been used for years in aircraft engine applications (Schilke, 2004). Superalloy 718 has been used for manufacture of discs in aircraft engines for more than 25 years (Schilke, 2004). Both these alloys have been produced through the conventional ingot metallurgy route.

Powder Metallurgy (PM) processing is being extensively used in production of superalloy components for gas turbines. PM processing is essentially used for Nickel-based superalloys. It is primarily used for production of high strength alloys used for disc manufacture such as IN100 or Rene95 which are difficult or impractical to forge by conventional methods. LC Astroloy, MERL 76, IN100, Rene95 and Rene88 DT are the PM superalloys where ingot metallurgy route for manufacture of turbine discs was replaced by the PM route.

The advantages of PM processing are listed in the following:

- Superalloys such as IN-100 or Rene95 difficult or impractical to forge by conventional methods. P/M processing provides a solution
- Improves homogeneity / minimizes segregation, particularly in complex Ni-base alloy systems
- Allows closer control of microstructure and better property uniformity within a part than what is possible in cast and ingot metallurgy wrought products. Finer grain size can be realized.
- Alloy development flexibility due to elimination of macro-segregation.
- Consolidated powder products are often super-plastic and amenable to isothermal forging, reducing force requirements for forging.
- It is a near net shape process; hence significantly less raw material input required and also reduced machining cost, than in case of conventional ingot metallurgy.

Several engines manufactured by General Electric and Pratt and Whitney are using superalloy discs manufactured through PM route.

Table 4 gives the details of disc superalloys for aircraft engines.

Grade	Chemical composition	Remarks
A286	Fe15Cr25Ni1.2Mo2Ti0.3Al0.25V 0.08C 0.006B	Iron-base superalloy; ingot metallurgy route
718	Ni19Cr18.5Fe3Mo0.9Ti0.5Al5.1Cb 0.03C	Nickel-iron-base superalloy; ingot metallurgy route
IN 100	60Ni10Cr15Co3Mo4.7Ti5.5Al0.15C 0.015B 0.06Zr1.0V	Nickel-base superalloy; powder metallurgy route
Rene 95	61Ni14Cr8Co3.5Mo3.5W3.5Nb2.5Ti3.5Al 0.16C0.01B0.05Zr	Nickel-base superalloy; powder metallurgy route
LC Astroloy	56.5Ni15Cr 15Co5.25Mo3.5Ti4.4Al 0.06C0.03B0.06Zr	Nickel-base superalloy; powder metallurgy route
MERL-76	54.4Ni12.4Cr18.6co3.3Mo1.4Nb 4.3Ti5.1Al0.02C0.03B0.35Hf0.06Zr	Nickel-base superalloy; powder metallurgy route
Rene88 DT	56.4Ni16cr13Co4Mo4W0.7Nb3.7Ti 2.1Al0.03C0.015B0.03Zr	Nickel-base superalloy; powder metallurgy route
Udimet 720	55Ni18Cr14.8Co3Mo1.25W5Ti2.5Al0.035C 0.033B0.03Zr	Nickel-base superalloy; ingot metallurgy / powder metallurgy route
Udimet 720LI	57Ni16Cr15Co3Mo1.25W5Ti2.5Al0.025C0. 018B0.03Zr	Low C, low B variant of Udimet 720.

Table 4. Disc superalloys for aircraft engines

5.2 IGTs – Steels and superalloys

Turbine discs of most GE single shaft heavy duty gas turbines are made of 1%Cr-1.25%Mo-0.25%V steel in hardened and tempered condition (Schilke, 2004). 12%Cr steels such as M152 have higher rupture strength than Cr-Mo-V steel, in addition to outstanding fracture toughness and capacity to attain uniform and high mechanical properties in large sections. Use of A286 for IGTs started in 1965, when the technological advancements made it possible to produce large ingots of this material with required quality (Schilke, 2004).

With the advent of advanced of gas turbine engines with much higher firing temperatures and compressor ratios, it became necessary to utilize a nickel-base superalloy, alloy 706 for the rotors. The use of this material provides the necessary temperature capability required to also meet the firing temperature requirements in the future. This alloy is similar to the Alloy 718, an alloy that has been used for rotors in aircraft engines for more than 25 years. Alloy 706 contains lower concentrations of alloying elements known for their tendency to segregate. Consequently it is less segregation-prone than Alloy 718 and could be produced in large diameters unlike Alloy 718. Accordingly large sized rotors of Alloy 706 could be produced to serve large IGTs for land-based power generation (Schilke, 2004). Alloy 718, the most frequently used superalloy for aircraft gas turbines, because of its segregation tendency, could be produced, until the turn of the century, to a maximum ingot size of 500mm. Developments made with reference to remelting techniques, together with very close control on chemical composition have enabled production of ingots of Alloy 718 as large as 750 mm in diameter. This has resulted in the ability to process Alloy 718 to the large disk sizes needed in modern IGTs.

The importance of Alloy 718 and Alloy 706 can be seen from the fact that several international conferences have been devoted to developments related to these alloys (Loria, 1989, 1991, 1994, 1997, 2001, 2005; Caron et al., 2008).

Grade	Chemical composition	Remarks
CrMoV steel	Fe1Cr0.5Ni1.25Mo0.25V0.30C	Medium carbon low alloy steel
M152	Fe12Cr2.5Ni1.7Mo0.3V0.12C	12% Cr steel
A286	Fe15Cr25Ni1.2Mo2Ti0.3Al0.25V 0.08C 0.006B	Iron-base superalloy
706	Ni16Cr37Fe1.8Ti2.9Cb0.03C	Nickel-iron-base superalloy
718	Ni19Cr18.5Fe3Mo0.9Ti0.5Al5.1Cb 0.03C	Nickel-iron-base superalloy
Udimet 720	55Ni18Cr14.8Co3Mo1.25W5Ti2.5Al0.035C0.033B0.03Zr	Nickel-base superalloy
Udimet 720LI	57Ni16Cr15Co3Mo1.25W5Ti2.5Al0.025C0.018B0.03Zr	Nickel-base superalloy

Table 5. Disc materials for IGTs

Udimet 720 also evolved as an advanced wrought alloy for land based gas turbines. Reductions in Cr content to prevent sigma phase formation and in carbon and boron levels

to reduce stringers and clusters of carbides, borides or carbonitrides have led to the development of the Alloy 720LI. Both these alloys have been of considerable interest to land based gas turbines. They have also been incorporated in some aircraft gas turbines (Furrer & Fecht, 1999). Table 5 gives details of special steels / superalloys used for production of discs for land-based gas turbines.
The reader is referred to an overview by Furrer and Fecht on nickel-based superalloys for turbine discs for land based power generation and aircraft propulsion (Furrer & Fecht, 1999).

6. Turbine blades and vanes – Cast superalloys

Recognition of the material creep strength as an important consideration for the gas turbine engines, understanding generated between age hardening, creep and γ' volume fraction and the steadily increasing operating-temperature requirements for the aircraft engines resulted in development of wrought alloys with increasing levels of aluminum plus titanium. Component forgeability problems led to this direction of development not going beyond a certain extent. The composition of the wrought alloys became restricted by the hot workability requirements. This situation led to the development of cast nickel-base alloys. Casting compositions can be tailored for good high temperature strength as there was no forgeability requirement. Further the cast components are intrinsically stronger than forgings at high temperatures, due to the coarse grain size of castings. Das recently reviewed the advances made in nickel-based cast superalloys (Das, 2010).
Buckets (rotating airfoils) must withstand severe combination of temperature, stress and environment. The stage 1 bucket is particularly loaded, and is generally the limiting component of the gas turbine. Function of the nozzles (stationary airfoils) is to direct the hot gases towards the buckets. Therefore they must be able to withstand high temperatures. However they are subjected to lower mechanical stresses than the buckets. An important design requirement for the nozzle materials is that they should possess excellent high temperature oxidation and corrosion resistance.

6.1 Conventional equiaxed investment casting process

6.1.1 Aircraft engines

Cast alloy IN-713 was among the early grades established as the materials for the airfoils in the most demanding gas turbine application. Efforts to increase the γ' volume fraction to realize higher creep strength led to the availability of alloys like IN 100 and Rene 100 for airfoils in gas turbine engines. Increased amount of refractory solid solution strengtheners such as W and Mo were added to some of the grades developed later and this led to the availability of grades like MAR-M200, MAR-M246, IN 792 and M22. Addition of 2 wt% Hf improved ductility and a new series of alloys became available with Hf addition such as MAR-M200+Hf, MAR-M246+Hf, Rene 125+Hf.
General Electric pursued own alloy development with Rene 41, Rene 77, Rene 80 and Rene 80+Hf having relatively high chromium content for improved corrosion resistance at the cost of some high temperature strength. Other similar alloys with high chromium content are IN738C, IN738LC, Udimet 700, Udimet 710.
Table 6 gives details of superalloy compositions of airfoils produced by conventional equiaxed investment casting process.

Grade designation	Chemical composition
IN 713	74.2Ni12.5Cr4.2Mo2Nb0.8Ti6.1Al0.1Zr0.12C0.01B
IN 100	60.5Ni10Cr15Co3Mo4.7Ti5.5Al0.06Zr0.18C0.014B
Rene 100	62.6Ni9.5Cr15Co3Mo4.2Ti5.5Al0.06Zr0.15C0.015B
MAR-M200	59.5Ni9Cr10Co12.5W1.8Nb2Ti5Al0.05Zr0.15C0.015B
MAR-M246	59.8Ni9Cr10Co2.5Mo10W1.5Ta1.5Ti5.5Al0.05Zr0.14C0.015B
IN 792	60.8Ni12.7Cr9Co2Mo3.9W3.9Ta4.2Ti3.2Al0.1Zr0.21C0.02B
M 22	71.3Ni5.7Cr2Mo11W3Ta6.3Al0.6Zr0.13C
MAR-M200+Hf	Ni8Cr9Co12W2Hf1Nb1.9Ti5.0Al0.03Zr0.13C0.015B
MAR-M246+Hf	Ni9Cr10Co2.5Mo10W1.5Hf1.5Ta1.5Ti5.5Al0.05Zr0.15C0.015B
Rene 41	56Ni19Cr10.5Co9.5Mo3.2Ti1.7Al0.01Zr0.08C0.005B
Rene 77	53.5Ni15Cr18.5Co5.2Mo3.5Ti4.25Al0.08C0.015B
Rene 80	60.3Ni14Cr9.5Co4Mo4W5Ti3al0.03Zr0.17C0.015B
Rene 80+Hf	59.8Ni14Cr9.5Co4Mo4W0.8Hf4.7Ti3Al0.01Zr0.15C0.015B
IN 738	61.5Ni16Cr8.5Co1.75Mo2.6W1.75Ta0.9Nb3.4Ti3.4Al0.04Zr0.11C0.01B
Udimet 700	59Ni14.3Cr14.5Co4.3Mo3.5Ti4.3Al0.02Zr0.08C0.015B
Udimet 710	54.8Ni18Cr15Co3Mo1.5W2.5Ti5Al0.08Zr0.13C
TMD-103	59.8Ni3Cr12Co2Mo6W5Re6Ta0.1Hf6Al

Table 6. Conventionally cast nickel-base superalloys for gas turbine blading applications in aircraft gas turbines

6.1.2 Land-based gas turbine engines

6.1.2.1 Bucket materials for land based gas turbines

Many of the GE engines used U-500 for stage 1 buckets in mid1960's. It is being used for later stages of buckets in selected gas turbine models (Schilke, 2004). IN738 has been used as stage 1 bucket material on several GE engines during 1971-1984. In recent years it has been also used as stage 2 bucket material in some GE engines (Schilke, 2004). The alloy has an outstanding combination of elevated temperature strength and hot corrosion resistance and this makes it attractive for heavy duty gas turbine applications. Developments in processing technology have enabled production of the alloy in large ingot sizes. The alloy is used throughout the heavy duty gas turbine industry. Subsequently GE has developed the alloy GTD-111, with higher strength levels than 738, but maintaining its hot corrosion resistance. GTD-111 has replaced IN738 as bucket material in different GE engine models (Schilke, 2004).
Table7 gives details of conventionally cast superalloys for blading applications in IGTs.

Grade	Chemical composition
Udimet 500	Ni18.5Cr18.5Co4Mo3Ti3Al0.07C0.006B
Rene 77	Ni15Cr17Co5.3Mo3.35Ti4.25Al0.07C0.02B
IN738	Ni16Cr8.3Co0.2Fe2.6W1.75Mo3.4Ti3.4Al0.9Cb0.10C0.001B1.75Ta
GTD 111	Ni14Cr9.5Co3.8W1.5Mo4.9Ti3.0Al0.10C0.01B2.8Ta

Table 7. Conventionally cast nickel-base superalloys for blading applications in IGTs

6.1.2.2 Nozzle materials for land based gas turbines

GE engines use FSX 414, a GE-patented cobalt base alloy for all stage 1 nozzles and some later stage nozzles. Cobalt base alloys possess superior strength at very high temperatures compared to nickel base superalloys - hence the choice of cobalt base superalloy. It has a two-three fold oxidation resistance compared to X40 and X45, also cobalt based superalloys used for nozzle applications. Use of FSX 414 over C40/C45 hence enables increased firing temperatures for a given oxidation life (Schilke, 2004).
Later stage nozzles must also possess adequate creep strength and GE developed a nickel base superalloy GTD222 for some stage 2 and stage 3 applications. The alloy has significantly higher creep strength compared to FSX414. N155, an iron-based superalloy, has good weldability and is used for later stage nozzles of some GE engines (Schilke, 2004). Table 8 gives the details of materials used for nozzles in IGTs.

Grade	Chemical composition	Remarks
X40	Co-25Cr10Ni8W1Fe0.5C0.01B	Cobalt-base superalloy
X45	Co-25Cr10Ni8W1Fe0.25C0.01B	Cobalt-base superalloy
FSX414	Co-28Cr10Ni7W1Fe0.25C0.01B	Cobalt-base superalloy
N155	Fe-21Cr20Ni20Co2.5W3Mo0.20C	Iron-base superalloy
GTD-222	Ni-22.5Cr19Co2.0W2.3Mo1.2Ti0.8Al0.10V 0.008C1.0B	Nickel-base superalloy

Table 8. Nozzle materials for IGTs

6.2 Directionally solidified (DS) castings

6.2.1 Aircraft engines

The major failure mechanism for gas turbine airfoils involved nucleation and growth of cavities along transverse grain boundaries. Elimination of transverse grain boundaries through directional solidification of turbine blades and vanes made an important step in temperature capability of these castings. Use of DS superalloys could improve the turbine blade metal temperature capability by about 14 °C relative to the conventionally cast superalloys.

Grade designation	Chemical composition	Remarks
DS MAR M-200+Hf	59.5Ni9Cr10Co12.5W2Hf1.8Nb2Ti5Al0.05Zr0.15C0.015B	First generation
CM247LC	61.7Ni8.1Cr9.2Co0.5Mo9.5W3.2Ta1.4Hf0.7Ti5.6Al0.01Zr0.07C0.015B	First generation
PWA1422	59.2Ni9Cr10Co12W1.5Hf1Nb2Ti5Al0.1Zr0.14C0.015B	First generation
DMD4	66.8Ni2.4Cr4Co5.5W6.5Re8Ta1.2Hf0.3Nb5.2Al0.07C0.01B	Third generation

Table 9. DS nickel-base superalloys for blading applications in aircraft engines

By early 1980s, DS superalloys became available and were operating in gas turbines. DS MAR-M-200+Hf became available. Another DS grade CM247LC is the outcome of extensive efforts to optimize the chemical composition to improve carbide microstructure, grain boundary cracking resistance, to minimize the formation of deleterious secondary phases and to avoid HfO_2 inclusion problem. Pratt and Whitney developed an equivalent DS grade PWA 1422.
Table 9 gives details of DS superalloy compositions for aircraft engines.

6.2.2 Land-based gas turbine engines

GE has been using the DS version of DTD-111 for stage 1 buckets of different engines. It is same as DTD-111 equiaxed, except tighter control on alloy chemistry. DS version of DTD-111 is stated to possess improved creep life, improved fatigue life and higher impact strength, compared to equiaxed version (Schilke, 2004). Use of DS superalloys could improve the turbine blade metal temperature capability by about 14 °C relative to the conventionally cast superalloys. TMD-103 belongs to the recent advances in DS alloy castings for IGT airfoil castings. It has very attractive long term creep rupture strength and hot corrosion resistance. The alloy could be directionally solidified in the form of large hollow blades for 2000KW IGT. Alloy chemistry of IGT buckets/vanes differs greatly from that of aeroengine blade/vane alloys, both on account of different operating scenarios and DS processing difficulties due to the large size of IGT components. Table 10 gives details of DS superalloy compositions for airfoils in IGTs.

6.3 Single crystals

In single crystal (SC) castings all grain boundaries are eliminated from the microstructure and an SC with a controlled orientation is produced in an airfoil shape. SCs required no grain boundary strengtheners such as C, B, Zr and Hf. Elimination of these elements while designing the SC compositions helped in raising the melting temperature and correspondingly the high temperature strength. Figure 2 schematically shows the improvement in creep strength of a cast superalloy by switching over from equiaxed polycrystalline investment casting to DS casting to SC casting.

Grade designation	Chemical composition	Remarks
DTD 111	Same as DTD 111 for equiaxed version, but with tighter control on alloy chemistry	
TMD-103	59.8Ni3Cr12Co2Mo6W5Re6Ta0.1Hf6Al0.07C0.015B	Third generation superalloy

Table 10. DS nickel-base superalloys for application as rotating blades in IGTs

6.3.1 Aircraft engines

The early SC superalloys included RR2000, RR2060 of Rolls Royce, PWA1480 of Pratt and Whitney, CMSX2 and CMSX3 of Cannon Muskegon and ReneN4 of GE. These SC alloys provided about 20 °C metal temperature advantage over the existing DS alloys.
Attempts to further improve the metal temperature capability of SC superalloys by way of increasing the refractory alloying elements, prominently Rhenium, led to the development

of SC superalloy grades PWA 1484, CMSX4, Rene N5, TUT92. These grades gave about 30 °C metal temperature improvement over the early SC superalloys.
Development of SC superalloys continued with the target of achieving another 30 °C improvement in metal temperature capability while maintaining the environmental resistance and freedom from appearance of deleterious phases in the microstructure. This led to emergence of grades CMSX10, Rene-6, TMS75, TMS80, MC-NG developed by Onera in France, DMS4 developed by DMRL, India, TMS-196, developed by NIMS, Japan. Detailed studies / evaluation have been carried out on these grades and they are potential candidate alloys for future gas turbine engines with enhanced performance.
Figure 3 schematically shows the improvement in stress rupture strength of superalloys, by moving over from DS (CM247) to first generation SC (CMSX2) to second generation SC (CMSX4) to third to fifth generation (CMSX10 and TMS 196) SCs, in the form of a Larson Miller plot.
Table 11 gives details of the superalloys used for blading applications in aircraft engines.

6.3.2 Land-based gas turbines

Development of SC castings has also benefited to improve the efficiency of combined cycle power plants by way of increasing the engine firing temperatures. GE has been applying the SC bucket technology for last several years. SC alloys such as CMSX11B, AF56, PWA1483 containing about 12%Cr for long term environmental resistance together with additions of C, B, Hf to enhance alloy tolerance to low angle boundaries have been developed as airfoil materials. SC alloys such as CMSX 11C and SC 16 have been developed with Cr >12% to increase resistance to hot corrosion and oxidation. Long term phase stability was an important consideration in design of these alloys. Gibbons reviewed the improvements that are taking place with reference to alloys and coatings for integrated gasification combined cycle systems (IGCC) (Gibbons, 2009)
Table 12 gives details of SC superalloys used for rotating blade application in IGTs.

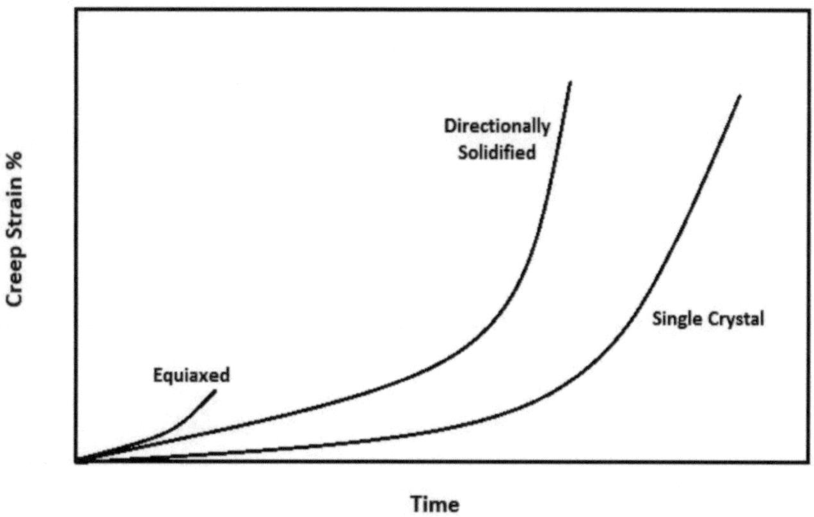

Fig. 2. Relative creep deformation of equiaxed, DS and SC superalloy castings (schematic)

7. Vanes - Oxide Dispersion Strengthened (ODS) superalloys

A limited use exists for ODS superalloys in gas turbine engines. ODS superalloys are advanced high temperature materials which can retain useful strength up to a relatively high fraction of their melting point. This advantage is due to the uniformly dispersed, stable oxide particles which act as barriers to dislocation motion. MA754 has been in production by General Electric as a vane material since 1980. Because of its high long time elevated temperature strength, it has been extensively used for aircraft gas turbine vanes.

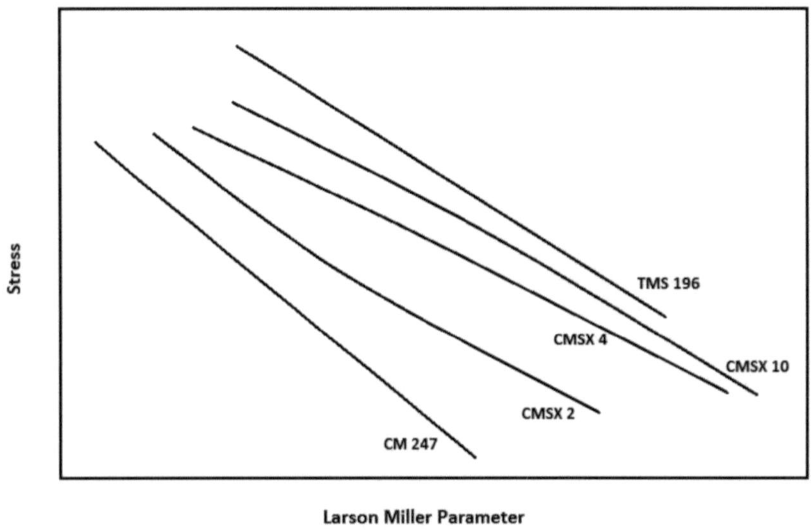

Fig. 3. Improvements made in the stress rupture strength of superalloys through advancements in processing technology (Schematic)

The modern high-performance gas turbine engines would not have been there but for the major advances made in superalloy development over the past 50 years, as outlined above. Excellent monographs / handbooks / technical guides are available on the subject of superalloys, covering different aspects – metallurgy, processing, properties and applications (Davis, 1997; Davis, 2000; DeHaemer, 1990; Donachie & Donachie, 2002; Durand-Charre, 1998; Reed, 2006). The reader is referred to them for more detailed information.

8. Reaction with the operating environment

During operation the turbine components are subjected to environmentally induced degradation. The reaction with the environment is essentially of two types – hot corrosion and high temperature oxidation.

Hot corrosion is a rapid form of attack generally associated with alkali metal contaminants - sodium and potassium – reacting with sulfur in the fuel to form molten sulfates. Two distinct forms of hot corrosion have been identified – high temperature hot corrosion occurring in the temperature range 850-950 °C and low temperature hot corrosion taking place in the range 593-760 °C. Macroscopic and microscopic characteristics and mechanisms of the two forms of hot corrosion in gas turbine components have been reviewed (Eliaz et

Generation	Grade designation	Chemical composition	Metal temperature capability °C
First	RR2000	62.5Ni10Cr15Co3Mo4Ti5.5Al1V	
	RR2060	63Ni15Cr5Co2Mo2W5Ta2Ti5Al	
	PW1480	62.5Ni10Cr5Co4W12Ta1.5Ti5Al	1060
	CMSX2	66.2Ni8Cr4.6Co0.6Mo8W6Ta1Ti5.6Al	
	CMSX3	66.1Ni8Cr4.6Co0.6Mo8W6Ta0.1Hf1Ti5.6Al	
	Rene N4	62Ni9.8Cr7.5Co1.5Mo6W4.8Ta0.15Hf0.5Nb3.5Ti4.2Al	
Second	PWA1484	59.4Ni5Cr10Co2Mo6W3Re9Ta5.6Al	1120
	CMSX4	61.7Ni6.5Cr9Co0.6Mo6W3Re6.5Ta0.1Hf1Ti5.6Al	
	Rene N5	63.1Ni7Cr7.5Co1.5Mo5W3Re6.5Ta0.15Hf6.2Al0.05C0.004b0.01Y	
	TUT 92	68Ni10Cr1.2Mo7W0.8Re8Ta1.2Ti5.3Al	
Third to fifth	CMSX10	69.6Ni2Cr3Co0.4Mo5W6Re8Ta0.03Hf0.1Nb0.2Ti5.7Al	1135
	ReneN6	57.3Ni4.2Cr12.5Co1.4Mo6W5.4R e7.2Ta0.15Hf5.8Al0.05C0.004B	1110
	TMS 75	59.9Ni3Cr12Co2Mo6W5Re6Ta0.1Hf6Al	1115
	TMS 80	58.2Ni2.9Cr11.6Co1.9Mo5.8W4.9Re5.8Ta0.1Hf5.8Al0.5B3.0Ir	
	MC-NG	70.3Ni4Cr<0.2Co1Mo5W4Re5Ta0.1Hf0.5Ti6Al4.0Ru	
	DMS4	67Ni2.4Cr4Co5.5W6.5Re9Ta0.1Hf0.3Nb5.2Al	1140
	TMS 196	59.7Ni4.6Cr5.6Co2.4Mo5.0W6.4Re5.6Ta0.1Hf5.6Al5.0Ru	1150

Table 11. SC nickel-base superalloys for blading applications in aircraft engines

Grade designation	Chemical composition
CMSX11B	62.1Ni12.5Cr7Co0.5Mo5W5Ta0.04Hf0.1Nb4.2Ti3.6Al
AF56	61.4Ni12Cr3Mo5Ta4.2Ti3.4Al
PWA1483	60.3Ni12.2Cr9Co1.9Mo3.8W5Ta0.5Hf4.1Ti3.6Al0.07C0.008B
CMSX11C	64.5Ni14.9Cr3Co0.4Mo4.5W5Ta0.04Hf0.1Nb4.2Ti3.4Al
SC16	70.5Ni16Cr3Mo3.5Ta3.5Ti3.5Al

Table 12. SC superalloys for advanced IGTs

al., 2002). When there is no presence of alkali metal contaminants and sulfur in the environment, high temperature oxidation dominates. Higher the temperature, more rapid is the oxidation. In this context, it is important to note the difference in the operating conditions of aircraft engines and heavy-duty gas turbines. Metal temperatures in the former are higher. The latter are subjected to excessive contamination through presence of sodium and sulfur in the operating environment and there is potential for extensive hot corrosion damage.

Alloying with elements which have a beneficial effect of cutting down the extent of hot corrosion has been adopted as an important approach to mitigate the problem. Similarly alloying with aluminum to enable the material form its own protective layer has been adopted to prevent high temperature oxidation. However these approaches do not take very far, as there are other functional requirements which also have to be taken care. Accordingly protective coatings have been developed to ward off the degradation. Most of the superalloys used for gas turbine components receive protection from specially engineered coatings.

9. Coatings

Having to perform under increasing firing temperatures and excessive contamination in the operating environment, it has become difficult to design superalloys which have the necessary creep strength on one side and the required resistance to corrosion / oxidation on the other side. It has hence become inescapable to bring coatings on to the surface of the blades to provide the necessary protection to the blades. The progress in coatings for gas turbine airfoils has been reviewed (Goward,1998). The function of the coating is to act as reservoir of elements which will form very protective and adherent oxide layers, thus protecting the underlying base material from oxidation, corrosion attack and degradation. Hot corrosion is distinctly different from the pure oxidation of an aircraft environment; it can therefore be readily appreciated that coatings for heavy duty gas turbines have different capabilities, compared to coatings for aircraft engines.

There are three basic types of coatings

- Aluminide (diffusion) coatings
- Overlay coatings
- Thermal barrier coatings (TBCs)

The diffusion coatings have been the most common type for environmental protection of superalloys. An outer aluminide layer (CoAl or NiAl) with an enhanced oxidation resistance is developed by the reaction of Al with the Ni/Co in the base metal. In recent years extremely thin layers of noble metals such as platinum have been used to enhance the oxidation resistance of aluminides. For most stage 1 buckets, GE used a platinum-aluminum diffusion coating until 1983 (Schilke, 2004). This coating offered superior corrosion resistance to straight aluminide coatings both in burner rig tests and in field trials. Their high temperature performance is however limited by oxidation behavior of the coatings.

GE has since switched over to overlay type coatings for stage 1 buckets (Schilke, 2004) . At least one of the major constituents in a diffusion coating (generally Ni) is supplied by the base metal. An overlay coating, in contrast, has all the constituents supplied by the coating itself. The advantage is that more varied corrosion resistant compositions can be applied to optimize the performance of the coating and thickness of the coating is not limited by process considerations. The coatings are generally referred to as MCrAlY, where M stands for Ni or Co or Ni+Co. Incorporation of yttrium improves corrosion resistance. The coatings are generally applied by vacuum plasma spray process. A high temperature heat treatment is performed (1040-1120 ^{o}C) to homogenize the coating and ensure its adherence to the substrate.

The TBCs provide enough insulation for superalloys to operate at temperatures as much as 150 °C above their customary upper limit. TBCs are ceramics, based on ZrO_2 - Y_2O_3 and produced by plasma spraying.

The ceramic coatings use an underlay of a corrosion protective layer e.g., MCrAlY that provides the oxidation resistance and necessary roughness for top coat adherence. Failures occur by the thermal expansion mismatch between the ceramic & metallic layers and by environmental attack on the bondcoat. This type of coating is used in combustion cans, transition pieces, nozzle guide vanes and also blade platforms. Improved efficiency of gas turbine engines is realized by adopting TBCs (Gurrappa & Sambasiva Rao, 2006)

9.1 R&D efforts in progress in the area of coatings

Development of even more corrosion resistant coating materials has been intensely pursued over the last few years. In particular, improvement of oxidation resistance and thermal fatigue resistance has been a focal theme for the R&D efforts in the recent years. Development work is also on advanced TBCs to tailor their structure in such a way that they withstand thermal fatigue conditions better and give a longer life (Schilke, 2005). Development of techniques to ensure uniformity of applied coating has been another important area of research. High-velocity plasma appears to be getting established as the technique for application of overlay coatings. Much stronger bond between the coating and the workpiece and much higher coating densities can be achieved using this technique.

10. Advanced materials under R&D

10.1 Ceramics

Increase in turbine inlet temperatures, beyond what is possible with superalloys, can be conceived if ceramic materials can be used in place of superalloys in gas turbine engine. Turbines would then operate at higher temperatures, yielding higher power with smaller engine sizes. Ceramic materials are known for their capability to withstand high temperatures. In addition they are quite tolerant to contaminants such as sodium and vanadium which are present in low cost fuels and highly corrosive to the currently used nickel-base superalloys. Ceramics are also up to 40% lighter than comparable high temperature alloys. They also cost much less - their cost is around 5% the cost of superalloys. Ceramic materials based on silicon carbide and silicon nitride were identified in 1960's as potential candidates for gas turbine application. Substantial efforts have subsequently been conducted worldwide to identify and seek solutions for key challenges: improvement in properties of candidate materials, establishing a design and life prediction methodology, generating a material database, developing cost-effective fabrication of turbine components, dimensional and non-destructive inspection, and validation of the materials and designs in rig and engine testing. Enormous technical progress has been made, but ceramic-based turbine components still have not found application in gas turbine engines, because of the problem of brittleness (Richerson, 2006). There have been efforts to improve their ductility, e.g., through addition of aluminum to ceramics. Unless the problem of brittleness is overcome satisfactorily, the use of ceramics in gas turbines will not be practical.

10.2 Intermetallics

During the last 30 years, extensive efforts have gone into development of intermetallic alloys for application in aircraft gas turbine engines. The primary driving force was to replace nickel based alloys with a density of 8-8.5 gm/cm^3 with lower density materials (4-7 gm/cm^3) and gain weight saving of the engine. Titanium and nickel based aluminides were the systems which received the maximum attention.

Excellent reviews are available on the subject of development of titanium based intermetallics for aero-engine applications (Gogia, 2005; Kumphert et al., 1998; Lasalmanie, 2006; Leyans & Peters, 2003)

The titanium aluminum system offers two possibilities - the Ti_3Al (α_2) based intermetallics and the γ TiAl based alloys. In the 1970's the research was centered mainly around the α_2 alloys. Although interesting results were obtained, these materials did not get into flying engines because their fracture toughness and resistance to growth of fatigue cracks was significantly inferior to high temperature titanium alloys processed through conventional methods. They offered little or no advantage with reference to temperature capability over the alloys such as IMI 834, Ti6242, Ti1100. In their present state of development, there is not enough justification for wide spread usage of Ti_3Al based intermetallics into aeroengines.

There is large volume of published work on second generation γ TiAl alloys, developed in 1990's. Considering the specific strength and oxidation resistance they are potential candidates to replace nickel alloys in the temperature range 650-750 °C. These alloys have also interesting properties at lower temperatures - high Young's modulus and resistance to fire and good HCF properties. All the main turbine engine manufacturers including General Electric Aircraft Engines, Pratt and Whitney and Rolls Royce have successfully gone through demonstration programmes for rotating and static engine components in the compressor, combustor, turbine and nozzle.

Stronger third generation γ TiAl alloys have been developed by GKSS with a wider temperature range of interest - room temperature to 850 °C, making them candidate materials also for LP compressor components (Lasalmanie, 2006).

The most recent alloy family within the titanium aluminides is represented by the orthorhombic titanium based intermetallics based on Ti_2AlNb. They appear to have better toughness, higher ductility, higher specific strength and lower coefficient of thermal expansion than TiAl base intermetallics. This property profile makes orthorhombic titanium aluminides attractive for compressor casings. Even compressor discs can be considered, if their damage tolerance can be improved.

Table 13 is a compilation of representative grades from different groups of titanium aluminides and their maximum service temperature.

Group of titanium aluminide	Chemical composition of a typical grade(s) (At%)	Maximum service temperature °C
Ti_3Al based	Ti-24Al-11Nb, Ti-25Al-10Nb-3V-1Mo	550
γ-TiAl based (2nd generation)	Ti-48Al-2Cr-2Nb	800
γ-TiAl based (3rd generation)	Ti-45Al-8Nb	850
Ti_2AlNb based	Ti-22Al-25Nb	700

Table 13. Representative grades of different groups of titanium aluminides

None of the potential applications talked above are expected to enter the production phase in the immediate future. There are serious hurdles, both technical and economic, on the way. The materials show large scatter in mechanical properties, with the result that the minimum property may be so low that the weight saving becomes negligible. They show very low tolerance to defects such as casting porosities, ceramic inclusions, machining cracks etc. There are a number of manufacturing difficulties. The production cost is much higher, compared to present technologies, particularly for complex components. Further research in alloy development and processing is required, it appears, before they get into flying aircraft engines.

β NiAl is better than current Ni alloys in high temperature oxidation and in creep at very high temperatures and was actively considered for turbine components in the range 1100-1650 °C. Serious obstacles to productionisation were faced - manufacturing difficulties, high cost of production, poor properties below 1000 °C, intrinsic brittleness. Large amount of research was done to improve the mechanical properties by developing complex multiphase intermetallic structures. The manufacturing difficulties and the brittleness still plague these materials and they are unlikely to be adopted as gas turbine materials (Lasalmanie, 2006).

Many other intermetallic aluminide systems have been studied, to a lesser extent, in the context of application in aircraft engines; the materials have not taken off; the problems faced with their development are similar to those enumerated above.

10.3 Composites

10.3.1 Polymer matrix composites

Substantial progress has been made with reference to development and use of polymer matrix composites in the cold section of jet engines. GE is producing its front fan blades out of epoxy resin-carbon fiber composites, resulting in substantial weight savings.

10.3.2 Titanium based metal matrix composites

Continuous fibre reinforced titanium metal matrix composites have been the subject of intense R&D activity, as they can lead to design changes from the conventional disc and dovetail arrangement to a bladed ring with a weight reduction of about 70%. A number of processes are being evaluated for production of these composites. One of the major factors limiting the use of these composites is the fibre-matrix interaction leading to the degradation of properties. There have been many studies on the interaction between the fibre and the titanium based matrix. Different types of coatings have been tried to minimize this interaction. Titanium aluminides have received interest as matrix materials, because they show much less reaction with the fibres. Cutting down the cost of production, achieving improved performance levels and amenability to mass production are the key factors for the introduction of titanium based composites into gas turbine engines. Good overviews on the subject of titanium-based composite materials are available for the reader to get more information.(Gogia, 2005; Kumphert et al., 1998; Leyans & Peters, 2003)

10.3.3 Ceramic matrix composites (CMCs)

The manufacturers of gas turbines are continually striving to increase the operating temperatures of their engines, leading to greater thermal efficiency, and reduced emission of harmful exhaust gases. These two drivers place an ever increasing burden on the materials used in, and the design of, hot gas path components. The introduction of CMCs into hot gas path components such as combustion chamber liners has long since been identified as a possible route to the achievement of increasing operating temperatures without incurring the penalties associated with increased cooling air use.

SiC-matrix composites appear to be highly tailorable materials suitable to gas turbine application at high temperatures. Melt infiltrated (MI) SiC/SiC composites are particularly attractive for gas turbine applications because of their high thermal conductivity, excellent thermal shock resistance, creep resistance, and oxidation resistance compared to other CMCs. They are tough, although their constituents are intrinsically brittle, when the fiber-matrix bonding is properly optimized through the use of a thin interphase deposited on the fibers prior to the infiltration of the matrix. They display good mechanical properties at high temperatures when prepared from stable fibers, as well as a high thermal conductivity if their residual porosity is low enough. The matrix composition can also be tailored to improve the oxidation resistance of the composites.
Replacing the superalloys by light, tough, refractory and creep resistant SiC-matrix composites will permit a significant increase of service temperature and hence an increase of the engine efficiency, a reduction of the NOx/CO emission (through an optimization of the fuel/air ratio), a simplification of the part design and a weight saving (typically, 30–50%). However, their use still raises a number of questions dealing with their durability, reliability, manufacture, design and cost. Presently, the development of SiC-matrix composites is limited mostly to non rotating parts including combustor liners, after-burner components (exhaust cone and flame holder) and exhaust nozzles (outer and inner flaps) in military aerojet engines, as well as combustor liners of large size in stationary gas turbine for electrical power/steam cogeneration. The GE Rolls-Royce Fighter Engine Team's F136 development engine for the Joint Strike Fighter (JSF) contains third-stage, low-pressure turbine vanes made by GE from CMCs.
In summary, although substantial progress has been made, significant risks and challenges still remain before these composites can be commercialized for gas turbine components. The reader is referred to an overview on design, preparation and properties of non-oxide CMCs for application in gas turbine engines (Naslain, 2004)

10.4 Chromium based alloys as gas turbine materials

The increased efficiency associated with higher operating temperatures in gas turbines has prompted designers to search for new materials that can be used at temperatures above the useful limit of nickel-based superalloys. Chromium based alloys have been considered as a possible base for alloy systems due to their high melting point, good oxidation resistance, low density (20% less than most nickel-based superalloys), and high thermal conductivity (two to four times higher than most superalloys). Considerable effort was made in the past, to explore the possibility of developing chromium-based alloys for high-temperature applications such as in jet engines. Two major disadvantages came in the way of their commercial exploitation. First, chromium alloys have a high ductile-to-brittle transition temperature. Second, chromium exhibits further embrittlement resulting from nitrogen contamination during high-temperature air exposure. Since the late 1970s, chromium alloys received very little attention. In recent years there is a revival of interest in these alloys (Gu et al., 2004). Aims of the recent researches have been (i) improvement in high temperature strength (ii) protection from nitridation / oxidation embrittlement (iii) improvement in the impact ductility at ambient temperature. There have been some encouraging findings - (i) trace additions of silver can improve room temperature ductility significantly (ii) strengthening with intermetallics can improve high temperature strength (iii) substantial progress has been made in the area of high temperature coatings (Gu et al.,2004). It is still a long way, however, before the engine manufacturers get interested in these alloys.

10.5 Molybdenum based alloys as gas turbine materials

Molybdenum alloys are currently used as components for ultra-high-temperature applications under protective atmosphere, taking advantage of their high melting point and, very good mechanical and creep strength. However, they suffer from severe oxidation in air above around 500 °C. Compositions based on Mo-9Si-8B have shown promise as structural materials for applications in excess of 1,100 °C in air (Heilmeier et al., 2009). The silicide and boride phases serve to provide oxidation resistance. Experiments to increase the oxidation resistance through alloying with Cr have shown promising results. Alloying with the reactive element Zr was also found to bring down the rate of oxidation. Creep resistance of the Mo-9Si-8B composition was found to be comparable or even superior to that of CMSX4 over the temperature range 1100-1200 °C; there is scope for R&D to further increase the creep strength.

10.6 Platinum based alloys

Platinum-based alloys possess the potential to be used at temperatures up to 1,700 °C. Despite their high prices, they are attractive for some gas turbine applications due to their exceptional resistance to oxidation, high melting points, ductility, thermal shock resistance, and thermal conductivity. They are envisaged to have potential for highly thermally loaded, but non-rotating parts in gas turbines (Alven, 2004). Work is in progress for development of platinum-based alloys with microstructures similar to those seen in commercial nickel-based alloys (Yamabe-Mitarai, 2004; Vorberg et al., 2004). Numerous research groups are currently involved in the study of these alloys. Significantly more data needs to be generated before the platinum-based alloys are ready for the designers of gas turbine engines.

11. Conclusion

Turbine entry temperature has increased by ~500 °C over last 6 decades and about 150 °C of that is due to improved superalloys and introduction of DS / SC technologies for blade casting. Advanced thermal barrier ceramic coatings on platform and full airfoil have contributed to another about 100 °C of this improvement. The developments in gas turbine materials and coatings have been largely due to increasing demands placed by the aircraft sector - higher engine thrust, thrust to weight ratio and fuel efficiency - necessitating higher operating temperatures and pressures. The land based industrial gas turbine industry has placed its own demands on materials, bringing in resistance to hot corrosion as an important requirement. Several SC superalloy compositions have been developed for aircraft gas turbines on one side and land based gas turbines on the other side. Partial γ' solutioning has been adopted in a number of SC IGT alloys to avoid incipient melting and control the extent of recrystallisation. Intense R&D is also going on development of advanced materials for gas turbine engine application - intermetallics, ceramics, composites, chromium / molybdenum / platinum based materials to improve the engine efficiency and bring down the harmful emissions. Major improvements in the coating technology have also been achieved. Present day coatings last 10-20 times longer than the coatings used in the late 90's. As much as 100% improvement is now being achieved in the blade life in the field through the process of coating. TBCs are being used in the first few stages in all advanced gas turbines. Intense R&D is underway to improve the thermal fatigue of the TBC's and thereby increase their life. This includes development of techniques for production of uniform and high density coatings.

12. Acknowledgment

The author is grateful to the Management of VIT University for their kind consent to publish this Chapter. He is also indebted to Ms. Brunda, his wife, for all the support he received from her in preparing this manuscript.

13. References

Alven, D.A. Refractory- and Precious Metal- Based Superalloys, JOM, Vol.56, No.9, pp 27 ISSN 1047-4838

Bayer, R.R. An Overview on the Use of Titanium in the Aerospace Industry, Materials Science and Engineering A, Vol.A213, (1996), pp103-114 ISSN 0921-5093

Boyce, M.P. (2006). Gas Turbine Engineering Handbook, (Third Edition), Gulf Professional Publishing, ISBN 0-88415-732-6, OxfordCaron, P.; Green, K.A. & Reed, R.C. (Eds.). (2008) Superalloys: Proceedings of the Eleventh International Symposium on Superalloys , TMS, ISBN 0873397282, Warrendale, PA

Coutsouradis, D.; Davin, A. & Lamberigts, M.(1987) Cobalt-based Superalloys for Application in Gas Turbines, Materials Science and Engineering, Vol.88, pp11-19, ISSN 0921-5093

Das, N. (2010). Advances in Nickel-base Cast Superalloys, Transactions of The Indian Institute of Metals, Vol.63, No.2-3, (April-June 2010), pp265-274, ISSN 0972-2615

Davis, J.R. (Ed.). (1997). Heat Resistant Materials, ASM Speciality Handbook, ASM International, ISBN 978-0-87170-596-9, Materials Park, OH

Davis, J.R. (2000). Nickel, Cobalt and their Alloys, ASM Speciality Handbook, ASM International, ISBN 0-87170-685-7, Materials Park, OH

DeHaemer, M.J. (1990). ASM Handbook, Volume 1, Properties and Selection: Irons, Steels, and High Performance Alloys, Wrought and Powder Metallurgy (P/M) Superalloys, ASM International, ISBN 0-87170-377-7, Materials Park, OH

Donachie, M.J. (August 2000). Titanium, A Technical Guide, 2 Edition, ASM International, ISBN 0871706865, USA

Donachie, M.J.; & Donachie, S.J. (2002). Superalloys A Technical Guide, (Second Edition), ASM International, ISBN 978-0-87170-749-9, The Materials Society, Materials Park, Ohio

Durand-Charre, M. (1998). The Microstructure of Superalloys, CRC Press, ISBN 9056990977

Eliaz, N.; Shemesh, G. & Latanision, R.M. (2002). Hot Corrosion in Gas Turbine Components. Engineering Failure Analysis, Vol.9, pp.31-43, ISSN 1350-6307

Furrer, D. & Fecht, H. (1999). Ni-Based Superalloys for Turbine Discs. JOM, Vol.51, No.1, (January 1999), pp. 14-17, ISSN 1047-4838

Gibbons, T.B. Superalloys in Modern Power Generation Application, Materials Science and Technology, Vol.25, No.2, pp129-135

Gogia, A.K. (2005). High Temperature Titanium Alloys, Defence Science Journal, Vol.55, No.2, (April 2005), pp149-173, ISSN 0011748X

Goward, G.W. (1998). Progress in coatings for gas turbine airfoils. Surface and Coatings Technology, Vol.108-109, pp. 73-79, ISSN 0257-8972

Gu, Y.F.; Harada, H. & Ro,Y. (2004). Chromium and Chromium-Based Alloys: Problems and Possibilities for High Temperature Service. JOM, Vol.56, No.9 (September 2004), pp. 28-33, ISSN 1047-4838

Gurrappa, I. & Sambasiva Rao, A. (2006). Thermal Barrier Coatings for Enhanced Efficiency of Gas Turbine Engines, Surface & Coatings Technology, Vol.201, (2006), pp3016-3029 ISSN 0257-8972

Heilmeier, M.; Krüger, M.; Saage, H. & Rösler, J. Metallic Materials for Structural Applications Beyond Nickel-based Superalloys JOM, Vol.61, No.7, pp61-67 ISSN 1047-4838

Kumphert, J.; Peters, M. & Keysser, W.A. The Potential of Advanced Materials on Structural design of Future Aircraft engines, Proceedings of RTO AVT Symposium on " Design, Principles and Methods for Aircraft gas Turbine engines", ISBN 92-837-0005-8, Toulouse, France, May 11-15, 1998

Lasalmanie, A. (2006). Intermetallics: Why is it so Difficult to Introduce them in Gas Turbine Engines?, In: Intermetalics, Vol.14, No.10-11, (October 2006), pp1123-1129, ISSN 0966-9795

Leyans, C & Peters, M. (2003). Titanium and Titanium Alloys: Fundamentals and Applications, Wiley-VCH Verlag GmbH & Co, ISBN 9783527305346, Weinheim, FRG

Loria, E.D. (1989). Proceedings of Conference on Superalloy 718 - Metallurgy and Applications, TMS, ISBN 0-87339-097-0, Warrendale, PA

Loria, E.D. (1991). Proceedings of Conference on Superalloy 718, 625 and various Derivatives, TMS, ISBN 0-87339-173X, Warrendale, PA

Loria, E.D. (1994). Superalloys 718, 625, and Various Derivatives, The Minerals, Metals & Materials Society, ISBN 0-87339-235-3, Warrendale, PA

Loria, E.D. (1997). Superalloys 718, 625, 706 and Various Derivatives, The Minerals, Metals & Materials Society, ISBN 0-87339-376-7, Warrendale, PA

Loria, E.D. (2001). Proceedings of the Fifth International Conference on Superalloys 718, 625, 706 and Various Derivatives, TMS, ISBN 0-87339-510-7, Warrendale, PA

Loria, E.D. (2005). Proceedings of Sixth International Symposium on Superalloys 718, 625, 706 and Derivatives, TMS, ISBN 978-0-87339-602-8, Warrendale, PA

Naslain, R. (2004) Design, preparation and properties of non-oxide CMCs for application in engines and nuclear reactors: an overview. Composites Science and Technology, Vol.64, pp155-170, ISSN 0266-3538

Reed, R.C. (2006). Superalloys: Fundamentals and Applications, Cambridge University Press, ISBN 9780521859042

Richerson, D.W. (2006). Historical review of addressing the challenges of use of ceramic components in gas turbines, Proceedings of 2006 ASME 51st Turbo Expo, ISBN 0791842371, Barcelona, May 2006

Schilke, P.W. (2004) Advanced Gas Turbine Materials and Coatings, 18.04.2011, Available from www.gepower,com/prod_serv/products/tech_docs/en/downloads/ger3569g.pdf

Vorberg, S.; Wenderoth, M.; Fischer, B.; Glatzel, U. & Völkl, R. Pt-Al-Cr-Ni Superalloys: Heat Treatment and Microstructure, JOM, Vol.56, No.9, pp40-43 ISSN 1047-4838

Wright, I.G. & Gibbons, T.B.(2007) Recent developments in gas turbine materials and technology and their implications for syngas firing. International Journal of Hydrogen Energy, Vol.32, pp 3610-3621, ISSN 0360-3199

Yamabe-Mittarai, Y. (2004). Platinum-Group-Metals-Based Intermetallics as High-Temperature Structural Materials, JOM, Vol.56, No.9, pp ISSN 1047-4838

14

Titanium in the Gas Turbine Engine

Mark Whittaker
Swansea University
UK

1. Introduction

The development of the gas turbine engine over the past 60 years has been mirrored by the success of the titanium industry, with a clear symbiant relationship existing between the two industries. Immediately apparent in the early days of the evolution of the gas turbine was the need for a material which could provide the strength required for component operation, whilst at the same time providing a low enough density to allow for successful flight applications. Whilst aluminium based alloys offer an excellent strength to weight ratio, their operation is limited to temperatures below approximately 130^{0}C, reducing possible applications within the gas turbine to a minimum. 300 series stainless steels offer a similar strength to most conventional titanium alloys, but come with a significant density penalty of over 50% and, whilst offering reductions in cost, do not provide significant benefits in terms of operating temperatures.

Titanium however, has long been viewed as having a desirable balance of properties for applications towards the front end of the gas turbine engine (i.e. fan discs/blades, compressor discs/blades, along with other smaller components). Titanium has a density of 4.5g/cm^3 (which, apart from a limited number of alloys such as Ti811, does not vary significantly in alloys considered for aerospace applications) which is higher than aluminium, but lower than nickel and steel alloys. Titanium is allotropic with a HCP lattice (α phase) stable to 882°C, transforming to a BCC (β phase) lattice above this temperature. Alloying elements act to stabilize either of these phases (Al, Sn for example stabilize the alpha phase, whereas Mo, V, Cr stabilize the beta phase) meaning that the transformation temperature can be altered, and subsequently the proportions of each phase existing at room temperature can be varied. The morphology of these phases may however vary, dependent on the process history, with alpha phase material being classed as primary alpha (persisting during heat treatment in the $\alpha+\beta$ phase field) or secondary alpha (structures arising from the $\beta\rightarrow\alpha$ phase transformation). This allows for the development of a range of bimodal microstructures which provide titanium alloys with inherent strength and also allows for further refinement of properties through various heat treatment and processing regimes. For example designers requiring creep strength and good elevated temperature properties may choose to opt for alloys with more alpha stabilizers (alpha or near alpha alloys), whereas metastable beta alloys, which are heavily beta stabilized offer improved forgeability. Alpha-beta alloys contain a more balanced mix of stabilizers and are widely used due their balance of properties. Ti6-4 (Ti-6Al-4V) for example has been a stalwart of the titanium industry since the 1950s due to its good weldability, relatively high strength and good fatigue properties.

The continued extensive use of titanium in the gas turbine, however, is constantly under threat, with potential new technologies showing promise at the lower temperature end of the gas turbine (fan blades) and at higher temperatures (HP compressor). Composite fan blades offer designers the opportunity for further weight reductions and, although concerns may still be raised over impact resistance, these materials offer great incentives to replace Ti6-4 as a fan blade material. Furthermore, in order to operate gas turbine engines more efficiently, it is necessary to continue to raise the temperature capability of the engine so that fuel can be burnt at temperatures closer to the stoichiometric value and also to achieve a higher compression ratio of the gas through the compressor stages. The second of these issues impacts significantly on the materials utilised within the compressor, since a higher pressure ratio will result in an increase in temperature where high temperature titanium alloys, such as Ti834 are utilized as disc materials. Ti834 was developed for high temperature applications and shows exceptional properties up to approximately 630°C. However, as operating temperatures at the disc rim increase, it clearly becomes necessary to consider viable alternatives, since Ti834 is limited by environmental degradation in the form of alpha case formation. Unfortunately however, no other titanium alloys show appropriate properties at temperatures higher than this, and designers are forced to consider other alloy systems. Nickel alloys are most commonly utilised as replacement alloys, although a significant density penalty is attached, with typical intermediate temperature alloys such as IN718 having a density of 8.19g/cm^3. Clearly on this basis designers will opt to utilise high temperature titanium alloys as far as possible to avoid these increases in weight. With pressure being applied to the use of titanium in the gas turbine from materials offering either lower density or a higher temperature capability, particularly in the case of $\alpha+\beta$ alloys, it is clear that it is necessary for the titanium industry to further develop the class of alloys for in service applications. However, it would be unrealistic to assume that advancements in terms of mechanical properties can be maintained through the development of new, improved alloys as has been the case in the past. Whilst alloy refinement and development may still contribute to these advances, it is clear that alternative, innovative approaches are necessary to extract further improvements in performance. The current work seeks to demonstrate three areas which may offer opportunities for advancement:-

i. The harnessing of crystallographic texture. The orientation of the HCP alpha phase lattice can result in significant anisotropy in mechanical properties. Whilst original research into titanium alloys sought to minimise this anisotropy, more recently it has been acknowledged that it should be possible to match the best properties from the textured material with the most demanding loading directions.
ii. The development of appropriate fatigue lifing methods for alloys operating in conditions where additional failure mechanisms such as creep and environmental damage interact with fatigue. Whereas the traditional view of engineers was to ensure that materials were operating at temperatures below where significant creep deformation would occur, improved lifing techniques enable accurate predictions to be made in these temperature regimes, allowing for safe operation.
iii. An improved understanding of the effect of processing on alloys which may prove extremely sensitive to issues such as residual stress. It is acknowledged that processing of components is likely to produce a less than perfect surface condition, and that the effects of surface roughness, residual stress and 'damage' to the material may cause

variations in the mechanical properties. These variations are rarely simple, with for example, the residual stress conditions being dependent on temperature and dislocation density and a lack of understanding may result in the requirement for overly conservative safety factors.

2. Generation of test data

The mechanical test data detailed in the following section was produced at Swansea University as part of a number of academic programmes funded either in part or fully by EPSRC, Rolls-Royce plc and TIMET UK.

3. Potential developments in titanium alloys

3.1 Crystallographic texture (Ti6-4)

The highly anisotropic nature of the HCP α phase lattice clearly gives rise to potential variations in mechanical properties of titanium alloys. This can be demonstrated by early work on the subject (Zarkades & Larson, 1970) which illustrates the variation of Young's modulus with crystal orientation, as shown by Figure 1.

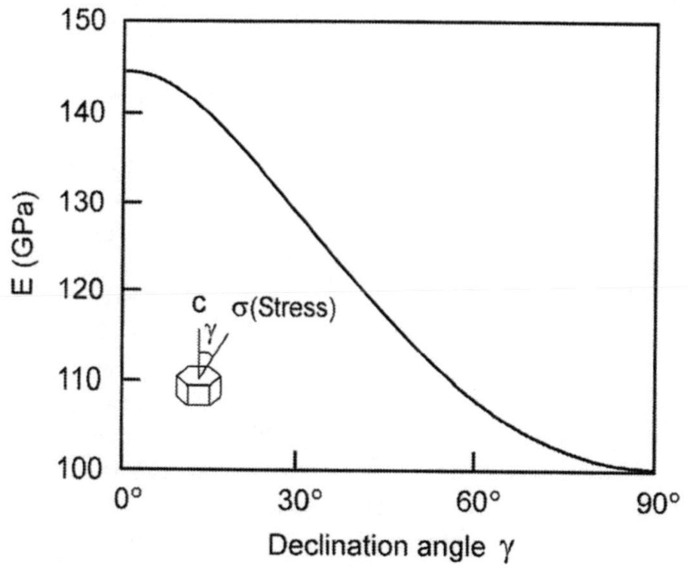

Fig. 1. The dependence of Young's modulus on increasing angle with c-axis.

It is clear that the due to the denser atomic packing along the c-axis (<0001> direction) a maximum value of the Young's modulus occurs in this direction, which shows a gradual decrease with increasing angle made with the c-axis. It is also apparent that this variation can account for differences of up to 40GPa, which significantly alters the mechanical response of the alloy. However, these changes do not only affect elastic deformation of these alloys, but also the plastic deformation, in terms of the orientation of slip planes with relatively low critical resolved shear stress (CRSS) values. As such it is necessary to understand the effects that this 'crystallographic texture' may impart to the alloy.

Clearly the anisotropy of the HCP lattice within a single grain demonstrates only the most extreme case. In terms of macroscopic properties, the averaging of these orientation effects over many thousands of grains will reduce the impact of crystallographic texture. However, as previously demonstrated (Lutjering, 1998) various forms of material processing tend to produce alignment of grain orientations, and macroscopic textures may occur which produce anisotropy in the mechanical properties of the alloy. As stated, these variations will affect not only the elastic deformation of the material, but also the plastic deformation through slip behaviour.

Fatigue accounts for approximately 80% of all in service failures and is an essential design criterion in any engineering application. LCF can be defined as failures in the region 10^3-10^5 cycles, and typically involves bulk plastic deformation within the material, whereas HCF failures (>10^5 cycles) usually exhibit only localized plastic deformation. Both however are affected by crystallographic texture, and it is important that designers understand the potential variations of this mechanical anisotropy. Whilst research in the 1970s sought to minimise the variations to produce a more isotropic product, more recently attention has turned to texture as a means of extracting improved properties from the alloy.

Many studies (Bowen, 1977; Peters et. al., 1984; Evans et. al., 2005a) have focussed on the type of variations in monotonic, axial fatigue and torsion-fatigue that can be expected in strongly textured alloys. However, for many design applications these conditions are overly simplistic. The presence of stress raising geometric features results in a non-linear stress state in components, which requires evaluation. In the study undertaken at Swansea, consideration was given to these features, which are represented in a laboratory environment by notched specimens. The geometrical discontinuities give rise to biaxial or even triaxial stress states which have not yet been thoroughly evaluated in textured materials. Furthermore, previous work has tended to focus on fully reversed (R=-1) or zero-maximum (R=0) loading conditions. The current work sought to further understanding by considering R ratios from R=-1 to R=0.8.

The titanium alloy Ti6-4 has long been regarded as the workhorse of the titanium industry and accounts for approximately 60% of total titanium production (Boyer, 1996). It is an $\alpha+\beta$ alloy which is used in all forms, including forgings, bar, castings, foil, plate and sheet. In the current study however, only the plate form is considered. Uni-directional (UD) rolling of the alloy within the $\alpha+\beta$ temperature field usually leads to the formation of a transverse or basal/transverse texture. The current material is no exception with a bimodal microstructure with an average primary alpha grain size of 15μm (some deviation from equiaxality was shown in the form of slight elongation towards the rolling direction), and a basal/transverse texture with a x3 random intensity as shown in Figure 2, recorded over an area of 0.5 x 0.5 mm. The crystallographic texture of the material is completely described by the three orthogonal pole figures shown in Figure 2.

A number of previous studies have also focussed on the behaviour of the material under load controlled conditions. However, a number of potential applications for textured alloys more closely simulate strain controlled deformation behaviour, mainly because of the constraint of surrounding material. It is also argued that material at the notch root in tested specimens (and therefore at stress raising features in components) undergoes an essentially strain control type of deformation (Evans, 1998). It is therefore critical that the effect of texture under this mode of deformation is accurately characterised.

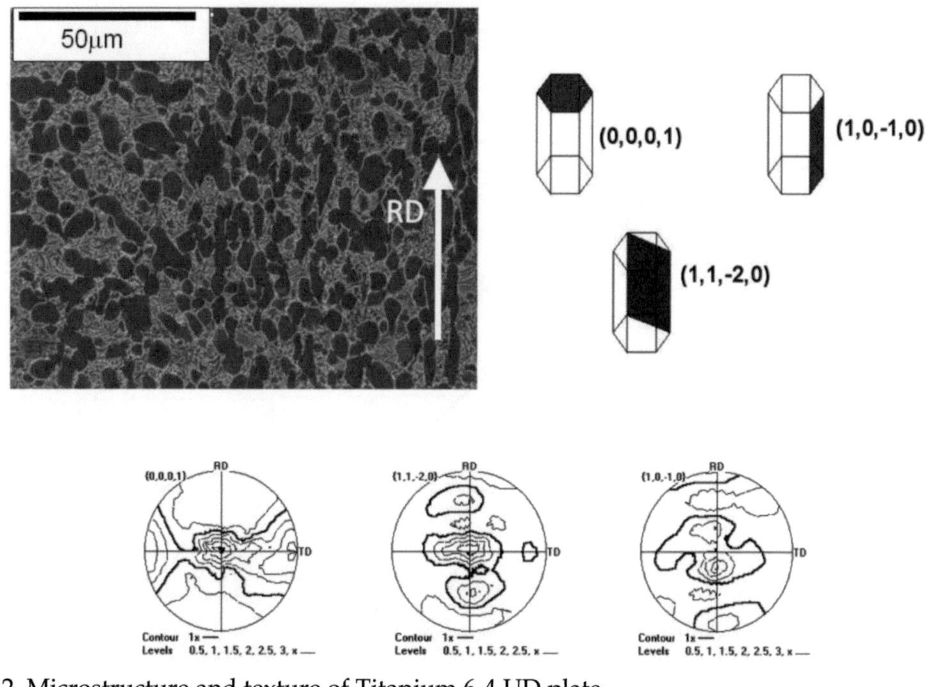

Fig. 2. Microstructure and texture of Titanium 6-4 UD plate.

3.1.1 Plain specimen behaviour

Figure 3 illustrates the type of effect that may be seen in textured alloys. Under strain control testing at R=0 (20°C, 1-1-1-1 second trapezoid cycle) it can be seen that specimens taken parallel to the rolling direction of the material show longer fatigue lives than counterparts taken from the transverse direction of the plate (TD), i.e. at 90° to the rolling direction. Initially this result may be considered counterintuitive; the increased density of basal planes perpendicular to the TD direction results in a higher modulus in the TD specimens, and also higher yield stress and UTS values. However, in order to understand this effect, it is necessary to consider the behaviour of the material under strain control, and the effect of the material texture.

Figure 4 illustrates the evolution of the maximum and minimum stress values achieved during a strain control test with a peak strain of 1.4%. It can be seen that there are significant differences between the RD and TD specimens, with considerably more stress relaxation in the early part of the test in the RD specimen. This is related to the availability of prismatic planes (which have a low critical resolved shear stress in titanium at room temperature) for slip in the early part of the test (see Figure 2). This leads to the TD specimen operating at a higher maximum stress. In combination with this, the TD specimen also has a higher modulus, and consequently a lower minimum stress during the test. The TD specimen therefore operates over a larger stress range for a given strain range and shows a reduced fatigue life. When considered only on a stabilised stress range basis, previous work has shown that no difference in fatigue life occurs between the RD and TD specimens in this material.

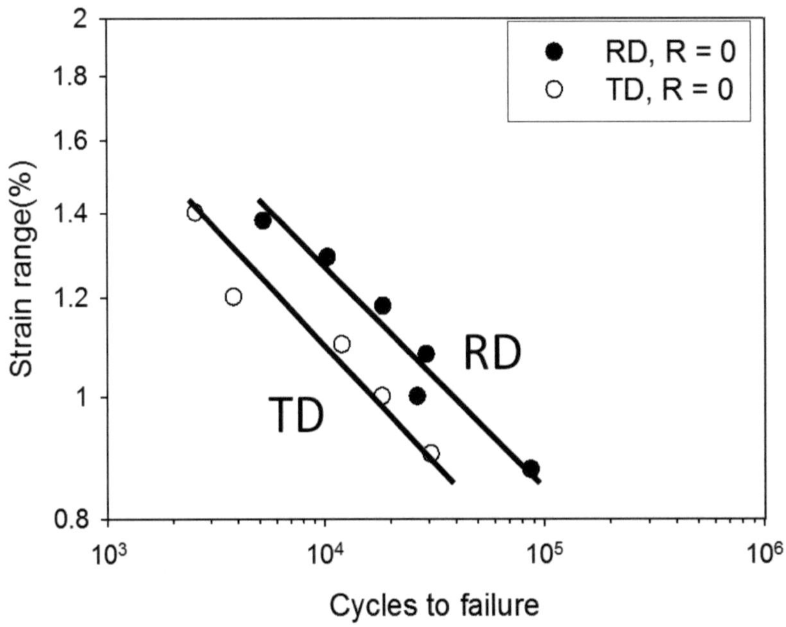

Fig. 3. The effect of orientation on the strain-life of Ti6-4.

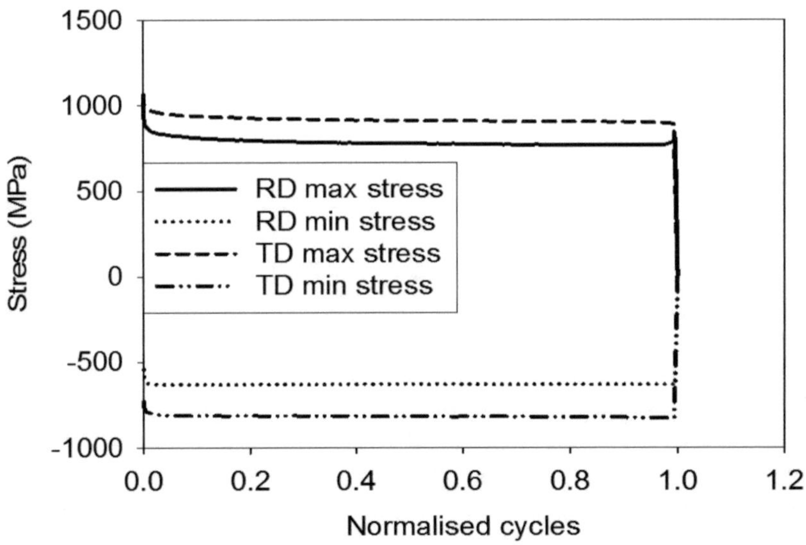

Fig. 4. The effect of orientation of the stress response of Ti6-4 under strain control loading (ε_{max}=1.4%, R=0).

3.1.2 Notched specimen behaviour

In considering notched specimen behaviour, it is important to acknowledge the requirement for a predictive methodology, to enable designers to extrapolate to conditions for which reliable test data does not exist. Previous work has shown that the Walker strain approach

(Walker, 1970) is an appropriate method for these types of predictions. The Walker strain relationship is an empirical method for correlating R values and involves correlating strain control data of different R ratios, allowing for the derivation of a 'master curve'. As stated earlier, the material at a notch root is assumed to experience strain control type conditions, due to restraint from material surrounding the critically stressed volume of material. Through application of Neuber's rule (Neuber, 1968), that the product of stress and strain is a constant, conditions at the notch root can be approximated allowing for the calculation of the individual Walker strain value for that specimen. Subsequently a predicted life can be inferred from the 'master curve' based on the strain control data. This approach has been found to be accurate for similar titanium alloys to Ti6-4 (Whittaker et. al., 2007), but has not previously been tested on a textured alloy.

During the course of the work, two notched specimen geometries were tested, both with cylindrical notches; the first was a V shaped cylindrical notch (VCN) which has a stress concentration factor, K_t, of 2.8, the second a round cylindrical notch (RCN) with a K_t of 1.4. Initially apparent from Figure 5 is the fact that no orientation effect appears to exist in the RCN specimen, with both RD and TD specimens showing similar fatigue lives to the plain specimen data. However, this is not the case with the VCN specimen, as shown in Figure 6, with the RD specimens showing longer fatigue lives than the TD specimens.

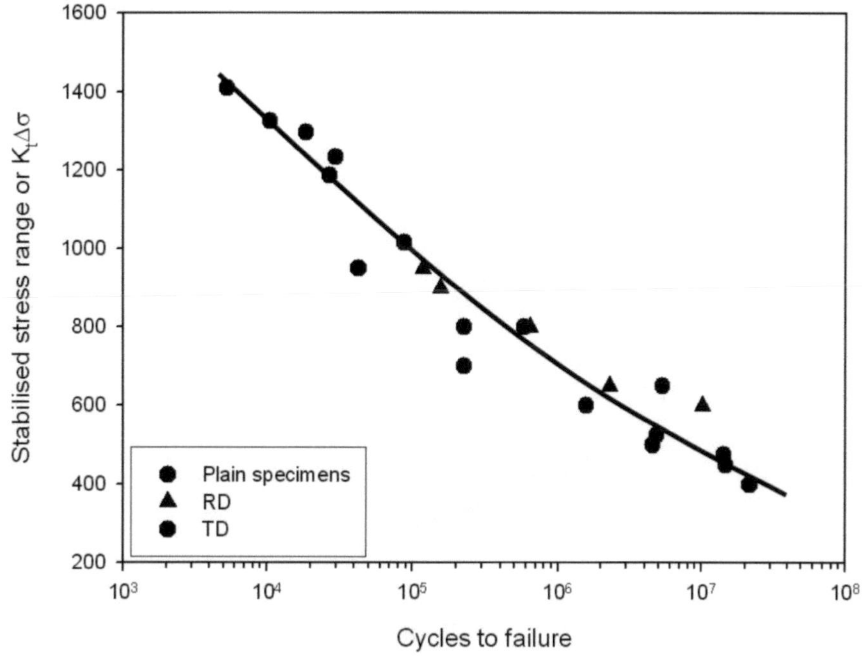

Fig. 5. Comparison of notched (RCN) and plain specimen response showing no orientation effect.

To interpret these results, it should be noted that these notched specimen tests are performed under load control; it is the geometry of the notch which imposes strain control type conditions on the material at the root of the notch. Figure 5, showing the results for the RCN specimen, is illustrative in a number of ways. Along with the fact that no orientation effect exists, it can also be seen that RD and TD specimens show similar fatigue lives. Furthermore, the notched specimen behaviour correlates well with the plain specimen response. The VCN specimens, however, do not follow either of these trends, Figure 6. Specimens in the RD orientation show longer fatigue lives than either plain or RCN specimens. This is consistent with previous experience since a lower volume of material is critically stressed in the VCN specimen. Since fatigue is essentially probabilistic in nature and relies on 'weak links' present in the material to initiate the fatigue process, a lower material volume infers a lower probability of a 'weak link' being present, and hence a longer fatigue life is statistically more likely.

The fact that the RCN specimen shows no orientation effect and correlated well with the plain specimen data when plotted on a stabilised stress basis indicates that a lack of constraint is occurring at the notch root. In this case a large volume of material is critically, or near critically stressed, similar to the plain specimens. Since the notch testing is performed under load control, the lack of constraint at the notch results in a shallow stress gradient, and hence the material at the notch root experiences conditions closer to load control than strain control. As a result these specimens behave like the plain specimens when considered on a stabilised stress basis, with no orientation effect. In the VCN specimens the stress gradient is far steeper, constraining material at the notch root, which then behave like the plain specimens, when considered on a strain range basis, and RD specimens show longer lives than TD specimens for the same reasons described in plain specimens (i.e. changes in relaxation behaviour and differences in modulus), as explained in the previous section.

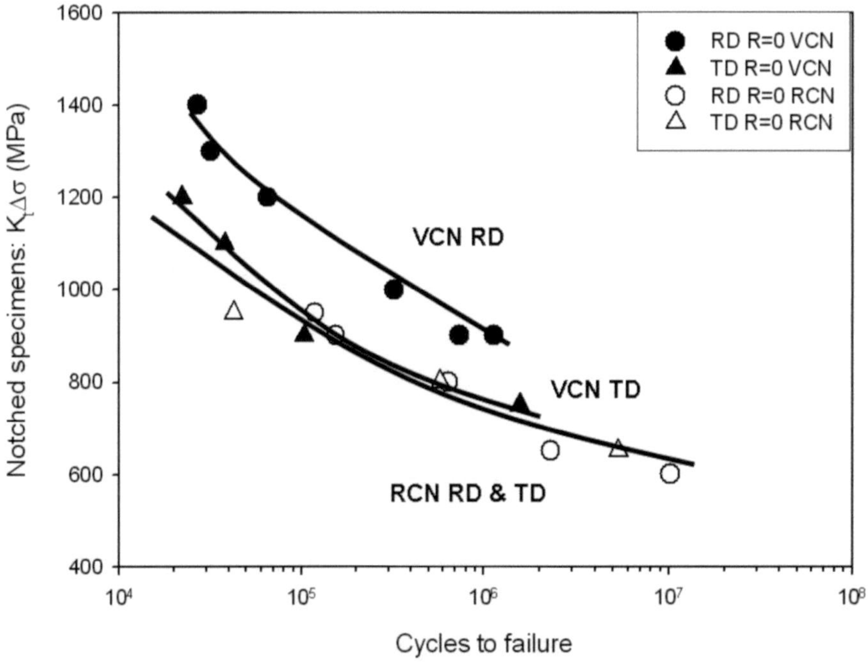

Fig. 6. Comparison of notched specimen fatigue lives showing an orientation effect in the VCN notch, whereas no such effect exists in the RCN notch.

In considering the ability of the Walker strain method to accurately predict fatigue lives, only RD specimens are currently considered, although similar calculations can be made for TD specimens (Evans & Whittaker, 2006). Although the Walker strain method is a relatively simplistic method, and does not compensate for notch type, it is a useful approach that has previously been shown to give excellent results in titanium alloys (Whittaker et. al., 2007). Figure 7 shows the type of predictions which can be made using this approach, over a wide range of R ratios. In order to consider a total life prediction methodology it should be recognised that this type of approach predicts only fatigue crack initiation in notched specimens. In strain control specimens, when a crack initiates, it will propagate quickly to failure. This is not the case in a notched specimen where the crack will grow more slowly through material away from the notch root. Previous crack monitoring work has shown that assuming a propagation phase of 50% of the total life allows for reasonable predictions (Whittaker et. al, 2010a).

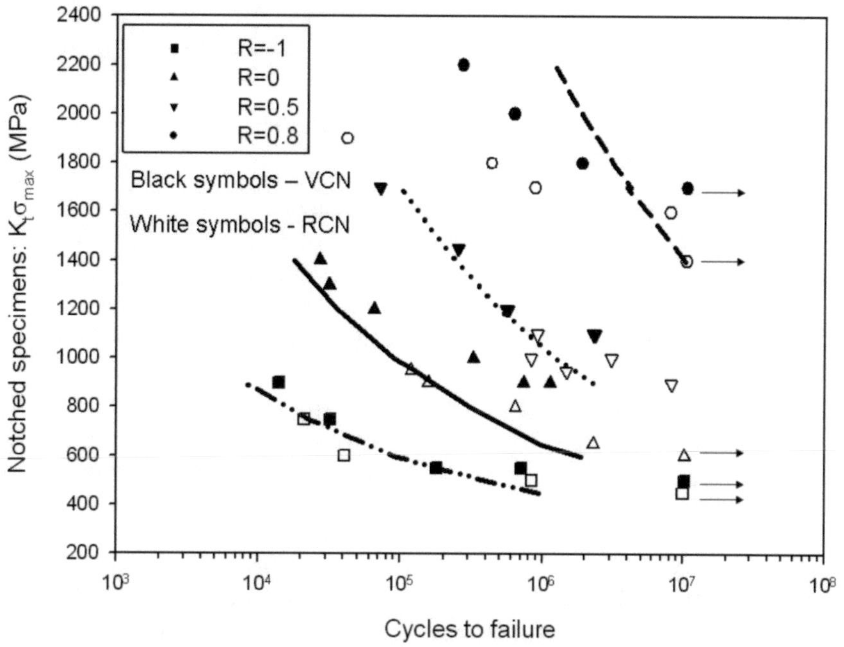

Fig. 7. Predictions of notched fatigue lives in RCN and VCN notches by the Walker strain method.

Based on these assumptions it is clear that excellent predictions are made for R ratios of -1, 0 and 0.5. However, significant over predictions are made at an R=0.8, particularly for the RCN specimens. The reason for this lies in the introduction of additional failure mechanisms. Strain accumulation at low temperatures has been widely reported in near α and $\alpha+\beta$ titanium alloys and is loosely termed 'cold dwell'. Particularly at high mean stresses, these failures are characterised by the formation of quasi-cleavage facets which form due to stress redistribution from so called 'soft' (suitably orientated for slip) grains onto 'hard' grains (unsuitably orientated for slip), as shown by the Evans-Bache model in Figure 8(a) (Bache & Evans, 1996). Clear evidence of these facets was found in both RCN and VCN R=0.8 specimens, although an increased density was found in the RCN specimens. The result of this is the reduction in fatigue lives (when compared with the Walker predictions) seen in Figure 7. The effect is more pronounced in the RCN specimens because of the larger amount of material being critically or near-critically stressed.

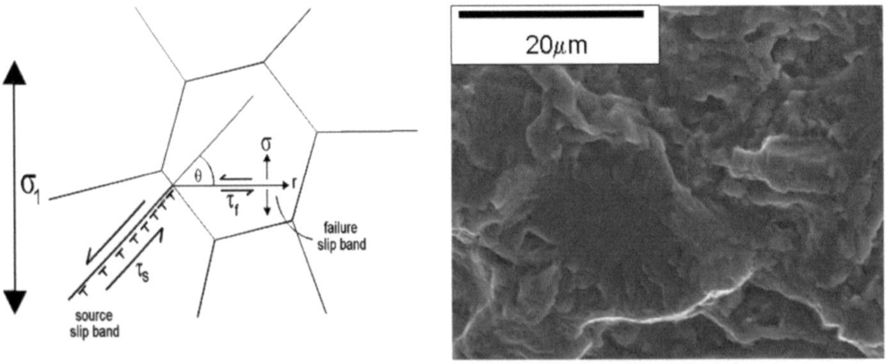

Fig. 8. The Evans-Bache model for facet generation in titanium alloys, with an example facet from an RCN, R=0.8 notched specimen.

Whilst it is clear that it is possible to accurately life notched specimens in a textured alloy, it is also evident that there are limitations. In the current work predictions have been made based on strain control data from the same orientation. Without this it is impossible to make accurate predictions. It is also apparent that for Ti6-4 there is a limited range of R ratios over which predictions can be made, with additional failure mechanisms playing a role.

3.2 High temperature lifing (Ti6246)

As temperatures rise in the gas turbine engine designers turn to titanium alloys with a higher temperature capability than Ti6-4, for which operation is limited to less than approximately 350°C. Ti6246 (Ti-6Al-2Sn-4Zr-6Mo) is such an alloy with good low cycle fatigue properties and improved creep resistance over Ti6-4, Figure 9. It is immediately apparent that the microstructure of Ti6246 differs significantly to Ti6-4, showing a fine Widmanstatten microstructure that would be typical of a material processed above the beta transus. The fine nature of the microstructure infers the high strength of the material and also offers good resistance to crack propagation.

Widely used as a compressor disc alloy, Ti6246 has traditionally been employed at temperatures where creep effects would not be considered significant. However, it is not necessary for the alloy to be limited in this way provided appropriate lifing techniques are employed. The following work describes the construction of a total life prediction capability for fatigue at high temperatures in the alloy. Again, the focus of the work is on notched specimens, due to the importance of the stress raising features within the gas turbine engine. Figure 10 demonstrates the importance of considering additional failure mechanisms to fatigue by considering crack propagation rates at 550°C in Ti6246. The vacuum 1Hz sinewave data (square symbols) represent solely the influence of fatigue on the crack propagation rate whereas the circular symbols indicate that as a dwell period is added to the waveform, by employing a trapezoidal 1-1-1-1 waveform, a significant increase is seen in the crack propagation rate. This is further increased by adding a 2 minute dwell period at peak

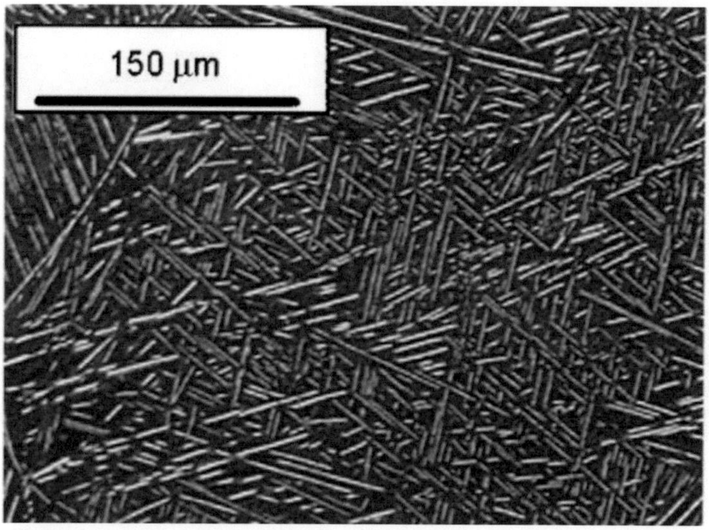

Fig. 9. Micrograph of Ti6246, showing a fine Widmanstatten type microstructure.

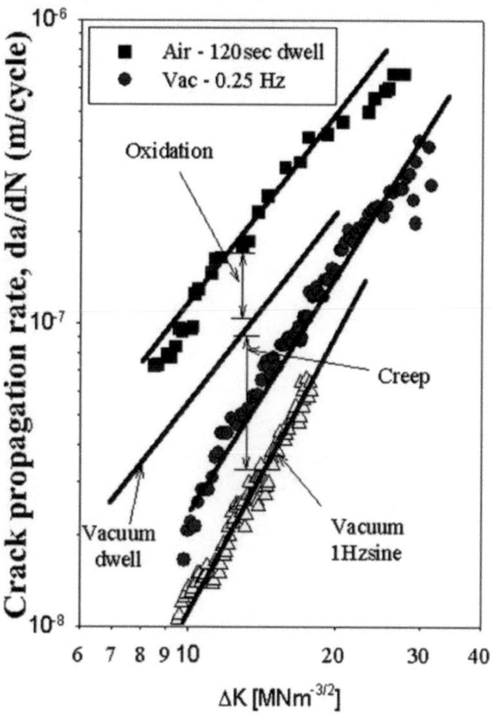

Fig. 10. Fatigue, creep and environmental effects in crack growth in Ti6246 (Evans et. al., 2005b).

load (1-1-120-1 waveform) as indicated. This increase in crack propagation rate is due to the effect of creep, with evidence seen of creep voids ahead of the crack tip. However it is also clear that at this temperature, creep and fatigue are not the only damage mechanisms in

operation. For tests conducted in air, rather than under high vacuum (10^{-6} mbar) conditions, a significant further increase in propagation rate is seen when the same 1-1-120-1 second trapezoid waveform is applied. This effect is environmental damage and as indicated by the graph, also requires consideration, since the increases in crack growth can be similar to, or even surpass those due to creep.
Whilst these results give an indication of the roles of fatigue, creep and environmental damage, it is clear that in order to build a total life prediction capability, their effects on fatigue crack initiation must be considered.

3.2.1 Fatigue modelling

As described previously the Walker strain method (Walker, 1970) has been shown to be a useful approach to the prediction of notched specimen behaviour, particularly in terms of predictions over a wide range of R ratios. However, the previous analysis was performed only at room temperature and it is necessary to investigate whether the Walker strain approach still offers accurate results at higher temperatures. In this work the notch considered is a double edged notch (DEN) with a $K_t = 1.9$.
Figure 11 illustrates predictions made using the Walker strain approach at 20°C and 450°C, with notch root conditions again approximated by use of Neuber's rule (Neuber, 1968). As described previously, these predictions do not account for the crack propagation phase of a notch test and assuming a propagation phase of approximately 50% of the total life has previously been shown to be a reasonable assumption (Whittaker, 2010a). Whilst predictions under R=-1 loading conditions are excellent, it can be seen that predictions for R=0 tests at 20°C and 450°C tend to be non-conservative when the propagation phase is added. This is obviously undesirable for designers of critical parts.

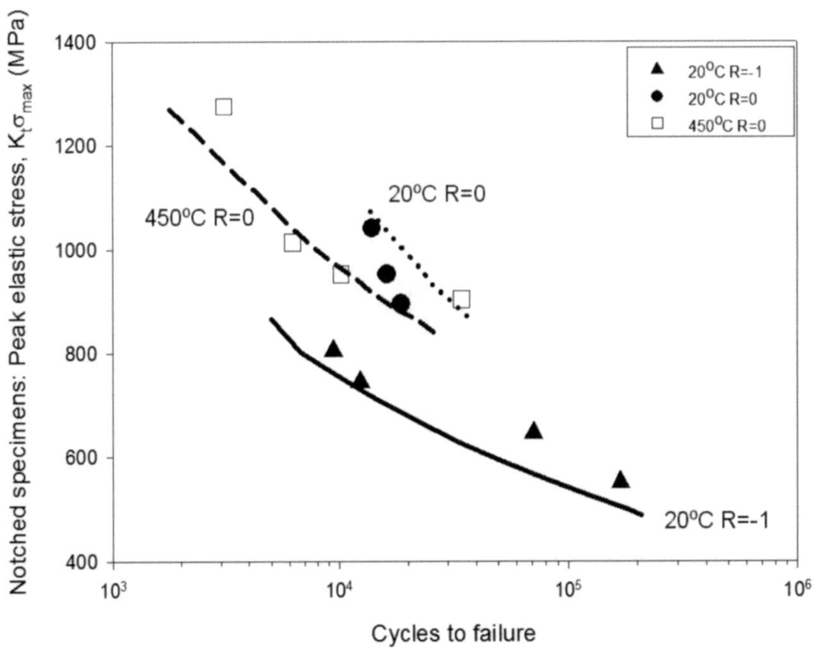

Fig. 11. Predictions of notched specimen behaviour at 20°C and 450°C using the Walker strain method.

The predictions made for R=-1 notch tests have improved accuracy over the R=0 tests simply for the reason that it is easier to predict the stress/strain state at the notch root for these tests. The highest load which was employed in fatigue testing of the R=-1 tests resulted in a peak elastic stress of 800MPa, which would be below yield for Ti6246 at room temperature, at a typical strain rate of 0.5%/sec. As such the stress/strain conditions at the notch root are simply 800MPa and 0.0067 (from strain = stress/modulus). However, in the R=0 tests, significant plasticity is induced at the notch root. Whilst in Ti6-4 this plasticity could be accurately approximated by Neuber's rule, clearly more accurate description is required in the current case.

3.2.2 Development of FEA model in ABAQUS

In order to achieve greater accuracy a model was developed in the modelling suite ABAQUS based upon open hysteresis loops generated under fully reversed strain control loading of Ti6246, over a range of temperatures. The loops were generated under laboratory air conditions so that fatigue/environment and subsequently fatigue/creep/environment interactions could be studied. The model was based around the Mroz multilayer kinematic hardening model (Mroz, 1969) which compared well with experimental observations that stress redistribution within the material allowed for the stabilization of the peak/minimum stress during the initial cycles of a strain control test. A typical stress-strain loop generated by the model is shown in Figure 12. It can be seen that the loop generated in ABAQUS accurately describes the test data generated for a strain control test with a peak strain of 1.5%.

Modelling of the double-edged notch specimen was achieved through the construction of a three dimensional 1/8 symmetrical FE model using 20-noded isoparametric rectangular elements (C3D20) with 18833 nodes and 4032 elements, with element size reduced near to the notch to improve accuracy. Calculations of the fatigue life were then based on the stabilised conditions of stress and strain at the node adjacent to the notch root.

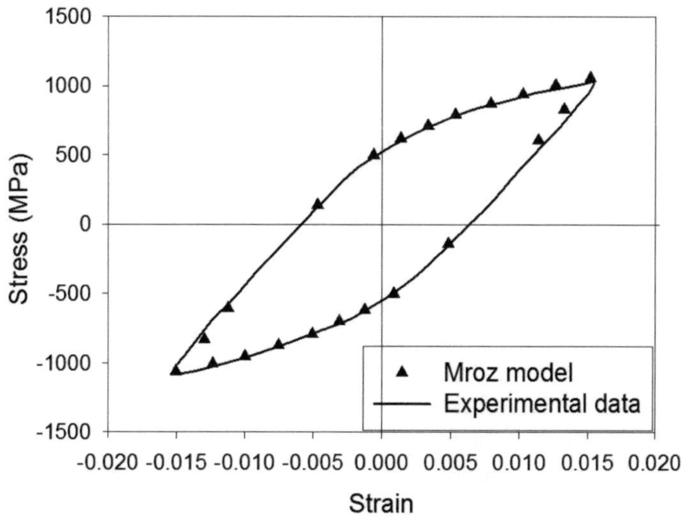

Fig. 12. ABAQUS modelling of a stress-strain loop at 20°C in Ti6246 (Whittaker et. al., 2010a).

3.2.3 Creep and environmental damage

Figure 13 shows the predictions made by the model under 20°C R=-1 loading conditions, and also 500°C R=0 loading conditions. It can be seen that the low temperature predictions of initiation life are again extremely accurate. At 500°C the predictions are slightly conservative, but clearly more acceptable than those previously demonstrated without the use of FEA. Previous work (Whittaker et. al., 2010a) has in fact shown that in this material, using DEN specimens, fatigue lives at 500°C are actually longer than at 450°C. This is due to the effect of creep within the vicinity of the notch root. At 450°C creep has a limited effect, whereas at 500°C it becomes more prevalent, and acts to decrease the stresses around the notch root, creating a shallower stress gradient and hence an improved fatigue life. Further increases in temperature to 550°C however, lead to a reduction in fatigue life as creep and environmental effects become more damaging.

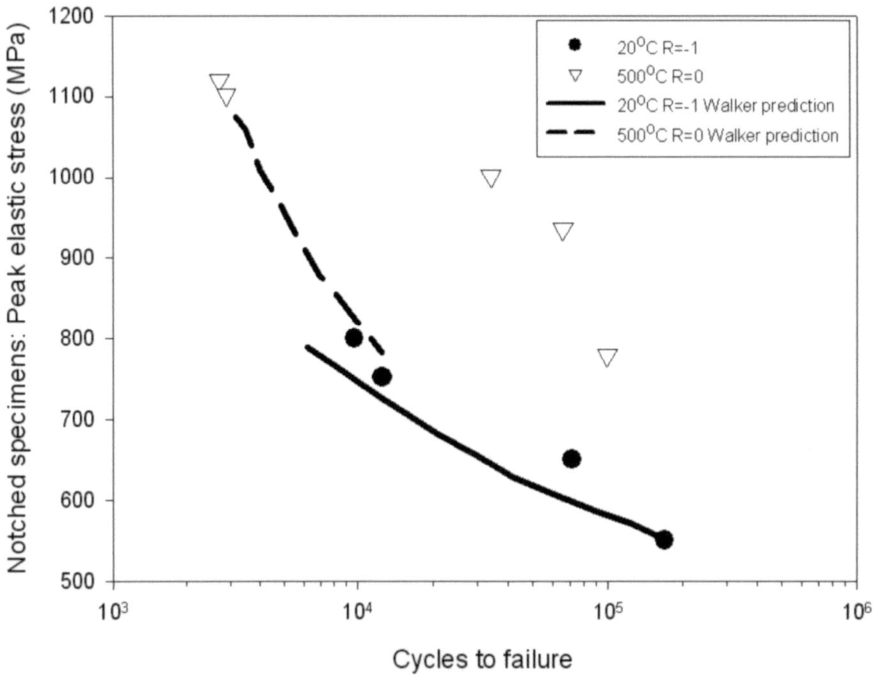

Fig. 13. Predictions of notched fatigue life made by ABAQUS model at 20°C and 500°C.

Further evidence of the significance of environment is demonstrated in Figure 14. Previous authors have described the development of a marked transition in the fatigue life curve of Ti6246 when tested under strain control (Mailly, 1999). Similar effects have been observed in the current work, where for lives greater than approximately 10^4 cycles the fatigue lives of the material may be highly variable as the curve becomes very flat. At this point the material is protected by an oxide layer which forms during the test, preventing further oxidation. However, as material strain increases as the applied stress is raised, the oxide layer cracks and allows further ingress of oxygen, causing damage to the material and resulting in a more typical fatigue curve. The effect is not observed at 20°C, but interestingly has been seen in strain control tests at temperature as low as 80°C.

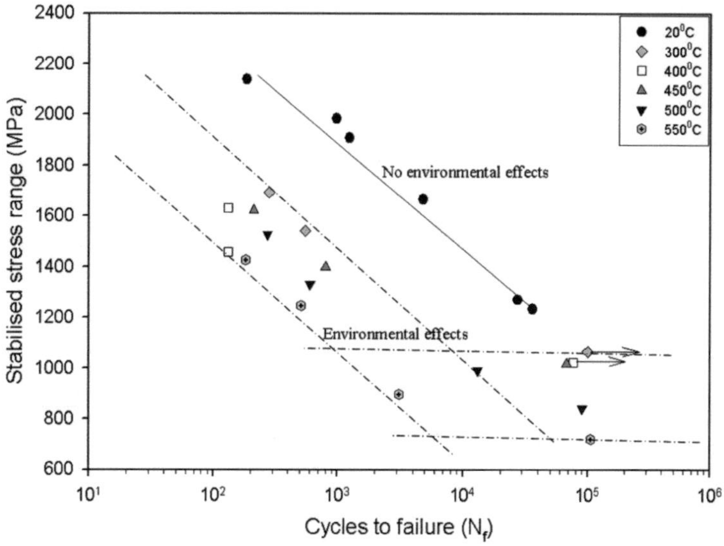

Fig. 14. Influence of environment on the fatigue lives of notched specimens in Ti6246 (Whittaker et. al., 2010a).

3.2.4 Combining fatigue, creep and environmental damage

Clearly the interactions of fatigue, creep and environment within the material are complex and offer a challenge to designers who wish to make accurate life predictions. However, some limited success has been achieved by the development of a fatigue-creep-environment model for crack growth. To construct the model it was necessary to combine fatigue crack predictions based on laboratory air conditions with damage due to creep effects, in the form

$$\left(\frac{da}{dN}\right)=\left(\frac{da}{dN}\right)_f+\int\left(\frac{da}{dN}\right)dt$$

where the first term on the right hand side represents fatigue damage and the second term represents creep damage.

To calculate the creep damage, a suitable creep model was required. Previous experience had shown that the theta projection method offered acceptable results of creep behaviour in Ti6246, so an ABAQUS subroutine for the relationship was compiled which included creep rupture based on a Kachanov type failure process. In order to predict fatigue crack growth at high temperatures fatigue and creep damage were then calculated separately and combined at each time increment to give the total growth rate. Figure 15 indicates the results of this approach for growth rates at 500°C, R=0.1. It is clear that a prediction based purely on fatigue significantly underestimates the growth rate, but when the combined fatigue-creep-environment prediction is made, predictions are accurate. The effect of further creep damage is represented by the growth rates under a waveform with a two minute dwell at peak stress, although currently predictions have not been made for this data.

Whilst it is acknowledged that there is still much work to be completed in developing a total life prediction methodology for fatigue performance at high temperatures, the results of the work are encouraging. It has been demonstrated that interactions between fatigue, creep and environment are complex and produce many non-linear effects which are difficult to model. However, some success has been achieved in the production of a fatigue crack growth model at high temperatures and the belief is that similar models could be produced to describe crack initiation lives based on suitable deformation data for the alloy. Clearly, to accurately build the model, all three damage mechanisms should be considered independently before coupling in a model which considers their effects. However, in order to achieve this, a greater proportion of vacuum data will be required, particularly under strain control conditions, which will be experimentally challenging.

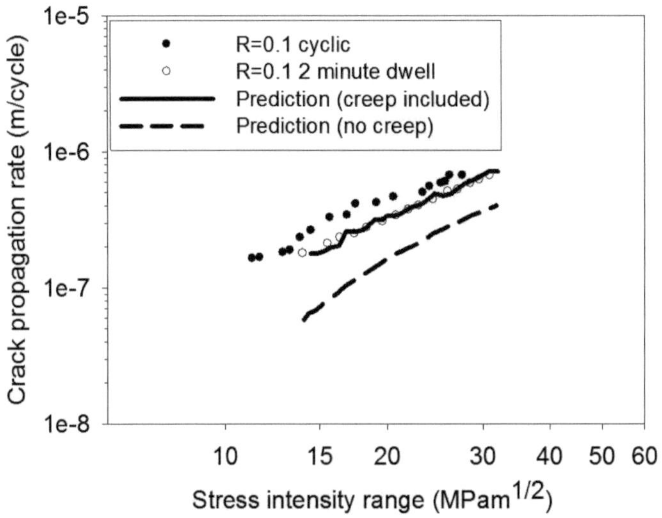

Fig. 15. Predictions of crack growth behaviour of Ti6246 at 500°C, R=0.1 (Whittaker et. al., 2010a).

3.3 Application of prestrain (Ti834)

Ti834 is a near α titanium alloy which was developed with a carefully controlled microstructure to enable exceptional mechanical properties at temperatures up to approximately 630°C. Combined with the high strength to weight ratio of the alloy, this excellent elevated temperature behaviour makes the alloy a popular choice for applications such as compressor discs and blades.

Clearly it is critical that the alloy is utilised under well understood conditions where any effect of the processing history can be accounted for. This may be a complex issue with different forging/machining/peening parameters influencing the surface condition of the material. During processing of components it is highly likely that surface roughness variations may occur, along with the possibility of further 'damage' to the material. Combined with the effect of residual stresses brought about by the peening process (commonly used to extend fatigue life), it is clear that significant variation may occur in the material mechanical properties. It is therefore necessary to understand these effects through a detailed investigation. In the current work, this was undertaken through a programme of mechanical testing aimed at detailing these variations in a range of prestrained Ti834 specimens.

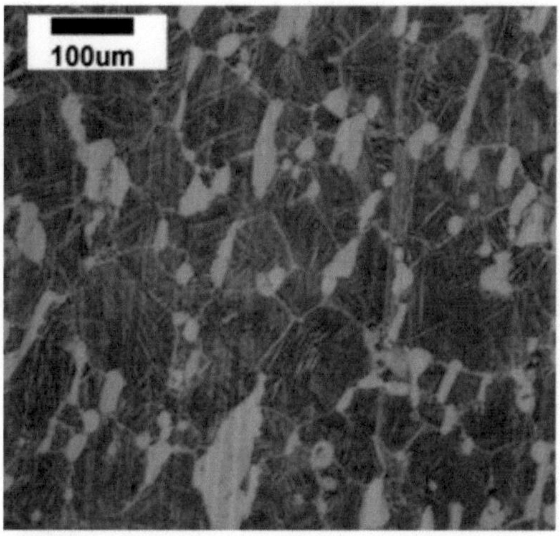

Fig. 16. Micrograph of Ti834, indicating a bimodal microstructure with primary alpha grains ranging in size from 20-200μm.

Figure 16 shows the microstructure of Ti834 tested, with a bimodal microstructure clearly evident encompassing primary α grains of 20-200μm in diameter. Total tensile prestrains of 2% and 8% (resulting in 1.25% and 7.25% plastic prestrains respectively) were applied to individual batches of specimens along with a compressive prestrain of 2%, denoted as -2% (plastic prestrain of -1.25%). These four different conditions, -2%, 0% (as received), 2% and 8% could then be compared under different loading conditions such as fully reversed strain control fatigue (20°C), stress relaxation at 1% strain (20°C) and creep (20°C and 600°C).

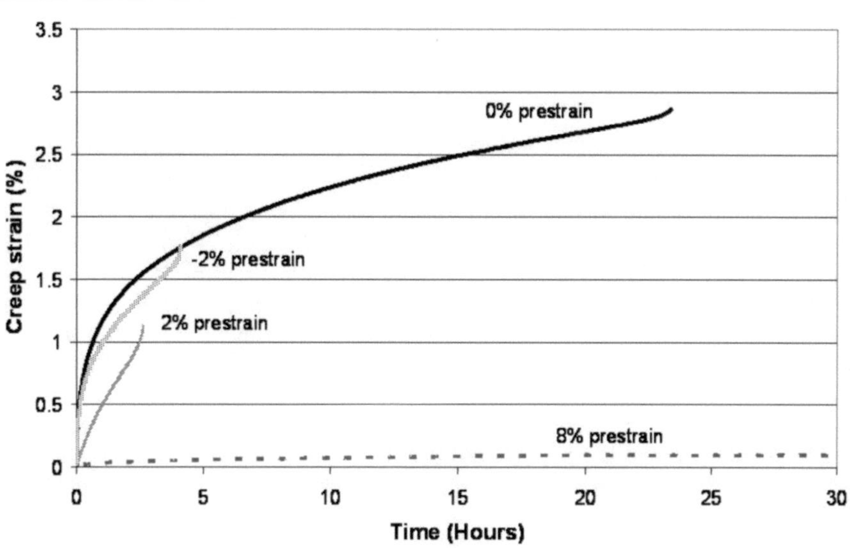

Fig. 17. Effect of prestrain on the creep rate of Ti834 at 20°C (Whittaker et. al. 2010b)

Fully reversed strain control loading at a peak strain of 1% was shown not to result in the formation of quasi-cleavage facets and as such was used as a control mechanism to produce eventual failure in test samples. In this way specimens could be separated and fracture surfaces investigated with the confidence that any facets generated would have been generated by the previous loading conditions and not the strain control fatigue.
Creep rates at room temperature (tested at 950MPa) were shown to be significantly affected by the application of prestrain, Figure 17. It can be seen that for 2% and -2% tests the primary creep is greatly reduced, although the creep strain rate increases and the strain at failure and creep life are markedly reduced. These effects offer only a limited improvement window for the material, in which creep strain is reduced over first few hours. However, the specimen which had undergone 8% prestrain showed a dramatic reduction in creep rate, eventually being removed from test after 250 hours, in which little creep was seen.

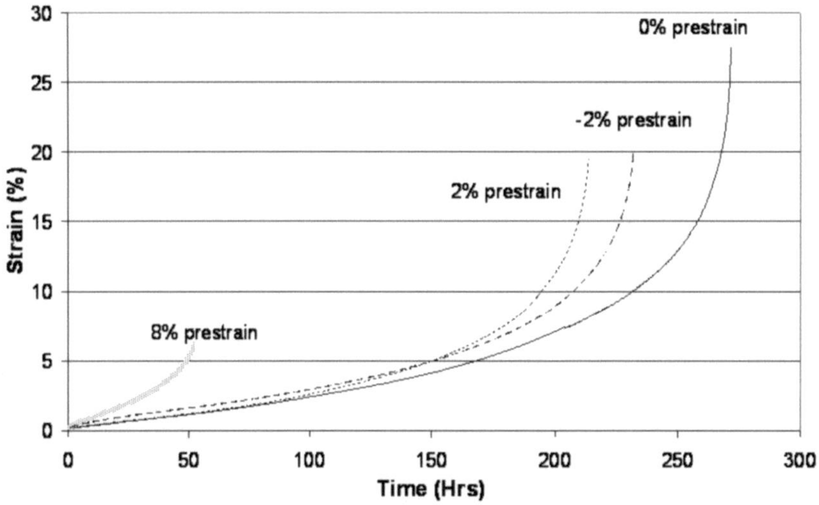

Fig. 18. Effect of prestrain on the creep rate of Ti834 at 600°C (Whittaker et. al., 2010b).

Clearly for designers interested in reduced creep rates at room temperature, this effect is attractive. Figure 18 illustrates though that these advantages will be temperature dependent. At 600°C the 8% prestrain specimen now shows a faster creep rate, shorter lives and reduced strain at failure. The reason for both of these effects will be related to the dislocation structure following prestrain. During the prestrain process, dislocations are generated as the yield stress is exceeded, which occurs at approximately 0.75% strain. It is clear that as the specimen continues to extend towards 8% strain, the dislocations will continue to multiply and as a result a high dislocation density occurs in the material. At room temperature, under creep conditions, dislocation mobility in the structure is significantly reduced, and the creep rate remains very low. However, at 600°C the increased thermal energy means that processes such as climb and cross slip become more prevalent, increasing dislocation mobility. This results in an increased creep rate when compared with the as received (0% prestrain) material.
Based on these results, it is clear that the effects of prestrain on the creep performance of the alloy vary significantly with temperature, and as such, dislocation mobility. At low

temperatures increased prestrain restricts further creep damage because of the high dislocation densities and apparent difficulty in processes such as climb and cross slip. Conversely, these processes occur more readily at 600°C and increases in prestrain lead to an acceleration in creep damage.

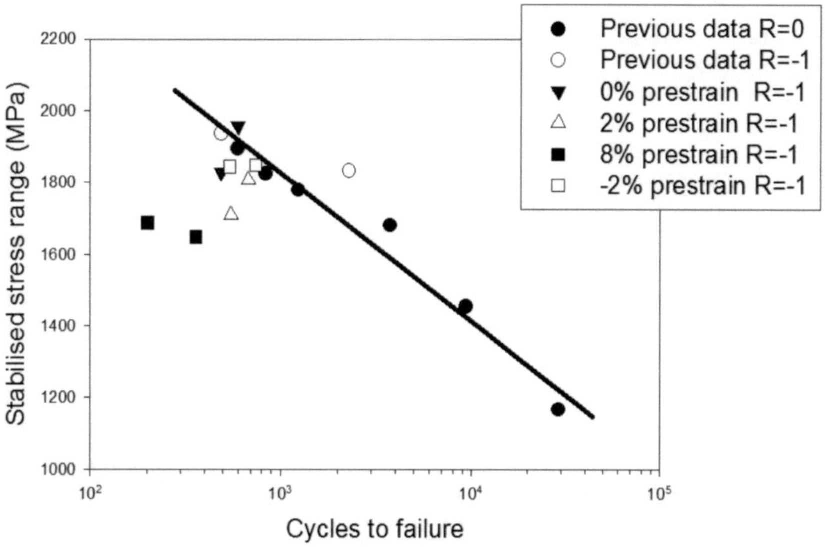

Fig. 19. Effect of prestrain on the fatigue properties of Ti834 at 20°C (Whittaker & Evans, 2009).

However, stress states in the gas turbine are rarely static and as such further consideration must be given to the effect on fatigue performance of the material, Figure 19. In the current work it was found that a small period of stress relaxation (<2 seconds) occurred at the end of the prestrain process before the specimen was unloaded. Previous work (Evans, 1998) has demonstrated that near α and $\alpha+\beta$ titanium alloys tend to form facets under stress relaxation, and fractographic analysis of failed specimens showed that this was indeed the case here. These facets offer initiation sites for fatigue cracks, which along with the increased dislocation density contributes to the 8% prestrain specimens showing significantly shorter fatigue lives.

4. Discussion

It is clear that whilst the rate of development of new titanium alloys has slowed in recent years, there are further areas which may be explored in order to achieve further improvements in mechanical properties. The research described here has shown that there is definite potential through the harnessing of texture, improved high temperature lifing techniques or improved understanding of processing effects.

Of these, perhaps improvements in high temperature lifing offer designers the greatest reward. Since the development of the gas turbine engine, increased efficiency has acted as the driver which has led to operation at higher and higher temperatures. Enabling components to operate at higher temperatures whilst retaining low density/low cost materials in their manufacture is obviously desirable and operation at temperatures where creep and environmental damage operate need not be ruled out. Indeed, provided a robust methodology is developed, it need not only apply to titanium alloys, and could be utilised throughout the engine.

It is useful, however, that the method is developed using Ti6246. The alloy is well understood and offers relatively few surprises to design engineers, particularly in its lack of susceptibility to cold dwell. The creep behaviour of the alloy is well detailed and suitably described by techniques such as the theta projection method. It also shows relatively good environmental resistance to approximately 500°C, giving a desirable combination of properties.

The research described here has shown that accurate predictions can be made for the fatigue behaviour of the alloy at high temperature, in the form of a model that described fatigue crack growth. It is recognised that this is only an initial step in the process of developing a total life prediction capability, but it is at least encouraging. Further work would seek to produce strain control deformation behaviour under vacuum conditions in order to isolate the effects of environment, which has been shown to be at least as significant as creep. It is also recognised that the model requires further refinement in order to describe the effects of the type that cause notch lives to be extended as the temperature is raised from 450°C to 500°C. To allow for this the creep deformation should be integrated more closely to fatigue damage in each cycle to describe the stress state of the notched specimen. There is no doubt that these requirements are challenging, but as described, the benefits are clear.

Textured alloys have been trialled for some applications previously, but the current research has demonstrated that it may be possible to widen the field of opportunity. By showing that textured alloys provide a predictable, consistent response, even in the presence of stress raising features, confidence can be gained towards applications in more complex components. Furthermore the work demonstrates that in this alloy, techniques such as the Walker strain approach are able to deal with a wide range of R ratios, although limitations do exist at high R ratios when cold dwell failure mechanisms become apparent. However, it is recognised that this may not be the case across all titanium alloys or process conditions.

The work in fact indicates how cold dwell still acts as a limiting factor in a number of titanium based applications. Essentially it can be seen that a threshold stress exists above which prediction becomes difficult and an extremely shallow S-N curve develops. Recent research at Swansea has shown that in attempting to model this type of behaviour, it is often more appropriate to base predictions on the creep behaviour of the alloy at room temperature, rather than its fatigue response, although further work is still required to characterise behaviour in this area. Such predictions, based on time at a high stress, rather than cyclic fluctuations, have been shown to capture the shape of the curve more accurately and to give reasonable estimates of life.

Indeed, one of the goals of the gas turbine industry is the increased understanding of cold dwell in order to further raise safe operating stresses. However, whilst a good understanding of the mechanisms of facet formation exists (Sinha et. al, 2009, Bache et. al., 1996) further work on the sensitivity of particular alloys would be useful. The three alloys considered here accurately demonstrate the range of effects seen in near $\alpha/\alpha+\beta$ titanium alloys. Ti834 has always shown a high sensitivity to cold dwell, particularly in disc form (Bache et. al., 1997) and as such has required designers to carefully consider operating conditions in components where it is utilised. Ti6246 on the other hand, has shown almost no sensitivity to cold dwell (Bache et al, 2007). Whilst this lack of sensitivity is possibly related to the fine Widmanstatten microstructure of the alloy, the mechanisms are still not fully understood. Ti6-4 sits between these two more extreme cases, showing cold dwell sensitivity which is often affected by microstructural form. However, since it is an alloy that tends to be used for more low temperature applications than Ti6246 or Ti834, it is clear that a good understanding of the effects is critical for safe utilisation of the alloy.

5. Conclusions

Whilst titanium alloys are under pressure to improve from either lighter, more complex materials (such as composite fan blades) or materials with higher temperature capability (polycrystalline nickel alloys in the HP compressor) it is clear that there are areas of development which have not yet been fully explored, which may offer significant opportunities for titanium alloys. In particular:-

a. The harnessing of crystallographic texture is capable of providing improved properties, providing the most effective orientations are aligned with direction of loading. It has been shown that despite this anisotropy, predictions of fatigue life can still be made accurately provided the input data for approaches such as the Walker strain method is from the same orientation and of a high enough quality.
b. High temperature lifing allows for extending the operational envelopes of alloys and is an area which requires further research. It has been shown that predictive models can be accurate provided all damage mechanisms (i.e. fatigue, creep and environment) are considered. Whilst the work here is only a start, it is clear that there is scope for further model development/refinement which may result in the safe operation of alloys such as Ti6246 at temperatures in excess of those currently used in service.
c. Improved understanding of the effect of processing conditions, and the resultant surface finish/damage and residual stress effects in alloys such as Ti834 can lead to increased minimum properties and hence a reduction in safety factors. It has been demonstrated that these effects, represented by prestrained specimens, can significantly alter mechanical properties, and show significant variation with temperature. Whereas room temperature creep rates may be reduced, the creep rates at 600°C are increased and would need to be accurately accounted for in deformation modelling of components Furthermore, fatigue properties at room temperature are reduced, as a result of the formation of quasi-cleavage facets under stress relaxation. This is the type of critical result which designers must account for when lifing such components.

6. Acknowledgements

The author would like to acknowledge funding and financial assistance from EPSRC, Rolls-Royce plc, TIMET UK, Cosworth Racing and QinetiQ during the course of this work.

7. References

Bache, MR; Evans, WJ. The effect of tension-torsion loading on low temperature dwell-sensitive fatigue in titanium alloys. *Mechanical Engineering publications* (1996) pp. 229-242.

Bache, MR; Cope, M; Davies, HM; Evans, WJ; Harrison, G. Dwell sensitive fatigue in a near alpha titanium alloy at ambient temperature. *International Journal of Fatigue* 19, Supp. 1, (1997), pp. S83-S88. ISSN: 0142-1123

Bache MR; Germain, L; Jackson, T; Walker ARM. Mechanical and texture evaluations of Ti6246 as a dwell fatigue tolerant alloy. *Proceedings of 11th world conference on titanium*, (2007), pp. 523-526. ISBN 978-4-88903-406-6

Boyer, RR. An overview on the use of titanium in the aerospace industry. *Materials Science & Engineering,* A213 (1996) pp. 103-114. ISSN 0921-5093

Bowen, AW. Texture stability in heat treated Ti6-4 alloys. *Material Science & Engineering,* 29 (1977) pp. 19-28.

Evans, WJ. Optimising mechanical properties in α/β Titanium alloys. *Materials Science & Engineering. A* 243 (1998) pp. 89-96. ISSN 0921-5093

Evans, WJ; Jones, JP; Whittaker, MT. Texture effects under tension and torsion loading conditions in titanium alloys. *International Journal of Fatigue* 27 (2005) pp. 1244-1250. ISSN: 0142-1123

Evans, WJ; Jones, JP; Williams, S. The interactions between fatigue, creep and environmental damage in Ti6246 and Udimet 720Li. *International Journal of Fatigue* 27 (2005), pp. 1473-1484. ISSN: 0142-1123

Evans WJ; Whittaker, MT. Prediction of notched specimen behaviour in textured Ti6-4. *Proceedings of 9th International Fatigue Congress*, Atlanta, June 2006.

Lutjering, G. Influence of processing on microstructures and mechanical properties of $\alpha+\beta$ Titanium alloys. *Material Science & Engineering* A243 (1998), pp. 32-45. ISSN 0921-5093.

Mailly, S. Effects de la temperature et d'l'environment sur la resistance a la fatigue d'alliages de titane. *PhD thesis, L'Universite de Poitiers*, (1999).

Mroz, Z. An attempt to describe the behaviour of metals under cyclic loads using a more general work hardening model. *Acta Mechanica.* 7, (1969), pp 199-212. ISSN 1619-6937.

Neuber, H. Theory of stress concentration for shear-strained prismatical bodies with Arbitrary Nonlinear Stress-Strain. Law Transactions ASME , *Journal Applied Mechanics,* 28 (1968), pp. 544-550. ISSN 0021-8936.

Peters, M; Gysler, A; Lutjering, G. Influence of texture on fatigue properties of Ti6-4. *Metallurgical Transactions,* A 15 (1984) pp. 1597-1605. ISSN 1073-5623.

Sinha, V; Mills, MJ; Williams, JC. Crystallography of fracture facets in a near alpha titanium alloy. *Metallurgical and Materials Transactions A,* 37, (2006), pp. 2015-2026. 1073-5623.

Walker, K. Effects of Environment and Complex Loading History on Fatigue Life, *ASTM STP 462,* (1970), pp. 1-14.

Whittaker, MT; Hurley, PJ; Evans, WJ; Flynn, D. Prediction of notched specimen behaviour at ambient and high temperatures in Ti6246. *International Journal of Fatigue* 29 (2007) pp. 1716-1725. ISSN: 0142-1123

Whittaker, MT; Evans, WJ. Effect of prestrain on the fatigue properties of Ti834. *International Journal of Fatigue* 31, (2009), pp. 1751-1757. ISSN: 0142-1123

Whittaker, MT; Harrison, W; Hurley, P; Williams, S. Modelling the behaviour of titanium alloys at high temperature for gas turbine applications. *Materials Science and Engineering,* A527, (2010), pp. 4365-4372. ISSN 0921-5093.

Whittaker, MT; Jones, JP; Pleydell-Pearce, C; Rugg, D; Williams, S. The effect of prestrain on low and high temperature creep in Ti834. *Materials Science and Engineering,* A527, (2010), pp. 6683-6689. ISSN 0921-5093.

Zarkades, A; Larson, FR (1970) Elasticity of Titanium sheet alloys, *Science, technology and applications of Ti 1970* pp 933-936.

15

Platinum-Based Alloys and Coatings: Materials for the Future?

Lesley A. Cornish and Lesley H. Chown
DST/NRF Centre of Excellence in Strong Materials,
and School of Chemical and Metallurgical Engineering,
University of the Witwatersrand,
South Africa

1. Introduction

Since platinum has a similar chemistry and atomic structure to nickel, platinum-based alloys are possible contenders for partial or complete substitution of nickel-based superalloys. Although the major disadvantages are high price and density, platinum-based alloys have many advantages, including excellent chemical and oxidation resistance and high strength at high temperatures. Since the melting point is higher than nickel, there is potential that Pt-based alloys can exhibit mechanical properties that surpass those of the nickel-based superalloys. Pt-modified coatings are already employed on turbine blades. These can be modified with the addition of different elements and various coating procedures can be used so that the coatings can complement different substrates. This chapter covers the basic properties of a range of Pt-based alloys and describes the different strengthening mechanisms that exist in these alloys, mainly through a structural approach. The oxidation resistance and corrosion resistance are also described. Further, new alloying additions and their effects on the structure and properties are identified.

Nickel-based superalloys (NBSAs) have excellent mechanical properties due to precipitation strengthening. The microstructure comprises many small, strained-coherent, particles of the γ' phase based on Ni_3Al, in a softer matrix of the γ phase, the solid solution (Ni) of nickel (Sims et al., 1987). The strengthening originates from dislocations being slowed down as they negotiate the small ordered γ' particles. There are several mechanisms whereby a unit dislocation has to split into partial dislocations to pass through the ordered precipitates and then re-associate to pass into the random matrix. Each stage requires energy, thus slowing the dislocation movement and providing strengthening. The strengthening depends on the amount of interfacial boundary between the two phases, and is highest when the amount of boundary to be negotiated is highest. This occurs when there are many small precipitates densely distributed, comprising at least 70 vol. % in the alloy (Vattré et al., 2009). For NBSAs, these are usually cubic γ' precipitates aligned on the {100} planes (Kear & Wilsdorf, 1962). Additionally, there is solid solution strengthening in the (Ni) matrix, as other elements are dissolved into the nickel, forming a random solid solution. The (Ni) strengthening depends mainly on the difference in elastic and bulk moduli between the Ni and solute atoms (Gypen & Deruyttere, 1981), as well as the size misfit, since each atom varies in size according to its atomic number. There are also bonding effects, all of which make it more difficult for the dislocations to pass.

Although NBSAs are used at high temperatures, significant precipitate coarsening does not readily occur, as the driving force for coarsening is low, due to the low surface energy between the matrix and precipitates (Hüller et al., 2005). Both matrix and precipitates are based on the face centred cubic structure: the γ matrix has a random fcc structure and the γ' particles have an $L1_2$ ordered structure. Aluminium atoms prefer the corners and nickel atoms prefer the faces of the face centred cube. The lattice misfit between these structures is very small and renders the surface energy negligible (Sims et al., 1987), leading to high stability of the fine structure at elevated temperatures, and hence reducing coarsening. Thus, NBSAs are the state-of-the-art material for high temperature, high stress and aggressive environment conditions, with good ductility at both room and high temperatures, and thermal stability.

Although the nickel-based superalloys have excellent mechanical properties, they have nearly reached their temperature limit for operation in turbine engines, unless extreme air-cooling is used, due to the relatively low melting point of nickel (1543°C) and dissolution of the strengthening γ' precipitates at ~1150°C. This limits the current operating temperature of NBSAs to ~1100°C (Chen et al. 2009). There have been many developments to increase the temperature capability of these alloys with complex alloying additions and improvements in processing technology and alloy design methodology (Davis, 1997), but in 24 years an increase of only ~100°C has been accomplished, with little scope for any further increases.

The need for increased application temperature arises because turbine engines are more efficient and provide greater thrust at higher temperature. This increased efficiency means less fuel burned, reduced cost as well as reduced CO_2 emissions. Since before 1990, Boeing has halved the mass of emissions by using higher temperature alloys and improved coatings (NIMS, 2007). A number of approaches can be used to obtain higher temperature alloys, including: increased alloying additions to the current nickel-based superalloys, addition of temperature-resistant coatings, or the use of entirely new materials. Since the increase in temperature capability in nickel-based superalloys is constrained by the melting point of nickel, there is interest in developing a whole new suite of similarly structured alloys. These alloys would have higher melting points than the NBSAs for use at temperatures of ~1300°C. Potential alloys systems include Mo-B-Si alloys, ceramics, or platinum group metal (PGM) based alloys which have already shown high temperature capability in the glass industry (Selman & Darling, 1973; Roehrig, 1981; Heywood, 1988). However, obstacles to using the PGM-based alloys are their high price and that the conventional alloys show comparably low mechanical resistance at elevated temperatures, although they have been designed with corrosion resistance to the harsh glass-producing environment as their primary attribute.

Platinum-based alloys have high melting points, good thermal stability and thermal shock resistance, and good corrosion- and oxidation-resistance. In several applications, the high electrical and thermal conductivity of Pt are important. Mechanically, Pt alloys combine high ductility with adequate creep strength. These properties give the alloys potential for applications in the chemical industry, space technology and glass industry (Fischer, 2001; Whalen, 1988; Lupton, 1990). In the spacecraft industry, Pt-materials are used to increase the heat resistance of rocket engine nozzles. High-purity optical glasses and glass fibres are manufactured using platinum-containing tank furnaces, stirrers and feeders to withstand high temperatures, mechanical loads and corrosive attack. In glasses, outstanding purity, homogeneity and the absence of bubble inclusions can be achieved only by using platinum. If ceramic melting vessels were used, ceramic particles would be loosened by erosion, contaminating the glass melts and compromising optical properties such as transmittance.

While pure platinum has low mechanical strength at high temperatures, alloying with iridium (Ir) or rhodium (Rh) significantly increases stress rupture strength (Fischer et al., 1997; 1999a and 1999b). These solid solution strengthened alloys have good ductility at high temperatures and can be welded. However, due to evaporation of oxides during annealing above 1100°C in air, Pt-Ir alloys show relatively high mass loss. Conversely, Pt-10%Rh and Pt-20%Rh alloys have a very low evaporation rate. However, grain coarsening in these solid solution strengthened alloys deteriorates the mechanical properties, promoting premature failure of components. To compensate, oxide dispersion strengthened (ODS) Pt-based alloys, with small amounts of finely distributed zirconium or yttrium oxides in the Pt matrix, were developed to improve the high temperature properties (Völkl et al., 1999). The reduction of dislocation mobility and grain boundary stabilisation by the stable oxide dispersoids increased the stress-rupture strength to ~1600°C.

Conventional Pt ODS alloys are manufactured by complicated and expensive powder metallurgical processes (Hammer & Kaufmann, 1982) and exhibit brittleness and susceptibility to cracking. Their low ductility also renders Pt ODS alloys unable to withstand stress concentrations caused by thermal expansion during frequent and rapid temperature changes. Difficulties in fabrication, especially decreased strength due to coagulation of oxide particles after welding, have discouraged the use of ODS platinum alloys (Fischer, 2001). Heraeus (2011) produces ODS Pt-alloys (dispersion hardened platinum DPH® alloys), in a process with internal oxidation, removing the disadvantages of conventional ODS Pt-alloys. These Pt DPH® alloys have good ductility, with comparable oxidation and corrosion resistance to solid solution strengthened Pt-alloys (Merker et al., 2003).

By assessing the applications of a range of Pt alloys, PGMs have been selected as potential base materials in the quest to provide a higher temperature alternative to the Ni-based superalloys. Fischer et al. (1997; 1999a; 1999b; 2001) and Völkl et al. (2000) stated that problems encountered in the aerospace industry could be solved by using Pt-based alloys because they perform exceptionally well in various high-temperature applications, including areas such as glass manufacturing and the handling of corrosive substances (Fischer et al., 2001; Coupland et al., 1980; Hill et al., 2001a). The idea of using an fcc PGM analogue has arisen several times (Bard et al., 1994): in NIMS, Japan with good properties for iridium-based and rhodium-based alloys (Yamabe et al., 1996; 1997; 1998a; 1998b and 1999; Yu et al., 2000), Pt-Ir alloys (Yamabe-Miterai et al., 2003) Pt-Al-Nb and Pt-Al-Ir-Nb alloys (Huang et al., 2004), and platinum-based alloys in South Africa (Wolff & Hill, 2000).

The PGMs and the NBSAs have similar structures (mostly fcc) and similar chemistry for the formation of similar phases. Advantages of PGM-based alloys over the NBSAs are the increased melting temperatures (e.g. 2443°C for iridium, 1769°C for platinum compared to 1455°C for nickel) and the excellent corrosion properties. Although platinum-based alloys are unlikely to replace all NBSAs on account of both higher price and higher density (Pt density of 21.5 $g.cm^{-3}$, compared to Ni density of 8.9 $g.cm^{-3}$), they have potential for use in components subjected to the highest temperatures. For Pt-based alloys, an increase in application temperature of at least 200°C could be gained (and more for Ir-based alloys). Although changes in engine design could be necessary, the higher application temperatures could offset the increased density and expense, and the alloys could be recycled.

Most PGMs are fcc structured, with the exception of ruthenium, which has a hcp structure. Iridium has a higher melting point than platinum, but has the disadvantage of brittleness (Panfilov et al., 2008) and is in short supply. Thus, platinum is the preferred alloy base among the PGMs in the most extreme environments in terms of elevated temperatures, aggressive atmospheres and higher stresses (Wolff & Hill, 2000; Hill et al., 2001a; Cornish et al., 2003). In terms of coatings, investigations have studied the possibilities that either no coatings would be necessary, or at least simpler coatings could be used than those currently used on nickel-based superalloys (Cornish et al., 2009a, 2009b; Douglas et al., 2009).

Currently, there are three major ranges of Pt-based alloys which have been developed. One is based on Pt-Al-Cr-Ni (Hüller et al., 2005; Wenderoth et al., 2005 & 2007; Rudnik et al., 2008; Völkl et al., 2005 & 2009), and two are based on Pt-Al-Cr-Ru (Cornish et al., 2009a; 2009b; Douglas et al., 2009). Of the latter range, one is more malleable but less resistant to extreme chemical environments, whereas the other has greater chemical resistance, but is more difficult to form. Although no alloys have yet been produced commercially, they are the subject of an ongoing project to develop them further. The major problems in finding a suitable application are that the alloys are too dense for current designs of turbine engines, and that they are extremely expensive. Within these restrictions, the alloys could have potential as coatings on other lighter, more affordable substrates.

2. Rationale for developing Pt-based superalloys

A rationale for the development of the Pt-based superalloys was derived, with the required material structure of fine precipitates in a matrix and using nickel-based superalloys (NBSAs) as the role model. Although Pt-based alloys are used in the glass industry, these did not have the required microstructure, as they were developed primarily for corrosion resistance, rather than strength (Selman & Darling, 1973; Roehrig, 1981; Hammer & Kaufmann, 1982; Heywood, 1988). This meant that there were few platinum alloy systems which could be used as a base for Pt-based superalloys.

Initially, several ternary alloys were produced to identify potential systems on which to base the alloys. The alloys were examined for the required two-phase structure with ordered fcc (cubic $L1_2$) phases, and were mechanically tested and tested for oxidation resistance. The first alloys were selected by examining binary phase diagrams (Massalski et al., 1990) containing platinum for the presence of suitable two-phase alloys. From pairs of such binary Pt-alloys, a number of ternary Pt-X-Z alloys were made, where X was a component for precipitate strengthening, and Z was for solid solution strengthening and/or chemical resistance. The alloys were to comprise two-phase microstructures: a γ fcc (Pt) matrix and γ' ordered fcc ($L1_2$) Pt_3X precipitates (Hill et al., 2001b, 2001c & 2001d; Cornish et al., 2009a). The candidates for X were: Al, Nb, Ta and Ti, and for Z were: Ni, Re and Ru. Each alloy was first analysed for the required two-phase γ/γ' structure, then was tested for hardness and oxidation resistance. At this stage, an alloy, and its candidate system, was rejected as soon as it failed one of the tests. Thus, alloys were abandoned if they were not two-phase, did not have good mechanical properties, or if they exhibited poor oxidation resistance (Hill et al., 2001b). Although NBSAs are based on ~Ni_3Al / (Ni) (γ'/γ), there were two reasons why the Pt-Al system was not chosen at the outset as the basis for the precipitation strengthened alloys. Firstly, Pt_3Al has at least two forms, making the system more complicated than Ni-Al (McAlister & Kahan, 1986; Oya et al., 1987). Secondly, a predisposition to Pt-Al was avoided, in case some other potentially beneficial candidate system was ignored.

Those systems which showed promise became the basis for more detailed studies, involving production of more alloys and further phase diagram studies where necessary. Large losses of niobium showed that the Pt-Nb-Z alloy was insufficiently stable, and the lath-like second

phase containing Nb was incoherent with the matrix, so Nb contents would be limited. Rhenium additions had to be limited to ≤3 at.% in order to avoid precipitation of the Re-rich needle-like phase. Two-phase microstructures with a γ fcc (Pt) matrix and γ' ordered fcc ($L1_2$) ~Pt_3X precipitates were achieved in Pt-Ti-X and Pt-Al-X systems, where X is a metal (Hill et al., 2001b). Since the Al-containing alloys had considerably better oxidation behaviour than other alloys, further work focused entirely on Pt-Al-Z alloys. Chromium was found to stabilise the $L1_2$ cubic form of ~Pt_3Al and Ru was added as a good solid solution strengthener (Hill at al., 2001d). High temperature oxidation behaviour was found to be suitable, as an external alumina film was formed which protected the alloys, since no internal oxidation occurred during long-term exposure (Hill et al., 2000; Süss et al., 2001a).
Simultaneously, work was undertaken on the microstructure and properties of platinum-hafnium-rhodium and platinum-rhodium-zirconium alloys (Fairbank et al., 2000; Fairbank, 2003). The binary Pt_8Hf and Pt_8Zr phases precluded a γ-γ' NBSA analogue in the Pt-Hf and Pt-Zr alloys, because these phases were between the fcc γ and ($L1_2$) γ' phases. However, the Pt_8Hf and Pt_8Zr phases did not penetrate far into the Pt-Hf-Rh and Pt-Rh-Zr systems, and γ-γ' regions were formed beyond their limits of penetration. Compressive proof stress results indicated better properties for $Pt_{74.5}$:Hf_{17}:$Rh_{8.5}$ (at.%) than the initial ternary alloys (Hill et al., 2000), but the oxidation resistance was much poorer.
From these beginnings, two series of Pt-based alloys were developed: Pt-Al-Cr-Ru (Douglas et al., 2009; Cornish et al., 2009b) and Pt-Al-Cr-Ni (Hüller et al., 2005; Wenderoth et al., 2005 & 2007; Rudnik et al., 2008; Völkl et al., 2005 & 2009). Both used the advantageous properties of aluminium (with the added benefit of chromium) for forming the ~Pt_3Al precipitates and protective alumina films. The former used ruthenium as a solid solution strengthener, whereas the latter alloys used nickel, and avoided ruthenium because of concerns over possible room temperature formation of RuO_4, a volatile, toxic oxide (Eagleson, 1993). The Pt-Al-Cr-Ni alloys were subjected to more rigorous mechanical testing by the researchers in Germany (Hüller et al., 2005; Wenderoth et al., 2005 & 2007; Rudnik et al, 2008; Völkl & Fischer, 2004; Völkl et al., 2005 & 2009). Further work was done by researchers in South Africa on the addition of alloying elements other than ruthenium to reduce the platinum content and thereby both the expense and the mass (Shongwe et al., 2009; 2010).
Testing of a valuable material requires different testing techniques to those conventionally used. Small samples were produced, usually by arc-melting, with masses of 2–50g, depending on the subsequent testing. Initial characterisation was done on small samples and mechanical properties were determined by performing microhardness tests. Young's modulus can be determined from the hardness values and fracture toughness can be gleaned from the deformation mode, *e.g.* planar or wavy slip, and cracking around the indentation. Initial oxidation tests could also be done on small samples.

3. Phases and microstructure of platinum-based alloys

As previously stated, platinum has a similar crystallographic structure and chemistry to nickel, and so has similar phase diagrams and hence phase relationships. Since the nickel-based superalloys (NBSAs) are based on the Ni-Al system, with the Ni_3Al precipitates being in a Ni-rich solid solution, a two-phase Pt-Al alloy of similar composition is a good starting point for a NBSA analogue. The Ni-Al and Pt-Al phase diagrams are similar for the Ni-rich and Pt-rich portions (Massalski et al., 1990). Both have fcc solid solutions based on nickel or platinum, and both have eutectic reactions forming Ni_3Al or Pt_3Al, with Ni_3Al being formed peritectically from NiAl, whereas Pt_3Al melts congruently. This indicates that similar processing in the region where the γ solid solution exists, albeit at higher temperatures, could be utilised to obtain the small γ' Pt_3Al precipitates necessary for the good mechanical properties.

However, there are two important differences between the Ni-Al and Pt-Al systems. One difference is that the limit of the platinum-rich solid solution is less temperature dependent than the nickel-rich solid solution in Ni-Al. This is shown by a comparatively vertical solvus in the binary Pt-Al system, which is a serious disadvantage, as a gradually sloping solvus is necessary for the development of a high proportion of precipitates and precipitation strengthening. However, this can be resolved by further alloying additions, considering that some NBSAs contain up to 15 alloying elements. Alloying additions to Pt-Al have already decreased the slope of the solvus and produced more precipitates. In Pt-Al-Cr-Ni alloys, Wenderoth et al. (2005) attained over 30 vol. % precipitates, thereby deducing that the solvus had to be less vertical than in the Pt-Al binary system.

At elevated temperatures, $L1_2$-Pt_3Al does not show an anomalous increase of the flow strength with increasing temperature as exhibited by $L1_2$-Ni_3Al (Takeuchi & Kuramoto, 1973), although Pt_3Al should be stronger than Ni_3Al at any temperature (Wenderoth et al., 2005). In the Ni-Al system, Ni_3Al has only one structure, whereas the Pt_3Al phase has at least two (McAlister & Kahan, 1986) if not three (Oya et al., 1987) forms. There are two conflicting phase diagrams regarding the transformation temperatures of γ' Pt_3Al. According to McAlister & Kahan (1986), there is a transformation of the high temperature Pt_3Al phase from $L1_2$ to a tetragonal low temperature variant (designated D0'c) at ~1280°C. However, Oya et al. (1987) showed an additional transformation at a lower temperature, with their transformations given as $\gamma' \rightarrow \gamma'_1$ at ~340°C and $\gamma'_1 \rightarrow \gamma'_2$ at 127°C.

The high temperature cubic $L1_2$ Pt_3Al allotrope is morphologically identical to $L1_2$ Ni_3Al. However, as the lower temperature Pt_3Al allotropes are tetragonal, it is necessary to stabilise the high temperature allotrope. Phase transformations between the cubic and tetragonal allotropes cause a change in volume, leading to local stresses and perhaps premature failure. The presence of the lowest temperature form has been confirmed (Oya et al., 1987), but conditions for its formation have not been fully explained, and may depend on impurities (Douglas et al., 2009) since separate research groups, using different source materials have given different, but reproducible results. Another Pt_3Al allotrope has been recognised in a binary Pt-Al alloy using transmission electron microscopy (TEM) where an unusual ordering phenomenon was found in the Pt_3Al precipitates (Douglas et al., 2007). A similar allotrope has also been calculated using *first principles* (Chauke et al., 2010), showing that the phase relationships are quite complex. This problem would be solved if the high temperature form was stabilised by alloying. The high temperature $L1_2$ allotrope has been stabilised and transformations to the lower temperature allotrope(s) inhibited, by small additions of Ti, Ta and Cr (Hill et al., 2001a, 2001e & 2002) and Zr, Hf, Mn, Fe and Co (Hüller et al. (2005).

Two-phase microstructures, leading to considerable precipitation-strengthening, were achieved in Pt-Ti-Z and Pt-Al-Z systems, where Z = Ni, Ru or Re (Hill et al., 2001b). Alloys in these systems showed promising mechanical properties at room temperature, with hardness values higher than 400 HV_1 and high resistance to crack initiation and propagation. However, during annealing at 1350°C for 96 hours, the Pt-Ti-Z alloys reacted with air, precluding further work on these alloys. Further studies were made on the phases and room temperature mechanical properties of Pt-Al-Z alloys after annealing the alloys at 1350°C for 96 hours, where Z = Ru, Re, W, Mo, Ni, Ti, Ta or Cr. Microstructures similar to

Ni- and cobalt- based superalloys were achieved in the Pt-based alloy Pt_{86}:Al_{10}:Z_4 (at.%), consisting of cuboidal ~Pt_3Al precipitates in a (Pt) matrix. Chromium was found to stabilise the cubic form of the ~Pt_3Al phase, while ruthenium acted as a solid solution strengthener (Biggs et al., 2001; Hill et al., 2001e & 2002). The lowest misfit between the (Pt) and ~Pt_3Al phases was found at 3-5 at.% Ru and over 20 at.% Al (Biggs, 2001). A lower temperature Pt_3Al form was found in both W- and Ni- containing alloys. In Mo-containing alloys, coarse microstructures were formed and Mo substituted for Pt in the Pt_3Al phase. All the Cr-, Ta- and Ti-containing alloys had favourable microstructures, and the cubic $L1_2$ form of ~Pt_3Al was stabilised.

Since no ternary Pt-based alloy had the required microstructure with a sufficiently high proportion of γ' precipitates, other alloying additions were needed. Several alloys were made with the objective to increase the γ' volume fraction, and original compositions were selected based on the studies of the ternary Pt-Al-Cr and Pt-Al-Ru systems (Cornish et al., 2009a & b; Douglas et al., 2009). A potential alloy, Pt_{84}:Al_{11}:Ru_2:Cr_3(at%), was composed entirely of a fine two-phase γ/γ' structure, with no primary phase, and its oxidation resistance was superior to the original ternary alloys (Süss et al., 2001b).

Other attempts have been made to improve the properties, and decrease the alloy cost and density by additional alloying elements. Nickel was added to improve the solution strengthening of the (Pt) matrix, although it was not as effective as expected. Cobalt was also added for solid solution strengthening, and extensive phase diagram work was done on the Pt-Al-Co (Chown & Cornish, 2003: Chown et al., 2004) and Pt-Al-Ni systems (Glaner & Cornish, 2003). The Co additions increased the formability. Currently, Nb and V additions are being studied in the hope that they can decrease the Pt content - and hence density and price - while simultaneously increasing the melting point (Shongwe et al., 2009 & 2010).

4. Mechanical properties of ternary pt alloys

4.1 Basis of assessment

As developmental alloys need a comparator, it was decided at the outset that two commercial alloys - MAR-M247 and PM2000 - would be tested. MAR-M247 was selected as a NBSA, and was therefore a representative of the alloys which the platinum alloys might replace. PM2000 was chosen as a comparator because of its advanced microstructure and high-temperature applications. PM2000 is ferrous-based (Fe-Cr-Al) with a ferritic matrix, and is mechanically alloyed with yttrium oxide (Y_2O_3) dispersion material.

4.2 Compressive testing

Ternary substitutional alloying additions to Pt-Al-Z alloys (where Z = Ti, Cr, Ru, Ta & Re) showed that Pt-Al-Z alloys had higher compressive strengths above 1150°C than the commercial NBSA MAR-M247 (Hill et al., 2001c & 2001e). High temperature compressive strength is a useful test for comparison, but it does not equate to creep strength, the latter being a crucial property for most high temperature applications. Using previous results (Biggs et al., 2001; Hill et al., 2001c & 2001e) to select the alloys, creep tests were done at 1300°C on PM2000 and standardised composition Pt_{86}:Al_{10}:Z_4 (at.%) alloys, where Z = Ti, Cr, Ru, Ta or Ir, as shown in Figure 1 (Süss et al., 2002). PM2000 had the highest strength of the alloys tested, but the shallow slope of the stress-rupture curve indicated high stress sensitivity and brittle creep behaviour. This indicated that PM2000 would be more likely to fail in the presence of stress concentrations or short overloads during usage. Pt_{86}:Al_{10}:Cr_4 exhibited the highest strength of the investigated Pt-based alloys.

Figure 2 shows selected creep curves tested at 30 MPa (Süss et al., 2002). No primary creep stage was seen for the Pt-based alloys within the measurement error. After secondary creep, the Pt-based alloys experienced substantial tertiary creep, leading to fracture strain values between 10 and 50% at 1300°C. It was not possible to resolve different stages of the creep curves for PM2000, because of very low creep rates and fracture strains below 1%.
Stress rupture lives of PM2000 and the most promising ternary Pt-based alloy, $Pt_{86}:Al_{10}:Cr_4$, are shown in Figure 3. Stress-rupture strength values after 10 hours at 1300°C of several conventional solid-solution strengthened Pt-based alloys (Lupton et al., 2000) and zirconia grain stabilised (ZGS) platinum, an oxide dispersion strengthened Pt alloy from Johnson Matthey Noble Metals (Bard et al., 1994) are also shown in Figure 3. The precipitation of γ' particles increased stress-rupture strength, by a factor of 8 from 2.2 MPa for the pure Pt matrix to 17 MPa for $Pt_{86}:Al_{10}:Cr_4$. The strength of the $Pt_{86}:Al_{10}:Cr_4$ alloy was also higher than the conventional solution strengthened Pt-based alloys, Pt-10 Rh (wt %) and Pt-20 Rh. By comparing strengths, it can be seen that a rhodium alloying addition of 30 wt% Rh was required for equivalent strength of the $Pt_{86}:Al_{10}:Cr_4$ alloy. This was encouraging, since the exceptionally high price of Rh and machining difficulties limit the practical use of Pt-30 Rh. At 1300°C, the creep strengths of the Pt-based alloys were higher than those of the Ni- and Co-based superalloys, whose precipitates dissolve at this high temperature, thus losing strength, and were comparable to a platinum ZGS alloy.

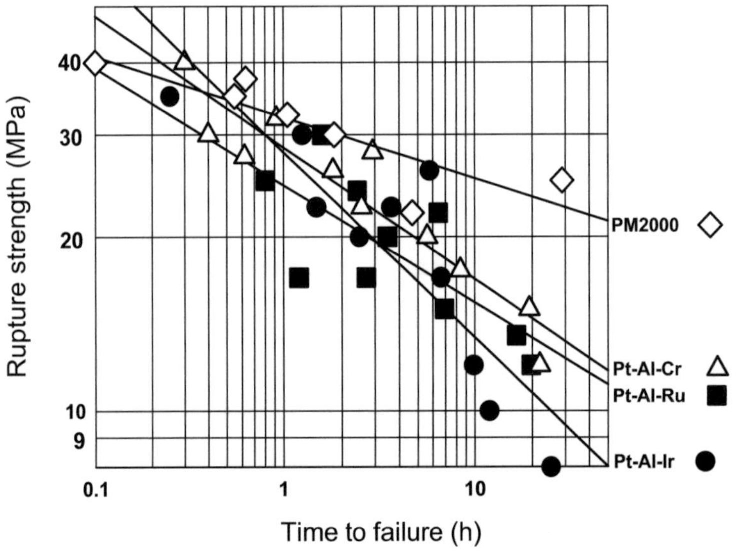

Fig. 1. Stress rupture curves at 1300°C in air of various $Pt_{86}:Al_{10}:Z_4$ alloys compared to PM2000 (Süss et al., 2002).

In summary, Pt-Al-Z alloys had higher compression strengths above 1150°C than the commercial Ni-based superalloy MAR-M247, and at 1300°C, the Pt-based alloys underwent pronounced tertiary creep leading to fracture strain values of 10-50%. $Pt_{86}:Al_{10}:Cr_4$ possessed the highest strength of the investigated Pt-based alloys, outperforming several conventional solid-solution strengthened Pt-Rh alloys, and was comparable to Pt ZGS and PM2000. These results were encouraging, as the precipitate volume fraction of ~40% was sub-optimal compared to the ~70% used in commercial NBSAs (Sims et al., 1987).

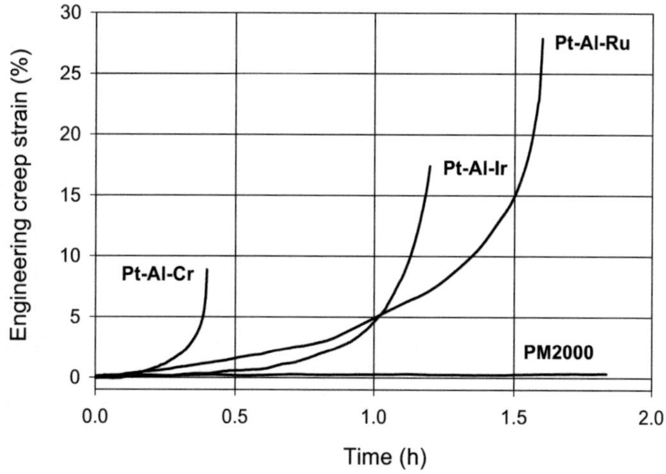

Fig. 2. Creep curves for $Pt_{86}:Al_{10}:Z_4$ and PM2000 alloys tested at 30 MPa and 1300°C (Süss et al., 2002).

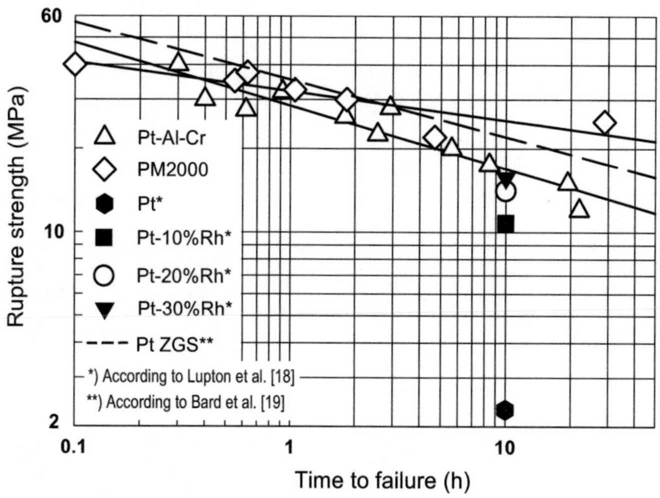

Fig. 3. Stress-rupture lives at 1300°C of $Pt_{86}:Al_{10}:Cr_4$ (at.%), PM2000 and Pt ZGS compared with stress-rupture strength values ($R_{m/10h/1300°C}$) of some conventional solid-solution strengthened Pt-base alloys (Süss et al., 2002.; Lupton et al., 2000; Ochiai, 1994).

Alloys annealed at 1300°C for 96 hours followed by helium quenching gave better all round compression results for $Pt_{80}:Al_{14}:Ru_3:Cr_3$ than $Pt_{86}:Al_{10}:Cr_4$ and $Pt_{86}:Al_{10}:Ru_4$. This was attributed to an increased volume fraction of the ~Pt_3Al phase, although the precipitates were coarser (Keraan & Lang, 2003a & 2003b; Keraan, 2004).

4.3 Tensile tests

The first tensile tests were undertaken on unoptimised alloys (Süss & Cornish, 2004; Cornish et al., 2009b), although the samples had been heat-treated to promote a homogeneous two-phase microstructure. Since mechanical properties become strain rate dependent at high

temperatures, only the room temperature tensile properties of Pt_{86}:Al_{10}:Cr_4 and Pt_{86}:Al_{10}:Ru_4 ternary alloys were evaluated, and compared with that of the best quaternary alloy, Pt_{84}:Al_{11}:Ru_2:Cr_3. The high temperature compressive strength of Pt_{84}:Al_{11}:Ru_2:Cr_3 was significantly higher than that of Pt_{86}:Al_{10}:Cr_4 (Keraan & Lang, 2003a & 2003b; Keraan, 2004).
Macro-scale tensile testing was excluded because of the high material cost. Smaller specimens than the ASTM standard sub-size specimen for tension testing (ASTM E8-93, 1993) were machined, utilising the small specimen test technology and experience from fusion materials development (Lucas et al., 2002). Yield stress is independent of specimen thickness for thicknesses greater than a critical thickness t_c and is not affected by specimen width w, while the ultimate stress is independent of the aspect ratio (t/w) above a critical aspect ratio $(t/w)_c$ (Panayotou, 1982; Kohyama et al., 1987). Specimens were made assuming that a thickness of 3 mm and an aspect ratio of 1 (i.e. a width of 3 mm) would ensure the tensile properties would be independent of the specimen dimensions (Lucas et al., 2002; Kohno et al., 2000). The specimen dimensions were: total length of 46mm and a thickness of 3mm, with a gauge width of 3 mm and gauge length of 18 mm.
After ageing in air at 1250°C for 100 hours followed by water quenching to retain the two-phase γ/γ' structure, flat mini-tensile specimens were machined from each 50g ingot by wire spark erosion. The tensile tests were performed with a cross-head speed of 5 $mm.min^{-1}$. The maximum ultimate tensile strength and elongation were determined and Vickers hardness tests (HV_{10}) were also performed, with the results given in Table 1. Also shown are the mechanical properties of pure platinum and some commercial high temperature alloys.
Minor cross-contamination of ~0.01 wt% Ru or Cr occurred in some ternary alloys which was probably from sputtering during arc melting. TEM and X-ray diffraction (XRD) showed that, except for the Ru alloys which were 95 vol.% (Pt), all samples comprised (Pt) and ~Pt_3Al (Figure 4). The volume fraction of precipitates varied between specimens.

Alloy composition (at.%)	Reference	Hardness (HV_{10})	UTS (MPa)	Elongation (%)
Pt_{86}:Al_{10}:Cr_4		317 ± 13	836	~4
Pt_{86}:Al_{10}:Ru_4		278 ± 14	814	~9
Pt_{84}:Al_{11}:Ru_2:Cr_3		361 ± 10	722	~1
Pure Pt - annealed	Johnson Matthey, 2011	40-50	124-245	35-40
Ferritic ODS alloy PM2000	Matweb, 2011	290	720	~14
γ-TiAl	Pather et al., 2003	~250	950	~1
CMSX-4	Maclachlan & Knowles, 2001	~370	870	~4

Table 1. Room temperature mechanical properties of selected Pt-based alloys compared to pure platinum and some commercial high temperature alloys.

Pt_{86}:Al_{10}:Cr_4 had a higher UTS and hardness than Pt_{86}:Al_{10}:Ru_4. Although Pt_{84}:Al_{11}:Ru_2:Cr_3 contained a significant amount of ~Pt_3Al and was the hardest, it had the lowest UTS (Süss & Cornish, 2004). The quaternary alloy failed intergranularly, while the ternary alloys failed mainly by intragranular cleavage, with some localised dimpling. The lower UTS and lower elongation were related to the intergranular failure. This more brittle failure could also have been due to the larger precipitates (Figure 4b), and the unfavourable ogdoadically-diced precipitate morphology (Westbrook, 1958) which is associated with inferior properties (Sims et al., 1987), especially when compared to $Pt_{86}:Al_{10}:Cr_4$. These results showed that the microstructure of $Pt_{84}:Al_{11}:Ru_2:Cr_3$ needed to be optimised (Keraan & Lang, 2003a & 2003b; Keraan, 2004). Since Ru is a better solid solution strengthener than Cr in these alloys (Hill et al., 2001), the higher strength of $Pt_{86}:Al_{10}:Cr_4$ over $Pt_{86}:Al_{10}:Ru_4$ was due to the low volume fraction (~5%) of ~Pt_3Al precipitates in $Pt_{86}:Al_{10}:Ru_4$, showing that it had been annealed above its solvus in the range 1250-1300°C (Süss et al., 2001b). The superior ductility was thus due to its nearly single phase (Pt) nature.

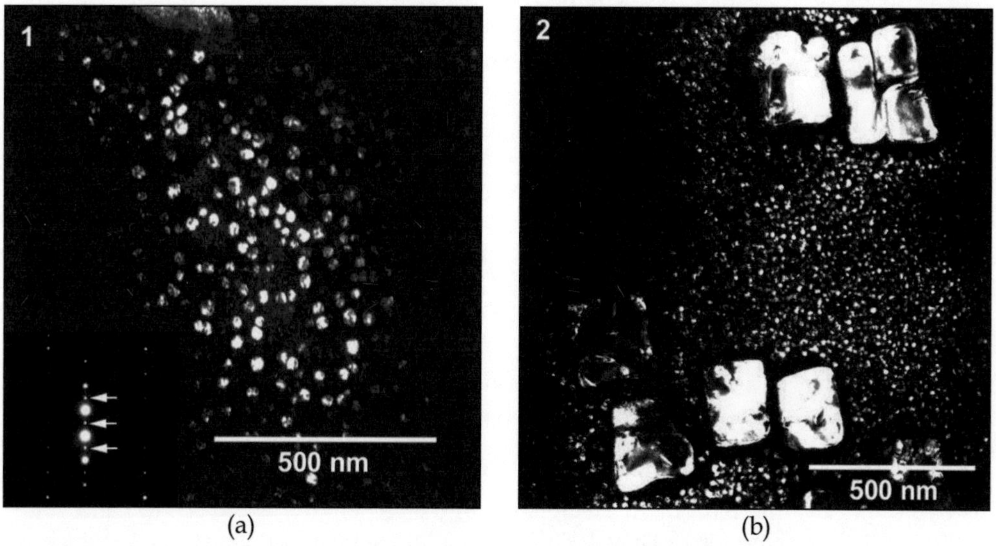

Fig. 4. (a) Dark field TEM image of $Pt_{86}:Al_{10}:Cr_4$ (Insert: SAD pattern); (b) Dark field TEM image of $Pt_{84}:Al_{11}:Ru_2:Cr_3$ (Süss & Cornish, 2004; Cornish et al., 2009b).

Although the spread and inconsistencies in the results were disappointing, testing was only undertaken on a small range of specimens due to material expense. As shown in Table 1, the mechanical properties of the tested Pt-based alloys were significantly higher than those of pure Pt in the soft state (Johnson Matthey, 2011). More importantly, the Pt-based alloys showed mechanical properties in a similar range to other high temperature alloys, e.g. ferritic ODS alloy PM2000 (Plansee, 1998), γ-TiAl (Pather et al., 2003) and CMSX-4 (Maclachlan & Knowles, 2001), even though the microstructures were not yet optimised.

Creep test results of $Pt_{84}:Al_{11}:Cr_3:Ru_2$ were inferior to those of $Pt_{86}:Al_{10}:Cr_4$, but it was deduced that this was because the former was done in air, whereas initial tests on the latter were done in argon (Cornish et al., 2009b). However, the results of $Pt_{84}:Al_{11}:Cr_3:Ru_2$ were slightly inferior to those of a commercial dispersion hardening (DPH) Pt alloy. The high temperature compressive strength of $Pt_{84}:Al_{11}:Cr_3:Ru_2$ was significantly higher than that of $Pt_{86}:Al_{10}:Cr_4$ (Keraan & Lang, 2003a & 2003b; Keraan, 2004).

Other additions, such as cobalt or nickel, have also been tested to improve the properties of the alloys and decrease their cost and density. Nickel was added to improve the solution strengthening of the matrix, although the intended solution strengthening was not fully achieved. Surprisingly, the melting temperature was increased by Ni additions. The work was not continued because of the disappointingly low hardness results (Glaner & Cornish, 2003), although other research has been more encouraging (Vorberg et al., 2004 & 2005; Hüller et al., 2005; Völkl, et al., 2005; Wenderoth et al., 2005 & 2006).

Pt-Al-Co and Pt-Al-Co-Cr-Ru alloys subjected to cold rolling yielded very interesting results (Chown and Cornish, 2003; Cornish et al., 2009b). Alloys with hardnesses below 400 HV_{10} showed good formability at room temperature (>75% total reduction in thickness), whereas the formability was poor (<40%) for hardnesses above 450 HV_{10}. The alloys with good formability were two-phase, comprising (Pt,Co) and ~Pt_3Al, with compositions of 5-20 at.% Co and <20 at.% Al. Excellent formability was obtained for alloys containing (Pt) and $CoPt_3$, whereas the alloys containing other intermetallic compounds showed extremely poor formability (<5% total reduction). The formability of the Pt-Al-Co alloys improved sigmoidally with increased Pt+Co content in the tested range of 60-90 at.% (Pt+Co). The cold formability of these alloys was far superior to other Pt-Al alloyed with Cr, Ru or Ni.

4.4 Developments at Fachhochschule Jena and the University of Bayreuth, Germany

4.4.1 Quaternary Pt-Al-Cr-Ni alloys

The University Bayreuth and the Fachhochschule Jena-University of Applied Sciences in Germany were already researching NBSAs in the 1990s. However, the basis for their development programme on precipitation hardened Pt alloys was the ternary Pt-Al-X alloys being investigated in South Africa (Süss et al., 2002). The Pt_{86}:Al_{10}:Cr_4 composition, with good properties (Figure 3), was the foundation for further work (Wenderoth et al., 2005; Vorberg et al., 2004 & 2005).

Nickel was added to a Pt-Al-Cr alloy in varying amounts (Hüller et al., 2005) since nickel has a good solid-solution strengthening effect on the (Pt) matrix (Zhao et al., 2002), but also to decrease the Pt content, and thus the density and price of the Pt-based alloys. A very promising microstructure was found in the alloy Pt_{79}:Al_{11}:Cr_3:Ni_7: a homogeneous distribution of $L1_2$-ordered ~Pt_3Al precipitates with edge lengths of 200-500 nm and volume fraction of 23%. Ageing for 120 h at 1000°C gave a lattice misfit of about -0.1 %, which is in the same range as in commercial Ni-based superalloys.

Based on these results, Pt-Al-Cr-Ni alloys with an increased Al content (Wenderoth et al., 2005), near the solubility limit of ~15 at.% Al (Oya et al., 1989) were made to increase the γ' volume fraction. A two-step heat treatment under flowing argon was used: homogenisation at 1500°C for 12 h, followed by water quenching; then a precipitation heat treatment (ageing) at 1000°C for 120 h, also followed by water quenching. Despite heat treatments being undertaken in flowing argon, there was 6 ppm by volume of residual oxygen, which was sufficient to form an oxide layer 10 μm thick on the sample surfaces.

Very fine γ' precipitates formed throughout the alloys from the supersaturated matrix during quenching after ageing, although much coarser γ' precipitates were found at a depth of 100 μm beneath the surface oxide layer (Wenderoth et al., 2005). The concentration profile showed a decrease of Al and an increase of Ni and Pt from the sample interior to the surface. Alloying the Pt-Al system with both Cr and Ni consistently stabilised the $L1_2$ high temperature allotrope of Pt_3Al at room temperature. The Pt_{86-76}:Al_{11}:Cr_3:Ni_{0-10} alloys were all single phase after solution heat treatment at 1450°C, and ageing at 1000°C produced microstructures analogous to Ni-based superalloys. In alloys with less than 6 at.% nickel, precipitates lost coherency after ageing, resulting in spherical particles. Pt_{80}:Al_{11}:Cr_3:Ni_6 had the highest γ' volume fraction (~23%) after ageing, and well-aligned cuboidal precipitates with 0.2–0.5 μm edge lengths and a misfit of -0.1%. The γ' shapes were due to a decreasing absolute misfit with increasing Ni content. Ageing at 1100°C produced coarse γ' particles and reduced the γ' volume fraction. The volume fraction was less affected by temperature in Ni-containing alloys than in Ni-free alloys.

4.4.2 Variation of the Al content for high γ' volume fraction

Pt-Al_x:Cr_3:Ni_{4-8} alloys were produced with 12–15 at.% Al (solubility limit is ~15 at.% Al) to increase the γ' precipitate volume fraction (Wenderoth et al., 2005). Alloys with up to 13 at.% Al were successfully homogenised in the single phase γ-region at 1500°C, but at higher Al contents the interdendritic eutectic formed, even after heat treatment at 1530°C.

Homogeneous distributions of ~Pt_3Al particles were achieved by ageing alloys with up to 13 at.% Al for 120 h at 1000°C. The absolute lattice misfit between γ and γ' decreased with increasing nickel content. Coherency between the γ and γ' phases was shown in alloys with >5 at.% Ni by slightly negative misfit values at room temperature, together with cubic or spherical particles (Figures 5b and c). Conversely, there was limited coherency and a high negative misfit of about -0.5% at nickel contents below 5 at.% Ni (Figure 5a).

Nickel has a smaller atomic radius than platinum (Schubert, 1964). When Ni additions partition to the γ matrix, the γ lattice parameter decreases more than the γ' lattice parameter, lowering the absolute lattice misfits. By increasing Ni, the γ' morphology changed from irregular shaped particles to almost perfect cubes, then to spherical particles, due to the decreasing absolute misfit. Low lattice misfit leads to spherical γ' particles (Qiu, 1996), and in the Pt_{77}:Al_{14}:Cr_3:Ni_6 alloy, medium lattice misfits led to cubic γ' particles. High lattice misfit, as in Pt_{79}:Al_{14}:Cr_3:Ni_4, caused loss of coherency, explaining the irregularly-shaped γ' particles.

As in the ternary alloys (Süss et al., 2002), there was a bimodal distribution of the precipitate sizes: coarse precipitates of ~400 nm length and smaller precipitates of ~100 nm length (Wenderoth et al., 2005). Mean γ' precipitate sizes were 520 nm in Pt_{77}:Al_{14}:Cr_3:Ni_6 and 660 nm in Pt_{75}:Al_{14}:Cr_3:Ni_8 - larger than in Pt-Al-Cr-Ni alloys with 11 at.% Al, which showed maximum sizes of 500 nm after equivalent ageing. (Hüller et al., 2005).

4.4.3 Variation of the Cr γ' volume fraction

The composition of Pt-$Al_{12.5}$:Cr_{0-6}:Ni_6 was selected to ensure coherency between γ and γ' (Wenderoth et al., 2005). Dendritic as-cast structures of Pt-$Al_{12.5}$:Cr_3:Ni_6, Pt-Al_{12}:Ni_6 and Pt-Al_{12}:Cr_6:Ni_6 were homogenised by heat treatment at 1500–1510°C, but γ' formation was almost completely suppressed in Pt-$Al_{12.5}$:Cr_3:Ni_6 after homogenisation for 12 h at 1500°C followed by water quenching. Air cooling produced homogeneous distributions of γ' Pt_3Al particles with 200nm average edge lengths and a volume fraction of about 30%, as shown in Figure 6b. Furnace cooling from 1500°C gave rise to a bimodal particle distribution, and a total γ' volume fraction of 34%. Increasing the Cr content to 6 at.% led to average edge lengths of 500nm and a volume fraction of 50% in Pt-Al_{12}:Cr_6:Ni_6 after homogenisation for 6 h at 1500°C + 6 h at 1510°C in Ar, followed by air cooling. Thus, with controlled air cooling after solution heat treatment, good microstructures were achieved.

Most precipitation strengthened alloys are generated from an eutectic system, and alloys are usually optimised by homogenisation (solution annealing) just below the eutectic temperature in the single phase field of the matrix. Highest strengths are attained with compositions near the solubility limit of the precipitate-forming elements in the matrix (Wenderoth et al., 2005). As-cast Pt-Al-Cr-Ni alloys with 11 at.% Al contained low volume fractions of γ/γ' eutectics, and were successfully solution treated at 1450°C, although annealing at 1400°C was insufficient to produce a single-phase microstructure. Ageing at 1000°C gave a maximum γ' volume fraction of ~23 % for Pt_{80}:Al_{11}:Cr_3:Ni_6 (Hüller at al., 2005). As-cast alloys with 13 at.% Al had dendritic structures with high volume fractions of the γ/γ'

eutectic. Rapid cooling (in the water-cooled copper hearth of the arc furnace) did not prevent secondary precipitation, indicating rapid γ' nucleation and initial growth.
The dendritic structures were converted to fine, homogeneous γ' particles in the γ matrix by solution heat treatment for 12 h at 1500°C and water quenching. Secondary precipitation of γ' occurred despite rapid cooling, which agreed with previous work on other Pt-based alloys (Wolff & Hill, 2000; Vorberg et al., 2004). Precipitation of fine γ' particles within the γ matrix channels has also been seen in NBSA CMSX-10 after multi-step ageing heat treatment (Erickson, 1995), improving the lower temperature tensile and creep strengths.

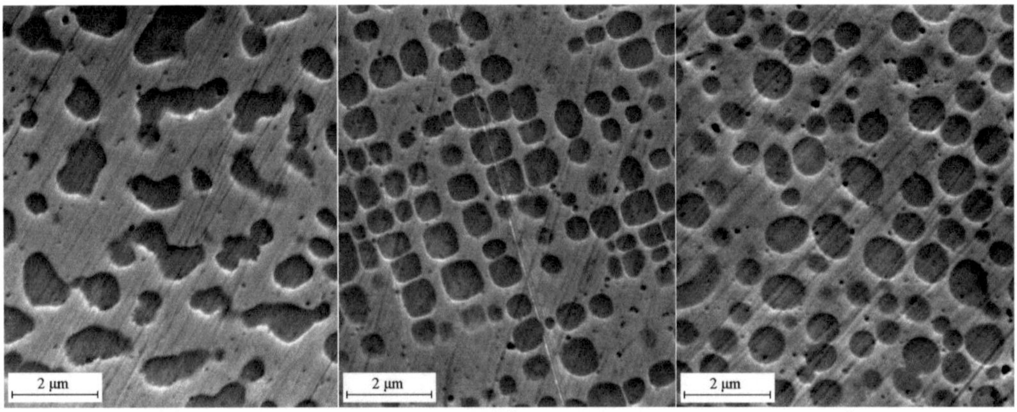

a) b) c)

Fig. 5. Secondary electron SEM micrographs of the Pt-based alloys, showing ~Pt_3Al γ′ precipitates (dark) in the (Pt) γ matrix (Hüller et al., 2005): a)~Pt_{79}:Al_{14}:Cr_3:Ni_4, b)~Pt_{77}:Al_{14}:Cr_3:Ni_6, and c)~$Pt_{7:5}$:Al_{14}:Cr_3:Ni_8 (at.%). [Annealed for 12 h at 1500°C, then 120 h at 1000°C in Ar. Electrolytically etched in 5% aqueous potassium cyanide (KCN) solution].

The simpler structure of the Ni-rich γ matrix (and by inference, the Pt-rich matrix) gives a much lower yield strength than for γ', but its misfit-induced compression stresses improve creep resistance during the early stages of tensile creep deformation, as long as coherent γ/γ' interfaces remain (Glatzel & Feller-Kniepmeier, 1989; Müller et al., 1993; Völkl et al., 1998). However, high negative misfit values increase the interfacial energy and thus boost the tendency for precipitate coarsening at high temperatures (Sims et al., 1987). Thus, the alloy Pt_{77}:Al_{14}:Cr_3:Ni_6 with an ambient temperature misfit of -0.3% in the solution heat treated condition and -0.1% in the aged condition, shows great potential (Wenderoth et al., 2005).

4.4.4 Substitution of Ni

Although nickel is beneficial in the reduction of misfit, it has a low melting point, and at least some substitution of high melting elements, such as Nb, Ta and Ti is required for improved high temperature properties (Völkl et al., 2009). XRD verified the fcc γ matrix and $L1_2$-ordered γ' Pt_3Al phases in Pt-Al_7:Cr_6:Nb_5, Pt-Al_7:Cr_6:Ta_5 and Pt-Al_7:Cr_6:Ti_5 (at.%). The lattice misfit ratios were in the order of -3 x 10^{-3} in all alloys, and homogenisation promoted bimodal γ' size distributions. Besides coarse and irregularly shaped particles with a volume fraction of 10-20%, there were small cuboids of ~300nm across. Ageing for 264 h at 1200°C with water quenching, resulted in Nb and Ta being almost equally partitioned to both γ and γ', whereas Ti partitioned to γ'. The γ' volume fractions after ageing were 34% in Pt-

$Al_7:Cr_6:Nb_5$, 33% in Pt-$Al_7:Cr_6:Ta_5$ and 35% in Pt-$Al_7:Cr_6:Ti_5$ (Völkl et al., 2009). Compression strengths of polycrystalline Pt-$Al_7:Cr_6:Nb_5$, Pt-$Al_7:Cr_6:Ta_5$ and Pt-$Al_7:Cr_6:Ti_5$ were higher than that of Pt-$Al_{12}:Cr_5$ (Süss et al., 2002). At 800°C, Pt-$Al_7:Cr_6:Ta_5$ was strongest, whereas at higher temperatures Pt-$Al_7:Cr_6:Nb_5$ was the strongest. Above 1200°C, Pt-$Al_7:Cr_6:Nb_5$ and Pt-$Al_7:Cr_6:Ta_5$ outperformed the single-crystal NBSA CMSX-4. The deformed CMSX-4 samples showed rupture on the outside surface, which was also observed in the Nb- and the Ti-containing Pt-based alloys.

Völkl & Fischer (2004) customised equipment to perform stress-rupture tests at high temperatures. At 1300°C, Pt-$Al_{12}:Cr_6:Ni_5$ had higher stress rupture strengths than both Pt-10%Rh and Pt-10%Rh DPH (Völkl et al., 2009). A Norton plot was generated by plotting minimum creep rate against stress in a double logarithmic plot. This shows data points on near straight lines, where the slope is the Norton exponent *n* of the Norton creep law:

$$\varepsilon_{min} = A\,\sigma^n \qquad (1)$$

where: A = a constant which depends on temperature, material and its condition
σ = stress in MPa, n = Norton exponent

The Norton exponent of Pt-10%Rh at 1600°C was calculated as n = 3.3 and A = 34, the values for pure Pt were n = 3.8 and A = 10.5 under similar conditions and the Norton exponent of $Pt_{77}:Al_{12}:Cr_6:Ni_5$ was 3.6 (Völkl et al., 2009). The low Norton exponents for pure Pt, Pt-10%Rh and Pt-$Al_{12}:Cr_6:Ni_5$ are typical for the viscous-drag controlled creep of single-phase solid solution alloys. However, the low value of Pt-$Al_{12}:Cr_6:Ni_5$ could also be explained by the intergranular fracture observed in the creep-deformed samples, indicating some brittleness and weakness of the grain boundaries, which has also been found in $Pt_{84}:Al_{11}:Ru_2:Cr_3$ (Süss & Cornish, 2004; Cornish et al., 2009b). To alleviate this weakness, very small amounts of boron were added, because B tends to segregate to grain boundaries and alters grain boundary adhesion. Rhenium is very beneficial for the creep strength of Ni-based super-alloys (Sims et al., 1987), so both B and Re were added to Pt-$Al_{12}:Cr_6:Ni_5$. Compression creep tests at 1200°C showed that minor B additions increased both creep strength and ductility considerably, and 0.3 at.% B with 2 at.% Re further increased creep strength. Rhenium was found to retard precipitate growth and to increase creep strength (Völkl et al., 2009).

5. Relation of deformation to microstructure

In order to understand the effect of composition on the γ' ~Pt_3Al precipitates and their deformation, TEM was done on $Pt_{86}:Al_{10}:X_4$ ternary alloys, where X was Ru, Cr, Ta, Ti and Ir (Hill et al., 2001c; Douglas et al., 2001; Santamarta et al., 2003). The precipitate distribution was bimodal, which could even have been interpreted as trimodal. Ti, Cr and Ta partitioned to ~Pt_3Al, and stabilised the higher temperature $L1_2$ structure of the Pt_3Al phase, resulting in cuboid precipitates (Figure 6a). Conversely, Ru and Ir stabilised a lower temperature modified $D0'_c$ type lath structure. This was recognised as originating from the displacive cubic ~Pt_3Al → tetragonal ~Pt_3Al reaction (McAlister & Kahan, 1986; Oya et al., 1987), since there were distinct bands (twins) in the precipitates (Douglas et al., 2001, 2003 & 2007; Santamarta et al., 2003; Douglas, 2004). These precipitates often had more complex appearances (Figure 6b), such as ogdoadically-diced shapes identified by Westbrook (1958). The amount of ruthenium in any alloy would need to be balanced to provide both beneficial solid solution strengthening and corrosion resistance properties (Potgieter et al., 1995; van der Lingen & Sandenbergh, 2001; Shing et al., 2001), without the disadvantage of stabilising the lower temperature ~Pt_3Al forms, although chromium could be used to offset the latter effect (Hill et al., 2001a, 2001e & 2002).

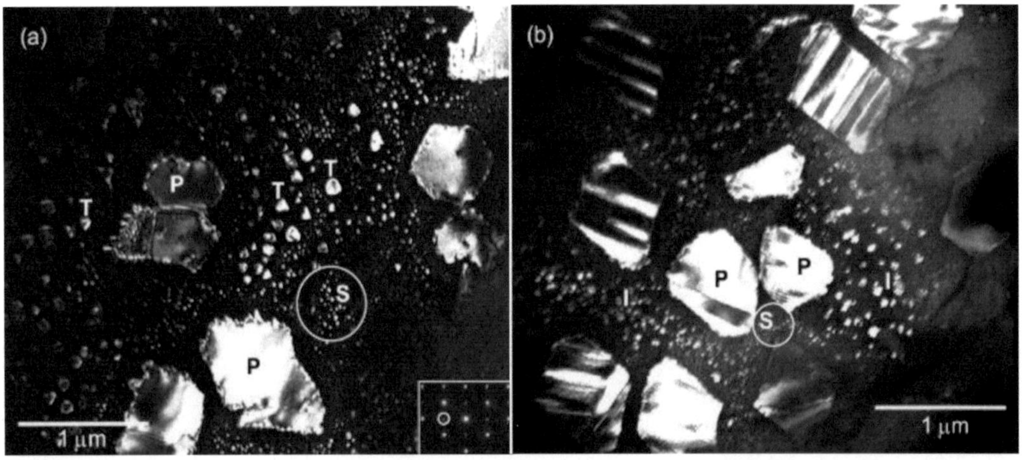

Fig. 6. a). Typical TEM micrograph of the $L1_2$ Pt_3Al precipitates stabilized by Cr, Ta and Ti addition. Inset shows the selected area diffraction (SAD) pattern, confirming the $L1_2$ structure. b). Typical TEM micrograph of the D0'c precipitates stabilised by Ru and Ir additions. P, T, I and S indicate the different size ranges of the precipitates (Hill et al., 2001c).

Lattice parameters were measured by XRD, using the (220) peak for (Pt) lattice parameters, (112) for tetragonal D0'c Pt_3Al, and (211) for cubic $L1_2$ Pt_3Al. The misfit δ was calculated (Sims et al., 1987) using the expression:

$$\delta = 2 \cdot \frac{(a_{ppt} - a_{matrix})}{(a_{ppt} + a_{matrix})} \tag{2}$$

where a is the lattice parameter.
Table 2 shows misfits at room temperature and 800°C between the matrix and precipitate phases (Hill et al., 2001c). All misfit values were negative, with the smaller misfits arising from the $L1_2$ precipitates, and the larger misfits belonging to the transformed D0'c structure. Since the interaction between dislocations in the matrix and precipitates determines the mechanical properties, dislocations were studied and compared to those in nickel-based superalloys (Douglas et al., 2001 & 2003). To simplify the process of understanding the above mechanisms, only ternary alloys were studied. The alloys were Pt_{86}:Al_{10}:X_4 (where X = Cr, Ru, Ti, Ir and Ta) compressed at different temperatures (21, 800, 1000 or 1300°C) (Douglas et al., 2004 & 2009). Lattice parameters of the (Pt) matrix were determined by using selected area electron diffraction (SAED), from <112> zone patterns. Slightly different results from Hill et al. (2001c) were attributed to calibration differences in the microscope cameras. The Burger's vectors were the same as the dislocations in the other alloys.
Pt-Al-Cr: The precipitates were octahedral in shape, not cuboid. With increasing compression temperature, the density of small precipitates decreased, while the dislocation density in the matrix increased. The structure of the dislocation system remained unchanged

and no other slip systems were activated at higher temperatures (Douglas et al., 2004 & 2009).

Alloy	Room temperature			
	$a_{(Pt)}$ (nm)	a_{Pt3Al} (nm)	δ	Pt_3Al type
Pt_{86}:Al_{10}:Ti_4	3.8921	3.8642	-0.0072	$L1_2$
Pt_{86}:Al_{10}:Cr_4	3.9022	3.8741	-0.0072	$L1_2$
Pt_{86}:Al_{10}:Ru_4	3.9001	3.8530	-0.0121	D0'c
Pt_{86}:Al_{10}:Ta_4	3.8941	3.8682	-0.0067	$L1_2$
Pt_{86}:Al_{10}:Ir_4	3.8983	3.8507	-0.0123	D0'c
	800°C			
	$a_{(Pt)}$ (nm)	a_{Pt3Al} (nm)	δ	Pt_3Al type
Pt_{86}:Al_{10}:Ti_4	3.9246	3.8961	-0.0073	$L1_2$
Pt_{86}:Al_{10}:Cr_4	3.9390	3.9103	-0.0073	$L1_2$
Pt_{86}:Al_{10}:Ru_4	3.9349	3.8967	-0.0098	D0'c
Pt_{86}:Al_{10}:Ta_4	3.9246	3.8961	-0.0073	$L1_2$
Pt_{86}:Al_{10}:Ir_4	3.9246	3.8747	-0.0128	D0'c

Table 2. Comparative lattice misfits of selected Pt_{86}:Al_{10}:X_4 alloys (Hill et al., 2001c).

Pt-Al-Ru: The precipitates were very different from those observed in Pt-Al-Cr, as they were ogdoadically-diced and twinned, originating from the martensitic transformation to tetragonal Pt_3Al. Precipitate interfaces were curved, indicating low surface energies, but at higher compression temperatures the interfaces became increasingly regular. The twin bands also became more developed at higher compression temperatures and dislocations were observed in alternating twin bands. The dislocations occurred in pairs, as expected from the ordering of the precipitates. This is indicative of superlattice dislocations analogous to those in γ' of NBSAs (Douglas et al., 2004 & 2009).

Pt-Al-Ti: After compression at 20°C, the precipitates had well-defined, straight borders, without dislocations, although the matrix contained many dislocation tangles. After compression at 800°C, there were no dislocations within the precipitates, and the dislocation density in the matrix was significantly lower than at ambient temperature. There was also a high density of smaller precipitates in the matrix, adding to the matrix strengthening. Dislocations were only present in the precipitates after compression at 1100°C (Figure 7). The low density of dislocations in the matrix at elevated temperatures indicated significant recovery. At 1300°C, the precipitates were once again dislocation-free, possibly due to recovery (Douglas et al., 2004 & 2009).

Pt-Al-Ir: The precipitates were similar to those in Pt-Al-Ru, as shown in Figure 8. No dislocations were seen in the precipitates, although some dislocations in the matrix extended to the precipitate/matrix interface. After compression at 800°C, the precipitate edges straightened (although their corners were still rounded) and the first dislocations appeared in the precipitates. Isolated dislocations traversed the matrix. At 1100°C, most of the ogdoadically-diced precipitates had become spherical. Twin bands were still visible, with dislocation pairs in some, and multiple twinning occurred. At 1300°C, the precipitates were more irregular. The dislocation density was high, with dislocations threading through each twin band and not only through alternating bands, implying that a second slip system had become operative at the higher temperature (Douglas et al., 2004 & 2009).

Pt-Al-Ta: After compression testing at ambient temperature, the precipitates had straight interfaces, with a low density of dislocations within the precipitates and a high density of interfacial precipitates. Isolated tangles of mixed screw and edge dislocations were observed in the matrix. After compression testing at 800°C, irregular precipitates were observed with no dislocations, while the matrix had a high dislocation density with small cubic precipitates. At 1100°C, matrix dislocations disappeared while there were dislocations observed in the more regular precipitates. At 1300°C, the precipitate interfaces were more defined, and the character of the dislocations inside the precipitates was different. From a dislocation viewpoint, this alloy would have the highest strength (Douglas et al., 2004 & 2009), although further testing of the alloy had earlier been abandoned due to poor oxidation resistance (Süss et al., 2001a).

High temperature TEM work: Since the alloys are being developed for high temperature application, the higher temperature cubic allotrope of Pt_3Al is preferred, as it has both higher strength and ductility. This required a study of the effect of temperature on the precipitates, especially the stability of the different Pt_3Al phases, and a study of the dislocations to understand the deformation mechanisms. This work was critical to the planning of other alloying additions to Pt-Al. The simplest alloys were chosen: binary Pt_{85}:Al_{15} and a ternary Pt-Al-Ir alloy. TEM with bright field imaging was performed at elevated temperatures using *in-situ* heating up to 1100°C (Douglas et al., 2004 & 2009), with a heating rate of 10°C.min^{-1} (5°C.min^{-1} was used for observing more detailed changes).

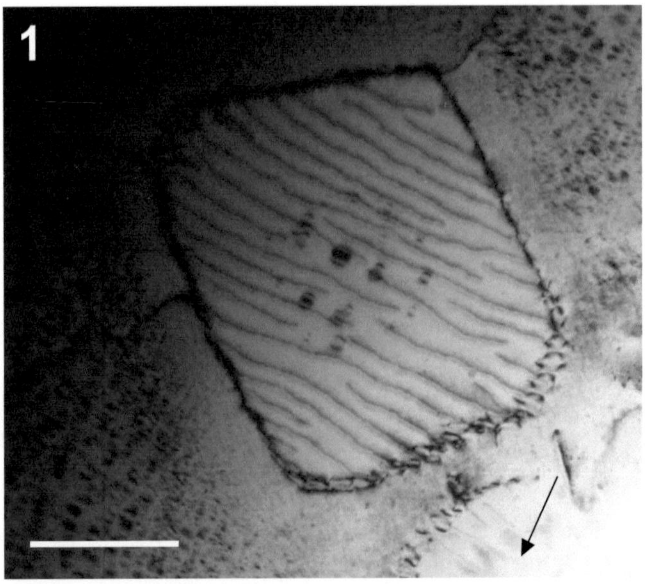

Fig. 7. Bright field TEM micrograph showing the dislocation distribution in a Pt_{86}:Al_{10}:Ti_4 alloy after deformation at 1100°C (scale bar indicates 300nm). (Douglas et al., 2004 & 2009).

At ambient temperature, the precipitates had martensite-like twin bands, with alternating bands having either no dislocations or a high density of dislocations, as would be expected considering the different orientations and dislocation invisibility criteria (Zhang & Zhang, 2001). The dislocation density decreased with increasing temperature and with increasing time. The twin bands disappeared as the crystal structure changed during the phase

Fig. 8. Dark field TEM micrograph of Pt_{86}:Al_{10}:Ir_4 alloy compressed at room temperature, with P denoting an ogdoadically-diced precipitate. The insert shows a SAED pattern with reflection used to obtain image circled. (Douglas et al., 2004 & 2009).

transformation. As a check, the transformation was also observed in diffraction mode (because the diffraction patterns of the two phases are different), and confirmed that the phase diagram of Oya et al. (1987) for the Pt-Al system is more accurate than that of McAlister and Kahan (1986) for these alloys. This was an important result, both because previous attempts to decide which was the most applicable phase diagram - McAlister & Kahan (1986) or Oya et al. (1987) - using SEM, XRD and DTA had been unsuccessful, and because knowledge of the transformation temperatures is vital for application of the alloys.

The strength of a precipitation strengthened alloy depends partially on the small precipitates which reduce dislocation mobility (Kear & Wilsdorf, 1962). To be effective, the precipitates must remain at the high application temperatures to provide a barrier to the dislocation movements. However, in the Pt_{85}:Al_{15} binary alloy, many of the strengthening precipitates dissolved at high temperatures. There was a high density of small precipitates (and a precipitate-free zone around the large precipitates) in the matrix surrounding the large precipitate at 580°C. At 810°C, the precipitates started to dissolve in the matrix, and between 810°C and 870°C the precipitate dissolution was more rapid, with most of the small precipitates disappearing by 870°C. All small precipitates had dissolved by 960°C, implying reduced precipitate strengthening at high temperatures, although the contributions to strengthening of the various precipitate sizes has not yet been established. The large precipitates were much smaller at 1170°C than at 1030°C, and the interfacial dislocation network had completely disappeared. Thus, other alloying additions are required to increase precipitate stability at high temperatures. This could potentially be achieved with the addition of high melting temperature elements.

The dislocations in the Pt-based alloys were more complex than those in NBSAs, mainly because of the lower temperature form of ~Pt_3Al (γ') and the varying misfits (Hill et al., 2001c; Douglas et al., 2007). However, in the Pt-Al-Cr-Ni system, the dislocations were more similar to those in NBSAs (Vorberg et al., 2005), indicating that the Pt_3Al precipitates were the fully cubic $L1_2$ allotrope. In studies on more complex alloys (Rudnik et al., 2008; Shongwe et al., 2008 & 2010), it was found that the relationships between mechanical properties, precipitate size and misfit were difficult to quantify, as different elements acted differently at different compositions, and Ta actually reduced the γ' amount at higher additions, even though it substituted for Al in ~Pt_3Al (Rudnik et al., 2008). The only relationship that was obeyed was that misfit was related to the precipitate morphology, as stated by Qiu (1996).

6. Corrosion studies

6.1 Oxidation of the ternary alloys

Although some alloys with potential for high-temperature applications had already been identified (Hill et al., 2001d), information on high temperature properties was needed. The oxidation resistance was ascertained by a stepped thermogravimetric test, with isothermal holding times of 10 000 s at 900°C, 1100°C, 1300°C and 1400°C. Internal grain boundary oxidation was seen in the Pt-Ti-Ru alloys, and there was extensive internal oxidation in the Pt-Nb-Ru and Pt-Ta-Re alloys. The alloys containing Al exhibited considerably better oxidation behaviour than the other alloys due to the formation of a protective alumina surface layer. Internal oxidation was observed in alloys containing Ti instead of Al, and this was presumed to be the cause of their inferior properties. Thus aluminium is an essential component of the alloys, not only to provide the basis for the precipitation strengthened precipitates, but also to develop the oxidation-resistant alloys (Hill et al., 2001b).

High temperature oxidation behaviour of Pt-Al-Z alloys (where Z = Re, Ta, Ti, Cr, Ir or Ru) was studied by isothermal oxidation tests at 1200°C, 1280°C and 1350°C for at least 100 hours (Hill et al., 2000; Süss et al., 2001a). Results showed increased thickness of the continuous alumina surface layer with time (Figure 10). A dispersion-strengthened alloy, PM2000 (Fe-Cr-Al with a fine dispersion of Y_2O_3 particles in a ferritic matrix) was used as a benchmark. The Pt_{86}:Al_{10}:Ti_4 and Pt_{86}:Al_{10}:Ru_4 alloys showed similar parabolic oxidation behaviour to the PM2000. The Pt_{86}:Al_{10}:Ir_4 and Pt_{86}:Al_{10}:Cr_4 alloys showed parabolic behaviour during the early stages of oxidation, with high initial oxidation rates, and subsequent logarithmic growth of the oxide layer, giving these two alloys the thinnest continuous oxide layers after 800 hours exposure.

Following an initial transient period when discontinuous alumina particles precipitated in a Pt matrix, an external alumina film was formed. This occurred because oxygen diffused through the scale more rapidly than the aluminium diffused in the alloy. Only when a critical volume of oxides was reached, did transition from internal oxidation to external scale formation occur (Wood & Stott, 1987). Once formed, the continuous film then provided protection for the alloy, since no further internal oxidation occurred even during long-term exposure. However, the ternary alloys did not perform as well as PM2000, which formed a perfectly continuous oxide layer from the start. Thus, accelerated formation of the continuous alumina layer was found to be necessary. The critical factor was deduced to be a specified minimum amount of aluminium, implying that the protection should increase by increasing the Al content.

The quaternary alloy Pt_{84}:Al_{11}:Ru_2:Cr_3 (at.%) showed improved oxidation resistance to the original ternary alloys (Suss et al., 2001a, 2001b & 2003). The Al content was increased to accelerate oxide scale formation (Süss et al., 2001b). One hour at 1350°C produced a thin continuous oxide layer, and after 10 hours exposure, the scale was already about three times

thicker than the scale layer on Pt_{86}:Al_{10}:Cr_4. No zone of discontinuous surface oxides or any other internal oxidation was observed, as had been seen in some of the earlier Pt-Al-X (where X = Re, Ta and Ti) ternary alloys (Hill et al., 2000; Süss et al., 2001a).
The increased Al content of the alloys accelerated the formation of a continuous alumina layer. Although good properties were achieved for the short test period, the oxidation rates are possibly too high for long application times. Ideally, a continuous oxide layer should form rapidly on the surface, but further mass increase should then follow logarithmic behaviour.
Wenderoth et al. (2007) showed that more complex alloying could promote internal oxidation. A beneficial effect was that Mo, Re and Ru additions reduced the width of the γ'-depleted layer, with Re have the greatest effect.

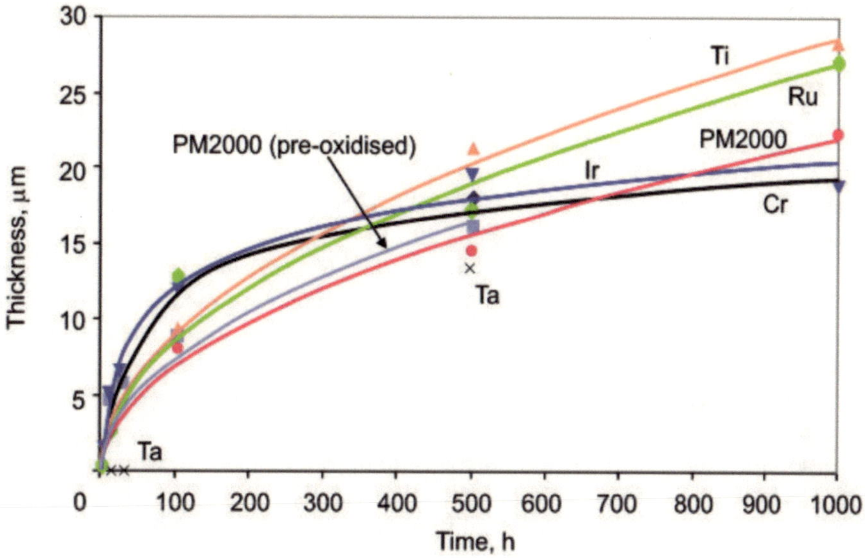

Fig. 10. Isothermal oxidation behaviour of Pt-Al-Z alloys at 1350°C, compared with PM2000 (Süss et al., 2001a).

6.2 Corrosion tests

In addition to high temperature strength, creep resistance and oxidation resistance, high-temperature corrosion resistance is also critical in the selection of high temperature materials (Pint et al., 2006). Increasing application temperatures imply that conditions become increasingly extreme, and materials work closer to their safe operating limits (Elliot, 1989). These conditions cause continued and exacerbated corrosion problems for the high temperature materials. As well as causing catastrophic failures, corrosion has a serious economic impact, as significant time and money is consumed by corrosion-related problems. Improved corrosion control is thus worthwhile, being more economic, safer and extending operational life (Sidhu & Prakash, 2006).
Increasing the high temperature strength of NBSAs was achieved partly by increasing the aluminium and reducing the chromium content, but this increased susceptibility to high temperature corrosion. This was combated by the introduction and development of coatings

(Gurrappa, 2001). Currently, thermal barrier coatings, overlay coatings and diffusion coatings are used in hot section components in the gas turbine as they can withstand temperatures exceeding the substrate melting temperature and protect the substrate from corrosion (Fritscher et al., 1995). However, coatings are expensive and are not completely reliable - catastrophic failure often occurs when coatings are breached. Thus alternatives to coatings are being sought, in particular bulk materials that require no coatings.

Hot corrosion is an aggravated and accelerated form of oxidation that occurs in the presence of sodium sulphate salt (Na_2SO_4) at high temperatures (Eliaz et al., 2002). In turbines, hot corrosion is observed in the low-pressure gas turbine where contaminants can easily accumulate, rather than in the high-pressure turbine. There are two types of hot corrosion: Type I or high-temperature hot corrosion and Type II or low-temperature hot corrosion. Type I hot corrosion usually occurs around 850-950°C, and Type II hot corrosion occurs in the 650°C to 800°C range (Mazur et al., 2005; Sidhu & Prakash, 2006). The main difference between them is their degradation morphology: Type I forms non-porous protective scale with internal sulphidation and chromium depletion, whereas Type II forms a pitted surface with no internal sulphidation. Hot corrosion can increase corrosion loss by up to 100 times (Tsaur et al., 2005), subsequently causing catastrophic failure (Deb et al., 1996). Parameters that affect hot corrosion attack include: material composition, thermomechanical properties, contaminant composition, flux rate, operating temperatures, temperature cycles, gas composition and gas velocity (Potgieter et al., 2010).

A typical test used in the glass industry is the crucible test (Fischer, 1992), where specimens are partially or wholly immersed in fused salt, such as Na_2SO_4 or mixtures of Na_2SO_4–NaCl, and the attack is rapid and severe. This is an advantageous test for materials in turbines, as Na_2SO_4 is an impurity in air or in fuels, where oxyanions in the molten salt act as the source of sulphur during corrosion attack (Eliaz et al., 2002). The presence of sodium chloride (NaCl) and vanadium pentoxide (V_2O_5) salts aggravates corrosion by forming low-melting eutectic compounds that are extremely corrosive to high-temperature materials (Tsaur et al., 2005; Sidhu & Prakash, 2006). Polarisation techniques are often used in hot corrosion testing of superalloys. Another method of evaluating corrosion is the burner-rig test which more closely simulates the operating environment of turbine engines, and so should be the subject of further work for potential turbine materials (Potgieter et al., 2010).

Four Pt-based alloys and a NBSA CMSX-4 ($Ni\text{-}Cr_{5.7}{:}Co_{11}{:}Mo_{0.42}{:}W_{5.2}{:}Ta_{5.6}{:}Al_{5.2}{:}Ti_{0.74}$) were subjected to a crucible test. Thin discs of the samples were placed in zirconium crucibles, covered in Na_2SO_4, held in a furnace at 900°C for 168 hours, and weighed periodically (Potgieter et al., 2010). After testing, the corroded surfaces were examined in an SEM, and the corrosion product was analysed using XRD. Results of the alloys are shown in Table 3.

Very little change in mass occurred for the Pt-based alloys. CMSX-4 showed a mass gain followed by a mass loss, and as the sample was almost entirely degraded, the test was discontinued. X-ray diffraction of the corrosion product on CMSX-4 showed a mixture of compounds based mainly on sodium and nickel. Conversely, XRD showed that the surface of the Pt-based alloys mainly consisted of α-alumina -the protective oxide coating that also forms naturally at high temperature.

Alpha-alumina acts as a diffusion barrier, minimising diffusion of substrate elements to the surface (Rhys-Jones,1989; Müller & Neuschütz, 2003), due to its high thermal stability at high temperatures (Zheng et al., 2006) and low solubility in molten salts (Chen et al., 2003). The presence of $\alpha\text{-}Al_2O_3$ was an indication that the Pt-based substrate was suitably supporting the alumina layer, as the sample surfaces showed no change in appearance to the naked eye. SEM did show pits in the samples. The alloy Pt_{86}:Al_{10}:Cr_4 was the least affected, and Pt_{84}:Al_{11}:Cr_3:Ru_2 and Pt_{86}:Al_{10}:Ru_4 were slightly pitted, with the latter showing more pits. Pt_{79}:Al_{15}:Co_6 showed a mixture of both more pitted and less pitted areas.

Sample (at.%)	Mass (g)						
	Initial	Day 1	Day 2	Day 3	Day 4	Day 5	Day 6
Pt_{86}:Al_{10}:Cr_4	0.611	0.613	0.612	0.611	0.612	0.612	0.612
Pt_{86}:Al_{10}:Ru_4	0.478	0.477	0.478	0.479	0.479	0.479	0.478
Pt_{84}:Al_{11}:Cr_3:Ru_2	0.522	0.524	0.523	0.524	0.522	0.522	0.522
Pt_{79}:Al_{15}:Co_6	0.600	0.600	0.600	0.600	0.601	0.601	0.601
CMSX-4	4.430	4.431	4.596	4.583	N/A	N/A	N/A

Table 3. Mass gain of the samples in the crucible test in Na_2SO_4 at 900°C for 168 hours.

A crucible test was then conducted at 950°C to increase the corrosion kinetics (Potgieter et al., 2010), as hot corrosion is more damaging at this temperature (Elliot, 1990; Yoshiba, 1993). Five Pt-based alloys and two single-crystal CMSX-4 superalloy samples were tested. A thin platinum aluminide coating (Pt_2Al - $Pt_{67}Al_{33}$, in at.%) of ~1.25µm thickness was deposited on one of the CMSX-4 samples, while the other was uncoated. Samples were covered by analytical anhydrous Na_2SO_4 salt, the corrosive electrolyte, inside a furnace with a static dry air environment. The test was performed for 540 hours, with an initial 60 cycles (1 hour of heating to 950°C, 20 minutes of cooling to room temperature), followed by long cycles of 72 hours of heating. Samples were washed free of salt residues, and were then weighed after every cycle.

SEM studies of the samples showed that both coated and uncoated CMSX-4 samples experienced much greater attack than the Pt-based alloys, forming a non-protective porous scale (Potgieter et al., 2010). The Pt_{73}:Al_{15}:Co_{12} and Pt_{79}:Al_{15}:Co_6 samples showed a disintegrated scale layer, indicating that the scale was not protective in this environment. Pt_{86}:Al_{10}:Cr_4, Pt_{86}:Al_{10}:Ru_4 and Pt_{84}:Al_{11}:Cr_3:Ru_2 had similar appearances, with a very thin oxide film on the surface of the alloys which was not visible using optical microscopy. These films were more tenacious and complete, although apparently porous, and provided more protection against hot corrosion than in the Pt-based superalloys with cobalt. Although this scale protected the substrate against high temperature corrosion, the porous nature allowed some internal attack to depths of ~15µm beneath the scale.

Five Pt-based alloys and CMSX-4 underwent potentiodynamic tests with exposure to Na_2SO_4 solutions of 20 or 80 mass % at 60°C. The rationale of the two solutions was that low concentrations favour the attack in turbines, whereas high concentrations are conventionally used to assess hot corrosion. The potentiodynamic curves were obtained at a polarisation scanning rate of 1 $mV.s^{-1}$ by ramping from -600 to 1 000 mV for the NBSA, and from -300 to 1 000 mV for the Pt-based alloys. The Tafel slopes were established, and i_{corr} was estimated where these tangents intersected E_{corr}. Since corrosion resistance is proportional to i_{corr}, these values were used as a measure of the corrosion resistance. The results are listed in Table 4, and show that the i_{corr} values were much lower for the Pt-based alloys than for the NBSAs, implying that the Pt-based alloys have higher corrosion resistance.

Alloy composition (at.%)	Solution			
	20% Na_2SO_4		80% Na_2SO_4	
	E_{corr} (mV)	I_{corr} (*10^{-7} A/cm^2)	E_{corr} (mV)	I_{corr} (*10^{-7}A/cm^2)
Pt_{86}:Al_{10}:Cr_4	51	2.4	20	28
Pt_{86}:Al_{10}:Ru_4	41	9.8	9	10
Pt_{84}:Al_{11}:Cr_3:Ru_2	73	2.6	2	25
Pt_{79}:Al_{15}:Co_6	214	2.7	131	34
CMSX-4	-387	2.4	329	113

Table 4. Polarisation results for Pt-based alloys and CMSX-4 in Na_2SO_4 solutions.

7. Comparison of the PGM-based alloys with other targeted materials for high temperature applications

The Pt-based alloys show excellent oxidation resistance with simple alloying, especially compared to other materials which are also being developed for high temperature applications (Figure 11), although the experiments were not undertaken under identical conditions (Zhao & Westbrook, 2003). Although the strength values are comparable with other competitors, once normalised against density they are less encouraging (Figure 12). However, before dismissing the Pt-based alloys, several other factors have to be taken into account. Firstly, the excellent corrosion resistance (Figure 11) shows potential that these materials will need simpler coatings, or possibly no coatings. This would reduce manufacturing cost and time. It is also a safety benefit, as it could avoid the potential catastrophic failures due to breaches in coating integrity. Additionally, the Pt-based alloys are formable, which could enable new component designs, thereby utilising the high strength, but with thinner sections (Zhao & Westbrook, 2003). Even though these properties have yet to be accurately determined, the microstructure and mechanical properties already measured indicate that fracture toughness, impact resistance and fatigue resistance is likely to be high for the Pt-based alloys.

The other materials in Figures 11 and 12 (Zhao & Westbrook, 2003) have other limitations which must also be considered. The silicon carbide composites have relatively low strengths, even on the density-normalised plot (Figure 12), although the impact resistance and the high temperature stability are favourable. There are still problems in combating evaporation of SiO_2 from the surface, although there has been good progress with environmental-barrier coatings. However, the manufacture of complex shapes needs to be optimised. The range of Nb silicide composites has good oxidation resistance, fatigue resistance, high-temperature strength and impact resistance. The fracture toughness is reasonable and the material is castable. Good coatings have also been developed for the composites. However, more work needs to be done to achieve both the high oxidation resistance and high strength for the same composition.

Molybdenum-silicon-boron composites have exceptional high temperature creep strength and yield strength, with good oxidation resistance above ~1000°C, although at intermediate temperatures, the oxidation resistance, fatigue resistance, impact resistance and fracture toughness are poor. Additionally, these composites are difficult to manufacture. The oxide-oxide composites (represented by Al_2O_3/$GdAlO_3$ in Figures 11 and 12) show excellent high temperature strength and oxidation resistance, with reasonable fracture toughness, but the thermal shock resistance is poor. Thus, taking the favourable properties into account and even considering the high density, overall the Pt-based alloys have potential for application at high temperatures. The price is high, but in view of the high formability, it has been estimated that the price of a Pt-alloy based turbine would be twice that of a current Ni-based superalloy (Glatzel, 2004).

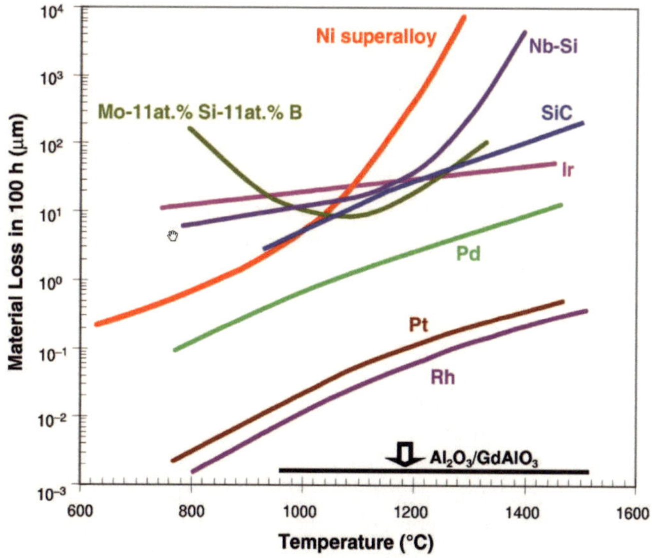

Fig. 11. Oxidation/recession rates of selected high temperature materials (Zhao & Westbrook, 2003). Material loss is usually by formation and spallation of a thermally grown oxide scale, or by evaporation of the metal and oxide. [N. B. The oxidation data were not obtained under identical conditions, so the graph can only be used as a comparison.]

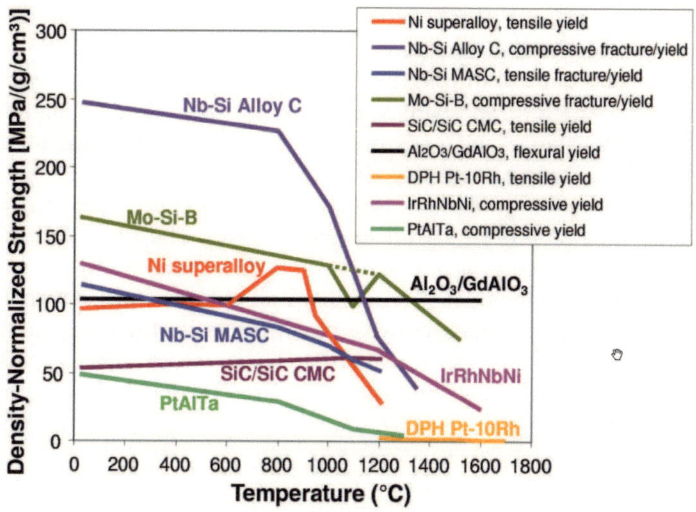

Fig. 12. Comparison of the density-normalised strength graph showing selected potential materials for high temperature applications (Zhao & Westbrook, 2003).

8. Coatings

A second niche is using the Pt-based alloys as coatings on a cheaper substrate, such as a stainless steel. Coatings are employed for a variety of reasons including attractiveness, corrosion resistance and/or wear resistance (although this unlikely for the current alloys). Coatings can be preferable for cheaper components, or where a long component life is not required. Thus, it is more economic to coat a cheaper substrate with a more expensive alloy than to make the whole component from the expensive alloy. Coatings can also be advantageous where low weight is important. The Pt-based alloys are dense, so using them as coatings with the reduced density of a lighter substrate would be beneficial. An added advantage is that the Pt-based alloys form their own enhanced protection in the form of the alumina scale layer.

Pt-modified coatings such as Pt aluminide are preferred for protection against Type I hot corrosion, and have better scale adherence and diffusion protection (Purvis & Warnes, 2001). The addition of other noble metals to Pt-modified coatings is likely to be beneficial, considering the use of such Pt-based alloys for handling molten glass (Fischer et al., 1999b).

Coatings are susceptible to thermal cycling and thermal shock. Variations in thermal expansion coefficients promote thermal stresses and possibly subsequent cracking and spallation of coatings. Even the most resistant alloys are susceptible to hot corrosion attacks (Sidhu et al., 2005 and 2006) and no coatings are totally reliable, which can result in catastrophic failure when they do fail. The design of coatings is becoming progressively difficult for increasingly higher temperature applications, due to the lack of compatibility of thermal expansion coefficients between coatings and high-temperature alloys. This further exacerbates the need for new high temperature materials, such as the Pt-based alloys (Wolff & Hill, 2000). However, the high density of platinum might mean that the most likely application would probably be as coatings on suitable substrate materials.

9. Conclusions

Potential NBSA analogues can be made from Pt-based alloys, having an (Pt) fcc solid solution (γ) matrix and $L1_2$ (γ') ~Pt_3Al precipitates. Although other systems, such as Pt-Ti, were initially tested, the Pt-Al system was identified as the most suitable base in terms of microstructure and mechanical properties. This was attributed to formation of ~Pt_3Al precipitates which provided strengthening in a softer, formable (Pt) matrix. Additionally, promising oxidation resistance and corrosion resistance were ascribed to the noble metal contents of the alloys and to the protective alumina layer.

Other lighter and cheaper materials are being researched, but all of these have disadvantages and most are very difficult to manufacture. The Pt-based alloys show great promise for high temperature applications despite their high cost and density, as they have good mechanical properties, including formability and oxidation resistance. However, more research needs to be undertaken, as there is much potential for other alloying additions, especially those that could increase the melting temperature and decrease the density, e.g. niobium and vanadium. Further mechanical testing, especially long-term creep testing, should also be undertaken.

10. Acknowledgements

In South Africa, the financial assistance of the South African Department of Science and Technology (DST), the Platinum Development Initiative (PDI) and the DST/NRF Centre of

Excellence in Strong Materials is gratefully acknowledged for the support of the work at Mintek and the University of the Witwatersrand.

11. References

ASTM Standards (1993). Designation E8-93: Standard Test Methods for Tension Testing of Metallic Metals, *ASTM Standards*, pp. 130-149.

Bard, J., Selman, G., Day, J., Bourne, A.A., Heywood A.E. and Benedek, R.A. (1994). Dispersion-Strengthened Materials - Platinum-Based Alloys, In: *Mechanical Properties of Metallic Composites*, ed. Shojiro Ochiai, Marcel Dekker Inc., New York, U.S.A., pp. 341-371.

Biggs, T. (2001). An Investigation into Displacive Phase Transformations in Platinum Alloys, Ph.D. Thesis, University of the Witwatersrand, Johannesburg, South Africa.

Biggs, T. Hill, P. Cornish L. A. and Witcomb, M. (2001). Investigation of the Pt-Al-Ru diagram to facilitate alloy development, *J. of Phase Equilibria*, Vol. 22, No. 3, pp. 214-215.

Chauke, H., Minisini, B, Drautz, R, Nguyen-Manh, D Ngoepe, P. and Pettifor D (2010). Theo-retical investigation of the Pt_3Al ground state, *Intermetallics*, Vol. 18, pp. 417-421.

Chen, Z., Wu, N., Singh, J. and Mao, S. (2003). Effect of Al_2O_3 overlay on hot-corrosion behavior of yttria-stabilized zirconia coating in molten sulfate-vanadate salt, *Thin Solid Films*, Vol. 443, No. 1, p. 46.

Chen, Z. B., Huang, Z., Wang, Z. and Zhu, S.J. (2009). Failure behavior of coated nickel-based superalloy under thermomechanical fatigue, *J. Mater. Sci.* , Vol. 44, pp. 6251–6257.

Chown, L.H. and Cornish, L.A. (2003). The Influence of Cobalt Additions to Pt-Al and Pt-Al-Ru-Cr Alloy Systems, in *Africa Materials Research Society Conference*, University of the Witwatersrand, Johannesburg, 8-11 December 2003, pp. 136-137.

Chown, L.H., Cornish, L.A. and Joja, B. (2004). Structure and Properties of Pt-Al-Co Alloys, in *Proc. Microsc. Soc. south. Afr.*, Vol. 34, Pretoria, 30 Nov. - 3 Dec. p. 11.

Cornish, L.A., Fischer B. and Völkl, R. (2003). Development of Platinum Group Metal Based Superalloys for High Temperature Use, *Materials Research Bulletin*, Vol. 28, No. 9, pp. 632-638.

Cornish, L.A., Süss, R., Douglas, A., Chown, L.H. and Glaner, L. (2009a). The Platinum Development Initiative: Platinum-Based Alloys for High Temperature and Special Applications: Part I, *Platinum Metals Review*, Vol. 53, No. 1, pp. 2-10.

Cornish, L.A., Süss, R., Chown, L.H. and Glaner, L. (2009b). The Platinum Development Initiative: Platinum-Based Alloys for High Temperature and Special Applications: Part III, *Platinum Metals Review*, Vol. 53. No. 3, pp. 155-163.

Coupland, D.R., Corti C.W. and Selman, G.L. (1980). The PGM Concept: Enhanced Resistant Superalloys for Industrial and Aerospace Applications, in *Behaviour of High Temperature Alloys in Aggressive Environments*, ed. I. Kirman, Proc. of the Petten International Conf., Petten, The Netherlands, 15-18 Oct. 1979, TMS London.

Davis, J. R. (1997) *Heat-resistant Materials*, ASM International, pp. 1-591.

Deb, D., Rama Krishna Iyer, S. and Radhakrishnan, V.M. (1996). Assessment of high temperature performance of a cast nickel base superalloy in corrosive environment, *Scripta Materiala*, Vol. 35, No. 8, pp. 947-952.

Douglas, A., Neethling, J.H., and Hill, P.J. (2001). Suppression of Martensite Phase in $L1_2$ Pt-Al Alloys, In *5th Multinational Congress on Electron Microscopy (MCEM5)*, Lecce, Italy, 20-25 Sep. 2001.

Douglas, A., Neethling, J.H., Santamarta, R., Schryvers D. and Cornish, L.A. (2003). TEM investigation of the microstructure of Pt_3Al precipitates in a Pt-Al alloy, in *Proc. Microsc. Soc. south. Afr.*, Vol. 32, Cape Town, 3–5 December 2003, p. 14.

Douglas, A. (2004). Microstructure and Deformation of Ternary Platinum Alloys as Superalloy Analogues, Ph.D. Thesis, University of Port Elizabeth.

Douglas, A., Neethling, J.H. and Cornish, L.A. (2004). Dislocation distribution in a Pt-based analogue of Ni-Based superalloys, in: *Proc. Microsc. Soc. south. Afr.*, Vol. 34, Pretoria, p. 12.

Douglas, A., Neethling, J.H, Santamarta, R., Schryvers D. and Cornish, L.A. (2007). Unexpected ordering behaviour of Pt_3Al intermetallic precipitates, *J.of Alloys and Compounds*, Vol. 432, pp. 96–102.

Douglas, A., Hill, P.J., Cornish, L.A. and Süss, R. (2009). The Platinum Development Initiative: Platinum-Based Alloys for High Temperature and Special Applications: Part II, *Platinum Metals Review*, Vol. 53, No. 2, pp. 69-77.

Eagleson, M. (1993). *Concise Encyclopaedia Chemistry*, Eds. H-D Jakubke and H Jeschkeit, De Gruyter, 1993, p. 960.

Eliaz, N., Shemesh, G. and Latanision, R.M. (2002). Hot corrosion in gas turbine components, *Engineering Failure Analysis*, Vol. 9, No. 1, pp. 31-43.

Elliott, P. (1989). "Catch 22" and the UCS Factor - Why Must History Repeat Itself?, *Materials Performance*, Vol. 28, No. 7, pp. 75-78.

Elliot, P. (1990). Practical Guide to High Temperature Alloys, *Materials Performance*, NACE International, Vol. 28, No. 8, pp. 57-66.

Erickson, G.L. (1995). A New Third Generation, Single Crystal, Casting Superalloy, *JOM*, Vol. 47, No. 4, pp. 36-39.

Fairbank, G.B., Humphreys, C.J., Kelly, A. and Jones, C.N. (2000). Ultra-high Temperature Intermetallics for the Third Millennium, *Intermetallics*, Vol. 8, No. 9-11, pp. 1091-1100.

Fairbank, G. B. (2003). The Development of Platinum Alloys for High Temperature Service, Ph.D. Thesis, University of Cambridge, U.K.

Fischer, B. (1992). Reduction of Platinum Corrosion in Molten Glass, *Platinum Metals Review*, Vol. 36, No. 1, pp. 14–25.

Fischer, B., Freund D. and Lupton, D.F. (1997). Stress-Rupture Strength and Creep Behaviour of Platinum Alloys, In *Precious Metals 1997, Proc. IPMI 21st Annual Conference on Precious Metals*, San Francisco, California, USA, 15-18 June 1997, Int. Precious Metals Institute, Pensacola, Florida, USA, pp. 307-322.

Fischer, B., Behrends, A. Freund, D., Lupton D.F. and Merker, J. (1999a). Dispersion Hardened Platinum Materials for Extreme Conditions, *Proc. of the 128th Annual Meeting and Exhibition of TMS*, San Diego, California, USA, 28 Feb.-4 Mar. 1999, pp. 321–331.

Fischer, B., Behrends, A., Freund, D., Lupton D.F. and Merker, J. (1999b). High Temperature Mechanical Properties of the Platinum Group Metals, *Platinum Metals Review*, Vol. 43, No. 1, pp. 18-28.

Fischer, B. (2001). New Platinum Materials for High Temperature Applications, *Advanced Engineering Materials*, Vol. 3, No. 10, pp. 811-820.

Fritscher, K., Leyens, C. and Peters, M. (1995). Structural materials: properties, microstructure and processing, *Mat. Sci. and Eng. A*, Vol. 190, No. 1-2, pp. 253-258.

Glaner, L. and Cornish, L.A. (2003). The Effect of Ni Additions to the Pt-Al-Cr-Ru System, in *Proc. Microsc. Soc. south. Afr.*, Vol. 33, Cape Town, 3-5 December 2003, p. 17.

Glatzel, U. and Feller-Kniepmeier, M. (1989). Calculations of Internal Stresses in the γ/γ' Microstructure of a Nickel-Base Superalloy with High Volume Fraction of γ'-Phase, *Scripta Metallurgica*, Vol. 23, pp. 1839-1844.

Glatzel, U. (2006). Private communication to R. Süss.

Gurrappa, I. (2001). Identification of hot corrosion resistant MCrAlY based bond coatings for gas turbine engine applications, *Surf. Coat. Tech.*, Vol. 139, No. 2-3, pp. 272-283.

Gypen, L. and Deruyttere, A. (1981). The combination of atomic size and elastic modulus misfit interactions in solid solution hardening, *Scripta Metal*, Vol. 15, No. 8, pp. 815-820.

Hammer, G. and Kaufmann, D. (1982). Degussa AG, *German Patent Appl.* 3,030,751 A1.

Heraeus (2011). Dispersion Hardened Platinum Materials, Date of access: 22 February 2011. Available from Heraeus website:
http://heraeusptcomponents.com/en/ downloads_1/technical_informations/publications_1/

Heywood, A.E. (1988). Johnson Matthey PLC, *German Patent* 3,102,342 C2.

Hill, P.J., Cornish, L.A. and Witcomb, M.J. (2000). The Oxidation Behaviour of Pt-Al-X Alloys at Temperatures between 1473 and 1623 K, In: *Proc. High Temperature Corrosion and Protection 2000*, pp. 185-190, 17-22 Sep. 2000, Sappora, Japan, ISBN1-900814-35-8.

Hill, P. J., Cornish L. A. and Fairbank G. B. (2001a). New Developments in High-Temperature Platinum Alloys, *JOM*, Vol. 53, No. 10, pp. 19-20.

Hill, P.J., Biggs, T., Ellis, P., Hohls, J., Taylor S. and Wolff, I.M. (2001b). An assessment of ternary precipitation-strengthened Pt alloys for ultra-high temperature applications, *Mat. Sci. and Eng. A*, Vol. 301, No. 2, 167-179.

Hill, P.J., Yamabe-Mitarai, Y., Murakami, Y., Cornish, L.A., Witcomb, M.J., Wolff, I.M. and Harada, H. (2001c). The Precipitate Morphology and Lattice Mismatch of Ternary (Pt)/Pt_3Al Alloys, In *3rd Int. Symp. on Structural Intermetallics*, TMS, Jackson Hole, Wyoming, U.S.A., Sept, 28 Apr. - 1 May, 2002, pp. 527-533.

Hill, P.J, Yamabe-Mitarai, Y. and Wolff, I (2001d). High-temperature compression strengths of precipitation-strengthened ternary Pt-Al-X alloys, *Scripta Mat.*, Vol. 44, No. 1, pp.43-48.

Hill, P.J., Cornish, L.A., Ellis, P. and Witcomb, M.J. (2001e). The Effect of Ti and Cr Additions on Phase Equilibria and Properties of (Pt)/Pt_3Al Alloys, *J. of Alloys and Compounds*, Vol. 22, pp. 166-175.

Hill, P.J., Adams, N., Biggs, T., Ellis, P., Hohls, J., Taylor, S.S., and Wolff, I.M. (2002). Platinum alloys based on Pt–Pt_3Al for ultra-high temperature use, *Mat. Sci. and Eng. A.*, Vol. 329-331, pp. 295-304.

Huang, C., Yamabe-Miterai, Y. and Harada, H. (2004). The stabilization of Pt_3Al phase with $L1_2$ structure in Pt-Al-Ir-Nb and Pt-Al-Nb alloys, *J. of Alloys and Compounds*, Vol. 366, pp. 217-221.

Hüller, M., Wenderoth, M., Vorberg, S., Fischer, B., Glatzel, U. and Völkl, R. (2005). Optimization of Composition and Heat Treatment of Age-Hardened Pt-Al-Cr-Ni Alloys, *Metall. and Mater. Trans. A*, Vol. 36, No. 13, pp. 681–689.

Kear B.H. and Wilsdorf, H.G.F. (1962). Trans TMS-AIME, Vol. 224, p. 382. Cited in: Vattré, A., Devincre, B. and A. Roos (2009) Dislocation dynamics simulations of precipitation hardening in Ni-based superalloys with high γ' volume fraction, Intermetallics, Vol. 17, Issue 12, pp. 988-994.

Keraan, T. and Lang, C.I. (2003a). High Temperature Investigation into Platinum-base Super-alloys, in *Proc. Microsc. Soc. south. Afr.*, Vol. 32, Cape Town, 3-5 Dec. 2003, p. 14.

Keraan, T. and Lang, C.I. (2003b). High Temperature Mechanical Properties and Behaviour of Platinum-base Superalloys for Ultra-high Temperature use, In: *Africa Materials Research Society Conference*, University of the Witwatersrand, Johannesburg, 8-11 December 2003, pp. 154-155.

Keraan, T. (2004). High Temperature Mechanical Properties and behaviour of Platinum-base Alloys, M.Sc. Dissertation, University of Cape Town.

Kohno, Y., Kohyama, A., Hamilton, M.L., Hirosi, T., Katoh, Y. and Garner, F.A. (2000). Specimen size effects on the tensile properties of JPCA and JFMS, *J. of Nuclear Materials* , Vol. 283-287, pp. 1014-1017.

Kohyama, A., Asano, K. and Igata, N. (1987). Influence of radiation on materials properties: *15th Int. Symposium (Part II), ASTM-STP 956*, p. 111; cited in: Cornish et al., 2009b.

Johnson-Matthey (2011). Accessed 3 Mar. 2011. Available from:www.platinum.matthey.com

Kamm, J.L. and Milligan, W.W. (1994). Phase stability in $(Ni,Pt)_3Al$ Alloys, *Scripta Metallurgica et Materialia*, Vol. 33, No. 11, pp. 1462-1464.

Lucas, G.E., Odette, G.R., Sokolov, M., Spätig, P., Yamamoto, T. and Jung , P. (2002). Recent progress in small specimen test technology, *J. of Nuclear Materials*, Vol. 307-311, pp. 1600-1608.

Lupton, D.F. (1990). Noble and Refractory Metals for High Temperature Space Applications, *Advanced Materials*, pp. 29–30.

Lupton, D. F., Merker, J., Fischer, B. and Völkl, R. (2000). Ductile High-Strength Platinum Materials for Glass Making, *Proc. on CD-ROM, 24th Int. Precious Metals Conference 2000*, Williamsburg, USA, 11–14 June, cited by Cornish et al., 2009a.

MacLachlan D.W. and Knowles, D.M. (2001). Modelling and prediction of the stress rupture behaviour of single crystal superalloys, *Mat.s Sci. and Eng.* A, Vol. 302, pp. 275–285.

Massalski, T.B., Okomoto, H., Subramanian, P.R. and Kacprzak, L. (Eds.) (1990). *Binary Alloy Phase Diagrams, 2nd Ed.*, ASM International, Ohio, USA.

Matweb (2011). Date of access: 6 March 2011. Available from: www.matweb.com/

Mazur, Z., Luna-Ramírez, A., Juárez-Islas, J. A. and Campos-Amezcua, A. (2005). Failure Analysis of a Gas Turbine Blade Made of Inconel 738LC Alloy, *Eng. Fail. Anal.*, Vol. 12, No. 3, pp. 474-486.

McAlister, A.J. and Kahan, D. J. (1986). The Al-Pt (Aluminium-Platinum) System, *Bulletin of Alloy Phase Diagrams*, Vol. 7, pp. 45-51.

Merker, J., Fischer, B., Völkl, R. and Lupton, D.F. (2003). Investigations of New Oxide Dispersion Hardened Platinum Materials in Laboratory Tests and Industrial Applications, *Materials Science Forum*, Vol. 426-432, pp. 1979–1984.

Müller, J. and Neuschütz, D. (2003). Efficiency of α-alumina as diffusion barrier between bond coat and bulk material of gas turbine blades, *Vacuum*, Vol. 71, No. 1-2, pp. 247-251.

Müller, L., Glatzel U. and Feller-Kniepmeier, M. (1993). Calculation of the Internal Stresses and Strains in the Microstructure of a Single-Crystal Nickel-Base Superalloy During Creep, *Acta Metallurgica et Materialia*, Vol. 41, No. 12, pp. 3401-3411.

NIMS (2007). Research and Development of Superalloys for Aeroengine Applications, Date of access: 5 May 2010. Available from: National Institute of Materials Science website: http://sakimori.nims.go.jp/topics/hightemp_e.pdf

Ochiai, S. (Ed.) (1994). Dispersion-Strengthened Materials- Platinum-Based Alloys, In: *Mechanical Properties of Metallic Composites*, Marcel Dekker Inc., New York, pp. 341-371. Cited by Cornish et al., 2009b.

Oya, Y., Mishima, U. and Suzuki, T. (1987). $L1_2 \leftrightarrow D0c$ Martensitic Transformation in Pt_3Al and Pt_3Ga, *Zeitschrift für Metallkunde*, Vol. 78, No. 7, pp. 485-490.

Panfilov, P., Pilugin, V.P. and Antonova, O.V. (2008). On Specific Feature of Plastic Deformation in Ir, in: *Creep 2008: 11th Int. Conf. on Creep and Fracture of Engineering Materials and Structures, Book of Abstracts*, Bayreuth, Germany, 4-9 May, p. CP-131.

Panayotou, N.F. (1982). The use of microhardness to determine the strengthening and microstructural alterations of 14 MeV neutron irradiated metals, *J. of Nuclear Materials*, Vol. 108, pp. 456-462.

Pather, R., Mitten, W.A., Holdway, P., Ubhi, H.S. and Wisbey, W.A. (2003). Effect of High Temperature Environment on High Strength Titanium Aluminide Alloy, *Proc. Advanced Materials and Processes for Gas Turbines*, TMS, pp. 309-316, 22-26 Sep. 2002, Copper Mountain, Colorado, USA. ISBN 0-97339-556-5.

Pint, B. A., DiStefano J. R. and Wright, I. G. (2006). Oxidation resistance: One barrier to moving beyond Ni-base superalloys, *Mat. Sci. and Eng. A*, Vol. 415, No. 1-2, pp. 255-263.

Plansee (1998). Dispersion-Strengthened High-Temperature Materials, Plansee brochure, Lechbruck.

Potgieter, J.H., van Bennekom, A. and Ellis, P. (1995). Investigation of the Active Dissolution Behaviour of a 22% Chromium Duplex Stainless Steel with Small Ruthenium Additions in Sulphuric Acid, *ISIJ Int.*, Vol. 35, pp. 197-202.

Potgieter, J.H., Maledi, N.B., Sephton, M. and Cornish, L.A. (2010). The Platinum Development Initiative: Platinum-Based Alloys for High Temperature and Special Applications: Part IV - Corrosion, *Platinum Metals Review*, Vol. 54, No. 2, pp. 112-119.

Purvis, A.L. and Warnes, B. M. (2001). The effects of platinum concentration on oxidation resistance of superalloys coated with single-phase platinum aluminide, *Surf. and Coat. Tech.*, Vol. 146, pp. 1-6.

Qiu, Y.Y. (1996). The effect of the lattice strains on the directional coarsening of γ' precipitates in Ni-based alloys, *J. of Alloys and Compounds*, Vol. 232, No. 10, pp. 254-263.

Roehrig, F.K. (1981). Owens-Corning Fiberglass Corp., *World Patent Appl.* 81/00,977.

Rhys-Jones, T.N. (1989). Coatings for blade and vane applications in gas turbines, UK Corrosion '87-High Temperature Materials, *Corrosion Science*, vol. 29, no. 6, pp. 623-646.

Rudnik, Y., Völkl, R., Vorberg, S. and Glatzel, U. (2008). The effects of Ta additions on the phase compositions and high temperature properties of Pt base alloys, *Mat.Sci. and Eng. A*, Vol. 479, pp. 306-312.

Saltykov, P., Fabrichnaya, O., Golczewskí, J. and Aldinger, F. (2004). Thermodynamic Modeling of Oxidation of Al-Cr-Ni Alloys, *J. of Alloys and Compounds*, Vol. 381, No. 1-2, pp. 99-113.

Santamarta, R. R. Neethling, R., Schryvers D. and Douglas, A. (2003). HRTEM investigation of the low temperature phase of Pt_3Al precipitates in (Pt), In *Proc. Microsc. Soc. south. Afr.*, Vol. 32, Cape Town, 3-5 Dec. 2003, p. 15.

Schubert, K. (1964). *Kristallstrukturen Zweikomponentiger Phasen*, Springer Verlag OHG, 1st edn., Berlin, Germany, p. 30.

Selman, G.L. and Darling, A.S. (1973). Johnson Matthey PLC, *British Patent* 1,340,076.

Shing T.L., Luyckx, S., Northrop, I.T. and Wolff, I. (2001). The Effect of Ruthenium additions on the hardness, toughness and grain size of WC-Co, *Int. J. of Refractory Metals and Hard Materials*, Vol. 19, pp. 41-44.

Shongwe, M.B., Cornish, L.A. and Süss, R. (2009). Effect of Misfit on the Microstructure of Pt Based Superalloys, *Proc. Microsc. Soc. south. Afr.*, Vol. 39, p. 59, Durban, South Africa, 8–11 Dec. 2009, ISSN 0250-0418.

Shongwe, M.B., Odera, B., Samal, S., Ukpong, A.M., Watson, A., Süss, R., Chown, L.H., Rading, G.O. and Cornish, L.A. (2010). Assessment of Microstructures in the Development of Pt-based Superalloys, *Light Metals Conference, SAIMM*, Paper 184-202 Shongwe on CD, Muldersdrift, Johannesburg, 27-29 Oct. 2010.

Sidhu, B. S. and Prakash, S. (2006). Studies on the behaviour of stellite-6 as plasma sprayed and laser remelted coatings in molten salt environment at 900 °C under cyclic conditions, *Materials Process Technology*, Vol. 172, No. 1, pp. 52-63.

Sidhu, T.S., Agrawal, R.D. and Prakash, S. (2005). Hot Corrosion of Some Superalloys and Role of High-Velocity Oxy-Fuel Spray Coatings - A Review, *Surf. Coat.s Tech.*, Vol. 198, pp. 441-446.

Sidhu, T. S., Prakash S. and Agrawal, R.D. (2006). Hot Corrosion and Performance of Nickel Based Coatings, *Current Science*, Vol. 90, No. 1, pp. 41-47.

Sims, C.T., Stoloff, N.S. and Hagel W.C. (1987). *Superalloys II*: High Temperature Materials For Aerospace and Industrial Power, Wiley-Interscience, New York, USA, 1987.

Süss, R., Hill, P.J., Ellis, P. and Wolff I.M. (2001a). The Oxidation Resistance of Pt-Base γ/γ' Analogues to Ni-Base Superalloys, In: *Proc. 7th European Conf. on Advanced Materials and Processes*, Rimini, Italy, 10-14 June, 2001. Paper No. 287, CD-ROM, ISBN 8885298397.

Süss, R., Hill, P.J., Ellis, P. and Cornish, L.A. (2001b). The oxidation resistance of Pt-Base superalloy Pt_{80}:Al_{14}:Cr_3:Ru_3 compared to that of Pt_{86}:Al_{10}:Cr_4, *Proc. Microsc. Soc. south. Afr.*, 2001, Johannesburg, Vol. 31, p. 21.

Süss, R., Freund, D., Völkl, R., Fischer, B., Hill, P.J., Ellis, P., and Wolff, I.M. (2002). The creep properties of Pt-base γ/γ' analogues to Ni-base superalloys, *Mat. Sci. and Eng.* A, Vol. 338, pp. 133-141.

Süss, R., Cornish, L.A., Hill, P.J., Hohls, J. and Compton, D.N (2003). Properties of a New Series of Superalloys Based on Pt_{80}:Al_{14}:Cr_3:Ru_3', in *Advanced Materials and Processes for Gas Turbines*, Ed. G. Fuchs, A. James, T. Gabb, M. McLean and H. Harada, TMS, 22-26 Sep. 2002, Copper Mountain, Colorado, USA, pp. 301-307.

Süss, R. and Cornish, L.A. (2004). Tensile Test Properties of Pt-based Superalloys, *Beyond Ni-based Superalloys*, TMS 2004 133rd Annual Meeting and Exhibition, p. 269, Charlotte, North Carolina, USA, 14-16 Mar. 2004.

Takeuchi, S. and Kuramoto, E. (1973). *Acta Met.*, Vol. 21, pg. 415, cited in Kamm, J.L. and Milligan, W.W. (1994).

Tsaur, C-C., Rock, J. C., Wang C-J. and Su, Y-H. (2005). The hot corrosion of 310 stainless steel with pre-coated $NaCl/Na_2SO_4$ mixtures at 750 °C, *Mat. Chem. and Phys.*, Vol. 89, No. 2-3, pp. 445-453.

van der Lingen, E. and Sandenbergh, R.F. (2001). The cathodic modification behaviour of Ru additions to titanium in hydrochloric acid, *J. of Corrosion Science*, Vol. 43, pp. 577-590.

Vattré, A., Devincre, B. and Roos, A. (2009). Dislocation dynamics simulations of precipitation hardening in Ni-based superalloys with high γ' volume fraction, Intermetallics, Vol. 7, No. 12, pp. 988-994.

Völkl, R. Glatzel U. and Feller-Kniepmeier, M. M. (1998). Measurement of the Lattice Misfit in the Single Crystal Nickebase Superalloys CMSX-4, SRR 99 and SC 16 by Convergent Beam Electron Diffraction, *Acta Materialia*, Vol. 46, No. 12, pp. 4395-4404.

Völkl, R., Freund, D., Fischer, B. and Gohlke, D. (1999). Comparison of the creep and fracture behaviour of non-hardened and oxide dispersion hardened platinum base alloys at temperatures between 1200°C and 1700°C, *Proc. 8th Int. Conf. on Creep and Fracture of Engineering Materials and Structures*, Vol. 171-174, pp. 77–84.

Völkl, R., Freund, D., Behrends, A., Fischer, B., Merker, J. and Lupton, D. (2000). Platinum Base Alloys for High Temperature Space Applications, In: *Euromat 99 Series: Materials for Transport*, ed. P.J. Winkler, Wiley-VCH Verlag GmbH, Weinheim.

Völkl, R. and Fischer, B. (2004). Mechanical Testing of Ultra-High Temperature Alloys, *Experimental Mechanics*, Vol. 44, No. 2, pp. 121–127.

Völkl, R. Yamabe-Mitarai, Y. Huang C. and Harada, H. (2005). Stabilizing the $L1_2$ structure of Pt_3Al(r) in the Pt-Al-Sc system, *Met. Mat. Trans. A*, Vol. 36, No. 11, pp. 2881-2892.

Völkl, R., Wenderoth, M., Preussner, J., Vorberg, S., Fischer, B., Yamabe-Miterai, Y., Harada, H. and Glatzel, U. (2009). Development of a precipitation-strengthened Pt-base alloy, *Mat. Sci. and Eng. A*, Vol. 510-511, pp. 328-331.

Vorberg, S., Wenderoth, M., Fischer, B., Glatzel U. and Völkl, R. (2004). Pt-Al-Cr-Ni superalloys: heat treatment and microstructure, *JOM*, Vol. 56, No. 9, pp. 40-43.

Vorberg, S., Wenderoth, M., Fischer, B., Glatzel U. and Völkl, R. (2005). A TEM investigation of the γ/γ' phase boundary in Pt-based superalloys, *JOM*, Vol. 57, No. 3, pp. 49-51.

Wenderoth, M., Cornish, L.A., Süss, R., Vorberg, S., Fischer, B., Glatzel, U. and Völkl, R. (2005). On the Development and Investigation of Quaternary Pt-Based Superalloys with Ni Additions, *Met. Mat. Trans. A*, Vol. 36, pp. 567–575.

Wenderoth, M., Völkl, R., Yokokawa, T., Yamabe-Mitarai Y. and Harada, H. (2006). High temperature strength of Pt-base superalloys with different γ' volume fractions, *Scripta Materialia*, Vol. 54, No. 2, pp. 275–279.

Wenderoth, M., Völkl, R., Vorberg, S., Yamabe-Miterai, Y., Harada, H. and Glatzel, U. (2007). Micro-structure, oxidation resistance and high temperature strength of γ' hardened Pt base alloys, *Intermetallics*, Vol. 15, pp. 539-549.

Westbrook, J.H. (1958). Precipitation of Ni_3Al from nickel solid solution as ogdoadically diced cubes, *Zeitschrift für Kristallographie,* Vol. 110, pp. 21-29.

Whalen, M.V. (1988). Space Station Resistojets, *Plat. Met.Rev.*, Vol. 32, No. 1, pp. 2–10.

Wolff, I.M. and Hill, P.J. (2000). Platinum metals-based intermetallics for high-temperature service, *Platinum Metals Review*, Vol. 44, No. 4, pp. 158-166.

Wood, G. and Stott, F. (1987). Oxidation of alloys, *Mat. Sci. and Tech.*, Vol. 3, No. 7, pp. 519-530.

Yamabe, Y., Koizumi, Y., Murakami, H., Ro, Y., Maruko, T. and Harada, H. (1996). Development of Ir-base Refractory Superalloys, *Scripta Met.,* Vol. 35, No. 2, pp. 211-215.

Yamabe-Mitarai, Y., Koizumi, Y., Murakami, H., Ro, Y., Maruko, T. and Harada, H. (1997). Rh-base Refractory Superalloys for Ultra-high Temperature Use, *Scripta Met.,* Vol. 36, No. 4, pp. 393-398.

Yamabe-Mitarai, Y., Ro, Y., Harada, H. and Maruko, T. (1998). Ir-base Refractory Superalloys for Ultra-High Temperature Use, *Met. Trans. A*, Vol. 29, No. 2, pp. 537-549.

Yamabe-Mitarai, Y., Ro, Y., Maruko, T. and Harada, H. (1999). Microstructure dependence of strength of Ir-base refractory superalloys, *Intermetallics*, Vol. 7, No. 1, pp. 49-58.

Yamabe-Mitarai, Y., Ro, Y., Maruko, T. and Harada, H. (1998). Precipitation hardening of Ir-Nb and Ir-Zr alloys, *Scripta Materiala*, Vol. 40, No. 1, pp. 109-115.

Yamabe-Mitarai, Y. and Aoki, H. (2003). An assessment of Pt-Ir-Al alloys for high-temperature materials, *J.of Alloys and Compounds*, Vol. 359, pp. 143-152.

Yoshiba, M. (1993). Effect of hot corrosion on the mechanical performances of superalloys and coatings systems, *Corrosion Science*, Vol. 35, No. 5–8, pp. 1115–1124.

Yu, X.H., Yamabe-Mitarai, Y., Ro, Y. and Harada, H. (2000). Design of quaternary Ir-Nb-Ni-Al refractory superalloys, *Met. and Mat.Trans. A*, Vol. 31A, No. 1, pp. 173-178A.

Zhang, X-F. and Zhang, Z. (Eds.) (2001). *Progress In Transmission Electron Microscopy: Concepts and Techniques*, Vol. 1, Springer-Verlag, Germany, pg. 263.

Zhao, J-C., Jackson, M., Peluso, L. and Brewer, L. N. (2002). A Diffusion Multiple Approach for the Accelerated Design of Structural Materials, *MRS Bulletin*, Vol. 27, pp. 324–329.

Zhao, J-C. and Westbrook, J.H. (2003). Ultra high temperature materials for jet engines, *MRS Bulletin*, Vol. 28, No. 9, pp. 622-627.

Zheng, D., Zhu, S. and Wang, F. (2006). Oxidation and hot corrosion behavior of a novel enamel-Al_2O_3 composite coating on K38G superalloy, *Surf. Coat. Tech.*, Vol. 200, No. 20-21, pp. 5931-5936.

16

Unidirectionally Solidified Eutectic Ceramic Composites for Ultra-High Efficiency Gas Turbine Systems

Yoshiharu Waku
Shimane University
Japan

1. Introduction

To help solve environmental problems, it is vital to develop a material for use in energy conservation technologies and for curbing the emission of pollutants such as CO_2. In the advanced power generation field, studies all over the world are seeking to develop ultra-high temperature structural materials that will improve thermal efficiency in aircraft engines and high-efficiency gas turbines. For example, to improve the thermal efficiency of gas turbines, operating temperatures must be increased and to achieve this, the development of ultra-high temperature resistant structural materials is necessary. Currently Ni-base superalloys are the main thrust in this field, but these have melting points of less than 1673 K, and their strength deteriorates sharply near 1200 K. In order to overcome the high temperature limitations of metals, the development of turbine technology using advanced ceramic matrix composites, has been vigorously pursued in recent years.
It has been reported that a unidirectionally solidified Al_2O_3/YAG eutectic composite has superior flexural strength, thermal stability and creep resistance at high temperature (Mah & Parthasarathy, 1990; Parthasarathy et al., 1990; Parthasarathy et al.,1993) and is a candidate for high-temperature structural materials. However, since the eutectic composite consists of many eutectic colonies, a fairly strong influence of colony boundaries may be predicted (Stubican et al., 1096). The recently developed MGCs are a new class of ceramic matrix composites made by melting and unidirectional solidification of raw material oxides using a eutectic reaction to precisely control the crystal growth. The MGCs have excellent high temperature characteristics due to their unique microstructure without eutectic colony boundaries (Waku et al., 1996; Waku et al., 1997; Waku et al., 2001). In this paper, the fundamental concept of the MGC is explained and compared with sintered polycrystalline monolithic ceramics and single crystal ceramics. High temperature strength, tensile creep properties, compressive deformation, and oxidation resistance and thermal stability as typical high temperature characteristics of the MGCs are introduced. MGC ultra-high efficiency gas turbine systems and the manufacturing process for near-net-shaped MGC component castings are also discussed.

2. Temperature dependence of fracture strength vs microstructure

In general, strength at high temperature depends primarily on the diffusion process, because diffusion-controlled processes dominate at high temperatures and grain boundaries play a major role in the deformation process (Courtright et al., 1992). The schematic of Fig. 1 illustrates the temperature dependence of fast-fracture strength on microstructure of sintered polycrystalline monolithic ceramics, single crystal ceramics and MGCs. In the case of sintered polycrystalline monolithic ceramics, plastic deformation occurs by micrograin superplasticity due to grain-boundary sliding or rotation at high temperatures, so the fast-fracture strength decreases noticeably with an increase in temperature. Oxide ceramics especially show a sudden drop in fast-fracture strength above 1300 K. In the case of single crystals, diffusion becomes active at high temperatures and there is no barrier (i.e. interface) to disturb the dislocation motion, so the high temperature fast-fracture strength decreases gradually with the increase in temperature.

In contrast to these materials, MGCs have the unique microstructure of a three-dimensionally continuous network of single crystal phases without eutectic colony boundaries. Therefore, MGCs display no grain-boundary sliding or rotation. Additionally, the existence of interfaces in MGCs prevents dislocation motion. Consequently, the MGCs because of their unique microstructure have excellent high temperature fast-fracture strength, superior to the high temperature strength of polycrystalline or single crystal ceramic materials.

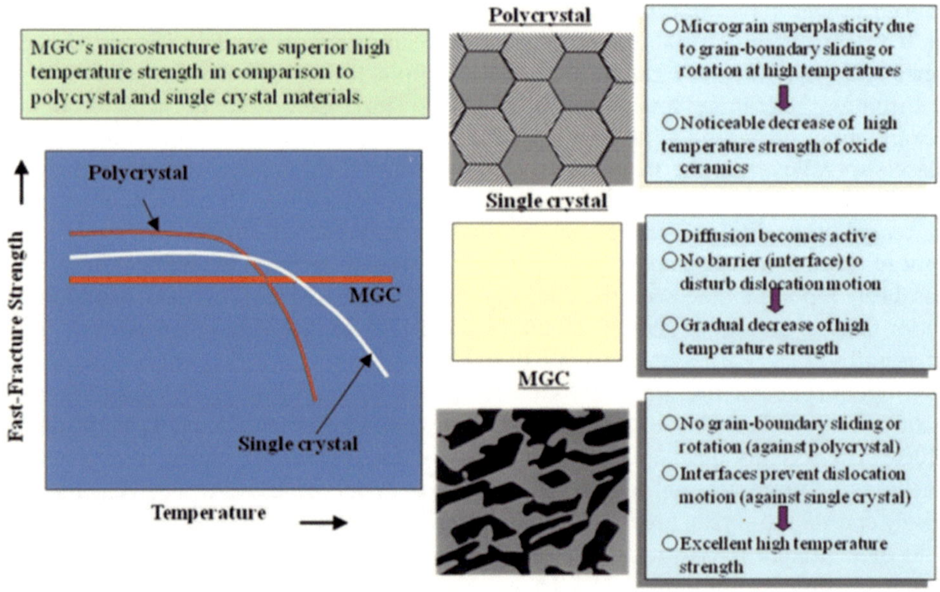

Fig. 1. Schematic illustration of temperature dependence of strength of polycrystal, single crystal and MGC materials.

3. MGC fabrication process

The MGC fabrication process involves unidirectional solidification of eutectic oxide composites as shown in Fig. 2. The preliminary melt was cast into a molybdenum (Mo) crucible (50 mm in outside diameter by 200 mm in length by 5 mm in thickness) placed in a vacuum chamber, and a graphite susceptor was heated by high-frequency induction heating. This heated the Mo crucible and facilitated the melting. After sustaining the melt of

2223 K (about 100 K above melting point) for 30 minutes, the Mo crucible was lowered at 5 mm an hour, completing the unidirectional solidification experiment.
The MGC fabrication process is a unidirectional solidification process that utilizes a eutectic reaction during melting. The MGC forming process is similar to that for single-crystal Ni-based cast superalloys. It is actually performed by lowering of a molybdenum crucible at constant speed using advanced-alloy crystalline-structure-controlling equipment. A detailed experimental procedure has been described in other manuscripts (Waku et al., 1997; Waku et al., 1998).

4. Microstructural characteristics of MGC

4.1 Microtructure

Fig. 3 shows SEM images of the microstructure of a cross-section perpendicular to the solidification direction of Al_2O_3/YAG binary, Al_2O_3/GAP binary and Al_2O_3/YAG/ZrO_2 ternary MGCs. For the Al_2O_3/YAG binary MGC (Fig. 3 (a)), the light area in the SEM microstructure is the YAG phase with a garnet structure, and the dark area is the Al_2O_3 phase with a hexagonal structure (identified by EPMA analysis), the dimensions of the microstructures are 20~30 μm (this dimension is defined as the typical length to the short axis of each domain seen in the cross-section perpendicular to the solidification direction) (Waku et al., 1998).

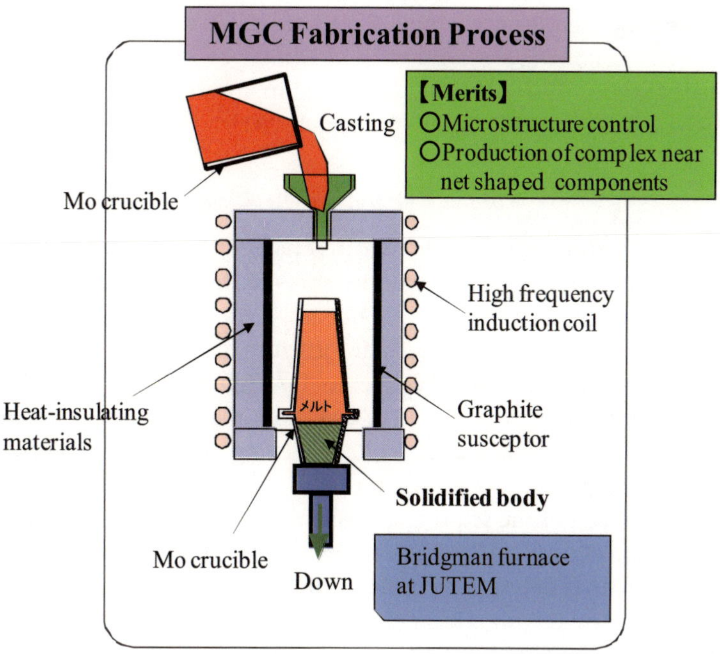

Fig. 2. MGC fabrication process.

For the Al_2O_3/GAP binary system (Fig. 3 (b)), the MGC consists of Al_2O_3 phases with the hexagonal structure and GAP phases with a perovskite structure. In the SEM microstructure, the light area is the GAP phase, the dark area is the Al_2O_3 phase from EPMA analysis. The dimensions of the microstructure of Al_2O_3/GAP binary MGCs are around 3-5

μm smaller than that of around 20-30 μm for the Al_2O_3/YAG binary MGC (Waku et al., 1997).
In the case of the Al_2O_3/YAG/ZrO_2 ternary system (Fig. 3 (c)), the microstructure consists of Al_2O_3, YAG phases and fully stabilized cubic-ZrO_2 phases ($Zr_{0.72}Y_{0.28}O_{1.86}$) with Y_2O_3 (determined from X-ray diffraction patterns). The gray area in the SEM micrograph is the YAG phase, the dark area is the Al_2O_3 phase and light area is the cubic-ZrO_2 (c-ZrO_2) phase (identified by X-ray diffraction and EPMA analysis). The dimensions of YAG phases in the present Al_2O_3/YAG/ZrO_2 ternary MGC are around 2-3 μm smaller in 1/10 than around 20-30μm of the Al_2O_3/YAG binary MGC (Fig. 3 (c)). Many of the c-ZrO_2 phases in the ternary MGC exist at interfaces between Al_2O_3 and YAG phases or in Al_2O_3 phases and seldom exist in YAG phases. Homogeneous microstructures with no pores or colonies are observed in all the binary and ternary MGC (Waku et al., 2001; Waku et al., 2002) .

4.2 Three-dimensional observation of MGC structure

Fig. 4 shows the three-dimensional image of the unidirectionally solidified Al_2O_3/YAG eutectic structure constructed from the reconstructed images (Yasuda et al., 2003). The growth morphology continuously changed, keeping the characteristic feature in the entangled structure. Entangling in the growth direction frequently occurred and the entangled domain was of the same order as the lamellar spacing. The three-dimensional image clearly indicates that the eutectic growth in the Al_2O_3/YAG system was far from the steady state. The specimen used for the CT consisted of α-Al_2O_3 and YAG single crystals, since the X-ray diffractions are consistently identified on the basis of the single crystal crystallographic domain. The following relationship for the crystallographic orientation between the two phases was obtained (Yasuda et al., 2005).

$$(0001)_{Al2O3}//(1\bar{1}2)_{YAG},[\bar{1}100]_{Al2O3}//[1\bar{1}\bar{1}]_{YAG}$$

This relationship coincides with an earlier work (Frazer et al., 2001). The lamellas tended to align in a certain direction. However, normal vectors of the interface between the Al_2O_3 and the YAG, which were evaluated from the CT images, were scattered over a wide range.
The entangled part in the Al_2O_3/YAG eutectic structure is shown in Fig. 5. A hole observed at the central part indicates that the Al_2O_3 pierces through the YAG. Since time evolution of the eutectic structure remains in the growth direction, the three-dimensional structure lets us know how the entangled part was formed. Branching of the YAG occurred at a position of A in Fig.5 (b). Namely, the Al_2O_3 grew over the YAG. At a position of B, branching of the Al_2O_3 occurred. As a result of the sequential branching of the YAG and the Al_2O_3, the hole in the YAG phase was produced. It should be emphasized that the branching frequently occurs and the entangle domain is of the same order as the lamellar spacing.
It is of interest to compare the branching observed in the Al_2O_3/YAG eutectic structure with that observed in the typical eutectic structures. The micro X-ray CT indicated that branching was rarely observed in regular eutectic structures (Sn-Pb alloys) (Yasuda et al., 2003). In irregular eutectic structures (Sn-Bi alloys), branching of the Bi phase is frequently observed whereas branching of the Sn phase was rarely observed (Yasuda et al., 2003) . The branching of the mate in the eutectic structure did not produce the entangled structure as shown in Fig.5 (b). The sequential branching of both phases results in the entangle structure.

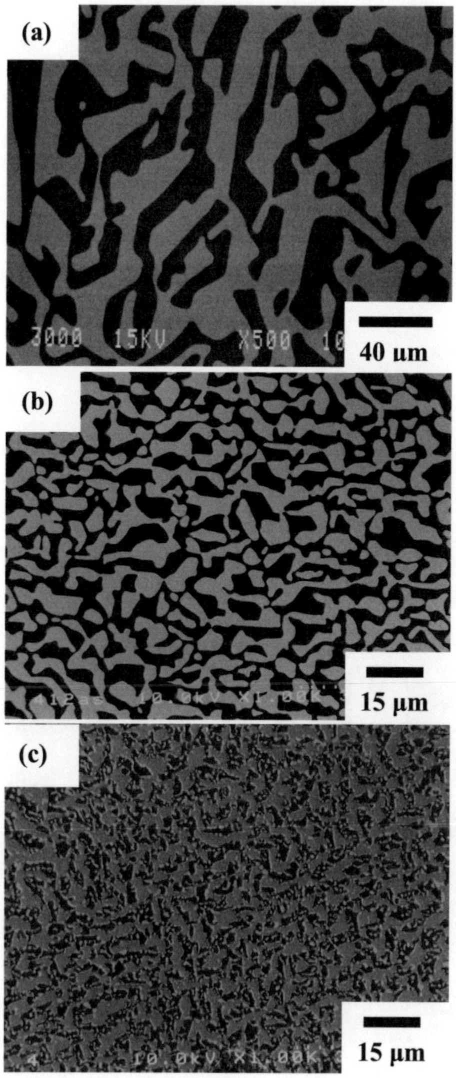

Fig. 3. SEM images showing the microstructure of a cross-section perpendicular to the solidification direction of the MGCs. (a) Al_2O_3/YAG binary system , (b) Al_2O_3/GAP binary system and (c) Al_2O_3/YAG/ZrO_2 ternary system.

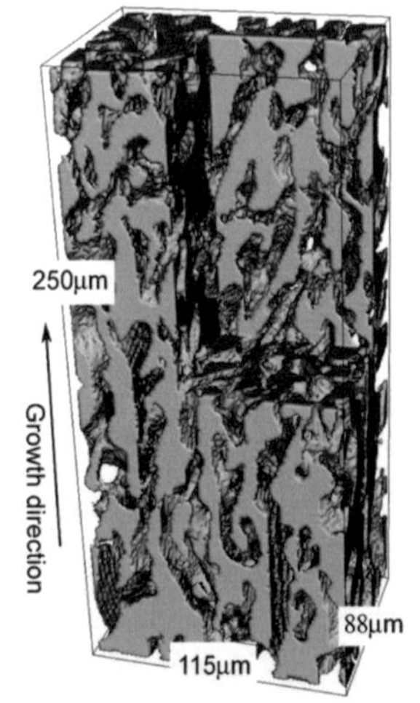

Fig. 4. Three-dimensional image of the unidirectionally solidified Al_2O_3/YAG eutectic structure. The α-Al_2O_3 phase was removed from image.

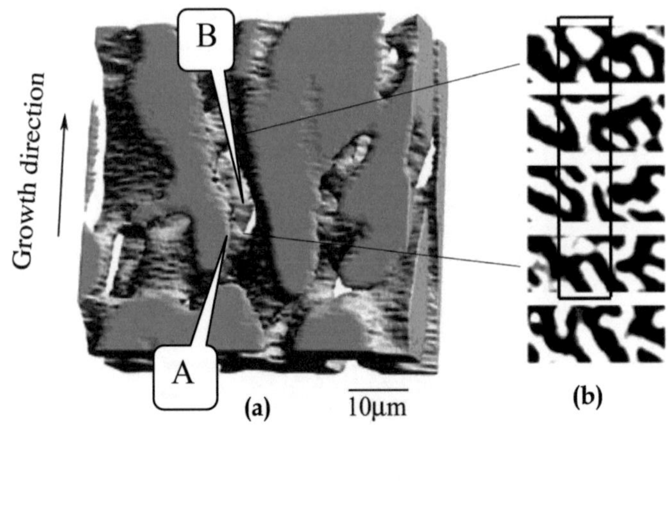

Fig. 5. (a) Three-dimensional image of the YAG phase in the entangled region and (b) sequence of the slice image perpendicular to the growth direction. Black and white phase are YAG and Al_2O_3, respectively.

5. High temperature characteristics of MGCs

5.1 Temperature dependence of flexural strength

The change in flexural strength of the binary and ternary MGCs as a function of temperatures is shown in Fig. 6 compared with those of superalloys (CMSX®-4) (Goulette, 1996), an a-axis sapphire and a Si_3N_4 advanced ceramic (Yoshida, 1998) which was recently developed for high temperature structural materials. Temperature dependence of the flexural strength of these materials is significantly different. With the exception of binary and ternary MGCs, the flexural strength of all other materials falls in a different style with a rise of temperatures. Superalloys are excellent high temperature structural materials in less than about 1300 K, but their strength decreases precipitously at more than about 1300 K. The Si_3N_4 advanced ceramics has the higher flexural strength than that of the other ceramic composite at room temperature, but its strength decreases gradually with an increase of temperatures above approximately 1000 K.

In the case of the a-axis sapphire, the flexural strength is almost the same as that of the Al_2O_3/GAP binary MGC at room temperature, but its strength decreases progressively with increases of temperatures until 1773 K. When the test temperature reaches above 1773 K, flexural strength of the a-axis sapphire drops sharply with rising temperatures.

In contrast, the binary MGCs maintain its room temperature strength up to very high temperature, with a flexural strength in the range of 300~400 MPa for the Al_2O_3/YAG binary MGC(Waku et al., 1996) and 500-600 MPa for the Al_2O_3/GAP binary MGC (Waku et al., 1997). Furthermore, the flexural strength of the Al_2O_3/YAG/ZrO_2 ternary MGC increases gradually with a rise in temperatures and its average flexural strength at 1873 K shows approximately 800 MPa, more than twice 350 MPa of the Al_2O_3/YAG binary MGC (Waku et al., 2002). This difference of the flexural strength's temperature dependence between binary and ternary MGCs is presumed to depend mainly on the dimensions of microstructure. Therefore, the higher high-temperature strength increases, the finer dimensions of microstructure become.

5.2 Tensile deformation

Fig.7 shows the nominal tensile stress-elongation curve obtained from tensile tests of an Al_2O_3/YAG binary MGC from room temperature to 2023 K. Above 1923 K a yield phenomenon occurs and the composites fracture after around 10-17%plastic deformation. The yield stress is about 200 MPa at 1923 K. Several cracks appeared in the microstructure at both the 1650 and 2023 K temperature levels. Nearly all of the cracks were in the YAG phase,with almost none observed in the Al_2O_3 phase. A SEM observation of the fracture surface at tensile testing reveals a constricted area in which a ductile fracture can be observed in the Al_2O_3 phase. In a part of the image, dimple-shaped fracture surface can also be observed. Also,the type of fracture is mixed; intergranular and transgranular fracture are both present.

Fig. 8 shows bright field TEM images of dislocation structures observed in the plastically deformed specimen in the tensile test at 1973 K for the Al_2O_3/YAG single crystal composite. Though the dislocation structures are to be observed in both single crystal Al_2O_3 and single crystal YAG, showing that the plastic deformation occurred by dislocation motion, dislocation densities and dislocation structures in both phases are largely different. Namely, many linear dislocations are observed in single crystal Al_2O_3. Meanwhile, low dislocationdensity is observed in single crystal YAG (Waku et al., 1998).

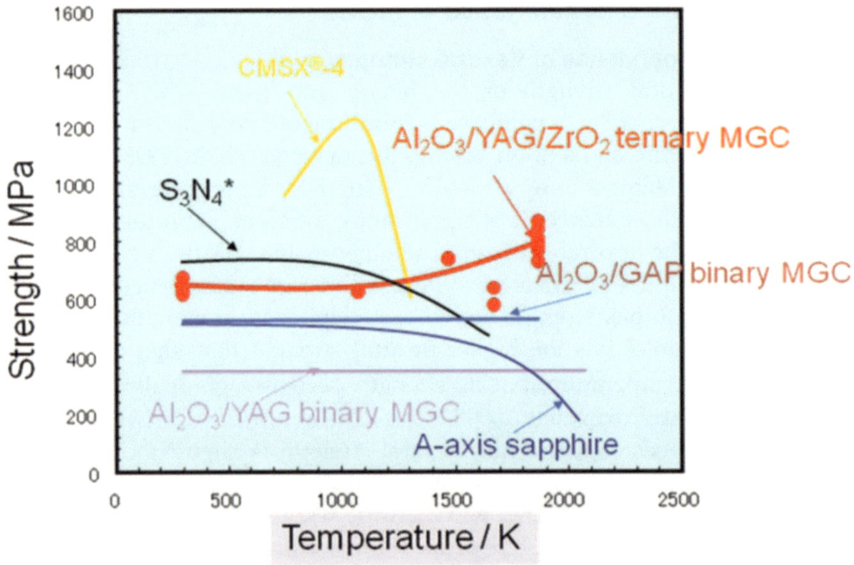

Fig. 6. Temperature dependence of flexural strength of the Al_2O_3/YAG, Al_2O_3/GAP binary and the Al_2O_3/YAG/ZrO_2 ternary MGCs in comparison with superalloys (CMSX-4), an a-axis sapphire and a Si_3N_4 sintered advanced ceramic.

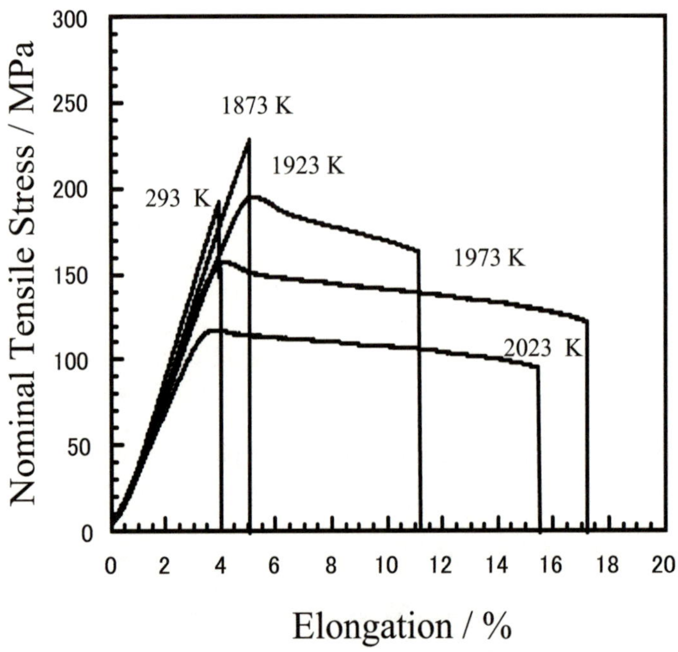

Fig. 7. Nominal tensile stress – elongation curves of an Al_2O_3/YAG single crystal composite from room temperature to 2023 K.

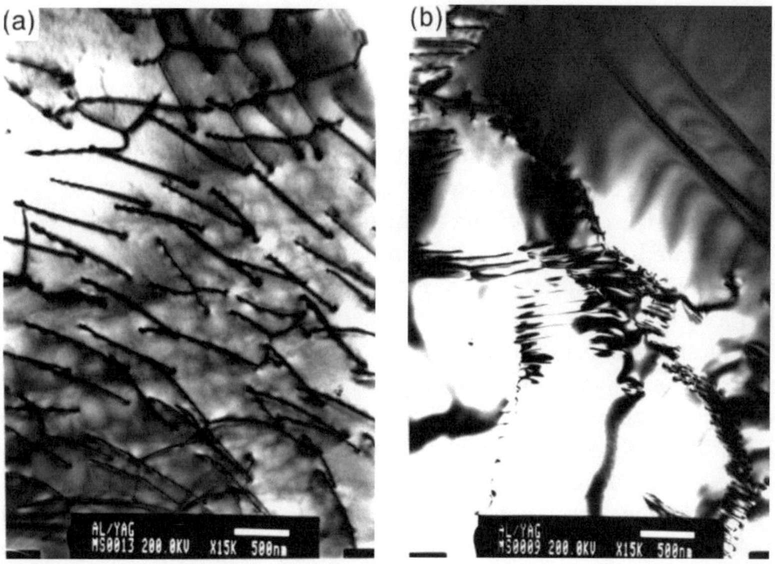

Fig. 8. TEM images showing the dislocation structures of (a) Al_2O_3 phases and (b) YAG phases of the plastically deformed specimens after the tensile test at 1973 K of the Al_2O_3/YAG single crystal composite.

5.3 Compressive deformation

Fig. 9 shows the relationship between compressive flow stress and the strain rate in an Al_2O_3/YAG single crystal composite and a sintered composite at test temperatures of 1773, 1873 and 1973 K. While the Al_2O_3/YAG single crystal composite and the sintered composite shared the same chemical composition and constitutional phases, their compressive deformation was markedly different. That is, the same strain rate of 10-4/s and test Al_2O_3/YAG single crystal composite's flow stress was approximately 13 times higher at 433 MPa. Moreover, as can be seen from the diagram, the Al_2O_3/YAG single crystal composite has creep characteristics that surpass those of a-axis sapphire and, as a bulk material, displays excellent creep resistance (Waku & Sakuma, 2000).

Fig. 10 shows the bright field TEM images of dislocation structure observed in the specimen plastically deformed around 14% in the compressive test at an initial strain rate of 10^{-5}/s and test temperature of 1873 K for an Al_2O_3/YAG single crystal composite and a sintered composite. Dislocation structure is observed in both Al_2O_3 phase and YAG phase for the Al_2O_3/YAG single crystal composite, showing that the plastic deformation occurred by dislocation motion (Wake & Sakuma, 2000; Waku et al., 2002). While dislocation was not observed in both Al_2O_3 phase and YAG phase for the sintered composite. The dislocation structures observed in the Al_2O_3/YAG single crystal composite also indicate that the plastic deformation mechanism of the present eutectic composite is essentially different from that of the sintered composite similar to the micrograin superplasticity of ceramics due to a grain-boundary sliding or a liquid phase present at grain boundary at a high temperature.

The steady state creep rate $\dot{\varepsilon}$ can be usually shown by the following equation:

$$\dot{\varepsilon} = A\sigma^n \exp(-Q/RT)$$

Here, A and n are dimensionless coefficients, σ is the creep stress, Q is the activation energy for the creep, T is the absolute temperature, while R is the gas constant.

In Fig. 9,the value of n is around 1-2 for sintered composites, and 5−6 for Al_2O_3/YAG single crystal composites. In sintered composites, it can be assumed that the creep deformation mechanism follows the Nabarro-Herring or Coble creep models,while in Al_2O_3/YAG single crystal composites,the creep deformation mechanism can be assumed to follow the dislocation creep models corresponding to the dislocation structure in Fig. 10. The activation energy Q is estimated to be about 700 kJ/mol from an Arrhenius plot, which is not so different from the values estimated from the high temperature creep in Al_2O_3 single crystal (compression axis is [110]) and YAG single crystal (compression axis is [110]). It is also reported that the activation energy for oxygen diffusion in Al_2O_3 is about 665 kJ/mol, which is not so far from the activation energy of Al_2O_3 single crystal for plastic flow even though that of A^+ diffusion is about 476 kJ/mol. This fact means that the deformation mechanism of the Al_2O_3 single crystal is the diffusion controlled dislocation creep. On the other hand,the activation energy for oxygen diffusion in YAG is about 310 kJ/mol, which differs significantly from the activation energy of YAG single crystal for plastic now. However,dislocation is always observed in both Al_2O_3 phase and YAG phase of compressively deformed specimens at 1773 K-1973 K and at strain rate of 10^{-4}/s - 10^{-6}/s. Therefore,the compressive deformation mechanism of the Al_2O_3/YAG single crystal composite must follow the dislocation creep models (Wake & Sakuma, 2000; Waku et al., 2002).

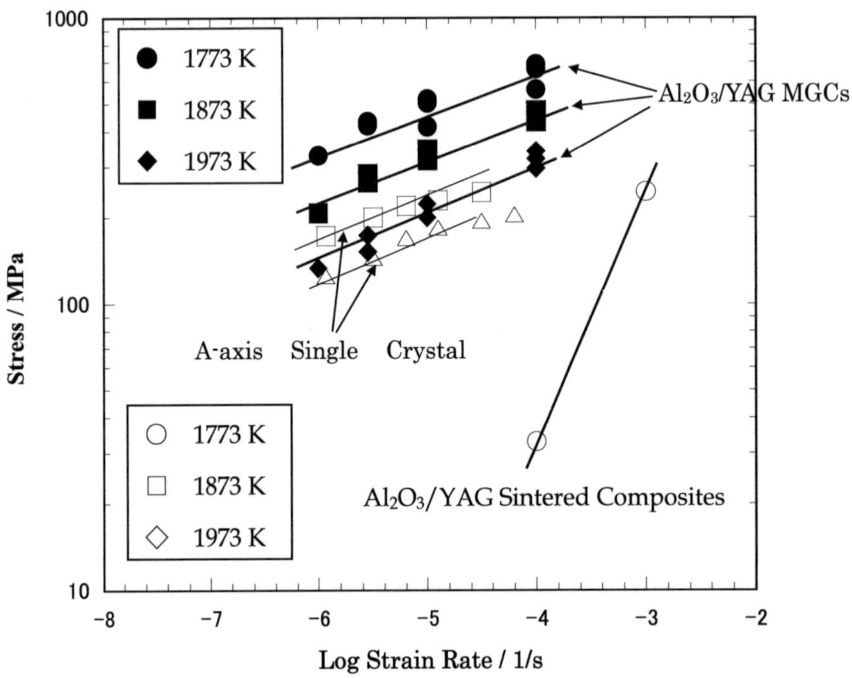

Fig. 9. Relationship between compressive flow stress and strain rate for an Al_2O_3/YAG single crystal composite, a sintered composite and an a-axis sapphire.

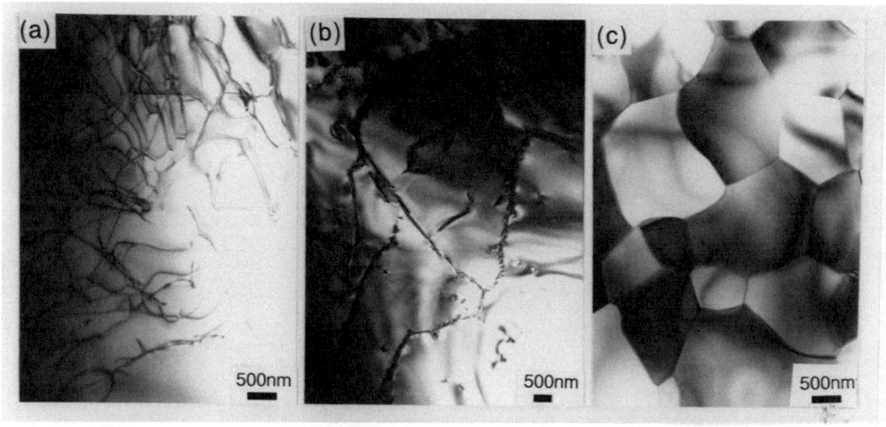

Fig. 10. TEM images showing the dislocation structure of (a) Al_2O_3 phases and (b) YAG phases in the Al_2O_3/YAG single crystal composite, and (c) the microstructure of Al_2O_3 and YAG phases in the sintered composite, of compressively crept specimens at 1873 K and strain rate of 10^{-5}/s.

5.4 Tensile creep rupture

To date a lot of isolated studies have been done on the creep behavior of various highly resistant structural materials. A direct comparison of creep results from different sources is not simple because they have usually been obtained under different test conditions; for instance, with different combinations of temperature and stress. To make a meaningful comparison of creep resistance, the creep data was evaluated here using a Larson-Miller parameter. Figure 3 shows the relationship between tensile creep rupture strength and Larson-Miller parameter, T(22+log t) (DiCarlo & Ynn 1999), for Al_2O_3/YAG binary MGC compared with that of polymer-derived stoichiometric SiC fibers: Hi-Nicalon Type S (Yun & DiCarlo, 1999), Tyranno SA (1, 2) (Yun & DiCarlo, 1999), and Sylramic (1) (Yun & DiCarlo, 1999) , those of silicon nitrides (Krause, 1999), an Al_2O_3/SiC nanocomposite (Ohji, 1994).

Here T is the absolute temperature; t is the rupture time in hours. For comparison, the Larson-Miller curve for a representative superalloy, CMSX®-10 (Erickson, 1996), is shown in Fig. 11 as well.

The relationship between tensile creep rupture strength and Larson-Miller parameter shows three broad regions. The Larson-Miller parameter for CMSX® -10 is 32 or less in region I. This material is already being used for turbine blades in advanced gas turbine systems at above 80% of its melting temperature, and its maximum operating temperature is approximately 1273- 1373 K. It is not envisioned that the heat resistance of this superalloy will be significantly improved in the future. On the other hand, advanced ceramics such as silicon nitrides, SiC fibers, and a nanocomposite are found in region II where the Larson-Miller parameter is between 33 and 42. These materials are promising candidates for high temperature structural materials. They have better high temperature resistance than the superalloys. The creep strength of SiC fibers is approximately coincident with that of silicon nitrides and significantly higher than that of the Al_2O_3/SiC nanocomposite.

In contrast, the Al_2O_3/YAG binary MGC is found in region III where the Larson-Miller parameter is between 44 and 48. The high temperature resistance of this MGC is superior to that of the silicon nitrides, the SiC fibers and the Al_2O_3/SiC nanocomposite. The creep deformation mechanisms for the MGC are believed to be essentially different from the grain boundary sliding or rotation of the sintered ceramics. We conclude that the network microstructure of MGC can be regarded as a suitable microstructure for super high temperature material (Waku et al., 2004).

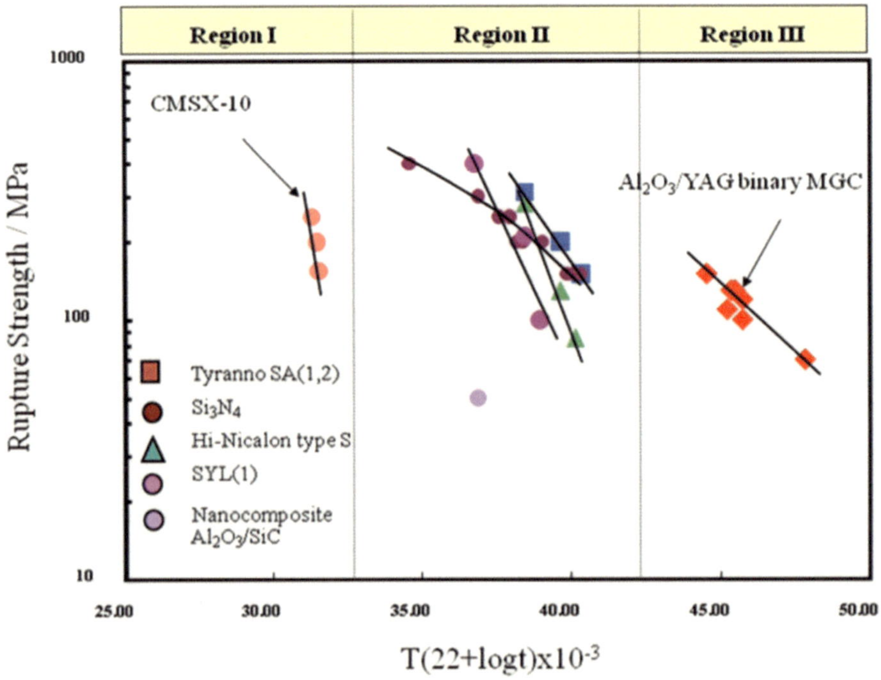

Fig. 11. Larson-Miller creep rupture strength of MGC compared to other heat-resistant materials.

5.5 Oxidation resistance and thermal stability

Fig. 12 shows the change in mass of eutectic composites manufactured by the unidirectional solidification method when these eutectic composites are exposed for a fixed period in an air atmosphere at 1973 K. For a comparison, Fig. 12 also shows the results of oxidation resistance tests performed under the same conditions on ceramics SiC and Si_3N_4. As the Fig. 12 shows, Si_3N_4 was shown to be unstable. When it was exposed to 1973 K for 10 hours in the atmosphere, the following reaction took place; $Si_3N_4 + 3O_2 \rightarrow 3SiO_2 + 2N_2$ and the collapse of the shape of the Si_3N_4 occurred. Likewise, when SiC was held at 1973 K for 50 hours, it was also shown to be unstable. The following reaction took place; $2SiC + 3O_2 \rightarrow 2SiO_2 + 2CO$ and the collapse of the shape also occurred (Waku et al., 1998).

On the other hand, when the unidirectionally solidified Al_2O_3/YAG eutectic composite was exposed in an air atmosphere at 1973 K for 1000 hours, the composite displayed excellent oxidation resistance with no change in mass whatsoever (Waku et al., 1998).

Fig. 13 shows the relationship between flexural strength and heat treatment time at 1973 K in an air atmosphere. For comparison, Fig. 13 also shows results for SiC and Si_3N_4. When the unidirectionally solidified eutectic composite was tested following exposure, there were no changes in flexural strengths both at room temperature and 1973 K, demonstrating that the composite is an extremely stable material. In contrast, when SiC and Si_3N_4 were heated to 1973 K in an air atmosphere for only 15 minutes, a marked drop in flexural strength occurred. Figure 9 shows changes in the surface microstructure of these test specimens before and after heat treatment. There was little difference in surface microstructure of the unidirectionally solidified eutectic composite following 1000 hours of oxidation resistance testing (Waku et al., 1998).

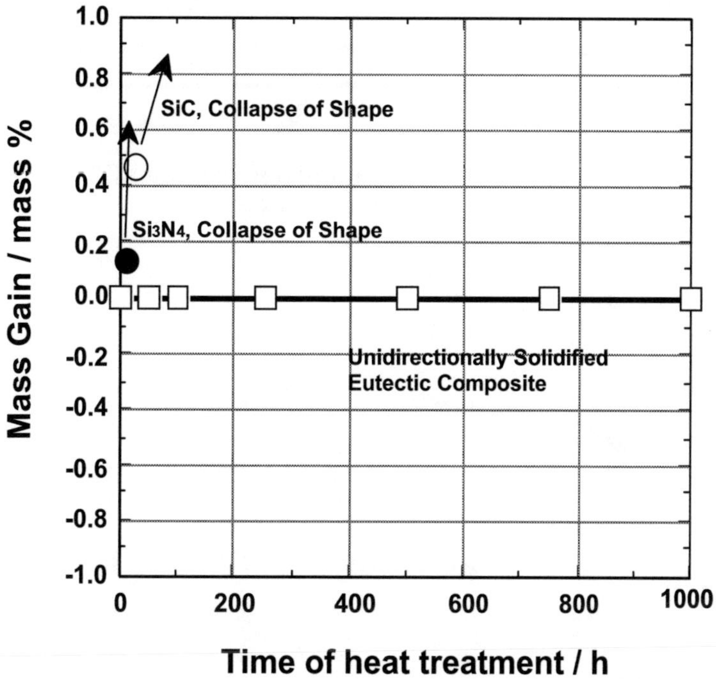

Fig. 12. Comparison of oxidation resistance characteristics of a unidirectionally solidified eutectic composites and advanced ceramics SiC and Si_3N_4 at 1973 K in an air atmosphere.

Al_2O_3/YAG and Al_2O_3/EAG binary MGCs have excellent oxidation resistance with no change in mass gain for 1000 hours at 1973 K in an air atmosphere(Waku et al., 1998). There were also no changes in flexural strength both at room temperature and 1973 K even after heat treatment for 1000 hours at 1973 K in an air atmosphere. In contrast, when advanced ceramic Si_3N_4 was exposed to 1973 K for 10 hours in the atmosphere, the collapse of the shape occurred. Likewise, when SiC was held at 1973 K for 50 hours, it was also shown to be unstable owing to the collapse of the shape also occurred.

Fig. 14 shows SEM images of the microstructure of an Al_2O_3/EAG binary MGC after 500 750, 1000 hours of the heat treatment at 1973 K in an air atmosphere. Even after 1000 hours of heat treatment no grain growth of microstructure was observed. The MGCs were shown to be very stable during lengthy exposure at high temperature of 1973 K in an air atmosphere. This stability resulted from the thermodynamic stability at that temperature of the constituent phases of the single-crystal like Al_2O_3 and the single-crystal EAG, and the thermodynamic stability of the interface. In contrast, a sintered composite shows grain growth and there are many pores lead to reduction of strength at 1973 K only for 100hr (Nakagawa et al., 1997; Waku et al., 1998).

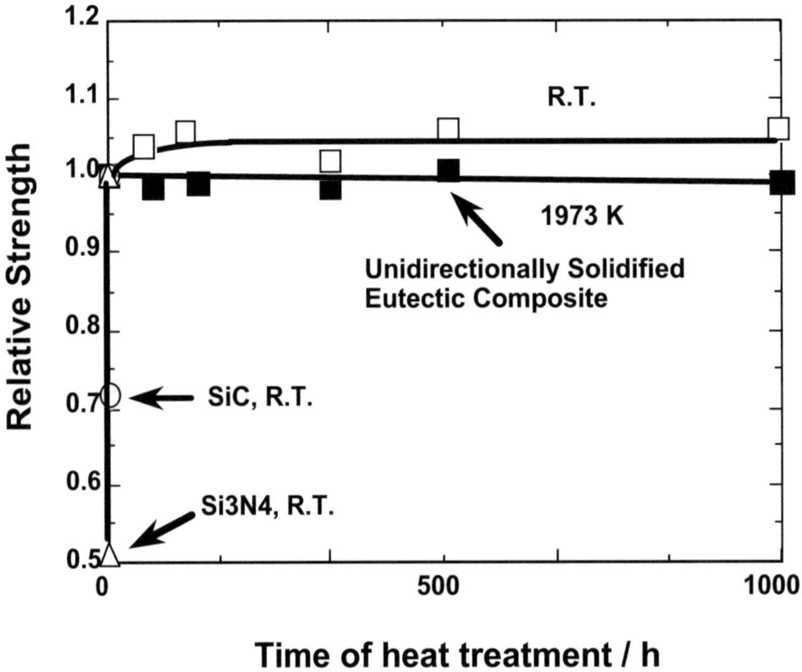

Fig. 13. Changes caused by length of heating in relative strength of unidirectionally solidified eutectic composites and advanced ceramics SiC and Si_3N_4 at room temperature at 1973 K. The relative strength is the ratio of flexural strength after a prescribed period of heating in an air atmosphere at 1973 K to as-received flexural strength.

6. MGC gas turbine systems

Feasibility studies were performed for a leading research project during 1988-2000 in Japan. Based on the results, work was conducted under a NEDO national project from 2001 to 2005. The objective of this project is the development of a 1973 K class uncooled, TBC/EBC-free gas turbine system using MGCs. A paper engine was designed to study component requirements and to estimate its performance. The size of the gas turbine chosen was a relatively small 5MW class. By increasing TIT from the conventional 1373 K to 1973 K, without cooling the nozzle vane and raising the engine pressure ratio from 15 to 30, the thermal efficiency of the gas turbine increased from 29% to 38%. Fig. 15 shows the estimated improvement compared with a current gas turbine. Both are simple cycle gas turbines, and the efficiency is defined at the electrical output (Kobayashi, K., 2002). The final targets of the national project for the MGC gas turbine system are: output power: 5MW class, overall pressure ratio: 30, turbine inlet temperature (TIT): 1973 K, and a non-cooled MGC turbine nozzle. The relationship between the thermal efficiency and the specific power depends strongly on the turbine inlet temperature and the overall pressure ratio. The current efficiency of a 5MW-class gas turbine is around 29%. In contrast, the efficiency of the MGC gas turbine with the uncooled turbine nozzle is higher than that of the conventional gas turbine. For a TIT of 1973 K and a pressure ratio of 30, the 29% efficiency of the conventional 5 MW-class gas turbine increases to 38%.

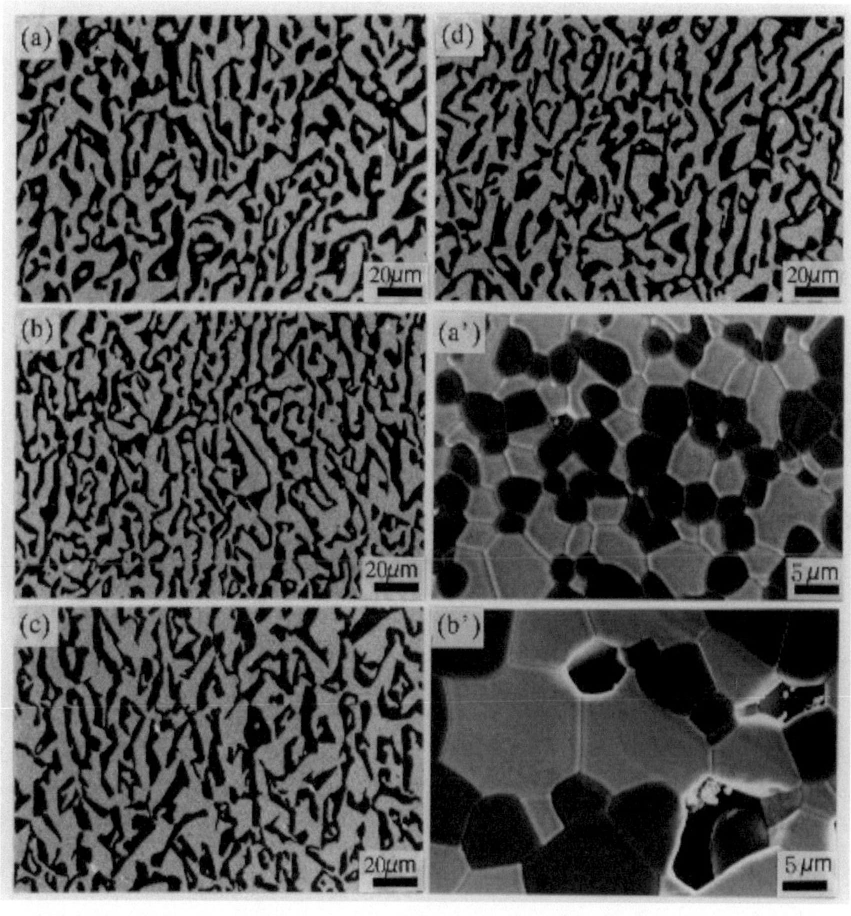

Fig. 14. SEM images showing thermal stability of the microstructures at 1973 K in an air atmosphere in Al_2O_3/EAG binary MGCs: (a) as-received, after heat treatment for (b) 500 h, (c) 750 h, (d) 1000 h and Al_2O_3/EAG sintered composites: (e) as-received and after heat treatment for (f) 100 h.

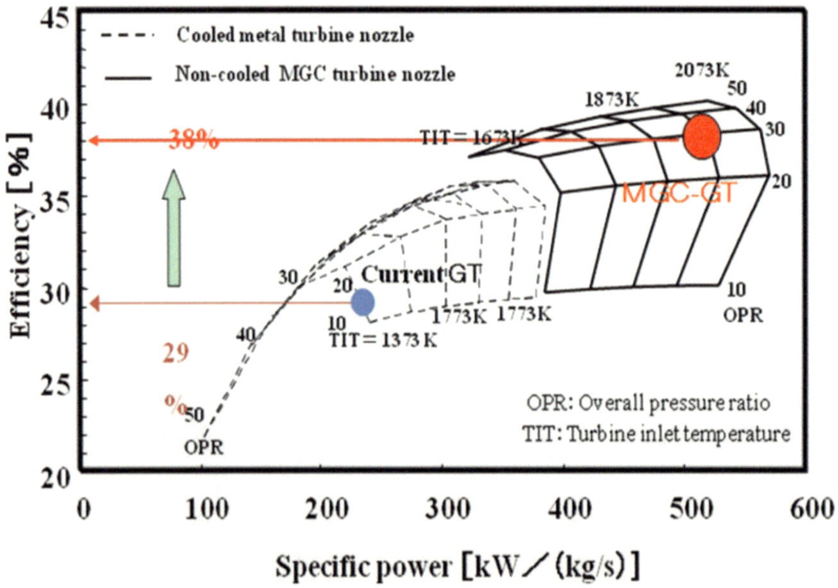

Fig. 15. Gas turbine performance curve as a function of specific power.

MGCs have outstanding high temperature characteristics up to a very high temperature, but the MGC has low thermal shock resistance. First, a hollow nozzle vane was tested at the maximum temperature of 1673 K which is the maximum allowable temperature for the current nozzle rig. The estimated maximum steady state stress using the measured temperature distribution was 211 MPa. To decrease the steady state stress more, a bowed stacking nozzle design is being developed.

An Al_2O_3/GAP binary MGC with high temperature strength superior to that of an Al_2O_3/YAG binary is being examined as a candidate material for the bowed stacking nozzle. Fig. 16 shows the external appearance of the bowed stacking nozzle machined from an Al_2O_3/GAP binary MGC ingot, 53 mm in diameter and 700 mm in length. The steady state temperature and thermal stress distribution at a TIT of 1973 K (see Fig 17) have been analyzed. The maximum temperature is around 1973 K, and it is observed along the central vane section from leading edge to trailing edge at the surface of the bowed stacking nozzle. The maximum steady state thermal stress, generated at the trailing edge of the nozzle, is estimated at 117 MPa. On the other hand, the maximum transient tensile stress in the bowed stacking nozzle during shut-down in one second from 1973 K to 973 K, generated at the leading edge near the mid-span location at 1373 K-1473 K, is estimated at 482 MPa (see Fig. 18). This value is smaller than the estimated ultimate flexural strength of 770 MPa at 1773K of the Al_2O_3/GAP binary MGC (Waku et al., 2003). A rig test at a gas inlet temperature of 1973 K is planned in order to ensure the structural integrity under steady state and thermal shock conditions. The bowed stacking nozzle in Fig. 16 was manufactured from an Al_2O_3/GAP binary MGC ingot by machining with a diamond wheel. Existing rig equipment is being improved for the 1973 K test to enable measurement of a continuous temperature distribution on the nozzle surface by using an infrared camera. It is feasible to verify the structural integrity of the MGC bowed stacking turbine nozzle using this equipment under these hot gas conditions.

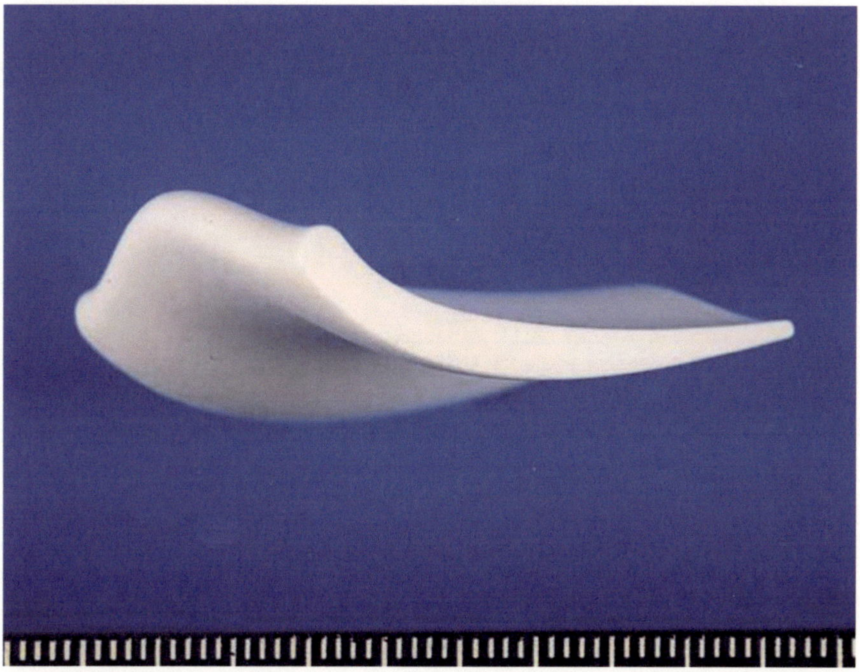

Fig. 16. A bowed stacking nozzle manufactured from an Al_2O_3/GAP binary MGC ingot by machining using a diamond wheel.

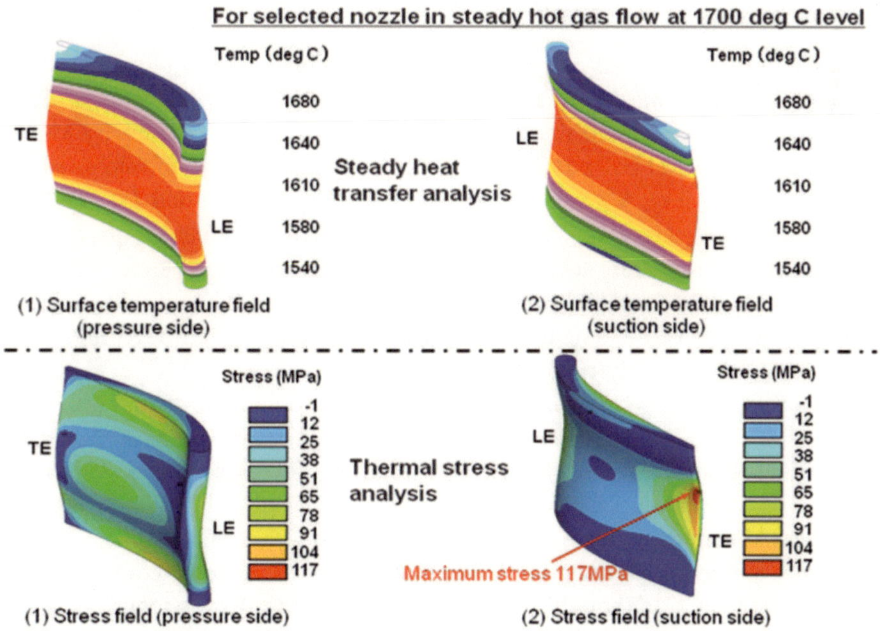

Fig. 17. Steady thermal stress generated during hot gas flow at 1700°C estimated by using numerical analysis.

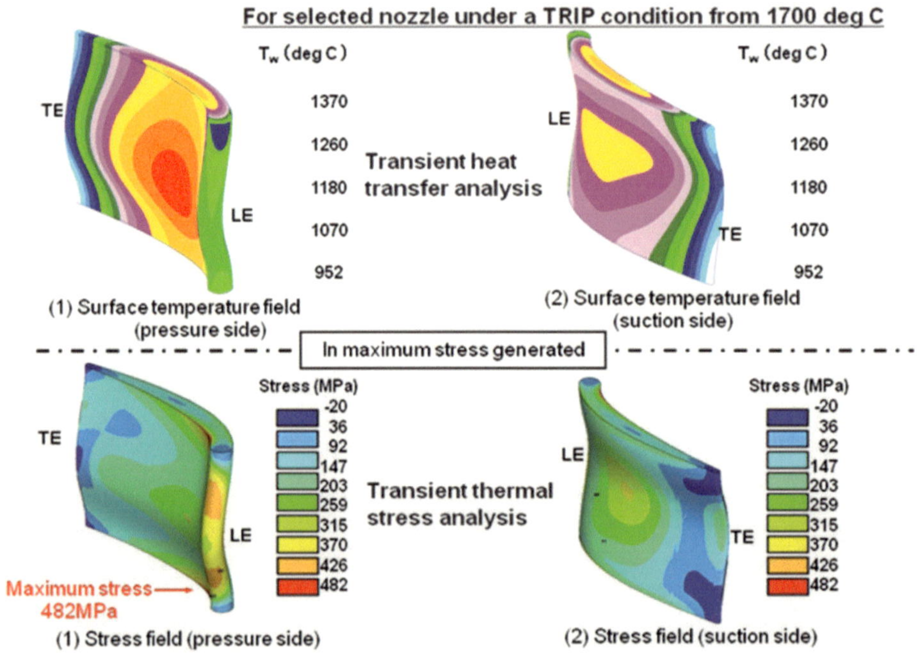

Fig. 18. Transient thermal stress under TRIP condition from 1973 K estimated by using numerical analysis.

7. MGC gas turbine component

Fig.19 shows the SEM images of microstructure of cross-section perpendicular to the solidification direction of the Al_2O_3/YAG and Al_2O_3/GAP binary MGCs after 0 - 1000 hours of heat treatment at 1700 °C in an air atmosphere. In case of Al_2O_3/YAG binary MGC (Fig.19 (a) and (b)) even after 1000 hours of heat treatment, no grain growth of microstructure was observed. While in case of Al_2O_3/GAP binary MGC (Fig. 19 6 (c) and (d)), a slight grain growth was observed. However, both MGCs were shown to be comparatively stable without void formation during lengthy exposure at high temperature of 1973 K in an air atmosphere. This stability resulted from the thermodynamic stability at that temperature of the constituent phases of the single-crystal Al_2O_3, the single-crystal YAG and the single-crystal GAP, and the thermodynamic stability of the interface.

Fig. 20 shows a relationship between flexural strength at room temperature and the time of heat treatment at 1973 K in an air atmosphere. The Al_2O_3/YAG binary MGC has about 300 - 370 MPa of the flexural strengths after the heat treatment for 1000 hours at 1973 K in an air atmosphere. This strength is the same value as the as-received. While, the flexural strength of the Al_2O_3/GAP binary MGC after heat treatment for 1000 hours at 1973 K in an air atmosphere has about 500 MPa slightly lower than that of the as-received. In the case of the Al_2O_3/GAP binary MGC, although the a little drop in the flexural strength in seen a shot time later of the heat treatment, the flexural strength after 200 hours of the heat treatment is independent of the heat treatment time. Both MGCs exhibited good thermal stability at very high temperature of 1973 K in an air atmosphere.

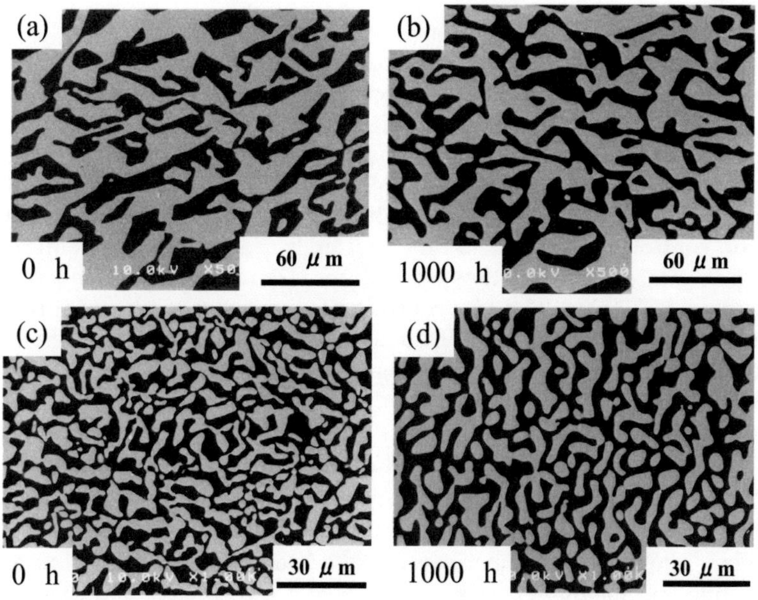

Fig. 19. SEM images showing microstructural changes of cross-section perpendicular to the solidification direction of binary MGCs before and after heat treatment until 1000 hours at 1973 K in an air atmosphere. (a) and (b): the Al_2O_3/YAG binary MGC. (c) and (d): the Al_2O_3/GAP binary MGC. (a) and (c) are for 0 hour. (b) and (d) are for 1000 hours.

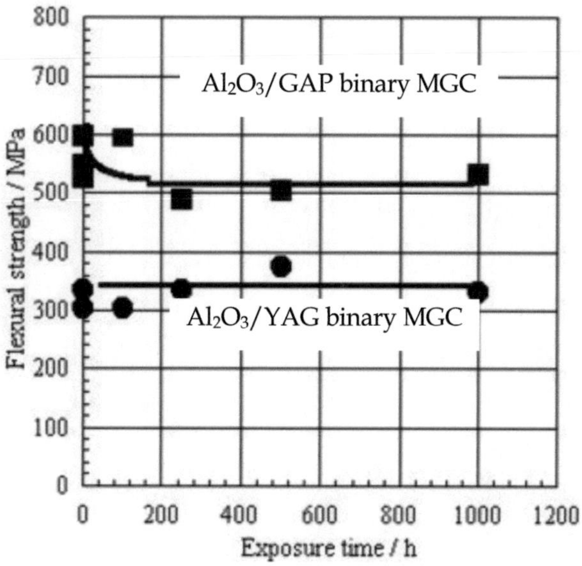

Fig. 20. Relationship between flexural strength at room temperature and time of heat treatment at 1973 K in an air atmosphere.

8. New Bridgman type furnace

MGCs are fabricated by unidirectional solidification from molten oxide eutectic compositions. The melting experiments are conducted at very high temperatures, hence a new Bridgman-type furnace, designed to accurately control many process parameters at super high temperatures, was acquired. Fig.21 is the schematic drawing of the new Bridgman-type furnace. The equipment consists of a casting chamber, a vacuum chamber and a driving device. The schematic on the right of Fig. 21 shows the casting chamber. The casting chamber consists of a heating and melting zone and a cooling zone. Both zones can independently control their temperatures by high frequency induction heating. The Bridgman type furnace has the following features: (1) measurement and control of high temperature around 2300 K, (2) precise control of temperature gradients at near the liquid-solid interface by controlling the melting zone and the cooling zone independently, and (3) the ability to fabricate large MGC components with a maximum size of 300 mm in diameter and 500 mm in length.

Fig.22 shows the external appearance of the new Bridgman-type furnace at JUTEM (Japan Ultra-High Temperature Materials Research Center). The equipment consists of a controller panel, a cooling water system, a casting chamber and a vacuum pump system. The upper right figure is the main controller panel. The lower right figure shows the inside of the casting chamber. The casting chamber consists of a heating and melting zone and a cooling zone. The heating and melting zone is used to heat a Mo crucible and then melt an oxide eutectic raw material in the Mo crucible. The cooling zone is used to control the temperature gradient close to an interface between solid and liquid. Both zones can independently control their temperatures by high frequency induction heating. The lower left figure shows the cooling water equipment.

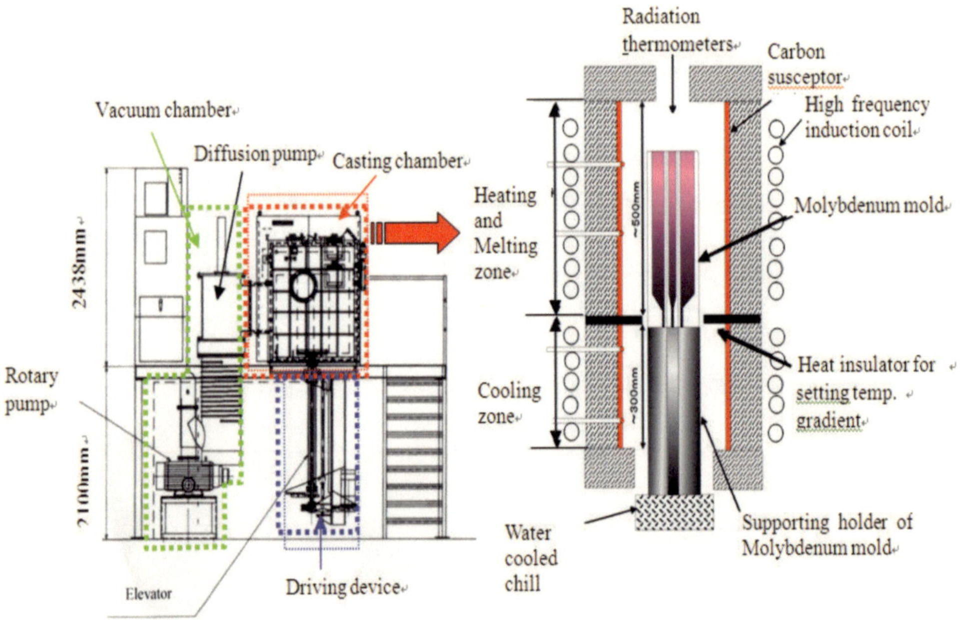

Fig. 21. Schematic drawing of new Bridgman-type furnace

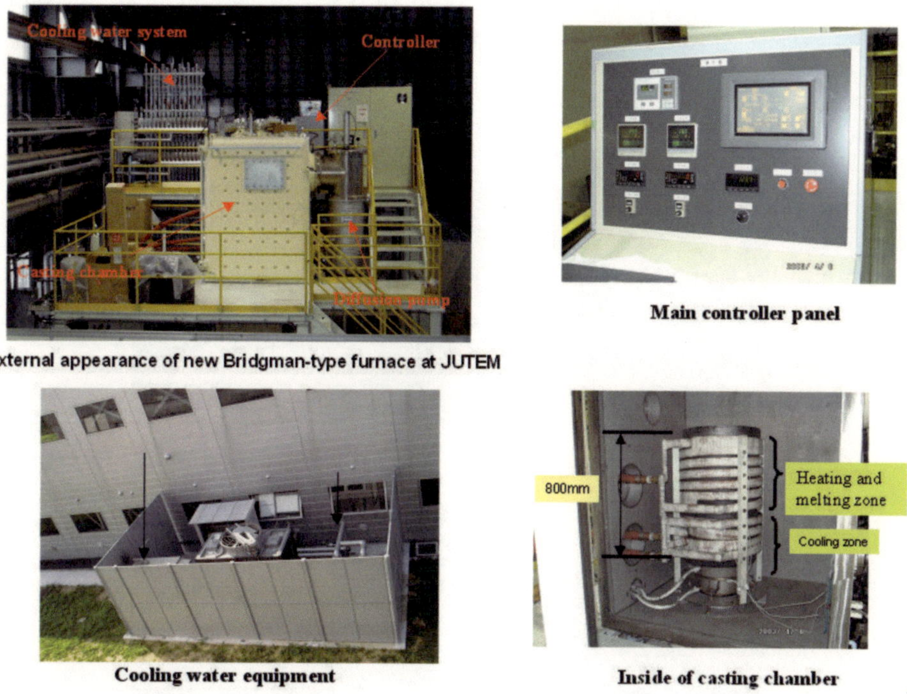

Fig. 22. New Bridgman-type furnace

9. Molybdenum crucible for near-net shape casting

9.1 Plasma sprayed molybdenum mold

Molybdenum is the most suitable material for fabricating MGC parts. Fabrication of a near-net-shaped crucible was attempted by plasma spraying of molybdenum powder on a copper model with a complex shape. The mold was obtained by plasma spraying the molybdenum powders on the copper model. Fig.23 shows plasma spraying of the molybdenum crucible for near-net-shape casting. The plasma spraying was performed using Mo powder with a particle diameter of 20~40 μm in a vacuum atmosphere (about 100 mmHg). The copper model was completely removed from the plasma sprayed mold by melting at 1473 K. Fig.24 shows the external appearance of the plasma sprayed molybdenum mold and its cross section at the middle of the longitudinal direction. The microstructure of the plasma sprayed Mo mold is relatively homogeneous, though it does include some pores. The thickness of wall of the mold is 2-3 mm.

Fig.25 shows the roughness of the internal surface of the plasma sprayed molybdenum mold, with or without shot peening to the Cu model, together with the relatively smooth surface of an extruded molybdenum mold for comparison. Shot peening causes the internal surface of the Cu model to be very rough. It is difficult to remove the MGC from the molybdenum mold after unidirectional solidification. Hence, it is necessary to improve the surface roughness of the mold. To achieve this, the molybdenum powder was plasmasprayed without shot peening to the Cu model. To improve the adhesion of the molybdenum powder, the temperature of the copper model was raised by about 150 K compared to the plasma spraying with shot peening. The surface roughness of the internal wall of the plasma sprayed mold without shot peening to the Cu model was found to be significantly improved compared with that of the plasma sprayed mold with shot peening. The surface roughness of the plasma sprayed molybdenum mold without shot peening to the Cu model appears to be comparable to that of the extruded molybdenum mold.

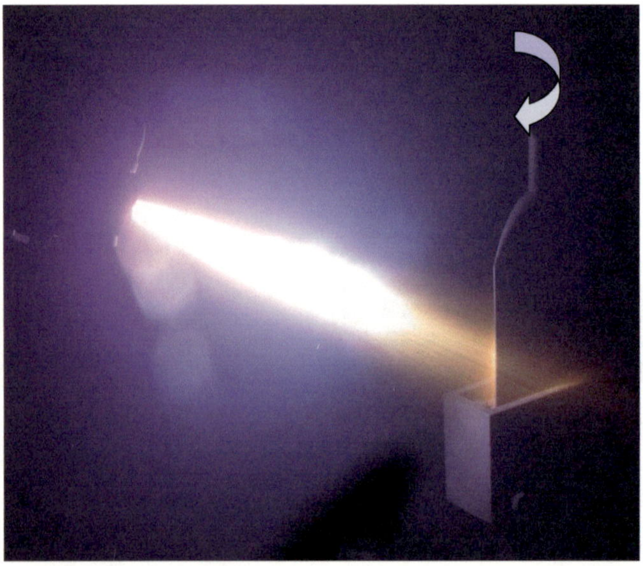

Fig. 23. A plasma spraying scene to produce the quasi-turbine nozzle mold for near-net shaped casting.

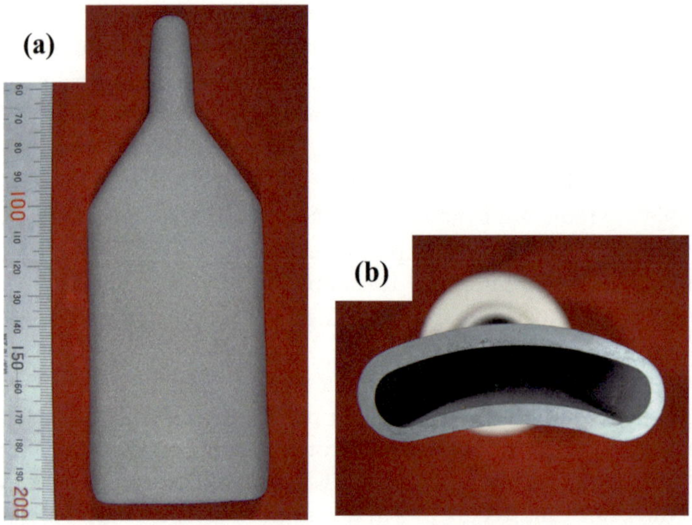

Fig. 24. Plasma sprayed crucibles. (a) External appearance of the crucible manufactured by plasma spraying of molybdenum powder on the copper master mold and (b) its cross-sectional diagram.

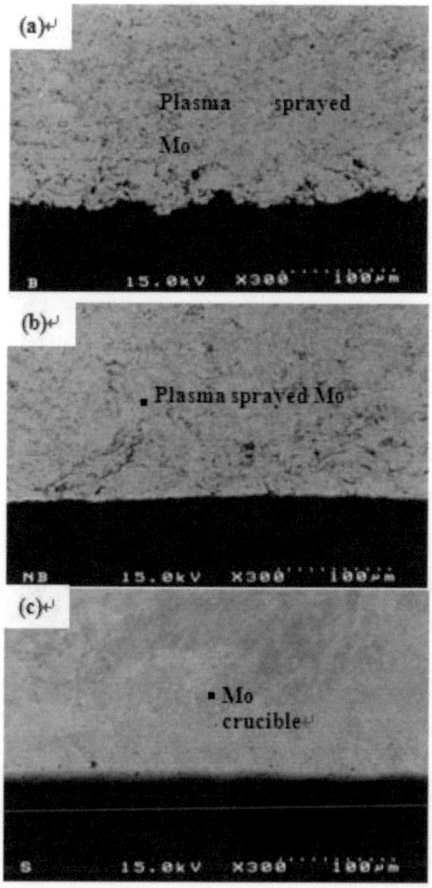

Fig. 25. Cross-sectional microstructure showing roughness on the internal surface of plasma sprayed crucible by Mo powder (a) with shot peening, (b) without shot peening, and (c) Mo crucible produced by extrusion.

9.2 Unidirectional solidification

Fig. 26 shows SEM images of the microstructure of a cross-section perpendicular to the solidification direction of an Al_2O_3/YAG and an Al_2O_3/GAP binary MGC fabricated using the plasma sprayed molybdenum mold and the new Bridgman type furnace. For the Al_2O_3/YAG binary MGC (Fig. 26 (a)), the light area is the YAG phase with a garnet structure, and the dark area is Al_2O_3 phase with a hexagonal structure in the same way as The dimensions of the microstructures are 15-20 μm, smaller than that of the Al_2O_3/YAG binary MGC produced using the extruded mold. Homogeneous microstructures with no pores or colonies are observed in the Al_2O_3/YAG binary MGC fabricated using the plasma sprayed Mo mold. However, the microstructure near the mold is bigger than that for the center.

For the Al_2O_3/GAP binary MGC (Fig. 26 (b)), the light area in the SEM micrograph is the GAP phase, and the dark area is the Al_2O_3 phase in the same way as Fig. 3(b). The dimensions of the microstructures are 2-4 μm, a little smaller than those of the Al_2O_3/GAP binary MGC produced using the extruded Mo crucible. Homogeneous microstructures with no pores or colonies are observed in the Al_2O_3/GAP binary MGC fabricated using the plasma sprayed Mo mold.

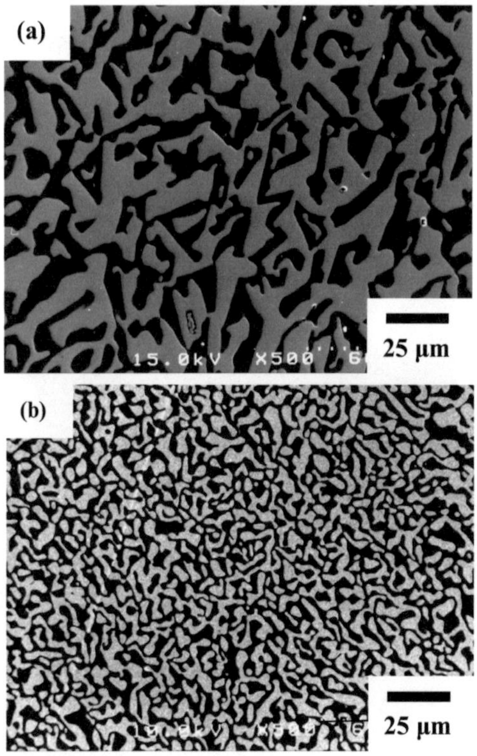

Fig. 26. SEM images of the microstructure of a cross-section perpendicular to the solidification direction of an Al_2O_3/YAG (a) and an Al_2O_3/GAP (b) binary MGC using manufactured using the plasma sprayed Mo mold.

9.3 High temperature strength

Fig. 27 shows the temperature dependence of the flexural strength from room temperature to 1973 K of Al_2O_3/YAG binary MGCs, which was produced by using the plasma, sprayed Mo mold and the extruded Mo crucible. Both MGCs maintain their temperature strength up to 1973 K, with a flexural strength in the range of 260~350 MPa. Namely, the temperature dependence of strength of the Al_2O_3/YAG binary MGC produced using the plasma sprayed Mo mold is almost the same as that of the Al_2O_3/YAG binary MGC produced using the extruded Mo crucible.

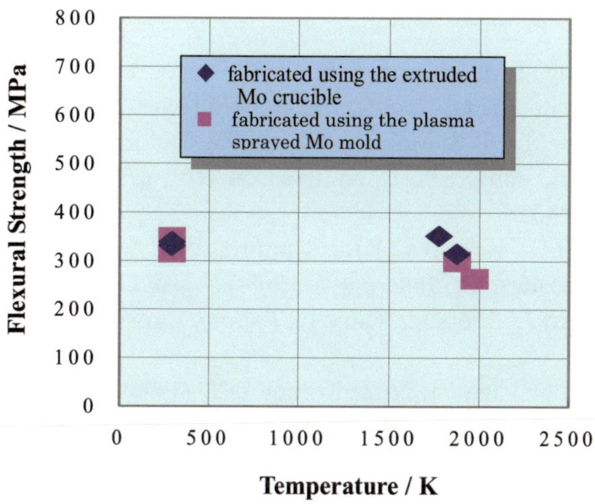

Fig. 27. Temperature dependence of the flexural strength from room temperature to 1973 K of different Al_2O_3/YAG binary MGCs fabricated using the plasma sprayed Mo mold and using the extruded Mo crucible.

10. Conclusions

MGCs have the unique microstructure of a three-dimensionally continuous network of single crystal phases without grain boundaries. Therefore, MGCs have many advantages such as excellent high temperature strength, creep resistance, superior thermal stability as ultra-high temperature structural materials. The NEDO project for a gas turbine system using MGCs has been briefly introduced along with current research topics for system integration and innovative process and manufacturing technology. The manufacturing process of a plasma sprayed molybdenum mold for near-net-shaped casting of the gas turbine component was also introduced. We have recently been successfully fabricated the

Al_2O_3/YAG and Al_2O_3/GAP binary MGCs using plasma sprayed Mo molds and a new Bridgman type furnace. Temperature dependence of strength of the Al_2O_3/YAG binary MGC fabricated using the plasma sprayed Mo mold is almost the same as that of the Al_2O_3/YAG binary MGC produced using the extruded Mo crucible.

11. Acknowledgements

The authors would like to express their thanks to the New Energy and Industrial Technology Development Organization (NEDO) and the Ministry of Economy, Trade and Industry (METI) for the opportunity to conduct "Research and Development of Ultra-high Temperature Heat-resistant Materials MGC".

12. References

Mah, T. & Parthasarathy, T.A. (1990). Processing and mechanical properties of $Al_2O_3/Y_3Al_5O_{12}$(YAG) eutectic composite. *Ceram. Eng. Sci. Proc.*, vol.11, No.9-10, pp. 1617-1627.

Parthasarathy, T.A.; Mah, T. & Matson, L.E. (1990). Creep behavior of an Al_2O_3-$Y_3Al_5O_{12}$ eutectic composite. *Ceram. Eng. Soc. Proc.*, vol.11No.9-10, pp. 1628- 1638.

Parthasarathy, T.A.; Mar, Tai-II & Matson, L.E. (1993). Deformation behavior of an Al_2O_3-$Y_3Al_5O_{12}$ eutectic composite in comparison with sapphire and YA," *J.Am.Ceram. Soc.*, 76[1], pp.29-32.

Stubican, V.S.; Bradt, R.C.; Kennard, F.L.; Minford, W.J. & Sorrel C.C. (1986). Ceramic Eutectic Composites. in *Tailoring Multiphase and Composite Ceramics*, Edited by Tressler, Richard E., Messing, Gary L., Patano, Carlo G. and Newnham Robert E., pp. 103-114.

Waku, Y.; Ohtsubo, H.; Nakagawa, N. & Kohtoku, Y. (1996). Sapphire matrix composites reinforced with single crystal YAG phases. *J. Mater. Sci.*, vol.31, pp. 4663-4670.

Waku, Y.; Nakagawa, N.,; Wakamoto, T.; Ohtsubo, H.; Shimizu, K. & Kohtoku, Y. (1997). A ductile ceramic eutectic composite with high strength at 1873 K. *Nature*, vol.389, pp. 49-52.

Waku, Y.; Nakagawa, N.; Wakamoto, T.; Ohtsubo, H.; Shimizu, K.& Kohtoku, Y. (1998). High-temperature strength and thermal stability of a unidirectionally solidified Al_2O_3/YAG eutectic composite. *J. Mater. Sci.*, vol.33, pp. 1217-1225.

Waku, Y.; Nakagawa, N.; Wakamoto, T.; Ohtsubo, H.; Shimizu, K. & Kohtoku, Y. (1998). The creep and thermal stability characteristics of a unidirectionally solidified Al_2O_3/YAG eutectic composite. *J. Mater. Sci.*, vol.33, pp. 4943-4951.

Waku, Y.; Nakagawa, N.; Ohtsubo, H.; Mitani, A. & Shimizu, K. (2001). Fracture and deformation behaviour of melt growth composites at very high temperatures. *J. Mater. Sci.*, vol.36, pp. 1585-1594.

Courtright, E.L.; Graham, H.C.; Katz, A.P. & Kerans, P.J. (1992). Ultrahigh temperature assessment study - ceramic matrix composites. Materials Directorate, Wright Laboratory, Air Force Materiel Command, Wright-Patterson Air Force Base,1.

Waku, Y.; Sakata, S.; Mitani, A. & Shimizu, K. (2001). A novel oxide composite reinforced with a ductile phase for very high temperature structural materials. *Materials Research Innovations,* vol.5, pp. 94-100.

Waku, Y.; Sakata, S.; Mitani, A.; Shimizu, K. & Hasebe, M. (2002). Temperature dependence of flexural strength of $Al_2O_3/Y_3Al_5O_{12}/ZrO_2$ ternary melt growth composites. *J.Mater. Sci.,* vol.37, No.14, pp.2975-2982.

Yasuda,H.; Ohnaka, I.; Mizutani, Y.; Morikawa, T.; Takeshima, S.; Sugiyama, A.; Waku, Y.; Tsuchiyama, A.; Nakano, T. & Uesugi,K. (2003). unpublished work, Osaka University, Osaka, Japan.

Yasuda,H.; Ohnaka, I.; Mizutani, Y.; Morikawa, T.; Takeshima, S.; Sugiyama, A.; Waku, Y.; Tsuchiyama, A.; Nakano, T. & Uesugi,K. (2005). *Journal of the European Ceramic Society,* vol. 25, pp. 1397-1403.

Frazer, C. S.; Dickey, E.C.; Sayir, A. (2001). *J. Cryst. Growth,* Vol.233, P. 187.

Yoshida, M.; Tanaka, K.; Kubo, T.; Terazone, H. & Tsuruzone, S. (1998). *Proceedings of the international Gas Turbine & Aeroengine Congress & Exibition* (The American Society of Mechanical Engineeers 1998).

Goulette, M. J. (196): *Proceeding of the eighth international symposium on superalloys* (TMS, Pennsylvania 1996).

Y. Waku, Y.; Sakata, S.; Mitani, A.; Shimizu, K.; Ohtsuka, A. & Hasebe, M. (2002). *J. of Mater. Sci.* vol. 37, p. 2975.

Waku, Y. & Sakuma, T. (2000). Dislocation Mechanism of Deformation and Strength of Al_2O_3-YAG Single Crystal Composite at High Temperature above 1700 K. *the Journal of the European Ceramic Society,* vol.20, pp. 1453-1458.

DiCarlo, J. A. & Yun, H. Y. (1999). Thermostructural performance maps for ceramic fibers," in *9th Cimtec World Forum on New Materials Symposium V* – Advanced Structural Fiber composites, Edited by P. Vincenzini, vol.22, pp29-42.

Yun, H.M. and DiCarlo, J.A. (1999). Comparison of the tensile, creep, and rupture strength properties of stoichiometric SiC fibers, *Ceram. Eng. Sci. Proc.,* vol.20, pp. 259-272.

Krause, R.F. Jr.; Luecke, W.E.; French, J.D.; Hockey, B.J. & Wiederhorn, S.M. (1999). Tensile creep and rupture of silicon nitride. *J. Am. Ceram. Soc.,* vol. 82, No.5, pp. 1233-1241.

Ohji, T.; Nakahira, A.; Hirano, T. & Niihara, K. (1994). Tensile creep behavior of alumina/silicon carbide nanocomposite. *J. Am. Ceram. Soc.,* vol.77, No.12, pp. 3259-3262.

Erickson, G.L. (1996). The development and application of CMSX-10. in *Proceedings of the Eighth InternationalSymposium on Superalloys,* September 22-26, 1996, Pennsylvania, USA, Edited by R. D. Kissinger, D. J. Deye, D. L. Anton, A. D. Cetel, M.V. Nathal, T. M.Pollock, and D. A. Woodford., pp.35-44.

Waku, Y.; Nakagawa, N.; Kobayashi, K.; Kinoshita, Y. & Yokoi, S. (2004). Innovative manufacturing Processes of MGC's Components for Ultra High Efficiency Gas Turbine Systems. *ASME TURBO EXPO 2004 – Power for Land, Sea & Air,* 14-17 June 2004, Vienna, Austria.

Nakagawa, N.; Waku,Y.; Wakamoto, T.; Ohtsubo, H., Shimizu, K. & Kohtoku, Y. (1997). The Creep, Oxidation Resistance Characteristics of a Unidirectionally Solidified $Al_2O_3/Er_3Al_5O_{12}$ Eutectic Composite. *6th International Symposium on Ceramic*

Materials & Components for Engines, October 19-24, 1997, Arita, Japan, (1997) pp701-706.

Waku, Y.; Nakagawa, N.; Kobayashi, K Kinoshita, Y. & Yokoi, S. (2003). unpublished work, HPGT Research Association, Tokyo, Japan.

17

Study of a New Type High Strength Ni-Based Superalloy DZ468 with Good Hot Corrosion Resistance

Enze Liu and Zhi Zheng
Institute of Metal Research, Chinese Academy of Sciences
China

1. Introduction

There is a great demand for advanced nickel-based superalloys, mainly for the application to industrial gas turbine blades. They should possess an excellent combination of hot corrosion resistance and high temperature strength. Despite the recent innovation of coating technology, hot corrosion resistance is still important for industrial turbines which are for a long term service. An increasing demand for the higher efficiency of gas turbines leads to the necessity of rising their operating temperatures and stresses, which requires a continued development of high strength superalloys for gas turbine components. Hot corrosion resistance is also important for industrial turbines, which are used for longer term than jet engines. Furthermore, oxidation resistance needs to be improved because of the general increase in the inlet-gas temperature of turbines [1, 2]. In order to improve high temperature strength, it is necessary to add Al, Ti, Nb, Ta, W, Mo, and so on. In order to gain good hot corrosion resistance property, Cr is indispensable alloying element in superalloys for maintaining hot corrosion resistance [3, 4]. However, the improvement in one property by adding one or more elements into the alloy may be accompanied by the deterioration of another property [5]. For example, the addition of Re improves both high-temperature creep strength and the hot corrosion resistance [6, 7]. However, increasing in the Re content in SC superalloys has the propensity to precipitate Re-rich topologically closed packed (TCP) phases which is known to reduce creep rupture strength [8, 9, 10].DZ125 alloy is one of using operating turbine blade with excellent mechanic property. IN738 alloy with excellent hot corrosion resistance was broadly using to produce industrial gas turbine blades. In this paper, we hope research a new alloy with the same mechanical property as that of DZ125 alloy and the same hot corrosion resistance as that of IN738 alloy on the basis of good phase stability. Based on DZ125 and IN738 alloys, a new alloy namely DZ468 was developed by institute of metal research, Chinese academy sciences. DZ468 show good mechanics properties, good environment properties and good phase stability.

2. Experiments and results

The DZ468 superalloy is a second-generation nickel-based directed solidified alloy developed by Institute of Metal Research; Chinese Academy of Sciences (IMR, CAS) based on DZ125 and IN738 alloys. Table 1 shows the compositions of DZ125, IN738 and DZ468 alloys. The alloy was melted in VZM-25F vacuum induction furnace. The directionally solidified specimens were made by the process of high rate solidification in ZGD2 vacuum induction directional solidification furnace. The temperature gradient was 80°C/cm and the withdrawal rate was 6 mm/min. The procedure of heat treatment was following: 1240°C/0.5 h +1260°C/0.5h +1280°C/2 h,AC+1120°C/4h, FC to 1080°C with 1h+1080°C /4h,AC+900°C /4h,AC (AC: air cooling, FC: fuel cooling).

Alloy	C	Cr	Mo	W	Co	Al	Ta	Ti	Re	Nb	Zr	Hf	B	Ni
DZ468	0.05	12	1	5	8.5	5.5	5	0.5	2.0	—	—	—	0.01	Bal.
IN738	0.05	16	1.8	2.6	8.5	3.5	1.8	3.2	—	0.8	0.1	—	0.01	Bal.
DZ125	0.08	9	2	7	10	5.2	3.8	1.0	—	—	—	1.5	0.015	Bal.

Table 1. Nominal composition of test alloys (mass fraction, %)

2.1 Microstructure

The microstructure of cast and heat treatment of DZ468 alloy were observed by scanning electron microscope(SEM) and optical microscope(OM).The specimens used for SEM were electrolyzed in a solution of 5ml HNO_3+10ml HCl+5ml H_2SO_4+100ml H_2O with a voltage of 7V. Rectangular specimens with dimensions of 10mm×10mm×8mm were cut by the electrical-discharge method. As shown in the Fig.1a, the microstructure of as-cast alloy are composed of γ, γ', carbides of MC type, (γ+γ') eutectic and a little boride at the edge of (γ+γ') eutectic. Fig.1b shows the size of γ' phase is large and the shape is roughly cubic. Most γ' phase particles show cube shape, but some reveal exaggerated octagonal form.

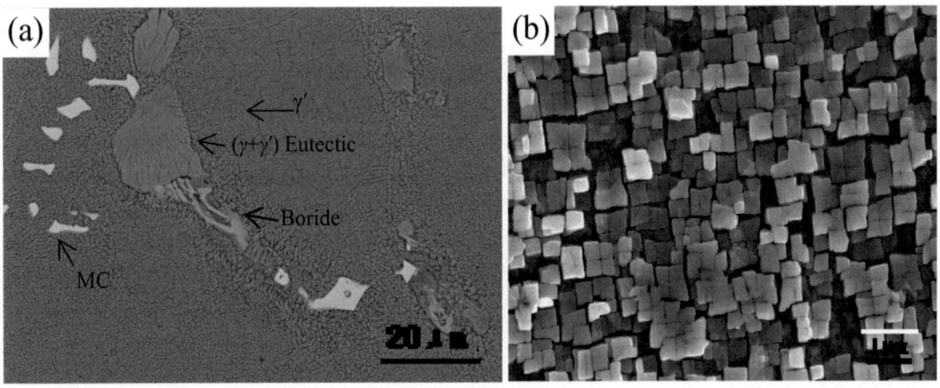

Fig. 1. Microstructure of cast DZ468 alloy (a) OM, (b) SEM

Microstructure of DZ468 alloy after heat treatment shows in the Fig.2a. After heat treatment, the microstructure of DZ468 alloy is composed of γ, γ' and carbides. The carbides are mainly MC and $M_{23}C_6$. There is no finding (γ+γ') eutectic and boride in the Fig.2a. After heated, γ' phase show good cubic shape and the variant size of γ' on inter-dendrite region and dendrite core is rather small as shown in the Fig.2b and Fig.2c. The microstructure of DZ468 alloy after aging at 900°C for 1000h was shown in the Fig.3. After prolong exposure, Coarsening of the γ' was observed and there is no finding TCP phase in the Fig.3. The types of carbide only are MC and $M_{23}C_6$ and there is a very small amount of acicular $M_{23}C_6$. It can be seen from Fig.2 and Fig.3 that DZ468 alloy displays excellent phase stability and uniform microstructure.

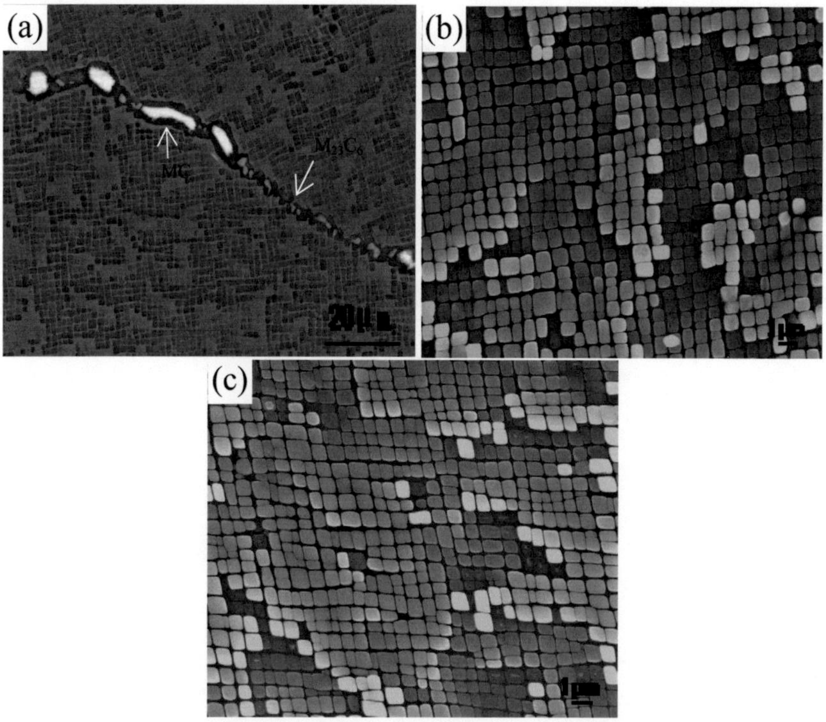

Fig. 2. Microstructure of DZ468 alloy after heat treatment (a) in the grain boundary (b) γ′ on inter-dendrite region, (c) γ′ on dendrite core

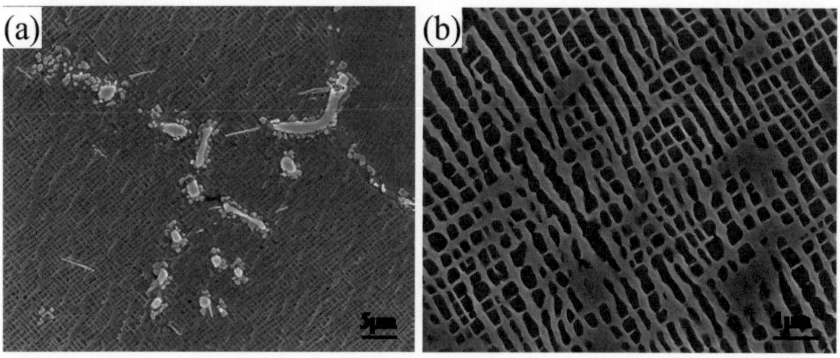

Fig. 3. Microstructure of DZ468 alloy after prolong exposure at 900℃ for 1000h (a) in the grain boundary (b) morphologies of γ′

2.2 Tensile properties

The tensile tests were performed at different temperatures from room temperature to 1000℃ a DCX-25T type universal test machine at a constant strain rate of $10^{-4}s^{-1}$. As shown in Fig.4, the change of tensile strength and yield strength of three alloys is similar. When temperature is lower than 760°C, the tensile strength(σ_b) and yield strength($\sigma_{0.2}$) of three alloys change slightly with increasing temperature. When the temperature is more than 760°C, the tensile strength and yield strength decrease sharply. The tensile strength and yield strength of DZ468 alloy is nearly the same as that of DZ125 alloy in the same condition, but its more than that of IN738 alloy.

The elongation (δ) and reduction of area (φ) are not without significant change from room temperature to 760°C in three alloys. When the temperature is more than 760°C, δ and φ quickly increase. As a whole, Ductility of DZ125 alloy displays better than that of DZ468 alloy in lower temperature, but difference of ductility between DZ125 alloy and DZ468 alloy is slightly in higher temperature.

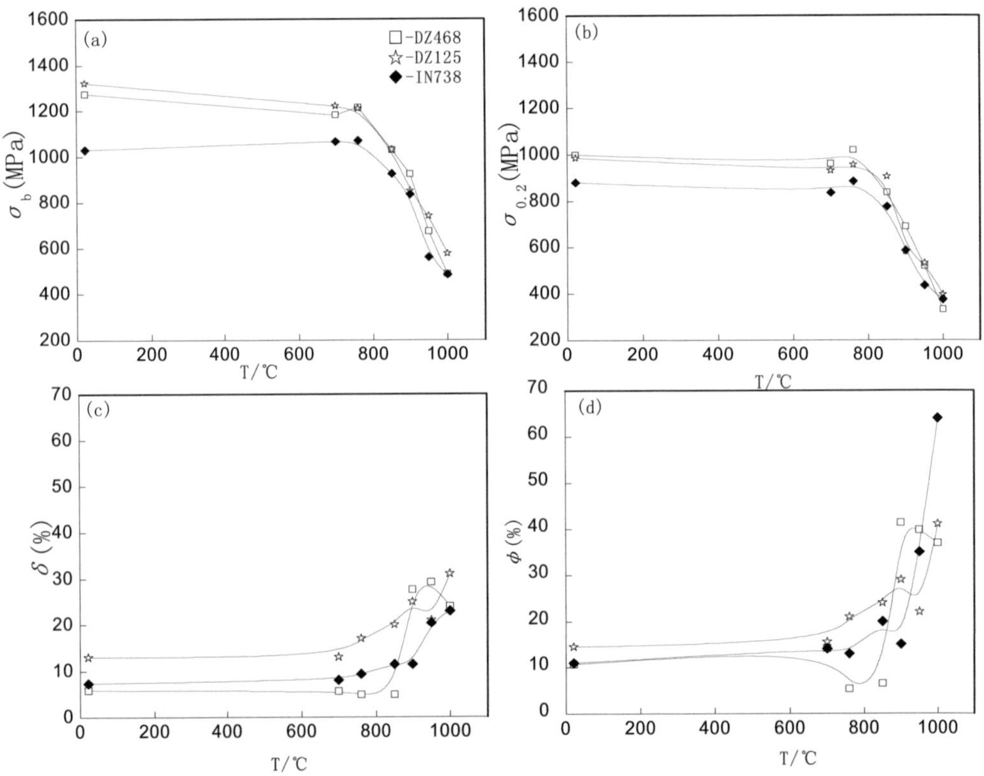

Fig. 4. Tensile properties of DZ468, DZ125 and IN738 alloys (a) the tensile strength, (b) the yield strength, (c) the elongation, (d) the reduction of area

2.3 The rupture properties

Constant load creep and rupture tests in air were carried out at different temperatures for specimens sampled from bars with normal heat treatments. Fig .5 shows the relationship between stress and time to rupture for specimens. The general trend of the rupture data was that the rupture life increased with decreasing test stress and test temperature, as is normally observed from other alloys. Fig.6 shows Larson- Miller curves of three alloys. It can be seen from Fig.6 the creep rupture life of DZ468 alloy is similar that of DZ125 alloy and observably more than that of IN738 alloy.

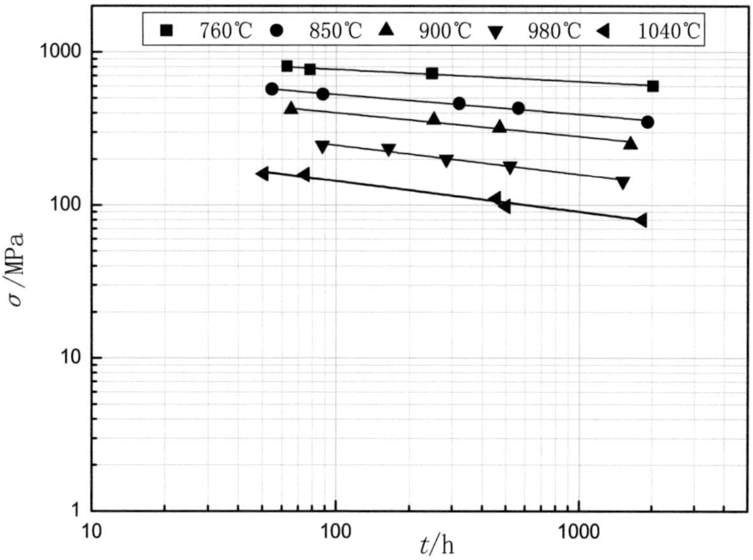

Fig. 5. Stress versus time to rupture in air for DZ468 alloy with different temperature and stress

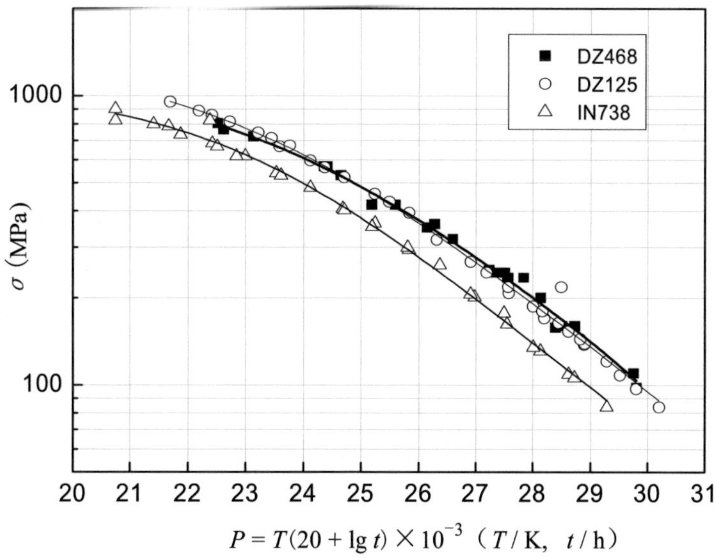

Fig. 6. Larson- Miller curves of DZ468, DZ125 and IN738 alloys

From Table 1, it can be seen the sum of Al, Ti and Ta respectively is 14.2 %(atom fraction) in DZ468 and13.8 %(atom fraction) DZ125 alloy. Hence, the γ' volume fraction of DZ468 and DZ125 is equivalent. In the DZ125 and DZ468 alloys, the total content of strengthening γ phase element W, Mo, Re is almost equivalent. Hence, creep rupture life of DZ468 alloy is similar to that of DZ125 alloy. It can be seen from Fig.7 the creep rupture life of DZ468 alloy to that of DZ125 alloy.

2.4 The creep properties

The creep curves of strain (ε) versus time (t) at different temperatures and stress levels are shown in three figures(fig.7, fig.8 and fig.9). It is indicated that the shape of the creep curve exhibits strong temperature and stress dependence, and the strain rate during steady-state creep is enhanced and creep lifetimes are obviously shortened with the increase of the applied stresses. The observed creep curves are similar and show a respective course at the same testing temperature. The creep curves show an obvious primary creep stage followed by an extended steady-state creep stage and then an accelerating creep stage leading to failure at 760℃ (Fig.7). The 850℃ creep curves demonstrate a very short primary stage, and a longer accelerating creep stage without steady-state creep stage(Fig. 8).It can be seen from fig.7, fig.8 and fig.9, with the increasing of test stress, the creep rate is obviously increasing. Table.2 and table.3 show the creep strength of DZ468 and DZ125 alloys. At the same condition of temperature, the creep strength of DZ468 is lower than that of DZ125 alloy.

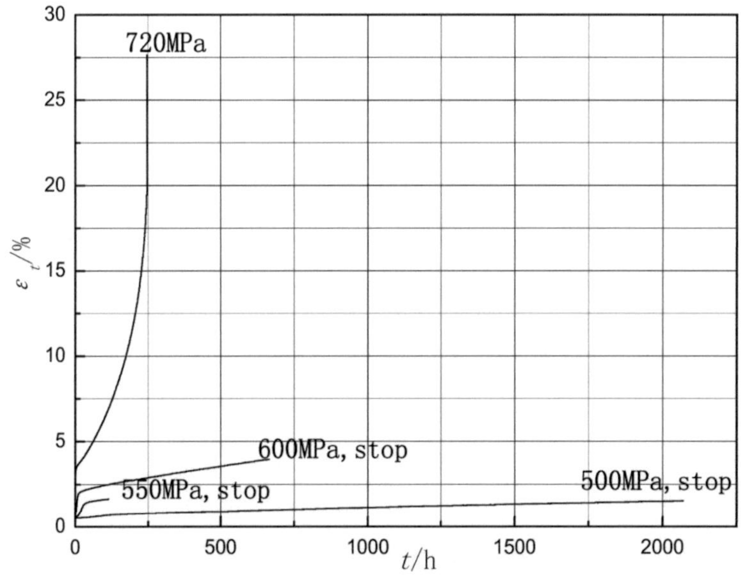

Fig. 7. Creep curves of DZ468 alloy at 760°C with different stress

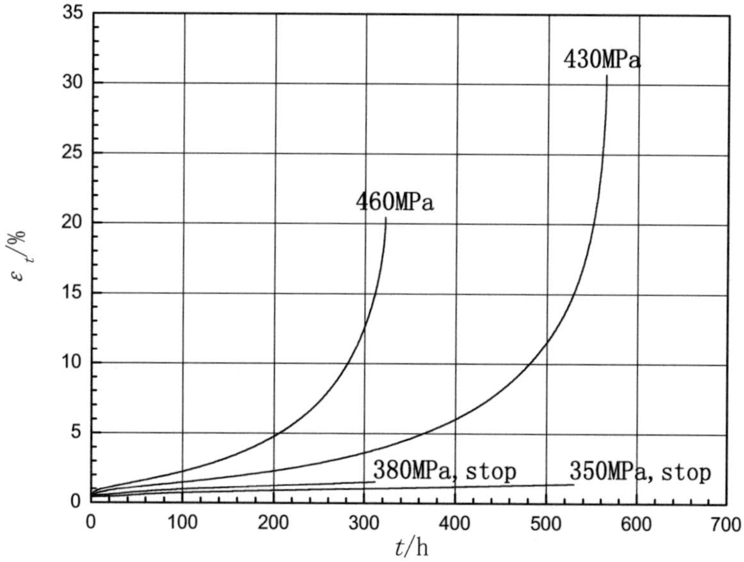

Fig. 8. Creep curves of DZ468 alloy at 980°C with different stress

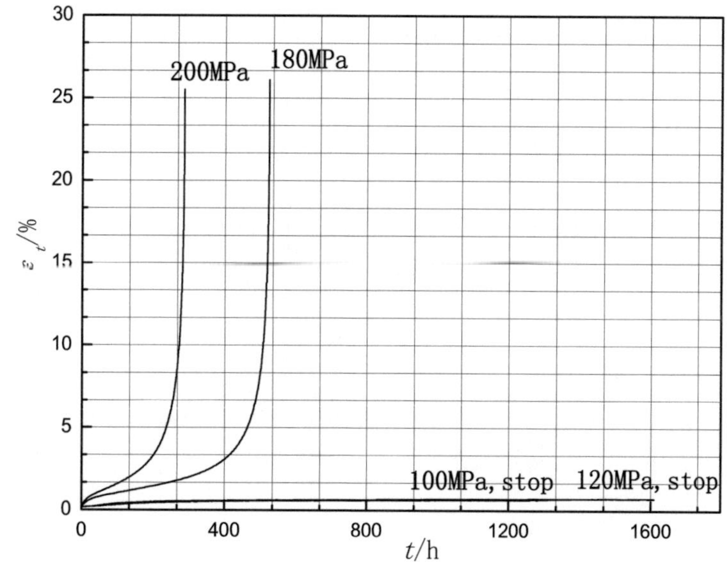

Fig. 9. Creep curves of DZ468 alloy at 850°C with different stress

T/℃	$\sigma_{0.1/100h}$/MPa	$\sigma_{0.2/100h}$/MPa	$\sigma_{0.5/100h}$/MPa
760	530	545	591
850	337	343	405
980	126	140	168

Table 2. The creep strength of DZ468 alloy with different plasticity strain at different temperature

T/℃	$\sigma_{0.1/100h}$/MPa	$\sigma_{0.2/100h}$/MPa	$\sigma_{0.5/100h}$/MPa
760	550	595	620
850	340	380	420
980	170	190	230

Table 3. The creep strength of DZ125 alloy with different plasticity strain at different temperature

2.5 High cycle fatigue

The airfoil sections of turbine blades in aircraft engines are subjected to very high temperatures, high stresses, and aggressive environments. These factors can lead to fatigue behavior that is quite complex, and dependent on stress level (both alternating and mean) and creep and environmental effects. High-cycle fatigue (HCF) tests were performed using smooth round-bar specimens on a high frequency MTS machine. The average test frequency is 120HZ and the R-ratio (R =minimum/maximum stress) is -1. The temperature is 760°C and 900°C. The results, plotted as test life in cycles to failure vs. stress amplitude, are shown in Fig.10. The general trend of the S-N data was that the fatigue life increased with decreasing maximum stress level, as is normally observed from fig.10. It can also be noticed that the fatigue limits of DZ468 alloy at 760°C was higher than that at 900°C.

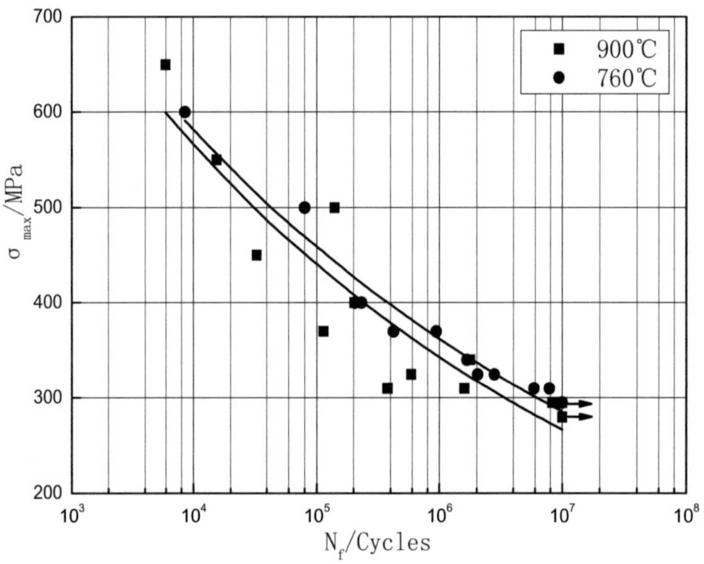

Fig. 10. The HCF S-N curves of DZ468 alloy at 760℃ and 900℃, the average test frequency is 120HZ and the R-ratio is -1.

2.6 Low cycle fatigue

The high temperature low cycle fatigue (LCF) failure is the major factor affecting the service life of the turbine blades. The type of fatigue tests and the experimental conditions were chosen in order to simulate the loading conditions of turbine blades knowing that these conditions are much more complex. LCF specimens were machined from solution treated bars with 15mm in diameter and 25mm in gage length. Before testing, non-destructive evaluation was used to check out the casting pores in specimens. A servo hydraulic testing machine was used to perform the fatigue tests at 800℃ in air. The total axial strain was measured and controlled by an extensometer mounted upon the ledges of specimens. The total strain range ($\Delta\varepsilon_t$) varied from ±0.15 to ±0.6% with a fully reversed strain-controlled push–pull mode, i.e., $R\varepsilon=\varepsilon min/\varepsilon max=-1$. The strain rate was $4\times10^{-3}s^{-1}$, applied in a triangular waveform with a frequency f=0.35 Hz. The temperature fluctuation over the gage length area was maintained within ±2 °C. Three specimens were prepared for each strain range at least. From the viewpoint of engineering applications, an important measure of a material LCF performance is the fatigue life as a function of total strain range, which is presented in Fig. 11 that shows the relationship curves of the total strain range versus number of cycles to failure. The fatigue life shows a monotonic decrease with increasing total strain range from 800℃. It can be seen from fig.11, the fatigue life of DZ125 is slightly longer than that DZ468 at the same total strain range.

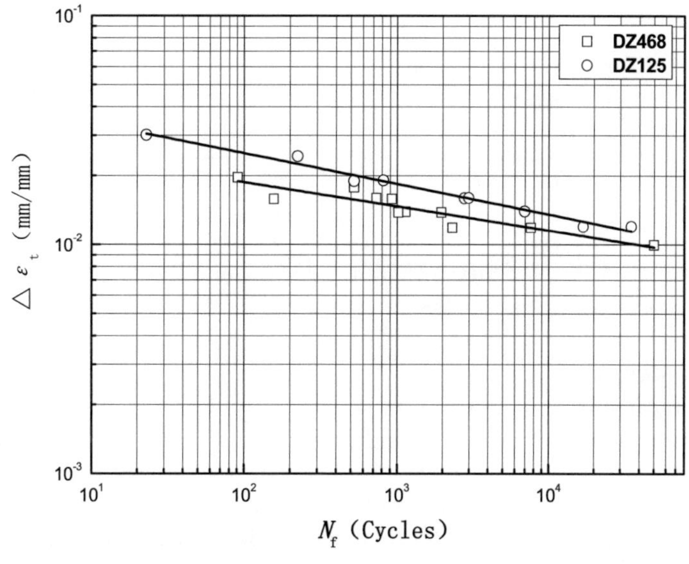

Fig. 11. Fatigue life of DZ468 and DZ125 alloys as a function of total strain range at 800°C

2.7 Hot corrosion resistance

The hot corrosion tests were conducted at 900°C. The surfaces were polished down to 1000-grit alumina paper. A mixture of 75% Na_2SO_4+25% (mass fraction) NaCl was used for hot corrosion experiment. The specimens and mixed salts contained in an Al_2O_3 crucible were placed in a muffle furnace after the furnace reached the desired temperature.

Fig.12 shows hot corrosion dynamics curves of DZ125, IN738 and DZ468 alloys. Both the DZ125 and IN738 alloys exhibit larger depth changes than DZ468 alloy. The absolute value of the depth change is the largest in the DZ125 alloy and the smallest in the DZ468 alloy. The element, Cr is well known to play an essential role in hot corrosion resistance, since it promotes the formation of a protective Cr_2O_3 scale [11]. Although the DZ468 alloy contains the middle Cr content among three experimental alloys, it shows the best hot corrosion resistance. This is due to Re content (2 Mass fraction %) in DZ468 alloy. As already reported, Re is effective in improving hot corrosion resistance as well as creep rupture strength [11, 12, 13]. Furthermore, DZ468 is a kind of low segregation alloy which has own uniform microstructure and chemical composition.

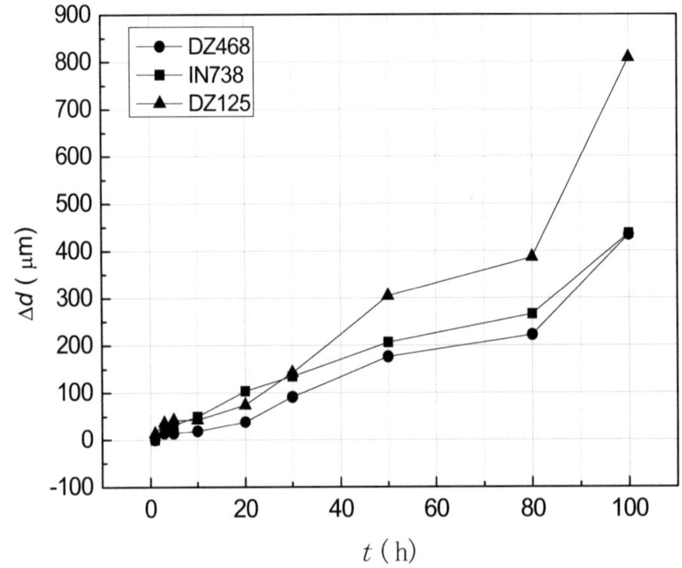

Fig. 12. Hot corrosion dynamics curves of DZ468, DZ125 and IN738 alloys in mixture of 75% Na_2SO_4+25% (mass fraction) NaCl at 900℃

2.8 Physics properties

The density of DZ468 alloy is about to 8.45g/cm^3. Fig.13 shows the mean linear thermal expansion coefficients (CTEs) of DZ468 and DZ125 alloy. It can be seen from fig.13, the l CTEs of DZ125 are larger than that of DZ468 alloy. Fig.14 shows the thermal conductivity of DZ468 and DZ125 alloys at different temperature. It can be seen from fig.14, when the temperature is more than 900°C, the thermal conductivity of DZ468 is higher than that of DZ125 alloy. The thermal conductivity of DZ468 alloy shows a monotonic increase with increasing temperature. Table.4 shows the Young's elastic modulus (*E*) of DZ468 and DZ125 alloy. The Young's elastic modulus (*E*) of DZ468 is decreasing with the increasing of test temperature. It is similar to the Young's elastic modulus DZ468 and that of DZ125.

T/°C	20	100	200	300	400	500	600	700	800	900	1000	1100
DZ468 /GPa	132.09	126.66	123.37	120.38	116.63	111.57	106.53	100.52	95.49	88.67	81.15	70.16
DZ125 /GPa	131.73	126.36	123.52	120.54	116.51	111.31	106.47	100.61	95.36	88.77	81.50	70.08

Table 4. The Young's elastic modulus of DZ468 and DZ125 alloys

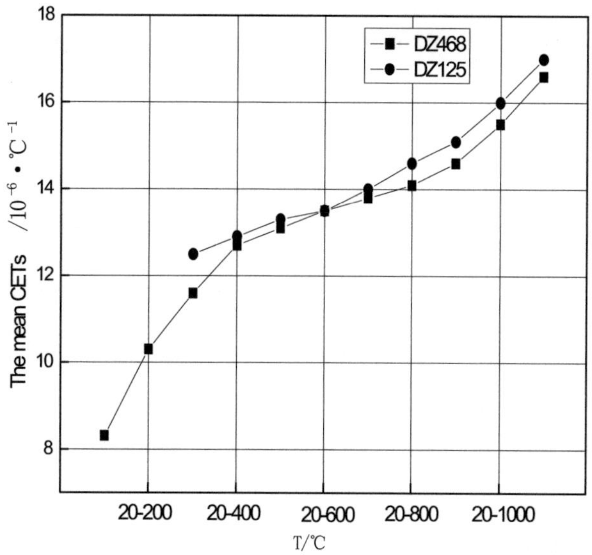

Fig. 13. The mean Linear Thermal Expansion Coefficient (CETS) of DZ468 and DZ125 alloys at different temperature interval

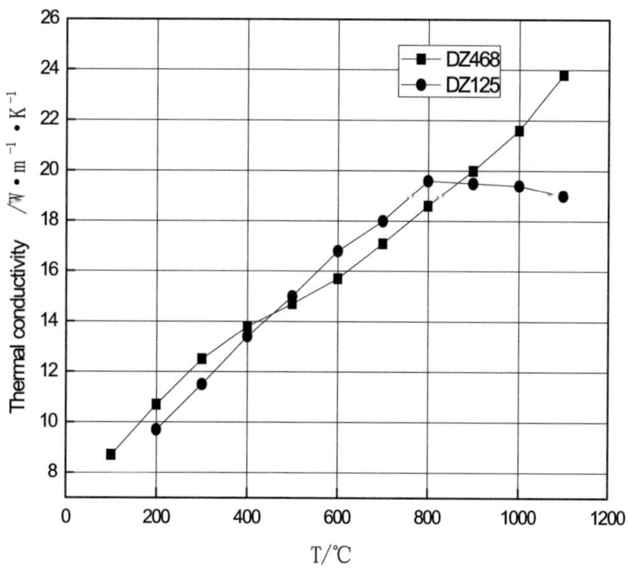

Fig. 14. The thermal conductivity of DZ468 and DZ125 alloys at different temperature

3. Conclusion

A new-typed directional solidification nickel-base superalloy that is named DZ468 was designed by low segregation technology. Microstructures of DZ468 as cast alloy are composed of γ, γ',(γ+γ') eutectics, MC type carbides and a few borides. After heat treatment, the microstructures of DZ468 alloy are composed of γ, γ′, MC and $M_{23}C_6$. DZ468 has excellent phase stability, good mechanics properties, physics properties and environment properties.

4. Acknowledgment

The great help of Mr. F. X. Yang from IMR National Laboratory on the temperature measurements during high-cycle fatigue testing is highly appreciated.

5. References

[1] Y.Murata, S. Miyazaki, et al., in: *Superalloys 1996*, edited by R. D. Kissinger, D. J, Deye, et al.,TMS (1996).
[2] Duhl David N, Chen Otis Y, GB Patent 2, 153, 848. (1985)
[3] Yamazaki Michio, Harada Hiroshi, U.S. Patent 4, 205, 985. (1980)
[4] Duhl, David N., Chen, Otis Y., U.S. Patent 4,597, 809. (1986)
[5] Sato Koji, Ohno Takehiro, Yasuda Ken, et al., U.S. Patent 5, 916, 382. (1999)
[6] Cetel Alan D., U.S. Patent 111,138. (2003)
[7] Cetel Alan D., Shah Dilip M., U.S. Patent 200,549. (2004)
[8] Sato Masahiro, Takenaka Tsuyoshi, et al., U.S. Patent 47,110(2010)
[9] T. Kobayashi, M. Sato, et al.in:*Superalloys 2000*,edited by T.M. Pollock, R.D. Kissinger, et al., TMS, (2000)
[10] Y. Murata, M. Morinaga, et al.: ISIJ International, Vol. 43(2003), p.1244
[11] K. Matsugi, Y. Murata, et al, in: *Superalloys 1992*, edited by S. D. Antolovich, R.D. Kissinger, et al., TMS, Warrendale, PA, (1992)
[12] K. Matsugi, M. Kawakami, et al.: Tetsu-to-Hagané, Vol.78 (1992), p.821
[13] T. Hino, Y. Yoshioka, K. Nagata, et al.in: *Materials for Adv. Power Eng.1998,* edited by J.Lecomte-Beckers et al., Forschungszentrum Julich Publishers, Julich, (1998)

18

BLISK Fabrication by Linear Friction Welding

Antonio M. Mateo García
CIEFMA - Universitat Politècnica de Catalunya
Spain

1. Introduction

Aircraft engines are high-technology products, the manufacture of which involves innovative techniques. Also, aero-engines face up to the need of a continuous improving of its technical capabilities in terms of achieving higher efficiencies with regard to lower fuel consumption, enhanced reliability and safety, while simultaneously meet the restrictive environmental legislations (External Advisory Group for Aeronautics of the European Commission, 2000). Technological viability and manufacturing costs are the key factors in the successful development of new engines. Therefore, the feasibility of enhanced aero-engines depends on the achievements of R&D activities, mainly those concerning the improvement of materials and structures.

Advanced compressor designs are critical to attain the purposes of engine manufacturers. Aircraft engines and industrial gas turbines traditionally use bladed compressor disks with individual airfoils anchored by nuts and bolts in a slotted central retainer. Nevertheless, an improvement of the component disk plus blades is the BLISK, a design where disk and blades are fabricated in a single piece. The term "BLISK" is an acronym composed of the words "blade" and "disk" (from BLaded dISK). BLISKs are also called integrated bladed rotors (IBR), meaning that blade roots and blade locating slots are no longer required. Both designs are illustrated in Figure 1.

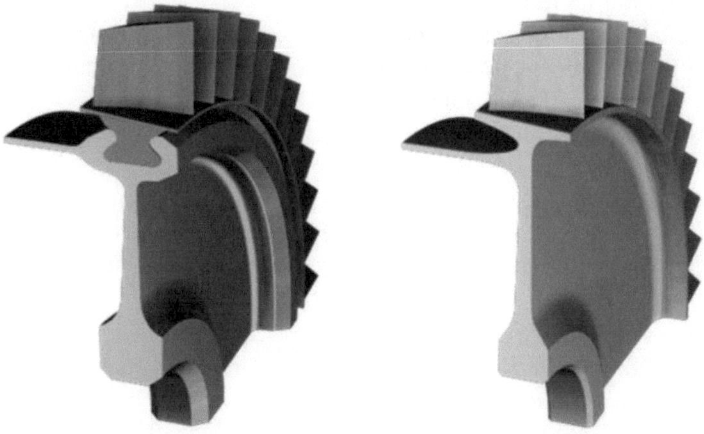

Fig. 1. Illustrations of the mechanical attachment blade-disk (left side) and of a BLISK (right side).

BLISKs can be produced by machining from a single forged part or by welding individual blades to a disk structure. Electron-beam and inertia welding have been used for this application (Roder et al., 2003). However, these techniques are generally not recommended in critical applications concerning fatigue (Broomfield, 1986). An interesting alternative technique is linear friction welding.

Hence, this chapter is devoted to this welding process and its application to manufacture BLISKs of titanium alloys. It is obvious that for such a critical application the integrity of linear friction welds must be totally demonstrated. For that reason, extensive experimental studies were carried out to find the optimum process parameters that assure the reliability of Linear friction welding for the manufacture of BLISK. Results concerning the characterisation of the monotonic and cyclic behaviour of linear friction welds on different titanium alloys are presented. These results demonstrate that linear friction welds may offer similar tensile and fatigue properties than the corresponding base materials.

2. Friction welding

Friction welding technologies convert mechanical energy into heat at the joint to be welded. Coalescence of metals takes place under compressive contact of the parts involved in the joint moving relative to one another. Frictional heating occurs at the interface between the workpieces, raising the temperature of the material to a level suitable for forging. Friction welding is a solid state process as it does not cause melting of the parent material (Messler, 2004).

Friction welding techniques have significant advantages:

- No additional filler material is used.
- Neither fluxes nor gases are required.
- Efficient utilisation of the thermal energy developed.
- The process can be used to join many similar or dissimilar metal combinations. Even dissimilar materials normally not compatible for welding can be friction welded.
- Joint preparation is minimal.
- Consistent and repetitive process.
- Suitable for quantities ranging from prototype to high production.
- Environmentally friendly process: no fumes, gases or smoke generated.
- Being a solid state process, porosity and slag inclusions are eliminated.
- Creates narrow heat- affected zones.
- Friction processes are at least two and even one hundred times faster than other welding techniques.

The relative movement between the workpieces to joint can be linear or in rotation, giving rise to the diverse friction welding processes, which are described in the following subsections. Special attention is paid to the linear friction welding process.

2.1 Rotary friction welding

Rotary friction welding was the first of the friction processes to be developed and used commercially. There are two process variants: direct drive rotary friction welding and stored energy friction welding. The first one is the most conventional technique and usually is simply known as "friction welding". It consists in two cylindrical bars held in axial alignment. The moving bar is rotated by a motor which maintains an essentially constant

rotational speed. The two parts are brought in contact under a pre-selected axial force and for a specified period of time. Rotation continues until achieving the temperature at which metal in the joint zone reaches the plastic state. Then, the rotating bar is stopped while the pressure is either maintained or increased to consolidate the joint. Figure 2 illustrates the stages of this process.
The other variant of rotary friction welding is the stored energy process, more often called "inertia welding". The rotating component is attached to a flywheel which is accelerated by a motor until a preset rotation speed is reached. At this point, drive to the flywheel is cut and the rotating flywheel, with stored energy, is forced against the stationary component. The resultant braking action generates the required heat for welding. Sometimes additional pressure is provided to complete the weld.

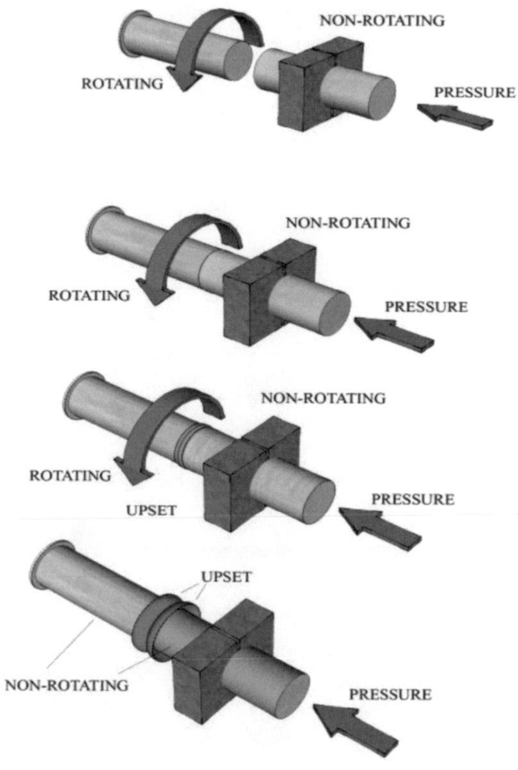

Fig. 2. Illustration of the stages of the direct drive rotary friction welding process.

The industrial acceptance of those benefits, together with the high quality obtained when using conventional rotary friction welding to produce joints in round section metallic components, led in the 1980's to the development of other welding techniques based on friction, such as friction stir welding and linear friction welding. These new friction welding processes allow joining non-round or complex geometry components.

2.2 Friction stir welding

Friction Stir Welding (FSW) is considered to be the most significant development in metal joining in the last decades of 20th century. Figure 3 shows the different stages of this process.

Essentially, a cylindrical non-consumable spinning tool is rotated and slowly plunged into the joint line between two pieces of sheet or plate material, which are butted together. The parts have to be clamped onto a backing bar in a manner that prevents the abutting joint faces from being forced apart. Frictional heat is generated between the wear resistant welding tool and the material of the workpieces. This heat causes the latter to soften without reaching the melting point. As the tool traverses the weld joint, it extrudes material in a distinctive flow pattern and forges the material in its wake. The resulting solid phase bond joins the two pieces into one. FSW can be regarded as a solid phase keyhole welding technique since a hole to accommodate the probe is generated, then filled during the welding sequence. Nowadays, FSW is used to join high-strength aerospace aluminium alloys with astounding success (Threadgill et al., 2009). For example, in the Eclipse 500 aircraft, now in production, 60% of the rivets are replaced by FSW. This fact has naturally stimulated exploration of its applicability to other alloys, such as copper (Won-Bae & Seung-Boo, 2004), titanium, magnesium and nickel (Mishra & Mahoney, 2007) and attempts have even been made to investigate it for the joining of polymers (Strand, 2003). In the particular case of steels, FSW tools would have to go through temperatures higher than 800°C in order to achieve a sufficiently plasticised steel to permit the material flow to enable a sound weld to be fabricated. Cost effective tool materials which survive such conditions for extended service remain to be developed (Bhadeshia & DebRoy, 2009).

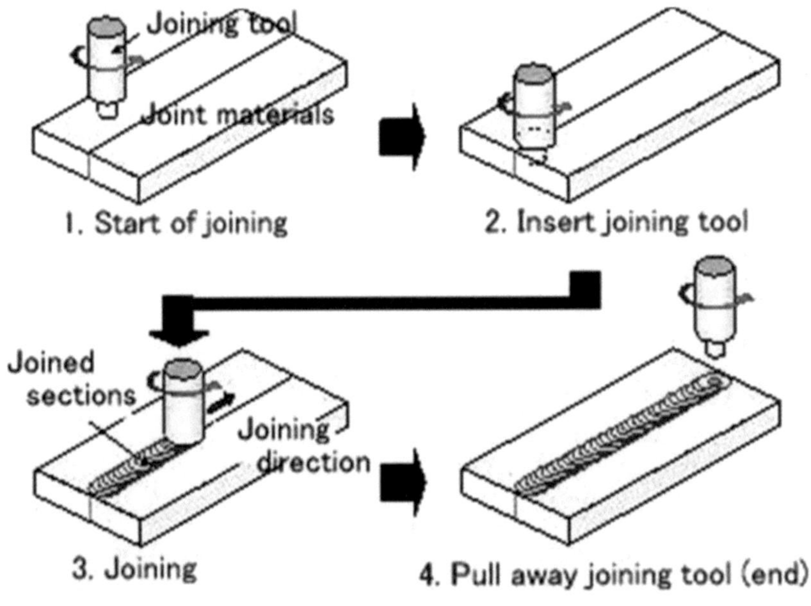

Fig. 3. Illustration of the stages of the friction stir welding process.

2.3 Linear friction welding

A British patent of The Caterpillar Tractor Co. described in 1969 a linear reciprocating equipment for welding steel (Kauzlarich et al., 1969), although no further information was published on this topic during the following decade. In the early 1980s, TWI (The Welding Institute) designed and built a prototype of electro-mechanical machine and demonstrated the viability of the Linear Friction Welding (LFW) technique for metals. Similar machines are now located at industrial plants of aircraft engine manufacturers in Europe and USA, such as MTU Aero Engines, Rolls Royce, Pratt & Whitney and General Electric, where it has proved to be an ideal process for joining turbine blades to disks. For this use, the elevated value-added cost of the components justifies the high price of a LFW machine. Nevertheless, the introduction of this welding technique to other more conventional applications requires novel solutions, which are still in development, principally to reduce the cost of the equipment (Nunn, 2005).

Like all the other friction welding techniques, LFW is able to join materials below their melting temperature. However, in LFW a linear reciprocating motion is the responsible of rubbing one component across the face of a second rigidly clamped part using an axial forging pressure, as depicted in Figure 4. The amplitude of the oscillating motion is small (1 to 3 mm) and the frequency uses to be in the range of 25 to 125 Hz. The maximum axial welding stress is around 100 MPa when titanium alloys are welded and it increases to 450 MPa for nickel pieces.

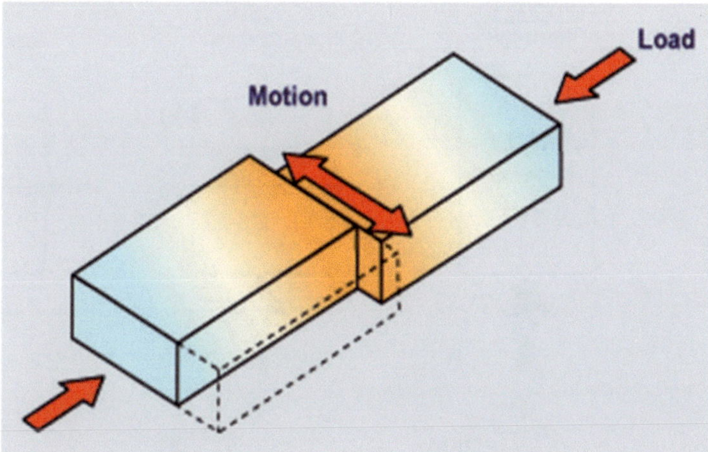

Fig. 4. Illustration of the motion of the parts in the linear friction welding process.

2.3.1 Linear friction welding stages

LFW process can be divided in four distinct stages, as shown in Figure 5. These stages were described in detail by Vairis & Frost (1998).

- Stage I: In the initial phase, both parts are brought in contact under pressure. The two surfaces rest on asperities and heat is generated from solid friction. The true contact area increases significantly throughout this phase due to asperity wear. There is no axial shortening of the specimens at this stage. If the rubbing speed is too low for a given axial force, insufficient frictional heat will be generated to compensate for the conduction and radiation losses, which will lead to insufficient thermal softening and the next phase will not follow.
- Stage II: In the transition phase, large wear particles begin to be expelled from the interface. The true contact area is considered to be 100% of the cross-sectional area. Both workpieces are heated by the friction and the material reaches a plastic state. The soft plasticised layer formed between the two materials is no longer able to support the axial load.

- Stage III: In the equilibrium phase, heat generated is conducted away from the interface and a plastic zone develops. The oscillatory movement extrudes material from the plasticised layer giving rise to flash formation. As a result, axial shortening of the parts takes place. If the temperature increases excessively in one part of the interface away from the centre line of oscillation, the plasticised layer becomes thicker in that section causing more plastic material to be extruded.
- Stage IV: In the deceleration phase, to complete the working cycle the oscillation amplitude decays until the total stop in times ranging from 0.2 to 1 seconds and the components are placed into perfect alignment. The decay rate is an important parameter because longer decay periods are less severe and assist bond formation. Finally, the axial welding pressure is maintained or increased to consolidate the joint. This pressure is usually called forge pressure.

The total cycle is very short, of the order of a few seconds.

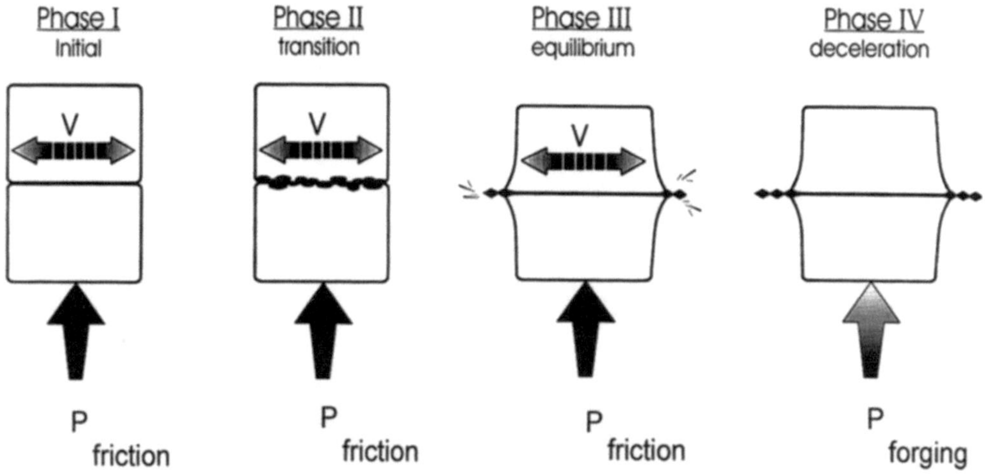

Fig. 5. Illustration of the four phases of the linear friction welding process.

2.3.2 Linear friction welding applications

Despite LFW is a relatively new welding process, it has demonstrated to be efficient to join many different metals, including steels, mainly high strength and stainless steels (Bhamji et al., 2010), aluminium (Ceschini, 2010), nickel (Mary & Jahazi, 2006) and titanium alloys (Wilhem et al., 1995). Even in the cases of intermetallic alloys (Threadgill, 1995), metal matrix composites (Harvey et al., 1995) and dissimilar joints, for example welding copper to aluminium for electrical conductors (Threadgill, 2011), LFW has been yet successfully employed.

LFW technique development has been always linked to aerospace industry. Its first important industrial use was for repairing damaged blades of aircraft engines made in nickel superalloys and titanium alloys. In this application, LFW process showed that it is particularly appropriate for welding titanium. The large affinity of titanium for oxygen, nitrogen and hydrogen makes that fusion welding of these alloys must be carried out under inert gas atmosphere. Conversely, LFW avoids the formation of liquid phase and can consequently be done in air. The next logical step was to expand LFW use to titanium BLISK production.

3. BLISK production

BLISK is one of the most original components in modern aero-engines. First used in small engines for helicopters, BLISK was introduced in the 1980's for military airplanes engines, and it is rapidly gaining position in commercial turbofan and turboprop engines. This is due to its advantages, such as:

- weight saving (usually as much as 20-30%): resulting from the elimination of blade roots and disk lugs;
- high aerodynamic efficiency: because BLISK diminishes leakage flows;
- eradication of the blade/disk attachment, whose deterioration by fretting fatigue is very often the life limiting feature.

Of course, BLISK has disadvantages too. The main one is the laborious, and then expensive, manufacturing and repairing processes. Also, an exhaustive quality control is required to ensure reliable performance. Development efforts are currently trying to mitigate these drawbacks.

As it was commented in the introduction of this chapter, BLISKs can be produced by machining from a single forging or by bonding single blades to a disk-like structure. Depending on the material and also on the design, factors that in turn depend on its location in the engine, each BLISK has its particularities that determine the selection of the manufacturing process. A complete description of the optimisation process for BLISK design and manufacture is given by Buβmann et al. (2005).

In the case of BLISKs produced by machining, there are also two possible paths: milling the entire airfoil or using electrochemical material removal processes. The first technique, illustrated in Figure 6, is used for medium and small size blades.

Fig. 6. Photograph of BLISK machining by high-speed milling (courtesy of MTU Aero-Engines).

In the case of low pressure compressor stages, where the length of the blades is a significant proportion of the diameter of the total component (disk + blades), machining the BLISK from a single forged raw part is a costly and inefficient way. Therefore, welding the blades to the disk becomes a more effective approach. Figure 7 shows the three first stages of the low pressure compressor of an EJ200 aero-engine. This engine is fabricated by using the BLISK technology.

Fig. 7. Low pressure compressor of the Eurojet EJ200 turbofan engine fabricated with the BLISK technology.

Full qualification for aero-engine application has been achieved for LFW manufacturing route. Design and manufacturing advantages derived of the fabrication of BLISKs by LFW in front of other processing routes are:

- High integrity welding technique;
- Low distorsion of the welded parts;
- Heat affected zone of very fine grain;
- Porosity free;
- Possible welding of dissimilar alloys for disk and for blades;
- Fabrication of large diameter BLISKs without the need for huge forged pancakes;
- Tolerances in position and angles of welded blades are very accurate.

4. Titanium alloys for BLISKs

A modern commercial aircraft is designed to fly over 60.000 hours during its 30-year life, with over 20.000 flights. This amazing capacity is for the most part a result of the high performance materials used in both the airframe and propulsion systems. One of those high performance materials is titanium.

The aerospace industry consumes 50% of the world's annual titanium production that was of almost 218.000 tonnes in 2010. Ti alloys make up 20% of the weight of modern Jumbos. For example, the new generation of huge commercial airplanes, i.e. Airbus A380 and Boeing 787 Dreamliner, include between 130 and 150 tons of titanium components per unit. Military aircraft demand also drives titanium usage. On the other hand, nowadays the range of Ti alloys available is very wide and this is a reflect of its growing use outside the aerospace sector, for example in chemical, marine, biomedical, automotive and other industrial applications.

In the case of the propulsion systems, selection of materials is based on their resistance to a combination of high loads and temperatures, together with extremely high safety levels. Typical engine materials are characterised by high specific strength values, i.e. strength divided by density, together with excellent reproducibility of mechanical properties. In this perspective, the development of modern gas turbine engines is mainly based on nickel-based superalloys, but Ti alloys also figure significantly. Around 33% of the weight for a commercial aircraft engine is due to the use of Ti alloys. For military engines this value approaches 50%. The main properties that justified the success of Ti alloys in aero-engines over the last five decades are their high specific strength (thanks to their low density of 4.5 g/cm^3), together with good corrosion resistance and weldability. However, titanium has a limited temperature capability, mainly due to oxidation constrains; therefore, Ti alloys are used for parts under moderate temperatures (i.e. fan and compressor) whereas nickel alloys are preferred for the high temperature regions (i.e. last stages of the high pressure compressor and turbines).

Titanium has two allotropic forms: alpha (α) and beta (β). α refers to hexagonal closed packed crystal structure, while β denotes cubic centred body structure. α and β are the basis for the commonly accepted classification of Ti alloys in four types: α, near-α, $\alpha+\beta$ and β. These categories denote the microstructure after processing and heat treatment. In general, α and near-α alloys have better creep and oxidation resistance, $\alpha+\beta$ alloys posses an excellent combination of strength and ductility, whereas β alloys have good formability and may be hardened to reach high strength levels (Donachie, 2000).

For low pressure compressors and the first stage of the high pressure compressor, where maximum operating temperature is 550°C, principally $\alpha+\beta$ Ti alloys are used. Typical titanium alloys in fan and compressor disks for civil aero-engines are Ti-6Al-4V (Ti-64) for applications up to 300°C and Ti-6Al-2Sn-4Zr-2Mo-0.15Si (Ti-6242) for service up to 480°C. The first alloy is the standard $\alpha+\beta$ alloy and the later one is a near-β alloy.

Ti-6Al-2Sn-4Zr-6Mo (Ti-6246) and Ti-5Al-2Sn-2Zr-4Mo-4Cr (Ti-17) are the only approved and certified (very important point in aero-engine business) high strength $\alpha+\beta$-titanium alloys. They are considered high strength alloys because offer 10-20% higher tensile strength than the typical Ti-64 and Ti-6242, and even higher values can be obtained when used in β-processed conditions. This high tensile resistance is maintained up to 300°C for Ti-17 and up to 450°C for Ti-6246.

On the other hand, the design constrains for disks and blades are different. Whereas high tensile strength and low cycle fatigue resistance are the most relevant properties for disk materials, high cycle fatigue and creep resistance are the main desired characteristics for blades. From this perspective, a possibility for optimisation of compressor performance would be to manufacture stages with a "disk-optimised" material condition for the disk combined with a "blade-optimised" material condition for the blades. Depending on the position in the compressor, a certain combination may be the optimum, whereas at another position another combination would be the right choice. For example, in the temperature regime from 430 to 520°C, an excellent combination would be a β-processed high strength Ti alloy (Ti-6246 or Ti-17) for the disk whit the typical Ti-64 or Ti-6242 alloys for the blades.

BLISKs produced by machining must comprise one single material, with the same microstructure for disks and blades, the specific condition being mostly optimised for the disks. In opposition, the use of the welding to manufacture BLISK opens this innovative possibility of joining dissimilar alloys, choosing the most convenient alloys and microstructures for each component, i.e. for disks and for blades.

5. DUTIFRISK project

It is expected that BLISKs with optimised material conditions, such as it was described in the section 4 of this chapter, would offer an improved in-service performance. But this will be true only if the mechanical properties of the weld-zone are good enough too. Nevertheless, these properties are difficult to predict because the microstructure produced by joining two different alloys by LFW may be very different of the base materials. Therefore, an ambitious European R&D Program was envisaged with the objective of characterizing the microstructure on the weld-zone and to determine the mechanical response of different combinations of titanium alloys selected to optimise BLISK performance. This research project, called "Dual Material Titanium Alloy Friction Welded BLISK", acronym DUTIFRISK, was carried out during 54 months, from April 2002 to September 2006. MTU Aero Engines (Germany) was the project coordinator, although another aero-engines producer, SNECMA Moteurs (France), contributed to the project too. Other participants in the project were: Böhler Schmiedetechnik (Austria), as a Ti alloys supplier; TWI (England), as the main LFW expert centre; ENSMA-CNRS (France) and CIEFMA-UPC (Spain), as research centres with large experience on materials science and technology.

5.1 Objectives

The following tasks, among others, were performed within the DUTIFRISK project:

- Production and characterisation in terms of relevant basic mechanical properties of the titanium alloys used in the project;
- Manufacture, testing and assessment of linear friction welded trial joints to optimise welding parameters;
- Manufacture, detailed testing and assessment of linear friction welded joints;
- Production and validation of a demonstrator BLISK.

The exploration for optimised linear friction welding parameters was developed on specimen scale size for various combinations of high strength titanium alloys (Ti-6246 and Ti-17) as disk-material and various $\alpha+\beta$−titanium alloys (Ti-6242, Ti-64, Ti-6246) as blade-materials. Post-weld heat-treatments were transferred from other welding processes for similar material combinations.

The exhaustive mechanical evaluation of the welds included different types of tests: standard tests (tensile, creep/creep rupture, low-cycle and high-cycle fatigue testing), standard tests adapted to evaluate the weld area properties (fracture toughness and fatigue crack propagation testing) and also specific "new" tests, such as micro-tensile test and Young's modulus measurement. A huge quantity of results was produced during the project. A few of them have been already published (Corzo et al., 2006, 2007; Mateo et. al., 2009; Roder at al., 2008). In subsection 5.2 some selected results, mainly concerning Ti-6246 alloy, are shown and the main conclusions are commented.

The final step of DUTIFRISK project was the production and testing of BLISK demonstrators. Their validation is the key point to prove the transferability of the results obtained on specimen scale to production scale.

5.2 Base materials

As previously explained in Section 4, $\alpha+\beta$ Ti-6246 is one of the titanium alloys which are particularly suitable for compressor disks, whereas the same alloy with $\alpha+\beta$ microstructure would be adequate for blades. Ti-6246 for DUTIFRISK project was produced by Böhler Schmiedetechnik following different fabrication processes depending whether the material was designated to produce the disk or it was for the blades, in order to achieve optimised microstructural characteristics for each part.

Ti-6246 for the disk was produced by die-forging. One of the die-forged disks is shown in Figure 8. It was forged in the β–region, i.e. at temperatures higher than the β-transus (945°C). Heat treatment consisted in a solution annealing at 915°C for 2 hours, with a forced air cooling, and finally an ageing at 595°C for 8 hours with air cooling.

Fig. 8. Ti-6246 die-forged disk in the as-forged condition (courtesy by BSTG)

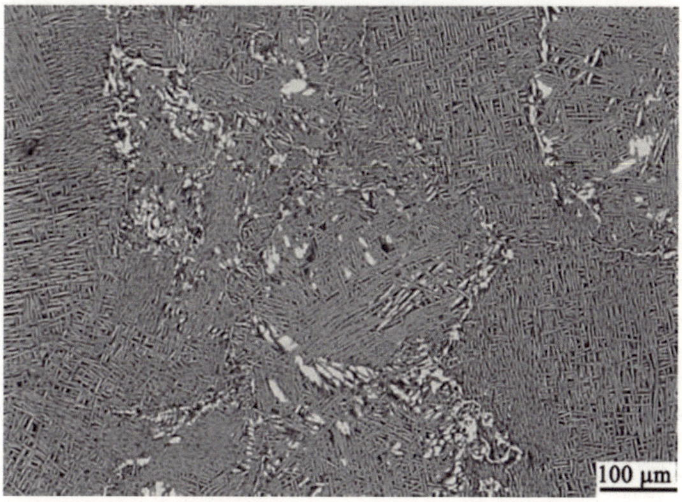

Fig. 9. Microstructure of β-forged Ti-6246.

Microstructural characterisation was carried out by optical microscopy and SEM (Scanning Electron Microscopy). Ti-6246 for the disks exhibits the typical aspect of a β–forged

microstructure, with platelet-like α_p-formation and the desired discontinuous α-layer along the grain boundaries (Figure 9). This type of microstructure is often designated as lamellar. The age hardening treatment produces α_{sec}-platelets in the β-matrix, between the α_p-plates, but they are only visible at high magnification.
In the case of Ti-6246 for the blades, slabs were forged in the α+β–field (around 900ºC), then annealed and aged following the same treatment than the disk material. The appearance of the (α+β)–forged alloy, with its typical bi-modal microstructure, can be observed in Figure 10. It is composed by globular α_p-particles embedded in a fine lamellar α+β matrix. The α_p content is around 28%vol. and the mean size of the nodules is 15 μm.

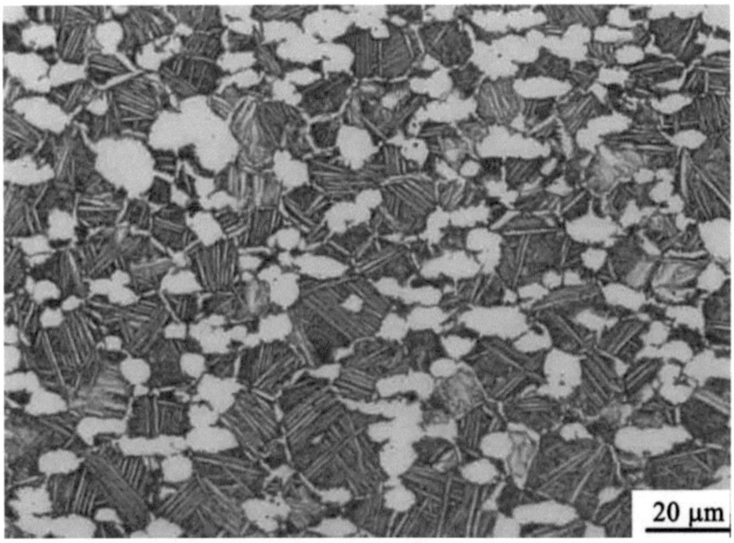

Fig. 10. Microstructure of α+β-forged Ti-6246.

5.3 Linear friction welds

Blocks of 60x36x15 mm were cut from the disks and slabs and welded using an electromechanical LFW machine instrumented to monitor and record the time dependant evolution of all significant process parameters. The appearance after welding is shown in Figure 11. A post-weld heat treatment at 620 ºC during 4 hours in vacuum was always performed. Those blocks were used for microstructural and mechanical characterisation of the different welds.
Cross sections of the welds were prepared for microstructural survey. An image corresponding to a transversal section is shown in Figure 12. The narrow weld-zone is clearly seen. Its apparent width is around 1 mm in the centre and wider at the extremes of the joint.
The microstructure of LFWs was analysed by SEM. Figure 13 is a global view of the weld centre and the heat affected zone where the evolution of both microstructures, i.e. α+β and β, when approaching the weld line is clearly appreciated.

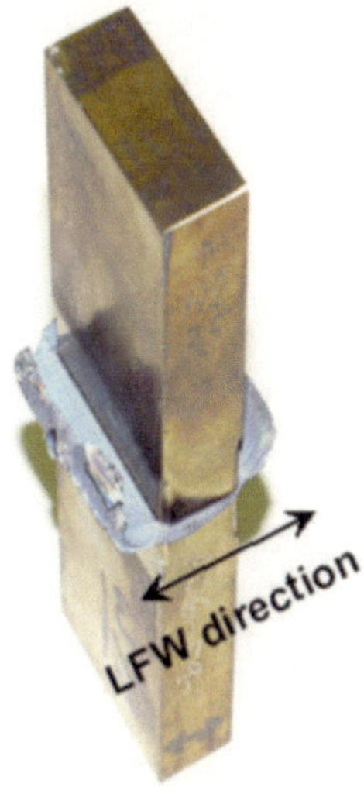

Fig. 11. Photograph of a welded block.

Fig. 12. Macrograph of the cross section of a weld (blade material in the upper part and disk material in the lower part).

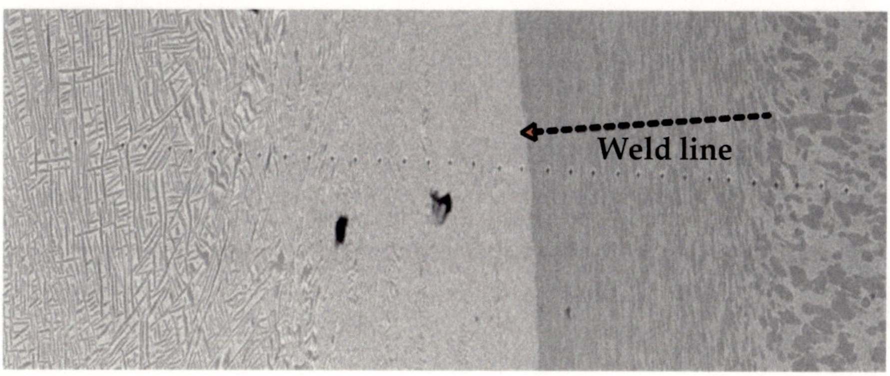

Fig. 13. Microstructure in the weld and heat affected zone. Blade ($\alpha+\beta$-Ti-6246) at the right side and disk (β-Ti-6246) at the left one.

Microstructural evolution was analysed at different positions of the weld line. In β–Ti-6246 alloy, a distortion of the α–lamellae is first observed and then the size of lamellas is progressively reduced as the weld zone is approached (Figure 14).

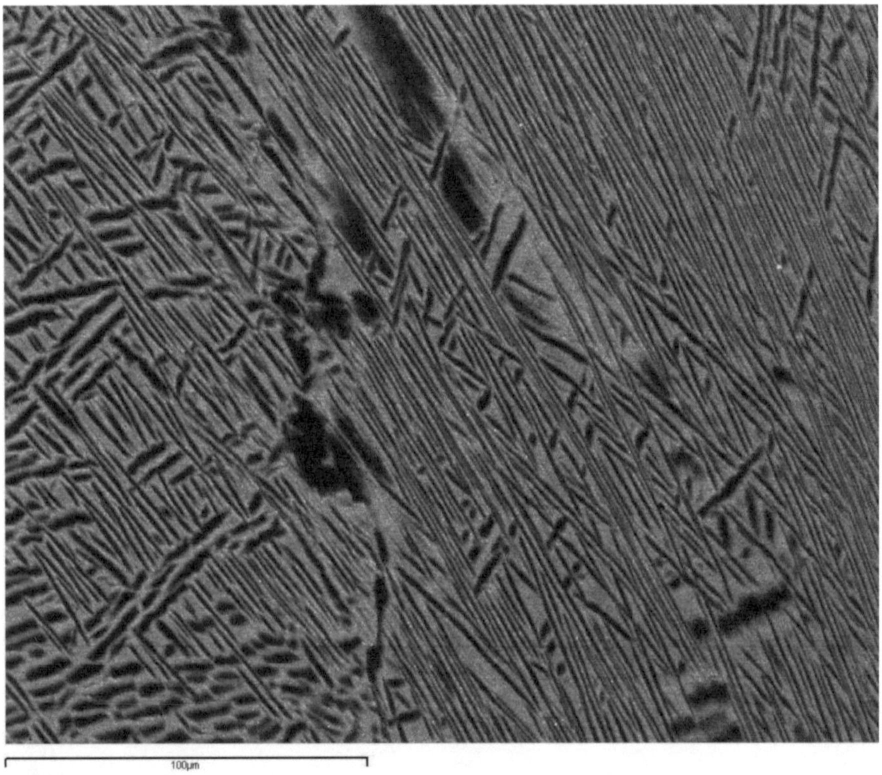

Fig. 14. Microstructure in the disk side when approaching the weld centre (situated to the right side).

Directly in the weld plane the microstructure is too fine to be resolved by conventional SEM. Therefore, more powerful techniques, such as FE-SEM (Field Emission) and TEM (Transmission Electron Microscopy), were used. Figure 15 shows the microstructure at both sides of the weld line, whereas Figure 16 is an even higher magnification picture of the weld centre where grains of nanometric size are observed. This very fine microstructure is characteristic of LFW and evolves from the fact that the temperature during the process exceeds the β-transus in a small volume of material and the cooling rate for this small volume is very high (Helm & Lutjering, 1999).
In the blade side, α+β Ti-6246 microstructure close to the weld zone shows highly deformed α_p nodules, elongated following the friction movement direction (Figure 17). Far away, the microstructures changes gradually towards that corresponding to the base materials, according to the decreasing temperature and plastic deformations reached at each point.

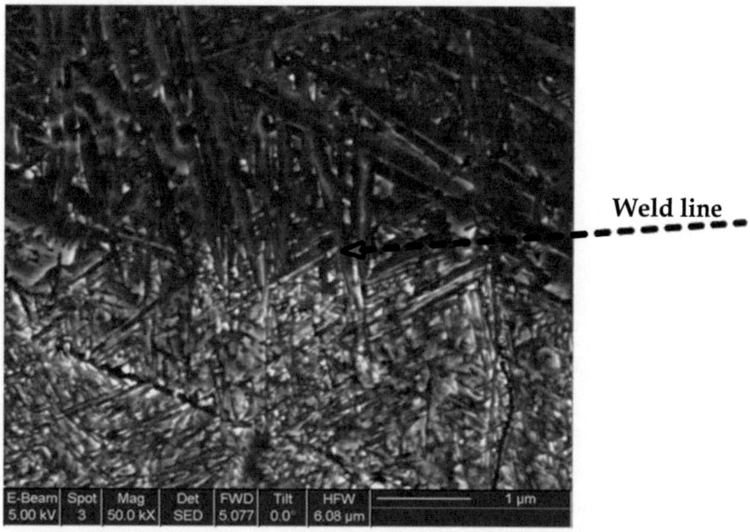

Fig. 15. Microstructure in the weld zone observed by FE-SEM. Disk material is in the upper part of the image and blade material in the lower one.

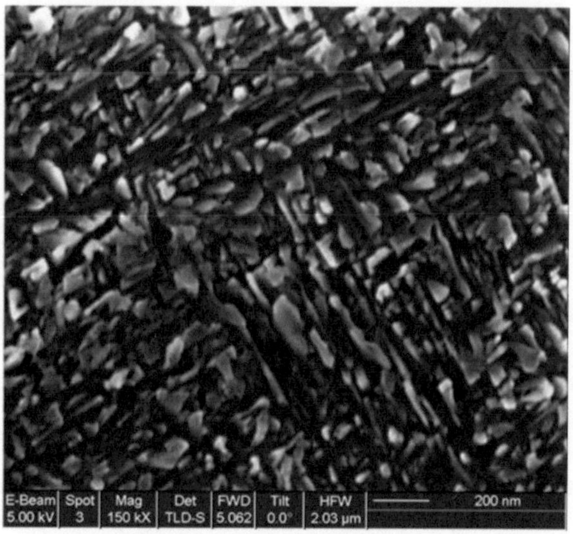

Fig. 16. Microstructure in the weld centre observed by FE-SEM.

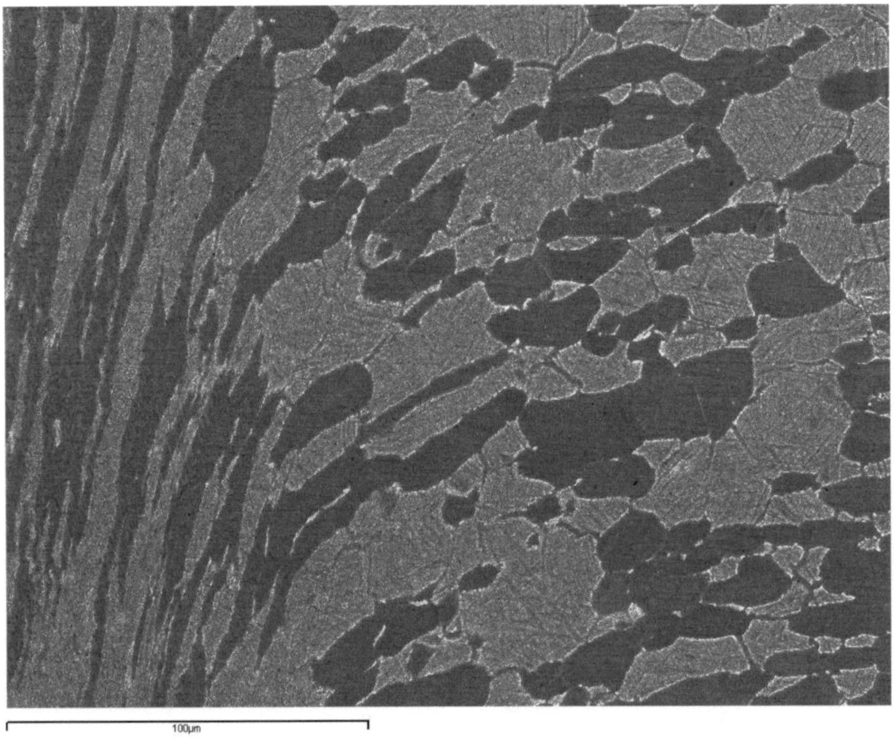

Fig. 17. Microstructure in the blade side when approaching the weld centre (situated to the left side).

5.4 Mechanical characterisation

The following subsections present the results of the different mechanical tests performed to evaluate both the monotonic and cyclic properties of LFW for BLISK manufacturing on titanium alloys.

5.4.1 Hardness profiles

The first step for the characterization of the mechanical response of the welds was measuring their hardness profiles. Loads of 0.5 kg were applied with a Vickers indenter. The profiles showed in Figure 18 correspond to the central part of the specimen, where the heat affected zone is narrower, and to a near flash zone.

5.4.2 Tensile properties

Concerning tensile testing of base materials, it was conducted on cylindrical specimens of 6 mm in diameter. The same geometry was used to machine specimens from the welded blocks, always with the load axis perpendicular to the joining plane of the weld. All tests were performed under displacement control at a rate of 2 mm/min and at least two specimens for condition were tested. Table 1 summarizes the main results of tensile tests.

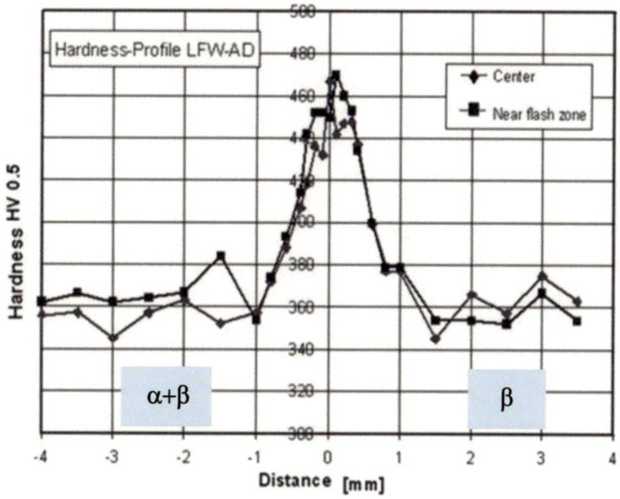

Fig. 18. Hardness profile of one of the welds.

Table 1 results evidence that both base material conditions have similar strength values, being slightly higher for the β-forged alloy. All of them are in the typical range reported for Ti-6246 (Donachie, 2000). Tensile properties of the welds are close to those obtained for bimodal α+β Ti-6246, which is the specimen side where the fracture occurred. It is important to note that fracture never took place by the weld zone. Moreover, it always began in the blade alloy side, at a distance of at least 10 mm from the centre of the weld, as shown in Figure 19.

Alloy	σ_{ys} (MPa)	σ_{uts} (MPa)	Elongation (%)
β Ti-6246	1051	1155	13
α+β Ti-6246	1044	1125	18
Welds	1030	1078	19

Table 1. Tensile testing results.

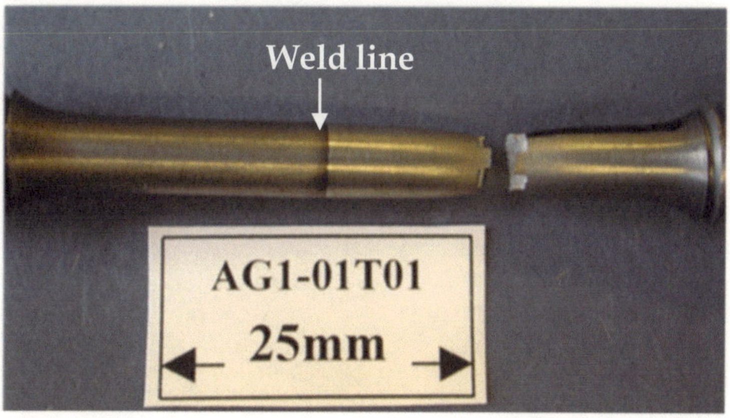

Fig. 19. Photograph of a tested tensile specimen. Note the fracture at the right side, far from the weld line.

5.4.3 Fatigue behaviour

The behaviour in the HCF (High Cycle Fatigue) regime of base Ti alloys and welded specimens was compared. Fatigue testing was carried out at a load ratio R of 0.1. Smooth specimens, carefully polished, with 4 mm in diameter and 35 mm of gauge length were used. The endurance limit was considered for $2x10^7$ cycles. Again, fracture never took place by the weld zone. Figure 20 plots the S-N curves obtained, in terms of maximum stress. Both Ti-6246 conditions, i.e. lamellar and bimodal microstructures, exhibit similar HCF behaviour. Red points corresponding to friction welded specimens are located at comparable and even higher stress values than base Ti alloys. However, the limited number of tested specimens, together with the high scatter on the results, makes difficult to fix a fatigue limit. Analogous fatigue behaviour appears in a published work dealing with different titanium alloys (Wisbey, 1999).

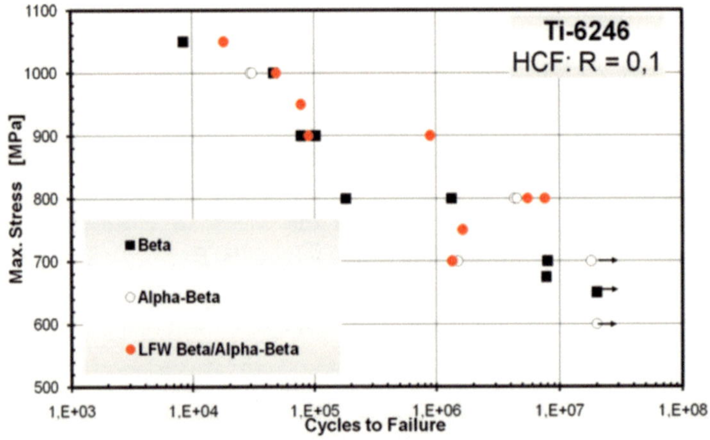

Fig. 20. HCF results for base materials and welds.

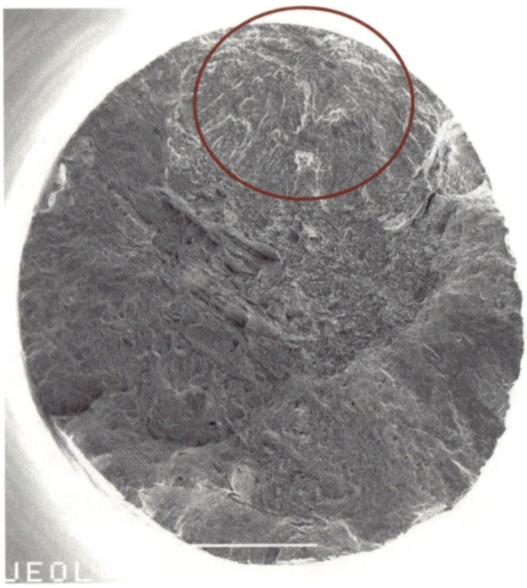

Fig. 21. Fracture surface of welded specimen tested at σ_{max} = 800 MPa and failed after $6x10^6$ cycles. Circle indicates the origin of fatigue fracture.

In order to study crack nucleation mechanisms, an exhaustive analysis of fracture surfaces of the specimens tested in HCF was carried out. This study revealed significant features:

- failures took place sometimes in the disk side and sometimes in the blade side, because both have similar fatigue resistance;
- for high stress levels, i.e. $\sigma_{max} \geq 800$ MPa, cracks nucleated at the surface or subsurface (Figure 21);
- internal crack nucleation (Figure 22) prevailed for low stress levels, i.e. $\sigma_{max} \leq 700$ MPa.

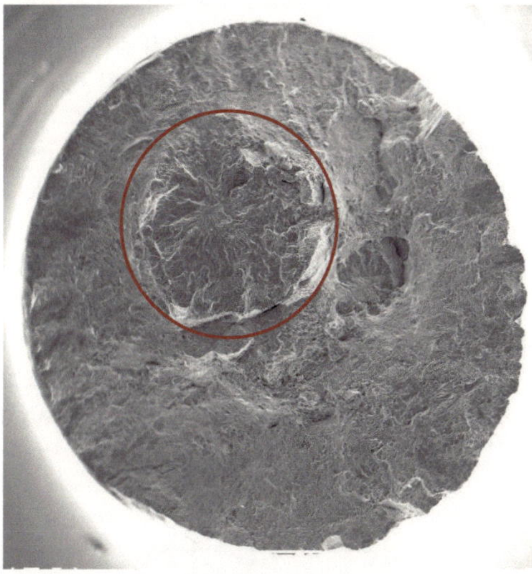

Fig. 22. Fracture surface of welded specimen tested at σ_{max} = 700 MPa and failed after $1,5x10^6$ cycles. Circle indicates the origin of fatigue fracture.

5.4.4 Fracture toughness

Fracture toughness (K_{Ic}) is a critical mechanical design parameter. Its determination was performed in both base materials and LFW samples according to the ASTM E-399 standard. Three-point-bending specimens had a section of 15x15 mm for base materials, whereas those taken from welded joints were 10x5 mm. For the pre-craking process, a sharp pre-crack was produced in the weld plane through application of cyclic compressive loads (Corzo et al., 2007).

Fracture toughness of Ti alloys is very dependent on microstructural and also crystallographic characteristics. Therefore, it is hard to establish a narrow range of "typical values". From literature data (Donachie, 2000), it can be realized that β processed alloys consistently have higher toughness than α+β processed ones. The same trend was appreciated for K_{Ic} values obtained for the two Ti-6246 conditions considered DUTIFRISK project, i.e. β–forged Ti-6246 specimens had higher toughness than α+β–processed Ti-6246 (Table 2). This is related to the fact that the lamellar microstructure induces crack deviations and bifurcations following prior β grain boundaries or colony boundaries (Saucer & Lütjering, 2001). Both deviations of the mode I serve to reduce the effective near tip stress intensity factor: in the case of deviations by inducing mixed mode near tip conditions, while bifurcations disperse the strain field energy among multiple crack tips.

AAlloy	FFracture Toughness, K_{Ic} (MPa.m$^{1/2}$)2)
β Ti-6246	67
α+β Ti-6246	54
Weld 1	15
Weld 2	31
Weld 3	45

Table 2. Fracture toughness testing results.

Three fracture toughness tests were performed on welded specimens. The scatter among the three K_{Ic} values obtained was very high, and all of them were clearly lower than those obtained for base materials. Two different behaviours can be separated. In one specimen both the fatigue pre-crack and the final fracture crack grew completely following the weld line, as shown in Figure 23. In this case, K_{Ic} measured was the lowest one, i.e. 15 MPa.m$^{1/2}$; in the other two tests, crack deviations towards the heat affected zones were observed (Figure 24) and toughness values were more than double than in the previous specimen, i.e 31 and 45 MPa.m$^{1/2}$. From these features, it can be concluded that the "real" toughness of the weld is around 15 MPa.m$^{1/2}$. This low value is not surprising, because the extremely fine microstructure on the weld zone has a very high hardness, which is usually associated with a loss in toughness. The results obtained for the other specimens are in fact evaluations of the toughness of the heat affected zone, because the crack deviated to this area. Values are higher than for the weld but lower than for the base material since the affected zone microstructure is a transition between both.

Fig. 23. Crack path for fracture toughness specimen corresponding to K_{Ic} = 15 MPa.m$^{1/2}$.

Fig. 24. Crack path for fracture toughness specimen corresponding to K_{Ic} = 45 MPa.m$^{1/2}$.

Summarizing the results of the exhaustive mechanical testing evaluation performed along DUTIFRISK project, the most relevant points are:

- Hardness values in the weld zone are very high as compared to base materials;

- Tensile tests on microspecimens taken from the welded zone confirmed its higher strength as compared to base materials (Corzo et al., 2006);
- Tensile properties of welded samples are equivalent to those of the base material with the lowest strength, and fracture always began in that part of the specimen, never in the weld zone;
- High cycle fatigue response of linear friction welds and base materials are qualitatively similar. Failure never took place in the weld zone.
- Fracture toughness evaluation depends on the crack path: when the crack grows along the weld line low values were obtained, however when the crack deviates from the weld centre higher values were attained. Therefore, it is essential to have inspection methods to reveal even the tiniest flaws in the welds.

5.5 Production and validation of BLISK demonstrator

Production of BLISK demonstrators was the last work package of DUTIFRISK project. The objective was assessed and validated the relevant linear friction welding process features on a real part scale. The geometry of the disk and the blades was simplified to reduce the machining cost. In the case of the blades, they were replaced by rectangular blocks, whereas Figures 25 is a photograph of a disk for one of the demonstrators.

Fig. 25. Disk for BLISK demonstrator.

Linear friction welded joints of BLISK demonstrators were validated by performing microstructural characterisation and mechanical testing and comparing the results with those obtained from laboratory specimens. Consequently, after welding the blades to the disk, demonstrators were cut in order to extract blocks for machining testing specimens. Figure 26 shows the pre-machined BLISK demonstrator after extraction of blade-blocks. Those blade-blocks were post-weld heat-treated prior to the machining of the specimen blanks. Tensile and HCF tests were carried out and mechanical properties values were similar to those obtained within the normal scatter of materials data.

Fig. 26. Pre-machined BLISK demonstrator, after extraction of blade-blocks.

6. Future research

BLISK is a critical part of an aero-engine, for that reason it is obvious that the integrity and reliability of linear friction welds must be totally verified. In order to guarantee in service welds performance, more research is needed in several fields:

- Optimisation of LFW parameters for each particular combination of alloys;
- Improving the understanding of the relationship between the very fine microstructure on the weld zone and its mechanical properties, especially concerning fracture toughness;
- Development of advanced inspection techniques adapted for BLISK inspection, such as computer tomography and ultrasonic inspection, is vital to reveal defects in the welds.

On the other hand, the final customers of aero-engines will accept BLISK technology only if suitable and affordable repairing technologies are available. Damages susceptible of being repairable are similar in a BLISK and in "normal" disk + single blades configuration, with the difference that an exchange of a blade in a BLISK is not as simple as for single blades. In DUTIFRISK project, a work-package was dedicated to this topic and the conclusion was that the most convenient repairing technique is the same LFW process (Mateo et al., 2009). However, further investigations should be committed to evaluate the effect on mechanical properties of successive welds in the same area and then to establish repairing procedures.

7. Conclusion

The overall conclusion drawn from the industrial experiences and research activities conducted on linear friction welding of Ti alloys for BLISK manufacturing, and in particular after all the tasks carried out during DUTIFRISK project, is that it is feasible to produce, by using linear friction welding, a dual alloy/dual microstructure BLISK with a disk optimised from the viewpoint of low cycle fatigue resistance and optimised blades for high cycle fatigue. It was further proven that it is possible to join different titanium alloys, thereby allowing customising the blades according to the required temperature capability of a certain stage.

8. Acknowledgement

The author of this chapter wants to recognize to all the colleagues that participated in DUTIFRISK project: J.-P. Ferte (SNECMA Moteurs), E. Gach (Böhler Schmiedetechnik), J. Méndez, J. Petit and P. Villechaise (ENSMA-CNRS), M. Nunn (TWI), M. Anglada and M. Corzo (CIEFMA-UPC) and especially to the project coordinator O. Roder (MTU Aero Engines). Also financial support of the European Commission (G4RD-CT-2001-00631, Acronym DUTIFRISK) and from the Spanish *Ministerio de Ciencia e Innovación* (MAT2009-014461) are acknowledged.

9. References

Bhadeshia, H.K.D.H. & DebRoy, T. (2009). Critical assessment: friction stir welding of steels. *Science and Technology of Welding and Joining,* Vol.14, No.3, pp. 193-196, ISSN 1362-1718

Bhamji, I.; Preuss, M.; Threadgill, P.L.; Moat, R.J.; Addison, A.C. & Peel, M.J. (2010). Linear friction welding of AISI 316L stainless steel. *Materials Science and Engineering A,* Vol.528, No.2, pp. 680-690, ISSN 0921-5093

Broomfield, R.W. (1986). Application of advanced joining techniques to titanium alloys, *Proceedings Conference on Designing with Titanium,* pp. 69-75, The Institute of Metals, Bristol, UK, 1986

Bußmann, M.; Kraus, J. & Bayer, E. (2005). An Integrated Cost-Effective Approach to Blisk Manufacturing, *Proceedings of 17th Symposium on Air Breathing Engines,* Munich, Germany, 4-9 September 2005

Ceschini, L.; Morri, A.; Rotundo, F.; Jun, T.S. & Korsunsky, A.M. (2010). A Study on Similar and Dissimilar Linear Friction Welds of 2024 Al Alloy and 2124Al/SiCP Composite. *Advanced Materials Research,* Vol.89-91, pp. 461-466, ISSN 1662-8985

Corzo, M.; Mendez, J.; Villechaise, P.; Rebours, C.; Ferte, J.-P.; Gach, E.; Roder, O.; Llanes, L.; Anglada, M. & Mateo, A. (2006). High-cycle fatigue performance of dissimilar linear friction welds of titanium alloys, *Proceedings of the 9th Int. Fatigue Congress,* Elsevier, paper O326, CD-ROM, Atlanta, USA, 14-19 May 2006

Corzo, M.; Casals, O.; Alcala, J.; Mateo, A. & Anglada, M. (2007). Mechanical evaluation of linear friction welds in titanium alloys through indentation experiments. *Welding International,* Vol.21, No.2, pp. 125-129, ISSN 0950-7116

Donachie, M.J. (2000). *Titanium: A technical guide,* ASM International, ISBN 0871706865, Ohio, USA

Esslinger, J. (2003). *Proceedings of 10th Titanium World Conference,* Ed. G. Lütjering and J. Albrecht, Wiley-VCH, Vol.V, pp. 2837-2844, ISBN 3527303065, Munich, Germany, 13-18 July 2003

External Advisory Group for Aeronautics of the European Commission (2000). Aeronautics for Europe, *Official Publications of the European Communities,* ISBN 92-828-8596-8, Belgium

Harvey, R.J.; Strangwood, M. & Ellis, M.B.D. (1995). Bond line structures in friction welded Al_2O_3 particulate reinforced aluminium alloy metal matrix composites, *Proceedings 4th Int. Conference on Trends in Welding Research,* ASM, pp. 803-808, ISBN 0871705672, Tennessee, USA, June 5-8, 1995

Helm, D. & Lutjering, G. (1999). Microstructure and properties of friction-welds in titanium alloys, *Proceedings of 9th World Conference on Titanium,* pp. 1726-1733, Saint Petersburg, Russia, 1999

Kauzlarich J.J. & Maurya Ramamurat R. (1969). *Reciprocating Friction Bonding Apparatus.* Patent Number US3420428, Caterpillar Tractor Co., 7th January 1969

Lee, W.B. & Jung, S.B. (2004). The joint properties of copper by friction stir welding, *Materials Letters*, Vol.58, Issue 6, pp. 1041-1046, ISSN 0167-577X

Mary, C. & Jahazi, M. (2006). Linear Friction Welding of IN-718 Process Optimization and Microstructure Evolution, *Advanced Materials Research*, Vol.15-17, pp. 357-362, ISSN 1662-8985

Mateo, A.; Corzo, M.; Anglada, A.; Mendez, J.; Villechaise, P.; Rebours, C. & Roder, O. (2009). Welding repair by linear friction in titanium alloys, *Materials Science and Technology*, Vol.25, No.7, pp. 905-913, ISSN 0267-0836

Messler, R.W. (2004). *Joining of materials and structures*, Elsevier, ISBN 0-7506-7757-0, Oxford, U.K.

Mishra, M. & Mahoney, M.W. (2007). *Friction stir welding and processing*, ASM International, ISBN 978087170848-9, Ohio, USA

Nunn, M. (2005). Aero Engine improvements through linear friction welding. *Proceedings 1st Int. Conference on Innovation and Integration in Aerospace Sciences*, Queen's University, Belfast, Northern Ireland, U.K., 4-5 August 2005

Roder, O.; Helm, D. & Lütjering, G. (2003). In: Titanium'03: Science and Technology, *Proc. 10th Titanium World Conference*, Ed. G. Lütjering and J. Albrecht, Wiley-VCH, Vol. V, pp. 2867-2874, ISBN 3527303065, Munich, Germany, 13-18 July 2003

Roder, O.; Ferte, J.-P.; Gach, E.; Mendez, J.; Anglada, M. & Mateo, A. (2008). Development and validation of a dual titanium alloy dual-microstructure BLISK, *Proceedings 1st EUCOMAS (European Conf on Materials and Structures in Aerospace)*, Ed. VDI BERICHTE, ISBN 0083-5560, pp. 309, Berlin, Germany, 26-27 May 2008

Saucer, C. & Lütjering, G. (2001). Influence of α layers and β grains boundaries on mechanical properties of Ti alloy, *Materials Science and Engineering A*, Vol.319-321, pp. 393-397

Strand, S.R.; Sorensen, C.D. and Nelson, T.W. (2003). *Proceeding Conference ANTEC 2003*, Society of Plastics Engineers, Vol.1, pp. 1078–1082; Nashville, USA, 4-8 May 2003

Threadgill, P.L. (1995). Joining of a nickel aluminide alloy. *Proceedings 4th Int. Conference on Trends in Welding Research*, ASM, pp. 317-322, ISBN 0-87170-567-2 Tennessee, USA, 5-8 June, 1995

Threadgill, P.L.; Leonard, A.J.; Shercliff, H.R. & Withers, P.J. (2009). *International Materials Reviews*, Vol.54, No.2, pp. 49-93, ISSN 0950-6608

Threadgill, P.L. (2011). Joining Linear friction welding, In: *TWI web page*, March 2001, Available from: http://www.twi.co.uk/content/ksplt001.html

Vairis, A. & Frost, M. (1998). On extrusion stage of linear friction welding of Ti-6Al-4V, *Wear*, Vol.217, pp. 117–131, ISSN 0043-1648

Wilhem, H.; Furlan, R. & Moloney, K.C. (1995). Linear friction bonding of titanium alloys for aeroengine applications, *Proceedings of the 8th World Conference on Titanium*, Ed. P.A. Blenkinsop et al., pp. 620-626, ISBN 1861250053, Birmingham, U.K., 22-26 October 1995

Wisbey, A.;Wallis, I.C.; Ubhi, H.S.; Sketchley, P.; Ward-Close, C.M. & Thereadgill, P.L. (1999). Mechanical properties of friction welds in high strength titanium alloys, In: Titanium'99, *Proceedings of the 9th World Conference on Titanium*, pp.1718-1725, Saint Petersburg, Russia, 1999

19

Damageability of Gas Turbine Blades – Evaluation of Exhaust Gas Temperature in Front of the Turbine Using a Non-Linear Observer

Józef Błachnio and Wojciech Izydor Pawlak
Air Force Institute of Technology (Instytut Techniczny Wojsk Lotniczxych-ITWL), Poland

1. Introduction

A turbine is a fluid-flow machine that converts enthalpy of the working agent, also referred to as the thermodynamic agent (a stream of exhaust gas, gaseous products of decomposition reactions or compressed gas) into mechanical work that results in the rotation of the turbine rotor. This available work, together with the mass flow intensity of the working agent, define power that can be developed by the turbine and subsequently used to drive various pieces of equipment (e.g. compressors of turbojet engines). The basic advantages of gas turbines include: possibility to develop high power at rather compact dimensions and low bare weight, relatively high efficiency of the energy conversion process, simple design and high reliability of operation (Błachnio, 2004, 2007; Kroes et al., 1992; Sieniawski, 1995). On the other hand, the drawbacks are: high operating temperatures of some components, sophisticated geometrical shapes of the components, e.g. blades and vanes, which makes the manufacturing process difficult, as well as high working speeds of rotors that impose the need to apply reduction gears, e.g. when turbine-power receivers show limited rotational speeds.

Because of the direction of flow of the exhaust gas, turbines are classified as axial-flow and radial-flow systems. Each turbine is made up of two basic subassemblies that compose the turbine stage.

- A stationary rim with profiled vanes fixed co-axially (axial-flow turbines) or in parallel (radial-flow turbines), i.e. the so-called turbine nozzle guide vanes, or shortly, the stator;
- A moving rim (one or several ones) with profiled blades fixed circumferentially (axial-flow turbines) or on the face surface (radial-flow turbines) of a rotating disk seated on the shaft, i.e. the turbine blade rim.
- Depending on the distribution of the inlet energy of exhaust gases among basic subassemblies, turbines are classified as:
  - action (impulse) turbines, - the exhaust gases are subject to decompression exclusively in turbine nozzle guide vanes,
  - reaction turbines - the exhaust gases are decompressed by both the guide vanes (stator vanes) and in the turbine rotor.

In industrial-type turbines a portion of the produced energy is used to drive a compressor, whereas the rest of it - to generate power transmitted then to power receivers. In aeronautical applications, the gas turbine is a structural component of a turbojet, a turboprop or a helicopter engine. The turbine power affects the engine performance; improvement in the turbine efficiency results in the engine thrust (power) increase and reduction in the unit fuel consumption.

Gas turbines, designed for industrial plants, vehicles, off-shore applications, power engineering systems, and aeronautical applications offer pretty high efficiency of 30 to 45% (Błachnio, 2004; Kroes et al., 19920). That efficiency depends, and to a very high degree, on temperature of exhaust gases at the turbine inlet. Over the recent years this temperature has increased by more than 450 K, which has resulted in substantial improvement in overall efficiency offered by turbines and made it possible to achieve even higher coefficient of unit power. Particular attention has been paid to increase heat resistance and high-temperature creep resistance of turbine components, especially of the 1st stage vanes and blades. For that purpose, dedicated systems are applied to cool down the vanes and blades in order to secure reliable operation of the machine under heavy-duty thermal and mechanical loads. This, in turn, reduces the working temperature of the material by as much as 625 K as compared to the exhaust gas temperature. It is much easier to develop a cooling system when designers deal with large vanes and blades. Application of similar cooling systems to compact turbines is associated with undesirable drop in the turbine efficiency. This is why further development of the turbine-blade production engineering processes, aimed at capabilities to increase temperature upstream the turbine, has been focused on spreading heat-resistant coatings showing good resistance to high-temperature corrosion, low thermal conductivity, and high stability of the material structure. The operating temperature of turbine blades and vanes can be maintained within the following intervals: (Błachnio, 2007; Kroes et al., 1992; Paton, 1997; Sieniawski, 1995; Taira and Ohtani, 1986), depending on structural materials used and cooling intensity:

- 1100 - 1200 K (when no dedicated cooling system is applied);
- 1200 - 1300 K (when blade/vane cooling system is applied);
- 1300 K and above (when an intense-cooling system is applied).

Moreover, vanes and blades are coated with materials that enable increase in blade/vane operating temperature. However, a highly sophisticated design and manufacture engineering processes increase the production overheads.

2. Description of failures to gas turbine vanes and blades

The process of gas turbine operation is associated with various failures to structural components of gas turbines, in particular blades. Condition of the blades is of crucial importance to reliability and lifetime of the entire turbine, and the 'parent' subassembly where it is installed. This is why the blades are subject to scrupulous checks, both during the manufacture and at the stage of assembly, when any deviations from the specification are detected and eliminated. Analysis of the literature and own experience (Błachnio, 2007; Błachnio & Bogdan, 2008; Hernas, 1999; Nikitin, 1987) show that only a small portion of damages/failures to turbine vanes and blades are caused by material defects, structural and/or engineering process attributable defects; most damages/failures are service-attributable (Fig. 1).

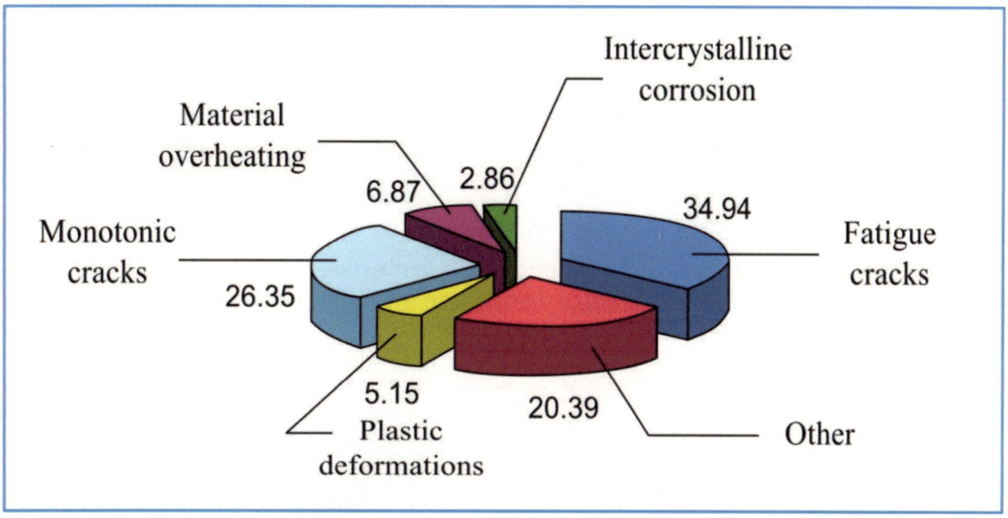

Fig. 1. Causes of failures to aircraft turbine engines during service (in percentage terms) (Błachnio, 2007)

Durability of turbine vanes and blades is a sum of a number of factors, where material quality is the matter of crucial importance. With respect to materials, durability can be defined as time of item operation when alloy properties developed during the manufacturing process remain steady (unchanged). Stability of the properties (the assumed service time) is defined at the design stage by selection of the desired characteristics (as compared to the expected loads and with account taken of the fact that the properties are subject to changes with time). High and stable strength properties of superalloys offer suitable microstructures that are resistant to any deterioration during the service. These structural features have been assumed a durability criterion.
During the service, gas turbine components may be subject to failures resulting from the following processes (Błachnio, 2007, 2009; Hernas, 1999; Poznańska, 1995; Swadźba, 2007; Szczepankowski et al., 2009; Taira & Otani, 1986; Tomkins, 1981):

1. Creeping;
2. Overheating and melting;
3. Low-cycle and high-cycle fatigue due to thermal and thermomechanical factors,
4. Corrosion and fatigue cracking
5. Chemical and intercrystalline corrosion,
6. Erosion
7. Other factors of less importance.

Failures to gas turbine vanes and blades most often are attributable to what follows

2.1 Mechanical failures

a. deformations due to foreign matter affecting the blade (Fig. 2).

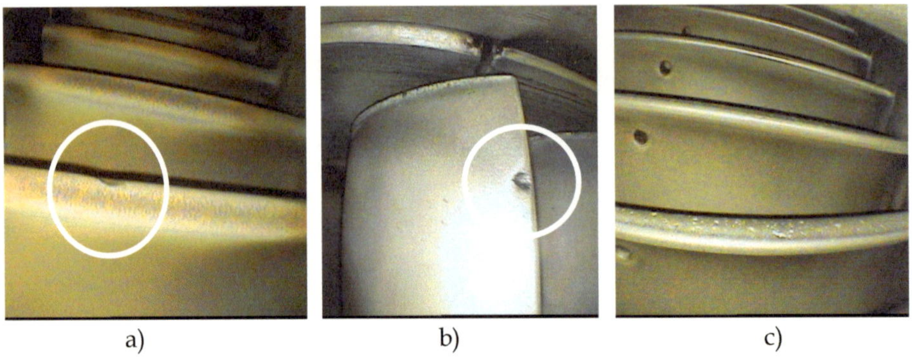

Fig. 2. Deformations of turbine blades in the form of dents caused by foreign matter (Reports, 2000-2010): a) - on the leading edge, b) - on the trailing edge, c) - on the on suction faces of blades

b. foreign-matter-attributable surface scratches (Fig.3).

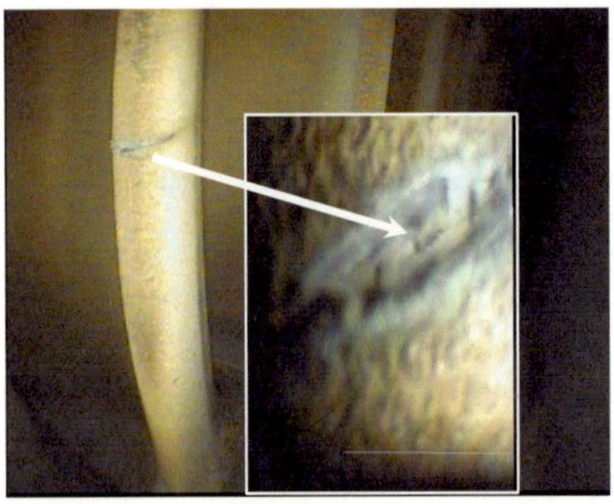

Fig. 3. Scratches on protective coating conducive to corrosion on the leading edge of a turbine blade (Reports, 2000-2010)

c. erosive wear (Fig. 4).

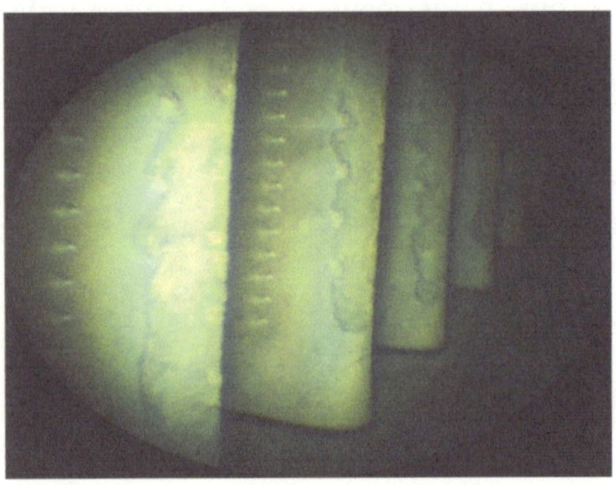

Fig. 4. Erosive wear of leading edges of rotor blades (Korczewski, 2008)

d. fatigue (Fig. 5).

a) b)

Fig. 5. Failures to gas turbine rotor blades caused by (Reports, 2000-2010):
a) - fatigue cracking of leading edge, b) - fatigue fracture located at the blade's locking piece

2.2 Thermal failures

a. creeping (Fig.6).

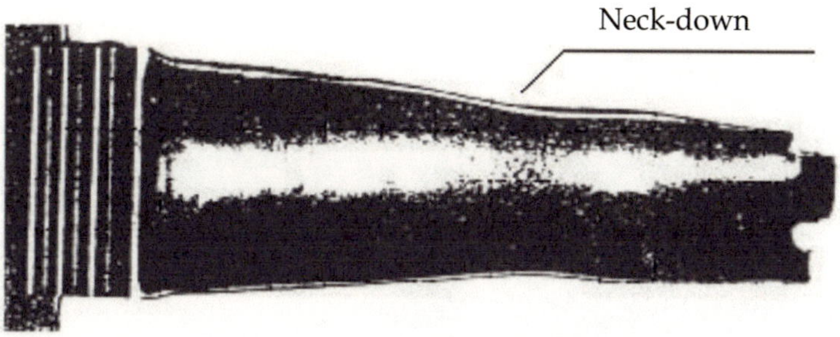

Fig. 6. Plastic deformation of the blade (Bogdan, 2009).

b. overheating of blade material (Fig. 7)

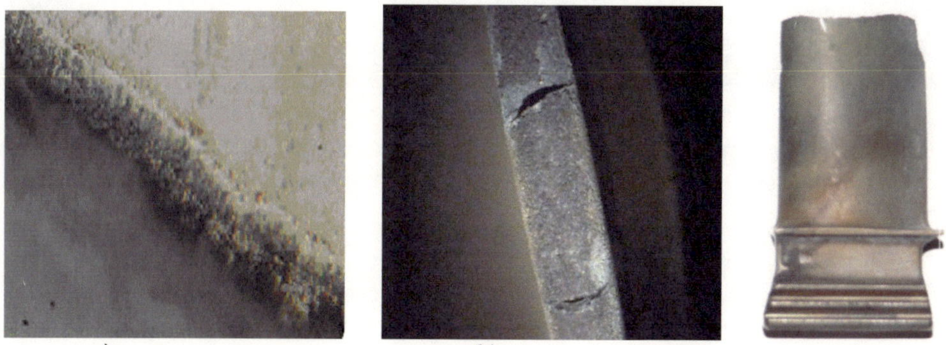

a) b) c)

Fig. 7. Characteristic forms of failures caused by overheating of blade material (Reports, 2000-2010): a– partial melting of blade's trailing edge, b) - cracks on blade's leading edge, c) - breakaway of the blade (Błachnio, 2010)

c. melting of the vane material (Fig. 8)

a) b)

Fig. 8. Characteristic forms of failures to gas turbine caused by long-lasting excessive temperature of exhaust gases (Reports, 2000-2010) : a) – burn-through of turbine rotor blades, b) – melting of a nozzle vane

2.3 Chemical failures

a. high-temperature corrosion (Fig. 9)

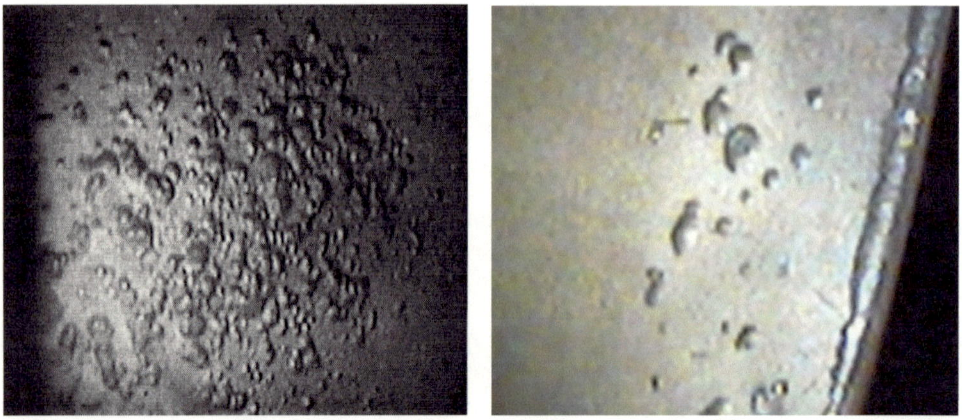

a) b)

Fig. 9. Failures to turbine blades operated in the seashore environment, caused by chemical impact of exhaust gases (Reports, 2000-2010): a) – on blade surface, b) – on blade leading edge

b. intercrystalline corrosion (Fig. 10).

Blade deformations in the form of dents (Fig. 2) are caused by a foreign matter ingested by the turbojet engine compressor and by particles of metal and hard carbon deposits from the combustion chamber. Such dents result in stress concentrations in blade material and prove conducive to the initiation of fatigue processes.

Scratches on blade surfaces (Fig. 3) due to the foreign matter impact are also reasons for local stress concentrations and, consequently, potential corrosion centers. What results is, again, material fatigue which, together with possible corrosion, prove conducive to fatigue fracture.

Fatigue of material of turbine rotor blades is caused by a sum of loads due to: non-uniform circumferential distribution of the exhaust gas stream leaving the combustion chamber and its unsteadiness in time, non-uniformity of the exhaust gas stream leaving the nozzle, and excitations from the structure of, e.g. the turbojet engine. The dynamic frequency of free vibration attributable to the rotor blade of variable cross-section depends on the centrifugal force, therefore, it is a function of rotation speed. It also depends on temperature of the working agent affecting the longitudinal modulus of elasticity (Young's modulus) of the material. The most hazardous are instances of turbine blade operation at resonance of the 1st form of vibration (single-node form). Such circumstances usually lead to fatigue cracking and finally, the blade breakaway Fig. 5).
Response of the gas turbine blade material to mechanical loads depends first and foremost on the blade operating temperature. Selection of material to manufacture a blade of specified durability should take account of mechanical properties in the area of maximum temperature. A typical temperature distribution along the blade is far from uniform (Fig. 10). Failures to first turbine stages are usually caused by exhaust gases of very high temperature, whereas blades of subsequent stages (i.e. the longest blades) suffer damages resulting mainly from mechanical loads (vibration, the centrifugal force).

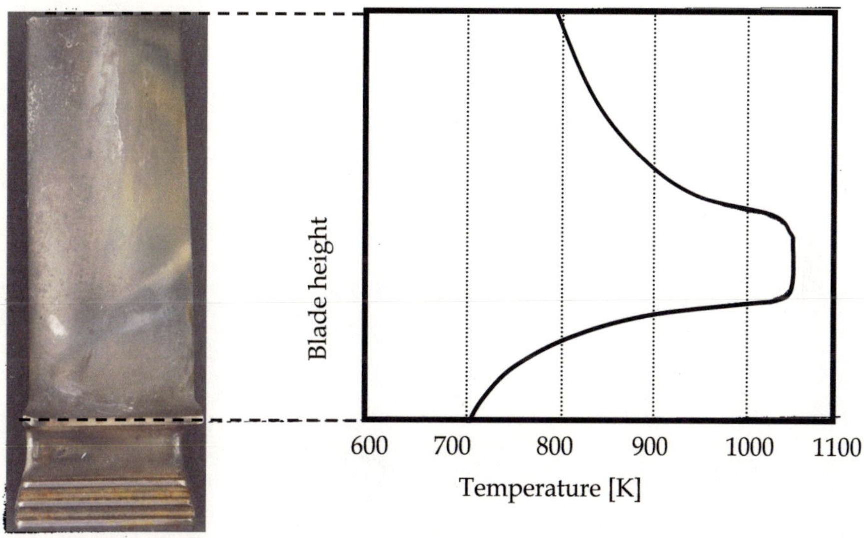

Fig. 10. Typical temperature distribution along the gas turbine blade

The predominant majority of failures to gas turbine blades are effected with inappropriate operation (misadjustment) of subassemblies mating with the turbine, first of all, the combustion chamber and, like with turbines of aircraft turbojet engines, the exhaust nozzle (in particular, the mechanism to adjust nozzle-mouth cross-section).
Quite frequent causes of failures are overheating of blade material and thermal fatigue of blades resulting from both the excessive temperature and the time the blade is exposed to high temperature. Overheating of vanes and blades takes place when the permissible average value of the exhaust gas temperature is exceeded. It may also result from the non-

uniform circumferential temperature distribution (Fig. 11). One of possible causes of non-uniform temperature distribution downstream the turbine lies in the improper fuel atomization due to excessive carbon deposit on fuel injectors (Fig. 12).

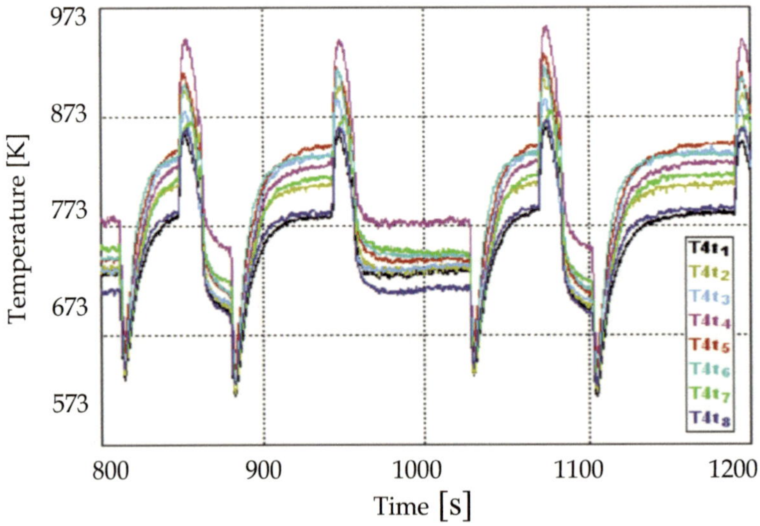

Fig. 11. Instantaneous circumferential non-uniform temperature T_4 distribution measured with 8 thermoelements ($T4t_1$ -$T4t_8$) located behind the turbine; measurements taken at increasing/decreasing rotational speeds

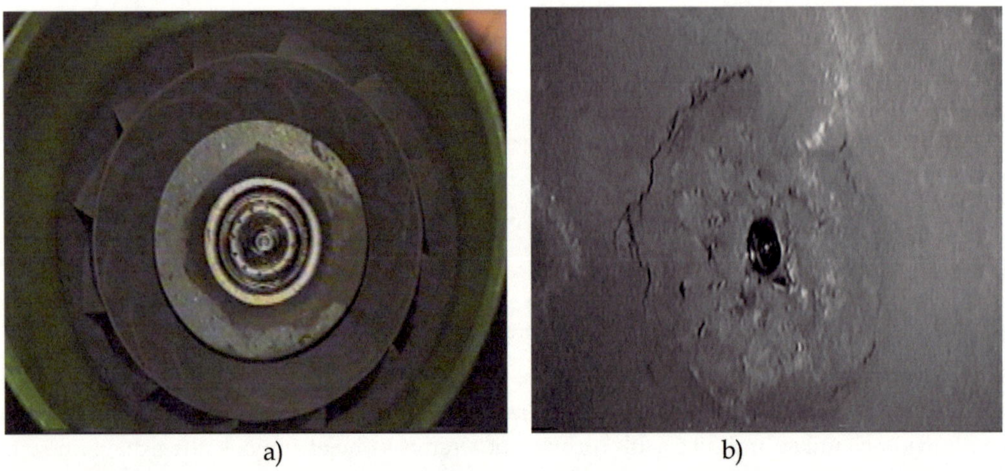

Fig. 12. Condition of combustion-chamber injectors: a) - clean, b) - polluted with carbon deposits from fuel

Elongation of the plasticized material of a rotor blade results from the blade being affected with overcritical temperature and centrifugal force. In such cases the rotor blade shows a characteristic 'neck-down' (Fig. 6). When it happens to a blade in the turbine nozzle blade row, it can suffer bending due to thermal extension of the material; the 'elongation capacity' of the blade is limited by the turbine's body.

Another very frequent cause of failures to vanes and blades is overheating of material combined with thermal fatigue caused by the excessive temperature and prolonged exposure time as well as by chemical activity of the exhaust gas (Fig. 7, Fig. 8). The high-temperature creep resistance of alloys for turbine vanes and blades is closely related with the strengthening γ' phase. The γ'phase is a component of the material's microstructure that has the strongest effect upon properties of supperalloys. The shape, size and distribution of γ' phase particles are factors of crucial importance to mechanical properties of the material.

Failures in the form of high-temperature corrosion of turbine vanes and blades are first and foremost caused by chemical compounds found in both the exhaust gas and the environment, e.g. moisture in seashore environment. Sulphur compounds in aircraft fuel, e.g. the Jet A-1 type (F-35) may contain not more than 0.3% of sulphur per a volume unit. This, in turn, may increase the content of SO_2 in the exhaust gas up to as much as approx. 0.014% (Nikitin, 1987; Paton, 1997; Swadźba, 2007). Hence the conclusion: the higher content of this element in aircraft fuel, the higher amount of SO_2 and SO_3 in the exhaust gas. It brings about the hazard of chemical corrosion on the surfaces of vanes and blades, which additionally may be caused by improper organization of the fuel combustion process. Chemical corrosion of turbine vanes and blades results in the formation of surface corrosion pits and, consequently, in the blade cracking and sometimes fracture.

Initiation and propagation of such failures is also affected by negligence in adhering to specified parameters while spreading protective coatings in the manufacturing or repair processes. The environment of operating the turbine, e.g. an aircraft turbine engine or a or turbojet is also of crucial importance to the system. Operating such engines in the seashore or offshore environments with elevated content of sodium chloride proves also conducive to chemical corrosion of turbine vanes and blades. Chemical corrosion considerably contributes to the formation of surface corrosion pits and, finally, to blade cracking and fracture when a substantial drop in mechanical properties occurs.

Another form of failures to vanes and blades of a gas turbine during operation thereof is the intercrystalline corrosion, which may result in changes to chemical composition of alloys at grain boundary. Propagation thereof is encouraged by environmental conditions under which the turbine is operated. The environment may contain aggressive compounds, such as sodium sulphite. If so, temperature above 1050 K is really conducive to the propagation of this type of corrosion (Antonelli et al., 1998; Swadźba, 2007). The intercrystalline corrosion usually attacks alloys with ferrous, nickel, or cobalt matrixes. The increased content of chromium in the alloy reduces the alloy susceptibility to intercrystalline corrosion, whereas the increased concentration of sodium chloride intensifies it, making the process proceed relatively fast. Author's experience proves that operation of the turbine under adverse conditions, i.e. at variable temperature, with permissible value thereof being periodically exceeded, substantially increases susceptibility of such alloys to intercrystalline corrosion. What results is a drop in the chromium content in the overheated region of the material, and the presence of relatively large carbides at grain boundary (Nikitin, 1987; Paton, 1997).

3. The assessment of condition of gas turbine vanes/blades throughout the operational phase

Throughout the operational phase of any gas turbine various forms of failures to turbine components may occur. These failures, different in intensity, may result in the malfunction of the turbine, and sometimes even in a notifiable accident, as e.g. in aviation. Failures/damages are always remedied by a major repair or overhaul of the turbine, both of which generate huge costs. The cost of engine major repair, not to mention an overhaul, are several thousand as high as unit price of a single vane or blade.

Any decision on whether the engine needs repair is taken by a diagnostic engineer who performs visual inspection with, e.g. a videoscope (Fig. 13) and is able to inspect and diagnose condition of difficult of access turbine components. The condition assessment is performed using a recorded image of the inspected component's surface and comparing it with pattern images of surfaces of serviceable and unserviceable (fit/unfit for use) components, e.g. analogous vanes and blades of the turbine. An experienced diagnostic engineer is capable of assessing the risk that failures such as dents, melting of materials, fatigue cracks or corrosion may pose. However, the assessment of, e.g. overheated material is much more difficult as it has to be based on the colour of the blade surface (Fig. 14).

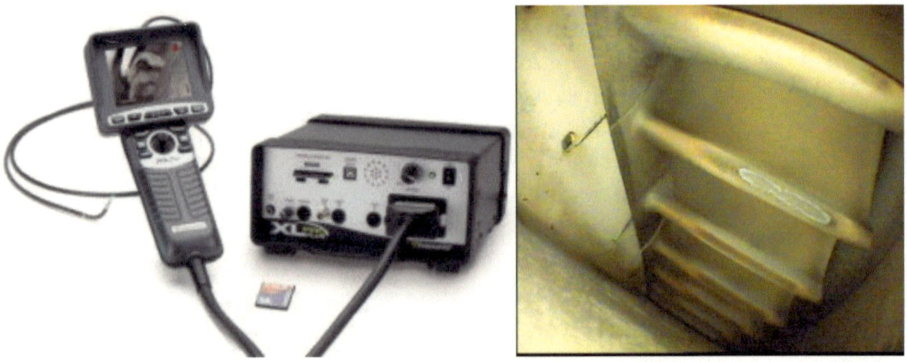

Fig. 13. An industrial videoscope and an image of gas turbine blades condition (Reports, 2000-2010)

Fig. 14. A gas turbine with visible changes in colour on surfaces of vanes - the evidence of different degrees of vane overheating (Reports, 2000-2010)

Such an assessment can be carried out using, e.g. a table of colours typical of the layer of oxides and corresponding temperatures upon a vane/blade fracture if the vane/blade is air-cooled (Table 1).

Temperature [K]	Colour of a layer of oxides upon vane fracture
~ 670	light gray
~ 770	light gray with a pale yellow shadow
~ 870	bright yellow
~ 920	yellow, dark yellow
~ 970	yellowish brown
~ 1020	yellowish brown with violet shadow
~ 1070	dark violet
~ 1120	blue, navy blue

Table 1. Colour of layer of oxides and corresponding temperatures upon vane/blade fracture if the vane/blade is air-cooled (Bogdan, 2009)

The trustworthiness of the condition assessment depends on a number of factors, i.e. skills and experience of the diagnostic engineer, the diagnostic method applied, condition of diagnostic instruments, external circumstances of the experiment, etc. To a large extent it is a subjective assessment by the diagnostic engineer, which always poses some risk R of the decision taken; the risk is expressed by the following formula (Błachnio & Bogdan, 2008).

$$\begin{aligned} R = {} & c_{11}p(w^1)\int_{-\infty}^{y_0} f(y_n/w^1)dy + c_{21}p(w^1)\int_{y_0}^{\infty} f(y_n/w^0)dy + \\ & + c_{12}p(w^0)\int_{-\infty}^{y_0} f(y_n/w^1)dy + c_{22}p(w^0)\int_{y_0}^{\infty} f(y_n/w^0)dy \end{aligned} \quad (1)$$

where:

$p(w^1)\int_{y_0}^{-\infty} f(y_n/w^0)dy$	probability of the 1st class error (a serviceable/fit-for-use object is assessed as an unserviceable/unfit-for-use one, probability of a false alarm, risk of placing an order),
$p(w^0)\int_{-\infty}^{y_0} f(y_n/w^1)dy$	probability of the 2nd class error (an unserviceable/ unfit-for-use object is assessed as a serviceable/fit-for-use one, contractor's risk),
c21 = w21	- cost (loss) in case of the 1st class error
cl2 = wl2	- cost (loss) in case of the 2nd class error
c11 = w11, c22 = w22	- right decision related cost (loss)

w^0 – status of serviceability,
w^1 – status of unserviceability,
y_0 – initial value of the status parameter,
y_n – final value of the status parameter.

Mistakes resulting from the subjective assessment carried out by the diagnostic engineer may lead to that the overheated vane is taken for a good one, and vice versa, the good one for an overheated one. In the first case, after a pretty short time of engine operation an air accident occurs, whereas the second-type mistake entails enormous cost of a major repair/overhaul of the engine. The assessment provided by the diagnosing engineer is verified with a destructive method, i.e. the microsection of the vane/blade in question is carefully analysed.

As already mentioned, the most difficult for type identification and for classification of vane/blade condition are failures in the form of material overheating, in particular of uncooled items. sometimes Apart from the strict bipolar classification 'serviceable/fit-for-use - unserviceable/unfit-for-use', in some instances of diagnosing vane/blade condition, the third, intermediate level of the component-condition assessment is used, namely the 'partly serviceable/fit-for-use'. This classification is applicable to, among other things, gas turbines installed, e.g. in aircraft turbojet engines, i.e. to very expensive systems expected (and required) to show the possibly maximum cost effectiveness (the 'durability to cost-of-operation' ratio). Therefore, if the diagnostic engineer delivers his subjective assessment with regard to the degree of overheating understood as a change in colour intensity, and to the size and location of the overheated area on the vane/blade, the three-grade assessment scale is applicable. If it is recognised that the degree of overheating suggests the vane/blade is classified to the 'partly serviceable/fit-for-use' category, the current assessment of the vane/blade condition is periodically carried out until the item reaches the 'unserviceable/ unfit-for-use' condition. Consequently, the turbine's life, i.e. its time of operation after a failure had occurred to a vane/blade (of an expensive aircraft engine) can be extended; the cost of engine operation is also reduced. Obviously, the flight-safety level of an aircraft with an engine furnished with a periodically diagnosed turbine cannot be compromised.

Currently, there are no unbiased criteria that enable unambiguous in-service assessment of the degree of overheating of vane/blade material with non-destructive methods. The case illustrated in Fig. 14 - there is no chance to unambiguously assess whether the surface of at least one vane exhibits symptoms of the material overheating, needless to say that nothing can be concluded about the degree of overheating if only the already existing criteria can be applied.

4. Examination of microstructures of damaged gas turbine blades

4.1 Object and methodology of the examination

Subject to examination were gas turbine blades with in-service damages (Fig. 15). Changes in the microstructure of a blade that has already been operated can be assessed on the basis of changes demonstrated by a new blade subjected to temperature within a specified range, and exposed to this temperature for sufficiently long time.

The examined blades were manufactured of the nickel-based superalloy EI 867-WD (HN62MWKJu-WD - to TC-14-1-223-72) intended for thermal-mechanical treatment, of he following chemical composition (% by weight): C = 0.03; Si = 0.14; Mn = 0.06; S = 0.005; P = 0.005; Cr = 9.69; Al = 4.65; W = 4.69; Mo = 9.29; Co = 4.84; Fe = 0.39; Ni = the rest. The manufacturing process comprises such processes as hot forging, surface machining by grinding, milling and polishing (Błachnio, 2009). The next step is thermal and chemical treatment of blades that consists in the introduction of aluminium to their surface layer in order to increase their resistance to thermal and chemical effect of exhaust gases. After the standard surface treatment, i.e. the solution heat treatment (1473 K/4 h/in air) and ageing (1223 K/8 h/in air) the material gains the Young's modulus E = 2.33×10^5 MPa and the Poisson coefficient $\nu = 0.3$ measured at the ambient temperature.

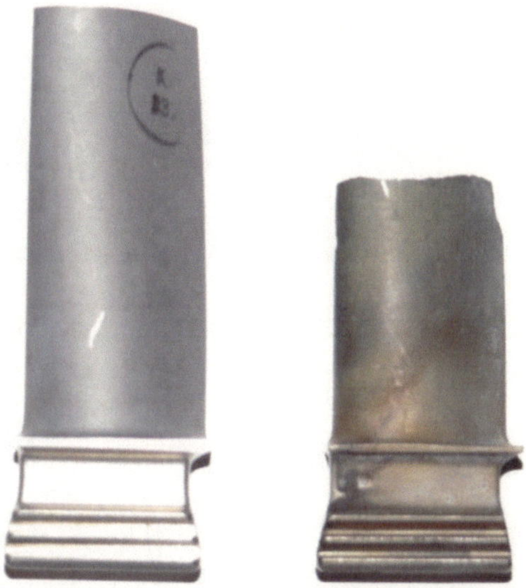

a) b)

Fig. 15. Gas turbine blades : a) – the new one, b) – the in-service damaged one, magn. x0.75

In order to investigate the kinetics of changes in the microstructure of the EI 867-WD alloy, new blades were subjected to soaking in a furnace with the application of: various times of thermal treatment at constant temperature, and various temperatures at constant time of soaking 1h. Further examination comprised preparation of metallographic microsections from specimens cut out of both the new blades and those damaged in the course of turbine operation. The specimens were subjected to etching with the reagent of the following composition: 30g $FeCl_3$; 1g $CuCl_2$; 0.5g $SnCl_2$; 100ml HCl; 500ml H_2O. The microstructures were analyzed with a scanning electron microscope (SEM).
Results of the examination of a new blade are presented in Fig. 16. One can see an aluminium coating (the bright part of the surface) and a part of it bound with the alloy structure by diffusion (Fig. 16a), also, the γ' phase precipitates cuboidal in shape (Fig. 16b).
The soaking at 1223 K results in the initiation of changes in precipitates of the strengthening γ' phase: the particles start changing their shapes from cuboidal to lamellar (Fig. 17b). On the other hand, the soaking at 1323 K results in evident changes in shapes of precipitates of the strengthening γ' phase to lamellar (Fig. 18b). At the same time, the surface roughness and thickness of the aluminium coating increase at both temperatures. These properties get intensified as the temperature growth. One can see the non-linear extension of the coating in function of the soaking time and temperature, both in the surface-adjacent area and in deeper layers where diffusion of aluminium had already occurred. The extension results in lower density of the material due to excessive porosity, which proves conducive to the penetration by the exhaust gases particles and leads to more intense destructive effects of both the thermal an chemical treatment upon the coating and the parent EI 867-WD alloy.

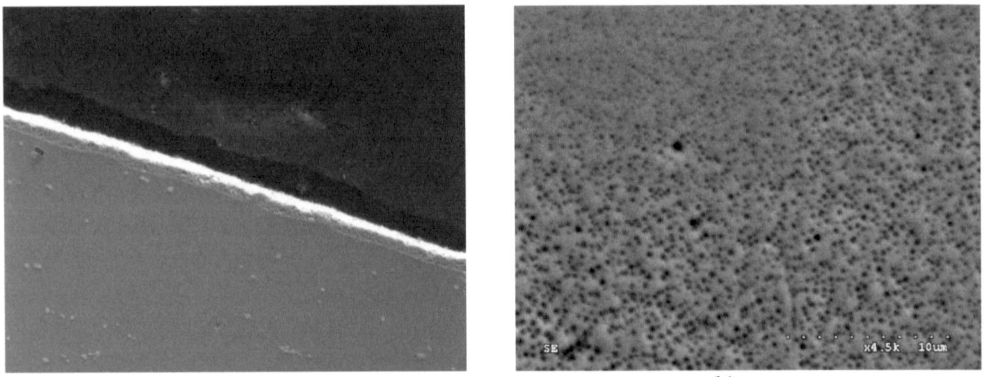

a) b)

Fig. 16. SEM microstructure of a new blade: a) – aluminium coating, magn. x450, b) - EI 867-WD alloy, magn. x4500

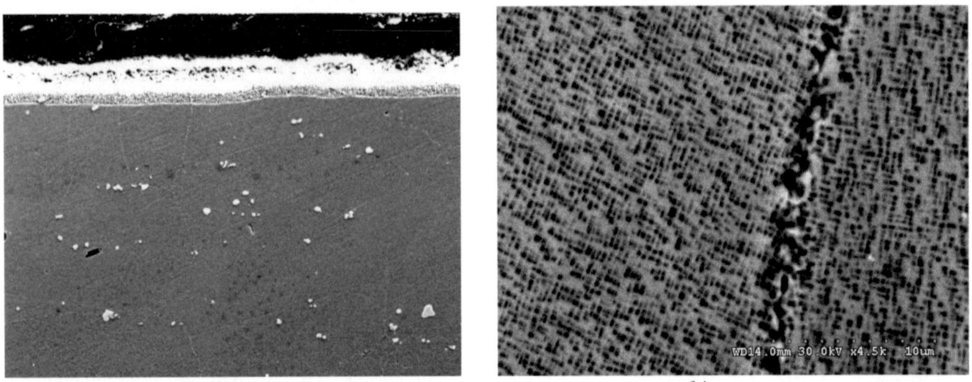

a) b)

Fig. 17. SEM microstructure of blade material subjected to soaking in a furnace at 1223 K: a)– aluminium coating, magn. x450, b) - EI 867-WD alloy, magn. x4500

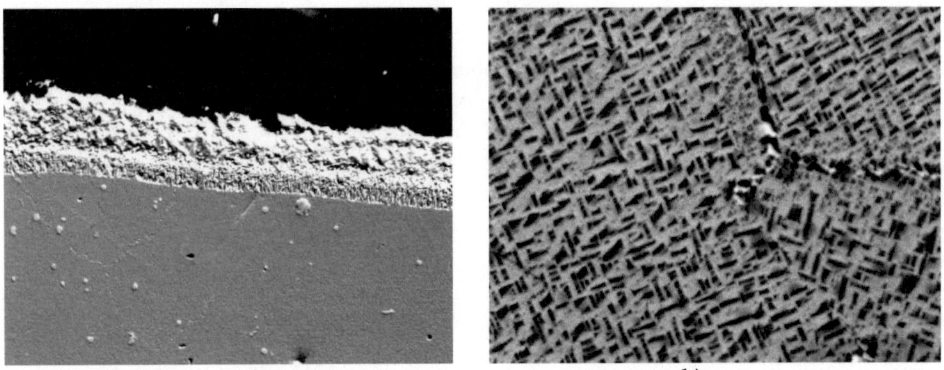

a) b)

Fig. 18. SEM microstructure of material subjected to soaking in a furnace at 1323 K: a) - aluminium coating, magn. x450; b) - EI 867-WD alloy, magn. x4500

4.2 Effect of operating conditions on material degradation of gas turbine blades

Examination results obtained for microstructure of the EI 867-WD alloy under laboratory conditions served as the basis for finding how turbine operating conditions affect degradation of the material used for the manufacture of gas turbine blades. As opposed to the laboratory conditions, extension of the heat resistant aluminium coating during the actual operation of engines entails a number of associated effects, such as erosion, oxidation and cracking, in particular on the leading edge of the blade profile (Fig. 19). Only the diffunded is durably bound to the parent metal, the rest of the coating was subject to decohesion, which resulted in the deterioration of heat resistance and high-temperature creeping resistance of the blade material. This, in turn allows of more intense penetration of the blade structure by exhaust gases and, consequently, to overheating of the alloy, initiation of cracks of thermal-fatigue nature and, quite probably, the break-away of the blade in the course of turbine operation (Fig. 15b) and finally, a gross failure to the gas turbine.

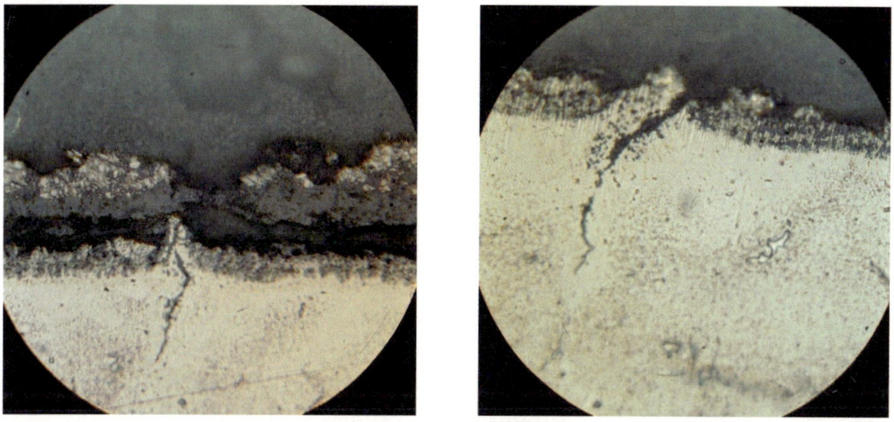

Fig. 19. Microstructures of exemplary in-service damages to gas turbine blades, magn. x500

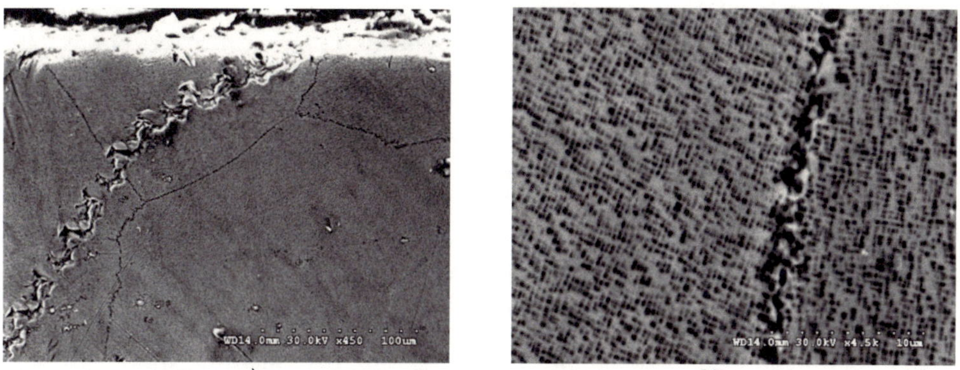

a) b)

Fig. 20. Results of examination of a gas turbine blade damaged due to long-lasting operation: a) - SEM microstructure for the EI 867-WD alloy at fracture, one can see a crack of the transcrystalline (TC) nature, magn. x450, b) – SEM microstructure of the alloy with visible changes in size and shape of the γ' phase, magn. x4500

Metallographic examination of specimens taken from an overheated blade (Fig. 15b) made it possible to find out degradation of the blade microstructure. Numerous microcracks along grain borundaries and transcrystalline ones were detected nearby the blade fracture (Fig. 20a). According to the studies (Okrajni & Plaza, 1995; Sieniawski, 1995; Tomkins, 1981), such decohesion results from the creeping and fatigue processes. The metallographic microsection enabled detection of the γ' phase coagulation. The coagulation and the precipitates dissolving effects intensify nearby the blade surface. In addition, fine-dispersion secondary precipitates are observed; the presence thereof increases susceptibility of the alloy to brittle cracking (Fig. 20a). Extension of the γ' phase in the alloy results in the change of phase shapes from cuboidal (Fig. 16b) to lamellar ones (Fig. 20b) as well as substantial extension of the size of this phase as compared to a new blade.
The morphology of particles within the γ' phase depends on the sign (direction) of the mechanical stress existing inside the blade. The tensile stress that acts along the blade axis in the course of turbine rotor's rotation is conducive to extension of the γ' phase within the plane that is perpendicular to the direction of stress. Consequently, the initially cuboidally-shaped particles (Fig. 16b) are converted into plates (Fig. 20b), with wider walls disposed perpendicularly to the stress direction whereas narrow walls are perpendicular to the remaining directions of the cube (Majka & Sieniawski, 1998; Paton, 1997).
Extension of particles within the γ' phase leads to loss of their stability, which leads to coagulation of some particles and dissolving of other ones (Majka & Sieniawski, 1998). That process takes place above some specific temperature typical of a given phase, and over the time of soaking. According to the results gained, when temperature of 1223 K is exceeded even for a very short time, a very intense extension of the γ'-phase precipitates takes place. This leads to the loss of shape stability and the formation of plates (Fig. 18b and Fig. 20b). This conclusion is also confirmed by results reported in (Majka & Sieniawski, 1998; Paton, 1997). Similar conclusion is outlined by authors of the studies (Poznańska, 1995; Taira & Otani, 1986). With the Udimet 700 alloy as an example one can find that at temperatures above 1093 K precipitates in the form of plates substantially deteriorate the yield strength. This effect one can see in Fig. 21 that presents changes in mechanical properties demonstrated by the EI-867 alloy as a function of temperature.
Kinetic characteristics of the γ' phase precipitates depend on the degree of saturation of the alloy matrix, i.e. the γ phase, with the admixture elements of the alloy. Shapes of precipitates depend on the degree of misfit between the lattice of alloy elements and the lattice of the basic material. The authors of (Nikitin, 1987; Paton, 1997) found out that for the misfit factor Δa = 0.2% the γ' phase is precipitated in the form of spheroid particles, for Δa = 0.5 – 1 % the particles of the γ' phase are of cuboidal shape whereas for Δa = 1.2 % the particles take lamellar shape. The theory of precipitate-based strengthening claims that crucial factors decisive to the degree of strengthening include diameters of the γ' phase particles and distances between them. These parameters depend on the extension rate (that is controlled by the volumetric diffusion) and coagulation of these particles.
Chemical composition of the γ' phase substantially affects the value of the lattice parameter $a\gamma'$ and the associated degree of misfit Δa to the matrix lattice $a\gamma$, where $\Delta a = (a\gamma - a\gamma') / a\gamma$. It influences morphology of the γ' and the range of its durability. It turns out that the degree of misfit between parameters of the phase lattices is the function of temperature. According to (Paton, 1997) the highest high-temperature creeping resistance is demonstrated by alloys, where the degree of misfit between the phase lattices is positive (>0) (Fig. 22). The chemical composition, morphology and distribution of the γ' phase

precipitates within the microstructure are crucial factors that decide mechanical properties of the alloy.

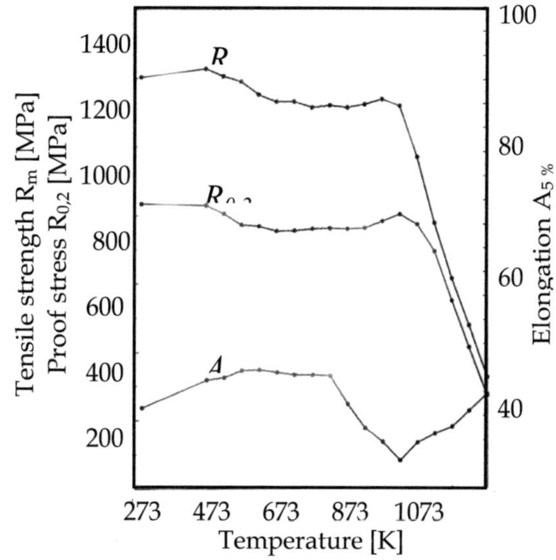

Fig. 21. Alterations in mechanical properties demonstrated by the EI-867 alloy vs. temperature (Poznańska, 1995)

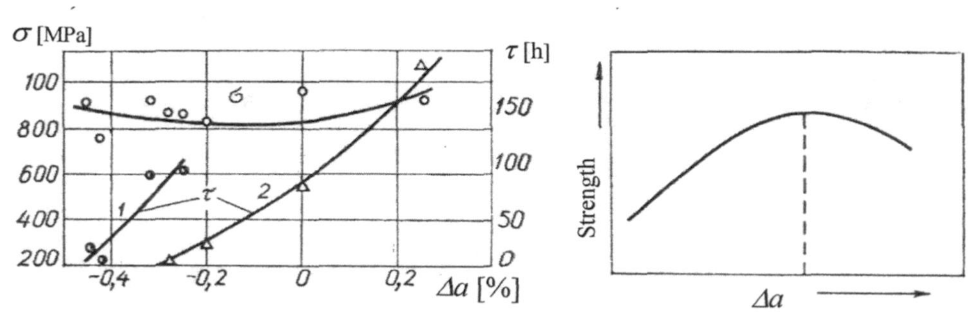

a) b)

Fig. 22. Effect of misfit between parameters of the crystallographic lattice for γ and γ' phases of a nickel alloy onto (Paton, 1997): a) strength limit (at 293 K) and durability limit (at 1373 K and $\sigma = 80$ MPa), b) – strength of a two-phase system

5. The method of assessment of temperature variations measured for exhaust gases upstream the gas turbine using a turbine state non-linear observer

The distinguishing peculiarity of low-cycle loads affecting the so called hot structural components of aircraft turbine reactive engines is superposition of adverse effects due to joint and simultaneous impact of both mechanical and thermal loads with high amplitudes.

The detrimental effect is particularly intensified when the engine is operated on a combat or a combined training and combat aircraft. It happens due to frequent and rapid operation of the engine control lever by a pilot when the aircraft is forced to make sophisticated manoeuvres. There are documented examples of substantial differences between low-cycle loads to engines installed on different aircrafts, for instance the ones that are used for group aerial stunts in a close line-up. It usually happens that the pilot of the guided aircraft, located at the line-up side changes the rpm range of the motor much more frequently, up to several dozens times during a single mission, as compared to the pilot of the guiding aircraft (Cooper & Carter, 1985). Consequently, the exact spectral measurements for low-cycle loads of a jet engine during its operation are the matters of crucial importance for unbiased assessment of its condition as a result of natural wear. To perform that task the researcher must be in possession of synchronous records for timings of momentary values for rotation speed of the turbine as well as for the average gas temperature at the outlet of the combustion chamber and downstream the turbine. In this study, the monitoring is focused on phenomena attributable to a turbojet engine with the longitudinal cross-section shown in Fig. 23.

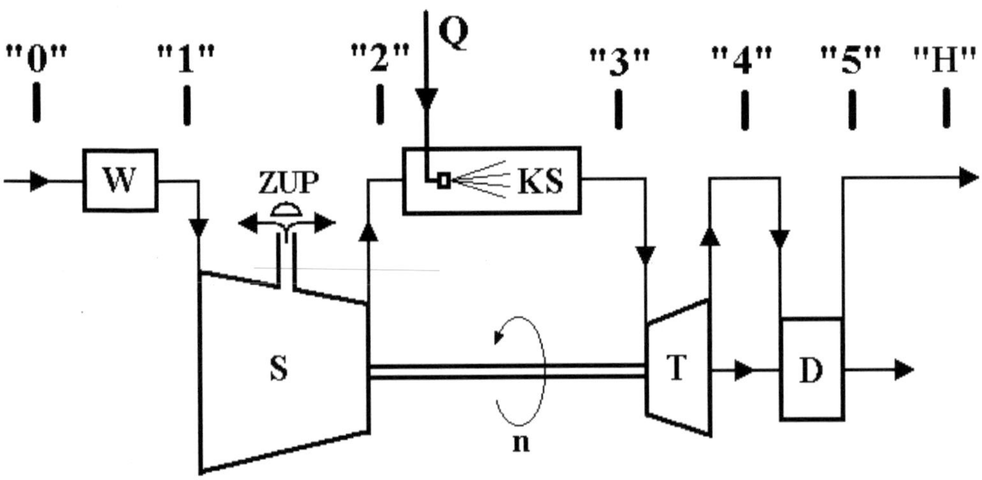

Fig. 23. The design configuration of the turbojet engine under tests with indication of calculation cross-sections of the flow path for the working medium

For the considered engine, temperature measurements for exhaust gases are carried out with use of a set made up of 8 thermoelements deployed in the channel downstream the turbine within the plane perpendicular to the flow velocity direction. Measurement results for the rotor rpm and the gas temperature are stored in the memory of the on-board digital recorder, along with a set of other parameters that are indispensable for further analyses, in particular flight parameters of the aircraft and ambient conditions.

List of symbols:

Cj – specific fuel consumption

Cp_{23} - average specific heat of the working medium inside the combustion chamber

D – convergent jet

G2 – mass flow intensity of the working medium at the compressor outlet

$G2_r$ – normalized mass flow intensity of the working medium at the compressor outlet
G3 – mass flow intensity of the working medium at the combustion chamber outlet
G3r – normalized mass flow intensity of the working medium at the combustion chamber outlet
h – increment for numerical integration
k_{34} – isentropic exponent average value for working medium in turbine
k_{45} – isentropic exponent average value for working medium in nozzle
KS – combustion chamber
n – rotational speed of the rotor (rpm)
n_{sr} – reduced rotational speed of the compressor
n_{tr} – reduced rotational speed of the turbin
P0 – total pressure of the working medium in the engine inlet
P1- total pressure of the working medium in front of the compressor
$P2_{start}$ – initial total pressure of the working medium downstream the compressor
P4 - total pressure of the working medium inside the convergent jet
$P4_{start}$ – initial total pressure of the working medium inside the convergent jet
PH – ambient pressure
Q – rate of fuel flow
Rg – gas constant
S - compressor
T – turbine
T1 – total temperature of the working medium upstream the compressor
T2 – total temperature of the working medium downstream the compressor
T3 – total temperature of the working medium upstream the turbine
T4 – total temperature of the working medium downstream the turbine
T4t – average temperature of exhaust gases measured with use of a thermoelements set
$T4t_{start}$ – initial average temperature of exhaust gases measured with use of athermoelements set
TH – ambient temperature
t_h – past service live in hours
u_1,u_2 – deviations of iterations
V5 – velocity of gas discharged from the convergent jet
w_1,w_2 – gain coefficients for iteration loops
Wo – calorific value of fuel
Wu - coefficient of air bleeding from the compressor for the needs of the airplane
ZUP – the system of valves for air bleeding from the compressor to prevent from the compressor stall
Π - pressure ratio of the working medium while flowing through the compressor
ε - pressure ratio of the working medium while flowing through the turbine
ϕ_ω - flow rate coefficient for the convergent nozzle
ϕ_π - total pressure preservation coefficient for the convergent nozzle
η_{ks} - efficiency of the combustion chamber
η_s – isentropic efficiency of the compressor
η_t - isentropic efficiency of the turbine

σ_{23} - the total-pressure preservation coefficient for the combustions chamber flow
σ_{01} - the total-pressure preservation coefficient for the compressor flow.
List of graphic symbols:

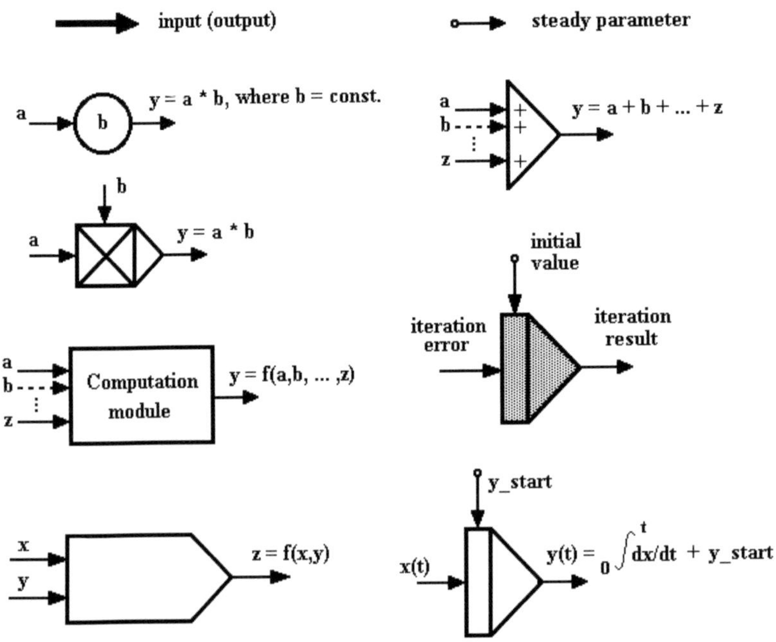

5.1 The problem of accurate measurements for momentary temperatures of gases that flow throughout the turbine

Measurement of the engine rotation speed is a relatively easy task, contrary to measurements of momentary temperature values that are associated with a number of hindrances. It is not enough to merely record temperature values indicated by a set of thermoelements. The reason for the hindrances is substantial and hard-to-define thermal inertia of measurements that entails considerable dynamic errors of the records. Such errors make the method useless for appropriate execution of the appointed task. In addition, a mathematical model capable to simulate operation of the thermoelements and potentially suitable for compensation of the dynamic errors has not been investigated yet in details.
The scope of research studies completed by the author made it possible to formulate, develop and test, with use of real-life data, a number of options for the so-called non-linear observers for singe rotor turbojet engines (Pawlak, 2005, 2006). It is worth to explain that the observer for the engine is a rather sophisticated iterative algorithm intended for calculation of the so-called hard-to-measure or non-measurable parameters of the engine operation under unsteady circumstances, where such values are calculated on the basis of other, easily measurable parameters of its operation. The example of such a parameter that is actually immeasurable in a direct way is a momentary temperature of built-up gases at the turbine inlet. On the other hand, the momentary average temperature of built-up exhaust gases in the engine jet or the thrust of a jet engine during the aircraft flight can serve as examples of hard-to-measure parameters. The particular reason for difficulties with direct measurements of temperature inside the engine jet is considerable and unpredictable with sufficient

accuracy thermal inertia of measuring thermoelement. The inertia results from sizeable dimensions of the thermoelement joint and very broad variation range of both the measured temperature and flow velocity of the exhaust gas stream. The thermoelements installed inside the jet of the engine under tests are made of wires with their diameters of 1.25mm. Such a large diameter of wires is a must due to the required durability of thermoelements that must be placed in the exhaust gas stream with variable flow velocity ranging up to several hundred meters per second and fast-changing temperature from 300 to 1100K. All in all, very high and hard to define with satisfying accuracy thermal inertia of thermoelements is the core reason for troubles with compensation of dynamic errors associated with temperature measurements when the engine operates in an unstationary mode, under various conditions in terms of the flight altitude and speed of the aircraft. On the other hand, exact compensation of measurement deviations for the specific temperature gauge needs perfect familiarity with the mathematical model of the latter.
Initial investigations of dynamic deviations for measurements of gas temperature inside the engine jet with use of thermoelements (Fig. 25) made it possible to find out that mathematical models in the form of an ordinary non-linear differential equation of 1st range that have already been used in referenced literature for thermoelements are inadequate (Michalski & Eckersdorf, 1986; Wiśniewski, 1983). The non-linear observers described in (Pawlak, 2005, 2006) and intended for scrutinizing operational parameters of turbojet engines make it possible to find out true characteristic curves for average temperatures of the built-up gas inside the engine jet (T4) (as well as the mass flow intensity under unsteady circumstances (G4)) with no need to measure the temperature in a direct way. When synchronous timings for the actual temperature (T4) determined with use of the observer and for the temperatures indicate by thermoelements (T4t) are available, one can verify whether the equality condition for the both temperatures is fulfilled, i.e. T4t = T4 at the moment when the value of the differential d(T4t)/dt equals zero (although the condition is obvious for the differential equation of 1st range). Meeting that obvious condition would serve as the proof for the mathematical model of a thermoelement that is shown in Fig. 24. in the form of an analog diagram.

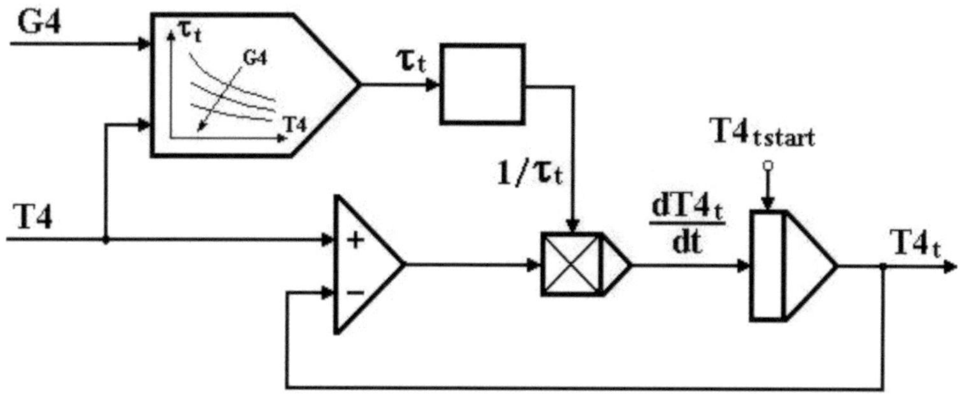

Fig. 24. The conventional mathematical model for arrangement of thermal components, reflecting the form of an ordinary non-linear differential equation of 1st range, here called in question

5.2 Applicability test for the existing model

Fig. 25 shows phase diagrams for the actual temperature (T4), measured indirectly with use of the observer and the temperature (T4t) measured directly with use of the set composed of 8 thermoelements installed inside the jet. The diagrams are plotted as functions of the rotation speed (n) of rotor of the unit made up of the turbine and the compressor. The diagram *T4t = f(n)* comprises points *a, b, c* and *d,* for which *dT4t/dt = 0.* Dark areas on the graph represent the dynamic error for temperature measurements with use of thermoelements.

The analysis of cycles plotted in Fig. 25 serves as the proof that the method that has already been used for mathematical modeling of dynamic properties attributable to thermoelements (Michalski & Eckersdorf, 1986; Wiśniewski, 1983) is inadequate. The diagrams for T4t and T4 temperatures intersect one another only for the phase of full acceleration, i.e. at the *'a'* and *'b'* points, whereas a slight discrepancy is observed for the *'c'* point.

Even still insufficient (for the current phase of developments) accuracy of the employed observer cannot serve as explanation for such a large quality divergence between diagrams for the T4t and T4 temperatures during the engine deceleration and for the fact that the diagrams do not intersect at the *'d'* point. Additionally, a slight phase displacement visible on both diagrams for the deceleration phase indicates herein that the predominant mechanism for heat exchange between measuring joints of the thermoelements and the ambient environment is probably radiation, whereas the mechanism of forced convection prevails on the acceleration area. Verification of the foregoing hypothesis should be included in the scope of intense research studies.

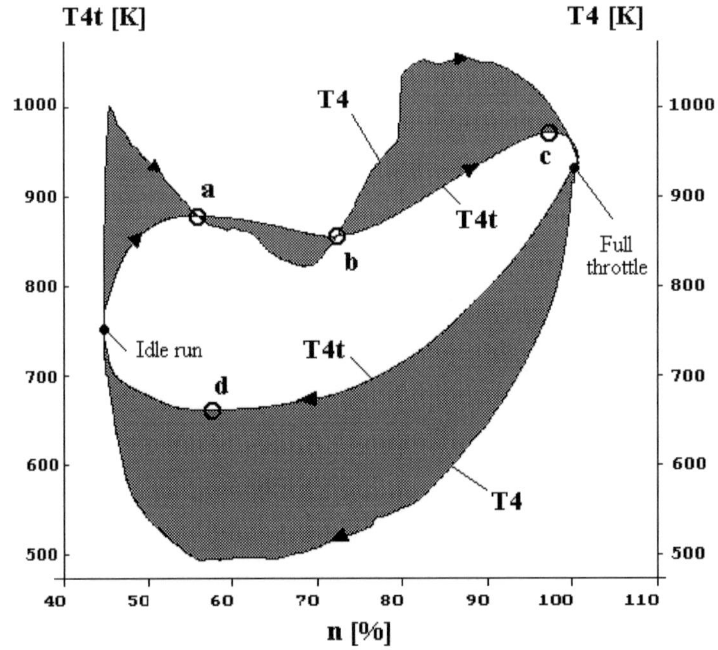

Fig. 25. Comparison between phase cycles of temperatures measured directly by means of a set of 8 thermoelements (T4t) and measured indirectly (T4) by a non-linear observer during full acceleration and deceleration of the engine under test at ground conditions

5.3 Non-linear observer for the K-15 engine

The structural diagram for one of numerous possible implementations developed for the non-linear observer for a single-rotor turbojet engine is shown in Fig. 26. The observer was designed owing to a simple transformation of a structural diagram for the simulation model of a single-rotor turbojet engine (Pawlak at all, 1996), where the rotation speed (*n*), obtained from integration of the equation for the rotor movement was substituted, for the observer with the actual rotation speed obtained from direct measurements for a real engine. The observer with its schematic diagram as shown in Fig. 26, is used for calculation of many other parameters for the engine operation (Pi,Ti,Gi,Vi), including also immeasurable or hard-to-measure operation parameters. These parameters can be found out on the basis of momentary values for appropriately composed set of other, easily measurable operation parameters of a real engine (PH, P0, T0, Q, n). Therefore, the observer can be used as a 'virtual gauge'. In particular, the observer is suitable for measurements of momentary values of the engine thrust (R), specific fuel consumption (C_j), temperature (T3) of the built-up working medium at the outlet from the combustion chamber.

The detailed algorithms for one of the possible options developed to build a non-linear observer for the engine under tests is shown in Fig. 27. The following simplifications were assumed for development of the observer:

- The working medium inside the engine channel is the ideal gas. Therefore, the parameters of the exponents for isentropic curves (k_{34}, k_{45}) as well as the specific heat of gas (Cp_{23}) are constant.
- The only accumulator that stores energy during non-stationary states of the engine is the rotor weight. It means that dynamic phenomena associated with accumulation of the working medium mass and enthalpy inside the volumes of the combustions chamber and the jet are omitted - which is justified in case of the engine under examination.

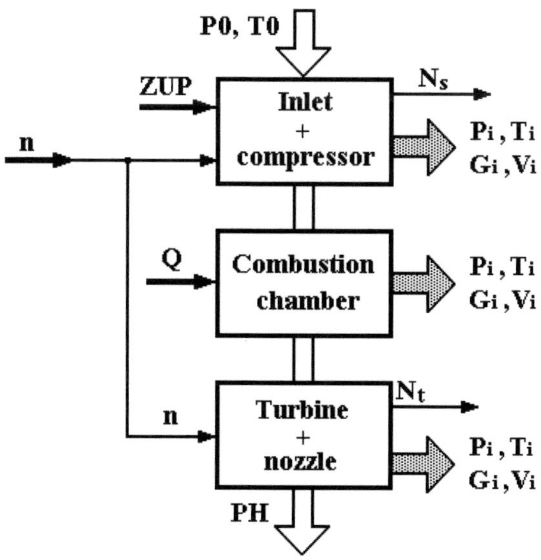

Fig. 26. The concept of a non-linear observer for a single-rotor turbojet engine (the parameters that are measured directly are: PH, P0, T0, Q, n, ZUP)

Adoption of the foregoing assumptions has demonstrated that the algorithm for the observer as shown in Fig. 27 has the form of a system of non-linear algebraic equations with the roots that represent momentary values for build-up pressures of the working medium downstream the compressor (P2) and inside the jet (P4). The input variables for the algorithm of the observer are momentary values of the following, directly measurable, parameters of the engine operation: ambient pressure (PH), pressure and temperature of air built-up at the inlet (P0, T0), fuel flow rate (Q), rotation speed of the rotor (n). The additional input variable that can be measured in the direct way is the status signal for valves that prevent the compressor from the stall effect (ZUP) that can report the 'open' status (ZUP=0) or the 'closed' status (ZUP=1).

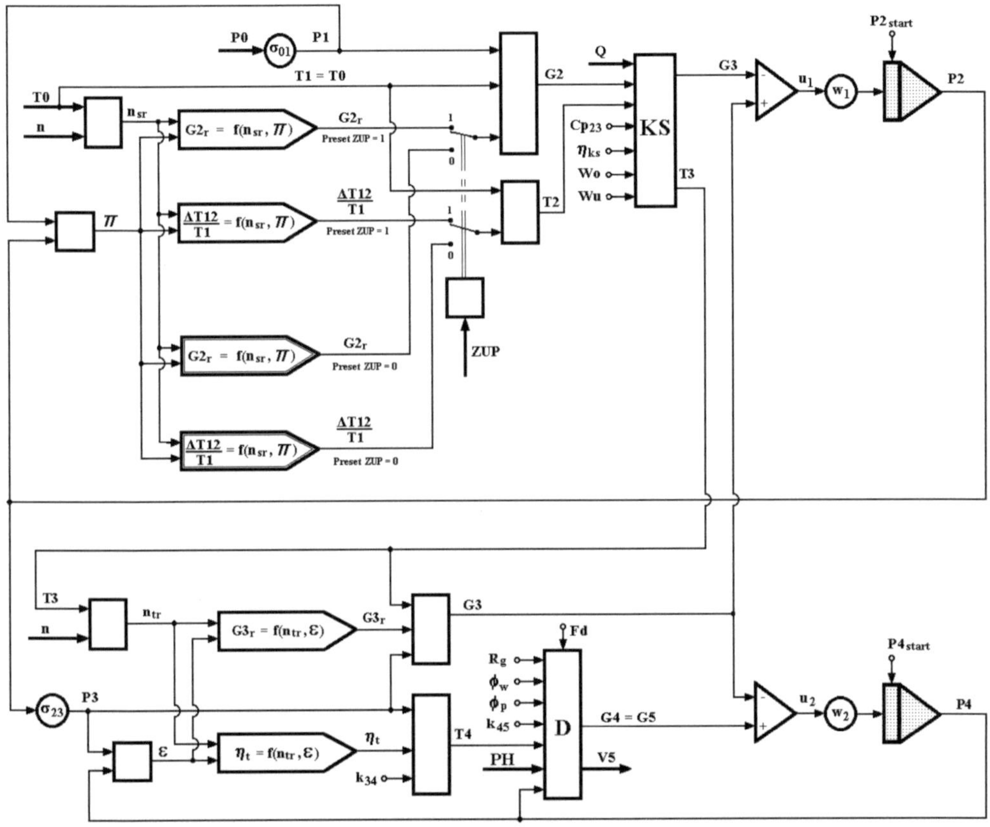

Fig. 27. Diagram of a non-linear observer for the engine under examinations (The ambient environmental and the engine parameters that are measured directly are: PH, P0, T0, Q, n ZUP)

The set of output variables includes all the other parameters of the engine operation that are shown in the diagram as well as possible additional variables that may result from processing them – such as already mentioned thrust and specific fuel consumption.

The major components of the algorithm for the observer that is shown in Fig. 27 include static characteristic curves of the compressor when it is operated with open (ZUP = 0) or closed (ZUP = 1) valves that prevent the compressor stall, i.e. the curves $G2_r = f(n_{sr}, \pi)$, $\Delta T12/T1 = f(n_{sr}, \pi)$. The next important components are static characteristic curves for the turbine: $G3_r = f(n_{tr}, \varepsilon)$ and $\eta_t = f(n_{tr}, \varepsilon)$. Accuracy of the calculation results obtained with use of the observer substantially depends on accuracy of the mentioned characteristic curves.

The observer diagram from Fig. 27 illustrates in the graphic way how to seek the roots of the system equations by a series of iterations. The method convergence, and, consequently, the time in which the algorithm can be completed on a digital computer, depends on how well the gain coefficients (w_1, w_2), also shown in the diagram, are selected. The iteration process is finished when values of deviations (u_1, u_2) are sufficiently low, which is illustrated in the diagram.

5.4 D spectrum of loads

Particular attention must be paid to spectra of totalized thermal and mechanical loads affecting the so-called hot parts of the engine design structure. The loads are caused by simultaneous variations of both the rotor rotation speed (*n*) and the average total temperature of the working medium at the outlet of the combustion chamber (T3) and inside the jet (T4). In order to illustrate spectra of these loads it was necessary to develop 3-D histograms (Fig. 28 - 32). Fig. 28 shows the 3-D histogram for the actual average temperature (T3) of the exhaust gas at the outlet of the combustion chamber and the rotation speed (*n*), whilst Fig. 29 contains the similar diagram for the actual average temperature (T4) of the exhaust gas inside the engine jet and the rotation speed (*n*). One has to pay attention that both monitored temperatures belong to the category of immeasurable parameters and could have been tracked only with use of the already mentioned non-linear observer for the engine. Consequently, the interesting phenomenon could be found out, namely stabilization of the histogram shapes after expiring of about 30-40h when the engine operation was monitored.

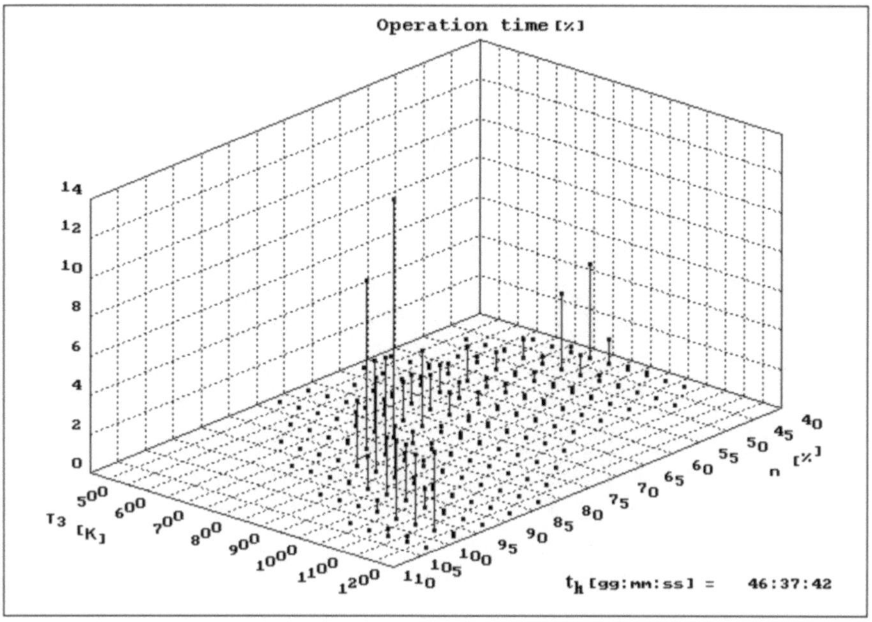

Fig. 28. Spectrum distribution for the average total temperature for the working medium at the outlet of the combustion chamber (calculated with use of the non-linear observer of the engine) and rotation speed

In addition, Fig. 30 shows the histogram for the rotation speed (n) and the gas temperature (T4t) inside the engine jet and measured directly with use of a set of thermoelements.

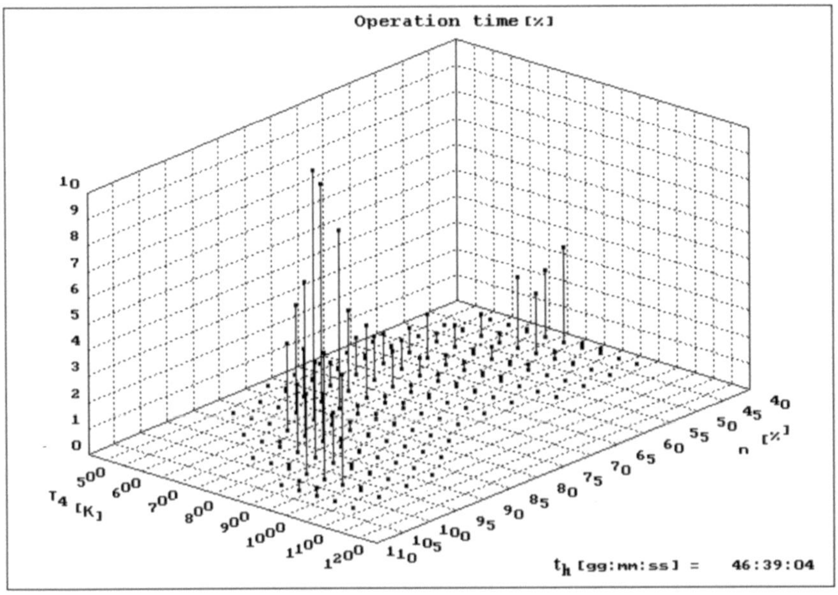

Fig. 29. Spectrum distribution for the average total temperature of the working medium inside the engine nozzle (calculated with use of the non-linear observer of the engine) and rotation speed

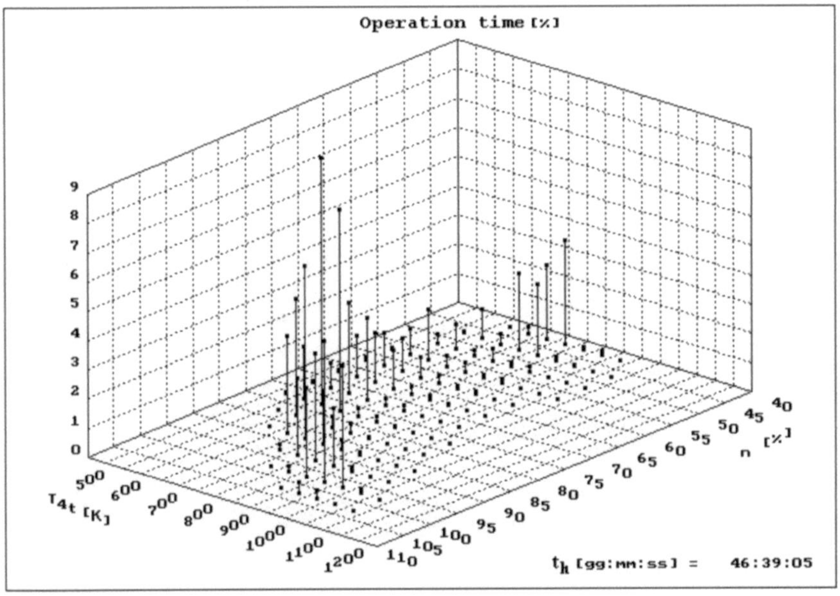

Fig. 30. Spectrum distribution for the average temperature of the working medium inside the engine nozzle (measured with use of set of thermoelements) and rotation speed

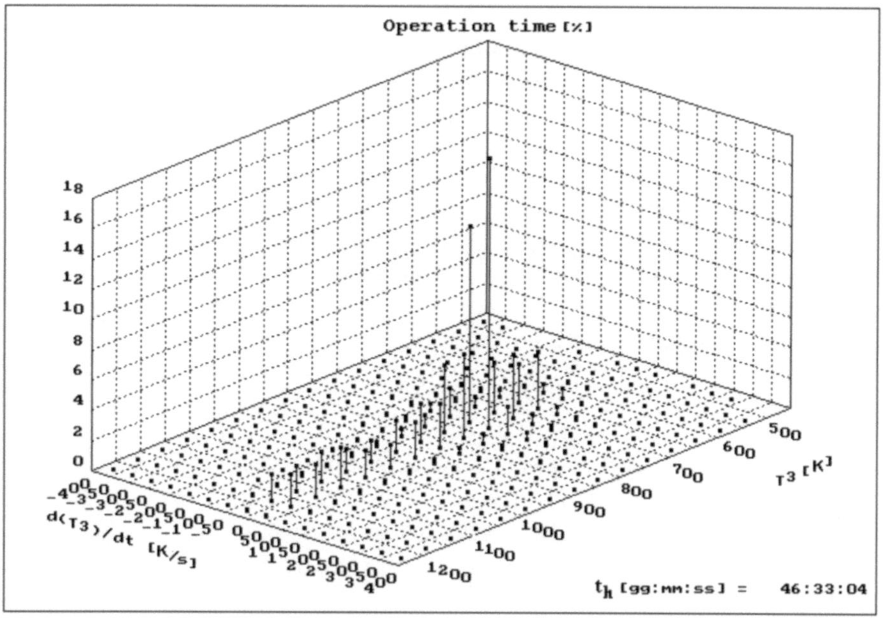

Fig. 31. Spectrum distribution for variations of the average total temperature of the exhaust gases at the outlet of the combustion chamber (calculated with use of the non-linear observer of the engine) and rotation speed

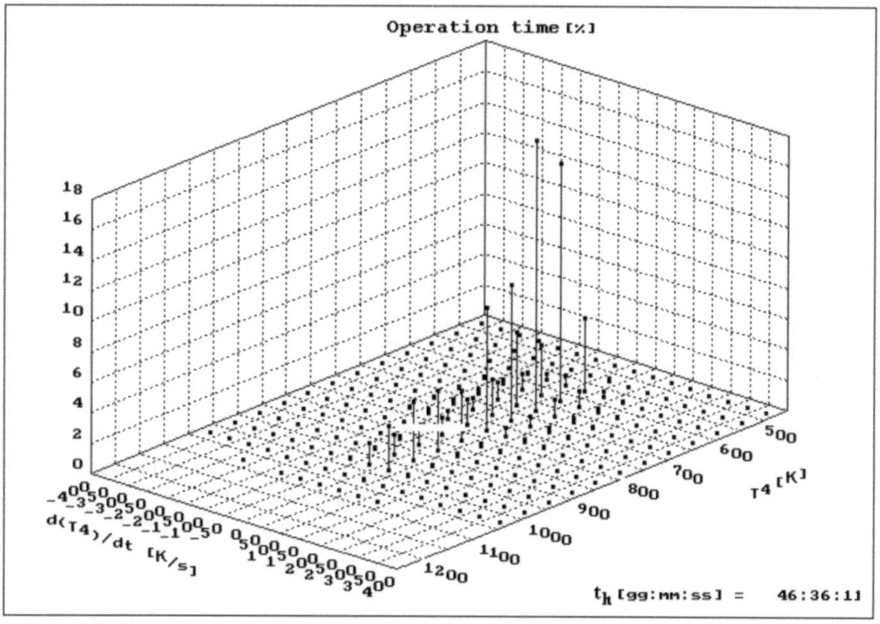

Fig. 32. Spectrum distribution for variations of the total temperature of the exhaust gases inside the engine nozzle (calculated with use of the non-linear observer of the engine) and rotation speed.

When thermal loads affecting the engine are subject to variations, it is crucial to know the rates of these variations. For that purpose the histograms presented in Fig. 31 and Fig. 32 were developed. These histograms show spectra for average temperatures of the working medium at the outlet of the combustion chamber (T3) and inside the engine nozzle (T4) together with their first derivatives (dT3/dt, dT4/dt). These histograms were also developed with use of the non-linear observer for the engine.
Availability of spectral diagrams for specific types of engines (e.g. in the form of the foregoing histograms) makes it possible to accurately determine their degree of low-cycle wear in pace with the operation time. It also enables to precisely schedule the fatigue tests for materials that are used for the engine design as well as to draw up plans and carry out the so-called rapid tests for critically loaded components of the engine structures.

6. Conclusions

The destructive processes affecting gas turbine vanes and blades begins with destruction of the aluminium coating. Consequently, the parent material of vanes and blades is exposed to the direct thermal and chemical effects of exhaust gases. Such circumstances lead, first and foremost, to the overheating of that material, which is manifested by adverse changes in the material microstructure. The factors that influence that effect include the supercritical temperature, the exposure time to the latter and chemical aggression of exhaust gases.
The comparative examinations of microstructures demonstrated by two blades, i.e. the new and overheated one and the one that has been in operation, have led to the conclusion that alterations of the damaged blade is a result of extension that happens to the intermetallic γ' phase. The size and shapes of that phase is comparable to the reciprocal parameters of the γ' phase for a new blade subjected to soaking to the temperature above 1223K for 1h. The achieved results make it possible to conclude that the damaged turbine blade was operated at the temperature of exhaust gases that exceeded the maximum allowed limit of 1013K for the time of about 1h (i.e. the time comparable to the heating time of a new blade) during air missions. Extension of the mentioned phase leads to coagulation of precipitates and changes of particle shapes from cubical to plate one. It leads to drop of the heat resistance and creep resistance parameters of the alloy.
Although the presented methodical approach to identification of reasons for failures of gas turbine blades during the process of the turbine operation ignores mechanical loads that occur during the turbine operation, it is sufficiently useful in case of engine defects when information of its operation conditions is incomplete.
Nevertheless, although the metallographic examinations enable determination of the material microstructure in an unbiased, trustworthy and dependable manner, it is still a destructive examination method. That is why there is a need to develop a set of non-destructive diagnostic tests aimed at assessing technical condition of gas turbine blades when the machine is still in operation. Each of non-destructive diagnostic methods that have already been applied to diagnostics of technical structures features a specific performance and applicability to assessment of that condition. Therefore, a combination of these methods, aided by computer technologies and with application of artificial intelligence, offers supreme research opportunities, much wider than any of the single method. Such an examination program shall enable really unbiased, trustworthy and dependable assessment of actual condition demonstrated by gas turbine blades during the process of operation

thereof against the three-threshold scale: applicability, partial (provisional) applicability and inapplicability.

The non-linear observer for jet engines pretty well performs the role of a 'virtual measuring gauge' for immeasurable or hard-to-measure parameters.

The basic hindrance that prevents the wider application of non-linear observers is the need to modify their structure as the time goes by in order to take into account slight variations of static characteristic curves for compressors and turbines of jet engines that occur in pace with the engine operation and result from natural wear and tear. It is the open problem that can become the topic of further research studies with a substantial practical impact.

Familiarity with dynamic timings for momentary temperature values of the working medium is the matter of crucial importance for determination of low-cycle thermal fatigue affecting the critically loaded 'hot' components of the engine design - turbine blades and disks.

Direct measurements of momentary temperature values for exhaust gases flowing via the turbine, when a set of thermoelements is applied, is burdened with a large and hard to compensate measurement error. The difficulty of compensation of such an error results from unavailability of a mathematical model that would be good enough to describe dynamic features of thermoelement operation. It is indispensable to undertake efforts in this area as the solutions that are available from literature sources are applicable only in a limited scope.

7. References

Antonelli, G.; Ruzzier, M. & Necci, F. (1998). Thickness measurement of MCrALY high-temperature coatings by frequency scanning eddy current technique. Journal of Engineering for Gas Turbines and Power, Transactions of ASME 120, pp. 537 - 542

Błachnio, J. (2004). Aircraft propulsion systems of the future. Journal of Transdisciplinary Systems Science, Vol. 9, pp. 29-37

Błachnio, J. (2007). Technical Analysis of Inefficiency and Failures of Aircrafts. Fundamentals of Aircraft Operation. Edited by J. Lewitowicz, J.: Publishing House of the Air Force Institute of Technology, Part. 4, pp. 181-264, ISBN 978-83-914337-9-X , Warsaw, Poland (in Polish)

Błachnio, J. & Bogdan, M. (2008) Diagnostics of the technical condition demonstrated by vanes of gas turbines. Diagnostics, No 1 (45), pp. 91-96, ISSN 641-6414

Błachnio, J. (2009). The effect of high temperature on the degradation of heat-resistantand high-temperature alloys. Journal Solid State Phenomena, Vol. 147-149, pp. 744 - 752, ISSN 1012-0394

Błachnio, J. (2010). Examination of changes in microstructure of turbine blades th the use of non-destructive methods. Journal of Polish CIMAC, Vol.5, nr 2, pp.17-28, ISBN 83-900666-2-9, ISSN 1231-3998

Bogdan, M. (2009). Diagnostic examinations of gas turbine blades by means of digital processing of their surface images. PhD Thesis, Technical University of Białystok, Białystok, Poland, (in Polish)

Hernas, A. (1999). Steel and Alloy High Temperature Creep Resistance, Part. 1. Publishing House of the Silesian University of Technology, Gliwice, Poland ISBN 83-88000-16-0 (in Polish)

Korczewski, Z. (2008). Archives of marine engines endoscopy, Publishing House of the Polish Navy University, Gdynia, Poland (in Polish)

Kroes, M. J.; Wild. T. W.; Bent, W. R. & Mc Kinley, J. L. (1992). Aircraft powerplants. Mc Graw- Hill

Majka, H. & Sieniawski, J. (1998). Examination of kinetic properties for extension and coagulation of the γ' phase in a nickel superalloy EI-867. Archives of Science on Materials, vol. 19, No. 4, pp. 237-254 (in Polish)

Nikitin, W. I. (1987). Corrosion and Protection of Gas Turbine Vanes. Leningrad, Russia (in Russian)

Okrajni, J. & Plaza M. (1995). Simulation of the fracture process of materials subjected to low-cycle fatigue of mechanical and thermal character. Journal of Materials Processing Technology, ELSEVIER, pp. 311 - 318

Paton, B. (1997). High-temperature creep resistance of cast nickel-base alloys; corrosion protection thereof, Naukowa Dumka, Kijew, Russia (in Russian)

Pawlak, W.; Wiklik, K. & Morawski, J.M. (1996). Synthesis and Examination of Control Systems for Aircraft Turbine Engines with Use of the Computer Simulation Methods, Scientific Library of the Aviation Institute, Warsaw, Poland (in Polish)

Pawlak, W. I. (2003). Nonlinear observer in control system of a turbine jet engine. The Archive of Mechanical Engineering, Vol. L, Warsaw, Poland

Pawlak, W. I. (2005). A non-linear observer in the warning system indicating faulty modes of operation of a turbine jet engine. The Archive of Mechanical Engineering, Vol. LII, Warsaw, Poland

Pawlak, W. I. (2006). The effect of convergent-nozzle volume on transient processes in a turbojet engine. The Archive of Mechanical Engineering, Vol. LIII, Warsaw, Poland

Pawlak, W. I. (2007). Computer simulation of transient processes in a turbojet engine, with special attention to amplitudes of thermal shocks in some selected fault modes of operation. The Archive of Mechanical Engineering, Vol. LIV, Warsaw, Poland

Pawlak, W. I. & Spychała, J. (2007). Performance of the advanced and simplified variants of a non-linear observer of a turbojet engine, comparison of results. Journal of Polish CIMAC, Gdańsk- Stockholm-Tumba, Poland-Sweden

Poznańska, A. (1995). Lifetime of blades made of the EI-867 alloy and used for aircraft engines at the aspect of non-uniform deformation and structural alterations, PhD Thesis, University of Technology in Rzeszów, Poland (in Polish)

Reports, (2000-2010) of Division of Aircraft Engines, Air Force Institute of Technology, Warsaw, Poland

Sieniawski, J. (1995). Assessment criteria and methods applicable to materials for components of turbine engines, Publishing House of the University of Technology in Rzeszów, Poland (in Polish)

Swadźba, L. (2007). Development of desired structures and properties demonstrated by protective coatings on selected alloys used for aircraft turbine engines, Publishing House of the Silesian University of Technology, Katowice, Poland (in Polish)

Taira, S. & Otani, R. (1986). Theories for high-temperature strength of materials, Moscow, Russia (in Russian)

Tomkins, B. (1981). Creep and fatigue in high temperature alloys, Ed. Bressers J., Applied Science Publishers, London,

20

New Non-Destructive Methods of Diagnosing Health of Gas Turbine Blades

Józef Błachnio[1,2], Mariusz Bogdan[2] and Artur Kułaszka[1]
[1]*Air Force Institute of Technology, Warszawa,*
[2]*Technical University of Białystok, Białystok,*
Poland

1. Introduction

The illuminated blade/vane surface can be recognized by a light-sensitive detector (a CCD matrix with an optical system - an optoelectronic device) owing to the fact that the surface becomes a secondary source of light (Zhang et al., 2004; Rafałowski, 2004; Tracton, 2006, 2007). Therefore, it is possible to diagnose the examined object in an indirect way by the processing and analysis of information acquired as digital images (Bogdan; 2008; Błachnio & Bogdan, 2010). Only a very tiny portion of an incident light beam is absorbed by metallic surfaces. Most of the light (90-95%) is reemitted from the reflecting surface as visible light with the same wavelength as the incident light. The remaining 5-10% of the absorbed energy is dissipated as heat (according to the rule of energy conservation). Chemical composition of the coating matter that covers surfaces of metallic objects decides the attenuation of some wavelengths in the spectrum of illuminating light, whilst the mixture of light that is selectively reflected from a specific surface is decisive for the perception of colours (some metals may exhibit specific colours due to selective reflection of light, e.g. gold, copper) (Zhang et al., 2004; Tracton, 2006). The CCD matrix is a set of many light-sensitive sensors, where each sensor is capable to record and then reproduce an electric signal that is proportional to the amount of light that has illuminated the surface (the photoelectric effect - emission of electrons as a consequence of their absorption of energy from photons that reach the matter; the diagram of such energy conversion and the associated losses are shown in Fig. 1).

The image-recording devices usually incorporate colour filters that are installed just upstream the light sensors and enable the intensity of a particular bandwidth within the light spectrum (a certain bandwidth of visible electromagnetic waves) to be recorded at a given location of the matrix. After processing by the central unit (CPU) the acquired information can be stored as digital images (graphic files) (Rafałowski, 2004).

The analysis of histograms of digital images is very useful since it provides a great amount of information about the image that has been processed, including information about the range of brightness levels and the number of levels represented in the images. In the case of images acquired under suitable illumination conditions the histogram includes also quantitative information about the brightness of the image acquired for the photographed object. On the other hand, the 2-D co-occurrence matrices presented by Haralick in his studies (Haralick & Shanmugam, 1973; Haralick & Shapiro, 1992) are used for analyses of textures as they are capable to extract spatial relationships from the image on the basis of the object brightness. Utilization of information acquired from histograms (statistical parameters such as: location of the maximum saturation, medians, quartiles, percentiles, average value, skewness, variance, kurtosis, the third and fourth central moments, excess) as well as that from co-occurrence matrices (such parameters as: energy, contrast, correlation, variance, homogeneity - the reverse differential moment, accumulated average, entropy) allows of the assignment of blade/vane surfaces to specific health levels. Based on the parameters found from histograms and from the co-occurrence matrices, with neural networks applied (the pattern recognition) it will be also possible to assign the blade image (recorded for a specific surface) to a defined class that corresponds with a given health level, i.e. the fit-for-use condition (Bogdan, 2009).

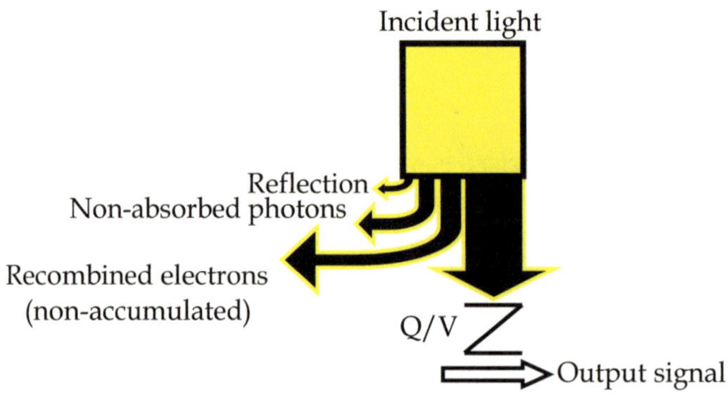

Fig. 1. Energy losses in the process of converting an optical signal into the electric one by means of a CCD cell (Rafałowski, 2004)

2. A method for the assessment of condition of new gas turbine blades (a laboratory experiment)

Subject to examination were new rotor blades of a gas turbine of aircraft jet engine. The blades are not cooled during the engine operation. They have been made from the nickel superalloy denoted as EI 867-WD (HN62MWKJu), suitable for the plastic working. This superalloy belongs to a sparse group of superalloys with no titanium admixture. The content of chromium is lower, therefore it is more susceptible to corrosion. For that reason various protective coatings are applied, e.g. aluminium ones. The TC 14-1-232-72 standard contains requirements for the chemical composition of the superalloy, thermal treatment and mechanical properties.

The research project assumed investigation of four new blades that were cut into four equal parts. To avoid structural alterations due to high temperature, the so-called ‘grinding burns', the cutting was carried out with a wire electric erosion machine. Pieces of blades were randomly selected and soaked (three pieces at a time) in a vacuum furnace at five various temperature levels, with the increment of 100 K, starting with the temperature of 1023 K. Then the specimens were cooled down inside the furnace until the ambient temperature was reached. Values of temperature applied while soaking the blade specimens correspond to temperature ranges that occur under both regular and failure in-service modes of rotor blades operation. Temperature of the working-agent stream (exhaust gas) at the outlet of the combustion chamber of the aircraft jet engine should remain in the range of 1173 K to 1223 K due to limitations imposed by thermal and chemical characteristics of materials used to manufacture solid and non-cooled blades of the gas turbine (Poznańska, 2000; Dżygadło et al., 1982; Kerrebrock, 1992; Hernas, 1999). Instances of short-term heating of blade material above the regular operating temperature quite often occur in the course of gas turbine operation. Therefore, such parameters as heat resistance and high-temperature creep resistance of the alloy to supercritical temperature are of crucial importance (Sieniawski, 1995; Sunden & Xie, 2010). The blades were cleaned in ultrasonic washers both before and after they were subjected to the soaking process.

The initial phase of metallographic investigation was focused on the determination of time the blade soaking process should take as it is the parameter that, along with the soaking temperature, affects the kinetics of growth and coagulation of the γ' phase particles. For that purpose the experiment was carried out, which consisted in that the specimens were soaked at temperature of T_{4max} (the maximum temperature downstream of the turbine, i.e. 1223 K, over 0.5 h, 1 h, 2 h, and 3 h). Metallographic microsections of these specimens were prepared with conventional methods to be then etched with the reactant of the following chemical composition:

$$30\text{ g } FeCl_2 + 1\text{ g } CuCl_2 + 0.5\text{ g } SnCl_2 + 100\text{ ml } HCL + 500\text{ ml } H_2O \qquad (1)$$

Both an optical microscope and a scanning electron microscope (SEM) were used to observe the microstructure. This enabled acquisition of information on structural changes in both the coating and the parent material of the blades, the changes being dependent on the soaking time and mainly including modifications of the size and distribution of the γ' phase (Fig. 2).

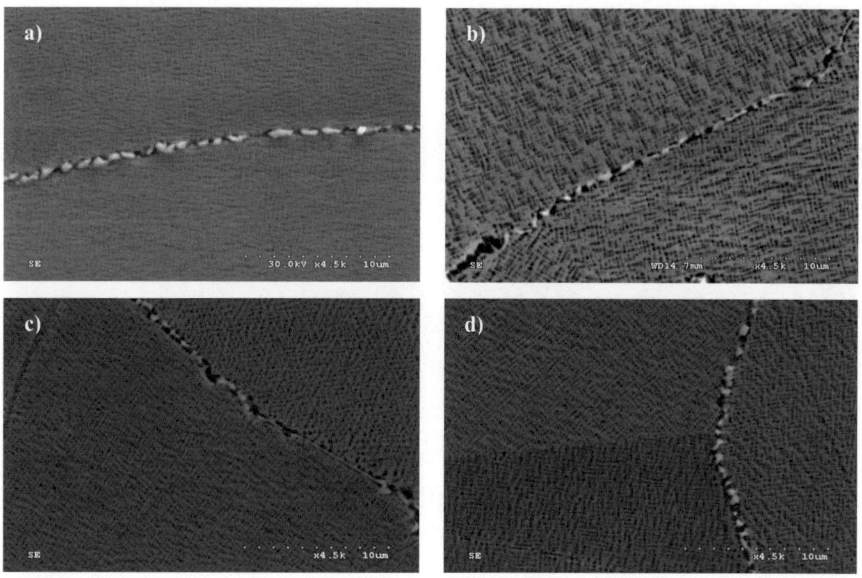

Fig. 2. Morphology of γ' precipitates - soaking at temperature 1223 K over: a) 0.5 h, b) 1 h, c) 2 h, d) 3 h (magn. x4500)

The dedicated software to analyze images of the material microstructures (Bogdan, 2009) in was used to determine alterations in size (surface area) of precipitates of the strengthening γ' phase against the soaking time (Fig. 3).

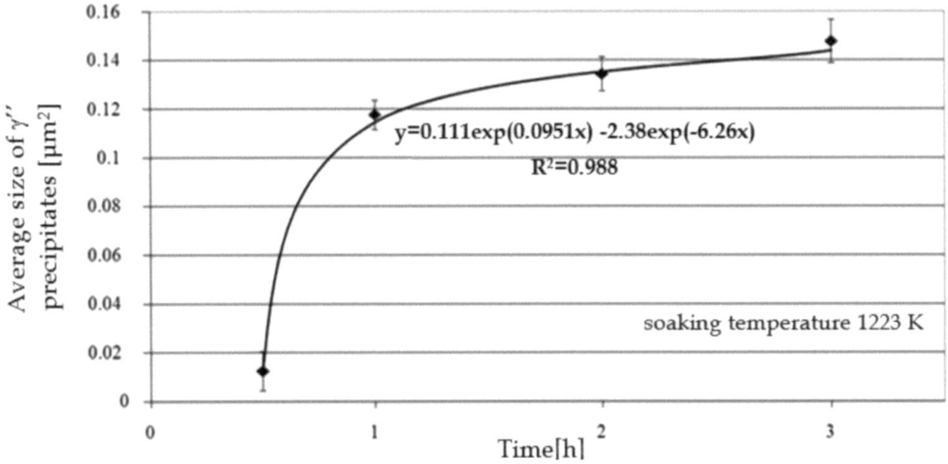

Fig. 3. Variations in the average size of γ′ precipitates against time of soaking the specimens

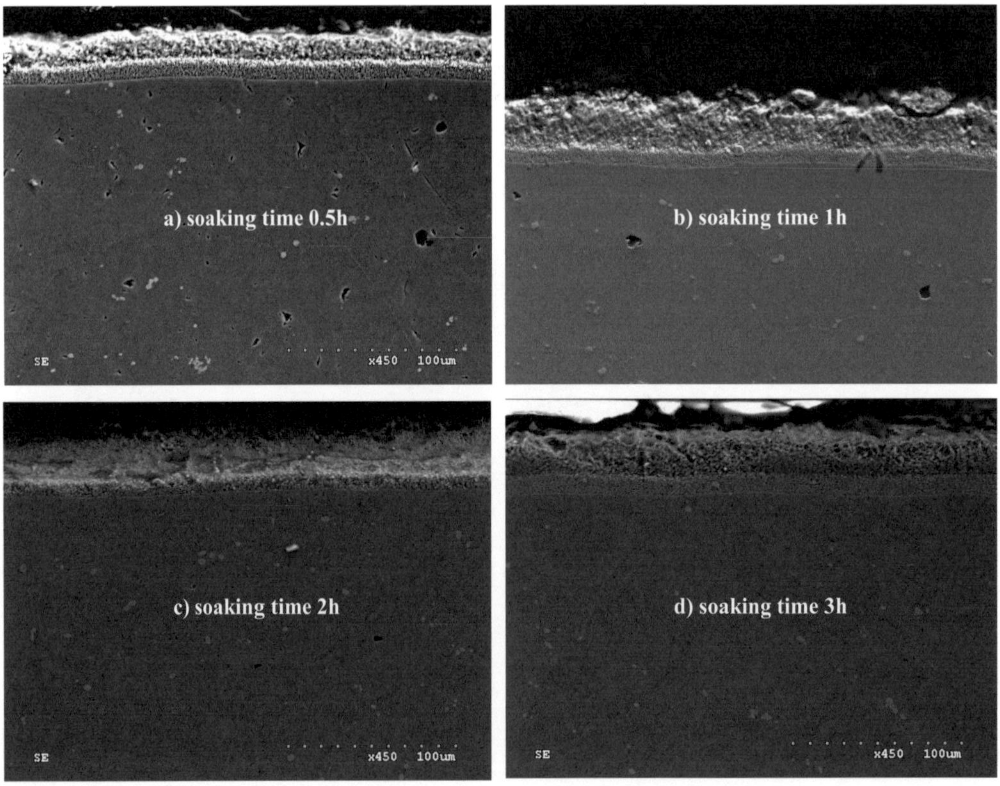

Fig. 4. Structure of the aluminium layer after soaking at temperature 1223 K over: a) 0.5 h, b) 1 h, c) 2 h, d) 3 h (magn. x4500)

The soaking time affected also the change in thickness of the aluminium protective coating (Fig. 4). The film thickness is calculated on the basis of ten distances measured in pixels (d_n, n = 1, 2, 3, ... 10) – Fig. 5. The obtained distances (in pixels) were then multiplied by the scale parameter, i.e. the size of one pixel in μm. In that way the value of average thickness for the aluminium coating was calculated for each of the recorded (and then analyzed) images. The coating thickness was measured at three locations, i.e. on the leading edge, in the centre, and on the trailing edge of the blade. Fig. 6 presents averaged values of the protective coating thickness for various soaking times. On the basis of graphs in Fig. 3 and Fig. 6, for the needs of examining the effect of high temperatures onto the blade material, the soaking time was assumed to be 1h at constant temperature, i.e. 1223 K. It was the time when rapid growth in the size of particles of the γ' phase occurred, with only slight increase in the coating thickness.

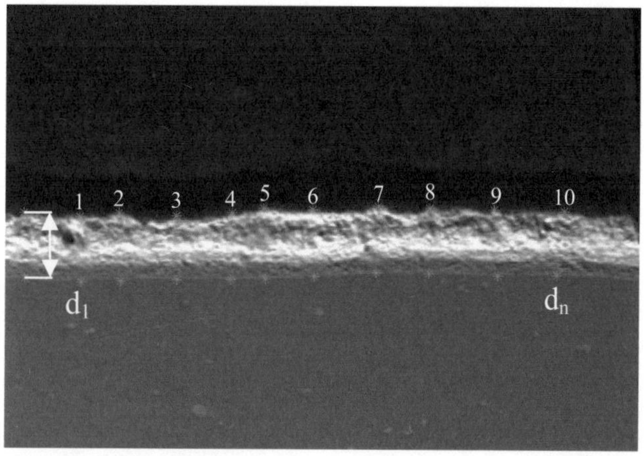

Fig. 5. Measurement of coating thickness

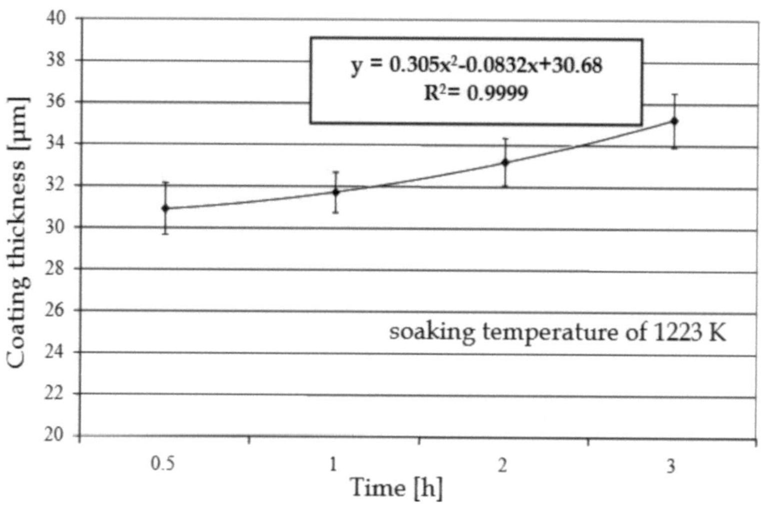

Fig. 6. Soaking-time dependent variation in thickness of the protective coating

Images of surfaces of blade specimens were acquired both before and after specimens soaking in the furnace. The photos were taken on a purpose-built workbench (Bogdan & Błachnio, 2007; Błachnio & Bogdan, 2008;) with a digital photo camera, while the surfaces were illuminated with scattered white light. Repeatability of the obtained results was proved by taking multiple photos of the same specimens, under the same conditions with appropriate settings of parameter of the digital photo camera. The soaking of blade specimens in the furnace led to alterations in colour of the surfaces. An exemplary set of images is shown in Fig. 7.

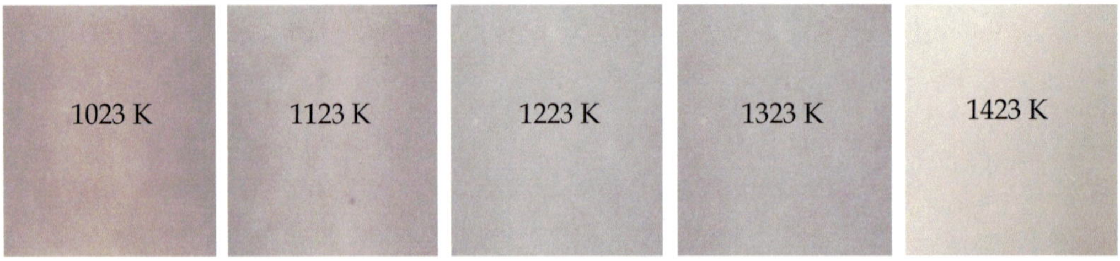

Fig. 7. Images of surfaces of specimens soaked at various temperatures

It was also determined how the temperature of blade soaking affects their microstructures. Examination was carried out using metallographic microsections and both an optical and a scanning electronic microscope (SEM). Fig. 8 shows the (new) blade structure before soaking. One can see the coating of the aluminium alloy (Fig. 8a) diffused in the blade parent metal as well as cuboidal precipitates of the γ' phase of the alloy (Fig. 8b).

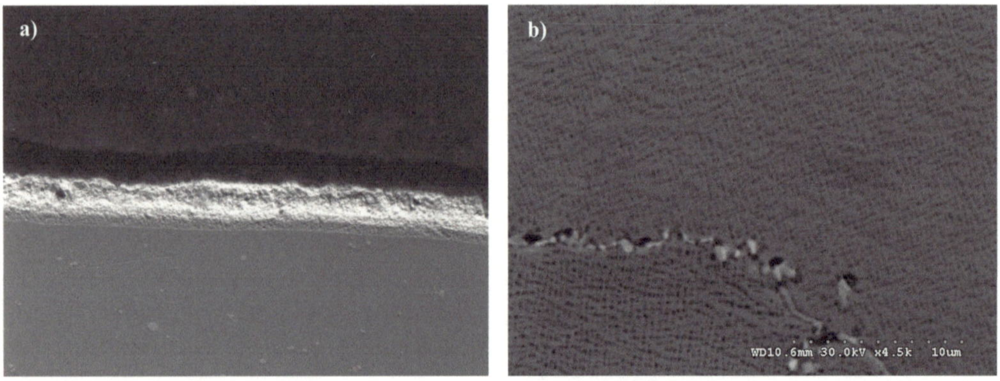

Fig. 8. Metallographic structure of the blade prior to soaking: a) coating (magn. x450); b) subsurface layer (magn. x4500)

The microstructures of high-temperature affected gas turbine blades were also observed. This provided detailed information about changes in the microstructures of both the coating layer (alteration in the coating thickness) and in the parent material. Changes in material parameters, mainly modifications in the size and distribution of the γ' phase, substantially affect mechanical properties of the material (Błachnio, 2009; Decker & Mihalisin, 1969; Dudziński, 1987; Mikułowski, 1997; Poznańska, 2000; Sims et al.,1987). Results of the examination of specimens subjected to soaking in the furnace at 1223 K and 1323 K are shown in Fig. 9 and Fig. 10, respectively.

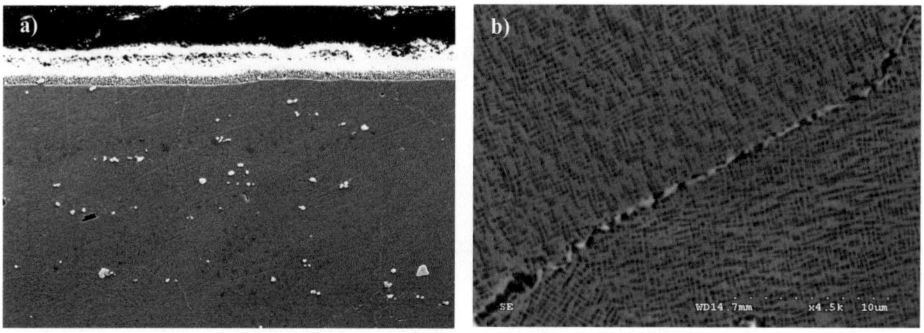

Fig. 9. Metallographic structure of the blade after soaking for 1 h at 1223 K: a) coating (magn. x450); b) subsurface layer (magn. x4500)

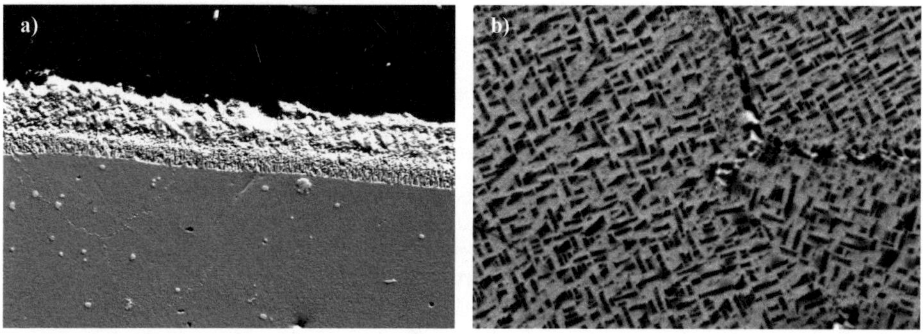

Fig. 10. Metallographic structure of the blade after soaking for 1 h at 1323 K: a) coating (magn. x450); b) subsurface layer (magn. x4500)

Relationship between the average thickness of the aluminium alloy coating and the soaking temperature of specimens is graphically shown in Fig. 11.

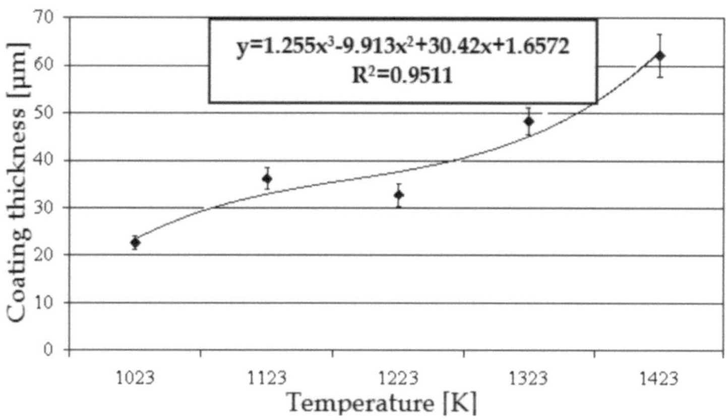

Fig. 11. Variation in the aluminium layer thickness against temperature

One can see the non-linear growth of the coating as a function of temperature, both nearby the surface and within the diffused layer. In consequence of that growth the layers exhibit less density (poorer tightness) and increased roughness that leads to amendments of the reflection parameters with regard to the incident light that illuminates the surface. In turn, the graphic relationship between the average value of the γ′ strain hardening phase emissions and the heating temperatures for the EI 867-WD alloy is plotted in Fig. 12 and demonstrates the exponential nature, but can be approximated with a polynomial.

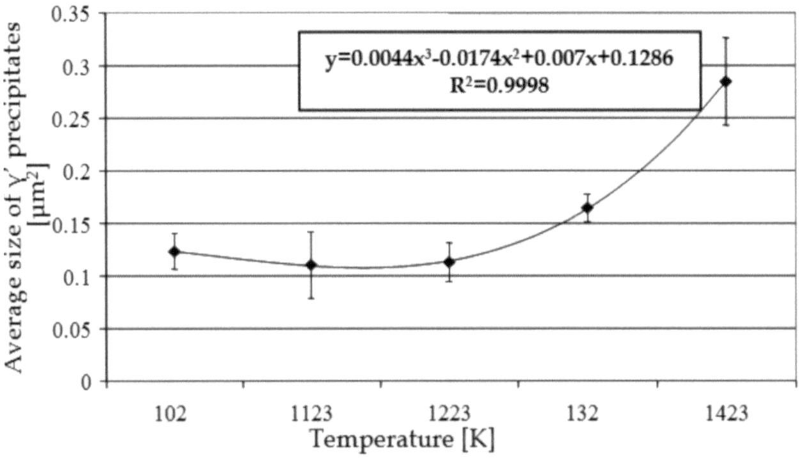

Fig. 12. Variation in γ′ particles of average size against temperature

Examination of the microstructure of blade specimens revealed that as early as at 1123 K there appeared the initial stage of coagulation of precipitates of the strengthening γ′ phase of relatively regular structure and very high density. As the temperature kept growing, the structure of the γ′ phase became less regular, and grain size was also growing. The initial period when cubic grains joined together to form plates started at 1223 K (Fig. 9b). It was found that as soon as the temperature reached 1323 K, the substantial growth and coagulation of γ′ phase precipitates followed; the γ′ precipitates adopted shapes of plates (Fig. 10b). Also, the number of particles was reduced but they were much larger than those at 1223 K.

To determine the blade serviceability (fit-for-use) threshold, it proved reasonable to develop a nomogram that presented correlation between the colour saturation in blade images and the size of the γ′ precipitates. The following assumptions resulting from the already described laboratory experiment were adopted:

1. Illumination - scattered white light;
2. No disturbing interferences of light reflected from other surfaces;
3. New gas turbine blades were used for tests;
4. Specimens cut out of blades were randomly selected and subjected to soaking (three pieces at a time) at five temperature values with the increment of 100 K, starting from the temperature of 1023 K;
5. Alteration in saturation (amplitudes of different wavelengths) of primary colours was adopted as the parameter that defines alterations in both chrominance and luminance of the examined surfaces.

To determine parameters that would enable description of the degree to which the microstructure of examined surfaces was changed (overheated), the technique of image analysis for the decomposition of primary colours, i.e. Red, Green and Blue (RGB) and shades of grey (parametric description of histograms) was employed. Due to the nature of the investigated phenomenon it was reasonable to only consider changes in the locations of maximum saturation amplitudes (for individual histograms representing distributions of brightness of digital images (Bogdan, 2008) – Fig. 13.

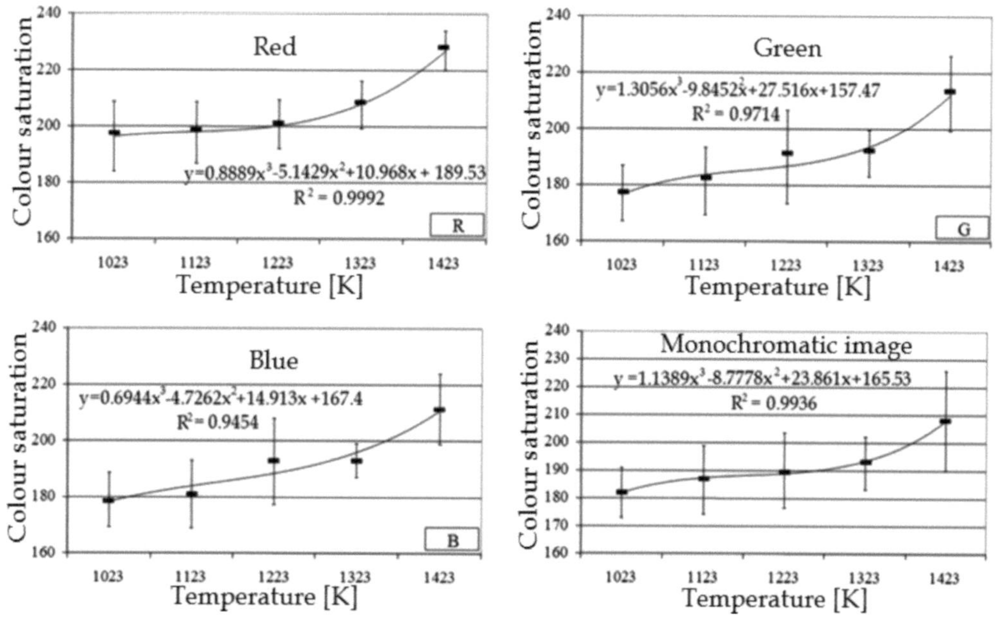

Fig. 13. Changes in locations of maximum amplitudes of saturation with RGB colours and shades of grey for various temperatures of specimen soaking

In order to find correlations between changes in colour of blade surfaces and the effect of temperature upon the blade microstructure the following nomograms were developed (Fig. 3.14 a, b) for the assessment of blade condition.

The assessment of blade condition is based on colour analysis of blade-surface images and is closely related with the material criterion (modification in the strengthening γ' phase , i.e. in both changes of shapes from cuboidal to plate-like and growth of precipitates), i.e. deterioration in high-temperature creep resistance and heat resistance after exceeding the temperature threshold of 1223 K. The nomogram that presents relationship between changes in colours of blade surface (in Red and greys) and temperature of blade soaking serves as the basis for the assessment of how much the microstructure of the EI 867-WD alloy was affected. When a mathematical description of the discussed phenomenon is introduced, the following regression curve equations result (the nomogram in Fig. 14b) for changes in:

- intensity of shades of grey (x_2):

$$x_2 = 0.2793e^{0.0189(z1-1150)} + 187.1 \tag{2}$$

  - the square of the correlation coefficient: R^2=0,9998
- average size of γ' precipitates (y_2):

$$y_2 = 0.0058e^{0.0142(z1-1150)} - 0.1 \tag{3}$$

  - the square of the correlation coefficient: R^2=0,9998

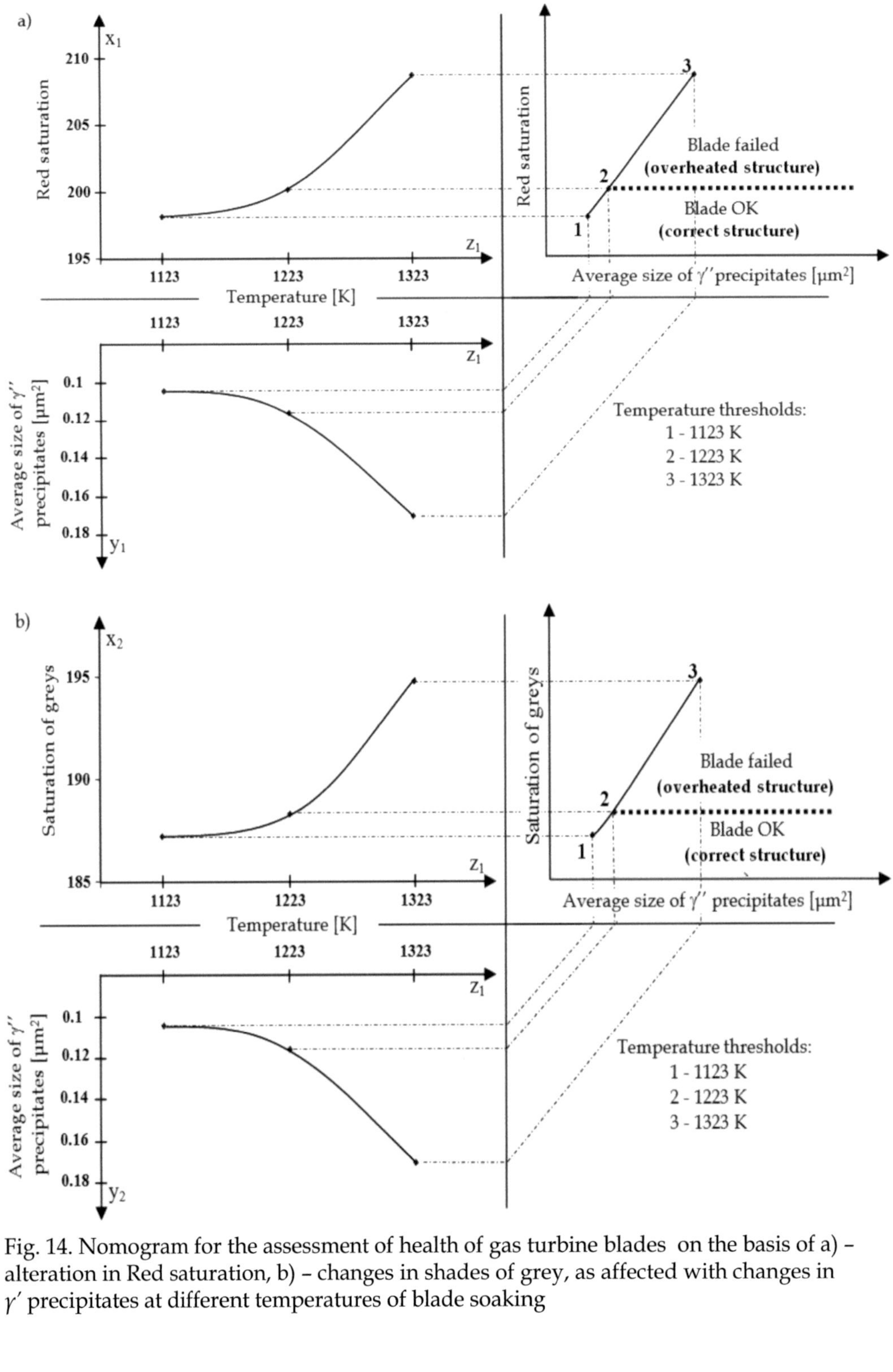

Fig. 14. Nomogram for the assessment of health of gas turbine blades on the basis of a) – alteration in Red saturation, b) – changes in shades of grey, as affected with changes in γ' precipitates at different temperatures of blade soaking

- average size of γ' precipitates as a function of greys intensity:

$$y_2 = 0.11512(x_2 - 187.1)^{0.7513} \quad (4)$$

where: z_1 – temperature [K].

Based on the foregoing functional relationship (equation 4) it is possible to assess condition of any blade (by its microstructure, i.e. the average size of the γ' precipitates) on the basis of the already calculated value of the degree of grey on the images of blade surfaces. Such an approach may prove useful, after taking account of disturbances and interferences, in formulating a mathematical model - the assessment of blade condition on the basis of changes in colours.

High temperature not only entails both changes in thickness of the aluminium coating (variable light-reflecting area) and modifications in the structure of γ' phase. In practice, alterations of the aluminium coating lead to variations of the luminance and chrominance of the surface that is recorded by the optoelectronic system furnished with the light-sensitive detector, i.e. the CCD matrix (digital images). The investigated microstructure of the subsurface layer reflects transformation of the EI 867-WD alloy and serves as the evidence for overheating of its structure (Fig. 10b, 11) after heating of the blade specimens at temperatures exceeding 1223K. When assuming the material criterion, i.e. size alterations of emissions for the γ' phase, as a criterion that is decisive for approval of blades for further operation, it is possible to find out the operability threshold that would qualify or disqualify blades for further use.

The soaking of blade specimens leads to structural changes in the superalloy. At the same time, roughness changes and thickness of the aluminum coating increases (Fig. 11). Changes in the coating's parameters (roughness, thickness) influence capability of the surface to reflect a luminous flux and its spectral composition (saturation in RGB). In addition, investigation into the chemical composition revealed that the soaking results in modification of the percentage weight-in-weight concentration of elements that make up the coating - Table 1. A substantial difference can be noted mainly in the content of such elements as W, Mo, Ni and Al.

Soaking temperature [K]	Elements by weight [%]							
	O	Al	Cr	Fe	Co	Ni	Mo	W
1423[K]	9.89	9.66	11.73	0.68	4.50	41.12	11.73	10.58
1023[K]	6.26	2.94	10.31	0.84	5.44	57.27	5.66	7.28

Table 1. Chemical composition of the aluminium coating subjected to soaking at 1023 and 1423 [K]

These are also the factors that affect conditions of reflecting the luminous flux to result in changes of colours of blade surfaces for particular soaking temperatures.

3. Diagnostic examination of operated stator vanes

The research program assumed examination of gas-turbine stator vanes of an aircraft jet engine. The vanes were manufactured of the ŻS6K alloy. The alloy in question has been strengthened with cubical γ' phase particles, the content of which amounts to approx. 64%. It is classified to the group of cast alloys. Figures below (Figs 15, 16 and 17) present exemplary sets of recorded images of turbine vanes with different degrees of overheating (according to the already applied classification of vane condition).

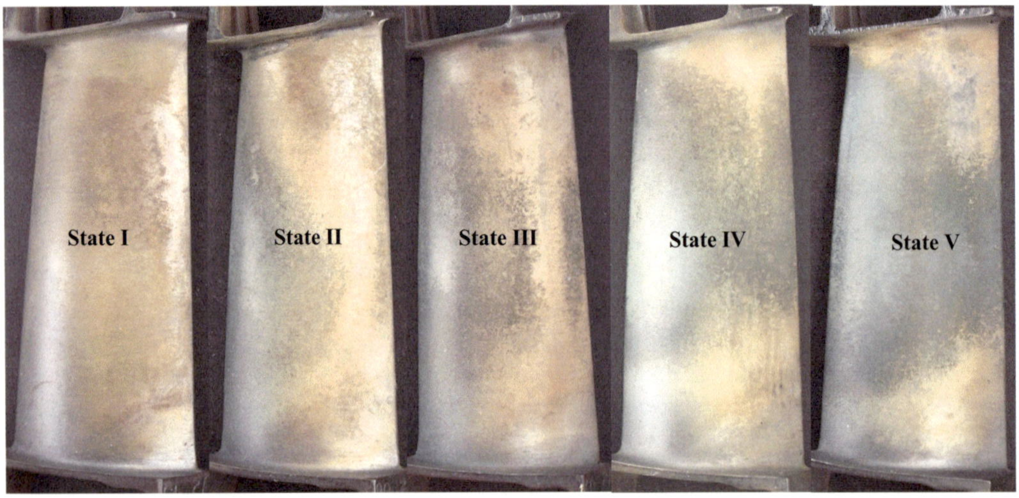

Fig. 15. Recording of vane surface images with a photo camera

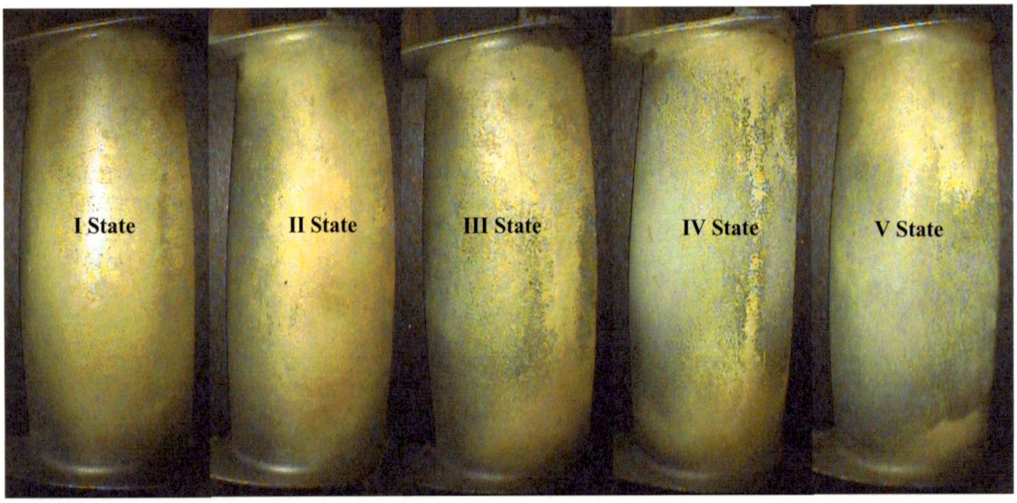

Fig. 16. Recording of vane surface images with a videoscope No. 1

Differences in colours of recorded images of turbine vanes surfaces result from properties of optoelectronic systems (chiefly, the CCD matrix) and variations in illumination (type of light) used in particular instruments. When images were taken with a photo camera, the illuminating light was uniformly scattered on entire surfaces of vanes, whilst the light emitted by videoguides was of focused nature.

The analysis of the collected vane-surface images in terms of estimation of changes in colours and shades of grey resulted in finding out the following changes in locations of maximum amplitudes for particular component colours:

- for images recorded with the digital photo camera (Fig. 18):
- for images recorded with use of the videoscope No 1(Fig. 19):

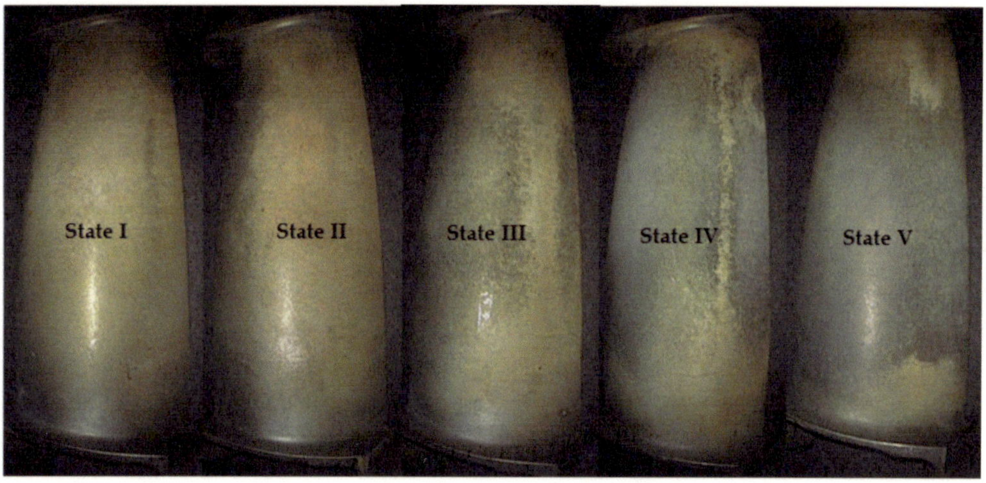

Fig. 17. Recording of vane surface images with a videoscope No. 2

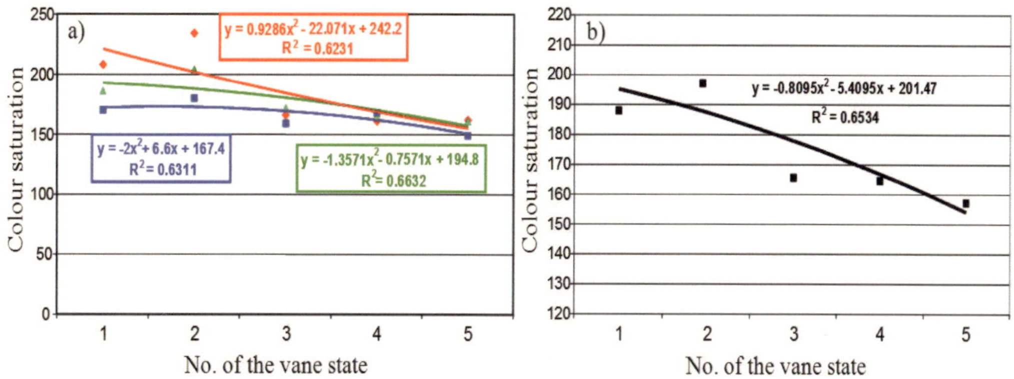

Fig. 18. Dislocation of the maximum saturation amplitudes of the image for various states of vanes: a) RGB components; b) grey shades

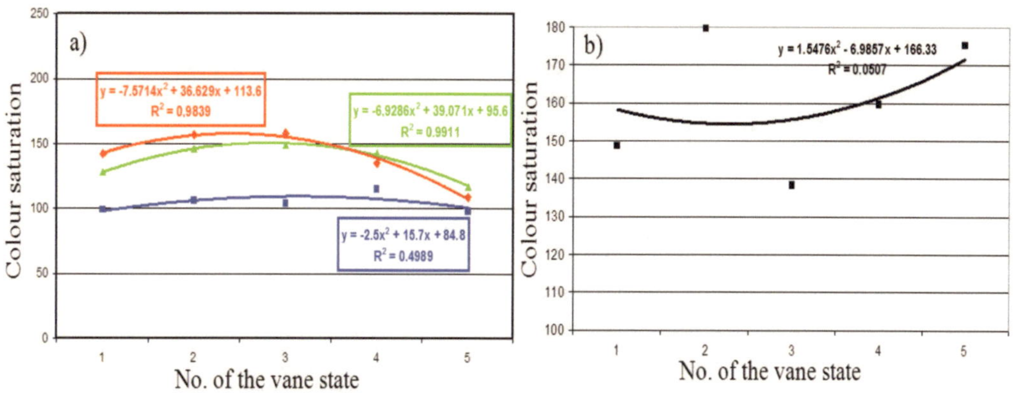

Fig. 19. Changes in locations of maximum amplitudes of image saturation for various states of vanes: a) RGB components; b) shades of grey

- for images recorded with the videoscope No 2 (Fig. 20)

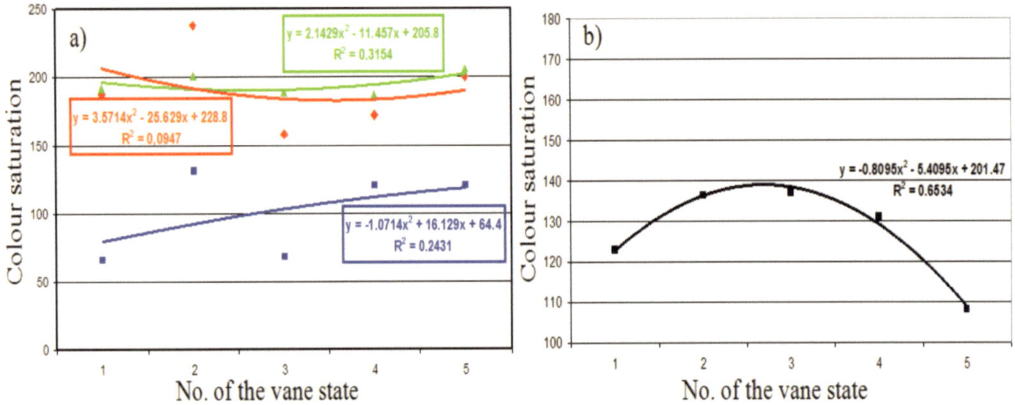

Fig. 20. Changes in locations of maximum amplitudes of image saturation for various states of vanes: a) RGB components; b) shades of grey

The curves (trend lines) demonstrate correlation coefficients much worse than those obtained from laboratory tests. It has been caused by the forms of histograms (the colour range of images is wider). However, for images recorded with a digital photo camera the surface colour represents changes due to the exposure of the material to high temperature (Fig. 18). To recognise microstructures of vanes that had already been in operation further metallographic examination was carried out under laboratory conditions. As in the experiment with new blades subjected to soaking, the examination was carried out using metallographic microsections. Two microscopes were used: optical and scanning (SEM) ones. After long-time operation the vanes manufactured of the ŻS6K alloy demonstrated different health conditions. On the basis of metallographic examination (Bogdan, 2009) it was found that initially, after some time of operation, the vane coating suffers no degradation and its thickness is nearly the same as that of a new vane. Later on, it starts to suffer swelling, which after a pretty short time may result in crack nucleation due to thermal fatigue. Since the working agent (exhaust gas) of high kinetic energy keeps affecting the vane material (the surface layer), successive changes in thickness of this layer follow. The coating is getting thinner and thinner and, therefore, loses its protective properties. Consequently, temperature of vane material grows by approx. 100 K and it is no longer protected against chemical effect of the exhaust gas. The vane becomes much more vulnerable to the exhaust gas, which results in complete deterioration of the protective coating or even the parent material. Furthermore, morphology of the γ' phase has been found to prove that after critical temperature is exceeded the alloy becomes overheated. The turbine vane cannot be then considered serviceable (fit for use). Therefore, on the basis of findings of vane microstructure analysis it is possible to state that vane no. 1 (i.e. State I) exhibits correct microstructure, whilst the structure of vane no. 5 (i.e. State V) is overheated. When these results are compared to those of the analysis of blade surface images, it is possible to infer that vanes no. 1 and 2 are in sound condition, since parameters of image properties are comparable. On the other hand, vanes no. 4 and 5 are overheated, as values calculated from the histogram (as well as from the co-occurrence matrix) are much different from those for earlier discussed items. Thus, it is feasible to demonstrate correlation between images of surfaces of turbine vanes in service and condition of microstructures of these vanes made of the ŻS6K alloy, covered with protective coatings. Metallographic examination of vanes in service has also allowed of the development of two methods for scanning surface images, i.e. one based on colour profiles, and another based on the value of plane. The subsequent stages of the first method are listed below:

- acquisition of images with a digital photo camera (laboratory conditions) or two videoscopes (real operating conditions),
- cutting of vane no. 5 (according to the earlier assessment, considered as overheated);
- plotting of averaged colour profiles down the cutting lines with account taken of the width of cutting (the model adopted to represent digital images - the RGB model);
- determination of variations in the coating thickness and changes in both sizes of precipitates and distribution of the strengthening γ' phase (the SEM microscope - the computer-aided analysis of metallographic images);
- on the basis of alterations in microstructure parameters - determination of colour lines that represent overheated and non-overheated structures;
- scanning of images of conditions I - V against the selected colour profiles.

According to the already applied classification, the fifth condition (state) denotes an overheated vane. To verify this judgement, further metallographic examination was carried out along two cutting lines. The changes in the coating thickness (on the aluminium matrix) were measured and changes in size of precipitates and shape of the strengthening γ' phase (the scanning (SEM) microscope, the computer-aided analysis of metallographic images). Alterations in these two parameters are of crucial importance for the heat resistance and high-temperature creep resistance of turbine vanes. Thus, it was possible to plot an averaged colour profile (taking account of the width of cutting) that represents an overheated structure (the selected range along line *1* - Figs 21a, b, and a non-overheated structure (the selected range along line 2 – Figs 22 a, b).

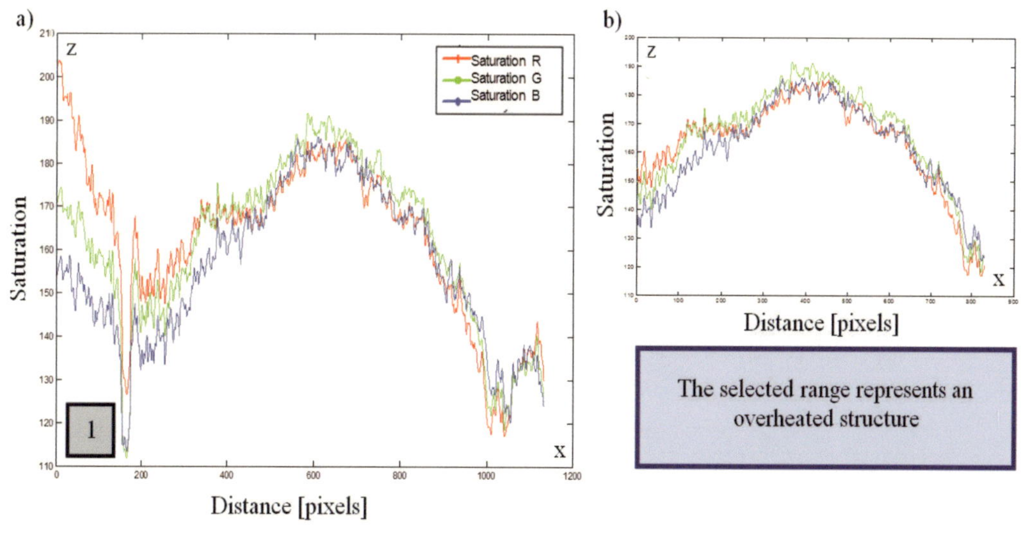

Fig. 21. Averaged RGB profiles: a) along line no. 1 – parallel to the normal direction (KN); b) the selected range that represents an overheated structure

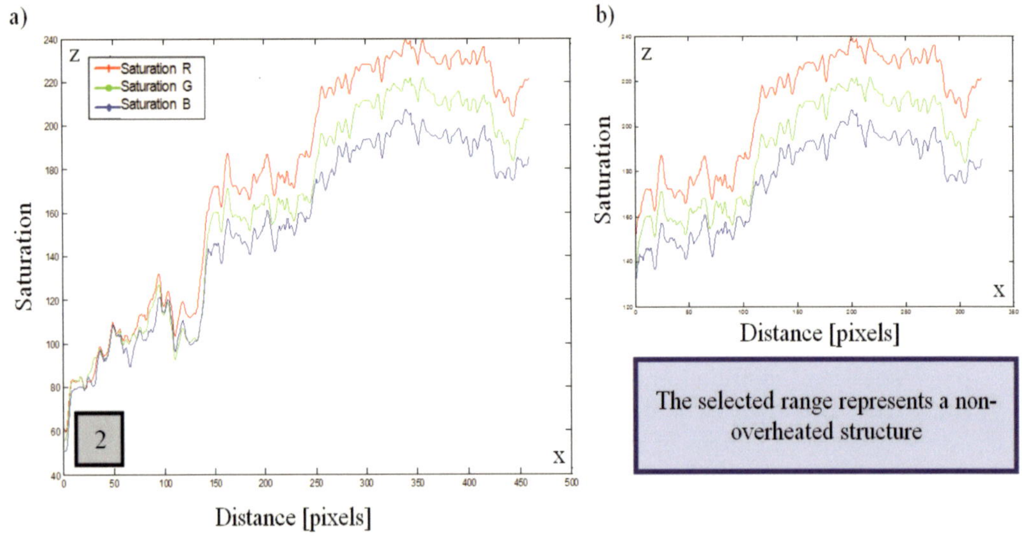

Fig. 22. Averaged RGB profiles: a) along the line no. 2 - perpendicular to the normal direction (KN); b) the selected range that represents a non-overheated structure

Next, on the basis of two ranges of colour profiles (Fig. 21b, Fig. 22b) each component (one pixel after another) of surfaces from states (I-V) of vanes was examined with regard to the occurrence of colour dots (RGB) that correspond to either overheated or non-overheated structure. Finally, the ratio of the overheated surface area to the overall surface area of the vane was obtained (Fig. 23).

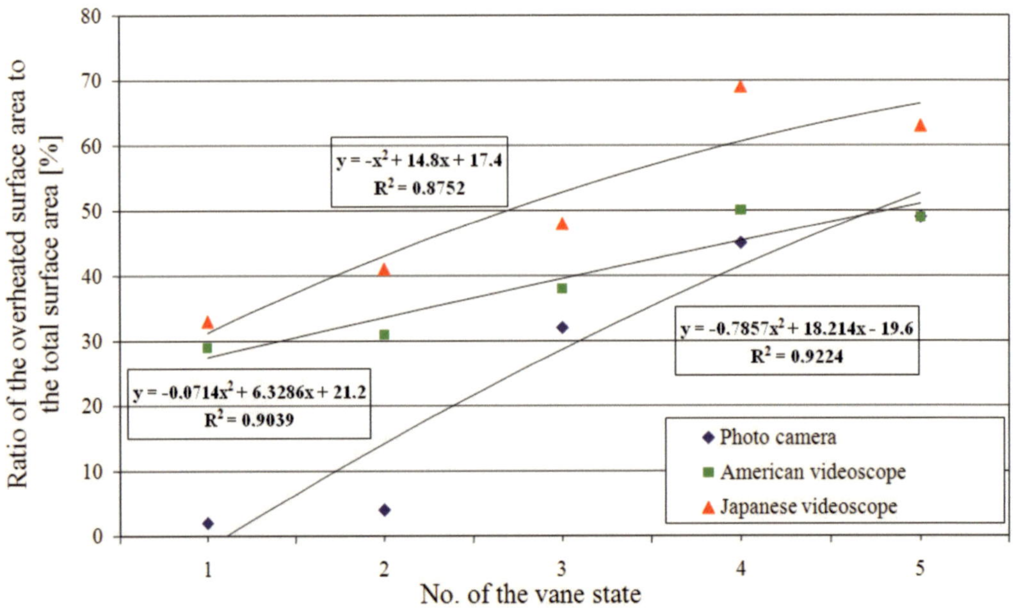

Fig. 23. Ratio of the overheated surface area to the total surface area for particular states of the vane - detection of images with a photo camera and two types of videoscopes

In the second method (Bogdan & Błachnio, 2009), the surface colour of the State V vane (the overheated structure according to metallographic examination) was assumed the overheating criterion. Based on the developed histograms, the criterion threshold was determined, where the threshold value was calculated on the basis of saturation (location of the maximum amplitude) for individual RGB components (R+G+B/3=162). The criteria threshold (value of the plane) was then referred to 3-D distributions of colours on surfaces of individual vanes from states I to V. Points with values below the determined plane were deemed the overheated surface points (pixels). Instances of estimating overheated surfaces for vanes from states I and V have been graphically shown in Figs 24 and Fig. 25, with images recorded with the digital photo camera. To make the image more clear, the Cartesian coordinate system was adopted (where: x, y - dimensions of the vane image in pixels, z - RGB saturation).

The dashed lines (Fig. 24a, Fig. 25a) represent the non-uniform effect of temperature on the vanes under examination, caused by faulty operation of injectors - irregularities in the combustion process inside the combustion chamber.

The area of overheated surface (the set of image points) extends as condition of vanes deteriorates - Fig. 24c, Fig. 25c. Introduction of the threshold plane (criterion of vane material overheating) in the 3-D charts of RGB distribution in images of surfaces of the turbine component under examination allows of the determination of the ratio of the overheated surface to the overall surface.(Fig. 26).

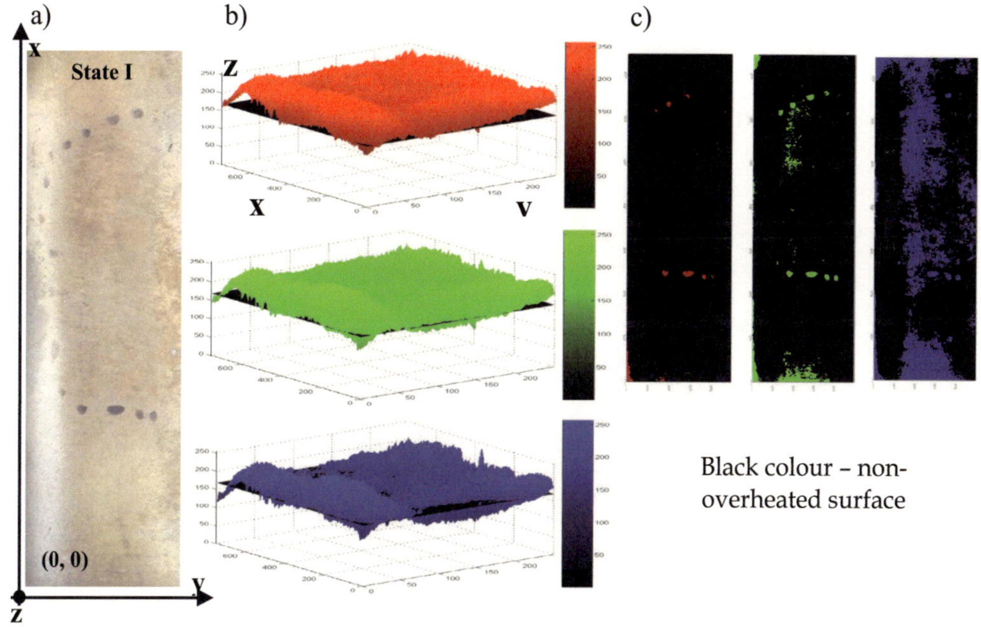

Fig. 24. Vane representing State I: a) surface image; b) 3-D distribution of RGB primary colours; c) vane surface viewed from below - the result of introduction of the criterion plane

The best results were gained for images of vane surfaces recorded under laboratory conditions with the digital photo camera. On the basis of the plotted curves (Fig. 23, Fig. 26 and Fig. 27) one can conclude that changes in colours of vane surfaces reflect health/maintenance status of the examined turbine components. Application of one of the two proposed methods , or both of them, of scanning the vane surfaces, i.e. the method based on the already determined colour.

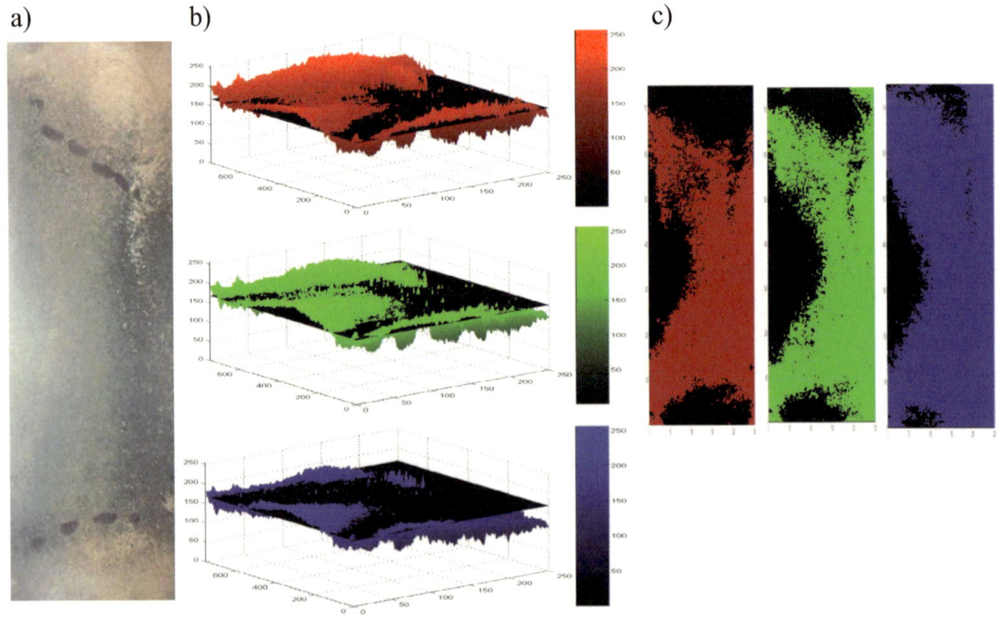

Fig. 25. Vane representing State V: a) surface image; b) 3-D distribution of RGB primary colours; c) vane surface viewed from below – the result of introduction of the criterion plane

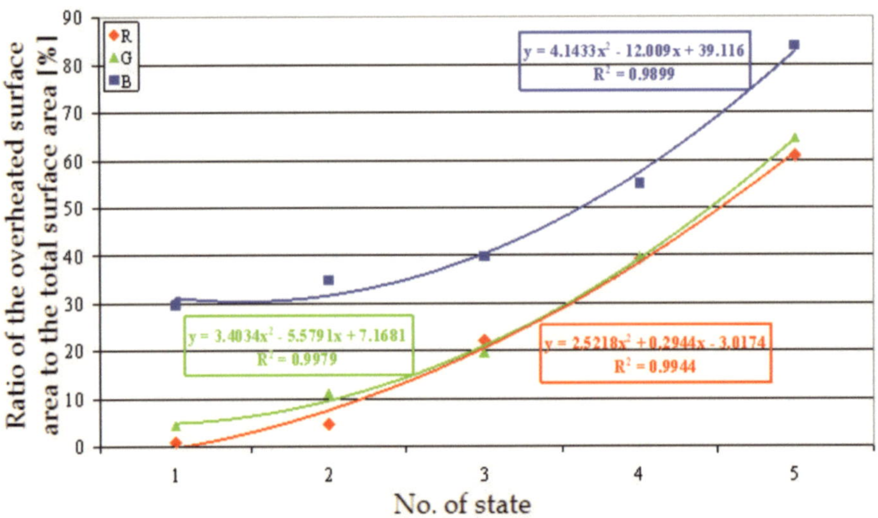

Fig. 26. Ratio of the overheated surface area to the overall surface area –images recorded with a digital photo camera

Identical relationships were determined for images recorded with two videoscopes (Fig. 27).

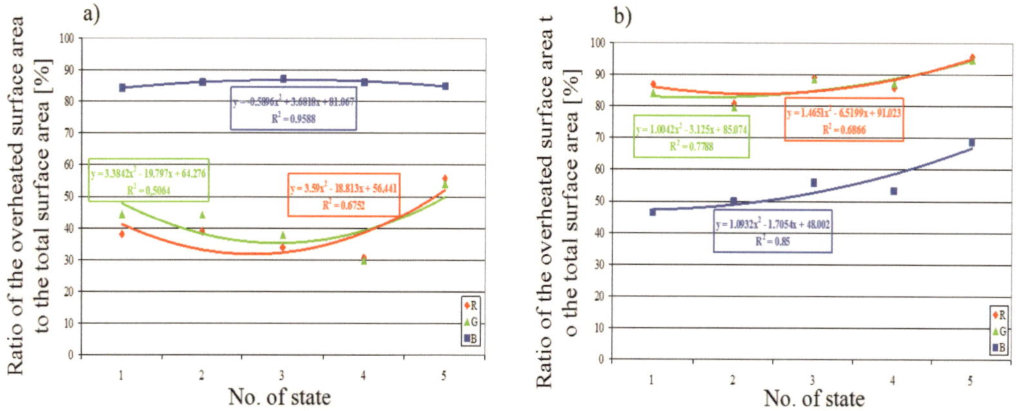

Fig. 27. Ratio of the overheated surface area to the overall surface area, a) images recorded with videoscope No. 1, b) images recorded with videoscope No. 2

profiles that respectively represent the correct and the overheated structures, and the method of the criterion plane improves likelihood (objectivity) of the assessment of the vane condition. The computer-aided acquisition of images together with dedicated software for image recognition will improve the process of the assessment itself and contribute to more trustworthy analyses than it used to be in past. Percentage differences between particular states result from the applied type of light and the way of illuminating the examined vanes. Under laboratory conditions only white scattered light was used, whilst videoscopes incorporate sources of light focused in other colour. Capability to recognize and record colours may also be different due to light-sensitive CCD matrices installed in various detection instruments. Nevertheless, it must be noted here that the application of endoscopes (videoscopes) for the acquisition of images may be used to track (monitor) changes in vane condition (development of failures, health/maintenance status of components under examination) in the course of periodical inspections with no need to dismantle the entire gas turbine.

4. Application of neural networks in diagnostics of vanes

The subsequent paragraphs present the opportunities to apply artificial neural networks to diagnostic examination of vanes, both new ones (after heating) and those that have already been in operation. The major objective was to develop such a neural network that would be capable of diagnosing the technical status of the turbine component under test on the basis of parameters for images of their surfaces. The metallographic examinations were carried out to assess technical condition of the turbine component in question. Alterations of the metal structure were taken into account, such as thickness alteration of the protective aluminium coating and changes in average size of emissions for the γ' phase (the strain hardening phase of the alloy, which is the phase that predominantly decides on creep resistance properties). The metallographic examination made it possible to classify vanes according to their technical condition. Fig. 28 explains an example of such classification of vanes that demonstrate various technical conditions (wear degree) – the material criterion.

On the basis of conclusions related to assessment of the overheating degree (vanes applicable and inapplicable for further operation) and drawn from microstructure examinations, the pattern images were adopted for vane surfaces representing various degrees of deterioration (the neuronal pattern classification). Nowadays a great number of supervised networks are available, although, in fact, they are merely options or variants of a limited number of models. For this study only models that offered the best results (verification of classification correctness on a set of test benchmarks) were taken into account, i.e. the Multi-Layer Perceptron (MLP) and the network with Radial Basis Function (RBF). Examples for structures of such networks are shown in Fig. 29. Each of the networks is made up of three layers (one input layer, one hidden and one output) with the same number of neurons per each layer.

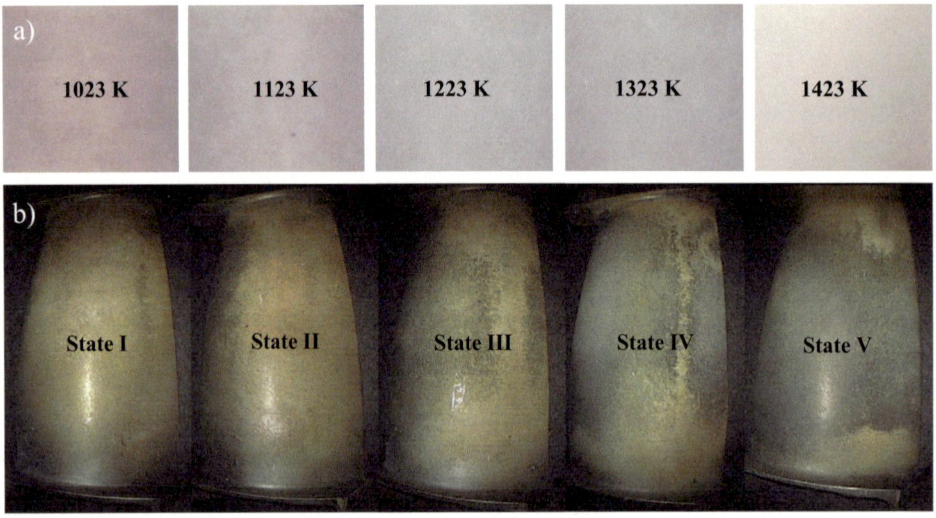

Fig. 28. Acquisition of surface images of blades/vanes: a) heated (images recorded with a photo camera); b) operated (images recorded with the videoscope no. 2 Three-state)

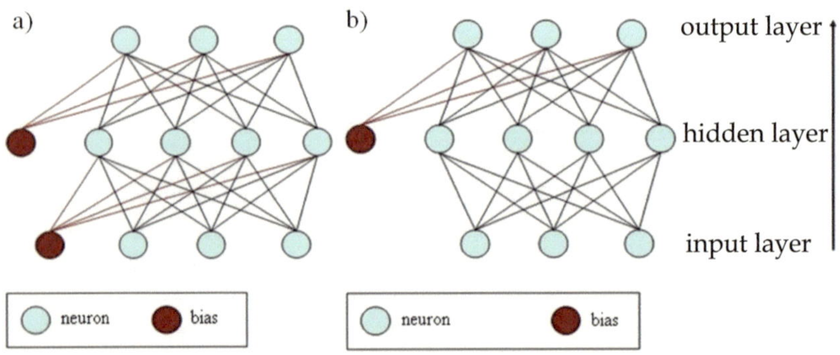

Fig. 29. Diagram of the network: a) the Multi-Layer Perceptron (MLP); b) the Radial Basis Function (RBF) network

For the perceptron network the neurons are deployed exclusively between subsequent layers and signals propagate to only one direction (the unidirectional network). The number of hidden layers is actually unlimited, but it has been proved that two layers are perfectly sufficient for any transformation of input data into output ones. Learning of networks of

such a type is usually carried out in the mode with a teacher, by means of the gradient method of the first of second order, by minimization of the error function. In case of a multi-layer perceptron the excitation level of a neuron is the weighted sum of inputs (plus the threshold value that is added as a bias). When a auxiliary bias input is added to a neuron the networks acquires higher ability to learn owing to the possibility to shift the activation threshold depending on the weight of the bias input. For Radial Basis Function (RBF) networks the bias input is added exclusively to neurons within the output layer. Moreover, the network type that uses the radial basis functions has usually one hidden layer that comprises neurons with a radial function of activation. Output neurons usually represent the weighted sum of signals coming from radial neurons deployed in the hidden layer. Learning of that type of networks consists in selection of weight coefficients for the output layer and parameters of the Gaussian radial basis functions.
The objective of the newly designed neuronal classifier was to develop a (computer-aided) method that would enable to recognize technical condition of a specific vane on the basis of its surface image (its properties). Two following cases were considered:

1. Two-state classification (for blades and operated vanes):
   - - class 1: operable status (non-overheated blade/vane);
   - - class 2: inoperable status (overheated blade/vane);
2. 2. Three-state classification (only for operated vanes):
   - - class 1: operable status (non-overheated vane);
   - - class 2: partly operable status (the vane suspected to be overheated);
   - - class 3: inoperable status (overheated vane);

The first phase of the development consisted in acquisition of data that were subsequently used to model the network (input data) and for further tests (verification of ability to correct classification). In order to reduce the amount of information, the colour images were converted into black and white ones (8 bit encoding of grey shades, 0-255). Then 10 input parameters were selected (image parameters). Six first parameters (P1-P6) describe the histogram, i.e. distribution of pixel brightness. Four subsequent parameters (P7-P10) were found out on the basis of the co-occurrence matrix (for the distance of 1 and angle of 0°) - Table 2.

Designation	Specification
P1	value of maximum saturation
P2	value of average brightness
P3	fluctuation of brightness distribution
P4	histogram skewness
P5	histogram kurtosis
P6	histogram excess
P7	contrast
P8	correlation
P9	energy
P10	homogeneity

Table 2. Input data – the feature vector

The preliminary metallographic investigations (the material criterion) made it possible to state that the new blades heated at the temperatures of 1023K and 1123K exhibit correct metallographic structure whilst the ones that are heated at 1323K or 1423K are overheated (Fig. 28a). In case of vanes that have already been in operation, the vanes of correct metallographic structure are those of the I and II state, whilst overheated vanes are from the IV and V state (Fig. 28b). Owing to such classification it was possible to embark on the network modelling. The modelling phases were the following:

1. Standardization of data and encoding of outputs (classes);
2. Subdivision of data into the learning pattern and test pattern (at the shares of 50% to 50%);
3. Determination of parameters for the neuronal network, such as minimum and maximum number of hidden layers (for MLP and RBF networks), types of activation functions, both for hidden and output neurons (for the MLP network), minimum and maximum values for reduction of weight coefficients, for both hidden and output neurons (for the MLP network).

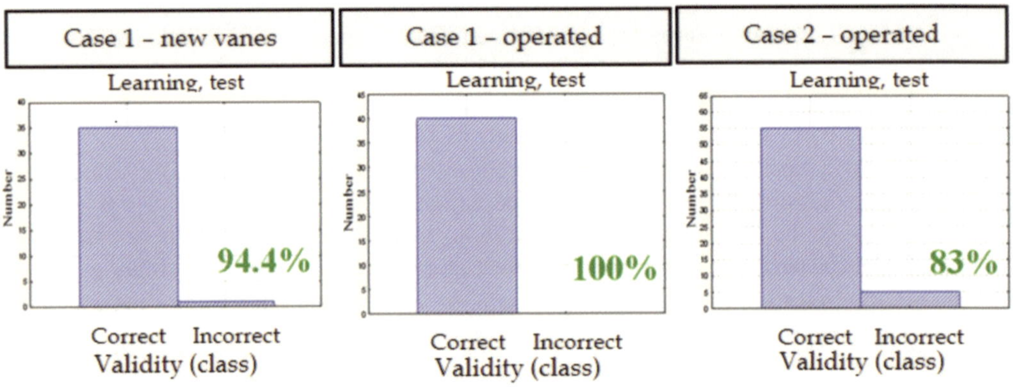

Fig. 30. Comparison between validity of classification

The network was subjected to the learning process with use of the set of input data. As a result of the simulation process, both in the learning and the test modes, the optimum models of neuronal networks were developed for each case (two-state and three-state classifications). It was confirmed that neuronal networks offer a useful tool to assess status of vanes, both new ones (after heating) and those that have already been in operation (Fig. 30).

The developed neural classification models (networks with a defined architecture) make it possible to determine technical condition of vanes on the basis of features (parameters) attributable to their images with satisfying dependability. The additional advantage of such an approach is the possibility to carry out the diagnosis process under conditions of continuous operation of the turbine (with no need to have it dismantled), where images of specific parts of turbines are acquired with use of a videoscope and then transmitted to the control computer, where the dedicated software extracts the required features (parameters) of the images. Finally, the 'modelled network' (well learned) indicates whether the vane is suitable for further operation. The three-state classification enhances the diagnostic process by the possibility of approving a vane for further, but supervised operation i.e. until the date of scheduled assessment.

5. The thermographic method for technical condition assessment of gas turbine vanes and blades

5.1 The passive infrared thermography

Thermographic methods represent relatively new but rapidly developing approach to non-destructive diagnostic examination of materials. Thermography not only enables measurements of temperature, but also determination of temperature distribution on the basis of the detection of infrared radiation emitted by examined surfaces. In the literature, the thermography is frequently referred to as 'thermovision' (Oliferuk, 2008).

For the entire spectral range, power of electromagnetic (EM) radiation emitted by surfaces of materials depends on temperature of a given surface, and peaks of the radiation of power fall within the infrared (IR) range. The infrared radiation fits within the range of electromagnetic waves from 0.75 to 100 μm, i.e. remains outside the very narrow interval of light visible to the human eye (0.4 to 0.7 μm). With a suitable infrared detector available, and with both the relationship between radiation power and temperature of the emitting surface and dependencies of the signal at the detector output on this power known, it is possible to determine temperature of the surface in a non-contact way. The infrared thermography method based on detection of infrared radiation (like all non-destructive testing methods) can be split into passive and active techniques. A research method based on detection of infrared radiation without the need to additionally stimulate the examined object (supplying with energy) is referred to as the passive infrared thermography.

Emissivity of materials is expressed by the following formula (Oliferuk, 2008):

$$E(\lambda, T) = \frac{dW_e}{d\lambda} \tag{5}$$

where: dW_e is the energy of electromagnetic radiation emitted in a time unit by a unit of surface of the material within the range of wavelengths λ to $\lambda + d\lambda$.

Among a number of fields, the thermographic method is also widely applied to technical diagnostics, owing to advanced thermographic systems offering the possibility of determining the temperature distribution on the examined surface with temperature resolution better than 0.1K. The range of applications of the method includes inspection of electric circuits and systems, integrated circuits, mating parts of machinery and structures, civil engineering, power engineering, diagnostics of high-temperature structures.

The diagnosing of gas turbines with the passive thermographic techniques consists in the recording of images of temperature distribution of turbine components at the exhaust nozzle's outlet. The starting point for any efforts intended to assess condition of the turbine components is development of pattern thermograms for correct operation of the turbine. Then, in the course of routine inspection carried out during regular operation of the turbine components the generated thermograms are compared with available patterns. If only a slight anomaly appears, it is considered a signal to initiate searching for any reason for the discrepancies. Owing to such an approach, it is possible to detect defects such as erosion of the turbine, failures to vanes/blades, incorrect operation of the combustion chamber, etc., i.e. ones hard to detect with other non-destructive inspection methods (Korczewski, 2008; Lewitowicz, 2008)

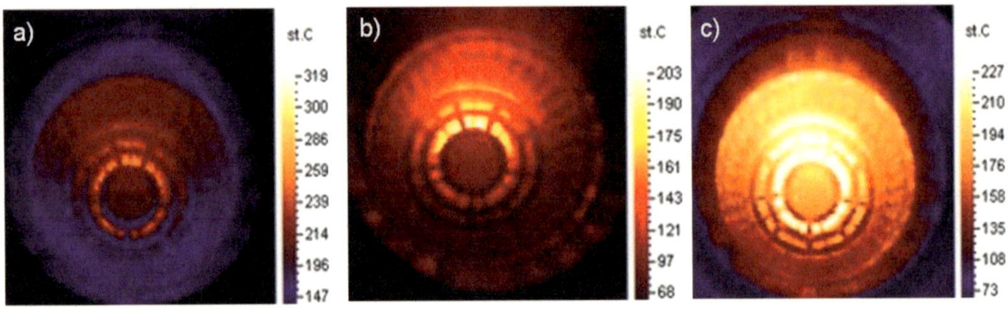

Fig. 31. Infrared radiation emitted by the turbine during the engine start-up: a) - start-up, b) - pre-heat, c) - operation (Haralick R. M. et al, 1973)

5.2 Active infrared thermography

The essence of the active infrared thermography consists in determination of thermal response of the examined material to stimulation by means of an external pulse of heat. Nowadays, research into the application of the active infrared thermography to detection of defects in surface layers of materials experience flourishing development. When a specific quantum of heat is delivered to the material surface, e.g. in the form of a heat pulse, the temperature of the material surface will be changing rapidly after the pulse termination. Owing to thermal diffusion, a thermal front moves deeper into the material. The presence of areas that differ in thermal properties (i.e with defects) from defect-free areas provokes some change in the rate of diffusion.

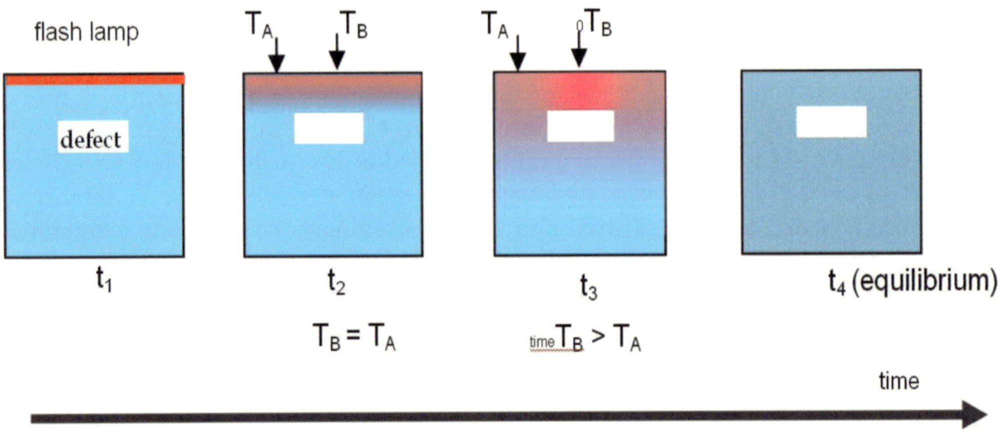

Fig. 32. Change in temperature of the specimen's surface after stimulation with an external heat pulse (Thermal Wave Imaging, Inc., 2009)

Thus, the monitoring of the temperature field on the surface of a specimen subjected to cooling provides capability to show locations of defects. The simpliest method of processing the signal recorded with an infrared thermograph while cooling the examined surface down consists in calculating the temperature contrast. The contrast is defined by the following relationship (Oliferuk, 2008):

$$C_a(t) = T_p(t) - T_{pj}(t), \tag{6}$$

where: $C_a(t)$ – absolute contrast, $T_p(t)$ – temperature at any point of surface of the examined material, $T_{pj}(t)$ – temperature at the point of surface above the homogeneous (i.e. defect-free) material.

Values of the absolute contrast are higher than zero at the points of surface right above the area of the material where a discontinuity exists. Depending on the stimulation method, several types of the active thermography are distinguished, namely: the pulsed thermography, the lock-in thermography with modulated heating and the pulsed phase thermography (Maldague, 2001; Thermal Wave Imaging, Inc., 2009).

The pulsed thermography is deemed as a rather simple variation of the active thermography and consists in the determination and analysis of temperature distribution on the examined surface when the surface is being cooled down after having been uniformly heated with a thermal pulse (Fig. 33). For the one-dimensional model and homogenous material, the equation that describes the change in temperature while the surface is being cooled down after heating with a short thermal pulse takes the following form (Oliferuk, 2008; Luikov, 1969):

$$T(t)-T(0)\approx Q\alpha^{-\frac{1}{2}}t^{-\frac{1}{2}} \tag{7}$$

where: Q stands for energy of the thermal pulse per each surface unit, t - time when the surface is being cooled down, α- thermal diffusivity, $T(0)$ – temperature at a selected point or on the area of the heated surface, just after termination of the thermal pulse and $T(t)$ - temperature at any moment of the cooling process.

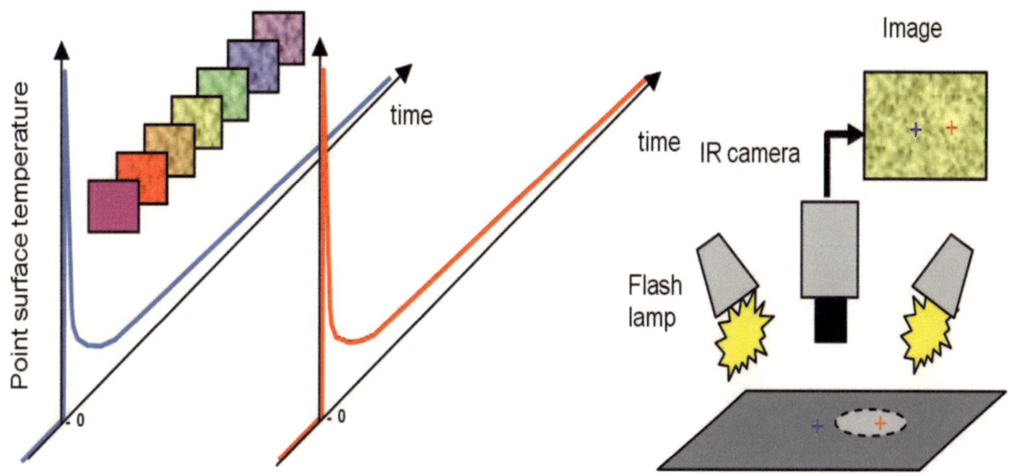

Fig. 33. Diagram to explain the application of pulse thermography (Thermal Wave Imaging, Inc., 2009)

When any flaws occur in the examined material, diffusion rates are reduced, which makes the temperature of the area of the surface above the flaw be different from the temperature of the area of the surface below with no defect – the nature of the foregoing interrelationship changes. Various physical properties of materials facilitate specialized diagnostic examination, e.g. determination of materials health, constitution of their structures, or identification of the material under examination, etc. The pulsed thermography enables one to distinguish temperature values in the course of cooling the specimen's surface down after preliminary treatment with a thermal pulse. The acquired information contained in thermal response from the examined surfaces enables detection of other types or grades of materials present in the specimen (Fig. 34)

Obviously, this method (as other ones) has its limitations, since it enables one to detect flaws only in subsurface layers of materials due to the fact that the temperature contrast rapidly fades out with the depth of penetration. The research work has demonstrated that flaws located in deeper layers reveal themselves, however, later and with a poorer contrast. Time t_d, from the termination of the stimulating pulse to the flaw revealing itself is proportional to the square of the flaw depth z (Oliferuk, 2008):

$$t_d \approx \frac{z^2}{\alpha} \tag{8}$$

whilst contrast C substantially fades as the depth of flaw increases (Oliferuk, 2008):

$$C \approx \frac{1}{z^3} \tag{9}$$

The experiments have also demonstrated that the radius of the smallest detected flaw must be at least twice as high as the depth where the flaw is located.

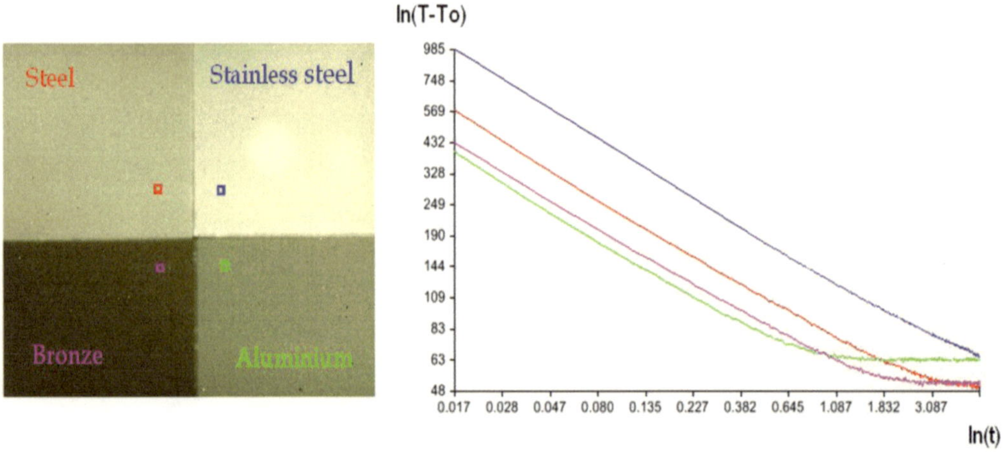

Fig. 34. Thermal response to a thermal pulse for selected grades of materials (Thermal Wave Imaging, Inc., 2009)

The drawback of the pulse thermography is that the examined surface has to meet requirements of the emission homogeneity, which is associated with the need to coat it before examination with a homogenous film, i.e. graphite.

On the other hand, methods based on another parameter, i.e. on the phase of the thermal wave, are free of the mentioned drawback. This is why these methods are widely applied in the active lock-in thermography with modulated heating and in the pulsed phase thermography (Oliferuk, 2008). The lock-in thermography with modulated heating, contrary to the pulsed one, not only allows of finding out surface distribution of power of infrared radiation emitted by the surface of the material under examination and distribution of the associated temperature, but also enables us to determine distribution of amplitudes and phases of thermal waves on the area in question. Amplitude of a thermal wave found on the basis of detected IR radiation emitted by the examined surface depends on the emissivity of the surface, whereas the phase is independent of this emissivity. These are properties deemed the most important advantage of the lock-in thermography with modulated heating. When the examined surface is not uniformly heated or the subsurface layer has been altered due to operating conditions, emissivity is locally affected. When the phase shifts have been mapped against the stimulating signal, one can infere the presence of flaws under the material surface (Oliferuk, 2008).

The pulsed phase thermography combines advantages of the pulsed thermograpy and the lock-in thermography with modulated heating. The response signal recorded with a thermovision camera represents the relationship between the surface temperature and time $T(t)$ for particular locations of the surface being cooled down after treating it with a thermal pulse. The signal is then subjected to discrete Fourier transformation (Oliferuk, 2008), which allows of finding particular waves for each point of the thermal image of the examined surface, and of development of phase maps. The phase maps, as in the method of the lock-in thermography with modulated heating, reveal locations of flaws in examined materials. The basic difference between the pulsed phase thermography and the lock-in thermography with modulated heating is that the pulsed phase thermography is focused on the analysis of a non-stationary process, i.e. the cooling of the surface of the object under examination, earlier treated with a thermal pulse. On the contrary, the lock-in thermography with modulated heating is applicable to stationary processes, i.e. stationary oscillations of the temperature field on the examined surface as a result of harmonic stimulation by heat (Oliferuk, 2008; Maldague, Matinetti, 1996; Maldague *et al*, 2002; Saenz *et al*, 2004).

5.3 Application of the thermographic method to assess condition of gas turbine vanes/blades

The pulsed thermography method was applied to a number of studies, including the project intended to determine the applicability of the method to assess flow capacity of internal cooling channels of turbine vanes/blades. Improvement in general efficiency of the turbine and increase in the power/weight ratio are directly associated with the exhaust-gas temperature. Increase in the exhaust gases temperature due to material problems has enforced application of turbine vanes/blades of more sophisticated geometrical shapes. It has, in turn, complicated vane/blade manufacturing processes and many other treatments, e.g. cooling the blades and vanes. Operational experience and examination of vanes and blades in repair workshops have demonstrated that, besides material defects, also disturbances in the internal cooling system caused by obstructions in cooling channels quite frequently cause defects of vanes and blades. Fig. 35 presents images of a damaged vane, taken with a conventional optical method, the raw pulsed thermography and the TSR (Thermographic Signal Reconstruction) technique employed in tomography devices. Application of the pulsed thermography method together with the dedicated software enables easy inspection of the internal system of cooling channels and flow capacity thereof. The advantages of the proposed method, as compared to the X-ray technique, are as follows: it keeps the operator safe from the hazardous X-ray radiation and, in consequence, does not require any dedicated, purposefully safeguarded rooms to carry out the examination; the unit cost of a test is reduced as there is no need to purchase expensive consumables; results are obtained in a very short time. The method based on the measurement of the amount of fluid flowing via the cooling channels within the blade offers much less accuracy and is more time- and labour-consuming than the thermographic technique.

Results of examining turbine vanes and blades with the pulsed thermography methods while investigating into discontinuities in the subsurface layer of the material became the inspiration to embark upon further research on the feasibility of this thermographic technique to assess alterations in microstructures of gas turbine blades and vanes using available devices and instruments. The examination involved specimens from new blades made of the EI 867-WD alloy and subjected to thermal ageing in a furnace at various temperatures. What resulted were distinct changes in the relationships between parameters of the thermal response from specimen materials and stimulation by a thermal pulse (Fig. 36).

Fig. 35. Images of high-pressure turbine blades (aircraft engine), acquired with various methods: optical, raw, TSR (Thermal Wave Imaging, Inc., 2009)

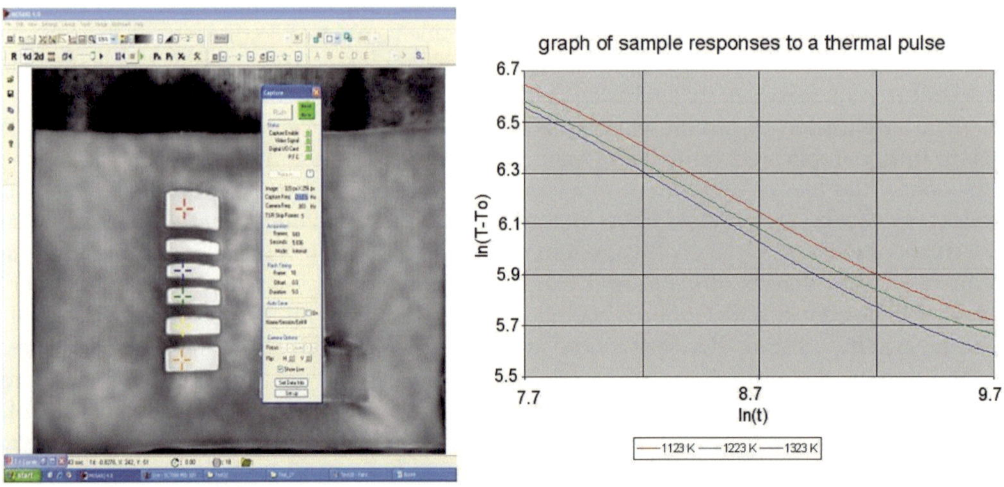

Fig. 36. Images of specimens cut out from blades (EI 867-WD alloy) subjected to soaking at 1123 K, 1223 K, 1323 K; graph of their responses to a thermal pulse

After completion of metallographic examination, the assessment of changes in microstructures of the specimens was carried out, mainly of change in the strengthening γ' phase -

$Ni_3(Al,Ti)$. Findings of this examination are presented in Fig. 37 as a nomogram. The relationship between the thermal response of the specimen's material, represented as the value of ln(T-To) against the average size of the γ′ precipitates allows of the assessment condition/health of the specimen material. This relationship, in conjunction with the knowledge on permissible changes in the microstructure, serves as a basis to judge whether the specimen's material remains fit for further service, or not.

High temperature results in both changes in thickness of the aluminium coating and modification of the γ′ phase structure. The examined microstructure of the subsurface layer reflects changes in the EI 867-WD alloy and proves the alloy structure suffered overheating as soon as the specimens were subjected to soaking at 1223 K (Figs 9 and 10). When the material criterion is adopted, i.e. a change in the size of γ′ precipitates, a threshold value of their remaining serviceable (fit for use) is considered the criterion that determines suitability of the blades for further operation. Results from metallographic examination confirm that the vane/blade material loses its high-temperature creep resistance at temperatures above 1223 K due to the clustering of fine-grain (Fig. 9) cubical particles of the γ′ phase and formation of plates (Fig. 10).

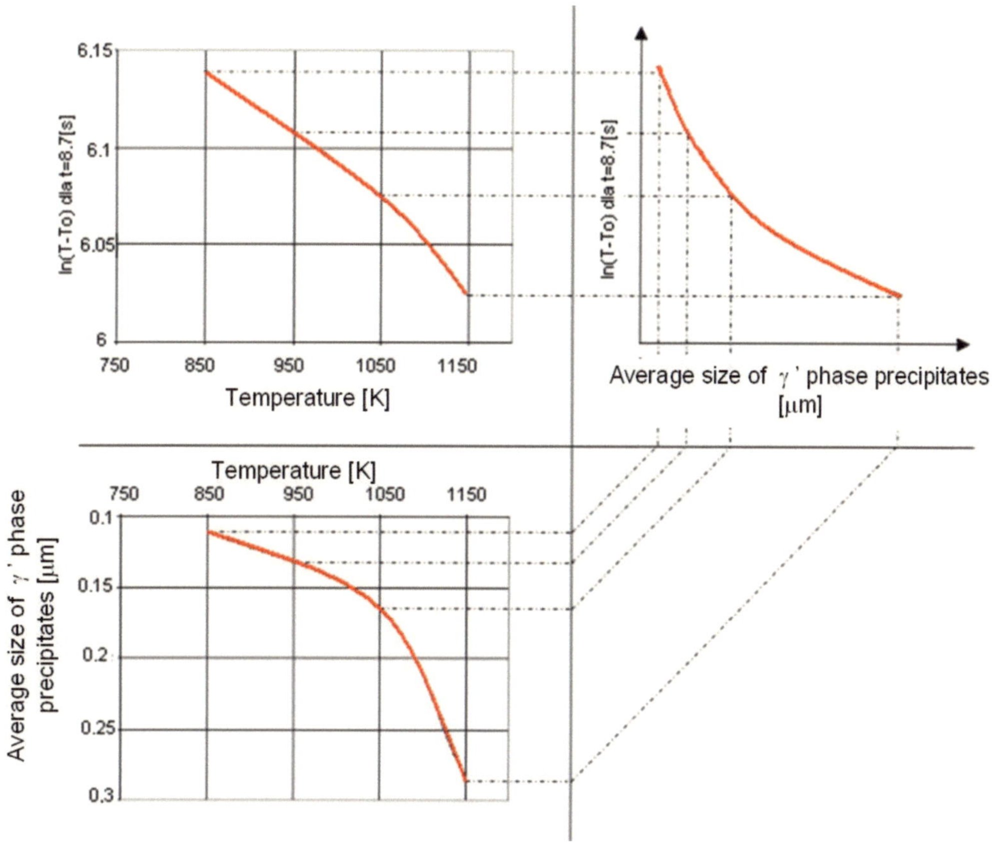

Fig. 37. Nomogram for the assessment of microstructures of specimens from gas turbine blades made of EI 867-WD alloy on the basis of relationship between change in *ln(T-To)* parameter and that in size of γ′ precipitates at different soaking temperatures

Further on, cognitive examination of items in service was carried out with the thermographic method. The examination was focused on the gas turbine stator vanes from the aircraft jet engine, the items in question being made of the ŻS6K alloy.
The vanes were initially classified according to the degree of overheating (from I to V) on the basis of visual criteria used in the course of engine inspection/overhaul. Results of the examination performed with the pulsed thermography have demonstrated that thermal response of the material of a vane deemed 'fit for service' (1st category) is rather uniform over the entire vane surface (Fig. 38). On the other hand, based on the analysis of response of a vane classified as 'unfit for service' to a thermal pulse one could easily find vane-surface areas that explicitly differed from the average. These areas overlapped the regions earlier classified, by visual inspection, as overheated. These findings, together with results of examining specimens cut out from vanes, serve as a basis to draw inferences on possible change in the material structure.
The already completed analyses of results obtained from the examination of vanes remaining in service and classified to the 1st and 5th category, and of new vanes, prove the pulsed thermography is suitable to judge about changes in the structure of material of turbine vanes remaining in service.

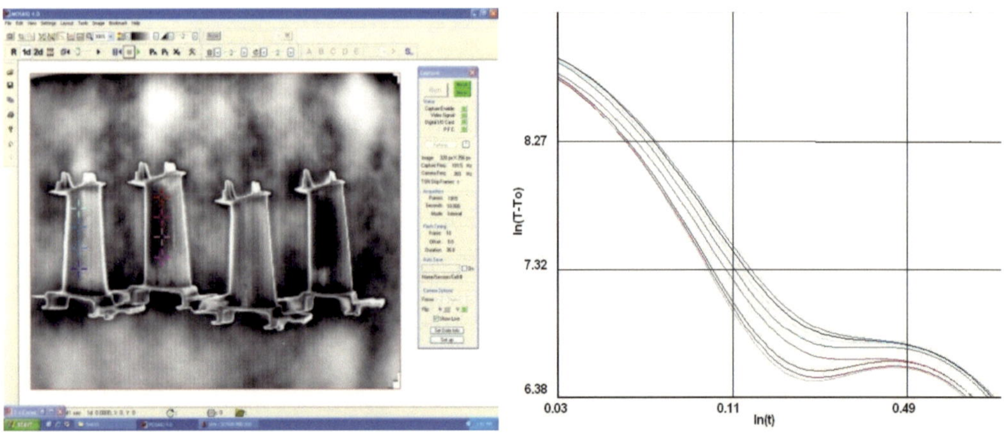

Fig. 38. Images of turbine vanes remaining in service , classified to the 1st and 5th categories (right); plotted are responses of vane materials to a thermal pulse

6. Conclusion

The assessment of health/condition of turbine blades and vanes is carried out by a diagnostic engineer on the basis of the recorded image of the component to be diagnosed, by comparing this image with pattern images of surfaces of turbine blades/vanes of the same type taken for items fit/unfit for use. However, such assessment criteria are very inaccurate since the diagnostic examination of blades/vanes is burdened with the element of subjective assessment (the organoleptic assessment of item condition). Moreover, colours depend on a number of physical and psychological factors. Therefore, the assessment performed by a diagnostic engineer is associated with a significant risk of a human error. The destruction process in a gas turbine blade/vane begins with a failure to the protective aluminium coating (which is visible on the blade/vane image as a change in colour of the surface). Application of digital image-recording technology, together with the computer-aided analysis of images and the computer-aided decision-making process (neural networks), significantly contribute to the improvement in the diagnosis of that turbine component. Additionally, the use of an instrument capable of recording images of blade/vane surfaces (e.g. videoscopes) allows of the diagnosing with no need to have the turbine disassembled. Diagnostic information collected in that way is gained pretty fast and the cost of data acquisition is really low as compared to other inspection procedures.

This study presents fundamentals of the new method dedicated to the assessment of condition of gas turbine blades/vanes, applicable to both new vanes (the laboratory experiment) and already operated ones (actual operating conditions). The method involves the digital processing and analysis of surface images. Results obtained from the diagnostic examination of vanes already in service are confirmed by two methods developed to scan surface images, i.e. the scanning method based on colour profiles and one that involves the value of criteria plane. Both methods enable determination of the size (dimensions) of local areas of overheating (the percentage ratio of overheated surface area to the overall area of the vane) and thus, allow of the assessment of the overheating degree.

The application of artificial neural networks clearly demonstrates the feasibility of a fully automated (computer-aided) decision-making process, where parameters of images, earlier obtained from histograms and the co-occurrence matrix, are classified and sorted according to their applicability during the neural-network teaching process. Therefore, the application of an artificial neural network enabled clear and complete mapping of sophisticated relationships between images of a blade/vane surface and condition thereof. The presented results for the two-state classification (fit-for-use status - the non-overheated blade/vane and the unfit-for-use status - the overheated blade/vane) are highly promising. Moreover, the three-state classification is presented as well, with the diagnostic process enhanced by the possibility to approve the blade/vane for further, however supervised operation, i.e. until the date of scheduled periodic inspection and condition assessment.

However, for further thorough investigation into the impact of the working agent (the exhaust gas) on the condition of surfaces of blades/vanes, it is necessary to pay more attention to the effect the operating conditions have upon turbine blades/vanes: to 'isolate' the most damaging periods of the engine operation (the start-up and acceleration) since thermal shocks are the most destructive factors affecting each engine. They are caused by massive variations of the exhaust gas temperature, the components of the hot section of the engine are exposed to. Other characteristics of the engine operation should be also taken into account, including the rotational-speed increase during the start-up of the engine, values and fluctuations of the exhaust gas temperature upstream the turbine.

The presented results of thermographic examination of gas turbine blades and vanes, both new ones and those that already in operation, explicitly prove that the method is perfectly suitable for the diagnosticing of the vane condition. These results serve also as evidence that relationships between thermal loads applied during the turbine operation, changes in signals of thermal responses attributable to blade/vane material and the microstructure

status of the turbine items in question really exist. They also give grounds for developing the essentials of a non-destructive thermographic method to assess the degree of the gas-turbine blade's/vane's material overheating.
The method of pulsed thermography offers also a number of benefits as compared to other techniques of non-destructive investigation, including short time required for obtaining the results and low unit cost of of the examination. The proposed method is expected to enable comprehensive analysis of the entire population of vanes and blades from a specific batch when they are submitted for verification to the repair workshop, instead of hardly reliable method that consists in the inferring on the condition of the entire set of items in question on the basis of destructive examination of one or two randomly selected pieces.

7. References

Błachnio, J. & Bogdan, M. (2008). Diagnostic procedures for the assessment of condition of gas turbine vanes in operation. *Diagnostics*, No.1(45), 2008, pp. 91-96, ISSN 641-6414.

Błachnio, J. (2009). The effect of high temperature on the degradation of heat-resistant and high-temperature creep resistant alloys. *Solid State Phenomena*, Vol. 147-149, pp. 744 - 752, ISSN: 1662-9779.

Błachnio, J. & Bogdan, M. (2010). A non-destructive method to assess condition of gas turbine blades, based on the analysis of blade-surface images, *Russian Journal of Nondestructive Testing*, Vol.46, No.11, pp. 860-866, ISSN 1061- 8309.

Bogdan, M. & Błachnio, J. (2007). The assessment of gas-turbine blade condition as based on the analysis of light reflected from the blade surface. *Archives of Transport*, Vol.19, No.4, pp. 5-16, ISSN 0866-9546.

Bogdan, M. (2008). An attempt to evaluate the overheating of gas turbine blades. *Journal of Polish CIMAC*, Vol.3, No.1, pp. 25-32, ISBN 83-900666-2-9, ISSN 1231-3998.

Bogdan, M. (2008). Computer processing of some surface images of engineering objects affected by high temperature conditions. *Acta Mechanica et Automatica*, Vol.2, No.3, pp. 19-23, ISSN 1898-4088.

Bogdan, M. & Błachnio, J. (2009). The assessment of condition of gas-turbine nozzle guide vanes, with digital analysis of images of guide-vane surfaces applied. *Journal of Polish CIMAC*, Vol.4, No.3, pp. 23-30, ISBN 83-900666-2-9, ISSN 1231-3998.

Bogdan, M. (2009). Diagnostic Examination of Gas Turbine Blades/Vanes by Means of Digital Processing of Surface Images. PhD Thesis, Technical University of Białystok, Białystok, Poland, (in Polish).

Decker, R. F. & Mihalisin, J. R. (1969). Coherency strains in γ hardened nickel alloys. *Trans. ASM*, Vol. 62,No. 2, pp. 481 - 489.

Dudziński, A. (1987). *The X-ray Structural Analysis of the EI-929 Alloy Subjected to Long-time Soaking*, PhD Thesis. Military University of Technology, Warsaw, Poland, (in Polish).

Dżygadło, Z. (et al.). (1982). *Rotor systems of turbine engines.* Communication and Communications Publishing, Warsaw, Poland, ISBN 83-206-0217-3 (in Polish).

Haralick, R. M. & Shanmugam K. (1973). Textural Features for Image Classification. *IEEE Transactions on Systems*, Vol. Smc-3, No.6, (November 2007), pp. 610-621, ISSN 0018-9472.

Haralick, R. M. & Shapiro L. G. (1992). *Computer and robot Vision*, Addison-Wesley, ISBN 0201569434, Addison-Wesley Longman Publishing Co., Inc. Boston, MA. USA.

Hernas, A. (1999). *Creep resistance of steel and alloys. Part 1.* Publishing House of the Silesian University of Technology, Gliwice, Poland, ISBN 83-88000-16-0 (in Polish).

Kerrebrock, J. L. (1992). *Aircraft engines and gas turbines (2nd edition)*, Massachusetts Institute of Technology (MIT), The MIT Press, ISBN-10 0262111624, ISBN-13 978-026211162.

Korczewski, Z. (2008). *Archives of marine engines endoscopy*, Publishing House of the Polish Navy University, Gdynia, Poland (in Polish).

Lewitowicz, J. (2008). *Fundamentals of Aircraft Operation*. Publishing House of the Air Force Institute of Technology, vol 4, Warsaw, Poland (in Polish).

Luikov, A. V. (1969). *Analytical Heat Diffusion Theory*, Academic Press, New York, USA.

Maldague,.B & Xavier. P.V. (2001). *Theory and Practice of Infrared Technology for Nondestructive Testin,* Wiley Interscience, ISBN 0-471-18190-0, USA.

Mikułowski, B. (1997). *Creep resistant and heat resistant alloys - superalloys,* Publishing House of the AGH University of Technology, Cracow, Poland (in Polish).

Oliferuk, W. (2008). *Infrared Thermography in Non-Destructive Tests of Materials and Equipment*, Gamma Office, ISBN 978-83-87848-61-3 Warsaw, Poland (in Polish).

Poznańska, A. (2000). *Lifetime of vanes made of the EI-867 alloy and operated in aircraft engines from the aspect of non-uniform deformations and structural alterations.* PhD Thesis. Technical University of Rzeszów, Poland (in Polish).

Rafałowski, M. (2004). *Integrated Image Analyzers for Lighting Technology based Measurements and Evaluation of Object Shapes.* Publishing House of the Technical University of Białystok, ISBN 0867-096X, Białystok, Poland (in Polish).

Sieniawski, J. (1995). *Criteria and methods for the assesment of materials for components of turbine engines.* Publishing House of the Technical University of Rzeszów, Poland (in Polish).

Sims, C. T.; Stoloff N.S & Hagel, W. C. (1987). *Superalloys II. High temperature materials for aerospace and industrial power,* Wiley & Sons, New York, USA.

Sunden, B. & Xie, G. (2010). Gas Turbine blade tip heat transfer and cooling: A Literature Survey. *Heat Transfer Engineering,* Vol. 31, Issue 7, Taylor & Francis, pp. 527-554, ISSN 1521-0537 (electronic), ISSN 0145-7632 (paper).

Thermal Wave Imaging, Inc.(2009). "EchoTherm User Manual".

Tracton, A. A. (Ed.). (2006). *Coatings Technology: Fundamentals, Testing, and Processing Techniques*, CRC Press, ISBN 978-1-4200-4406-5, Bridgewater, New Jersey, USA.

Tracton, A. A. (Ed.). (2007). *Coatings Materials and Surface Coatings*, ISBN 978-1420-04404-1, CRC Press, Bridgewater, New Jersey, USA.

Zhang, D.; Kamel, M. & Baciu, G. (2004). *Integrated Image and Graphics Technologies*, Kluwer Academic Publishers, ISBN 1-4020-7774-2, Norwell Massachusetts,

21

Repair of Turbine Blades Using Cold Spray Technique

Kazuhiro Ogawa and Dowon Seo
Fracture and Reliability Research Institute, Tohoku University
Japan

1. Introduction

Hot section parts of combined cycle gas turbines are susceptible to degradation due to high temperature creep, crack formation by thermal stress, and high temperature oxidation, etc. Thus, regularly repairing or replacing the hot section parts such as gas turbine blades, vanes, and combustion chambers is inevitable. For this purpose, revolutionary and advanced repair technologies for gas turbines have been developed to enhance reliability of the repaired parts and reduce the maintenance cost of the gas turbines. The cold spraying process, which has been studied as not only a new coating technology but also as a process for obtaining a thick deposition layer, is proposed as a potential repairing solution. The process results in little or no oxidation of the spray materials, so the surfaces stay clean, which in turn enables superior bonding. Since the operating temperature is relatively low, the particles do not melt and the shrinkage on cooling is very low. In addition, this technique is based on high velocity (300-1200 m/s) impinging of small solid metallic particle (generally 5-50 μm in diameter) to the substrate. In this spray process, the particles are accelerated by the subsonic gas jet which is usually lower than melting temperature of feedstock. Consequently, this process has solved the problem of thermal spraying, i.e. oxidation and phase transformation. A cold spray system is also simpler than a than low pressure plasma spraying (LPPS) system. Therefore, it has a possibility to apply cold spray technique instead of welding to repair the cracks. In this chapter, it is described that the possibility of applying cold spray technique for repairing the Ni-base turbine blades and its characteristic.

2. Cold spray

Cold spray is a process whereby metallic powder particles are utilized to form a coating or thick deposition by using ultra-high speed impingement upon a substrate (see Fig. 1). The metallic particles range in size from 5 to 50 μm and are accelerated by injection into a high velocity stream of gases. The high velocity gas stream is generated through the expansion of a pressurized, preheated gas through a de Laval (converging-diverging) nozzle. The pressurized gas is expanded to supersonic velocity, with an accompanying decrease in pressure and temperature. The powder particles, initially carried by a separate gas stream, are injected into the nozzle either prior to the throat or downstream of the throat. The particles are then accelerated by main nozzle gas flow and are impacted onto a substrate after exiting the nozzle. Upon impact, the solid particles deform and create a bond with the substrate. As the process continues, particles continue to impact the substrate and form bonds with the deposited particles, resulting in a uniform coating with very less pores and high bond strength. The term "cold spray" has been used to describe this process due to the relatively low temperatures of the expanded gas stream that exits the nozzle. Cold spray as a coating technique was initially developed in the mid-1980s at the Institute for Theoretical and Applied Mechanics of the Siberian Division of the Russian Academy of Science in Novosibirsk. The Russian scientists successfully deposited a wide range of pure metals, metallic alloys, and composites onto a variety of substrate materials, and they demonstrated that very high coating deposition rates are attainable using the cold spray process.

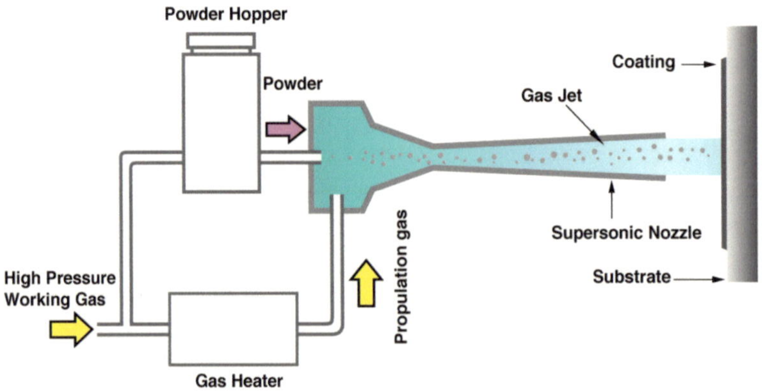

Fig. 1. Schematic illustration of cold spray apparatus

The temperature of the gas stream is always below the melting point of the particulate material during cold spray, and the resultant coating and/or freestanding structure is formed in the solid state. Since adhesion of the metal powder to the substrate, as well as the cohesion of the deposited material, is accomplished in the solid state, the characteristics of the cold spray deposit are quite unique. Because particle oxidation is avoided, cold spray produces coatings that are more durable with better bonding. One of the most deleterious effects of depositing coatings at high temperatures is the residual stress that develops, especially at the substrate-coating interface. These stresses often cause debonding. This problem is compounded when the substrate material is different from the coating material. This problem is minimized when cold spray is used. In addition, interfacial instability due to differing viscosities and the resulting roll-ups and vortices promote interfacial bonding by increasing the interfacial area, giving rise to material mixing at the interface and providing mechanical interlocking between the two materials.

A key concept in cold spray operation is that of critical velocity. The critical velocity for a given powder is the velocity that an individual particle must attain in order to deposit after impact with the substrate. Small particles achieve higher velocities than do larger particles, and since powders contain a mixture of particles of various diameters, some fraction of the powder is deposited while the remainder bounces off. The weight fraction of powder that is deposited divided by total powder used is called the deposition efficiency, and several parameters including gas conditions, particle characteristics, and nozzle geometry, affect particle velocity. And the quality of the cold sprayed coating is affected by not only particle velocity, but also the particle size and size distribution. What seem to be lacking, however, are investigation of influence of particle size distribution. In next section, the influence of the particle size distribution is explained.

2.1 Materials used and spray conditions

A nickel-based superalloy Inconel 738LC (IN738LC) was used in this study. This alloy was solution treated and then subjected to a typical aging treatment. Chemical composition is shown in Table 1 and heat treatment (HT) was applied as following step; first aging at 843°C/24h with air cooling and the second solution treatment at 1121°C/2h with air. Then, the alloy was to form 5.0 mm-thick sheets and vertically sprayed on with a high pressure cold-spray apparatus (PCS-203, Plasma Giken Co., Japan). The thickness of the deposited layer was approximately 800 μm. The powder particles used for cold spraying were prepared from IN738LC (same solution number, gas atomized). The sprayed particles had diameters of less than 25 μm, under 45 μm, and in the range of 25-45 μm. The effect of particle size variation on the strength of the sprayed layer was evaluated. The particle size distribution is shown in Fig. 2. And the spray conditions are displayed in Table 2.

Co	Cr	Mo	W	Al	Ti	Nb	Ta	C	Ni
8.25	15.95	1.7	2.6	3.43	3.42	0.95	1.74	0.11	61.85

Table 1. Chemical compositions of IN738LC (wt.%)

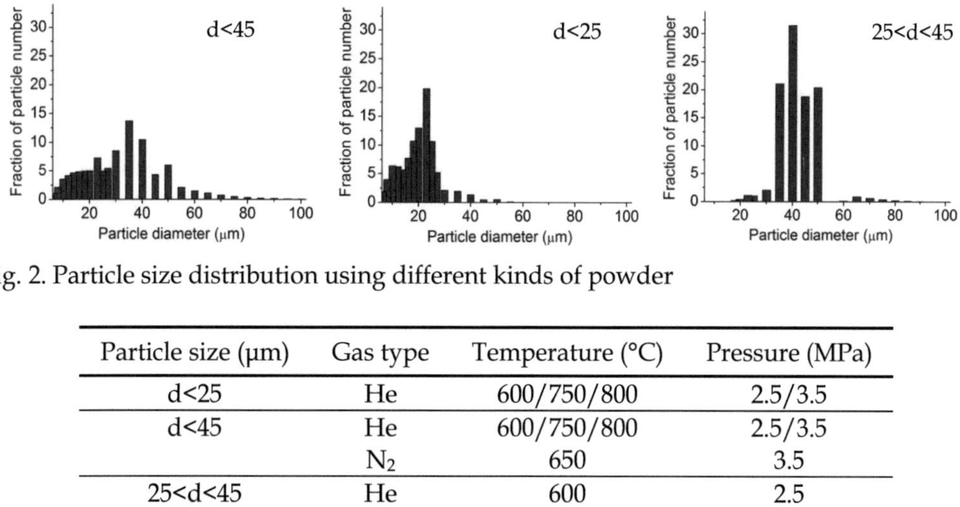

Fig. 2. Particle size distribution using different kinds of powder

Particle size (μm)	Gas type	Temperature (°C)	Pressure (MPa)
d<25	He	600/750/800	2.5/3.5
d<45	He	600/750/800	2.5/3.5
	N_2	650	3.5
25<d<45	He	600	2.5

Table 2. Cold spray conditions

2.2 Microstructures of cold sprayed Ni superalloy coatings

Typical scanning electron microscopy (SEM) images are shown in Fig.3. As shown in these images, it can be made it possible to form thick and dense deposition by cold spray technique. And it is clear that denseness of the cold spray coatings depend on spray conditions. In the case of using 25<d<45 μm powder, in spite of lower gas temperature compared to the others, coating density was high.

Normally, it has been widely accepted that particle velocity prior to impact is one of the most important parameters in cold spraying. It determines whether deposition of a particle or erosion of a substrate occurs on the impact of a spray particle. Generally, there exists critical velocity for materials such that a transition from erosion of the substrate to deposition of the particle occurs, as previously explained. Only those particles achieving a velocity higher than the critical one can be deposited to produce a coating. The critical velocity (ref. Fig. 4) is associated with properties of the feedstock (Alkimov et al., 1990; Van Steenkiste et al., 1999) and the substrate (Stoltenhoff et al., 2002; Van Steenkiste et al., 1999; Zhang et al., 2003). On the other hand, the particle velocity is related to the physical properties of the driving gas, its pressure and temperature, as well as the nozzle design in the spray gun (Dykhuizen & Smith, 1998; Gilmore et al., 1999; Li & Li, 2004; Van Steenkiste et al., 2002). Ordinarily, higher gas pressure and temperature cause higher particle velocity on cold spraying. Accordingly, by using higher temperature, it can be easy to deposit the particles on the substrate and already deposited particles, and to form dense coatings. However, from the Fig. 3, cold spraying at 600°C has better quality rather than that at 750°C. This means particle size and size distribution are also important for cold sprayed deposition. In the next section, it is described that influence of particle size distribution of used powder focusing on kinetic energy and rebound energy of cold sprayed particles.

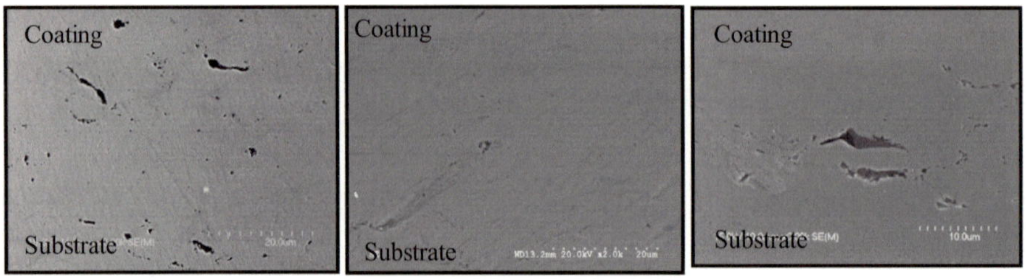

d<25 750 °C 2.5 MPa   25<d<45 600 °C 2.5 MPa   d<45 750 °C 2.5 MPa

Fig. 3. Examples of typical SEM images of cold sprayed Ni base superalloy coatings

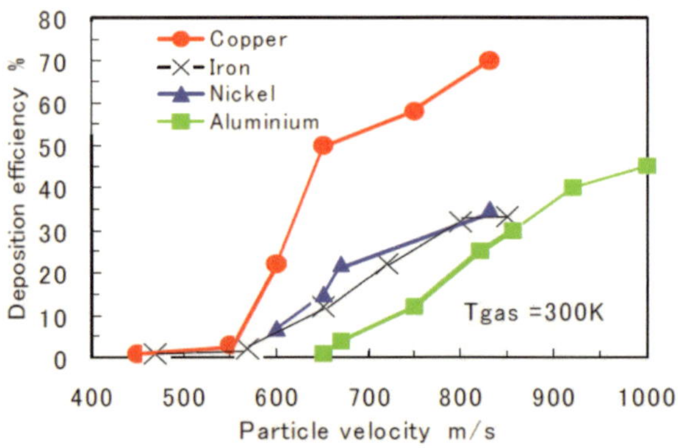

Fig. 4. Critical velocity of cold sprayed depositions

3. Kinetic energy and rebound energy

The cold spraying conditions were optimized by taking into account the particle kinetic energy and the rebound energy for application in repairing gas turbine blades. A high quality cold-sprayed layer is that which has lowest porosity; thus the spraying parameters were optimized to achieve low-porosity layer, which was verified by SEM.

The details on the coating formation mechanism and properties of the cold sprayed layers have not been elucidated thus far. Fukumoto et al. reported that by this technique, high deposition efficiency was achieved under the conditions of high velocity and high temperature of spraying particles (Fukumoto, 2006; Fukumoto et al., 2007). High velocity particles which have high kinetic energies tend to be more oblate and facilitate deposition. Moreover, erosion behavior can be observed when the particle velocity is low, like grit blasting. Plastic deformation of particles occurs at high kinetic energies of the particles having a high velocity; this plastic deformation induces the formation of a deposited layer. This critical velocity at which deposition begins depends on the mechanical properties of the substrate and the particles, the presence of an oxide layer, and the diameter of the particles. Vlcek et al. have conducted studies on cold spraying of aluminum, copper, and stainless steel on mirror-polished iron and steel to investigate in detail the critical velocities of each metal (Vlcek et al., 2001). In this section, it was suggested that particle impulse (particle mass × velocity) strongly affects deposition efficiency. From the results of the above studies, it can be concluded that the factors that influence the deposition efficiency in cold spraying are: 1) gas temperature, 2) particle mass, and 3) particle velocity. All of these factors depend on the kinetic energy of particles. Therefore, kinetic energy of particles can influence deposition efficiency and influence the strength of the deposited layer.

3.1 Rebound energy during deposition

During cold spraying, all the particles are accelerated by the working gas such as helium and nitrogen. The kinetic energy generated by the working gas induces the deposition. However, part of this kinetic energy is not utilized for the deposition of particles but gets converted to rebound energy. This rebound energy of the particles is calculated by Eq. 1 as follows (Johnson, 1985; Papyrin et al., 2003),

$$R = \frac{1}{2} e_r m_p V_p^2 \tag{1}$$

Here, e_r is the coefficient of rebound, and for spherical particles, its value is expressed as,

$$e_r = 11.47 \left(\frac{\overline{\sigma_Y}}{E^*} \right) \left(\frac{\rho_p V_p^2}{\overline{\sigma_Y}} \right)^{-\frac{1}{4}} \tag{2}$$

Here, $\overline{\sigma_Y}$ and E^* are the yield stress of the particle and the elastic modulus of the substrate, respectively; in this study, the parameters of the alloy IN738LC were determined by conducting proof strength (950 MPa) and indentation tests (201 GPa). ρ_p, V_p, and m_p are the particle density, particle velocity, and particle mass, respectively. In this study, both the particle diameter and mass are sufficiently small. V_p is considered to have the same value as the working-gas flow rate U_g. The U_g is evaluated by the following equation (Eq. 3);

$$U_g = \sqrt{2 \frac{\lambda}{\lambda - 1} R T_i \left[1 - \left(\left(\frac{P_e}{p_i} \right)^{\frac{\lambda - 1}{\lambda}} \right) \right] + U_{gi}^2} \tag{3}$$

Here, U_g and U_{gi} denote the nozzle outlet and inlet rates, respectively; λ and R denote the specific heat ratio and the gas constant, respectively; and P_e and P_i denote the nozzle outlet and inlet pressures, respectively. The U_g values calculated at different U_{gi} values are summarized in Table 3; during the calculations, P_i was set as atmospheric pressure. Obtained gas flow rates by Eq. 3 are listed in Table 3. From the table, the gas flow rate depends on gas temperature and gas pressure, but is affected by kind of gases in particular.

Gas	Spraying conditions	Gas flow rate at nozzle inlet, U_{gi} (m/s)	Gas flow rate at nozzle outlet, U_g (m/s)
He	800°C, 3.5 MPa	33.04	2910.54
	750°C, 2.5 MPa	23.77	2775.23
	650°C, 3.5 MPa	33.04	2699.48
	600°C, 3.5 MPa	33.04	2625.35
	600°C, 2.5 MPa	23.77	2563.72
N_2	650°C, 3.5 MPa	32.53	1105.50

Table 3. Gas flow rate at nozzle inlet and outlet under different conditions

3.2 Threshold diameter of adhered particle and rebound particle

Fig. 5 shows a schematic illustration of the effect of rebound energy of particles on collision with the substrate. During cold spraying, high velocity particles of various diameters impinge on the substrate. Consequently, the rebound energy of one particle is transferred to the other on collision. Let us that all the kinetic energy of the small particles is converted to adherent energy; Then, if the rebound energy of the coarse particle exceeds the kinetic energy of the small particles, then the rebound energy of coarse particles cause the coarse particles to delaminate into smaller particles, as illustrated Fig. 6. Here, the well-adhered particles are considered to be particles of average diameter. Under this consideration, Eq. 4 was obtained, and the threshold diameter of adherent particle is deduced from Eq. 5, where ρ, σ, and E are parameters characteristic to the particles. Therefore, D_{th} has a unique value.

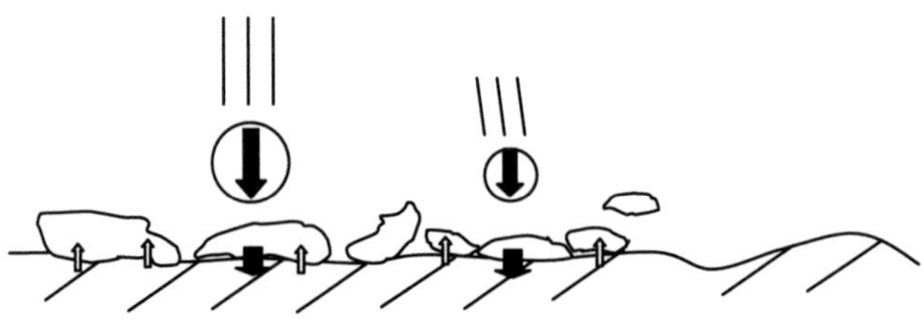

Fig. 5. Schematic illustration of effect of particle rebound energies on substrate

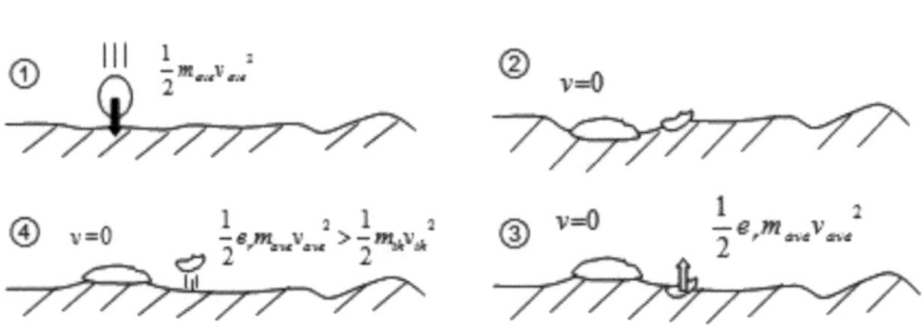

Fig. 6. Schematic illustration of rebound energy to surfaces' particle

$$\frac{1}{2}e_r m_{ave} V_{ave}^2 = \frac{1}{2} m_{th} V_{th}^2 \tag{4}$$

where, $m_{ave} = \frac{4}{3}\pi r_{ave}^3 \rho_{ave}$, $m_{th} = \frac{4}{3}\pi r_{th}^3 \rho_{th}$, and ρ is particle density.

$$D_{th} = 2r_{th} = 2r_{ave}\left[11.47\left(\frac{\overline{\sigma_Y}}{E^*}\right)\left(\frac{\rho_p V_p^2}{\overline{\sigma_Y}}\right)^{-\frac{1}{4}}\right]^{\frac{1}{3}} \tag{5}$$

If the rebound energy of a coarse particle exceeds the kinetic energy of well adhered particles, then the average-diameter particles can also be delaminated by coarse particle. The diameter of the resultant negative coarse particle can be calculated from Eq. 6.

$$D_{coa} = 2r_{coa} = 2r_{ave}\left[11.47\left(\frac{\overline{\sigma_Y}}{E^*}\right)\left(\frac{\rho_p V_p^2}{\overline{\sigma_Y}}\right)^{-\frac{1}{4}}\right]^{-\frac{1}{3}} \tag{6}$$

3.3 Equation for optimization of cold spray deposition

The current theoretical principles for cold spray deposition are in Fig. 7. In this figure, it was considered that d<45. The particle distribution result was obtained on the basis of fundamental assumptions. In this result, the kinetic energy used to achieve deposition is that between D_{th} and D_{coa}. Let α represent the number of particle of each diameter; then, the kinetic energy of the deposited particles can be evaluated by Eq. 7. Here, the rebound energy of the coarse particle has a negative effect on the kinetic energy. Therefore, the rebound energy of coarse particle is subtracted from Eq. 7 to give Eq. 8. Thus, $E_{deposit}$ in Eq. 8 represents the effective kinetic energy utilized to achieve deposition. A high $E_{deposit}$ value may imply high deposition efficiency and an improvement in the strength of the adhered layer. Fig. 8 shows the adhesion strength at different spray conditions, as calculated by Eq. 8. The porosity ratios are determined by carrying out SEM observations. The particle diameter and spray conditions are found to affect the quality of the deposited layer. In particular, small particle size can result in the formation of high quality deposited layer.

$$E_{area:M} = \frac{1}{2} * \frac{4}{3}\pi \rho v^2 \sum_{r_i = \frac{D_{th}}{2}}^{\frac{D_{coa}}{2}} \alpha_i r_i^3 \tag{7}$$

$$E_{deposit} = \frac{1}{2} * \frac{4}{3}\pi \rho v^2 \left(\sum_{r_i = \frac{D_{th}}{2}}^{\frac{D_{coa}}{2}} \alpha_i r_i^3 - \sum_{r_i = \frac{D_{coa}}{2}}^{\infty} \alpha_i r_i^3 \right) \tag{8}$$

Optimal particle ranges of each spray condition are listed in Table 4. From the result of Table 4, in the case of condition of d<25, 600°C, He, 2.5 MPa, optimal particle size can be

4.50 to 57.3 μm. The d<25 particle includes less than 4.50 μm particles. These smaller particles can induce formation of porosity or the other defects.

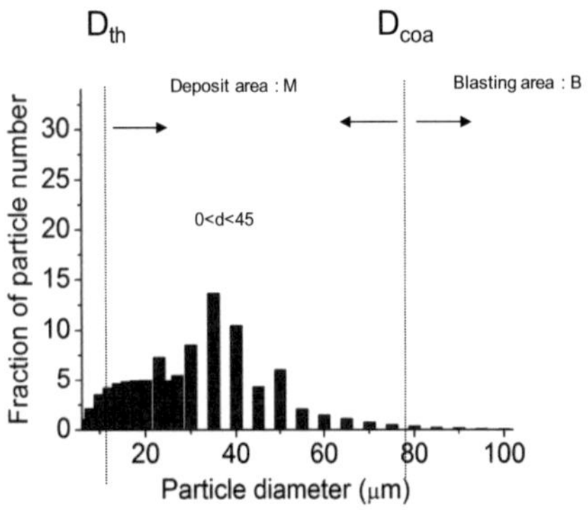

Assumptions;
- Initial particles are spherical.
- All kinetic energy is converted to adhesion energy.
- A lots of particles inpinge at the same time.
- All particle velocities are same with gas velocities.
- Fragmentation does not occur.

Fig. 7. Particle size distribution and range of particle deposition

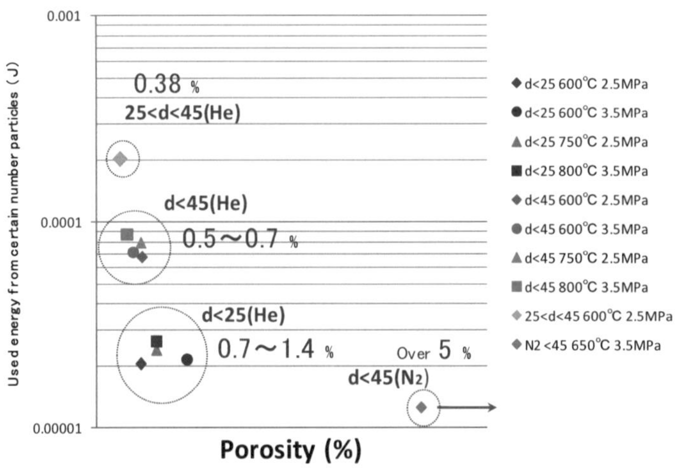

Fig. 8. Evaluation of adhesion strength from Eq. 8

Powder size (μm)	Gas	Temp. (°C)	Pressure (MPa)	D_{th} (μm)	D_{coa} (μm)
d<25	He	600	2.5	4.50	57.3
			3.5	4.47	61.5
		750	2.5	4.51	63.2
		800	3.5	4.47	63.7
d<45	He	600	2.5	6.56	91.0
			3.5	6.49	90.1
		750	2.5	6.58	92.2
		800	3.5	6.74	93.0
	N_2	650	3.5	11.97	123.6
25<d<45	He	600	2.5	10.20	142.7

Table 4. Optimal particle ranges of each spray condition

4. Microstructure and mechanical properties of as-sprayed coatings

Small punch tests were carried out for as-sprayed cold spray coatings as evaluation test of mechanical property. Spraying conditions were particle size of d<25 μm, gas temperature of 650°C, He and N_2 gas with 3.5 MPa. And cross-sectional SEM images of both samples are shown in Fig. 9. The nitrogen gas used coating had many pores, due to lower impinge velocity (ref. Table 3).

Schematic of small punch (SP) test is illustrated in Fig. 10. The samples for SP tests were taken from the cold sprayed deposition. The geometry of the SP specimen was Ø8 mm × 250 μm. The SP specimen was received compressive load by Ø1.0 mm alumina ball. The displacement was measured by Linear Variable Differential Transducer (LVDT). From the SP tests, maximum load and SP energy were evaluated. The schematic of the SP energy is illustrated in Fig. 11. The SP energy was estimated from the area of load-displacement curve. Relationship between applied load and displacement is shown in Fig. 12. And, SP energy is shown in Fig. 13. From Fig. 12, the He gas used specimens are 3 times higher maximum load than that of the N_2 gas used ones. And also, the SP energy of the He gas used specimens was 5 times higher than that of N_2 gas used ones. From these results, mechanical property of the cold sprayed Ni base superalloy coatings depends on coating quality, such as porosity ratio, cohesive force etc.

SEM images of the SP specimens after SP tests are shown in Fig. 14. In the case of the He gas used specimen (see Fig. 14b), radially-propagated cracks were observed. On the other hand, in the case of the N_2 gas used specimen, brittle fracture at the corner of the die was generated. The He gas used specimens, which has higher partcile velocity during spraying, have higher mechnical property than that of N_2 gas used one. However, the maximun load of bulk Ni base superalloys was approximately 1.0 kN in SP tests. This means that the mechnical property of the as-aprayed Ni base superalloy coatings are not enough. It is thought that HT for as-sprayed coatings can be effective for improvement of the mechanical property. And also, it is expected that the HT can control microstructures of cold sprayed coatings. In next section, effects of HT is introduced.

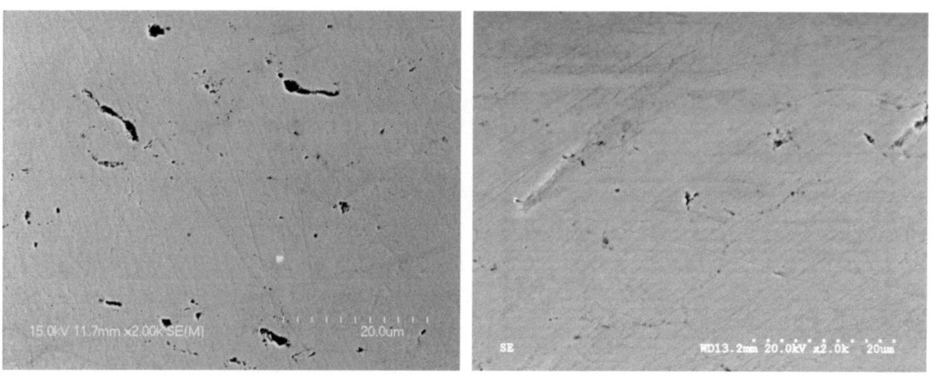

(a) N_2 gas used (b) He gas used

Fig. 9. Typical cross-sectional SEM images of cold sprayed Ni base superalloy coatings

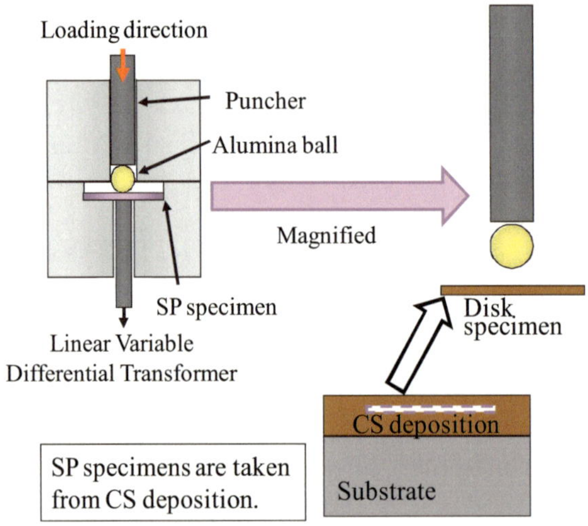

Fig. 10. Schematic illustration of small punch test

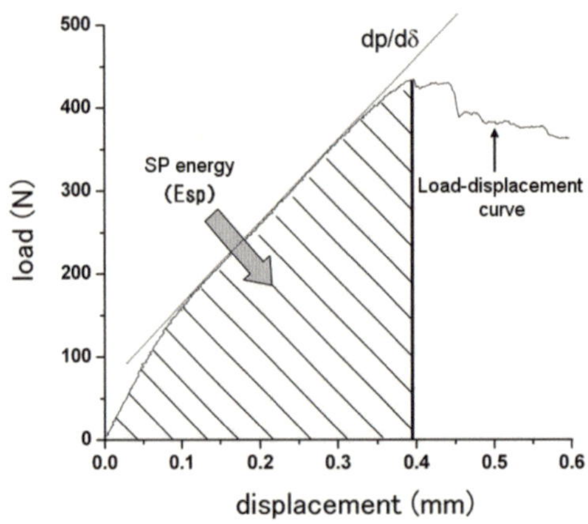

Fig. 11. SP energy

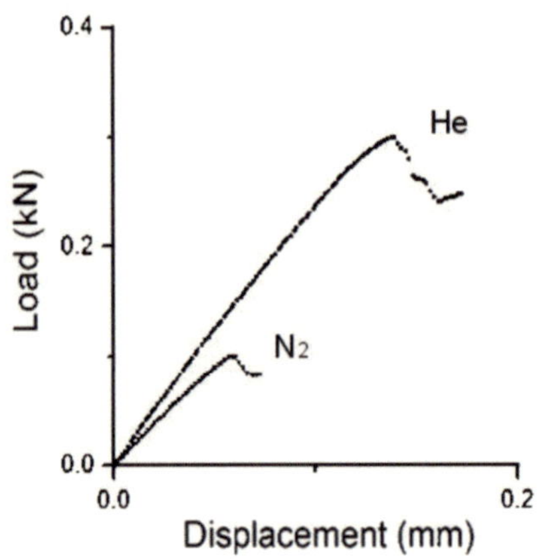

Fig. 12. Results of SP test

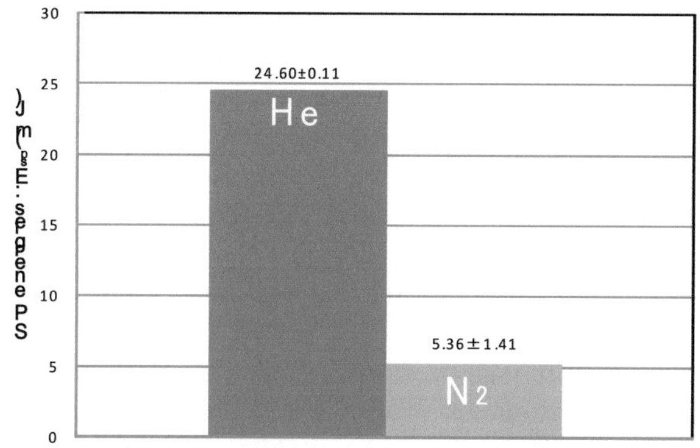

Fig. 13. SP energy of as-sprayed cold spray coatings

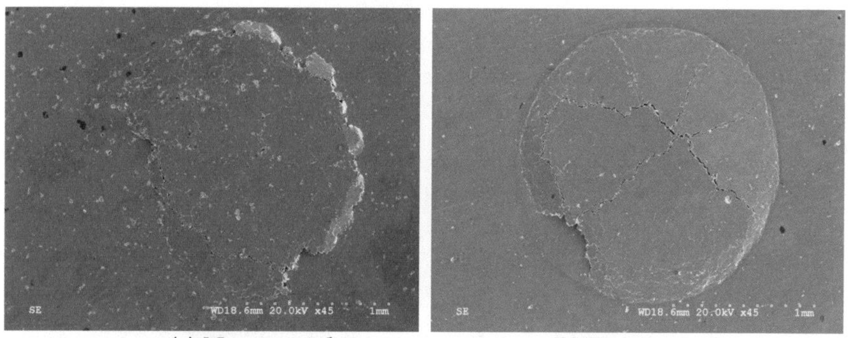

(a) N_2 gas used (b) He gas used

Fig. 14. SEM images of the SP specimens after SP tests

5. Post spray heat treatment (PSHT)

For the repair of gas turbine blades, polycrystalline Ni-based superalloy, IN738LC has been studied on the optimization of cold spray process, strength evaluation, and so on. But the mechanical properties of as-sprayed coatings are low compared with bulk materials. In contrast, their properties can be improved by applying standard HT (solution treatment and aging treatment) (Niki, 2009). It was also reported that the mechanical strength, ductility, and adhesion between coating/coating and coating/substrate can be improved by post treatment which causes atomic diffusion between the coating and substrate and the generation of intermetallic compounds (Li et al., 2006; Li et al., 2009; Spencer & Zhang, 2009). Thus many researches relating the post HT are necessary for the cold sprayed coatings. Therefore the following chapter focuses on the metallic structure of the coatings after post HT including the change of γ'-phase before and after HT, precipitation, grain, and so on.

5.1 Microstructure after HT

Fig. 15 shows the sectional micrographs of a) the as-sprayed CS coating and b) one after a standard HT. The existence of pores was confirmed at both samples. From comparing the two micrographs, the improved adhesive interfaces were observed at some location which revealed that adhesion can be improved by the post HT. Both porosities are shown in Fig. 16 calculated by image processing of cross sectional images. Decrease in porosity is confirmed from the results after HT. Decrease in porosity of yttria-stabilized zirconia coatings after post treatment also was reported (Renteria & Saruhan, 2006; Zhao et al., 2006). It might be resulted from the powder sintering effect by HT over 1000°C. Figs. 17 and 18 show the cross sectional micrographs of the as-sprayed coating and the heat treated coating respectively. In case of as-sprayed coatings, the distorted splats are observed but the grain boundary and intermetallic precipitation of γ′ phases are not observed as shown in Fig. 17b. In case of the coatings applied the post HT, on the other hand, two different grains which have diameter of about 800 nm and 250 nm was observed. It reveals the existence of γ and γ′ phase precipitation. They are irregular in shape, but in case of substrate, two phases are homogeneous in shape as shown in Fig. 19. High temperature strength of Ni-based superalloys such as IN738LC is highly dependent on containing the γ′ phase precipitating. Caron reported that the creep rupture strength of CMSX-2 having the regularly aligned γ′ phase is more higher than one having the irregularly shaped γ′ array (Caron & Khan, 1983) and it was also reported that Alloy143 having the smaller γ′ phase showed higher creep rupture strength at 982°C. From above reasons the appropriate post HT are essential for the cold sprayed coatings including IN738LC to improve their properties.

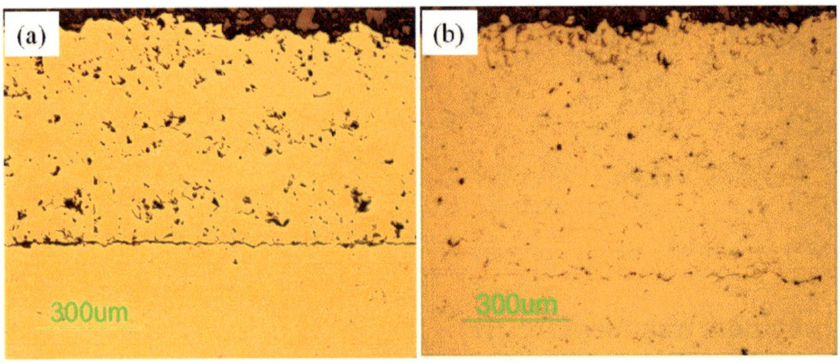

Fig. 15. Cross-sectional micrographs of the CS coatings, a) as-sprayed and b) applied the post HT

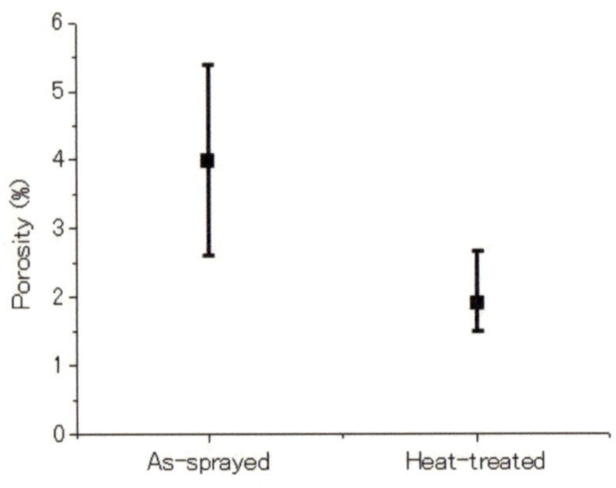

Fig. 16. Porosity measurement before and after HT

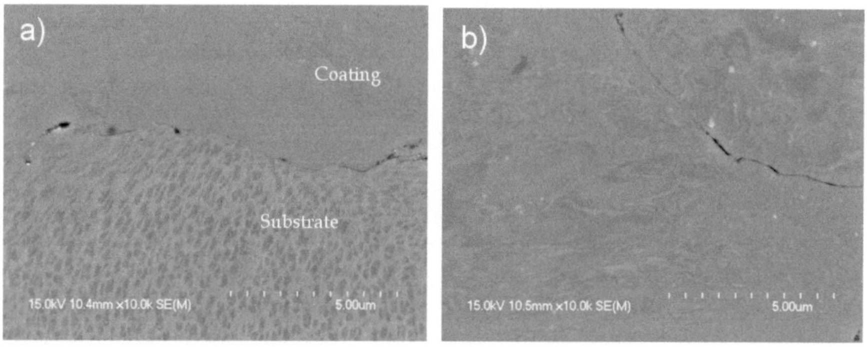

Fig. 17. Cross-sectional micrographs of the as-sprayed CS coatings showing a) the coating/substrate interface and b) near the interface

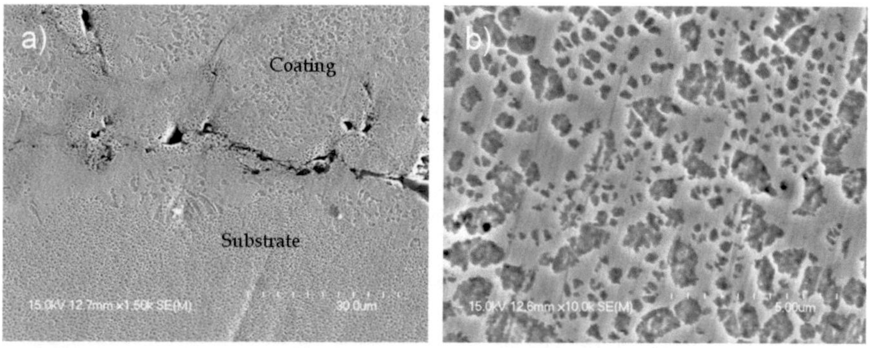

Fig. 18. Cross-sectional micrographs of the CS coating applied the post HT, showing a) the coating/substrate interface and b) the closed view of coating

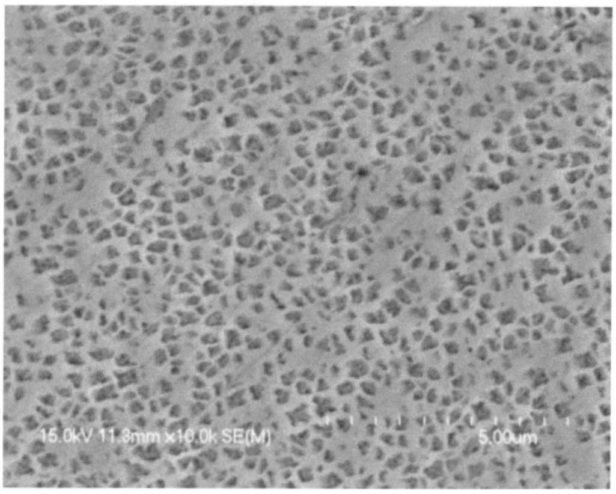

Fig. 19. Micrograph of IN738LC substrate after HT

5.2 Grain structure study via electron back scattering diffraction (EBSD)

From SEM observations, the grains in the as-sprayed coating were not clearly observed. Therefore, in order to observe the detailed structures of the as-sprayed CS coatings before and after HT, the EBSD analysis was carried out which is one of the useful crystallography methods. The EBSD patterns (Kikuchi) were produced when an electron beam incident on the sample surface which represent the crystal orientation. Fig. 20 shows a schematic diagram of the EBSD technique showing the principle of Kikuchi pattern generation and electron beam irradiation. The Kikuchi lines are appeared when the electron beam is reflected via the Bragg reflection angles by the inelastic dispersion of electron within a depth of 50 nm from the sample surface. It is possible to analyze the crystal structure and crystal orientation at the exact point of sample surfaces by electron irradiation. Strain is expected to be introduced inside the coatings during the deposition process. The dislocations might be caused by plastic strain in the coatings. The Kernel Average Misorientation (KAM) can be effective to analyze these dislocation and plastic strain comparing before and after HT.

Image Quality (IQ) map and Inverse Pole Figure (IPF) map representing the distribution of plastic strain were also used together with KAM map. The IQ is displayed in the map as the gray shade which quantified in the sharpness of the diffraction pattern generated from the sample. High value of IQ represents on the map by bright color close to white, and does more clearly where the diffraction pattern is strongly obtained. In general, if the grain boundaries and the distortion are large, or the superposition of patterns from several adjacent crystals, IQ values are decreased and it appears as black lines on the map. IPF is classified by color according to the crystal orientation for each grain. In addition, KAM map is classified by five colors showing the magnitude of plastic strain via the difference in the orientation of each pixel. Blue color indicates the smallest plastic strain and the red color indicates the biggest.

Figs. 21 and 22 shows the IQ map and KAM of the coating and substrate interfaces of the as-sprayed coating and the coating after applied the standard HT. Upper region of each image is the coating and the lower is the substrate. Dark parts on the IQ and IQ + IPF map, reveal that the crystal diffraction corresponding a FCC structure was not detected. In addition, from the EBSD results shown in Fig. 21, most of the diffraction pattern of crystal orientation cannot be seen in case of the as-sprayed coatings. Region that the crystal orientation is not observed is reached over dozens of micrometer from the interface. From the result after HT shown in Fig. 22, the fine grains of 1-20 um in diameter are observed on the coating region.

Fig. 23 shows the EBSD analysis of the IN738LC powder prior to the deposition. The grains are clearly observed from the feedstock powder, indicating that the crystal change was occurred after deposition.

The cold sprayed coatings are deposited by large plastic deformation which resulted from the acceleration and collision with the supersonic feedstock powder. In this case, the grain refinement is occurred. Zou et al. reported that the submicron grains were observed after the cold spray deposition with Ni powders via the EBSD analysis (Zou et al., 2009). And Borchers et al. reported that the deformed grains, sub-grains and fine grains are observed in the cold sprayed Ni coating by transmission electron microscopy (TEM) analysis (Borchers et al., 2004). In this study, the amorphous or nano-crystallized grains at the substrate and coating side near the interface are observed which caused from the plastic deformation of feedstock and substrates, and then they recrystallized and grew up to micron size by HT.

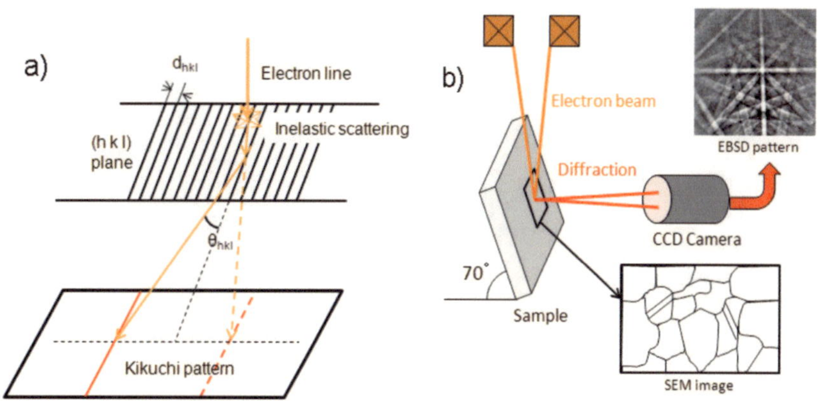

Fig. 20. Schematics presenting a) principle of Kikuchi pattern and b) EBSD technique

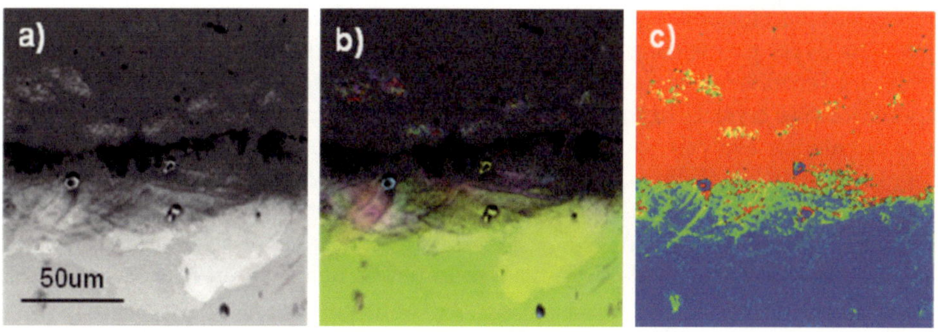

Fig. 21. EBSD analysis of as-sprayed samples; a) IQ map, b) IQ + IPF, and c) KAM map

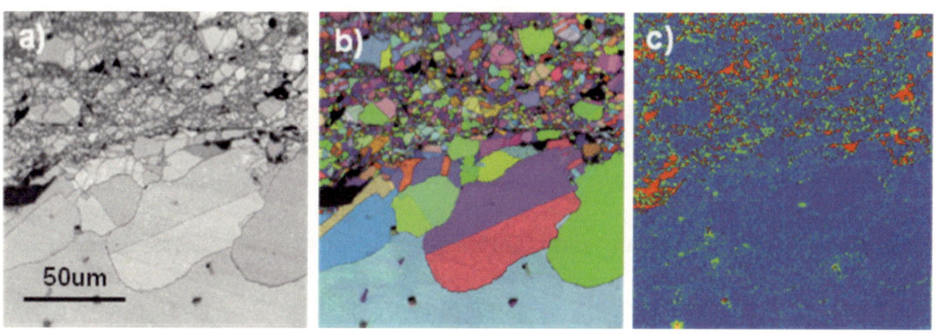

Fig. 22. EBSD analysis of the standard HT sample; a) IQ, b) IQ + IPF, and c) KAM map

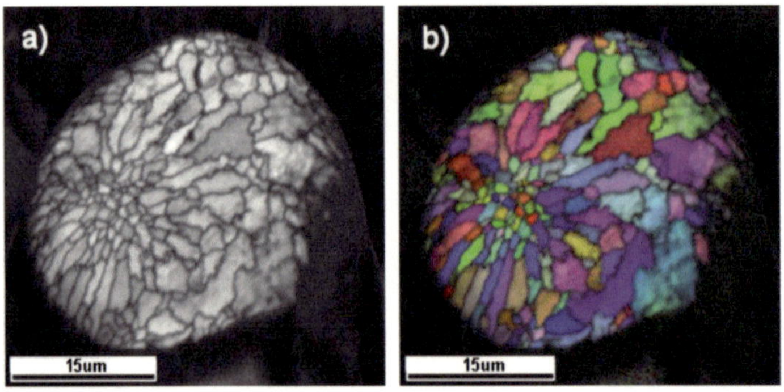

Fig. 23. EBSD analysis of the feedstock powder before spraying; a) IQ and b) IQ + IPF map

5.3 Nanostructure via transmission electron microscopy (TEM) analysis

XRD analysis results described in the preceding paragraph, CS in the deposition of CS μ m depth near the interface and the film was supposed to be because of nano crystals over the substrate, TEM observation was carried out because the nano-sized crystals were observed by XRD analysis. Sampling was made by the focused ion beam (FIB) apparatus, which possesses the coating/substrate interface shown in Fig. 24. The as-sprayed and heat treated samples were prepared for analysis.

Fig. 25 shows the TEM micrographs of the coating/substrate interface of the as-sprayed CS sample. The upper part is the coating area and the lower is substrate, and the white arrows indicate the coating/substrate interface. From this image, the as-sprayed sample is composed of fine crystals from several tens to hundreds of nm order. Fig. 26 shows the crystal orientation analysis in the as-sprayed sample. It was confirmed that the crystal orientation of the six places up to 500 nm in diameter was mixed in disorder at many regions. Especially near the interface, the points of c, d, and e are remarkable. The curved grain boundary was observed from Fig. 25. When the feedstock is attached to the substrate during the deposition by high speed, the severe plastic deformation is occurred in a large scale, and then the curved interfaces are formed from spherical shape of powder as shown in Fig. 27. From the expanded view of the splat boundary shown in Fig. 27, the columnar grain boundaries were observed following a line. Meanwhile, a number of small grains, tens of nm level, have become a gathering at the bottom of the splats. This difference of grain shape may come from the different magnitude and direction of plastic strain during the particle collision. From these results, the schematic diagram of changes in grains of the powder particles and the substrate during deposition can be inferred as Fig. 28. Nano-sized isotropic crystalline was formed at near the bottom surface of the particles as compression direction by the applied strain. In case of side region of coating, the shear strain in the direction parallel to the substrate surface was occurred and the columnar grains deformed along the direction of the strain. From the repeat of particle collision, the deposited coating is composed of columnar crystals and the isotropic nanoscale crystals.

Fig. 29 shows the TEM images at the coating/substrate interface after standard HT. Comparing with the as-sprayed sample in Fig. 25, the growth of large grains can be observed. The coating/substrate interfaces are indicated by white dotted lines and white arrows. As can be seen from this result, the grain was recrystallized and grew in grain size up to micro order, caused by HT. As shown in red and blue lines in the figure, in particularly, the interface to grow to the consolidated form. It can improve the adhesion strength of the coating/substrate interface by post HT, and such a partial recrystallization at the interface is considered likely to contribute to improve the adhesion of the coating (Zou et al., 2009). In addition, Fig. 30 shows the crystal orientation of four points, a, b, c, and d, around interface in the heat-treated sample. It was confirmed that the grain size of nano-order still remained at point b which close to the surface, even though the points of a, c, and d are single grains. Such nano-grain could be a starting point of coating delamination from the substrate. The post HT with optimal condition are necessary to cover such flaw and then to improve the interface adhesive.

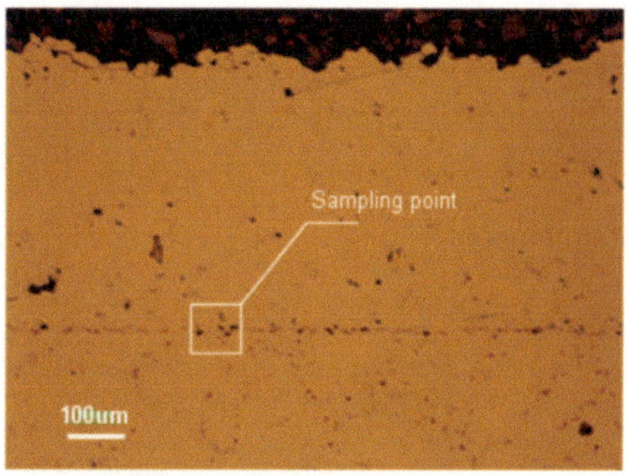

Fig. 24. TEM sampling position

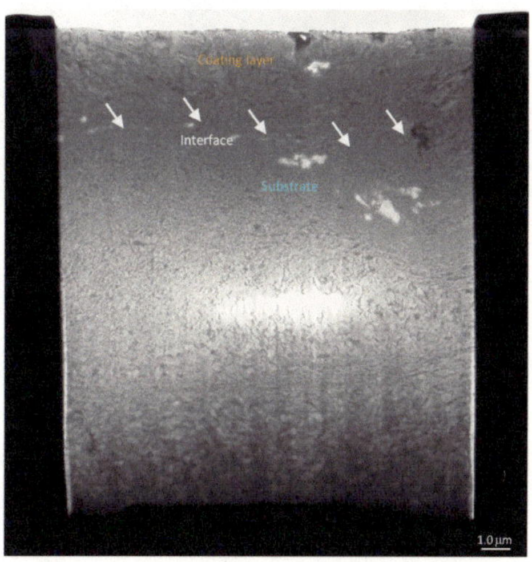

Fig. 25. TEM image at coating/substrate interface of the as-sprayed CS sample

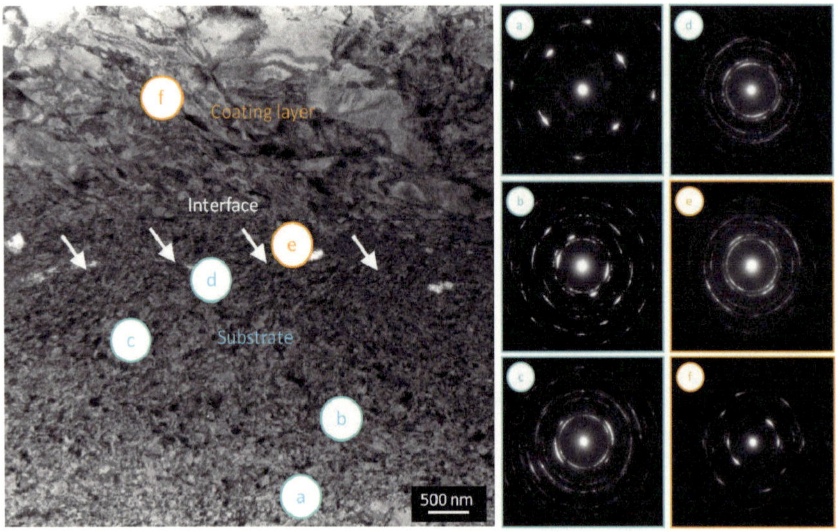

Fig. 26. TEM analysis of crystal orientation at coating/substrate interface of as-sprayed CS sample

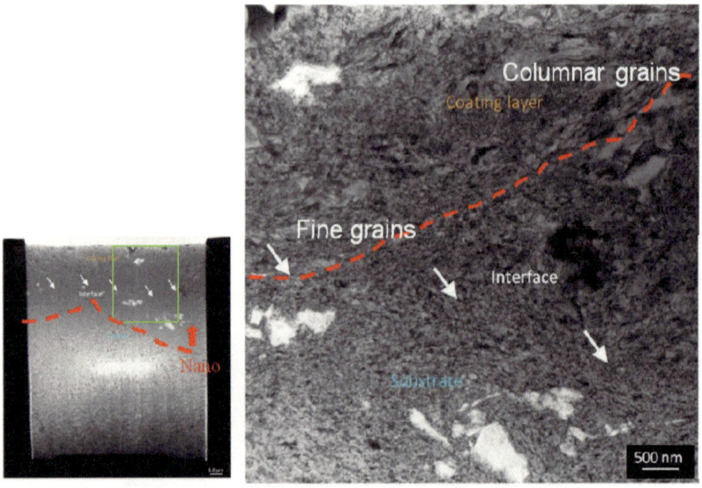

Fig. 27. Magnified image at the coating/substrate interface

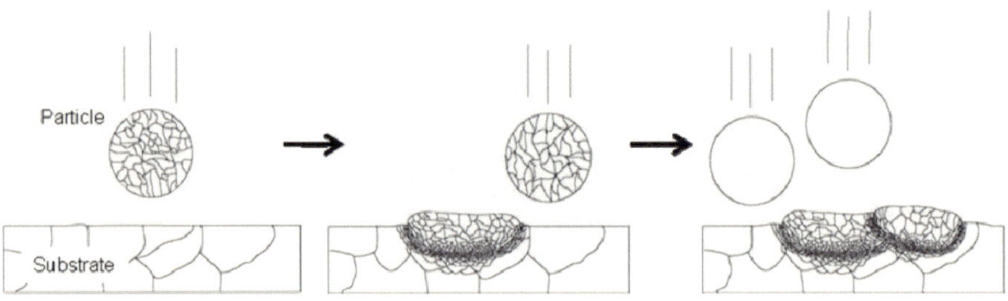

Fig. 28. Schematic of the powder grain changes by cold spraying

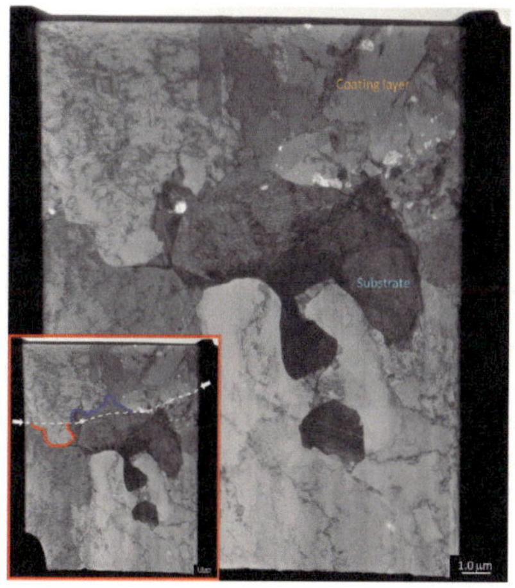

Fig. 29. TEM image at coating/substrate interface after post HT

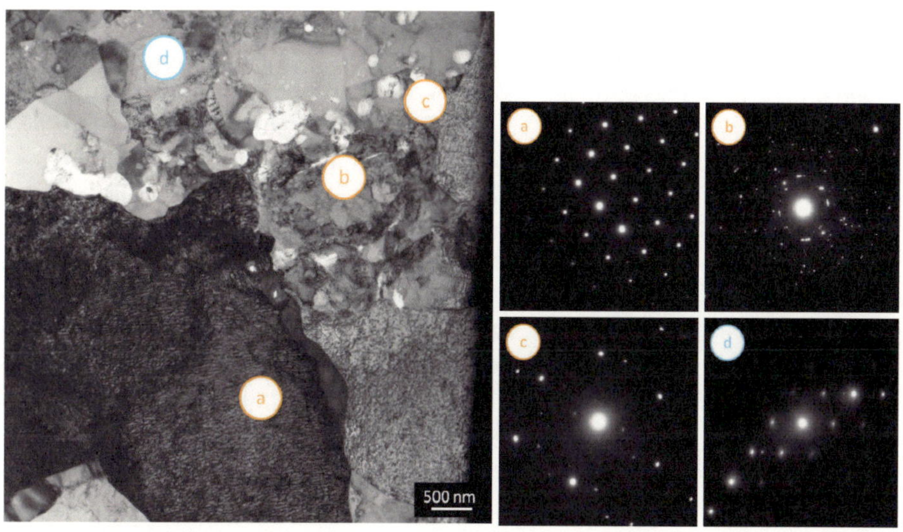

Fig. 30. TEM analysis of crystal orientation at coating/substrate interface after post HT

5.4 High temperature oxidation resistance after post heat treatments

Ni-base superalloys which used as gas turbine blade materials are commonly exposed to high temperature corrosive environment during operation. Among the turbine blade arrays, TBC on the first stage blades is designed to prevent oxidation and corrosion of blade materials from high temperature environment. After long-term use, however, it is frequently reported that the delamination at the interface of TBC system or the weakening of grain boundaries caused by oxidation proceeds from intergranular corrosion in materials, and the cause of crack initiation (Ejaz & Tauqir, 2006; Viswanathan, 2001). As previously mentioned, the grain size in the CS coating is smaller compared to the substrate material (substrate: several hundred microns, CS coating: several microns), and there are also boundaries between the splats. Therefore, it can be considered that oxygen and corrosive components are easily penetrated to grain boundary compared to the substrate. Therefore In this section, the oxidation resistance of IN738LC CS coatings under high temperature conditions was evaluated together with the effect of post HT on the oxidation resistance property. To evaluate the effect of HT on oxidation properties and to compare, the standard HT (NHT; treated at 1121°C/2h + 843°C/24h) and the high temperature HT (HHT; treated at 1171°C/2h + 843°C/24h) were applied. And the four condition of the exposure time such as 100h, 200h, 500h, and 1000h at temperature of 900°C was applied.

As shown in the Fig. 31, it consisted of double internal oxide layer, i.e., Cr_2O_3 oxide on the coating surface was mainly observed on Ni-base superalloy IN738LC at 900°C, and the scattered γ′ phase of Al_2O_3 also can be categorized into. Thus the evaluation of oxidation characteristics was performed by measurement of the external and inner oxide layer thickness and the average value of them. Figs. 32 and 33 show the thickness of external oxide and inner oxide layer according to the exposure time respectively. And Fig. 34 shows the cross-section images showing the oxide layer of each sample. The surface oxidation was progressed significantly at all samples even though they were exposed to only 500h. It could be resulted from the pore defects in the coatings. For each exposure time and condition, both of the external and internal oxide layers of as-sprayed state were thickest being followed by that of most thick, and following the post HT of NHT and HHT. As a result, each CS sample is considered to be related to grain size in the coating. In case of the external oxide layer, the substrate had the thickest layer than others, but in case of the internal oxide layer, the substrate and the NHT sample show a similar level of thickness.

To investigate what caused these differences, the cross-sectional specimen of each oxide by EDX elemental analysis was performed. From Figs. 35 to 38, they show the EDX analysis for each sample. From the element mapping of each, as noted in previous, the oxide layer on the surface is mainly Cr_2O_3, but the inner consisted of Al_2O_3 and TiO_2. Several elements presented a significant difference was also observed. Comparing the mapping images, Ti element was observed only at the substrate sample. In case of the substrate, Cr_2O_3, TiO_2, and Al_2O_3 were observed as the main oxides. In case of the cold sprayed coating, in contrast, Ti penetrated up to below Al layers. Because the coatings consist of smaller grains compared to the substrate and also the grain boundaries, the oxygen can penetrate and spread easily into the layer, and finally from the oxides.

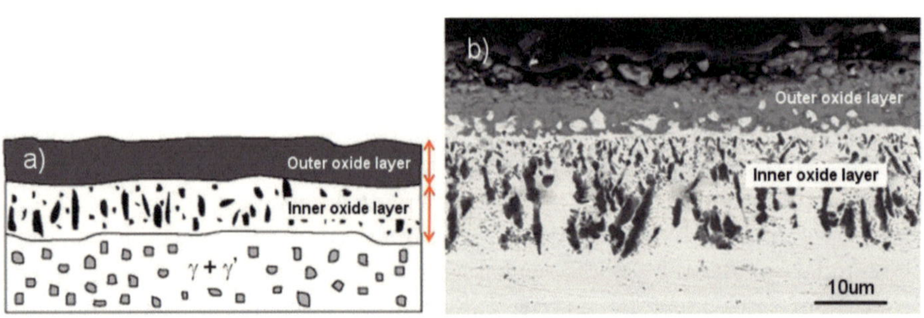

Fig. 31. Oxidation of Ni-based superalloy IN738LC; a) schematic of the double oxide layers and b) SEM image

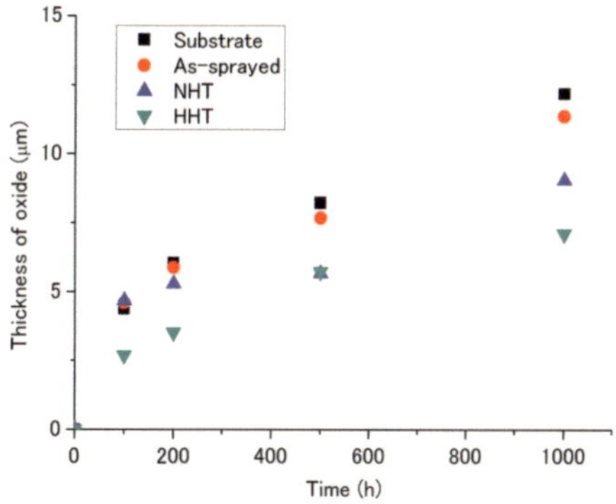

Fig. 32. External oxide thickness of each sample according to exposure time

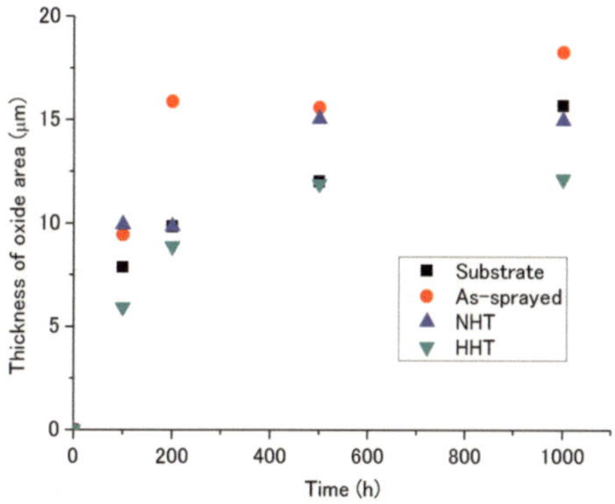

Fig. 33. Inner oxide thickness of each sample according to exposure time

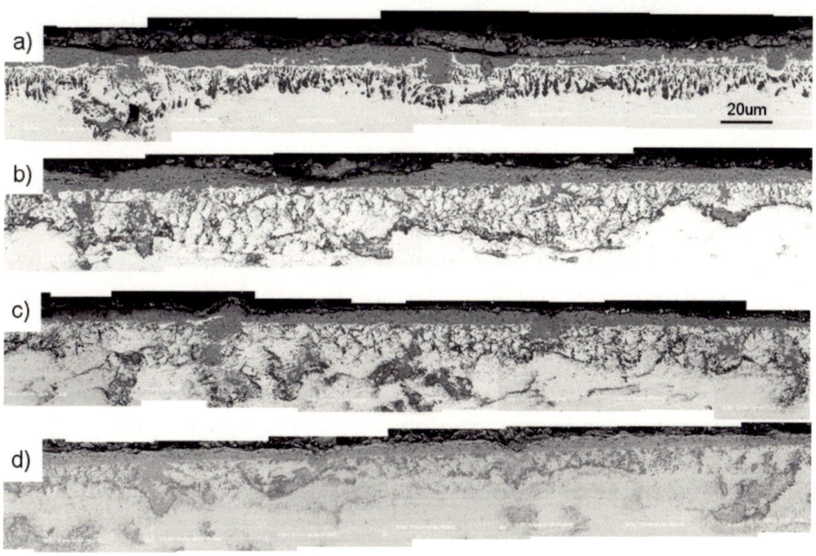

Fig. 34. Cross-sectional images of oxide layer formed at 900°C for 500h; a) substrate, b) as-sprayed, c) NHT, and d) HHT

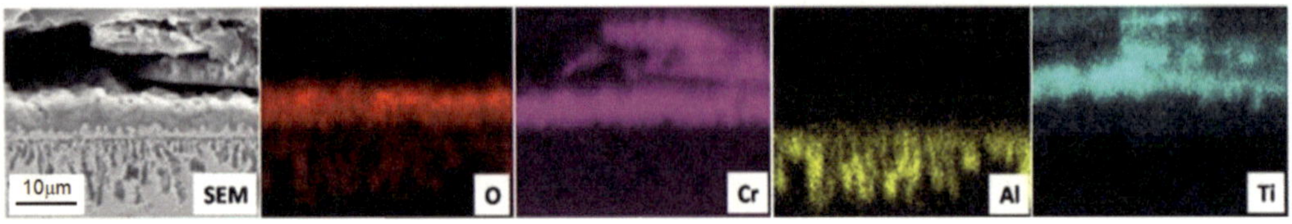

Fig. 35. EDX mapping of the substrate exposed at 900°C for 200h

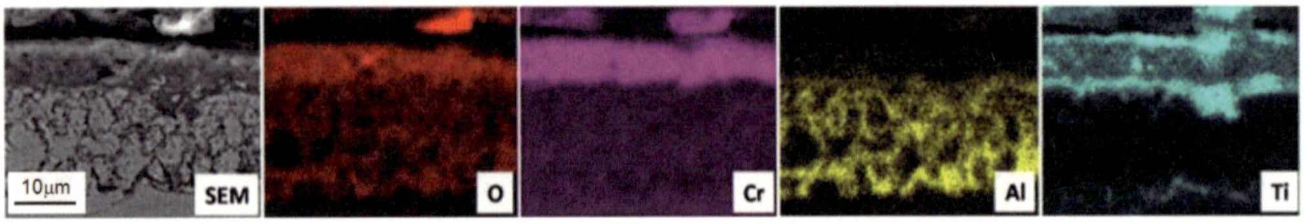

Fig. 36. EDX mapping of the as-sprayed coating exposed at 900°C for 200h

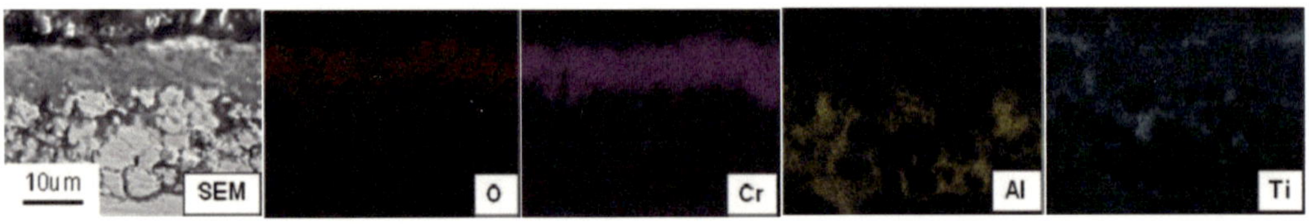

Fig. 37. EDX mapping of the NHT sample exposed at 900°C for 200h

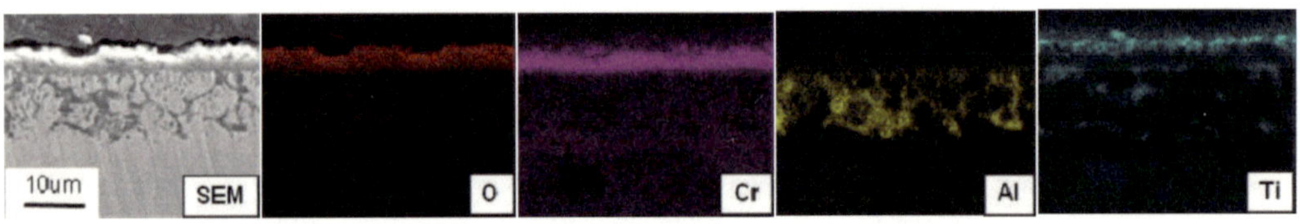

Fig. 38. EDX mapping of the HHT sample exposed at 900°C for 200h

5.5 Small punch tests after post heat treatments

Changes in mechanical properties after HT were evaluated by the maximum load and the absorption energy via the small punch test. Fig. 39 shows the p-δ curves and the SP absorbing impact energy E_{sp} before and after HT respectively. It is clear that the maximum load increased about four times and SP energy also increased about 34 times more after HT. It reveals that the interface adhesion between the coating and the substrate can be significantly improved by applying HT after cold spray deposition of IN738LC. The HT sample in Fig. 39 shows the discontinuous fracture phenomenon with segregation. It may come from the non-homogeneous adhesion between the deposited splats, i.e., some interfaces are improved their adhesive strength via the chemical bonding by the post HT but some part of them cannot bond together due to the gaps between deformed splats. When the cracks progress at the weak region consisted of more gaps, the mechanical load may be released temporarily and result in the segregated load flow.

Figs. 40, 41 and 42 show the SP p-δ curve, SP $dp/d\delta$ curve and SP impact absorbing energy Esp respectively. From Fig. 40, the optimal condition for adhesion layer was to withstand

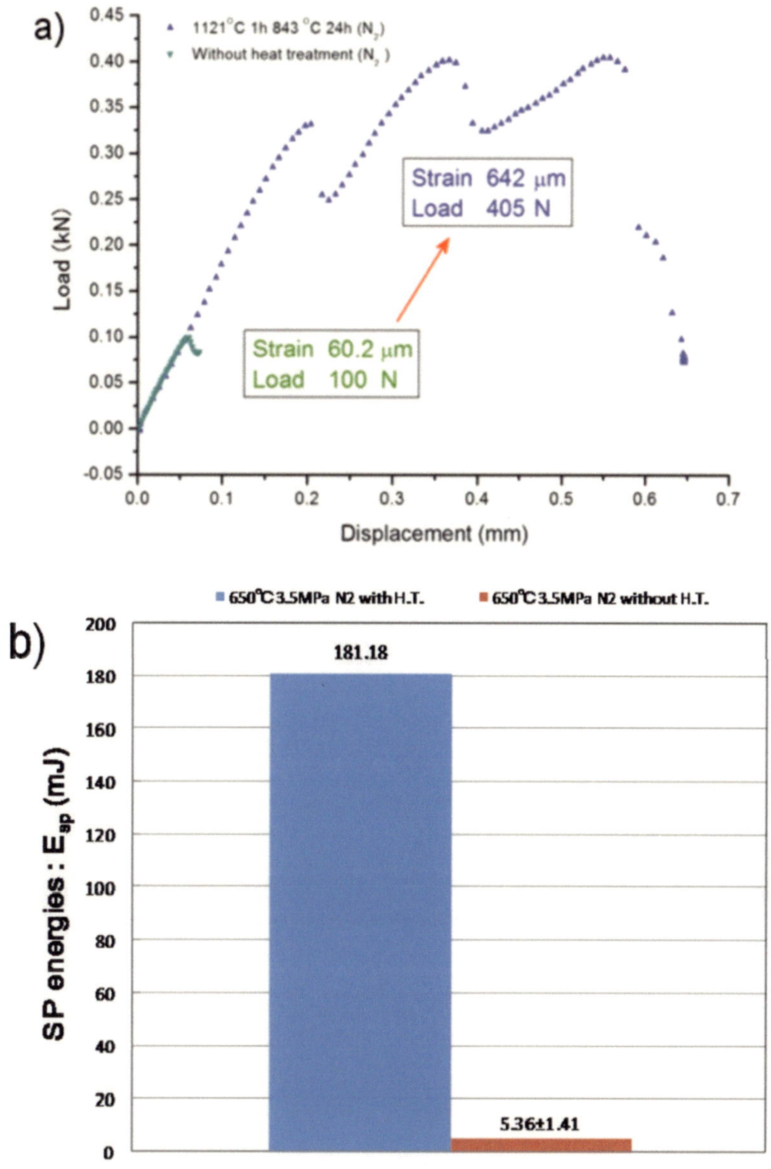

Fig. 39. Comparison of a) p-δ curves and b) SP energy with and without PSHT

about 1.5 times of maximum load after the standard HT. Also refer to Fig. 41, the $dp/d\delta$ value increased about 2-3 times at same condition. From this, even for good adhesion layer, it was found that post spray HT (PSHT) could improve the resistance to deformation or fracture. For a long time thermal aging material which intended for actual use, the maximum load was slightly less compared to the substrate, but SP absorbed energy value of the E_{sp} show about three times of the value of the substrate. Then it became clear that the ductile behavior of the strain to failure increases. Fig. 43 shows the cross-section of SP sample after etching. When the ball was in contact with the substrate/coating interface, it stopped after crack propagation of about 300 μm. It was found that the crack stopped after propagating into layer shown in figure. It revealed that the interfacial bonding strength among the splats is considered to be sufficiently high after HT. Thus, for repair of the gas turbine blades by cold spray deposition, the PSHT could be admitted the possibility of applying this technique layer from the viewpoint of the better mechanical properties of IN738LC.

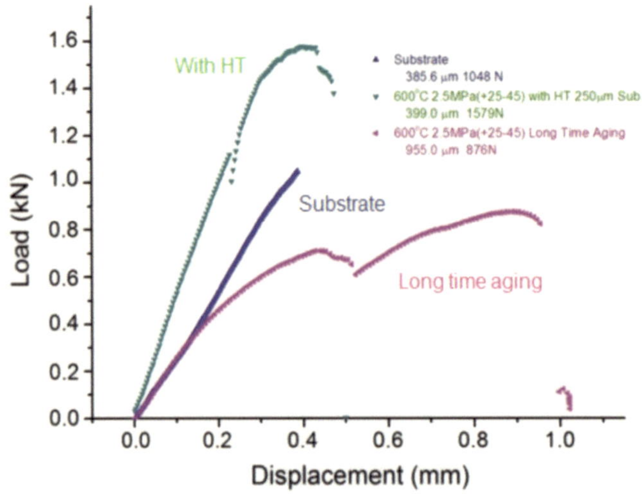

Fig. 40. Comparison of p-δ curves with short and long heat treatment

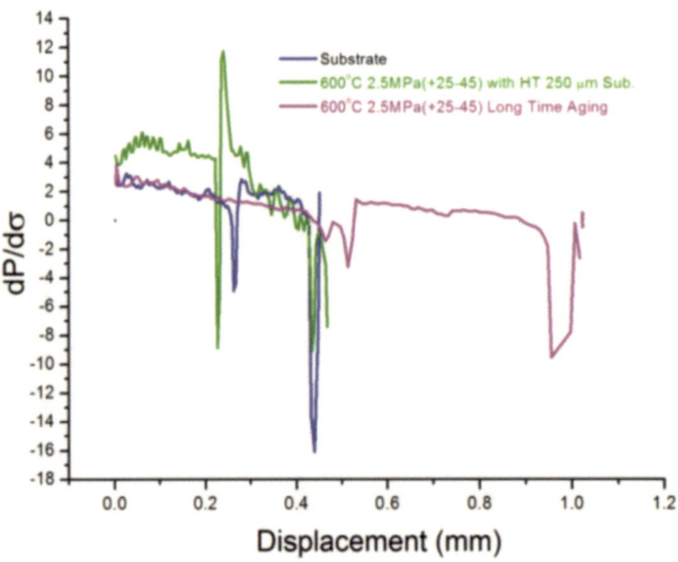

Fig. 41. Derivation of SP p-δ curve for each point

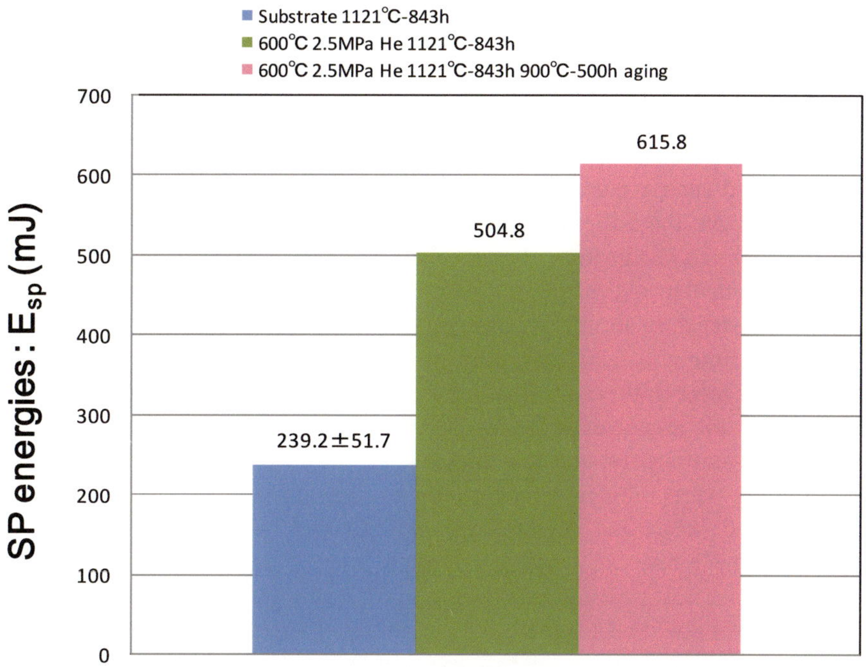

Fig. 42. Comparison of SP energies with short and long heat treatment

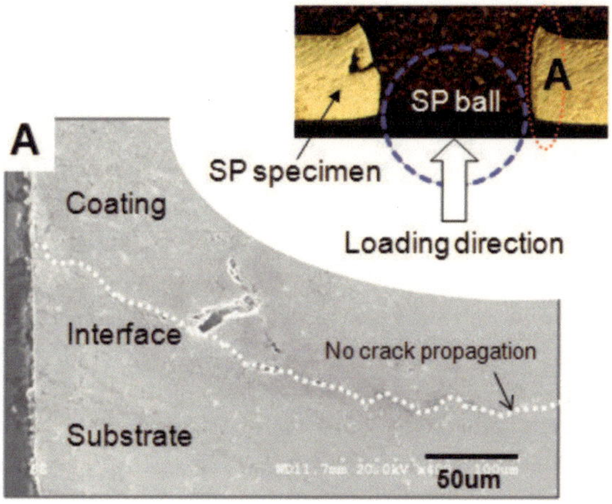

Fig. 43. Cross sectional SEM images of SP specimen

6. Summary

The optimization of cold spraying conditions based on the kinetic energy and rebound energy of the particles was carried out. The particle diameter and spray conditions were found to affect the quality of the diameter deposited layer. In particular, small particle size can effectively improve the quality of the deposited layer. In small punch tests, the He gas used specimens showed are higher maximum load and SP energy than that of the N_2 gas

used ones. However, He gas used coatings are not enough mechanical property compared to bulk Ni base superalloy.
Post heat treatment applied to the cold sprayed coatings was shown to be able to recover their crystalline structure and mechanical properties close to the bulk materials such as substrates. It can improve the adhesion strength of the coating/substrate interface by post heat treatment, and such a partial recrystallization at the interface is considered likely to contribute to improve the adhesion of the coating. It also causes the reduction of pores present inside coatings, the growth of grain size and more γ' phase formation from the as-sprayed state. Particular optimization of the heat treatment conditions is necessary to improve the microstructure and properties.
In each exposure time and condition, both of the external and inside oxide layers of as-sprayed state were most thick, and following the post heat treatment of NHT and HHT. As a result, each CS sample is considered to be related to grain size in the coating. Comparing to the bare substrate sample, the average thickness of the oxide layer of the both NHT and HHT CS samples were thinner at high temperature. This is because the dense grain boundaries near the surface and the formation of Al_2O_3 and TiO_2 in internal layer oxidation.
It revealed that the interfacial bonding strength among the splats is considered to be sufficiently high after heat treatment. Thus, for repair of the gas turbine blades by cold spray deposition, the PSHT could be admitted the possibility of applying this technique layer from the viewpoint of the better mechanical properties of IN738LC.

7. Acknowledgments

The authors would like to express thanks to Dr. T. Niki and Mr. S. Onchi who have contributed to a lot of experiments.

8. References

Alkimov, A.P., Kosarev, V.F., & Papyrin, A.N. (1990). A Method of Cold Gas Dynamic Deposition, *Dokl. Akad. Nauk SSSR*, 1990, Vol.318, No.5, (Sep 1990), pp. 1062-1065, ISSN 0002-3264

Borchers, C., Gartner, F., Stoltenhoff, T., & Kreye, H. (2004). Microstructural Bonding Features of Cold Sprayed Face Centered Cubic Metals, *J. Appl. Phys.*, Vol.96, No.8, (Oct 15 2004), pp. 4288-4292, ISSN 0021-8979

Caron, P. & Khan, T. (1983). Improvement of Creep Strength in a Nickel-base Single-crystal Superalloy by Heat Treatment, *Mater. Sci. Eng.*, Vol.61, No.2, (Nov 1983), pp. 174-184, ISSN 0025-5416

Dykhuizen, R.C. & Smith, M.F. (1998). Gas Dynamic Principles of Cold Spray, *J. Therm. Spray Technol.*, Vol.7, No.2, (Jun 1998), pp. 205-212, ISSN 1059-9630

Ejaz, N. & Tauqir, A. (2006). Failure due to Structural Degradation in Turbine Blades, *Eng. Fail. Anal.*, Vol.13, No.3, (Apr 2006), pp. 452-463, ISSN: 1350-6307

Fukumoto, M. (2006). New Trend in Particle Deposition, PD Process and Its Future, *J. Jpn. Weld. Soc.*, Vol.75, No.8, (Dec 2006), pp. 7-11, ISSN 0021-4787

Fukumoto, M., Tanabe, K., Yamada, M., & Yamaguchi, E. (2007). Clarification of Deposition Mechanism of Copper Particle in Cold Spray Process, *Q. J. Jpn. Weld. Soc.*, Vol.25, No.4, (Nov 2007), pp. 537-541, ISSN 0288-4771

Gilmore, D.L., Dykhuizen, R.C., Neiser, R.A., Roemer, T.J., & Smith, M.F. (1999). Particle Velocity and Deposition Efficiency in the Cold Spray Process, *J. Therm. Spray Technol.*, Vol.8, No.4, (Dec 1999), pp. 576-582, ISSN 1059-9630

Johnson, K.L. (1985). *Contact Mechanics*, Cambridge University Press, ISBN 0-521-34796-3, Cambridge, UK

Li, C.J. & Li, W.Y. (2004). Optimization of Spray Conditions in Cold Spraying based on the Numerical Analysis of Particle Velocity, *Trans. Nonferrous Met. Soc. China*, Vol.14, No.2, (Oct 2004), pp. 43-48, ISSN 1003-6326

Li, W.Y., Li, C.J., & Liao, H.L. (2006). Effect of Annealing Treatment on the Microstructure and Properties of Cold-Sprayed Cu Coating, *J. Therm. Spray Technol.*, Vol.15, No.2, (Jun 2006), pp. 206-211, ISSN 1059-9630

Li, W.Y., Zhang, C., Liao, H., & Coddet, C. (2009). Effect of Heat Treatment on Microstructure and Mechanical Properties of Cold Sprayed Ti Coatings with Relatively Large Powder Particles, *J. Coat. Technol. Res.*, Vol.6, No.3, (Sep 2009), pp. 401-406, ISSN 1547-0091

Niki, T. (2009). *Study of Repairing for Degraded Hot Section Parts of Gas Turbines by Cold Gas Dynamic Spraying and its Durability Evaluation*, PhD thesis, Tohoku Univ., Japan, March, 2009

Papyrin, A., Klinkov, S.V., & Kosarev, V.F.; Marple, B.R. & Moreau, C. (2003). Modeling of Particle-Substrate Adhesive Interaction under the Cold Spray Process, *Proceedings of ITSC 2003*, pp. 27-35, ISBN 0-87170-785-3, Orlando, FL, USA, May 5-8, 2003

Renteria, A.F. & Saruhan, B. (2006). Effect of Ageing on Microstructure Changes in EB-PVD Manufactured Standard PYSZ Top Coat of Thermal Barrier Coatings, *J. Eur. Ceram. Soc.*, Vol.26, No.12, (Jun 2006), pp. 2249-2255, ISSN 0955-2219

Sharghi-Moshtaghin, R. & Asgari, S. (2004). The Influence of Thermal Exposure on the γ' Precipitates Characteristics and Tensile Behavior of Superalloy IN-738LC, *J. Mater. Process. Technol.*, Vol.147, No.3, (Apr 2004), pp. 343-350, ISSN 0924-0136

Spencer, K. & Zhang, M.X. (2009). Heat Treatment of Cold Spray Coatings to Form Protective Intermetallic Layers, *Scr. Mater.*, Vol.61, No.1, (Jul 2009), pp. 44-47, ISSN 1359-6462

Stoltenhoff, T., Voyer, J., & Kreye, H.; Berndt, C.C., Khor, K.A., & Lugscheider, E.F. (2002). Cold Spraying - State of the Art and Applicability, *Proceedings of ITSC 2002*, pp. 366-374, ISBN 3-87155-783-8, Essen, Germany, March 4-6, 2002

Van Steenkiste, T.H., Smith, J.R., Teets, R.E., Moleski, J.J., Gorkiewicz, D.W., Tison, R.P., Marantz, D.R., Kowalsky, K.A., Riggs, W.L., Zajchowski, P.H., Pilsner, B., McCune, R.C., & Barnett, K.J. (1999). Kinetic Spray Coatings, *Surf. Coat. Technol.*, Vol.111, No.1, (Jan 1999), pp. 62-71, ISSN 0257-8972

Van Steenkiste, T.H., Smith, J.R., & Teets, R.E. (2002). Aluminum Coatings via Kinetic Spray with Relatively Large Powder Particles, *Surf. Coat. Technol.*, Vol.154, No.2-3, (May 2002), pp. 237-252, ISSN 0257-8972

Viswanathan, R. (2001). An Investigation of Blade Failures in Combustion Turbines, *Eng. Fail. Anal.*, Vol.8, No.5, (Oct 2001), pp. 493-511, ISSN 1350-6307

Vlcek, J., Huber, H., Voggenreiter, H., Fischer, A., Lugscheider, E., Hallen, H., & Pache, G.; Berndt, C.C., Khor, K.A., & Lugscheider, E.F. (2001). Kinetic Powder Compaction Applying the Cold Spray Process: A Study on Parameters, *Proceedings of ITSC 2001*, pp. 417-422, ISBN 0-87170-737-3, Singapore, May 28-30, 2001

Zhang, D., Shipway, P.H., & McCartney, D.G.; Marple, B.R. & Moreau, C. (2003). Particle-Substrate Interactions in Cold Gas Dynamic Spraying, Proceedings of ITSC 2003, pp. 45-52, ISBN 0-87170-785-3, Orlando, FL, USA, May 5-8, 2003

Zhao, X., Wang, X., & Xiao, P. (2006). Sintering and Failure Behaviour of EB-PVD Thermal Barrier Coating after Isothermal Treatment, *Surf. Coat. Technol.*, Vol.200, No.20-21, (May 2006), pp. 5946-5955, ISSN 0257-8972

Zou, Y., Qin, W., Irissou, E., Legoux, J.G., Yue, S., & Szpunar, J.A. (2009). Dynamic Recrystallization in the Particle/Particle Interfacial Region of Cold-sprayed Nickel Coating: Electron Backscatter Diffraction Characterization, *Scr. Mater.*, Vol.61, No.9, (Nov 2009), pp. 899-902, ISSN 1359-6462